THE OCEAN BASINS AND MARGINS

Volume 3
The Gulf of Mexico and the Caribbean

THE OCEAN BASINS AND MARGINS

Volume 1: The South Atlantic • 1973
Volume 2: The North Atlantic • 1974
Volume 3: The Gulf of Mexico and the Caribbean • 1975

In preparation:

Volume 4: The Mediterranean
Volume 5: The Arctic Ocean
Volume 6: The Indian Ocean
Volume 7: The Pacific Ocean

THE OCEAN BASINS AND MARGINS

Edited by

Alan E. M. Nairn

Department of Geology
University of South Carolina
Columbia, South Carolina

and

Francis G. Stehli

Department of Geology
Case Western Reserve University
Cleveland, Ohio

Volume 3
The Gulf of Mexico and the Caribbean

PLENUM PRESS · NEW YORK AND LONDON

Library of Congress Cataloging in Publication Data

Nairn, A E M
  The ocean basins and margins.

  Includes bibliographies.
  CONTENTS: V. 1. The South Atlantic—v. 2. The North Atlantic.—v. 3. The Gulf
of Mexico and the Caribbean.
  1. Submarine geology. 2. Continental margins. I. Stehli, Francis Greenough, joint
author. II. Title.
QE39.N27   551.4'608   72-83046
ISBN 0-306-37773-X (v. 3)

CONTRIBUTORS TO THIS VOLUME

Daniel D. Arden, Jr.
Georgia Southwestern College
Americus, Georgia

Philip O. Banks
Department of Earth Sciences
Case Western Reserve University
Cleveland, Ohio

Carl Bowin
Woods Hole Oceanographic Institution
Woods Hole, Massachusetts

J. E. Case
U. S. Geological Survey
Menlo Park, California

Gabriel Dengo
Instituto Centroamericano de Investigación y
    Technología Industrial (ICAITI)
Guatemala City, Guatemala

Thomas W. Donnelly
Department of Geological Sciences
State University of New York
Binghamton, New York

Paul J. Fox
Department of Geological Sciences
State University of New York at Albany
Albany, New York

Bruce C. Heezen
Lamont-Doherty Geological Observatory
    of Columbia University
Palisades, New York

James Helwig
Department of Geology
Case Western Reserve University
Cleveland, Ohio

Philip B. King
U. S. Geological Survey
Menlo Park, California

Ray G. Martin
U. S. Geological Survey
Corpus Christi, Texas

Georges Pardo
Gulf Oil Corporation
Pittsburgh, Pennsylvania

E . López Ramos
Petróleos Mexicanos
Mexico

Reginald Shagam
Department of Geology
Ben Gurion University of the Negev
Beer Sheva, Israel

John Frederick Tomblin
Seismic Research Unit
University of the West Indies
Trinidad

Elazar Uchupi
Woods Hole Oceanographic Institution
Woods Hole, Massachusetts

CONTENTS

Chapter 3. Geophysical Studies in the Caribbean Sea

J. E. Case

Chapter 4. **Basement Rocks Bordering the Gulf of Mexico and the Caribbean Sea**

Philip O. Banks

Chapter 5. **The Ouachita and Appalachian Orogenic Belts**

Philip B. King

Chapter 6. **Tectonic Evolution of the Southern Continental Margin of North America from a Paleozoic Perspective**

James Helwig

Chapter 7. **Geological Summary of the Yucatan Peninsula**

E. Lopez Ramos

Chapter 8.   Paleozoic and Mesozoic Tectonic Belts in Mexico and Central America

Gabriel Dengo

Chapter 9.   The Northern Termination of the Andes

Reginald Shagam

Chapter 10.   **Geology of the Caribbean Crust**

Paul J. Fox and Bruce C. Heezen

Chapter 11.   **The Lesser Antilles and Aves Ridge**

John Frederick Tomblin

Chapter 12. The Geology of Hispaniola

Carl Bowin

Chapter 13. **Geology of Cuba**

Georges Pardo

Chapter 14. **Geology of Jamaica and the Nicaragua Rise**

Daniel D. Arden, Jr.

Chapter 15. **The Geological Evolution of the Caribbean and Gulf of Mexico—Some Critical Problems and Areas**

Thomas W. Donnelly

Chapter 1

PHYSIOGRAPHY OF THE GULF OF MEXICO AND CARIBBEAN SEA*

Elazar Uchupi

Woods Hole Oceanographic Institution
Woods Hole, Massachusetts

I. INTRODUCTION

The Gulf of Mexico and the Caribbean Sea have attracted the attention of many geological investigators because of their topographic complexity. Early students suggested that both regions were sites of landmasses that had subsided to their present depths or been oceanized. Others have treated both areas as permanent oceanic basins with one (Gulf of Mexico) being surrounded by landmasses and the other (Caribbean) by island arcs and geosynclines (Paine and Meyerhoff, 1970; Meyerhoff and Meyerhoff, 1972). Recent investigators have attempted to explain their complex tectonics and stratigraphy, particularly the Caribbean, in the light of global tectonics. Proponents of sea-floor spreading have suggested that the Caribbean is a remnant of pre-Mesozoic sea floor trapped between the westerly drifting American continents and later sealed from the Pacific and Atlantic Oceans by subduction zones (Deuser, 1970; Edgar *et al.*, 1971; Malfait and Dinkelman, 1972). Other mobilists have interpreted the Caribbean as a young ocean basin formed during separation

---

* Contribution No. 3031 of the Woods Hole Oceanographic Institution.

1

of the American continents (Funnell and Smith, 1968; Ball and Harrison, 1969, 1970; JOIDES, 1971). The present report attempts to describe the physiography of both areas in the light of seismic reflection data collected by numerous investigators during the past decade, data from the Deep-Sea Drilling Project, and a new detailed bathymetric chart of the Gulf of Mexico and Caribbean. These new data greatly expand the fund of critical knowledge and have resulted in new insights into the origin and evolution of the Gulf of Mexico and Caribbean.

II. GULF OF MEXICO

A. General

The Gulf of Mexico is a small oceanic basin surrounded by continental masses (Ewing *et al.*, 1955; Ewing *et al.*, 1960; Antoine and Ewing, 1962; Ewing *et al.*, 1962; Fig. 1). This nearly landlocked basin is connected to the Atlantic by the Straits of Florida, and to the Caribbean by the narrow deep Yucatan Channel between Cuba and the Yucatan peninsula. Physiographically, the Gulf can be divided into two provinces, a terrigenous one to the west, and a carbonate one to the east. The terrigenous province consists of the Mississippi Delta and Cone, the northern, western, and southern continental terraces, and the abyssal gulf. Banco de Campeche and the shelf off west Florida comprise the carbonate province.

B. Terrigenous Province

1. *Mississippi Delta and Cone*

The eastern margin of the terrigenous province, partly overlapping the Florida carbonate platform, contains the Mississippi Delta and Cone. The delta and cone are the seaward end of a sedimentary apron that was initiated early in the Cenozoic and has prograded nearly 1000 km to the edge of the continental plateau (Russell and Russell, 1939). Detailed studies over the years have led to the identification and dating of many stages of the development of the Mississippi Delta. Changes in site of deposition appear to be due to progradation and lowering of gradient along the previous river course (Fisk, 1952). Breaching of the natural levee during flood of the river established a shorter and steeper course to the ocean and the initiation of new subdelta. During the past 6000 yr seven such deltas have been developed (Fig. 2). Rapid progradation of the modern delta over older shelf deposits and prodelta plastic clays has resulted in the vertical migration of the plastic clays into the over-

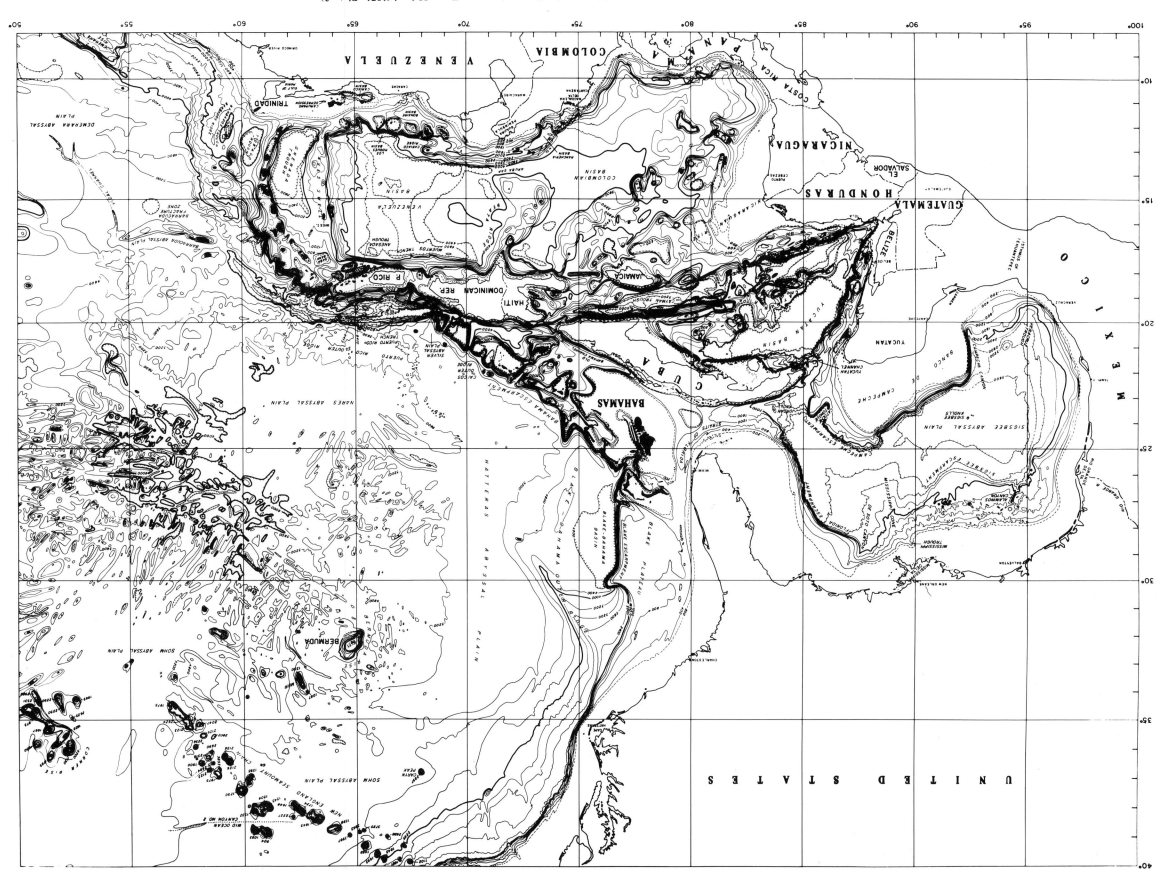

Fig. 1. Topography of the Gulf of Mexico and Caribbean and adjacent deep sea. From Uchupi (1971, Plate 3).

Fig. 2. Succession and ages of major subdeltas in the Mississippi Delta. (1) Sale–Cypremort, 5500 to 4400 yr ago; (2) Cocodire, 4600 to 3600; (3) Teche, 2900 to 2700; (4) St. Bernard, 2800 to 1700; (5) Lafourche, 1900 to 700; (6) Plaquemines, 1200 to 400; (7) Balize (modern), 500 to present. The black area represents the Pleistocene upland. Adopted by Emery and Uchupi (1972, Fig. 36) from Kolb and van Lopik (1966) and Ballard and Uchupi (1970). Published with permission of the American Association of Petroleum Geologists.

lying sediments. These diapirs or mudlumps have risen as much as 130 m (Morgan *et al.*, 1968). Between the diapirs are synclinal troughs filled with as much as 120 m of rapidly deposited near-strandline bar sand, silt, clay, and organic matter. From the topset to the foreset surfaces of the delta to a depth of about 60 m are many gullies, some of which terminate in small hills (Shepard,

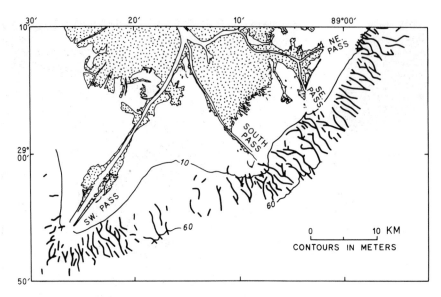

Fig. 3. Distribution of delta-front valleys of the Mississippi Delta. Adopted by Emery and Uchupi (1972, Fig. 61) from Shepard (1955; pl. 1). Published with permission of the American Association of Petroleum Geologists.

1955; Fig. 3). These gullies are believed to have been formed as a result of mass slumping along the rapidly prograding slope of the delta.

Seaward of the foreset slope of the Mississippi Delta is the Mississippi Cone, a thick mass of partly bedded Pleistocene and early Holocene sediments that grade into stratified hemipelagic sediments on the fringes of the cone (Wilhelm and Ewing, 1972). The eastern boundary of the cone is simple, whereas the western boundary is very irregular, probably due to diapiric structures. The cone slopes continuously southeastward between the Florida and Campeche escarpment to the western approaches of the Straits of Florida. Westward the Mississippi Cone grades into the continental rise and southwestward into the Sigsbee Abyssal Plain (Fig. 1). Pleistocene and early Holocene sediments of the cone are in sharp contact with the Florida and Campeche escarpments (Fig. 1), suggesting that sediment deposition from the scarps and adjacent shelves is minor. Several knolls characterized by strong magnetic anomalies rise above the southeastern fringe of the cone. These features may be related to the subsurface volcanic belt of the Florida carbonate platform (Gough, 1967) or the ultramafic belt in northwestern Cuba (Pyle *et al.*, 1969). The largest of these topographic highs, Jordan Knoll (Fig. 1), served as a foundation for a reef that flourished during Early Cretaceous. Rocks recovered from its crest indicate that the Jordan Knoll region was a shallow bank until latest Albian or early Cenomanian, that the Straits of Florida deepened gradually from the Cenomanian through Santonian, and that since the Santonian, bathyal conditions have prevailed on the western approaches to the Straits (Bryant *et al.*, 1969). That deep water conditions have prevailed since Late Cretaceous is also suggested by the sediments in JOIDES Hole 95 (Fig. 7), which consist of Late Cretaceous deep-water pebbly mudstones, black cherts, and Cenozoic pelagic oozes. Possibly the knoll may have been uplifted during the middle Eocene in response to the "Laramide" deformation of Cuba.

The apex of the Mississippi cone is bordered by two canyons, De Soto on the east and the Mississippi Trough (Fig. 1) on the west. De Soto Canyon can be traced landward to a depth of about 400 m. Seaward the canyon connects with a leveed channel that parallels the Florida Escarpment for 360 km (Jordan, 1951). To a large extent the canyon's shape is controlled by salt domes, with its eastern edge marking the eastern limit of the piercement structures that are so common farther west (Harbison, 1968; Antoine, 1972). This decrease in the number of salt structures in an easterly direction may result from thinning of the salt in that direction (Antoine and Bryant, 1969). Antoine (1972) has suggested that the eastern limit of the Triassic–Jurassic salt may be controlled by the basement structure trending southwest from the Ocala uplift (Fig. 17) and then westward just north of the Florida Escarpment. With the considerable volume of sediment transported to the Gulf of Mexico by the Mississippi River, it is difficult to see why De Soto Canyon has not been buried by now. The presence of a prograding apron on the western side indicates that part of the Mississippi Cone detritus is transported eastward. No drainage system from the north can be associated with the canyon, so its origin is enigmatic. The considerable evidence of erosion displayed by seismic profiles across the canyon led Antoine (1972) to suggest that it was formed by ocean currents, possibly the Loop Current (Nowlin, 1971). If the canyon has been cut and maintained by ocean currents since early Tertiary, then the present circulation of the Gulf has not changed appreciably for the last 65 million yr, and the Yucatan Channel, Straits of Florida, and the Gulf Stream have existed since the Cretaceous.

On the western margin of the Mississippi Cone is the Mississippi Trough, a U-shaped valley that connects with a filled channel that has been traced on the shelf and delta to Houma, Louisiana (Carsey, 1950). Seaward the trough extends to a depth of 2200 m where it terminates in a fan. This trough probably was the main conduit not only for sediments of the Mississippi Cone, but for much of the sediments of the abyssal gulf. Continuous seismic profiles indicate that the lower part of the cone consists of well stratified sediment and shows at least 14 major Cenozoic depositional cycles (Huang and Goodell, 1970). The upper cone, however, has many irregularities that are the result of slumping, gravitational sliding, and salt tectonics (Uchupi and Emery, 1968; Huang and Goodell, 1970; Walker and Massingill, 1970; Fig. 4, profile 4).

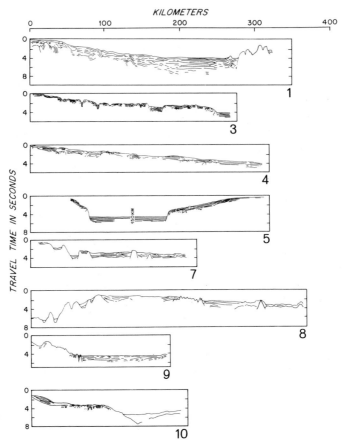

Fig. 4. Continuous seismic reflection profiles from the Gulf of Mexico and the Caribbean. For locations of profiles, see Fig. 16. The profiles were redrawn to a common scale from the following sources: (1) United States Geological Survey (1972*a*, line 10); (3) Uchupi and Emery (1968, line 249); (4) Uchupi and Emery (1968, line 245); (5) Uchupi and Emery (1968, line 233); (7) Uchupi (1973, line 403); (8) Edgar *et al.* (1971, Fig. 18); (9) Edgar *et al.* (1971, Fig. 17); (10) Krause (1971, Fig. 9).

2. *Northern Continental Terrace*

The most important structural element of the northern continental terrace is the Gulf Coast geosyncline extending from the Tamaulipas Carbonate Platform (Fig. 17) in northeast Mexico to the Florida Carbonate Platform. The northern limit of this huge prism of Cenozoic terrigenous sediments nearly 20 km thick is 320 km inland along the line of Mesozoic–Cenozoic outcrops. The geosyncline's southern boundary is along the Sigsbee Escarpment, and its axis is located immediately south of the present shoreline. Strata in the geosyncline dip and thicken southward and are modified by faults and flexures.

Both the structural axis and depocenters migrated with time, the axis southward and the depocenters eastward, with the Rio Grande River predominating as a sediment source during early Tertiary, and the Mississippi River drainage, during the late Tertiary–Pleistocene. Evolution of the geosyncline appears to have been controlled by structural weaknesses inherited from the Paleozoic Ouachita orogenic belt to the north, presence of the Gulf of Mexico to the south, uplift of the Rocky Mountains during the Paleocene which supplied a high volume of sediment to the geosyncline, and Late Triassic–Middle Jurassic evaporites which imparted structural mobility to the geosyncline's sedimentary apron (Lafayette and New Orleans Geological Society, 1968).

The 120–260 km wide continental shelf south of Texas and Louisiana has a depth of 120 m along its seaward edge. Topographic highs atop the shelf consist of cemented remnants of Pleistocene shorelines with relief of less than 4 m, a cluster of ridges north of the Rio Grande delta at a depth of 28 m which represent distributary ridges formed when the Rio Grande delta (Fig. 5) was located farther north (Mattison, 1948; Rusnak, 1960), and circular highs with reliefs greater than 4 m concentrated along the shelf's edge (Fig. 5). The circular highs are believed to be surface expressions of salt domes (Shepard, 1937; Nettleton, 1957; Neumann, 1958; Curray, 1960; Moore and Curray, 1963; Ewing and Antoine, 1966; Uchupi, 1967; Uchupi and Emery, 1968; Walker and Ensminger, 1970), although some may represent pre-Holocene biocherms (Matthews, 1963). These topographic prominences appear to have been modified in early Holocene by the carbonate accretion of coral and calcareous algae and by wave erosion during the Holocene transgression (Stetson, 1953). A topographic survey by Parker and Curray (1956) indicates that the topographic highs commonly have their tops at 17, ˙59, and 88 m. These levels may represent sea-level stands during the last transgression. Another tectonic element of the shelf's edge is growth faults which appear to be related to the flowage of salt away from the prograding shelf.

Topographic lows appear to be of two types. On the outer shelf, channels at right angles to the contours were probably carved by stream erosion during Pleistocene low-sea-level stands. In the middle of the shelf the channels trend obliquely to the contours and are associated with ridges. These channels and ridges probably represent barrier spits and bays formed during the Holocene transgression (Curray, 1960). Some of the bays are open to the east, opposite to present-day bays, which suggests that in the past the littoral drift was toward the east. Curray (1960) believed that this reversal in littoral drift was a result of minor glacial advances during the Holocene that forced the easterly winds to retreat southward. With this retreat, westerly winds became dominant, producing easterly longshore drift. Renewal of ice retreat caused the easterly winds to migrate northward, resulting in the present wind regime and westerly longshore drift.

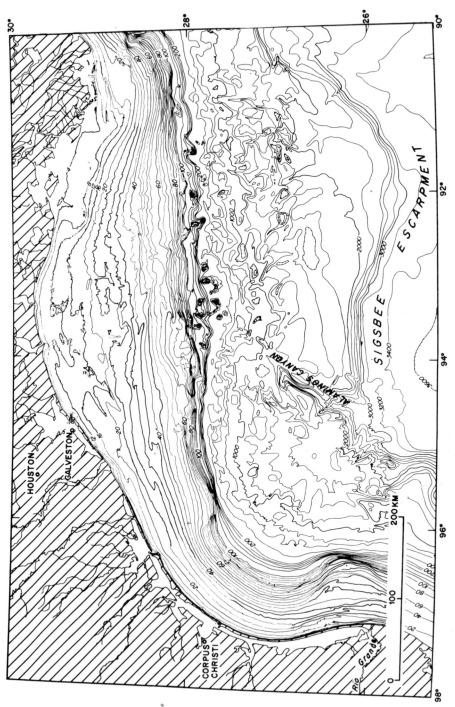

Fig. 5. Topography of the continental terrace off Texas and Louisiana. Adopted from Ballard and Uchupi (1970, Plate 1) and by Emery and Uchupi (1972, Fig. 39). Published with permission of the American Association of Petroleum Geologists.

The seaward and growing edge of the Gulf Coast geosyncline is the continental slope, a ridge formed of salt that has been forced southward by pressure from the thick sediment prism to the north (Fig. 4, profile 3). The Sigsbee Escarpment (Fig. 1) represents the front of this southward-moving salt wedge (Amery, 1969). Alaminos Canyon, a north–south system of valleys and basins, cuts the western end of this salt front (Fig. 1; Bouma *et al.*, 1968, 1972). This canyon may have been formed by coalescing salt fronts produced from separate basins (Garrison and Berryhill, 1970) or may be a channel cut by turbidity currents and modified by contemporaneous and later salt diapirism. A northeast–southwest valley also intersects the Sigsbee Escarpment at its eastern end (Bergantino, 1971). This valley bifurcates at 26° N with both branches continuing northward to 26°40′ N.

The northern section of the upper slope is disrupted by numerous topographic highs that rise as much as 1000 m above their surroundings. Relief of these features is believed to be due to the vertical migration of salt (Moore and Curray, 1963; Ewing and Antoine, 1966; Uchupi and Emery, 1968; Lehner, 1969; Antoine and Pyle, 1970; Ewing *et al.*, 1970; Antoine, 1972; Wilhelm and Ewing, 1972). The smoothness of the southern part of the upper slope just north of the Sigsbee Escarpment is due to salt diapirs that have risen to a common elevation. The vertical migration of salt has brought to shallow depths

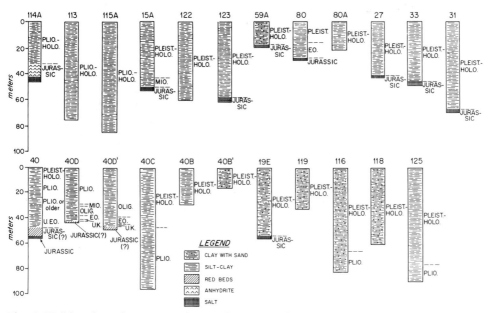

Fig. 6. Well logs from the upper continental slope south of Texas. See Fig. 16 for locations of wells. Compiled by Emery and Uchupi (1972, Fig. 80) from data from Lehner (1969). Published with permission of the American Association of Petroleum Geologists.

sediments that occur at considerable depth farther north, thus permitting a study of the early history of the geosyncline. Red beds of possible Triassic–Jurassic age and deposits of Cretaceous and Eocene age have been recovered from some of these structural highs (Fig. 6). Drill data indicate that throughout much of the slope the sediment blanket consists mainly of Pleistocene deposits (Fig. 6). This sediment blanket is very thick, and only in the steep-sided basins do sediment thicknesses exceed 3 km (Lehner, 1969). Thus, stratigraphically the slope is the outer edge of the Pleistocene clastic wedge, a major sedimentary unit of the Gulf Coast geocyncline. Much of the sediment in structural lows on the upper slope appears to have been emplaced by slumping, gravitational sliding, and turbidity currents (Lehner, 1969). Sediments atop the Sigsbee scarp (hole 92, Fig. 7) consist of overconsolidated Pleistocene mudstones fractured and, in some, cases rewelded (Scientific Staff, 1970).

3. *Western Continental Terrace*

The shelf of the western continental terrace was formed by progradation of terrigenous sediment over the Tamaulipas Platform, a Mesozoic carbonate platform east of the Mexican outer geosyncline that extends from the Gulf Coast geosyncline in the north to the Isthmian embayment (Fig. 17) in the south. This shelf is generally less than 50 km wide with its edge lying at a depth of about 100 m. In the southern half, off Tampico and Veracruz, coral reefs occur atop the shelf (Heilprin, 1890; Moore, 1958; Emery, 1963). These reefs consist mainly of common West Indian corals with some calcareous algae. Topographically they resemble the patch reefs that rise from the floors of the shallow lagoons within atolls in the Pacific and the bioherms of Paleozoic epicontinental seas.

The continental slope east of Mexico is characterized by linear ridges parallel to the shoreline (Fig. 8; Fig. 4, Profile 1; Bryant *et al.*, 1968; Bergantino, 1971). These ridges have wave lengths of about 12 km and an average relief of 400 m. Relief is greatest between the outer slope and the edge of the abyssal gulf. Although present farther west, they are buried by sediments derived from the west. The ridges can be divided into three zones. In the southernmost section (19°–22° N) the folds are concave toward shore and well developed, while the sediment blanket is conformable over ridges and troughs (Ewing *et al.*, 1970; Antoine, 1972). The folds in this zone lie adjacent and parallel to the offshore edge of the Golden Lane, an atoll on the eastern edge of the Tamaulipas Platform (Mina, 1965; Meyerhoff, 1967). The folds in zone 1 are separated from those of zone 2 by a series of east–west trending scarps that have a maximum relief of 2000 m over a horizontal distance of 5 km (Ewing *et al.*, 1970). These scarps may be a seaward extension of the Zacatecas Fracture Zone (Fig. 17) that extends across Mexico from Baja California to Tampico (Murray, 1961, p. 11). In zone 2 (22°–24° N) the folds trend northeast obliquely

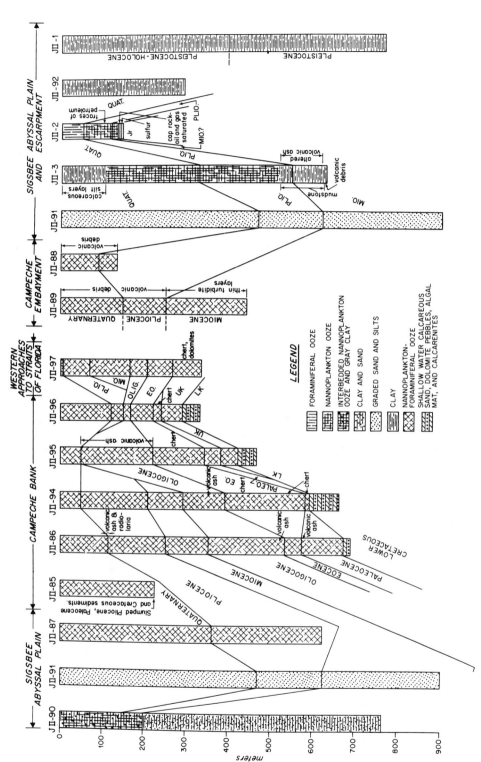

Fig. 7. Well logs from the Gulf of Mexico and western approaches to the Straits of Florida. Based on data from Ewing *et al.* (1969) and Scientific Staff (1970). See Fig. 16 for locations of wells.

to shore, with outer ones plunging beneath the sediments of the continental rise. The folds within this zone become less and less distinct northward, and their crests more diapiric, until in zone 3 (24°–26° N) the topography becomes more rugged, with most of the highs apparently due to diapirism. Based on data available, most geologists believe that the ridges consist of evaporites that were folded as a result of gravitational sliding on a décollement surface. Sediment loading has accentuated the initial folds, and in some cases, particularly in zone 3, loading has resulted in the formation of piercement structures along the ridge crests.

4. *Southern Continental Terrace*

The southern continental terrace borders the Isthmian embayment. This embayment is part of the Northern Central America orogen, an important tectonic element that extends from the southern Gulf of Mexico to the Caribbean Sea (Vinson and Brineman, 1963; McBirney and Bass, 1969a; Dengo, 1969; Dengo and Bohnenberger, 1969; Paine and Meyerhoff, 1970; Bateson, 1972; Hall and Bateson, 1972). Important structural features of the Northern Central American orogen include the east–west normal faults at the Gulf of Mexico end, the numerous wrench faults, Laramide folds, salt anticlines, salt domes, post-orogenic folds south of Yucatan, and northeast trending normal faults along the east coast of Yucatan. The most prominent transcurrent fault on the west side of the orogen is the north–south oriented Salina Cruz dextral fault (Fig. 17). This structure, which is more than 350 km long, crosses the Isthmus of Tehuantepec (Fig. 1) to the Gulf of Mexico. This together with the Zacatecas Fracture Zone and the Mexican Volcanic Belt (Fig. 17) are three structures that extend across Mexico bisecting the structural grain of the region. The other transcurrent faults arc around the southern edge of the Yucatan peninsula. Movements along these structures have been counterclockwise, with estimates of displacements ranging from several kilometers to possibly as much as 100 km (McBirney, 1963; Anderson, 1968; Dengo and Bohnenberger, 1969; McBirney and Bass, 1969a; Kesler, 1971).

Three geosynclinal cycles can be distinguished in the Northern Central American orogen. The earliest cycle consists of the Chuacus and Palacagüina metamorphics that were deformed and intruded in mid-Paleozoic. The ultramafic rocks of the Central Guatemala Cordillera have been interpreted as oceanic crust atop which the Chuacus metamorphics were deposited. A second cycle is represented by the Pennsylvanian–Permian strata resting unconformably on the older metamorphic rocks. These orthogeosynclinal and platform deposits (eugeosynclinal to the south, miogeosynclinal to the north, and platforms north and south of the geosyncline) were deformed and intruded during the latter part of the Paleozoic. The latest geosynclinal cycle which began in

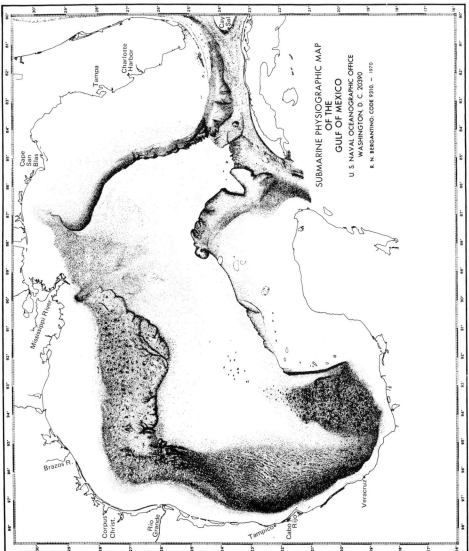

Fig. 8. Physiographic diagram of the Gulf of Mexico. From Bergantino (1971, Fig. 1). Note ridges east of Mexico. These features are known as the Mexican Ridges. Published with permission of the author and the Geological Society of America.

SUBMARINE PHYSIOGRAPHIC MAP
OF THE
GULF OF MEXICO

U. S. NAVAL OCEANOGRAPHIC OFFICE
WASHINGTON, D. C. 20390

R. N. BERGANTINO, CODE 9310. — 1970

mid-Mesozoic was deformed in mid-Cretaceous and mid-Eocene. The orogen was the site of evaporite deposition in Late Jurassic (the Isthmian basin which extended from the Gulf of Mexico to the Guatemalan border) and in latest Jurassic to possibly earliest Cretaceous (Chiapas salt basin which extended from the Gulf of Mexico through southern and central Guatemala to possibly western Belize) (Viniegra, 1971). In the Eocene the southern margin of the orogen was uplifted, folded, faulted, and intruded. This uplift resulted in the deposition of a large volume of terrigenous debris within the Isthmian embayment (Viniegra, 1971). As the orogeny continued, the sediment debris that reached thicknesses of about 8 km spread northward into the Gulf of Mexico. Deposition of this sediment has continued as recently as Pliocene and Pleistocene. Salt tectonism was concurrent with deposition, resulting in the formation of numerous salt domes, salt ridges, and down to the basin faults. The southern terrace with its complex upper continental slope due to salt diapirs (known as the Tabasco–Campeche Knolls; see Fig. 8), down to the basin faulting, and a thick Tertiary section has a geologic setting similar to that of the terrace off Texas and Louisiana. In contrast to the northern terrace, however, the salt tectonics here may have been triggered in part by uplift of the basement rather than solely by squeezing of the salt northward by the thick sediments to the south (Ensminger and Matthews, 1972; Scientific Staff, 1971a).

The salt diapirs on the slope display their greatest relief near the central part, decreasing eastward, westward, and northward (Fig. 8; Fig. 9, profile 2; Creager, 1953, 1958a; Bergantino, 1971). North of 22°45′ N diapirs are present only in the subsurface. Southward diapiric structures decrease in relief, but increase in number and complexity. The diapiric structures on the slope are flanked on the east by Campeche Canyon (Creager, 1958b) and on the west by the Veracruz Trough (Fig. 1). The floor of the canyon, a structural low between the knolls and Campeche Carbonate Platform, is somewhat irregular due to sediment creep and diapirism. The Veracruz Trough is a structural low separating the Tabasco–Campeche Knolls from the Mexican Ridges farther west (Fig. 8). Some of the structural features from both the Tabasco–Campeche Knolls and the Mexican Ridges extend into the trough, but have little surface expression except in the south where isolated diapirs have relief of about 60 m.

The youngest structural unit present in the southern margin is the Mexican Volcanic Belt that extends across Mexico, cutting the Cordillera System (Guzman and de Cserna, 1963). This late Cenozoic volcanic feature may follow a deep seated structure that appears to deflect the trend of the Cordillera system (King, 1969b). Seismic reflection profiles indicate that the volcanic belt terminates abruptly on the shelf, joining the volcanic belt in Guatemala in a sharp cusp (Scientific Staff, 1971a). Both of these volcanic arcs probably are related to the Acapulco and Guatemala segments of the Middle American Trench located on the Pacific side of Mexico (Scientific Staff, 1971a).

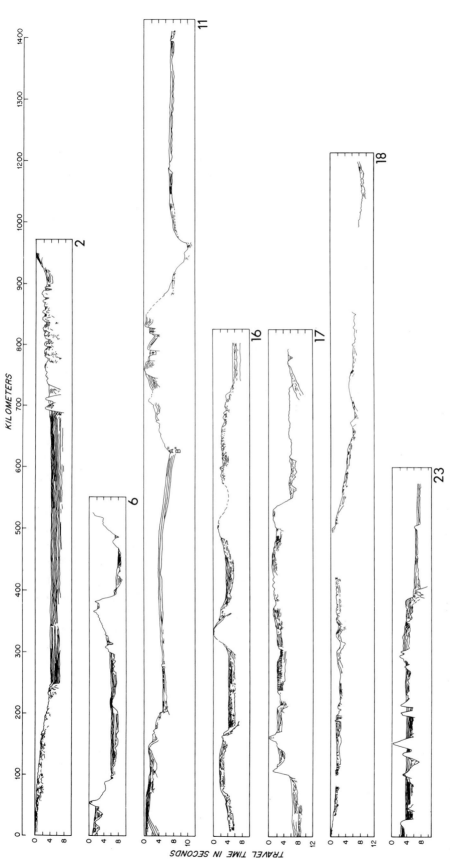

Fig. 9. Seismic reflection profiles from the Gulf of Mexico and Caribbean. For locations of profiles, see Fig. 16. These profiles were redrawn to a common scale from the following sources:: (2) United States Geological Survey (1972a, lines 1 and 2); (6) United States Geological Survey (1972b, line 32); (11) Bunce et al. (1974, line 3) and United States Geological Survey (1972d, lines 1 and 2); (16) Bunce et al. (1970, lines III-3, V-2, and VII-3); (17) United States Geological Survey (1972c, line 46); (18) Collette et al. (1969, 60° west section); (23) Uchupi et al. (1971, line 334).

5. *The Abyssal Gulf*

Seismic refraction data indicate that the western abyssal gulf has an oceanic crust and that this crust is overlain by as much as 9 km of sediment (Ewing *et al.*, 1962). The lower 5 km of this sedimentary blanket is believed to consist of latest Paleozoic and Triassic sediments (post-Ouchita orogenic sediments), and the upper 5-km sequence is made up of Jurassic to Holocene sediments (Wilhelm and Ewing, 1972). The approximately 2 km thick, 4–5 km/sec layer on the upper part of the lower sequence has been interpreted as evaporites, a seaward extension of the Louann salt, by Ewing *et al.* (1962). Salt tectonics may have uplifted the older sediments in the southern part of the Gulf. Pequegnat *et al.* (1971) have described violet siltstones containing Carboniferous (318×10^6 yr old) glauconite from the crest of one of the Sigsbee Knolls (Fig. 1). Possibly the siltstone is younger, as the glauconite may be reworked; an Early-Middle Mesozoic age for the siltstone may be more reasonable. The nature of the siltstone appears to be more typical of a paralic or continental rather than an abyssal environment. If the siltstone is in place, it may indicate that the Gulf was much shallower in the past. Similar deposits in the Isthmus of Tehuantepec, however, are characterized by flysch-like graded bedding suggestive of deep-water turbidity current deposition (Scientific Staff, 1971*a*).

The continental rise at the base of the Sigsbee Escarpment consists of a large depositional wedge about 800 m thick that lacks stratification. This wedge, known as the Sigsbee Wedge (Wilhelm and Ewing, 1972) appears to consist primarily of Quaternary massive to laminated clays with a variable admixture of sand and silt (Fig. 7, Hole JII-I). These sediments appear to have been transported to their present site by gravity slides originating on the scarp to the north. In places, the wedge ends rather abruptly against the Sigsbee Escarpment where it is in contact with the Pleistocene of the lower slope, but in other locations the wedge has been uplifted by recent movements along the scarp. Similar movements of the eastern edge of the Mexican Ridges have also affected the upper rise off Mexico. Southward the Sigsbee Wedge pinches out between bedded sediments with the strata above the pinchout correlating with the thin post—Wisconsin on top of the Mississippi Cone (Wilhelm and Ewing, 1972). The surface layers of the northern rise dip southward, but this dip first decreases, then reverses with depth to produce northward dip (Bergantino, 1971). On the rise east of the Mexican Ridges the Pleistocene strata dip eastward and thicken westward. In contrast, the older strata thicken and dip northwestward. Sediments on the southern rise consist primarily of pelagic oozes containing variable amounts of volcanic sediment (Fig. 7, JOIDES Holes JII-88 and JII-89). As a whole, strata of this segment of the rise are undisturbed except in the vicinity of piercement structures common in the southern abyssal gulf.

The sediments of the Sigsbee Abyssal Plain down to a depth of 300 m beneath the sea floor are horizontal, but deeper layers dip northwestward with the dip increasing with depth. The northwest dipping layers appear to be continuous with the deeper strata on the western rise and to be older than the Sigsbee Escarpment (Bergantino, 1971). Near the base of the Campeche Escarpment, exotic blocks displaced from the scarp occur on the abyssal plain. These slides appear to be pre-Pleistocene, as sediments of that age abut the erratic blocks (Wilhelm and Ewing, 1972).

The most prominent topographic feature of the Sigsbee Abyssal Plain are the Sigsbee Knolls that extend across the southern rise in a northeasterly direction nearly to the northern margin of the abyssal plain (Ewing *et al.*, 1958; Ewing *et al.*, 1962; Nowlin *et al.*, 1965). The knolls have reliefs of 100 to 200 m and diameters up to 10 km. Twenty-five such highs have been reported from the abyssal plain. Associated with the knolls are 21 diapiric structures without any surface expression (Worzel *et al.*, 1968). A similar structure has also been reported between the Florida and Campeche Escarpments (Wilhelm and Ewing, 1972). Recent drilling has clearly established that these features are salt diapirs (Fig. 7, JOIDES Hole JII-2). The well drilled atop one of the knolls penetrated a typical salt dome caprock mineral assemblage of calcite, gypsum, and free sulfur (Burk *et al.*, 1969). A middle to late Jurassic palynomorph assemblage occurs in the caprock (Kirkland and Gerhard, 1971). Thus, the Sigsbee salt is of about the same age as the Louann–Werner of southeastern United States (Murray, 1961), the Minas Viejas and Salina of Mexico (Viniegra, 1971), and the Punta Alegre of Cuba (Meyerhoff and Hatten, 1968). Sediments above the caprock consisted of a thin pelagic sequence ranging in age from late Miocene to Pleistocene. In contrast, the sediments on the adjacent abyssal plain contain a thick turbidite sequence (Fig. 7, JOIDES Holes JII-3, JII-87, JII-90, and JII-91). The major problem yet to be resolved is whether the evaporites that gave rise to the piercement structures were (1) squeezed out by the bordering margins toward the center of the basin, (2) deposited in deep water in the manner suggested by Schmalz (1969), or (3) deposited in a much shallower isolated Gulf that later subsisded to its present depth. That the salt may have been deposited in shallow depth is suggested by the reddish sediments associated with it.

C. Carbonate Province

1. *General*

Within the carbonate province of the eastern Gulf are the Florida and Campeche platforms (Fig. 17). Both platforms consist of a thick section of Late Jurassic–Cenozoic carbonate and evaporite deposits resting on Paleozoic and Triassic rocks (Paine & Meyerhoff, 1970). Along the northern edge of the Florida Platform the sediment blanket is about 2 km thick grading north-westward into terrigenous deposits. Southward the carbonates and evaporites thicken considerably to exceed 11 km (Paine and Meyerhoff, 1970). The Campeche Platform appears to be tilted westward, displaying the greatest sediment thickness (>3 km) on the western side of the platform.

2. *Shelves*

On the southern end of the shelf off west Florida, the submerged top of the Florida Platform is Florida Bay (Fig. 1). This broad area less than 3 m deep is subdivided by mud banks and separated from the open sea by the Florida Keys. The honeycomb pattern of shallow elongate basins on the bay's floor is believed to have been inherited from drainage systems of fresh-water swamps whose vegetation was destroyed during the Holocene transgression (Price, 1967). The upper Florida Keys south of Miami are floored by the Key Largo Limestone, and the irregular lower keys farther west by the Miami Oolite. The limestone has been interpreted by Hoffmeister *et al.* (1967) as a patch reef that was formed behind a reef belt similar to that fluorishing now along the seaward edge of the Florida Platform. The Miami Oolite is believed to be a lagoonal sediment that was deposited behind the reefs. Radioactive dating of some of the corals from the Key Largo Limestone indicates that the patch reefs were formed 97,000 yr ago when sea level was higher than now (Broecker and Thurber, 1965).

Atop the broad shelf to the north are a number of relict shorelines that mark sea-level stands during the Holocene transgression (Ballard and Uchupi, 1970, Fig. 10). Between Cape San Blas and Mobile is a zone of large sand waves that Hyne and Goodell (1967) believe originated by late Holocene wave erosion of two sand-barrier islands that mark sea-level stands at 21 and 27 m. Along a relict shoreline at a 60 m depth is a series of six ridges trending diagonally to the general contours. These ridges, which lie in water depths between 50 and 70 m, are about 80 km long. The 160-m shoreline is delineated by a prominent split that Jordan and Steward (1959) named Howell Hook (Fig. 10). Associated with these ancient shorelines are small prominences that are believed to be algal reefs (Jordan, 1952; Gould and Stewart, 1955; Ludwick and Walton, 1957; Jordan and Stewart, 1959). These algal reefs are also relict and flourished during low sea-level stands during late Pleistocene and early Holocene. At present these organic prominences support coral and other calcareous or-ganisms.

The shelf north of the Yucatan peninsula consists of a smooth inner zone that terminates near 60 m and an outer zone extending to the top of the slope at about 130 m (Logan *et al.*, 1969). The surface of the outer zone is quite

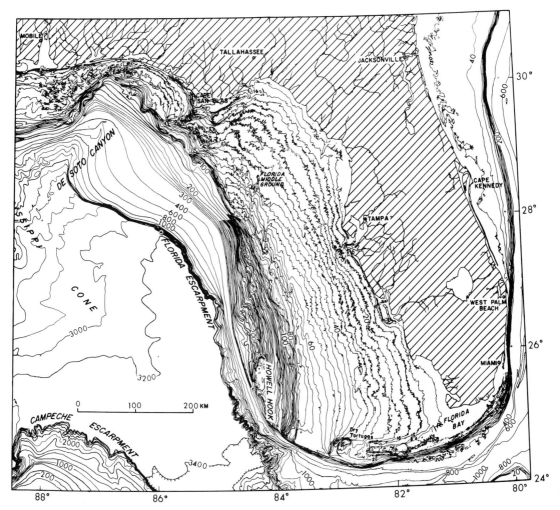

Fig. 10. Bathymetry of the continental terrace off the west coast of Florida. Contours
are in meters. Adopted from Ballard and Uchupi (1970, plate 1) by Emery and Uchupi
(1972, Fig. 34). Published with permission of the American Association of Petroleum
Geologists.

irregular, being characterized by numerous highs and intervening flats. The
greatest concentration of these features is on the shallow end of the terrace,
where they are flanked by a depression on the landward side. As in the West
Florida Shelf the erosional features atop the Campeche Platform are believed
to have been formed during the Holocene transgression. A series of elongate
rocky hills capped by Holocene biostromal deposits separate the two zones
of the shelf. The hills which show evidence of terracing at 31–37 m and planation at 31 m are believed to be sand dunes deposited along shore during the

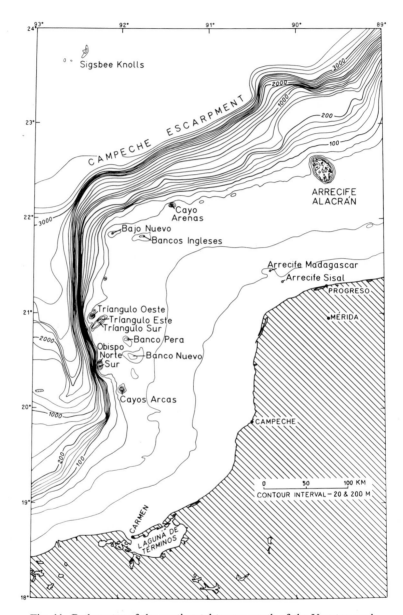

Fig. 11. Bathymetry of the continental terrace north of the Yucatan penin-
sula. Redrawn by Emery and Uchupi (1972, Fig. 41) from Uchupi (1968).
Published with permission of the American Association of Petroleum
Geologists.

Holocene transgression. The most prominent of the reefs rising above the relict dune hills is Alacran Reef (Fig. 11) with its large lagoon containing many irregular reef patches (Kornicker *et al.*, 1959; Kornicker and Boyd, 1962; Bonet, 1967; Folk, 1967; Logan, 1969). Madreporian coral is the chief reef builder, with calcareous algae a minor contributor.

3. *Continental Slope*

The continental slope of both carbonate platforms consists of two segments—a gentle, smooth upper segment and a steep lower one. In some areas the upper segment is disrupted by ridges that may be fault controlled (Jordan and Stewart, 1959; Bergantino, 1971). Seaward of the upper smooth sections are the Florida and Campeche Escarpments which in places have gradients that exceed 30° (Jordan, 1951). North of Latitude 27° the Florida Escarpment is smooth and has a simple northwest–southeast trend, but farther south the scarp is cut by short broad canyons and offsets that may be fault controlled. The Campeche Escarpment trends nearly north–south on the western side of the platform. North of the platform the scarp is cut by numerous valleys and troughs, and between longitudes 89° and 90°30′ W a terrace divides the scarp into an upper 400-m-high and a lower 2400-m segment. The terrace appears to be a down-dropped portion of the upper slope, with the 400-m-high inner scarp being a seaward dipping monocline (Bergantino, 1971). On its east side the scarp is cut by numerous valleys, the largest of which is Catoche Tongue (Fig. 1; Feden *et al.*, 1972). Gravity data (Krivoy and Pyle, 1972) suggest that a similar feature may be present beneath the shelf on the Florida platform. Along the southern flank of this subsurface basin is the Middle Ground Arch (Fig. 17) mapped by Winston (1969), and on the northern flank the seaward extension of the Ocala Arch (Antoine, 1972). Burial of this basin was probably due to reefal progradation.

Seismic reflection profiles across the shelf and upper slope of both carbonate platforms show a sequence of young Tertiary strata resting unconformably on unstratified reef platform deposits (Ewing and Ewing, 1966; Antoine *et al.*, 1967; Uchupi and Emery, 1968; Antoine, 1972; Wilhelm and Ewing, 1972; Fig. 4, Profile 5). On the Florida Platform this unconformity has been traced across the shelf to the Florida Escarpment where sampling and seismic data indicate that the unconformity is between the Miocene and Cretaceous (Antoine *et al.*, 1967; Uchupi and Emery, 1968; Antoine, 1972; Wilhelm and Ewing, 1972). The unconformity is believed to be a result of Laramide disturbance which caused the platforms to be uplifted above sea level and eroded at different times (Wilhelm and Ewing, 1972). Holes drilled along the outer edge of the upper slope encountered Upper Cretaceous and Cenozoic bathyal chalks and oozes (Scientific Staff, 1970; Fig. 7, JOIDES Holes JII-85, JII-86,

JII-94, JII-95, and JII-96). Parts of the Cenozoic are missing from the section which suggests considerable slumping along the scarp. Such an interpretation is verified by rubble encountered in hole 85 (Fig. 7) and exotic blocks observed at the base of the escarpment on some of the seismic profiles.

Seismic reflection profiles across the Florida and Campeche Escarpments show a domal structure associated with the tops of the scarps (Fig. 4, profile 5). Core and dredge samples, and borings (Fig. 7) indicate that this structure represents a Lower Cretaceous reef. The presence of a buried Lower Cretaceous reef from northeast of New Orleans to south Texas (Cullum *et al.*, 1962; Murray, 1961; Coogan, 1972) and in eastern Mexico indicates that a Lower Cretaceous reef probably surrounded the Gulf of Mexico except for a narrow opening at the Isthmus of Tehuantepec. The Lower Cretaceous reef atop Jordan Knoll and the Golden Lane Reef along the eastern margin of the Tamaulipas Platform appear to have been atolls separated from the main reef. The presence of a reef or bank edge in northern Cuba (Furrazola-Bermúdez *et al.*, 1964) indicates that the reef probably extended across the Straits of Florida. From there the reef may have extended to the western tip of the Puerto Rico Trench and then around the Bahamas to the Blake Escarpment where Lower Cretaceous reefal sediments and structures have been reported (Ewing *et al.*, 1966; Heezen and Sheridan, 1966; Emery and Zarudzki, 1967; Sheridan *et al.*, 1969). At present there are no data available on the Caribbean extension of the reef, but the occurrence of possible Mesozoic reefal sediments on the islands off Belize (Stoddart, 1962) suggests that the reef may extend into the Caribbean. Movements associated with the "Laramide" disturbance may have resulted in uplift and exposure of the reef and bank edge deposits from south Florida to north-central Cuba (Bryant *et al.*, 1969). Such uplift and subsequent erosion could account for the numerous canyons on the Florida Escarpment south of 27°, talus deposits present at the base of the Florida Escarpment, the apparent erosion of Jordan Knoll, and the reef, near reef, and bank carbonate clasts of Albian through Maestrichtian age present in the Eocene Cascarajicara Formation along the north side of the Sierra del Rosario in Cuba (Bryant *et al.*, 1969).

III. CARIBBEAN SEA

A. General

The Caribbean is a small inland sea bounded on the south and west by the American continents, and on the north and east by the Caribbean Island Arc. The topography of the sea floor enclosed by these land masses is highly varied. A striking structural feature of the region is a discontinuous belt of seismic activity that extends from the northern coast of South America to the Gulf of Honduras (Schulz and Weyl, 1959; Sykes and Ewing, 1965; Molnar and Sykes, 1969). Within the Caribbean Sea itself seismic activity is nearly absent. The distribution of seismic activity suggests that at present the Caribbean is a relatively rigid plate being deformed along its edges. The boundaries of this rigid plate are the Middle American Trench (a zone of underthrusting) on the west, the Cayman Trough, Old Bahama Channel, and Puerto Rico Trench (a zone of translation) on the north, and the seismic belt through the Lesser Antilles (a zone of underthrusting) on the east. A minor belt of activity extending from Panama to South America (a zone of translation) marks the tectonic boundary of the Caribbean plate on the southwest (Molnar and Sykes, 1969). Another important tectonic element of the area is the unusual crustal structure of the Venezuela and Colombia basins (Fig. 1) which consists of two layers having velocities of 6.1 and 7.2 km/sec atop a mantle having a normal velocity of 8.1 km/sec (Officer *et al.*, 1959; Edgar *et al.*, 1971). Topographically and structurally this complex region can be subdivided into the following physiographic provinces: the northern margin, eastern margin, southern margin, Venezuela Basin–Beata Ridge–Colombia Basin, Nicaraguan Rise–Cayman Trough–Cayman Ridge, and the western Caribbean which consists of the Yucatan Basin and the eastern Yucatan continental margin (Fig. 1).

B. Northern Boundary

1. *The Greater Antillean Ridge*

Along the northern margin of the Caribbean Sea are the islands of Cuba, Hispaniola, Puerto Rico, and the Virgin Islands. These islands are the emergent segments of the Greater Antillean Ridge, a deformed and uplifted Jurassic through early Tertiary orthogeosyncline. The western end of the geosyncline is located near Yucatan Channel, and the eastern end northeast of the Virgin Islands. From the Virgin Islands the geosyncline may turn southeastward (Meyerhoff and Meyerhoff, 1972), since possible pre-Late Jurassic and Late Jurassic eugeosynclinal rocks present in the Greater Antilles have been reported from La Désirade and the escarpment north of the island (Fink, 1970, 1971; Johnston *et al.*, 1971). Only in Cuba are all the elements of an orthogeosyncline present, with a eugeosyncline on the southern side of the island, a median welt, and a miogeosyncline in northern Cuba. East of Cuba only the eugeosyncline is exposed. The western (Cuban) end of the geosyncline may have a continental crust (Khudoley and Meyerhoff, 1972), whereas the rest of the geosyncline is underlain by oceanic crust (Donnelly, 1964, 1967; Bowin, 1966).

Up to 12 km of silicic to mafic igneous rocks and clastics accumulated in the geosyncline (Khudoley and Meyerhoff, 1972). Orogenies in Middle Cretaceous and Middle Eocene led to the breakup of the geosyncline, with the islands assuming their present form by uplift during the Miocene. Holocene deformation of a Pleistocene erosional surface off the Virgin Islands (Donnelly, 1965, 1968) indicates that this uplift may still be taking place. Donnelly has estimated a maximum rate of 5 mm/yr and an average of 1 mm/yr for the Holocene uplift. The Greater Antillean Ridge is nearly bisected by the Cayman Trough between Cuba and Hispaniola, and on the east end the ridge is separated from the Lesser Antilles by a deep graben, the Anegada Trough (Fig. 1). Seismic reflection profiles across the ridge between Hispaniola and Puerto Rico show considerable evidence of tensional faulting, whereas north of the crest the ridge's strata are undisturbed (Garrison and Buell, 1971, Fig. 9, profile 11). Other topographic modifications of this and other ridges are due to Quaternary carbonate deposition, particularly on the shelf south of Cuba (Daetwyler and Kidwell, 1959), and late Pleistocene, early Holocene algal reefs (Macintyre, 1972).

2. *Old Bahama Channel and Puerto Rico Trench*

Immediately north of the Greater Antillean Ridge are the Old Bahama Channel and the Puerto Rico Trench. Both of these topographic features follow the northern margin of the Caribbean plate as delineated by earthquake epicenters. Old Bahama Channel is a topographic extension of the Puerto Rico Trench that can be traced westward to the Straits of Florida (Fig. 1). As a whole, the sediments within the channel are undisturbed except northeast of Hispaniola, where they are folded and faulted (Fig. 12, profile 21). These structures probably are related to the seismicity of the region because focal mechanisms of seismic events in the area indicate underthrusting toward the south and southwest along a plane dipping 50° to 60° (Molnar and Sykes, 1969). East of Old Bahama Channel is the Puerto Rico Trench with a maximum depth of 8414 m (Fig. 1). This, the deepest depression in the Atlantic Ocean, is a structural puzzle. Seismic reflection profiles clearly show that the southeast end of the trench was formed by westward underthrusting of the ocean floor (Chase and Bunce, 1969). North of Puerto Rico, however, no evidence of such underthrusting has been demonstrated to date. Some have proposed that this segment of the trench is the result of the eastward motion of the Caribbean (see, for example, Malfait and Dinkelman, 1972). The undisturbed nature of the trench's sediment fill, the elongate ridges and depressions on the north wall of the trench that resemble fault slices, the elevated plains and the grabens and horsts of the south wall, and the fault block topography along the crest of the Greater Antillena Ridge indicate that vertical motion has been more

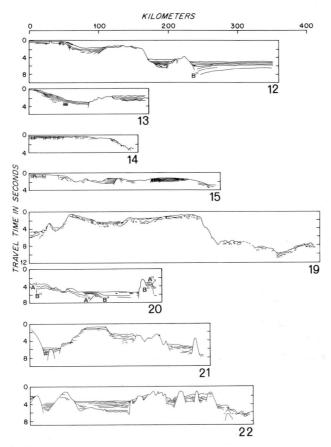

Fig. 12. Seismic reflection profiles from Bahamas and Caribbean. For locations of profiles, see Fig. 16. These profiles were redrawn to a common scale from the following sources: (12) United States Geological Survey (1972*d*, line 34); (13) United States Geological Survey (1972*d*, line 49); (14) Lattimore *et al.* (1971, line AA′); (15) Bassinger *et al.* (1971, line DD′); (19) Bunce *et al.* (1970, lines IV-1 and VI-1); (21) Uchupi *et al.* (1971, line 340); (22) Uchupi *et al.* (1971, line 335).

important than horizontal translation in the formation of the trench (Hersey, 1962; Ewing *et al.*, 1968). The transition from underthrusting to block faulting appears to occur near Long. 62° where the trend of the trench becomes south-easterly [Scientific Staff, 1972*a*; compare profiles 17 and 18 (Fig. 9) with profiles 11 and 19 (Figs. 9 and 12)]. Seismic refraction measurements suggest that the trench may be displaced downward at least 2 km relative to the Puerto Rico Outer Ridge farther north (Bunce and Fahlquist, 1962). As the trench parallels post-Eocene structural trends on Puerto Rico, the depression probably formed concurrently with the uplift of the island in late Miocene (Monroe, 1968).

3. *Bahamas*

Seaward of Old Bahama Channel are the Bahamas, a 1400-km-long archipelago extending from the Florida plateau to the western tip of the Puerto Rico Trench (Fig. 1). These banks are parts of a carbonate platform or para-geosyncline located north of the Greater Antillean geosyncline. Toward the northwest the Jurassic to Holocene carbonates and evaporites appear to rest on a continental crust. East of Great Bahama Bank the banks rest on oceanic crust. Throughout much of the area the sedimentary sequence, as shown by seismic profiles, consists of a poorly stratified, strongly reflecting lower unit and a well stratified upper unit. The lower unit, which at present is restricted to topographic highs, may represent reefal-lagoonal sediments, and the upper unit may represent deep-water carbonate oozes and turbidites (Fig. 9, profile 23; Fig. 12, profiles 21 and 22). Except for a small area north of Cuba and in Exuma Sound, the upper unit is generally undeformed. The steep normal faults in the upper unit north of Cuba (Fig. 9, profile 23) probably reflect the Cuban middle Eocene orogeny (Meyerhoff and Hatten, 1968), and deformation of the unit in Exuma Sound is a result of salt diapirs (Ball *et al.*, 1968). The north-west-trending Bahama Escarpment (Fig. 1) separating the Bahamas from the deep-sea floor appears to be the result of reefal deposition during the Mesozoic. This reef is part of the organic high that extends from the Caribbean to the Blake Plateau (Figs. 1 and 17). The foundation of the reef is a structural high formed during the early opening of the Atlantic (Uchupi *et al.*, 1971). This ridge and similar ridges between the escarpment and Old Bahama Channel served as the foundation for the reefal-lagoonal sediments of the southeastern Bahamas. Continued subsidence of the oceanic basement has restricted reefal deposition to the high, and the lows have become sites of open ocean deposition.

C. Eastern Margin

1. *Barbados Ridge*

Three north–south ridges that are slightly to markedly convex to the east and separated by deep troughs form the eastern margin of the Caribbean Sea (Fig. 9, profiles 16 and 17; Fig. 12, profile 19). The easternmost of these topographic highs is the Barbados Ridge or Anticlinorium, a double ridge with a central syncline (Bassinger *et al.*, 1971; Weeks *et al.*, 1971, Fig. 1). This structural high, which apparently was folded and faulted as a unit, can be traced from north of Barbados Island across the shelf off South America where it terminates against the east–west trending Pilar Block. Geophysical measurements suggest that the basement of the southern section of the ridge consists of rocks similar to the pre-Late Jurassic basement of the Araya–Paria peninsulas of Venezuela and the Cretaceous volcanic sequence of the Venezuela–Netherlands Antilles and Tobago (Meyerhoff and Meyerhoff, 1972). Thus, the southern end of the Barbados Anticlinorium structurally and stratigraphically belongs to the Venezuela–Netherlands Antilles. Sediments atop the basement are folded and faulted, indicating that the processes that formed the ridge were active during the early period of sedimentation. Seismic profiles show that the deformed sediments plunge under the flat-lying deposits of the Tobago Trough and depressions within the anticlinorium.

Barbados Island located 225 km south of the northern terminus of the 10-km-thick sedimentary high is the only emergent segment of the highly deformed sedimentary ridge east of the Lesser Antilles (Fig. 1; Fig. 9, profile 18). The oldest strata exposed on the island, the Paleocene to early Eocene Scotland Formation, consist of terrigenous marine turbidites and synorogenic flysch. The mineral composition and sediment thickness indicate that these deposits were derived from a terrain composed at the southwest of low to high grade metamorphosed acidic rocks and basic volcanic rocks cut by diorite and milky quartz veins (Meyerhoff and Meyerhoff, 1972). The basement, which nears the surface along profile 18 (Fig. 9), probably represents this terrain. On this profile, basement extends to within a short distance of Barbados. Strata of the Scotland Formation are folded, faulted, contain glided olistoliths and olistostomes, and strike east–northeast–west–southwest parallel to the structures of Tobago, Trinidad, and northern South America (Meyerhoff and Meyerhoff, 1972). The nature of the structure of the Scotland Formation led Baadsgaard (1960) to suggest that the sediments were emplaced by gravity sliding from a ridge to the south–southeast. According to Daviess (1971), the block came to rest in a deep trough in which the Oceanic Formation was being deposited. The east–northeast structural trend could be a consequence of this gravitational sliding, although Meyerhoff and Meyerhoff (1972) have postulated that such deformation is due to the north–northwest compressive stress that deformed and uplifted the Caribbean orthogeosyncline in northern South America. Once the block came to rest, the upper portions were subjected to erosion by bottom currents. As the mass slowly subsided, quiet seas prevailed, and the foraminiferal-radiolarian oozes of the middle Eocene–early Miocene Oceanic Formation were deposited. Ash beds within the latest Eocene strata of the Oceanic Formation reflect the volcanism of the Lesser Antilles. The present north–south trend of the ridge is the result of the westward underthrusting of the oceanic crust (Fig. 9, profiles 17 and 18) in the manner described by Chase and Bunce (1969) or the eastward motion of the Caribbean plate (Meyerhoff and Meyerhoff, 1972). Uplift has continued until recently, as attested by the uplifted Pleistocene reefs (Mesollela et al., 1970). Gradual tectonic uplift and eustatic change in sea level resulted in the formation over a period of 500,000 to 1,000,000 yr of a terraced coral cap on Barbados.

2. *Tobago Trough*

East of the Barbados Ridge is the Tobago Trough, an elliptical, asymmetrical depression probably formed as the result of minor adjustments that took place as the sediment apron east of the Lesser Antilles was uplifted and displaced eastward (Bowin, 1972). Seismic profiles across the trough show two sedimentary sequences—a folded lower one and a younger relatively flatlying one. Irregularities of the lower unit which plunges beneath the trough's upper horizontal sediments probably are due to the uplift of the sedimentary Barbados Ridge. Tobago Trough appears to terminate on the north against an east–west structural feature (Weeks *et al.*, 1971). North of the warp the sediments, which are about 1 km thick, are highly disturbed and tilted to the east, and the Barbados Ridge is nearly buried under the sediment blanket (Bunce *et al.*, 1970).

3. *Lesser Antilles Ridge*

The Lesser Antilles Ridge east of Tobago Trough is about 750 km long and about 25–40 km wide at its southern end. North of 16° N the ridge separates into two ridges with a total width of more than 100 km. Atop the outer ridge are the Limestone Caribs, and atop the inner ridge the Volcanic Caribs. The Limestone Caribs are limited to the northern half of the ridge, whereas the volcanic islands extend to within 150 km of Venezuela. The limestone islands consist of middle and late Eocene tilted andesitic tuffs and tuff-breccias intruded by hypabyssal basalt, andesite, and quartz diorite and capped by Oligocene–Miocene limestones (Christman, 1953). Volcanism in the islands, which was initiated in the Eocene, appears to have shifted westward with time and to have terminated in late Eocene or early Oligocene. Volcanism on the inner northern arc was initiated in the Oligocene and has continued to the present. The islands on the southern half of the arc have been sites of volcanism since the Eocene. The volcanic rocks on La Désirade, an east–west trending island at 16° N, are quite different from those on the other islands. On La Désirade the igneous basement consists of easterly striking pillowed spilite, quartz keratophyre, and bedded chert of unknown age that are intruded by Late Jurassic (142.2 ± 9.7 m.y.) trondhjemite (Fink, 1970, 1971). In addition, greenschists and related rocks crop out on the escarpment north of the island (Johnston *et al.*, 1971). Thus, the Lesser Antilles are not solely the result of Eocene–Holocene volcanism as previously believed. As rocks similar to these are common in the Greater Antilles geosyncline (Khudoley and Meyerhoff, 1971), their presence in the Lesser Antilles could imply that the two ridges were a continuous tectonic feature in pre-Eocene time.

4. *Grenada Trough and Aves Swell*

Grenada Trough is a convex eastward depression between the Lesser Antilles Ridge and the Aves Swell that contains over 7 km of sediment (Fig. 1). As in Tobago Trough, the flank sediments of the positive areas on either side dip beneath the flat-lying sediments of the trough. Toward the north the trough's sediment fill becomes thinner and rests on a deformed basement. These strata are more continuous than farther south and can be traced from Aves Swell, across the trough to and up the western slope of the Lesser Antilles (Bunce *et al.*, 1970).

West of Grenada Trough is Aves Swell, a 450-km-long probable volcanic ridge that extends from the Venezuela continental margin to the junction of the Greater and Lesser Antilles (Hurley, 1966). Along the swell's summit, which ranges in depth from 500 to 1300 m, are steep sided north–south trending pedestals having several hundred meters of relief, one of which, Aves Island, rises above sea level. Dredge samples indicate that these pedestals consist of diabase, porphyritic basalt, and metamorphosed basalt (Fox *et al.*, 1971). Potassium–Argon dating of volcanic material from one of the pedestals on the southern end of the swell indicates that the rocks are earliest Tertiary in age (Fig. 13). The extreme southern end of the swell has a granodiorite core

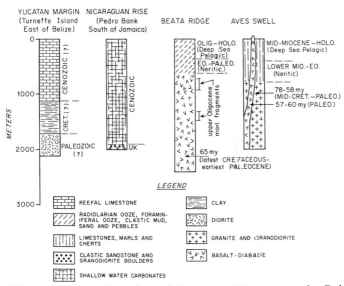

Fig. 13. Stratigraphic sections of the eastern Yucatan margin, Pedro Bank in the Nicaraguan Rise, Beata Ridge (northern end), and Aves Swell (extreme southern end). Based on data from D. R. Stoddart (1971, written communication), Neff (1971), Fox *et al.* (1970), and Fox *et al.* (1971).

(Fox *et al.*, 1971; Fig. 13). This granitic material, which is mid-Cretaceous to Paleocene in age (Fig. 13, Fox *et al.*, 1971), more properly belongs to the east–west structural high forming the foundation of the Netherlands–Venezuela Antilles rather than the swell itself. Thus, the swell is probably a volcanic structure much like the Lesser Antilles Ridge, and the 6.2–6.8 km/sec material forming the core of the high is not granitic, as suggested by Fox *et al.* (1971), but is of the same composition as the oceanic crust.

Data on the sedimentary blanket atop the swell has come from boreholes, cores, dredge samples, and seismic profiling. JOIDES Hole 148 along the western flank of the swell penetrated 272 m of Quaternary–Pliocene deep-water oozes resting unconformably on Miocene, Paleocene, and Cretaceous volcanic sands and minor foraminiferal sands (Scientific Staff, 1971*b*, Fig. 14). Within the pelagic sequence are ash beds indicative of six volcanic episodes, with the middle to late Pleistocene episodes reflecting the formation of the more recent volcanic centers in the Lesser Antilles. The mineral composition of the lower sedimentary unit indicates that it was deposited on the flanks of an emergent volcanic ridge. Separating the two sequences is a thin layer that may have been formed by subaerial erosion. This transition from bathyal environment to subaerial erosion and back indicates that the swell has undergone extensive vertical movement in the recent geologic past. JOIDES Hole 30 near the crest of the swell bottomed in middle Miocene and showed pelagic oozes containing appreciable quantities of volcanic ash and glauconite in the lower part of the section. Indistinct cross-bedding suggests current activity during deposition of the sequence (Scientific Staff, 1971*b*). Dredge samples from one of the pedestals (Fig. 13) consisted of limestones, marls, and cherts. The mid-Eocene to lower Miocene limestones appear to have been deposited in a shallow environment, whereas the mid-Miocene and younger sediments were deposited in a bathyal environment (Fox *et al.*, 1971). Lithified limestone from the crest of the swell and limestone samples from depths as great as 550 m contain concentric stromatoloid structures and other algal features—further evidence that the swell was much shallower in the past (Marlowe *et al.*, 1971). Thus, the following tectonic events can be deduced from the rocks present on the swell. Volcanic activity in earliest Tertiary led to the formation of the swell, which at that time probably extended above sea level. Subaerial erosion was followed by shallow water deposition from late Eocene to middle Miocene. In mid-Miocene the swell began to subside and a deep-water environment was initiated.

The three north–south ridges and associated troughs that form the eastern margin are believed to be the result of the interaction of the Caribbean plate and the ocean basin to the east. Meyerhoff and Meyerhoff (1972) proposed that the ridges were formed by the eastward motion of the Caribbean plate between the Greater Antilles and the Caribbean geosynclines. The first arc to

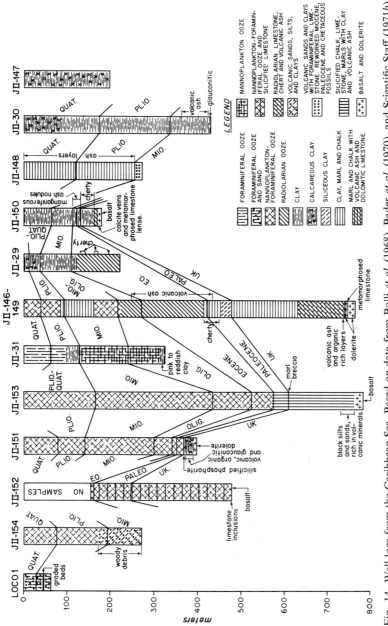

Fig. 14. Well logs from the Caribbean Sea. Based on data from Bolli *et al.* (1968), Bader *et al.* (1970), and Scientific Staff (1971*b*). For locations of wells, see Fig. 16.

be formed was the Aves Swell which was deformed and intruded during the early part of the "Laramide" orogeny. The Lesser Antilles Ridge was formed in Paleocene–Eocene time and superimposed partly on the southeastern inactive end of the Greater Antilles geosyncline (La Désirade Island). Proponents of the sea-floor concept believe that the arcs were formed by the westward under-thrusting of the Caribbean plate by the oceanic plate. According to this concept the east–west ridges were formed as a consequence of shifting of the zone of underthrusting. The first phase of subduction, lasting from Jurassic to Eocene, is reflected by the igneous rocks of the Aves Swell and La Désirade Island of the outer island arc. An eastward shift of the northern edge of the Caribbean plate in latest Eocene initiated volcanic activity along the inner arc. A shift of the whole eastern edge of the plate in late Miocene led to the uplift and deformation of the Barbados Ridge (Bowin, 1972). Thus, the north–south compressional trends and igneous activity in the region could readily be ex-plained by westward underthrusting of the oceanic crust. Not so easily explained under such a tectonic history, however, are the structural trends present in the area that are more or less at right angles to the arcs. These include not only folds, but also thrust faults (Weeks *et al.*, 1971). Among these cross trends are the east–west trend of Aves Island at the northern end of Aves Swell (Weeks *et al.*, 1971), the folds and thrust faults in Barbados that are oriented more or less east–west (probably formed during the emplacement of the Scot-land Formation by gravitational sliding), the east–west St. Lucia cross-warp (Weeks *et al.*, 1971), the east–west folds in Grenada, the southernmost island of the Lesser Antilles, and the east–northeast folded schist and phyllite in northern Tobago Island. Meyerhoff and Meyerhoff (1972) ascribed these structural trends to north–south stresses that deformed the Caribbean geo-syncline prior to the formation of the north–south ridges by east–west stresses.

D. Southern Margin

1. *Caribbean–Andean Geosynclines*

Along the southern boundary of the Caribbean bordering the Precambrian shield is a Mesozoic–Cenozoic orthogeosyncline that extends from Trinidad–Tobago on the east to Colombia on the west (Mencher, 1963; Alvarez, 1971; Bell, 1971, 1972; Case *et al.*, 1971). The orthogeosyncline consists of an outer geosyncline supposedly underlain by oceanic crust and an inner mio-geosyncline underlain by continental crust. Continuity of the geosyncline is disrupted in western Venezuela by the Maracaibo Platform (Fig. 17). Super-position of the western segment of the trough on an earlier Paleozoic geo-syncline has added further structural complications. The presence of Mesozoic eugeosynclinal rocks in the Venezuelan–Netherlands Antilles indicate that the geosyncline extends beyond the present coast to form the foundation of the continental terrace. The tectonic evolution of this trough is briefly described below. (For more details see other articles in this volume.)

A thick sequence of marine and volcanic sediments accumulated in an east–west trough along the present coast of Venezuela during Late Jurassic (?) and Cretaceous. By latest Cretaceous the northern part of this sedimentary-igneous section had been regionally metamorphosed and uplifted. Uplift of the basin resulted in emplacement of large masses in a basin to the south by gravitational sliding. In mid-Eocene this southerly trough was also uplifted, and another basin was formed farther south during the Oligocene. Tilting of the landmass concurrent with downwarping of the basin again resulted in gravitational sliding, this time on a much larger scale.

Cyclic uplift and gravitational sliding gave rise to the structural configuration of the southern coastal zone of Venezuela. Recently Bell (1972) ascribed these tectonic events to sea-floor spreading and suggested the following tectonic history: (1) underthrusting which formed the coastal trough with metamorphism due to frictional heating along a southward dipping Benioff Zone, (2) cessation of underthrusting resulting in the uplift of basinal sediments by isostatic rebound, with the uplift leading to the initial gravitational sliding, and (3) east–west right lateral motion between the Caribbean and American plates along the El Pilar and Oca faults in northern Venezuela, and Bocono fault in western Venezuela (Fig. 17) which began in mid-Eocene and initiated the latter phase of uplift, gravitational sliding, and strike slip faulting. Detailed field mapping by Metz (1965), however, indicates that maximum post-Cretaceous lateral movement along the El Pilar was 15 km and possibly could be less than 5 km.

Tertiary deformation of the eugeosyncline in Colombia gave rise to two smaller depressions. This progressive Tertiary tectonism reached maximum intensity during the Miocene–Pliocene when considerable folding and intrusive and extrusive igneous activity took place. Case *et al.* (1971) suggested that this deformation, which becomes progressively younger westward, is due to underthrusting from the Pacific which migrated westward, reaching its present position along the South American Pacific coast in late Cenozoic. Farther northeast, in the Guajira peninsula, deformation and metamorphism took place from Late Cretaceous to Eocene or early Oligocene. Two phases of deformation separated by deposition of late Eocene sediments can be recognized. Deformation during the first phase suggests uplift of at least two linear belts parallel to the east–west axis of the mountain system and compression perpendicular to it. During the Oligocene this segment of the geosyncline was broken into northwest-trending blocks, and Oligocene and later sediments were deposited in irregular depressions surrounded by uplifted blocks. The younger deformation phase is characterized by gentle folding with axes parallel to earlier structures.

2. *Southern Central American Orogen*

The other structural element of the southern margin of the Caribbean is the Southern Central American orogen (Fig. 17) that extends from eastern Panama to southern Nicaragua to form a link between the South American and the Northern Central American orogen. This tectonic element has an oceanic basement and was initiated during Middle to Late Mesozoic either as a ridge, a volcanic island arc, or a combination of both (Terry, 1956; Lloyd, 1963; Dengo, 1967). Erosion of the ridge, volcanism, and pelagic deposition led to the formation of a complex association of volcanics, cherts, siliceous limestones, and graywackes. Some of the reddish cherty sediments are probably deep-water sediments, and the older highly altered basalts probably represent oceanic basement. This sedimentary sequence was deformed, mildly metamorphosed, and uplifted in mid-Cretaceous. Faulting associated with this deformation resulted in uplift of serpentinized peridotite (possibly mantle) in northern Costa Rica. In Late Cretaceous–Eocene a deep trough was formed on the Caribbean side of the ridge. Deformation during the "Laramide" in mid-Eocene led to the fragmentation of this basin. Igneous activity which lasted from the Jurassic–late Miocene into the Pliocene culminated with granodiorite intrusions along the axis of the Isthmus of Panama in the Miocene. Continuing growth and uplift resulted in the formation in the Pliocene of the first uninterrupted connection between South America and nuclear Central America. This complex tectonic history has been ascribed to the northerly drifting Coco Ridge (Fig. 17) fragments that clog the subduction zone in the Middle American Trench (van Andel *et al.*, 1971), and to underthrusting of the Caribbean plate by the Nazca plate (Bowin and Folinsbee, 1970).

3. *South American–Southern Central American Continental Terrace*

The South American–Southern Central American Continental Terrace consists of a series of linear ridges and troughs parallel to the coast (Athearn, 1965; Lidz *et al.*, 1968; Avdeev and Beloussev, 1971; Ball *et al.*, 1971; Civrieux, 1971; Maloney, 1971; profiles 11–15, Figs. 9 and 12). Seismic reflection data indicate that these structural highs, together with those on land, were formed by vertical movement as a result of northwest–southeast compression and that horizontal translation has been minimal. Compressional stresses probably resulted from the north extension of the Caribbean. Such a movement appears to have been initiated during the Late Cretaceous and to have lasted into the present. The country rock forming the ridges varies considerably. Some of the ridges have cores made of Mesozoic igneous and metamorphic rocks similar to those along the coast; in others both igneous-metamorphic and sedimentary rocks form the foundation of the ridges; and in still others, like the Cariaco Ridge, only sedimentary rocks are involved in their formation. The structural highs have served as dams behind which sediments are being deposited. Some of the sediments, however, do reach the open sea by way of gaps in the ridges after first being transported along the axes of the troughs. The southernmost structural high is the El Pilar Block, a narrow horst bounded on the south and north by fault zones (Weeks *et al.*, 1971, Fig. 12, profile 15; Fig. 17). The block can be traced from 58°30′ W through the Araya–Paria peninsulas to west of Cariaco Basin (Figs. 1 and 17) where it loses its identity. On the east the block appears to have acted as a dam south of which over 4 km of sediment have accumulated.

Seismic refraction measurements indicate that vertical motion along the northern fault zone has been on the order of 2 km, which is comparable to an estimate of 1.8 km along the El Pilar fault in the southern fault zone (Bassinger *et al.*, 1971). The presence of collapse features along the block's crest indicates that the horst has not behaved as a rigid body. Cariaco Basin, the most prominent depression atop the block, is 192 km long, 32 km wide, and 1427 m deep (Maloney, 1971). Cores from the basin display three units, a 5–11 m thick upper dark greenish-gray organic lutite containing foraminifera and pteropods, a 1-m thick unit of firm clay poor in microfossils, and a 1–3 m thick bed of oxidized lutite (Heezen *et al.*, 1958). The lower unit is believed to have been deposited when the basin was ventilated in late glacial time, and the dark upper unit when the basin became stagnated. Radioactive dating indicates that this stagnation began 11,000 years ago. JOIDES Hole 147 (Fig. 14) drilled on the north slope of the basin consists of about 200 m of dark green calcareous clay with high organic content. A change in the planktonic foraminiferal assemblage indicates that the Holocene–Pleistocene boundary is at 7 m below the sea floor. Another change in the planktonic fauna at 42 m has been correlated with the end of the last interglacial period (Scientific Staff 1971*b*). The sedimentation rate for the last 140,000 to 200,000 yr has been estimated at 50 cm per thousand years. Details of the more recent sedimentary history of the Gulf of Paria, the broad reentrant west of Trinidad, have come from high-resolution echo sounding. Within the upper sedimentary sequence are three regionally correlative erosional surfaces separated by well-stratified sediments. The lowest and middle erosional surface meet at 43 m below present sea level, and the middle and upper surfaces at 20 m. Van Andel and Sachs (1964) believe that these surfaces were produced during the late Pleistocene regression and two minor regressions during the Holocene transgression.

North of the eastern end of the El Pilar Block is the north–south trending Barbados Ridge (Fig. 17). This structural high, which apparently was folded and faulted as a single unit, can be traced across the shelf to the northern boundary of the block. The basement rocks and folded sediments that support the positive elements of the ridge plunge seaward and beneath the flat-lying sediments of the structural lows within the ridge. The only emergent segment of this megastructure on the shelf is Tobago Island. Basement on the island is made up of metavolcanics folded in Late Cretaceous and strongly sheared ultramafic and dioritic intrusives and andesitic volcanics emplaced and deformed in earliest Tertiary. Resting unconformably on this basement are Miocene–Pliocene sands and coral limestone (Maxwell, 1948).

The north–south trending Tobago Trough located west of the Barbados high also extends across the shelf depressing it under a volume of sediments over 7 km thick (Edgar *et al.*, 1971). On approaching the El Pilar Block, the trough turns westward, paralleling the horst. The Curapano Sea Valley or Depression located farther west may be a structural extension of the Tobago Trough (Fig. 12, profile 14). This structural low loses its identity westward as it merges with the Cariaco Basin. The Tobago Trough–Curapano Depression is flanked on the north by a structural high that appears to extend from Grenada to about 70° W where it too loses its identity. Grenada Island, the southern terminus of the Lesser Antilles Ridge, is separated from the structural high along the outer edge of the shelf north of Venezuela by a fault-controlled, narrow, V-shaped, northwest-trending channel which widens eastward in the direction of the Tobago Trough. Although topographic continuity between the Lesser Antilles Ridge and the outer shelf's structure cannot be demonstrated, continuity of a 60 to 100 mgal free-air-gravity anomaly suggests that the two features are related (Bassinger *et al.*, 1971). Segments of the ridge along the outer shelf rise above sea level to form islands. Two of these, Los Testigos and Los Frailes, consist of diabasic diorite, quartz diorite, and tuff which are highly fractured and trend parallel to the northwest trend of the island. Isla de Margarita north of the Araya peninsula has a metamorphosed basement similar to the rocks exposed in northern Venezuela (Hess and Maxwell, 1949). The southern portion of the island contains folded Eocene and Miocene sediments. Deformation appears to have occurred at the end of the Cretaceous and between Eocene and Miocene.

La Tortuga Island on the western end of the ridge consists of Pleistocene coral limestone with a thin strip of Pliocene marls cropping out on the southern part of the island (Maloney, 1971). The island was apparently uplifted and tilted northward in late Pleistocene, an indication that the processes that formed the ridges are still active. North of this structural high is the Bonaire Basin, an elongated east–west depression 390 km long and 2050 m deep (Fig. 1). This depression may connect eastward with the Grenada Trough. Westward it may be separated from the Falcon Basin in northern Venezuela by the Pilar Block. Irregularities along the south wall of the basin appear to be the result of gravitational sliding (Scientific Staff, 1972*b*).

The Netherland Antilles are located along the crest of the ridge north of the Falcon–Bonaire–Grenada depression (Fig. 17). These islands have volcanic and metamorphic marine sediments intruded by a Late Cretaceous granitic batholith and capped by limestones and marls of middle Tertiary to Pleistocene age. The Late Cretaceous granite on the extreme southern tip of Aves Swell is probably part of this structural unit. The abundance of pillow basalts, diabase, and cherts on the islands suggest that this structural high is uplifted oceanic crust (Scientific Staff, 1972b). These rocks bear striking similarities to the eugeosynclinal rocks of the Guajira peninsula along the Venezuelan–Colombian border, but they are not quite what one would expect simply by projecting the rocks on the peninsula eastward (Alvarez, 1971). Los Roques Basin (Fig. 1) north of the Netherland Antilles and bounded on the north by Cariaco Ridge has a maximum depth of over 4800 m and is structurally similar to the Bonaire Basin. The Cariaco Ridge together with the Los Roques Basin and the Netherland Antilles Ridge can be traced to the southern end of the Aves Swell where they terminate against a northwest-trending trough that is probably fault controlled. The broad 300-km-long and 50 to 80 km-wide east-west trending Cariaco Ridge is the most prominent structural feature of the area. Seismic refraction measurements indicate that as much as 14 km of low-velocity sedimentary rocks underlie the ridge (Edgar et al., 1971). East of 68° W the entire ridge is deformed, but to the west only the northern slope shows any evidence of deformation (Scientific Staff, 1972b). West of Guajira peninsula the Cariaco Ridge becomes quite subdued, consisting of a fold belt at the base of the continental slope (Fig. 4, profiles 9 and 10, Fig. 15). Off Panama, the folded sequence at the base of the slope is bisected by seaward extensions of the north–south faults extending across Panama. Similar displacements also occur north of Colombia (Krause, 1971). These fold structures appear to be young, since they are penecontemporaneous with turbidites being deposited landward of the fold belt (Krause, 1971). Rapid progradation by the Magdalena Delta off Colombia has added further topographic irregularities to the slope in the area. Such rapid encroachment has resulted in massive slumping and formation of mudlumps similar to those off the passes of the Mississippi Delta (Shepard et al., 1968).

E. Venezuela Basin–Beata Ridge–Colombia Basin

The major structural lows of the eastern Caribbean are the Venezuela and Colombia Basins (Heezen, 1959). Aruba Gap (Fig. 1) south of the Beata Ridge serves as a deep-water connection between them. An unusual feature of these topographic lows is their crustal structure which consists of two crustal layers having velocities of 6.3 and 7.3 km/sec with a total thickness of 8 km underlain by a normal mantle with a velocity of 8.1 km/sec (Officer et al.,

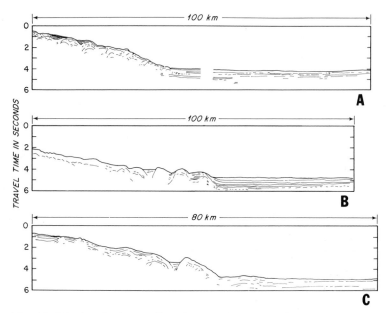

Fig. 15. Seismic reflection profiles of the continental margin and adjacent deep
sea off Colombia and Panama, which display folded structures on the slope.
For locations of profiles, see Fig. 16. Profiles courtesy of Dr. C. O. Bowin of
Woods Hole Oceanographic Institution.

1959; Ewing *et al.*, 1970; Edgar *et al.*, 1971). This unusual crust is probably
not a product of Mesozoic–Cenozoic sea-floor spreading, but is an old oceanic
crust that has been trapped between the drifting American continents (Edgar
et al., 1971). Sediments above this crust range in thickness from less than 2 km
in Venezuela Basin and less than 4 km in Colombia Basin to well over 5 km
along the basin edges. Maximum depths within the Venezuela Basin occur
in the Muertos Trench south of the Greater Antilles Ridge, where they exceed
5200 m (Fig. 1). The southern part of Colombia Basin is dominated by an
extensive deep-sea fan that grades northward into the Colombia Abyssal
Plain. As in the Venezuela Basin, maximum depths in Colombia Basin occur in
small depressions at the base of the Greater Antilles Ridge. Turbidity currents
initiated in the areas of the Magdalena River in Colombia and the Sixaola
River in Panama have deposited appreciable quantities of sand and vegetal
debris within the fan and on the abyssal plain (Heezen *et al.*, 1955). JOIDES
Hole 154 (Fig. 14), drilled on a high above the abyssal plain, is made up of
two sedimentary units—a calcareous upper unit and a volcanic–terrigenous
lower unit. The pelagic upper unit of early Pliocene to Holocene age is a
foraminiferal–nannoplankton ooze containing minor amounts of volcanic
minerals. The lower volcanic sequence of early Pliocene to late Miocene age
is dominated by volcanic turbidite sands, silts, and clays with minor amounts

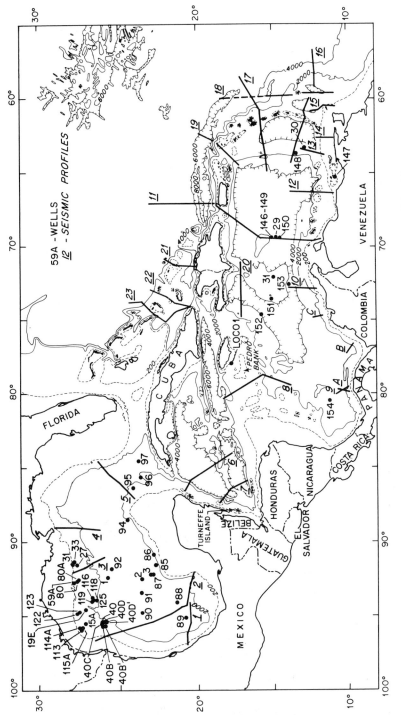

Fig. 16. Chart showing locations of seismic profiles and boreholes described in this report.

of woody debris and reworked fossils (Scientific Staff, 1971*b*). This sharp change from volcanic–terrigenous sediments to pelagic deposition in early Pliocene is probably the result of uplift of the site of deposition.

Seismic reflection profiles in the eastern Caribbean show two prominent reflectors. The lower one, known as horizon B″, is relatively smooth and reflects a coherent echo; and the upper one, horizon A″, is made up of two or three closely spaced horizons (profile 11, Fig. 9). These two horizons define two sedimentary layers—an upper one 0.55 sec thick and a lower one 0.4 sec thick that collectively are known as the Carib beds (Ewing *et al.*, 1967, 1968; Edgar *et al.*, 1971). The reflector between these two beds, A″, is generally stronger and the lower beds are acoustically more opaque than the upper one. Drilling data indicate that the sediments above A″ consist of Cenozoic pelagic oozes, that horizon A″ is made up of early Tertiary (early Eocene–late Paleocene) cherty sediments, that the beds beneath A″ are made up of early Cenozoic–latest Cretaceous pelagic sediments, and that Horizon B″ is the surface of a basalt–diabase layer of latest Cretaceous age (JOIDES Holes 29, 146–149, 150, Fig. 14). Volcanic debris occurs in appreciable quantities on the lower part of the Cenozoic section, with the cherts being the alteration product of the volcanic detritus. It was first suggested that horizon B″ was oceanic basement formed by a spreading process as the American continents drifted apart (JOIDES, 1971), but the baked contacts in holes 146–149, 150, and 152 and the presence of reflecting horizons beneath B″ in the northeast corner of the Venezuela Basin (United States Geological Survey, 1972*c*) indicate that horizon B″ is the top of a sill.

The Carib beds have been traced throughout most of the Venezuela Basin and the northern part of Colombia Basin (profile 20, Fig. 12), but have not been identified in the southern part of Colombia Basin. Throughout much of the area the beds are uniform in acoustic characteristics and do not increase in thickness toward the basins' margins as one would expect if the marginal landmasses had contributed sediments to the basins. Thus, the terrigenous debris eroded from surrounding lands must have been trapped in coastal troughs, allowing formation of the Carib beds in the central parts of the basin. Continuity between the deformed reflectors of the Cariaco Ridge and the horizontal beds of the Venezuela Basin (United States Geological Survey, 1972*d*) suggests that the ridge represents a trough that has been deformed recently. Within the basins themselves the Carib beds are arched up in mid-basin and dip toward the basin edges. This warping has resulted in the formation of narrow troughs along the margins of the basins which have been filled with turbidites to form small abyssal plains. In the Muerto Trench south of the Greater Antilles Ridge, the Carib beds can be traced for a distance of 18 km beneath the Greater Antilles Ridge which forms the north wall of the trench (United States Geological Survey, 1972*c*). Horizons A″ and B″ appear to be offset by small normal faults beneath the trench's sediment fill, and part of the fill itself is folded, uplifted, and incorporated in the Greater Antilles Ridge. Thus, the Muerto Trench could be interpreted as a zone of under-thrusting similar to that of the eastern Puerto Rico Trench (Scientific Staff, 1972a). Along the southern margin of the Venezuela basin the Carib beds also dip beneath young turbidites in the marginal trench, and in some profiles horizon B″ can be traced beneath the Cariaco Ridge for several kilometers (Scientific Staff, 1972b). In places the trench appears to be the result of normal faulting of A″ and B″. Such faulting, where present, parallels the long axis of the trench and is probably the result of upwarping of the beds. In some areas the Carib beds thin as they plunge beneath the wedge of turbidites. If this thinning is real and not the result of masking by strong reverberation by the turbidites (Edgar *et al.*, 1971), it may be the result of erosion by turbidity currents before the deposition of the young turbidites. The structural rela-tionship between the Carib beds and the sediment wedges along the margins of the basins suggests uplift of the basins' center and underthrusting of the basins' margins resulting in the deformation of the rise sediments off South America to form the Cariaco Ridge, uplift of the other southern ridges, uplift of the Greater Antilles Ridge, and possibly uplift of the ridges along the Carib-beans' eastern margin.

The Beata Ridge between the Venezuela and Colombia Basins is a horst tilted to the east and has a core of basalt–diabase (Fox *et al.*, 1970; Scientific Staff, 1971b). Potassium–argon analyses of whole rock samples gave an age of 65 m.y., and of feldspar separates 64 m.y. (Fox *et al.*, 1970). Thus, the basalt in Beata Ridge, Aves Swell, and that forming Horizon B″ are of about the same age and probably reflect the same igneous episode. Sediments atop the ridge's basement consist of Tertiary–Quaternary foraminiferal–nannoplankton oozes and early Tertiary silicified chalks, marls, and limestones (JOIDES Holes 31, 151, and 153, Fig. 14; Fig. 13). The occurrence of shallow-water Eocene limestone atop the ridge suggests that the ridge was uplifted by normal faulting in early Tertiary, and the presence of Oligocene–Holocene deep-water oozes indicates that by Oligocene the ridge had subsided to deep water. Uplift of Beata Ridge, like the Aves Swell, probably is related to the east–west com-pressive stress that formed the other north–south Caribbean ridges.

F. Nicaragua Rise–Cayman Trough–Cayman Ridge

Directly eastward of the Northern Central American orogen are the Nicaragua Rise, Cayman Trough, and Cayman Ridge. Meyerhoff (1966, 1967) has suggested that this rise–trough–ridge system is part of the northern Central American geosynclinal complex which originated in Paleozoic time. Arden (1969), however, believes that the Nicaragua Rise had its origin in early

Mesozoic as a result of rifting and separation of the American–Afro–European continents. Sediment thickness also indicates that the Nicaragua Rise–Cayman Trough–Cayman Ridge system is young, was formed by recent geologic events, and is not relict a feature that has survived since the Paleozoic. The Nicaragua Rise is a broad topographic high that extends from Central America to the western tip of Hispaniola. Its northwest flank is linear and steep, dropping directly into the Cayman Trough. In contrast, its southeast flank is gentle and quite irregular, with slope continuity disrupted by numerous topographic highs. Topographically the rise can be divided into two segments, a south-westerly one which is relatively smooth, and a northeasterly segment made up of highs and lows whose orientation appears to be structurally controlled. Along the northern edge of the smooth segment on the crest of a ridge parallel-ing the Cayman Trough are the Bay Islands. The westernmost of the Bay Island (Utila) consists of alkaline basalts erupted through Quaternary coral reefs (McBirney and Bass, 1969b). In Roatan, the central island, the basement consists of gently folded, probably Paleozoic metamorphic rocks capped by pre-Tertiary limestones and conglomerates. The easternmost island, Guanaja, has a basement of metamorphosed graywacke, silty shale, and chert. In this island, hornblende gabbro is associated with serpentine and aplite dikes. Faults on the islands are oriented both at a high angle to and parallel with Cayman Trough. Those parallel to the trough show no evidence of lateral displacement, but recent vertical movement has resulted in uplift and southward tilting of the islands. Exposures of probable Paleozoic rocks on the islands suggests that the smooth segment of the Nicaragua Rise structurally belongs to the Northern Central American orogen. East of the Bay Islands are the Swan Islands, which also are located along the crest of a ridge that parallels the Cayman Trough. The oldest rocks on these islands consist of Oligocene or early Miocene deep-water carbonate oozes deformed along a northwest structural axis. These deformed sediments are capped by post-early Miocene bank limestones, reef limestones, and beach deposits that strike east–west parallel to the Cayman Trough (United States Geological Survey, 1967). Scattered throughout the smooth segment of the Nicaraguan Rise are other highs, some bearing reefs that appear to be structurally controlled (Stewart et al., 1961, Milliman, 1969). Other tectonic features of the rise include salt (?) piercement structures south of the Bay Islands (Pinet, 1972), a possible eastern extension of the Jurassic Chiapas salt basin, and the northwest dextral faults and associated ultramafic dikes off Honduras (Pinet, 1971).

Within the topographically complex eastern segment of the Nicaragua Rise is the Island of Jamaica. A Cretaceous foredeep trough occupied the northeast of the island, and a Paleozoic (?) "backland" its western end. This "backland" is believed by Khudoley and Meyerhoff (1972) to be an eastward extension of the Northern Central American orogen. Physiographically the island can be divided into: (1) an eastern mountainous region consisting mainly of Mesozoic igneous and metamorphic rocks that mark the site of the Mesozoic foredeep; (2) the northwest-trending Wegwater belt, a trough containing lower-middle Eocene shales, sandstone, and conglomerates that is separated from the mountainous region to the east by the Blue Mountain thrust; and (3) the Cornwall–Middlesex block comprising four-fifths of the island in which are found Mesozoic–Cenozoic terrigenous and carbonate deposits (Zans *et al.*, 1962). This last physiographic province, basically an east–west anticline bounded on the north and south by faults and separated from the Wegwater belt to the east by the Wegwater thrust, is the site of the Paleozoic (?) "backland." The Late Cretaceous deformation of the foredeep and "backland" was accompanied by intrusions of ultramafic rocks and later granodiorites (Chubb and Burke, 1963). Deformation during early–middle Eocene was much more intense, involving strong folding, faulting, and thrust faulting (Zans *et al.*, 1962). In middle Miocene the whole island was uplifted and assumed approximately its present outline. Erosion following this uplift resulted in the development of the relatively wide shelf south of the island by sediment progradation atop a subsiding early Miocene surface (Goreau and Burke, 1966). Tectonic movements did not cease during the Miocene but were renewed during the Pliocene and appear to be continuing today. Recent tectonic activity appears to be concentrated along the southeast end of the thrust faults.

The sea floor from the vicinity of Jamaica to the western tip of Hispaniola is dominated by a series of highs and lows, some of which shoal to less than 50 m (Robinson and Cambray, 1971). The topographic highs have been interpreted as constructional features formed by reefal accretion atop a subsiding early Miocene surface (Burke, 1968), or as due to fracturing related to movements along the Cayman Trough (Meyerhoff, 1966). Pedro Bank, the largest of them, has a Late Cretaceous (?) granodiorite basement capped by about 10 m of granodiorite boulders and calcareous cemented sandstones which in turn are overlain by about 2000 m of shallow-water Tertiary carbonates (Fig. 13). The other topographic highs which trend northeastward can be divided into two types—those that reach to within 50 m of sea level, which are flat-topped and have smooth steep slopes, and those whose rugged crests are 600–800 m deep and whose side slopes are less steep. Differences between these structural highs are believed to be due to reefal buildup and Pleistocene planation by wave erosion. Sediments in the depressions between these structural highs consist of alternating turbidites and pelagic sediments (LOCO 1, Fig. 14). Taxonomic analyses of the samples indicate that the level at 23.54 m correlates with the "Nebraskan–Aftonian" boundary of the Gulf of Mexico (Bolli *et al.*, 1968). Oxygen-isotopic analysis also indicates that at least three glacial and interglacial intervals are recorded in the sediments.

Cayman Trough is a structural low 1700 km long and over 100 km wide extending from the Gulf of Honduras to the Gulf of Gonave in Hispaniola (Banks and Richards, 1969). Its seismicity and rugged topography make this depression one of the major tectonic units of the Caribbean. The trend of the trough ranges from northeast at its western end to about east–west on its eastern end (Fig. 1). This change in trend has been ascribed to a change in drift direction of the Caribbean plate from a northeast direction in Late Cretaceous to easterly in the Eocene (Malfait and Dinkelman, 1972). This change in trend supposedly led to tension in the western end of the trough and upwelling in the mantle, thus accounting for the shallowness of the mantle in the region (Ewing et al., 1960; Bowin, 1968). Geophysical measurements available to date indicate that the trough does not extend across the Greater Antilles Ridge but turns eastward and enters the Enriquillo Cul-de-Sac in southern Hispaniola. The structural low may continue even beyond Hispaniola, with the Anegada Trough south of the Virgin Islands being located along its eastern terminus.

The eastern end of the Cayman Trough's floor is dominated by three deeps, Oriente located south of Cuba, Bartlett south of the Cayman Islands, and Misteriosa Deep located south of the bank of that name. All three deeps reach depths in excess of 6000 m. Farther west the deepest segments of the Cayman Trough occur along the south wall with Swan Deep north of the islands and Bonacca Deep north of the Bay Islands reaching depths in excess of 5200 m. Dredge samples from the eastern part of the trough indicate that granodiorite ranging from fresh to highly sheared and hydrothermally altered crops out on the north and south wall of the trough from 3400 to 600 m (Fox and Schreiber, 1970). This granodiorite, which appears to be overlain in places by sheared chlorite–epidote metabasalt, is probably correlative with the Late Cretaceous granodiorite of Jamaica. Seismic profiles show that the western end of the trough is covered by sediments nearly 2 km thick. The sediments are deformed by numerous normal faults, probably due to recent seismic activity within the trough (Ericson et al., 1972). The sediments thin rapidly eastward until they occur only in isolated depressions. Isolated sediment ponds also occur in places along the side walls of the troughs behind linear spurs that parallel the trough's axis.

The morphology of the trough has been interpreted as due to tension associated with eastward translation of the Caribbean plate (Hess and Maxwell, 1963). Supposedly the Old Bahama Channel and the Puerto Rico Trench, located along the northern boundary of the Caribbean plate, were formed in the same manner. Total eastward migration of the Caribbean plate has been estimated to be from 800 km (Malfait and Dinkelman, 1972) to 1100 km (Pinet, 1972), with post-Eocene motion on the order of 180 to 200 km. Lack of evidence of great lateral motion in the Northern Central American orogen, the northern coast of South America (McBirney, 1963; Metz, 1965; Anderson, 1968; McBirney and Bass, 1969a; Dengo and Bohnenberger, 1969; Kesler, 1971), or the offshore islands (McBirney and Bass, 1969b), together with the blocky topography and the undisturbed sediment fill in the troughs, led Meyerhoff (1966, 1967), Meyerhoff and Meyerhoff (1972), and, more recently, Uchupi (1973) to suggest that tension has been important in formation of the troughs. Lithofacies maps (Mills et al., 1967; Bowin, 1968) suggest that formation of the Cayman Trough began in the Paleocene or possibly earlier. The small amount of sediment fill and the fact that the trough cuts across the Oligocene–early Miocene structural grain of the Swan Islands suggest that the trough was initiated later, probably in the Miocene concurrent with late Tertiary tectonic activity in Guatemala (Vinson, 1962). The Miocene and the post-Miocene tectonism in eastern Cuba, Jamaica, and Hispaniola is probably related to the initial opening of the Cayman Trough.

Northwest of the Cayman Trough is the Cayman Ridge, an uplifted block extending from eastern Cuba to the Gulf of Honduras. Its western end is buried by a sedimentary fan debouching into the Gulf of Honduras. Topographically this horst with its steep slopes can be separated into two segments; an eastern one trending nearly east–west is very narrow and capped by several islands, while a wider western segment is made up of several highs bounded by north–south faults. The Cayman Islands on the ridge's eastern segment are cored by middle Oligocene to Miocene Bluff Limestone (Matley, 1926). Surrounding this core is coral sand and marl of the Pleistocene Ironshore Formation forming a low coastal terrace that ends against raised marine cliffs cut into the Bluff Limestone. Dredge samples indicate that this segment of the ridge may have a granodioritic foundation capped by basalt (Fox and Schreiber, 1970). West of 82° W the ridge is made up of highs and intervening lows which are probably structurally controlled. Two of the highs shoal to less than 200 m (Fig. 1). According to Fahlquist and Davies (1971), early Pleistocene volcanic ash blankets the western end of the ridge.

G. Western Caribbean

The topographic provinces within the western Caribbean include the Yucatan Basin and the continental terrace off the east coast of Yucatan. An abyssal plain occupies much of the floor of the Yucatan Basin, a 4500-m depression north of the Cayman Trough. Southeastward the basin grades into a sedimentary fan that has buried the southwest ends of the Cayman Trough and Ridge. Strata within the basin are undeformed except near the western end of the Cayman Ridge where they are folded and faulted probably as a result of the uplift of the ridge (Fahlquist and Davies, 1971). Seismic reflection profiles in the Yucatan Basin show no evidence of horizons A'' and B''. In this low, the sedimentary sequence consists of a poorly stratified pelagic lower unit and a well-bedded turbidite upper unit. The sedimentary sequence is underlain by a strong, somewhat irregular reflector that is probably the top of the 3.9 km/sec layer delineated by seismic refraction (Ewing *et al.*, 1960). Although the velocity is somewhat low, this layer is probably lithologically similar to basaltic oceanic basement (layer 2) in the deep sea.

Off Yucatan the continental terrace consists of northeast-trending fault ridges that converge southeastward toward the Gulf of Yucatan (Baie, 1970; Dillon and Vedder, 1973; Scientific Staff, 1971*c*; Uchupi, 1973; profile 6, Fig. 9, and profile 7, Fig. 4). The ridge along the edge of the continental terrace north of 21° N turns eastward toward Cuba. Yucatan Channel, the deepwater passageway to the Gulf of Mexico, is cut in this ridge. Continuity of the ridge system is disrupted by east–west troughs that appear to be fault controlled. These gaps serve as conduits for sediments to reach the deep sea. Dredge and drill samples indicate that the marginal ridges are composed of igneous and metamorphic rocks resembling the possible Paleozoic rocks in the Isle of Pines, central and eastern Cuba, and the Paleozoic rocks in the Northern Central American orogen (Hill, 1959; Meyerhoff *et al.*, 1969; Stoddart, 1962). The ridges have served as foundations for reef buildup ranging from 959 to 1219 m. Parts of these reefal structures are probably Caribbean extensions of the Lower Cretaceous reefs that surround the Gulf of Mexico. Sediments ponded behind the ridges are generally undisturbed except in the Gulf of Honduras where they are broken by many small faults. This deformation is probably related to seismic activity in the Cayman Trough.

It has been suggested by Uchupi (1973) that the igneous and metamorphic rocks on the eastern Yucatan margin, the intensely metamorphosed rocks in Cuba, the rocks in the Nicaragua Rise, and the Paleozoic rocks in the Central American orogen are parts of the same geosyncline. Counterclockwise rotation of the Yucatan peninsula (probably due to the northwest migration of the North American continent) beginning in the Jurassic led to the fragmentation of this geosyncline and the formation of the Yucatan margin ridges, Yucatan Basin, and the Cayman Ridge–Cayman Trough–Nicaragua Rise complex.

IV. SUMMARY

The Gulf of Mexico and the Caribbean Sea with their geosynclines, salt piercement structures, gravitational slides, island arcs, active volcanoes, and carbonate platforms probably have the most diverse topography of any segment of the Atlantic Basin (Fig. 17). The terrigenous province of the Gulf of Mexico is dominated by the Gulf Coast geosyncline on the north and the Isthmian embayment to the south. Late Triassic–Middle Jurassic evaporites

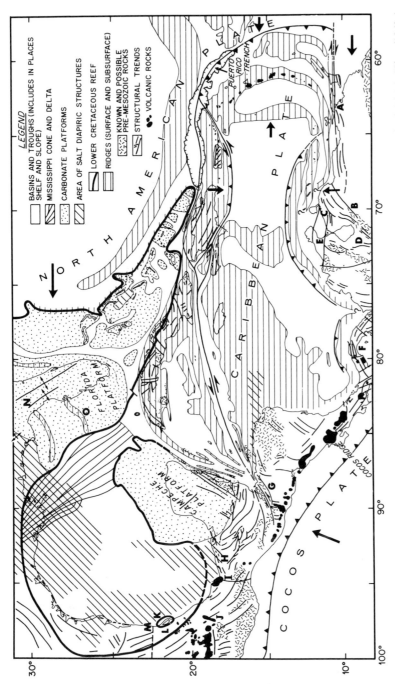

Fig. 17. Structural units of the Gulf of Mexico and the Caribbean. Based on data described in this report, Goddard (1965), King (1969a), and Ball et al. (1968). Major structural features are labeled as follows: (A) El Pilar fault; (B) Bonaco fault; (C) Falcon Basin; (D) Maracaibo Platform; (E) Oca fault; (F) Southern Central America orogen; (G) Northern Central America orogen; (H) Isthmian Embayment; (I) Salina Cruz fault; (J) Mexican Volcanic Belt; (K) Golden Lane; (L) Tamaulipas Platforms; (M) Zacatecas Fracture Zone; (N) Ocala Uplift; (O) Middle Ground Arch.

have given the terrigenous fill of both of these geosynclines considerable mobility. The Mexican Ridges on the western side of the terrigenous province were formed by the seaward migration of the evaporite sequence. Sediments in the abyssal gulf between the geosynclines are about 10 km thick and rest on oceanic crust. These strata, like those on the bordering continental terraces, are, in places, intruded by salt diapirs. The salt in these piercement structures may have migrated basinward from the edges, or it may be in place. The eastern edge of the terrigenous province is formed by the Mississippi Cone, a sedimentary fan deposited by the Mississippi River during the Pleistocene. Salt structures are found at the head of the cone, indicating that the eastern margin of the Triassic–Jurassic evaporite basin is along the western edge of the Florida Platform. The eastern Gulf of Mexico is dominated by the Florida and Campeche Carbonate Platforms. Reef deposition during the Lower Cretaceous and possibly later formed the steep seaward slopes of these carbonate platforms. In the past the carbonate structures extended farther westward (Fig. 17), but terrigenous progradation during the Cenozoic buried them. One buried structure, the Tamaulipas Carbonate Platform, forms the foundation of the continental terrace off eastern Mexico. During the Cretaceous, when the reefs were active throughout the periphery of the Gulf of Mexico, channels in the Isthmus of Tehuantepec and possibly the Straits of Florida connected the nearly enclosed Gulf of Mexico with the open sea.

The northern and southern margins of the Caribbean Sea consist of Mesozoic–Cenozoic geosynclines that were deformed from mid-Cretaceous to mid-Eocene and uplifted during the Miocene. In western Colombia and the Isthmus of Panama, tectonism reached its maximum intensity during the Miocene–Pliocene. These late orogenic movements also may have affected the continental rise off South America, resulting in the formation of the Cariaco Ridge. North of the Greater Antilles Ridge are the Bahama Banks, a carbonate platform atop a fracture zone formed during the opening of the Atlantic, and the Old Bahama Channel and Puerto Rico Trench, collapse features formed concurrently with the uplift of the Greater Antilles Ridge in the Miocene. The north–south ridges along the eastern margin of the Caribbean were formed by either eastward overthrusting of the Caribbean plate or westward underthrusting of the oceanic basin that may have begun as early as Jurassic. The northeast–southwest trough ridge–basin topography of the western Caribbean appears to have been formed by the fragmentation of the Northern Central American orogen. Such fragmentation, which probably began in the Jurassic, was a result of the separation of the Caribbean and North American plates. Present-day seismic and volcanic activity indicates that the Caribbean plate is still being deformed along its edges. Absence of such activity in the Gulf of Mexico, the westernmost Caribbean, and the Bahamas indicates that all three areas are stable today.

V. ORIGIN AND EVOLUTION OF THE GULF OF MEXICO AND THE CARIBBEAN SEA

Of the tectonic models that have been proposed to explain the structure and stratigraphy of the Gulf of Mexico and the Caribbean, only two receive any serious consideration. One group (staticists) believe that the Caribbean plate is an ancient oceanic structure and the presence of Late Jurassic structural elements in Barbados and La Désirade, geologic data which suggest that the position of Barbados relative to South America has not changed for the last 140 m.y., and lack of major translation along the edges of the plate are indicative that this plate has been in its present position relative to the surrounding continental masses for some time (Meyerhoff and Meyerhoff, 1972). The tectonic fabric of this ancient oceanic plate is explained by Meyerhoff and Meyerhoff (1972) as due to Mesozoic–Cenozoic compression between the American continents which deformed the Greater Antilles and Caribbean geosynclines, and to post-Eocene eastward overthrusting of the plate which formed the north–south ridges on the eastern Caribbean. These newer structures are superimposed on the eastern salients of the Greater Antilles and Caribbean geosynclines. The Nicaragua Rise–Cayman Trough–Cayman Ridge system of the western Caribbean is interpreted by Meyerhoff (1966) as eastward extensions of the Northern Central American orogen that originated in the Paleozoic and was also the site of deposition and orogenesis until the Tertiary. According to Paine and Meyerhoff (1970), subsurface and surface data indicate that the Gulf of Mexico has been oceanic since at least the Jurassic and possibly since Late Mississippian–Pennsylvanian.

Proponents of the mobilist concept of global tectonics believe that the tectonic fabric of the Caribbean and possibly the Gulf of Mexico is due to large horizontal motions of the surrounding continents. In the last few years numerous paleogeographic maps of the Caribbean have been compiled by combining sea-floor spreading and paleomagnetic information (Bullard et al., 1965; Funnell and Smith, 1968; Freeland and Dietz, 1971; Phillips and Forsyth, 1972). These maps show considerable overlap of the continents prior to their dispersion. Overlaps in southern Central America, Bahamas, Hispaniola, Puerto Rico, Virgin Islands, and the Lesser Antilles can readily be explained since these features appear to have an oceanic crust and were probably formed during the opening of the Atlantic. The overlap of northern Central America, an area underlain by Paleozoic rocks, by South America is a much more serious problem. Freeland and Dietz (1971) resolved this problem by rotating the overlap fragments into the Gulf of Mexico. However, there is no geologic justification for such a solution. Rodolfo (1971) states that such overlaps are probably due to plastic deformation of the original Pangean configuration during the breakup of this megacontinent. No evidence has been found to

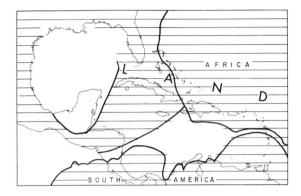

Fig. 18. Paleogeographic map of the Caribbean at the end of the Paleozoic.

verify this. An alternative solution to the overlap problem is that proposed by Drake *et al.* (1968) where the continental fit is made along the seaward margins of the magnetic quiet zone rather than along the continental slope. The proto-Atlantic shown in the Drake *et al.* reconstruction is almost surrounded by Paleozoic orogenic belts. In the paleogeographic reconstruction shown in Fig. 18 I have resolved the overlap by rotating South America to a somewhat greater degree than has been done previously.

Three origins have been proposed by the mobilists to explain the origin of the Caribbean. Some believe that the region was formed by sea floor spreading during the Late Cretaceous (JOIDES, 1971). However, there is no evidence of a spreading center in the region, and horizon B'' appears to be the top of an intrusive body, not oceanic basement formed by a spreading process. Others have stated that the Caribbean is an ancient oceanic plate which was trapped between the drifting American continents (Edgar *et al.*, 1971; Malfait and Dinkelman, 1972) and sealed off on the southwest and east by subduction zones. The Mesozoic–Cenozoic tectonic history of the Caribbean is believed to be related to the northeastward and eastward movements of this plate and to the convergence of the Americas. From Late Cretaceous to Eocene the margins of the northeastward moving Caribbean plate were a complicated trench system. Cuba and Hispaniola were sites of underthrusting by the Caribbean Plate, and farther east along the Puerto Rico Trench and the Aves Swell the Caribbean was a zone of overthrusting. These two zones of over- and underthrusting were separated by a hinge fault located along the Beata Ridge. Another underthrust region was present along the northern coast of South America. As the Caribbean plate began to move eastward in latest Eocene–earliest Oligocene, underthrusting became restricted to the site of the Lesser Antilles, and the northern and southern boundaries were converted from subduction zones to transform faults. At the same time the plate began moving eastward, underthrusting along the Middle American Trench, and decoupled the Caribbean plate from the East Pacific plate. As the Caribbean plate drifted eastward, crustal failure along the northern boundary fault resulted in the formation by tension of the fault block topography of the eastern Yucatan margin.

Another explanation for the origin of the tectonic fabric of the Caribbean is that proposed by Funnell and Smith (1968), Ball and Harrison (1969, 1970) and Ball et al. (1971). They suggested that the Atlantic west of the Mid-Atlantic Ridge consists of two plates—the North American plate north of 10° and the South American plate south of that latitude. The Caribbean region is believed to have been formed by north–south extension and left lateral shear due to (1) divergence of the plates during their westward migration and (2) a faster drift of the North American plate. The faster drift of the North American plate resulted in the underthrusting of the Caribbean, which became attached to South America during its formation. The huge normal faults with rotated blocks and the slumps and gravity slides with resultant thrusts along the southern Caribbean are explained as a result of subsidence into the zone of crustal extension (Ball et al., 1971). Anegada Trough, Puerto Rico Trench, Old Bahama Channel, Enriquillo Cul-de-Sac in southern Hispaniola, Cayman Trough, and deformation of the Greater Antilles geosyncline are products of north–south crustal extension and the interaction between the Caribbean and the North American plate as the North American plate drifted by.

Below I attempt to describe the origin of the Gulf of Mexico and Caribbean in the light of a model somewhat similar to the one above. I will first describe the evolution of the Gulf of Mexico and then that of the Caribbean. When the Atlantic Ocean closed in Late Paleozoic, it was reduced to two small basins, one in the Gulf of Mexico and another in the eastern Caribbean (Fig. 18). These two seas were separated by a landmass formed by the Paleozoic rocks in Yucatan, Cuba, Cayman Ridge, and Nicaraguan Rise. Possibly neither one of these basins existed prior to the closing of the Atlantic, but were formed by rifting of the sialic crust during the deformation and consolidation of the Paleozoic orogens in northern South America around the Gulf of Mexico (Fig. 17). If the basins were formed by rifting, then possibly the simatic layers solidified at shallow depth (Wilhelm and Ewing, 1972). Thus, the pre-Jurassic Gulf region may have been a relatively shallow and restricted epicontinental sea. The Atlantic probably began to open in latest Triassic–earliest Jurassic, as attested by the wide distribution of Late Triassic rift structures in the Americas. Counterclockwise rotation of Florida during the early phase of rifting may have resulted in the formation of the subsurface trough on the West Florida Shelf. During the early phase of the rifting, the Gulf of Mexico, northern Cuba, the Bahamas, and the Nicaragua Rise were sites of evaporite deposition. As the Atlantic widened, the Gulf of Mexico began to subside. The amount of subsidence, much of which occurred during the Cenozoic, ranges from 5 km in the eastern carbonate platforms, 7 km north of the Isthmian embayment, possibly over 10 km in the abyssal Gulf, and 15–16 km in the Gulf Coast geosyncline. This subsidence was slow enough for reefal structures to keep pace with it. Except for gaps in the Isthmus of Tehuantepec and the western approaches to the Straits of Florida, these Jurassic–Cretaceous organic structures encircled the Gulf. The reefs extended beyond the Gulf to the Blake Plateau by way of Cuba and the Bahamas (Fig. 17). Possibly the reefs off eastern Yucatan represent Caribbean extensions of the Gulf of Mexico structure. The Late Cretaceous–middle Eocene deformation of the Rocky Mountains, Northern Central American geosyncline, and Greater Antilles geosyncline had a great impact on the development of the sedimentary framework of the Gulf of Mexico. Erosion of these structures and deposition of vast quantities of Cenozoic detritus along the margins of the Gulf led to the burial of the western carbonate structures and to reduction of the Gulf to its present size. The Mesozoic evaporite sequence beneath this terrigenous embankment gave it considerable mobility. Increase in the rates of subsidence toward the end of the Cretaceous also led to the extinction of the reef structures throughout much of the area.

Tectonic features within the Caribbean are believed to have originated in the following manner. As the continents drifted past the Caribbean plate complex, rift structures, partly atop continental crust (Cuba and western Bahamas) but mainly on oceanic crust, were formed along the plate's northern and southern margins. These features grew eastward with the drifting continents and slowly subsided as the Atlantic widened. The northern structure formed the foundation for the Greater Antilles geosyncline and the Bahama Carbonate Platform. Initially reefal accumulation occurred throughout much of the edge of the rift zone, but with continued subsidence, reef accretion became restricted to topographic highs, and the lows became sites of open ocean deposition. Similarly the southern rift became the site of the Caribbean–Andean geosynclines. Detritus, eroded from the landmass in the western Caribbean, filled the northern rift, and detritus from the pre-Mesozoic orogens filled the southern rift. Continued separation led to fragmentation of the western Caribbean landmass by counterclockwise rotation of the Yucatan peninsula in the Jurassic. In early Tertiary, the northern and southern geosynclines were deformed and the Caribbean became welded to South America. Deformation of the Caribbean edges is probably due to north–south extension as the Americas drifted westward and the Atlantic widened. Periodic right lateral translation along northern South America probably results from intermittent failure of the Caribbean–South American weld, as suggested by Ball *et al.* (1971). Shortly after this deformational phase, the Caribbean began to be underthrust by the faster-moving North American plate. Initiation of this east–west stress appears to have led to the emplacement of the sill whose top forms horizon B″ and the uplift of the Beata Ridge. Crustal failure as the North American plate drifted by the Caribbean led to formation of the tensional features along the Caribbean northern margin.

In the Miocene the Caribbean plate was overthrust by the eastern end of the Greater Antilles Ridge and the South American margin (Fig. 17). This overthrust caused the upwarping of the Venezuela and Colombia basins, deformation of the rise off Venezuela to form the Cariaco Ridge, formation of the narrow depressions along the northern edge of the eastern basins, and continued subsidence of the Puerto Rico Trench and Old Bahama Channel. The basinward migration of the northern and southern margins forced the Caribbean's eastern margin and the subduction zone to shift eastward. This eastward shift led to the uplift of the Barbados Ridge. The Caribbean was isolated from the Pacific by the Pliocene when underthrusting of the Caribbean by the Nazca plate formed the Southern Central American orogen connecting the Andean geosyncline with the Northern Central American orogen.

ACKNOWLEDGMENTS

Financial support for this study came from the National Science Foundation Grant 28193 and the Office of Naval Research Contract N00014-66-CO-241. I wish to thank R. D. Ballard, K. O. Emery, and A. L. Peirson for their suggestions during the preparation of the report.

REFERENCES

Alvarez, W., 1971, Fragmented Andean belt of northern Colombia, in: *Caribbean Geophysical, Tectonic, and Petrologic Studies*, Connelly, T. W., ed.: Geol. Soc. Am. Mem. 130, p. 77–96.

Amery, G. B., 1969, Structure of Sigsbee Scarp, Gulf of Mexico: *Am. Assoc. Petr. Geol. Bull.*, v. 53, p. 2480–2482.

Anderson, T. H., 1968, Pre-Pennsylvanian and later displacements along the Chixoy–Polochic fault trace, northwestern Guatemala: Geol. Soc. Am. Ann. Meet., Abstr., p. 6–7.

Antoine, J. W., 1972, Structure of the Gulf of Mexico, in: *Contributions on the Geological and Geophysical Oceanography of the Gulf of Mexico*, Rezak, R., and Henry, V. J., eds.: Texas A. & M. Univ. Oceanogr. Stud., Gulf Publishing Co., Houston, Texas, v. 3, p. 1–34.

Antoine, J. W., and Bryant, W. R., 1969, Distribution of salt and salt structures in Gulf of Mexico: *Am. Assoc. Petr. Geol. Bull.*, v. 53, p. 2543–2550.

Antoine, J., and Ewing, J., 1962, Seismic refraction measurements of the margins of the Gulf of Mexico: *J. Geophys. Res.*, v. 68, p. 1975–1996.

Antoine, J., Bryant, W. R., and Jones, B. J., 1967, Structural features of continental shelf, slope, and scarp, northeastern Gulf of Mexico: *Am. Assoc. Petr. Geol. Bull.*, v. 51, p. 257–262.

Antoine, J. W., and Pyle, T. E., 1970, Crustal studies in the Gulf of Mexico; *Tectonophysics*, v. 10, p. 477–494.

Arden, D. D., Jr., 1969, Geologic history of the Nicaraguan Rise: *Trans. Gulf Coast Assoc. Geol. Soc.*, v. 19, p. 295–309.

Athearn, W. D., 1965, Sediment cores from Cariaco Trench, Venezuela; Woods Hole Oceanogr. Inst. Tech. Rep. No. 65-37, 20 p.

Avdeev, A. I., and Beloussev, I. M., 1971, On the geomorphology of the Caribbean Sea and the Gulf of Mexico, in: *Symposium on Investigations and Resources of the Caribbean Sea and Adjacent Regions*: UNESCO, Paris, p. 215–224.

Baadsgaard, P. H., 1960, Barbados, W. I.: Exploration results 1950–1958, in: *Structure of the Earth's Crust and Deformation of Rocks*, Kvale, A., and Metzger, A., eds.: 21st Intern. Geol. Congr. Rep., pt. 18, p. 21–27.

Bader, R. G., Gerard, R. D., Benson, W. E., Bolli, H. M., Hay, W. W., Rothwell, W. T., Ruef, M. H., Riedel, W. R., and Sayles, F. L., 1970, Initial reports of the deep sea drilling project. Volume IV, covering Leg 4 of the cruises of the drilling vessel "Glomar Challenger," Rio de Janeiro, Brazil, to San Cristobal, Panama, February–March, 1969, 753 p.

Baie, L. F., 1970, Possible structural link between Yucatan and Cuba: *Am. Assoc. Petr Geol. Bull.*, v. 54, p. 2004–2007.

Ball, M. M., Gaudet, R. M., and Leist, G., 1968, Sparker reflection seismic measurements in Exuma Sound, Bahamas: *Am. Geophys. Union Trans.*, v. 49, p. 196.

Ball, M. M., and Harrison, C. G. A., 1969, Origin of the Gulf and Caribbean and implications regarding ocean ridge extension, migration, and shear, *Trans. Gulf Coast Assoc. Geol. Soc.*, v. 19, p. 287–294.

Ball, M. M., and Harrison, C. G. A., 1970, Crustal plates in the central Atlantic: *Science*, v. 167, p. 1128–1129.

Ball, M. M., Harrison, C. G. A., Supko, P. R., Bock, W., and Maloney, N. J., 1971, Marine geophysical measurements on the southern boundary of the Caribbean Sea, in: *Caribbean Geophysical, Tectonic, and Petrologic Studies*, Donnelly, T. W., ed.: Geol. Soc. Am. Mem. 130, p. 1–33.

Ballard, R. D., and Uchupi, E., 1970, Morphology and Quaternary history of the continental shelf of the Gulf Coast of the United States: *Mar. Sci. Bull.*, v. 20, p. 547–559.

Banks, N. M., and Richards, M. L., 1969, Structure and bathymetry of the western end of the Bartlett Trough, Caribbean Sea, in: *Tectonic Relations of Northern Central America and the Western Caribbean*—The Bonacca Expedition, McBirney, A. R., ed.: Am. Assoc. Petr. Geol. Mem. 11, Tulsa, Oklahoma.

Bassinger, B. G., Harbison, R. N., and Weeks, L. A., 1971, Marine geophysical study of northeast Trinidad–Tobago: *Am. Assoc. Petr. Geol. Bull.*, v. 55, p. 1730–1740.

Bateson, J. H., 1972, New interpretation of geology of Maya Mountains, British Honduras: *Am. Assoc. Petr. Geol. Bull.*, v. 56, p. 956–963.

Bell, J. S., 1971, Tectonic evolution of the central part of the Venezuelan Coast Ranges, in: *Caribbean Geophysical, Tectonic, and Petrologic Studies*, Donnelly, T. W., ed.: Geol. Soc. Am. Mem. 130, p. 107–118.

Bell, J. S., 1971, Geotectonic evolution of the southern Caribbean area, in: *Studies in Earth and Space Science*, Shagam, R., ed.: Geol. Soc. Am. Mem. 132, p. 369–386.

Bergantino, R. N., 1971, Submarine regional geomorphology of the Gulf of Mexico: *Geol. Soc. Am. Bull.*, v. 82, p. 741–752.

Bolli, H. M., Boudreaux, J. E., Emiliani, C., Hay, W. W., Hurley, R. J., and Jones, J. I., 1968, Biostratigraphy and paleotemperatures of a section cored on the Nicaraguan Rise, Caribbean Sea: *Geol. Soc. Am. Bull.*, v. 79, p. 459–470.

Bonet, F., 1967, Biogeologia subsuperficial del Arrecife Alacran, Yucatan: *Univ. Nacional Autonoma de Mexico, Inst. Geologico Bol.* 80, 192 p.

Bouma, A. H., Bryant, W. R., and Antoine, J. W., 1968, Origin and configuration of Alaminos Canyon: *Trans. Gulf Coast Assoc. Geol. Soc.*, v. 18, p. 290–296.

Bouma, A. H., Chancey, O., and Merkel, G., 1972, Alaminos Canyon area, in: *Contributions on the Geological and Geophysical Oceanography of the Gulf of Mexico*, Rezak, R., and Henry, V. J., eds.: Texas A. & M. University Oceanographic Studies, Gulf Publishing Co., Houston, Texas, v. 3, p. 153–179.

Bowin, C. O., 1966, Geology of central Dominican Republic (a case history of part of an island arc), in: *Caribbean Geological Investigations*, Hess, H. H., ed.: Geol. Soc. Am. Mem. 98, p. 11–84.

Bowin, C. O., 1968, Geophysical study of the Cayman Ridge: *J. Geophys. Res.*, v. 73, p. 5159–5173.

Bowin, C., 1972, The Puerto Rico Trench negative gravity anomaly belt, in: *Studies in Earth and Space Science*, Shagam, R., ed.: Geol. Soc. Am. Mem. 132, p. 339–350.

Bowin, C. O., and Folinsbee, A., 1970, Gravity anomalies north of Panama and Colombia: *Trans. Am. Geophys. Union*, v. 51, p. 317.

Brœcker, W., and Thurber, D. L., 1965, Uranium-series dating of corals and oolites from the Bahaman and Florida Key limestones: *Science*, v. 149, p. 58–60.

Bryant, W., Antoine, J., Ewing, M., and Jones, B., 1968, Geology and geophysics of the Mexican continental shelf and slope, Gulf of Mexico: *Am. Assoc. Petr. Geol. Bull.*, v. 52, p. 1204–1228.

Bryant, W. R., Meyerhoff, A. A., Brown, N. K., Jr., Furrer, M. A., Pyle, T. E., and Antoine, J. W., 1969, Escarpments, reef trends, and diapiric structures, eastern Gulf of Mexico: *Am. Assoc. Petr. Geol. Bull.*, v. 53, p. 2506–2542.

Bullard, E., Everett, J. E., and Smith, A. G., 1965, IV. The fit of the continents around the Atlantic, in: *A Symposium on Continental Drift*, Blackett, P. M. S., Bullard, E., and Runcorn, S. K., organizers: *Phil. Trans. Roy. Soc. London, Ser. A*, v. 258, p. 41–51.

Bunce, E. T., and Fahlquist, D. A., 1962, Geophysical investigation of the Puerto Rico Trench and Outer Ridge: *J. Geophys. Res.*, v. 67, p. 3955–3972.

Bunce, E. T., Phillips, J. D., Chase, R. L., and Bowin, C. O., 1970, The Lesser Antilles arc and the eastern margin of the Caribbean Sea, in: *The Sea. Volume 4. New Concepts of Sea Floor Evolution. Part II. Regional Observations Concepts*, Maxwell, A. E., ed.: Wiley–Interscience, New York, p. 359–385.

Bunce, E. T., Phillips, J. D., and Chase, R. L., 1974, Geophysical study of Antilles Outer Ridge, Puerto Rico Trench, and northeast margin of Caribbean Sea: *Am. Assoc. Petr. Geol. Bull.*, v. 58, p. 106–123.

Burk, C. A., Ewing, M., Worzel, J. L., Beall, A. O., Jr., Berggren, W. A., Bukry, D., Fischer, A. G., and Pessagno, E. A., Jr., 1969, Deep-sea drilling into the Challenger Knoll, central Gulf of Mexico: *Am. Assoc. Petr. Geol. Bull.*, v. 53, p. 1338–1347.

Burke, K., 1968, Drowned dwarf atolls south of Jamaica: *Am. Assoc. Petr. Geol. Bull.*, v. 52, p. 348–350.

Carsey, J. B., 1950, Geology of the gulf coastal area and continental shelf: *Am. Assoc. Petr. Geol. Bull.*, v. 34, p. 361–385.

Case, J. E., Duran S. L. G., Lopez, R. A., and Moore, W. R., 1971, Tectonic investigations in western Colombia and eastern Panama: *Geol. Soc. Am. Bull.*, v. 82, p. 2685–2712.

Chase, R. L., and Bunce, E. T., 1969, Underthrusting of the eastern margin of the Antilles by the floor of the western North Atlantic Ocean, and origin of the Barbados Ridge: *J. Geophys. Res.*, v. 74, p. 1413–1420.

Christman, R. A., 1953, Geology of St. Bartholomew, St. Martin, and Anguilla, Lesser Antilles: *Geol. Soc. Am. Bull.*, v. 64, p. 65–96.

Chubb, L. J., and Burke, K., 1963, Age of the Jamaican granodiorite: *Geol. Mag.*, v. 100, p. 524–532.

Civrieux, J. M. Sellier de, 1971, Cañones submarinos frente al la cordillera de la costa de Venezuela, in: *Symposium on Investigations and Resources of the Caribbean Sea and Adjacent Regions*: UNESCO, Paris, p. 291–295.

Collette, B. J., Ewing, J. I., Lagaay, R. A., and Truchan, M., 1969, Sediment distribution in the oceans: The Atlantic between 10° and 19° N: *Mar. Geol.*, v. 7, p. 279–345.

Coogan, A. H., Debout, D. G., and Maggio, C., 1972, Depositional environments and geologic history of Golden Lane and Poza Rica trend, Mexico, an alternative view: *Am. Assoc. Petr. Geol. Bull.*, v. 56, p. 1419–1447.

Creager, J. S., 1953, Submarine topography of the continental slope of the Bay of Campeche: Texas A. & M. Tech. Rep. Ref. 53-10, 23 p.

Creager, J. S., 1958a, Bathymetry and sediments of the Bay of Campeche: Texas A. & M. Tech. Rep. 58-12F, 188 p.

Creager, J. S., 1958b, A canyon-like feature in the Bay of Campeche: *Deep-Sea Res.*, v. 5, p. 169–172.

Cullum, M., Granata, W., Gayer, S., Heffner, R., Pike, S., Herrmann, L., Meyertons, C., and Siegler, G., 1962, The basin frontiers and limits for exploration in the Cretaceous System of central Louisiana: *Trans. Gulf Coast Assoc. Geol. Soc.*, v. 12, p. 97–115.

Curray, J. R., 1960, Sediments and history of Holocene transgression, continental shelf, northwest Gulf of Mexico, in: *Recent Sediments, Northwest Gulf of Mexico,* Shephard, F. P., Phleger, F. B., and van Andel, Tj. H., eds.: Am. Assoc. Petr. Geol., Tulsa, Oklahoma, p. 221–226.

Daetwyler, C. C., and Kidwell, A. L., 1959, The Gulf of Batabano, a modern carbonate basin: Fifth World Petroleum Congress, Section 1, Paper 1, p. 1–21.

Davies, S. N., 1971, Barbados: A major submarine gravity slide: *Geol. Soc. Am. Bull.*, v. 82, p. 2593–2602.

Dengo, G., 1967, Geological structure of Central America, in: *Studies in Tropical Oceanography, No. 5*: Inst. Mar. Sci., Univ. Miami, Miami, Florida, p. 56–73.

Dengo, G., 1969, Problems of tectonic relations between Central America and the Caribbean: *Trans. Gulf Coast Assoc. Geol. Soc.*, v. 19, p. 311–320.

Dengo, G., and Bohnenberger, O., 1969, Structural development of northern Central America, in: *Tectonic Relations of Northern Central America and the Western Caribbean—The Bonacca Expedition*, McBirney, A. R., ed.: Am. Assoc. Petr. Geol. Mem. 11, Tulsa, Oklahoma, p. 203–220.

Deuser, W. G., 1970, Hypothesis of the formation of the Scotia and Caribbean seas; *Tectonophysics*, 10, p. 391–401.

Dillon, W. P., and Vedder, J. G., 1973, Structure and development of the continental margin of British Honduras: *Geol. Soc. Am. Bull.*, v. 84, p. 2713–2732.

Donnelly, T. W., 1964, Evolution of eastern Greater Antilles island arc: *Am. Assoc. Petr. Geol. Bull.*, v. 48, p. 680–696.

Donnelly, T. W., 1965, Sea-bottom morphology suggestive of post-Pleistocene tectonic activity of the eastern Greater Antilles: *Geol. Soc. Am. Bull.*, v. 76, p. 1291–1294.

Donnelly, T. W., 1967, Some problems of island-arc tectonics, with reference to the northeastern West Indies, in: *Studies in Tropical Oceanography, No. 5*: Inst. Mar. Sci., Univ. Miami, Miami, Florida, p. 74–87.

Donnelly, T. W., 1968, Caribbean island-arcs in light of the sea-floor spreading hypothesis: *Trans. N.Y. Acad. Sci.*, v. 30, p. 745–750.

Drake, C. L., Ewing, J. I., and Stockard, H., 1968, The continental margin of the eastern United States: *Can. J. Earth Sci.*, v. 5, p. 993–1010.

Edgar, N. T., Ewing, J. I., and Hennion, J., 1971, Seismic refraction and reflection in Caribbean Sea: *Am. Assoc. Petr. Geol. Bull.*, v. 55, p. 833–870.

Emery, K. O., 1963, Coral reefs off Veracruz, Mexico: *Geofis. Intern., Rev. Union Geofis. Mex.*, v. 3, p. 11–17.

Emery, K. O., and Uchupi, E., 1972, The western North Atlantic—its topography, rocks, structure, water, life, and sediments: Am. Assoc. Petr. Geol. Mem. 17.

Emery, K. O., and Zarudzki, E. F. K., 1967, Seismic reflection profiles along the drill holes on the continental margin off Florida: U.S. Geol. Surv. Prof. Paper 581A, 8 p.

Ensminger, H. R., and Matthews, J. E., 1972, Origin of salt domes in Bay of Campeche, Gulf of Mexico, *Am. Assoc. Petr. Geol. Bull.*, v. 56, p. 802–807.

Ericson, A. J., Helsley, C. E., and Simmons, G., 1972, Heat Flow and continuous seismic profiles in the Cayman Trough and Yucatan Basin: *Geol. Soc. Am. Bull.*, v. 83, p. 1241–1260.

Ewing, J., Worzel, J. L., and Ewing, M., 1962, Sediments and oceanic structural history of the Gulf of Mexico: *J. Geophys. Res.*, v. 67, p. 2509–2527.

Ewing, J., Antoine, J., and Ewing, M., 1960, Geophysical measurements in the western Caribbean Sea and in the Gulf of Mexico: *J. Geophys. Res.*, v. 65, p. 4087–4126.

Ewing, J., Ewing, M., and Leyden, R., 1966, Seismic-profiler survey of Blake Plateau: *Am. Assoc. Petr. Geol. Bull.*, v. 50, p. 1948–1971.

Ewing, J., Talwani, M., Ewing, M., and Edgar, L., 1967, Sediments of the Caribbean, in: *Studies in Tropical Oceanography, No. 5*: Inst. Mar. Sci., Univ. Miami, Miami, Florida, p. 88–102.

Ewing, J., Talwani, M., and Ewing, M., 1968, Sediment distribution in the Caribbean Sea, *Trans. Fourth Caribbean Geol. Conf., Trinidad*, p. 317–323.

Ewing, J. I., Edgar, N. T., and Antoine, J. W., 1970, Structure of the Gulf of Mexico and Caribbean Sea, in: *The Sea. Volume 4. New Concepts of Sea Floor Evolution. Part II, Regional Observations Concepts*, Maxwell, A. E., ed.: Wiley–Interscience, New York, p. 321–358.

Ewing, M., and Antoine, J., 1966, New seismic data concerning sediments and diapiric structures in Sigsbee Deep and upper continental slope, Gulf of Mexico: *Am. Assoc. Petr. Geol. Bull.*, v. 50, p. 479–504.

Ewing, M., and Ewing, J., 1966, Geology of the Gulf of Mexico, *2nd Marine Tech. Soc. Conf. Trans.*, June 27–29, 1966, Washington, D.C., p. 145–164.

Ewing, M., Worzel, J. L., Ericson, D. B., and Heezen, B. C., 1955, Geophysical and geological investigations in the Gulf of Mexico. Part I: *Geophysics*, v. 20, p. 1–18.

Ewing, M., Ericson, D. B., and Heezen, B. C., 1958, Sediments and topography of the Gulf of Mexico, in: *Habitat of Oil*, Weeks, L., ed.: Am. Assoc. Petr. Geol., Tulsa, Oklahoma, p. 995–1053.

Ewing, M., Lonardi, A. G., and Ewing, J. I., 1968, The sediments and topography of the Puerto Rico Trench and Outer Ridge, *Trans. Fourth Caribbean Geol. Conf., Trinidad*, p. 325–334.

Ewing, M., Worzel, J. L., Beall, A. O., Berggren, W. A., Bukry, D., Burk, C. A., Fischer, A. G., and Pessagno, E. A., Jr., 1969, Initial reports of the deep sea drilling project. Volume 1, covering Leg I of the cruises of the drilling vessel "Glomar Challenger," Orange, Texas, to Hoboken, New Jersey, August–September, 1968, 672 p.

Fahlquist, D. A., and Davies, D. K., 1971, Fault-block origin of the western Cayman Ridge, Caribbean Sea: *Deep-Sea Res.*, v. 18, p. 243–253.

Feden, R. H., Ensminger, H. R., and Massingill, J. V., 1972, Geophysical investigations of the Catoche Tongue, Gulf of Mexico: *Geol. Soc. Am. Bull.*, v. 83, p. 1157–1162.

Fink, L. K., Jr., 1970, Evidence for the antiquity of the Lesser Antilles arc: *Trans. Am. Geophys. Union*, v. 51, p. 326–327.

Fink, L. K., Jr., 1971, Evidence in the eastern Caribbean for mid-Cenozoic cessation of sea floor spreading: *Trans. Am. Geophys. Union*, v. 52, p. 251.

Fisk, H. N., 1952, Geological investigation of the Atchafalaya Basin and the problems of the Mississippi River diversion: U.S. Army, Corps of Engineers, Waterways Expt. Sta., v. 1, 145 p.; v. 2, 36 p.

Folk, R. L., 1967, Sand cays of Alacran reef, Yucatan, Mexico: Morphology: *J. Geol.*, v. 75, p. 412–437.

Fox, P. J., and Schreiber, E., 1970, Granodiorites from the Cayman Trench: Geol. Soc. Am. Ann. Meet., Abstr., p. 553.

Fox, P. J., Ruddiman, W. F., Bryan, W. B. F., and Heezen, B. C., 1970, The geology of the Caribbean crust, I: Beata Ridge: *Tectonophysics*, v. 10, p. 495–513.

Fox, P. J., Schreiber, E. S., and Heezen, B. C., 1971, The geology of the Caribbean crust: Tertiary sediments, granitic and basic rocks from the Aves Ridge: *Tectonophysics*, v. 12, p. 89–109.

Freeland, G. L., and Dietz, R. S., 1971, Plate tectonic evolution of Caribbean-Gulf of Mexico region; *Nature*, v. 232, p. 20–23.

Funnell, B. M., and Smith, A. G., 1968, Opening of the Atlantic Ocean: *Nature*, v. 219, p. 1328–1333.

Furrazola-Bermúdez, G., Judoley, C. M., Mijailóvskaya, M. S., Miroliúbov, Y. S., Novojatsky, I. P., Jiménez, A. N., and Solsona, J. B., 1964, *Geologia de Cuba*: Ministerio de Industrias, Inst. Cubano de Recursos Minerales, La Habana Cuba, 239 p. plus map folio volume.

Garrison, L. E., and Buell, M. W., Jr., 1971, Sea-floor structure of the eastern Greater Antilles, in: *Symposium on Investigations and Resources of the Caribbean Sea and Adjacent Regions*: UNESCO, Paris, p. 241–245.

Garrison, L. E., and Berryhill, H. L., 1970, Possible seaward extension of the San Marcos Arch: Geol. Soc. Am., South-Central Section, 4th Ann. Meet., Abstr., p. 285–286.

Goddard, E. N. (Chairman), 1965, Geologic map of North America: U.S. Geol. Surv., scale 1:5,000,000, 2 sheets.

Goreau, T., and Burke, K., 1966, Pleistocene and Holocene geology of the island shelf near Kingston, Jamaica: *Mar. Geol.*, v. 4, p. 207–225.

Gough, D. I., 1967, Magnetic anomalies and crustal structure in eastern Gulf of Mexico, *Am. Assoc. Petr. Geol. Bull.*, v. 51, p. 200–211.

Gould, H. R., and Stewart, R. H., 1955, Continental terrace sediments in the northeastern Gulf of Mexico, in: *Finding Ancient Shorelines*, Hough, J. L., ed.: Spec. Publ. 3, Soc. Econ. Paleontologists and Mineralogists, Tulsa, Oklahoma, p. 2–20.

Guzman, E. J., and de Cserna, A., 1963, Tectonic history of Mexico, in: *Backbone of the Americas*, Childs, O. E., and Beebe, B. W., eds.: Am. Assoc. Petr. Geol. Mem. 2, Tulsa, Oklahoma, p. 113–129.

Hall, I. H. S., and Bateson, J. H., 1972, Late Paleozoic lavas in Maya Mountains, British Honduras, and their possible regional significance: *Am. Assoc. Petr. Geol. Bull.*, v. 56, p. 950–963.

Harbison, R. N., 1968, Geology of De Soto Canyon, *J. Geophys. Res.*, v. 73, p. 5175–5185.

Heezen, B. C., 1959, Some problems of Caribbean submarine geology, *Second Caribbean Geol. Conf., Mayaguez, Puerto Rico*, p. 12–16.

Heezen, B. C., and Sheridan, R. E., 1966, Lower Cretaceous rocks (Neocomian–Albian) dredged from Blake Escarpment; *Science*, v. 154, p. 1644–1647.

Heezen, B. C., Ewing, M., and Menzies, R. J., 1955, The influence of submarine turbidity currents on abyssal productivity: *Oikos*, v. 6, p. 170–182.

Heezen, B. C., Menzies, R. J., Broecker, W. S., and Ewing, M., 1958, Date of stagnation of the Cariaco Trench, southeast Caribbean: *Geol. Soc. Am. Bull.*, v. 68, p. 1579.

Heilprin, A., 1890, The corals and coral reefs of the western waters of the Gulf of Mexico: *Proc. Acad. Nat. Sci. Philadelphia*, v. 42, p. 303–316.

Hersey, J. B., 1962, Findings made during the 1961 cruise of *Chain* to the Puerto Rico Trench and Caryn sea mount: *J. Geophys. Res.*, v. 67, p. 1109–1116.

Hess, H. H., and Maxwell, J. C., 1949, Geological reconnaissance of the island of Margarita. Part 1: *Geol. Soc. Am. Bull.*, v. 60, p. 1857–1868.

Hess, H. H., and Maxwell, J. C., 1963, Caribbean research project: *Geol. Soc. Am. Bull.*, v. 64, p. 1–6.

Hill, P. A., 1959, Geology and structure of the northwest Trinidad Mountains, Las Villas province: *Geol. Soc. Am. Bull.*, v. 70, p. 1459–1478.

Hoffmeister, J. E., Stockman, K. W., and Multer, H. G., 1967, Miami limestone of Florida and its recent Bahamian counterpart: *Geol. Soc. Am. Bull.*, v. 78, p. 175–190.

Huang, Ter-Chien, and Goodell, H. G., 1970, Sediments and sedimentary processes of eastern Mississippi Cone, Gulf of Mexico: *Am. Assoc. Petr. Geol. Bull.*, v. 54, p. 2070–2100.

Hurley, R. J., 1966, Geological studies of the West Indies, in: *Continental Margins and Island Arcs*, Poole, W. H., ed.: Geol. Survey of Canada, Ottawa, Canada, p. 139–150.

Hyne, N. J., and Goodell, H. G., 1967, Origin of the sediments and submarine geomorphology of the inner continental shelf off Choctawatchee Bay, Florida: *Mar. Geol.*, v. 5, p. 299–313.

Johnston, T. H., Schilling, J. G., Oji, Y., and Fink, L. K., Jr., 1971, Dredged greenstones from the Lesser Antilles island arc: *Trans. Am. Geophys. Union*, v. 51, p. 246.

JOIDES, 1971, Scientists find the Caribbean not as ancient as believed: Scripps Inst. Oceanogr. Release No. 162, 5 p.

Jordan, G. F., 1951, Continental slope off Apalachicola Bay, Florida, *Am. Assoc. Petr. Geol. Bull.*, v. 35, p. 1978–1993.

Jordan, G. F., Reef formation in the Gulf of Mexico off Apalachicola Bay, Florida; *Geol. Soc. Am. Bull.*, v. 63, p. 741–744.

Jordan, G. F., and Steward, H. B., Jr., 1959, Continental slope off southwest Florida: *Am. Assoc. Petr. Geol. Bull.*, v. 43, p. 974–991.

Kesler, S. E., 1971, Nature of ancestral orogenic zone in nuclear Central America: *Am. Assoc. Petr. Geol. Bull.*, v. 55, p. 2116–2129.

Khudoley, K. M. and Meyerholff, A. A., 1971, Paleogeography and geological history of Greater Antilles: Geo., Soc. Am. Mem. 129, 199 p.

King, P. B., 1969a, Tectonic Map of North America: U.S. Geol. Surv., scale 1:5,000,000, 2 sheets.

King, P. B., 1969b, The tectonics of North America—a discussion to accompany the Tectonic Map of North America: U.S. Geol. Surv. Prof. Paper 268, scale 1:500,000,000, 94 p.

Kirkland, D. W., and Gerhard, J. E., 1971, Jurassic salt, central Gulf of Mexico, and its temporal relation to circum-Gulf evaporites, *Am. Assoc. Petr. Geol. Bull.*, v. 55, p. 680–686.

Kolb, C. R., and van Lopik, J. R., 1966, Depositional environments of the Mississippi River deltaic plain—southeastern Louisiana, in: *Deltas in Their Geologic Framework*, Shirley, M. L., and Ragsdale, J. A., eds.: Houston Geol. Soc., Houston, Texas, p. 17–61.

Kornicker, L. S., and Boyd, D. W., 1962, Shallow-water geology and environments of Alacran Reef complex, Campeche Bank, Mexico: *Am. Assoc. Petr. Geol. Bull.*, v. 46, p. 640–673.

Kornicker, L. S., Bonet, F., Cann, R., and Hoskin, C. M., 1959, Alacran Reef, Campeche Bank, Mexico: *Publ. Inst. Mar. Sci. Univ. Tex.*, v. 6, p. 1–22.

Krause, D. C., 1971, Bathymetry, geomagnetism, and tectonics of the Caribbean Sea north of Columbia, in: *Caribbean Geophysical, Tectonics and Petrologic Studies*, Donnelly, T. W., ed.: Geol. Soc. Am. Mem. 130, p. 35–54.

Krivoy, H. L., and Pyle, T. E., 1972, Anomalous Crust beneath West Florida shelf, *Am. Assoc. Petr. Geol. Bull.*, v. 56, p. 107–113.

Lafayette and New Orleans Geological Society, 1968, Geology of natural gas in south Louisiana, Meyerhoff, A. A., ed., in: *Natural Gases of North America*, Beebe, B. W., and Curtis, B. F., eds.: Am. Assoc. Petr. Geol. Mem. 9, Tulsa, Oklahoma, p. 376–581.

Lattimore, R. K., Weeks, L. A., and Mordock, L. W., 1971, Marine geophysical reconnaissance of continental margin north of Paria peninsula, Venezuela: *Am. Assoc. Petr. Geol. Bull.*, v. 55, p. 1719–1729.

Lehner, P., 1969, Salt tectonics and Pleistocene stratigraphy on continental slope of northern Gulf of Mexico: *Am. Assoc. Petr. Geol. Bull.*, v. 53, p. 2431–2479.

Lidz, L., Ball, M. M., and Charm, W., 1968, Geophysical measurements bearing on the problem of the El Pilar fault in northern Venezuela offshore: *Mar. Sci. Bull.*, v. 18, p. 545–560.

Lloyd, J. J., 1963, Tectonic history of the South-Central-American orogen, in: *Backbone of the Americas*, Childs O. E., and Beebe, B. W., eds.: Am. Assoc. Petr. Mem. 2, Tulsa, Oklahoma, p. 88–100.

Logan, B. W., 1969, Coral reef and banks, Yucatan Shelf, Mexico, in: *Carbonate Sediments and Reefs, Yucatan Shelf, Mexico*, Logan, B. W., ed.: Am. Assoc. Petr. Geol. Mem. 11, Tulsa, Oklahoma, p. 129–198.

Logan, B. W., Harding, J. L., Ahr, W. M., Williams, J. D., and Snead, R. G., 1969, Late Quaternary sediments of Yucatan Shelf, Mexico, in: *Carbonate Sediments and Reefs, Yucatan Shelf, Mexico*, Logan, B. W., ed.: Am. Assoc. Petr. Geol. Mem. 11, Tulsa, Oklahoma, p. 5–128.

Ludwick, J. C., and Walton, W. A., 1957, Shelf-edge calcareous prominences in northeastern Gulf of Mexico: *Am. Assoc. Petr. Geol. Bull.*, v. 41, p. 2054–2101.

Macintyre, I. G., 1972, Submerged reefs of eastern Caribbean: *Am. Assoc. Petr. Geol. Bull.*, v. 56, p. 720–738.

Malfait, B. T., and Dinkelman, M. G., 1972, Circum-Caribbean tectonics and igneous activity and the evolution of the Caribbean plate: *Geol. Soc. Am. Bull.*, v. 83, p. 251–272.

Maloney, N. J., 1971, Continental margin off central Venezuela, in: *Symposium on Investigations and Resources of the Caribbean Sea and Adjacent Regions*: UNESCO, Paris, p. 261–266.

Marlowe, J. A., Kranck, K., and Mann, C. R., 1971, Research by the Atlantic Oceanographic Laboratory, Bedford Institute, in the Caribbean area during 1968, in: *Symposium on Investigations and Resources of the Caribbean Sea and Adjacent Regions*: UNESCO, Paris, p. 267–271.

Matley, C. A., 1926, The geology of the Cayman Islands (British West Indies) and their relation to the Bartlett Trough: *Q. J. Geol. Soc. London*, v. 83, p. 352–387.

Matthews, R. K., 1963, Continuous seismic profiles of a shelf-edge bathymetric prominence in northern Gulf of Mexico: *Trans. Gulf Coast Assoc. Geol. Soc.*, v. 13, p. 49–58.

Mattison, G. C., 1948, Bottom configuration in the Gulf of Mexico, *J. U.S. Coast Geod. Surv.*, v. 1, p. 76–81.

Maxwell, J. C., 1948, Geology of Tobago, British West Indies: *Geol. Soc. Am. Bull.*, v. 59, p. 801–854.

McBirney, A. R., 1963, Geology of part of the central Guatemala Cordillera: *Univ. Calif. Publ. Geol. Sci.*, v. 38, p. 177–242.

McBirney, A. E., and Bass, M. M., 1969a, Structural relations of pre-Mesozoic rocks of northern Central America, in: *Tectonic Relations of Northern Central America and the Western Caribbean*, McBirney, A. R., ed.: Am. Assoc. Petr. Geol. Mem. 11, Tulsa, Oklahoma, p. 269–290.

McBirney, A. R., and Bass, M. M., 1969b, Geology of the Bay Islands, Gulf of Honduras, in: *Tectonic Relations of Northern Central America and the Western Caribbean*, McBirney, A. R., ed.: Am. Assoc. Petr. Geol. Mem. 11, Tulsa, Oklahoma, p. 229–243.

Mencher, E., 1963, Tectonic history of Venezuela, in: *Backbone of the Americas*, Childs O. E., and Beebe, B. W., eds.: Am. Assoc. Petr. Geol. Mem. 2, Tulsa, Oklahoma, p. 73–87.

Mesolella, K. J., Sealy, H. A., and Matthews, R. K., 1970, Facies geometries within Pleistocene reefs of Barbados, West Indies: *Am. Assoc. Petr. Geol. Bull.*, v. 54, p. 1899–1917.

Metz, H. L., 1965, Geology of the El Pilar fault zone, state of Sucre, Venezuela, *Trans. Fourth Caribbean Geol. Conf.*, Trinidad, p. 1–8.

Meyerhoff, A. A., 1966, Bartlett Fault system: Age and offset, *Trans. Third Caribbean Geol. Conf., Jamaica*, p. 1–9.

Meyerhoff, A. A., 1967, Future hydrocarbon provinces of Gulf of Mexico–Caribbean region: *Trans. Gulf Coast Assoc. Geol. Soc.*, v. 17, p. 217–260.

Meyerhoff, A. A., and Hatten, C. W., 1968, Diapiric structures in central Cuba, in: *Diapirism and Diapirs*, Braunstein, J., and O'Brien, G. D., eds.: Am. Assoc. Petr. Geol. Mem. 8, Tulsa, Oklahoma, p. 315–357.

Meyerhoff, A. A., and Meyerhoff, H. A., 1972, Continental drift, IV: The Caribbean "plate": *J. Geol.*, v. 80, p. 34–60.

Meyerhoff, A. A., Khudoley, K. M., and Hatten, C. W., 1969, Geologic significance of radiometric dates from Cuba: *Am. Assoc. Petr. Geol. Bull.*, v. 53, p. 2494–2500.

Milliman, J. D., 1969, Carbonate sedimentation on four southwestern Caribbean atolls and its relation to the "oolite problem": *Trans. Gulf Coast Assoc. Geol. Soc.*, v. 19, p. 195–206.

Mills, R. A., Hugh, K. E., Feray, D. E., and Swolfs, H. C., 1967, Mesozoic stratigraphy of Honduras; *Am. Assoc. Petr. Geol. Bull.*, v. 51, p. 1711–1789.

Mina-U, F., 1965, Petroleum developments in Mexico in 1964: *Am. Assoc. Petr. Geol. Bull.*, v. 49, p. 1102–1111.

Molnar, P., and Sykes, L. R., 1969, Tectonics of the Caribbean and middle America regions from focal mechanisms and seismicity: *Geol. Soc. Am. Bull.*, v. 80, p. 1639–1684.

Monroe, W. H., 1968, The age of the Puerto Rico Trench: *Geol. Soc. Am. Bull.*, v. 79, p. 487–493.

Moore, D. G., and Curray, J. R., 1963, Structural framework of the continental terrace, northwest Gulf of Mexico, *J. Geophys. Res.*, v. 68, p. 1725–1747.

Moore, D. R., 1958, Notes on Blanquilla Reef, the most northerly coral formation in the western Gulf of Mexico: *Publ. Inst. Mar. Sci., Univ. Tex.*, v. 5, p. 151–155.

Morgan, J. P., Coleman, J. M., and Gagliano, S. M., 1968, Mudlumps, diapiric structures in Mississippi Delta sediments; in: *Diapirism and Diapirs*, Braunstein, J., and O'Brien, G. D., eds.: Am. Assoc. Petr. Geol. Mem. 8, Tulsa, Oklahoma, p. 145–161.

Murray, G. E., 1961, *Geology of the Atlantic and Gulf Coastal Province of North America*; Harper and Row, New York, 692 p.

Neff, C. H., 1971, Review of 1970 petroleum developments in South America, Central America, and Caribbean Sea: *Am. Assoc. Petr. Geol. Bull.*, v. 55, p. 1418–1482.

Nettleton, L. L., 1957, Gravity survey over a Gulf Coast shelf mound: *Geophysics*, v. 22, p. 630–642.

Neumann, A. C., 1958, The configuration and sediments of Stetson Bank, northwestern Gulf of Mexico; Texas A. & M. Univ. Tech. Rep. 58-5T, 125 p.

Nowlin, W. D., 1971, Water masses and general circulation of the Gulf of Mexico: *Oceanology*, v. 5, n. 2, p. 28–33.

Nowlin, W. D., Harding, J. L., and Amstutz, D. E., 1965, A reconnaissance study of the Sigsbee Knolls of the Gulf of Mexico: *J. Geophys. Res.*, v. 70, p. 1339–1347.

Officer, C. B., Ewing, J. I., Hennion, J. F., Harkrider, D. G., and Miller, D. E., 1959, Geophysical investigations in the eastern Caribbean: Summary of 1955 and 1956 cruises, in: *Physics and Chemistry of the Sea*, Ahrens, L. H., Press, F., Rankama, K., and Runcorn, S. K., eds.: Pergamon Press, London, v. 3, p. 17–109.

Paine, W. R., and Meyerhoff, A. A., 1970, Gulf of Mexico: Interactions among tectonics, sedimentation, and hydrocarbon accumulation; *Trans. Gulf Coast Assoc. Geol. Soc.*, v. 20, p. 5–43.

Parker, R. H., and Curray, J. R., 1956, Fauna and bathymetry of banks on continental shelf, northwest Gulf of Mexico: *Am. Assoc. Petr. Geol. Bull.*, v. 40, p. 2428–2439.

Pequegnat, W. E., Bryant, W. R., and Harris, J. E., 1971, Carboniferous sediments from the Sigsbee Knolls, Gulf of Mexico: *Am. Assoc. Petr. Geol. Bull.*, v. 55, p. 116–123.

Phillips, J. D., and Forsyth, D., 1972, Plate tectonics, paleomagnetism, and the opening of the Atlantic: *Geol. Soc. Am. Bull.*, v. 83, p. 1579–1600.

Pinet, P. R., 1971, Structural configuration of the northwestern Caribbean plate boundary: *Geol. Soc. Am. Bull.*, v. 82, p. 2027–2032.

Pinet, P. R., 1972, Diapirlike features offshore Honduras: Implications regarding tectonic evolution of Cayman Trough and Central America: *Geol. Soc. Am. Bull.*, v. 83, p. 1911–1922.

Price, W. A., 1967, Development of the basin-in-basin honeycomb of Florida and northeastern Cuban lagoon: *Trans. Gulf Coast Assoc. Geol. Soc.*, v. 17, p. 368–399.

Pyle, T. E., Antoine, J. W., Fahlquist, D. A., and Bryant, W. F., 1969, Magnetic anomalies in Straits of Florida: *Am. Assoc. Petr. Geol. Bull.*, v. 53, p. 2501–2505.

Robinson, E., and Cambray, F. W., 1971, Physiography of the sea floor east of Jamaica, in: *Symposium on Investigations and Resources of the Caribbean Sea and Adjacent Regions*, UNESCO, Paris, p. 285–289.

Rodolfo, S. K., 1971, Contrasting geometric adjustment styles of drifting continents and spreading sea floors, *J. Geophys. Res.*, v. 76, p. 3272–3281.

Rusnak, G. A., 1960, Sediments of Laguna Madre, Texas, in: *Recent Sediments, Northwest Gulf of Mexico*, Shepard, F. P., Phleger, F. B., and van Andel, Tj. H., eds.: Am. Assoc. Petr. Geol., Tulsa, Oklahoma, p. 153–196.

Russell, R. J., and Russell, R. D., 1939, Mississippi River Delta sedimentation, in: *Recent Marine Sediments*, Trask, P. D., ed.: Am. Assoc. Petr. Geol., Tulsa, Oklahoma.

Schmalz, R. F., 1969, Deep-water evaporite deposition: *Am. Assoc. Petr. Geol. Bull.*, v. 53, p. 789–823.

Schulz, R., and Weyl, R., 1959, Sismos y la estructura de la corteza terrestre en la parte norte de Centro America, *Bol. Sism. Ser. Geol. Nac.*, v. 5, p. 3–9.

Scientific Staff, 1970, Deep sea drilling project: Leg 10: *Geotimes*, v. 15, n. 6, p. 11–13.

Scientific Staff, 1971a, USGS-IDOE leg 1: Bahia de Campeche: *Geotimes*, v. 16, n. 11, p. 16–17.

Scientific Staff, 1971b, Deep sea drilling project: Leg 15: *Geotimes*, v. 16, n. 4, p. 12–15.

Scientific Staff, 1971c, USGS-IDOE leg 2: *Geotimes*, v. 16, n. 12, p. 10–12.

Scientific Staff, 1972a, USGS-IDOE leg 3: *Geotimes*, v. 17, n. 3, p. 14–15.

Scientific Staff, 1972b, USGS-IDOE leg 4: Venezuelan Borderland: *Geotimes*, v. 17, n. 5, p. 19–21.

Shepard, F. P., 1937, "Salt" domes related to Mississippi submarine trough: *Geol. Soc. Am. Bull.*, v. 48, p. 1349–1361.

Shepard, F. P., 1955, Delta-front valleys bordering the Mississippi distributaries, *Geol. Soc. Am. Bull.*, v. 66, p. 1489–1498.

Shepard, F. P., Dill, R. F., and Heezen, B. C., 1968, Diapiric intrusions in foreset slope sediments off Magdalena Delta, Colombia, *Am. Assoc. Petr. Geol. Bull.*, v. 52, p. 2197–2207.

Sheridan, R. E., Smith, J. D., and Gardner, J., 1969, Rock dredges from Blake Escarpment near Abaco Canyon: *Am. Assoc. Petr. Geol. Bull.*, v. 53, p. 2551–2558.

Stetson, H. C., 1953, The sediments of the western Gulf of Mexico, Part I. The continental terrace of the western Gulf of Mexico: Its surface sediments, origin and development: *Pap. Phys. Oceanogr. and Meteorol., Mass. Inst. Tech. and Woods Hole Oceanogr. Inst.*, v. 12, p. 5–45.

Stewart, H. B., Jr., Raff, A. D., and Jones, F. L., 1961, Explorer Bank—A new discovery in the Caribbean: *Geol. Soc. Am. Bull.*, v. 72, p. 1271–1274.

Stoddart, D. R., 1962, Three Caribbean atolls: Turneffe Island, Lighthouse Reef, and Glover's Reef, British Honduras: *Atoll Res. Bull.*, v. 87, p. 1–151.

Sykes, L. R., and Ewing, M., 1965, The seismicity of the Caribbean region: *J. Geophys. Res.*, v. 70, p. 5065–5074.

Terry, R. A., 1956, Geological reconnaissance of Panama: *Occas. Pap. Calif. Acad. Sci.*, v. 23, p. 1–91.

Uchupi, E., 1967, Bathymetry of the Gulf of Mexico: *Trans. Gulf Coast Assoc. Geol. Soc.*, v. 17, p. 161–172.

Uchupi, E., 1968, Map showing relation of land and submarine topography, Mississippi Delta to Bahia de Campeche: U.S. Geol. Surv. Misc. Geol. Invest. Map I-521, scale 1:1,000,000, 2 sheets.

Uchupi, E., 1973, The continental margin off eastern Yucatan and western Caribbean tectonics; *Am. Assoc. Petr. Geol. Bull.*, v. 57, p. 1074–1085.

Uchupi, E., and Emery, K. O., 1968, Structure of continental margin off Gulf Coast of United States: *Am. Assoc. Petr. Geol. Bull.*, v. 52, p. 1162–1193.

Uchupi, E., Milliman, J. D., Luyendyk, B. P., Bowin, C. O., and Emery, K. O., 1971, Structure and origin of the southeastern Bahamas: *Am. Assoc. Petr. Geol. Bull.*, v. 55, p. 687–704.

United States Geological Survey, 1967, Geological Survey research, 1967 oceanic islands, geology of Swan Island: U.S. Geol. Surv. Prof. Pap. 575, Chap. A: A127.

United States Geological Survey, 1972*a*, Acoustic-reflection profiles, Bay of Campeche. International Decade of Ocean Exploration. Leg 1, 1971 cruise, UNITEDGEO I, G. W. Moore, Chief Scientist: U.S. Geol. Surv. GD-72-002, 21 p. plus plates.

United States Geological Survey, 1972*b*, Acoustic reflection profiles, Yucatan peninsula. International Decade of Ocean Exploration. Leg 2, 1971 cruise, UNITEDGEO I, J. G. Vedder, Chief Scientist: U.S. Geol. Surv. GD-72-003, 15 p. plus plates.

United States Geological Survey, 1972*c*, Acoustic reflection profiles, eastern Greater Antilles. International Decade of Ocean Exploration. Leg 3, 1971 cruise, UNITEDGEO I, L. E. Garrison, Chief Scientist: U.S. Geol. Surv. GD-72-004, 19 p plus plates.

United States Geological Survey, 1972*d*, Acoustic-reflection profiles, Venezuela continental borderland. International Decade of Ocean Exploration, 1971 cruise, UNITEDGEO I, E. A. Silver, Chief Scientist: U.S. Geol. Surv. GD-72-005, 23 p. plus plates.

van Andel, Tj. H., and Sachs, P. L., 1964, Sedimentation in the Gulf of Paria during the Holocene transgression; a subsurface acoustic reflection study: *J. Mar. Res.*, v. 22, p. 30–50.

van Andel, Tj. H., Heath, G. R., Malfait, B. T., Heinrichs, D. F., and Ewing, J. I., 1971, Tectonics of the Panama Basin, eastern Equatorial Pacific, *Geol. Soc. Am. Bull.*, v. 82, p. 1489–1508.

Viniegra O., F., 1971, Age and evolution of salt basins of southeastern Mexico: *Am. Assoc. Petr. Geol. Bull.*, v. 55, p. 478–494.

Vinson, G. L., 1962, Upper Cretaceous and Tertiary stratigraphy of Guatemala: *Am. Assoc. Petr. Geol. Bull.*, v. 46, p. 425–456.

Vinson, G. L., and Brineman, J. H., 1963, Nuclear Central America, hub of Antillean transverse belt, in: *Backbone of the Americas*, Childs, O. E., and Beebe, B. W., eds.: Am. Assoc. Petr. Geol. Mem. 2, Tulsa, Oklahoma, p. 101–112.

Walker, J. R., and Ensminger, H. R., 1970, Effect of diapirism on sedimentation in the Gulf of Mexico: *Am. Assoc. Petr. Geol. Bull.*, v. 54, p. 2058–2069.

Walker, J. R., and Massingill, J. V., 1970, Slump features on the Mississippi Fan, northeastern Gulf of Mexico: *Geol. Soc. Am. Bull.*, v. 81, p. 3101–3108.

Weeks, L. A., Lattimore, R. K., Harbison, R. N., Bassinger, B. G., and Merrill, G. F., 1971, Structural relations among the Lesser Antilles, Venezuela and Trinidad–Tobago; *Am. Assoc. Petr. Geol. Bull.*, v. 55, p. 1741–1752.

Wilhelm, O., and Ewing, M., 1972, Geology and history of the Gulf of Mexico: *Geol. Soc. Am. Bull.*, v. 83, p. 575–600.

Winston, G. O., 1969, A deep glimpse of west Florida's platform, *Oil and Gas J.*, v. 67, n. 48, p. 128–133.

Worzel, J. L., Leyden, R., and Ewing, M., 1968, Newly discovered diapirs in Gulf of Mexico, *Am. Assoc. Petr. Geol. Bull.*, v. 52, p. 1194–1203.

Zans, V. A., Chubb, L. J., Versey, H. R., Williams, J. B., Robinson, E., and Cooke, D. L., 1962, Synopsis of the geology of Jamaica. An explanation of the 1958 provisional geological map of Jamaica: Bull. 4, Geol. Surv. Dept., Jam., 72 p.

NOTE ADDED IN PROOF

Since the manuscript on the physiography of the Gulf of Mexico and Caribbean Sea was written, the following papers have been published which will be of interest to readers:

Dillon, W. P., Vedder, J. G., and Graf, R. J., 1972, Structural profile of the Northwestern Caribbean: *Earth and Planet. Sci. Letters*, v. 17, p. 175–180.

Edgar, N. E. *et al.*, 1973, Initial reports of the deep sea drilling project, Vol. XV, covering leg 15 of the cruises of the drilling vessel "GLOMAR CHALLENGER," San Juan, Puerto Rico to Cristobal, Panama, December 1970–February 1971, 1137 p.

Garrison, L. E., and Martin, R. G., Jr., 1973, Geologic structures in the Gulf of Mexico basin: U.S. Geol. Survey Prof. Paper 773, 85 p.

Holcombe, T. L., Vogt, P. R., Matthews, J. E., and Murchison, R. R., 1973, Evidence for seafloor spreading in the Cayman Trough: *Earth and Planet. Sci. Letters*, v. 20, p. 357–371.

Moore, G. W., and Castillo, L. del, 1974, Tectonic evolution of the southern Gulf of Mexico: *Bull. Geol. Soc. Amer.*, v. 85, p. 607–618.

Pyle, T. E., Meyerhoff, A. A., Fahlquist, Antoine, J. W., McCrevey, J. A., and Jones, P. C., 1973, Metamorphic rocks from northwestern Caribbean Sea: *Earth and Planet. Sci. Letters*, v. 18, p. 339–344.

Silver, E. A., Case, J. E., and MacGillavry, H. J., 1975, Geophysical study of the Venezuelan Borderland: *Bull. Geol. Soc. Amer.*, v. 86, p. 213–226.

Weyl, R., 1973, El desarrollo paleogeografico de America Central: *Boletin Asociacion Mexicana de Geologos Petroleros*, v. 25, p. 375–425.

Worzel, J. L. *et al.*, 1973, Initial reports of the deep sea drilling project, Vol. X, covering Leg 10 of the cruises of the drilling vessel "GLOMAR CHALLENGER," Galveston, Texas to Miami, Florida, February–April 1970, 748 p.

Chapter 2

GEOPHYSICAL STUDIES IN THE GULF OF MEXICO

Ray G. Martin

U.S. Geological Survey
P. O. Box 6732, Corpus Christi, Texas

and

J. E. Case

U.S. Geological Survey
345 Middlefield Road, Menlo Park, California

I. INTRODUCTION

The Gulf of Mexico is a small ocean basin which, for the last 20 years, has been subject to intensive geological and geophysical investigation. Much has been learned about its framework, but a host of complex problems remains unsolved.

Quasi-oceanic crust, depressed substantially below normal oceanic levels and covered by an extremely thick sequence of sediments and sedimentary rocks, is known to underlie the central part of the basin. The formation of this crust, its antiquity, and its role in the overall evolution of the gulf basin are subjects of continuing speculative debate. Several models have been proposed. The gulf may represent (1) a foundered and oceanized continental mass; (2) a downwarp related to thermally controlled phase changes in the lower crust and upper mantle; (3) a gigantic tensional rift formed in relation to Mesozoic opening of the Atlantic Ocean, new crust being emplaced either by passive upwelling or from an active spreading center; or (4) a Paleozoic or older ocean basin.

The Gulf of Mexico abyssal plain is bordered on the south and east by massive carbonate platforms of Mesozoic–Cenozoic age, and on the north and west by thick embankments of terrigenous sediments built during Tertiary–Holocene time (Fig. 1). The vast wedge of clastic material that rims the gulf from Florida to Yucatan has been highly distorted by diapiric intrusions. The Continental Slope off eastern Mexico is a province of almost symmetrically folded sediments which is expressed by ridge and valley sea-floor topography.

The large volumes of salt, at least partly of Middle to Late Jurassic age (Kirkland and Gerhard, 1971), beneath the margins and the abyssal floor of the gulf have played a most important role in the morphological development of the basin. Diapiric salt masses, which dot the Coastal Plain, shelf, and slope from the De Soto Canyon in the northeastern gulf to northern Mexico, terminate at the foot of the Continental Slope along the Sigsbee and Rio Grande Escarpments. A belt of diapiric structures, the Campeche Knolls, reflected by steep hummocky topography of the Continental Slope, extends from onshore in southern Mexico to the Sigsbee Knolls in the abyssal plain. The folding of the Mexican ridges along the western gulf margin is, perhaps, related to salt tectonics.

The origin of the salt is of key importance in reconstructing the geological history of the gulf basin. From geophysical investigations and deep-sea drilling, it has been determined that the knolls in the Sigsbee Plain and the rugged hummocky topography of the Continental Slopes have resulted from salt deformation. However, the depositional environment, deformational history, and even the present areal extent of the salt are still matters open to debate. Several possibilities might explain the presence of evaporites in the gulf region: (1) deposition in shallow marginal basins peripheral to deeper water, parts of which later subsided to abyssal depths; (2) deep-water deposition of salt over an ancestral gulf whose margins were later uplifted; (3) contemporaneous deposition of salt in marginal and deep-water environments; or (4) deposition in shallow marginal basins peripheral to deep water, salt being squeezed into deep water by sediment overburden. Similar ages (Jurassic) for the Louann Salt in the Gulf Coastal Plain and the Sigsbee–Campeche salt in the southern gulf (Kirkland and Gerhard, 1971) suggest gulf-wide evaporite deposition.

Determination of the nature and ages of the rocks below the salt in the central Gulf of Mexico is a problem equally important to that of the origin of the salt. Seismic-refraction data indicate that sedimentary and/or volcanic type rocks lie between the salt layer and oceanic basement. These rocks presumably are pre-Jurassic in age and may well include Paleozoic and older rocks.

In the preceding chapter, Uchupi has ably summarized the structural and stratigraphic framework of the gulf, using data derived from thousands of

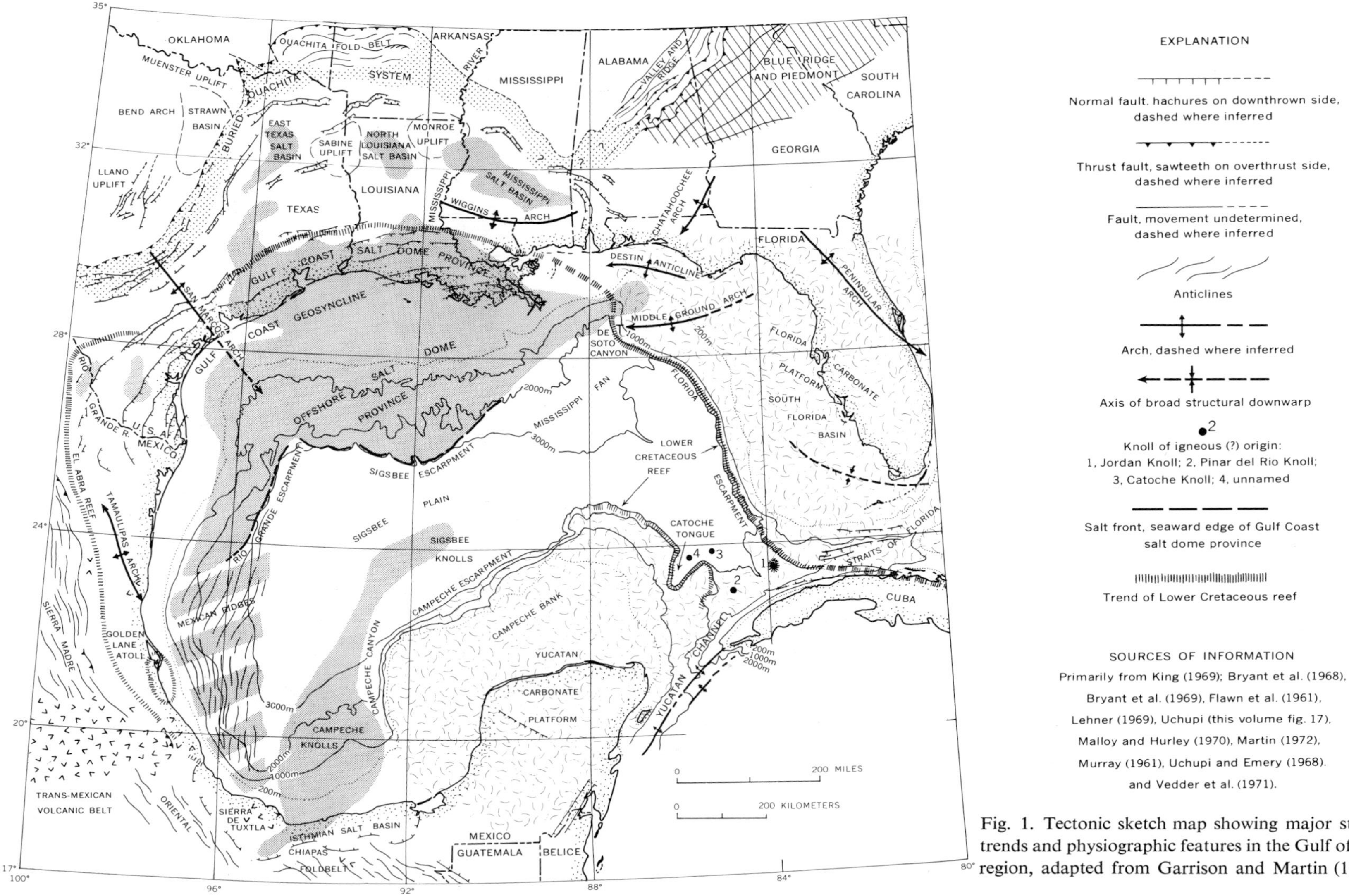

Fig. 1. Tectonic sketch map showing major structural trends and physiographic features in the Gulf of Mexico region, adapted from Garrison and Martin (1973).

track miles of seismic-reflection profiles and from the fundamental stratigraphic information obtained by the Deep Sea Drilling Project (DSDP). In this chapter we shall attempt to discuss the gulf's crustal structure, as deduced from seismic-refraction, magnetic, gravity, and heat-flow data, and to describe some of the main interpretations that bear on the evolution of the gulf basin. This chapter will concentrate, as far as possible, on findings from more recent geophysical investigations.

II. ABYSSAL GULF OF MEXICO

The Gulf of Mexico abyssal plain covers more than 350,000 km² of sea floor that includes the Sigsbee Plain in the west and the area of flat sea floor between the Campeche and Florida Escarpments in the east (Fig. 1). The plain is bounded by the steep Florida and Campeche Escarpments on the east and south and by the less abrupt slopes of the Sigsbee and Rio Grande scarps on the north and northwest. Elsewhere, the abyssal floor merges gently with the varied topographies of the Mississippi Fan and the Continental Slope of the western and southwestern gulf. The deep floor of the gulf is separated from the abyssal zones of the Atlantic Ocean and Caribbean Sea by the shallower depths of the Straits of Florida and the Yucatan Channel.

Seismic-refraction investigations across the Sigsbee Plain (Fig. 2) suggest that the central Gulf of Mexico is underlain by "oceanic" crust (layer 3) that is approximately 5.5 km thick (Fig. 3; Ewing *et al.*, 1955, 1960, 1962). The layer has an average P-wave velocity (Vp) of 6.8 km/sec, which is identical with typical oceanic crust (Fig. 4). Ewing *et al.* (1960, 1962) have reported mantle ($Vp = 8.0$–8.3 km/sec) depths of 16–17.5 km below sea level under the Sigsbee Plain (Fig. 3), which are about 5–6 km deeper than depths of mantle beneath typical ocean basins.

Sedimentary (and volcanic?) material ($Vp = 1.7$–5.1 km/sec), ranging in total thickness from about 4 km near the Campeche Escarpment to as much as 12 km adjacent to the Sigsbee Escarpment (Fig. 3), overlies the oceanic layer in the central gulf. This total thickness contrasts with the average thickness of 2.5 km for sedimentary rocks ($Vp = 1.7$–4.0 km/sec) and intermediate crustal material (layer 2) ($Vp = 4.5$–5.5 km/sec) found in typical oceanic provinces (Fig. 4).

Beneath the Sigsbee Plain, the intermediate layer has a velocity range of 4.5–5.1 km/sec and a thickness of 3–5 km, or about twice the average thickness of its counterpart in the typical ocean basin. Ewing *et al.* (1962) suggested that the upper part of the intermediate layer includes the Louann Salt (Jurassic) and possibly beds of Paleozoic age. The lower part was thought to be comparable to oceanic basement (layer 2) of other ocean basins. They also postulated that the surface of the layer (refractor 5, Fig. 3) corresponds

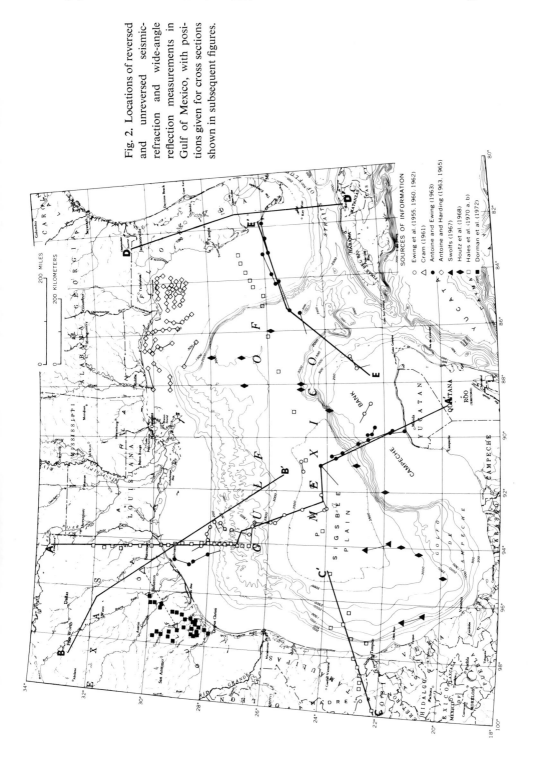

Fig. 2. Locations of reversed and unreversed seismic-refraction and wide-angle reflection measurements in Gulf of Mexico, with positions given for cross sections shown in subsequent figures.

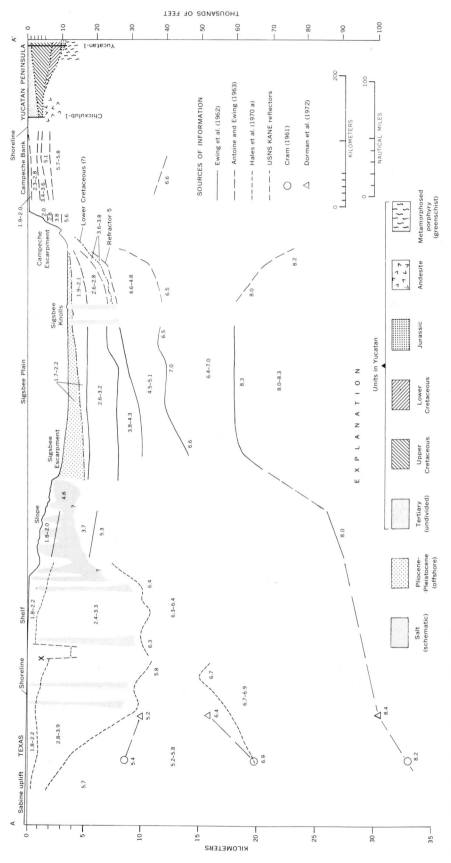

Fig. 3. Crustal section *AA'* across Gulf of Mexico from northeast Texas to central Yucatan. Location is shown in Fig. 2. Stratigraphic horizons in subsurface of Yucatan Peninsula, from Bryant *et al.* (1969) and Bass and Zartman (1969); in abyssal plain, scaled from seismic-reflection profiles and data of Worzel, Bryant *et al.* (1973). Refraction data from Ewing (1961) and Dorman *et al.* (1972) are projected northeast 100 and 250 km, respectively. Velocity of salt under Sigsbee Escarpment from Amery (1969). Velocities in km/sec.

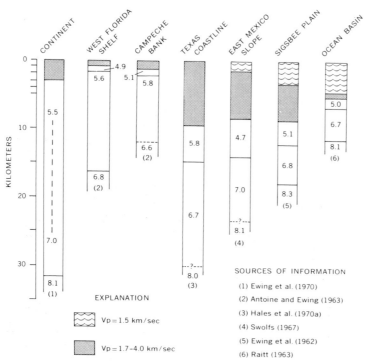

Fig. 4. Comparison of velocity structures of the margins and abyssal plain of the Gulf of Mexico with those of average ocean basin and continent. Velocities in km/sec.

to the top of the evaporite section from which the Sigsbee Knolls rise and concluded that salt is continuous from the knolls northward to the Sigsbee Escarpment and southwestward into the Isthmian salt basin (Fig. 1) in southern Mexico. Subsequent seismic-reflection surveys (Ewing *et al.*, 1962; Nowlin *et al.*, 1965; Worzel *et al.*, 1968; Bryant *et al.*, 1968; Ballard and Feden, 1970; Garrison *et al.*, 1972; Ensminger and Matthews, 1972; and Garrison and Martin, 1973) have demonstrated continuity of salt between the Sigsbee diapirs and those in southern Mexico. Reflection surveys, however, have not provided any definite evidence of a salt layer between the Sigsbee Knolls and the Texas–Louisiana slope (Fig. 1).

Oceanic and intermediate crustal layers under the Sigsbee Plain dip northward beneath near-horizontal layers of low-velocity material ($Vp = 1.7$–4.3 km/sec). These near-horizontal layers form a sedimentary prism which is 7 km thick south of the Sigsbee Escarpment but which thins to less than 3 km at the foot of Campeche Bank (Fig. 3). Both sedimentary prism and underlying crustal layers thicken westward into the Mexican continental margin (Fig. 4; Swolfs, 1967). Seismic-reflection profiles from the USNS *Kane* (Garrison and Martin, 1973) and DSDP drill-hole data (Worzel, Bryant, *et al.*, 1973) indicate

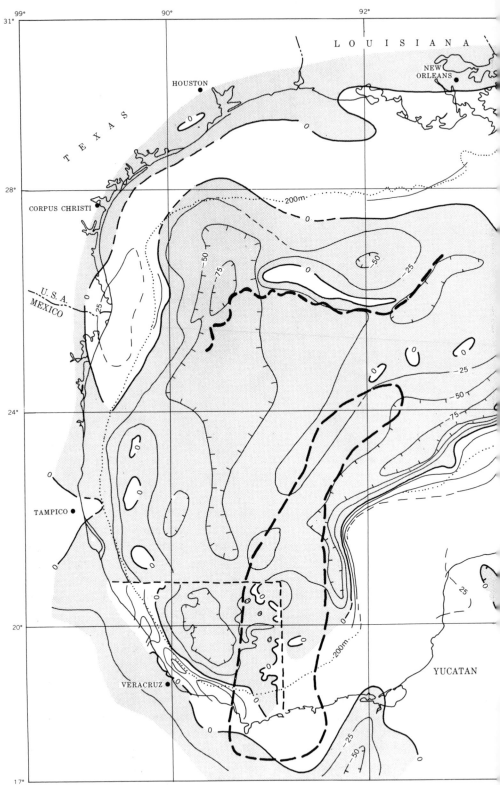

Fig. 5. Free-air gravity anomaly map of the Gulf of Mexico. Area 1 adapted from P. Rabinowitz (1962); and area 4, Dehlinger and Jones (1965). Heavy dashed lines outline Sigsbee Escarpment and

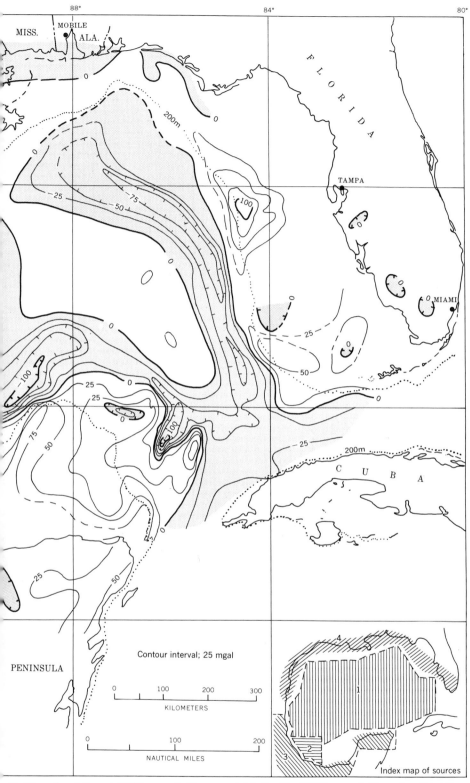

(*in* Worzel and Bryant, 1973); area 2, Moore and del Castillo (1974); area 3, Yungul and Dehlinger
Sigsbee–Campeche Knolls province. Shaded area represents negative free-air gravity anomaly field.

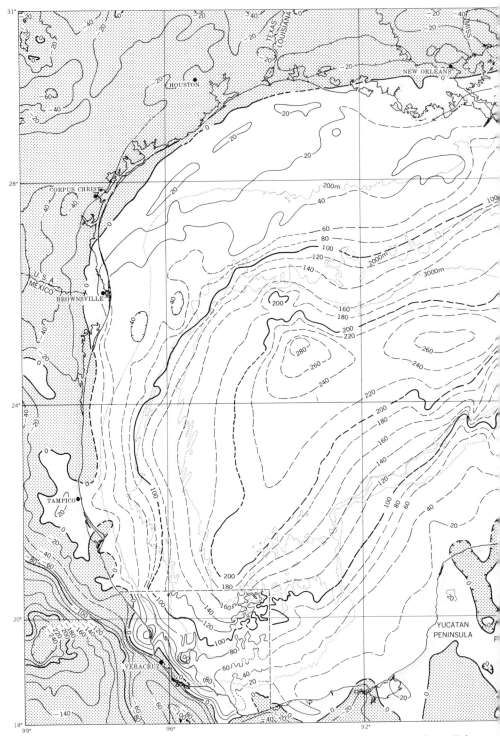

Fig. 6. Bouguer gravity anomaly map of the Gulf of Mexico. Area 1, adapted from Krivoy and area 4 from American Geophysical Union (1964). Arrows show trend of gravity

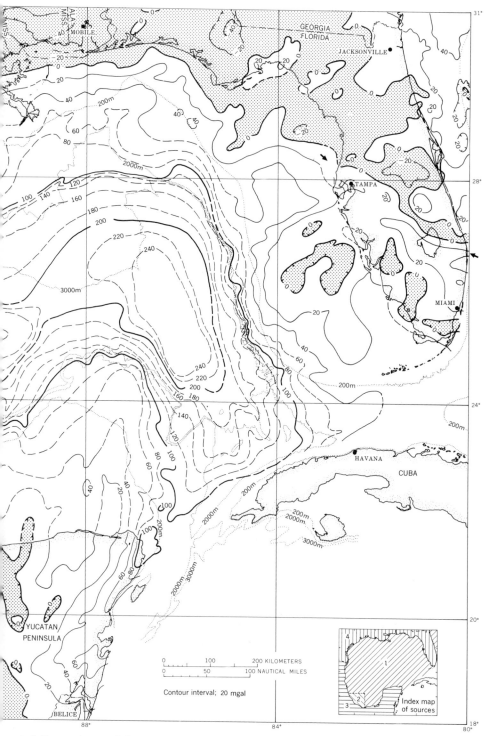

et al. (in press); area 2 from Moore and del Castillo (1974); area 3 from Woollard *et al.* (1969); gradient discussed in text. Shaded area represents negative Bouguer gravity anomalies.

that as much as 2 km of the low-velocity layer consists of beds of Miocene age or younger, and as much as 0.5 km of older sediments (Jurassic–Lower Cretaceous?) may lie between the top of the salt (*refractor* 5; Ewing *et al.*, 1962) and the deepest horizon detected on the *Kane* profiles (Fig. 3).

The free-air gravity field (Fig. 5) over the abyssal plain indicates that the basin is in approximate isostatic equilibrium, anomalies generally ranging between ± 25 mgal. Local mass anomalies having values more negative than -50 mgal occur along the bases of the Florida and Campeche Escarpments. Dehlinger and Jones (1965) suggested that these result from a normal variation caused by changes in crustal thicknesses and densities between oceanic and continental crust. They did not, however, totally rule out crustal faulting beneath the escarpments as a possible cause for the anomalies.

Oceanic crust beneath the abyssal plain is suggested by both the free-air (Fig. 5) and Bouguer gravity anomaly maps (Fig. 6). Bouguer anomalies over the abyssal gulf average about $+200$ mgal, which, allowing for differences in water depth, is comparable to Bouguer-anomaly values over the Caribbean basins and the western North Atlantic Ocean (Worzel, 1965; Emery and Uchupi, 1972a, Fig. 117). Steep gradients of Bouguer-anomaly contours along the Florida and Campeche Escarpments can be accounted for by the effects of abrupt change in water depth and normal transition from oceanic to continental crust.

Ewing *et al.* (1962) calculated that the observed gravity field of the gulf is about 200 mgal more positive than expected. They suggested that this may be due to (1) a greater density of the sediments, (2) a greater density of both sediments and oceanic crust, or (3) a denser mantle than predicted by the Nafe–Drake velocity–density curve. They preferred the idea of greater sediment and crustal density because it required the least deviation from data points on the Nafe–Drake curve. Antoine and Pyle (1970) favored a greater density than normal in the sedimentary prism alone, because the Gulf of Mexico contains a greater mass of turbidites than of pelagic deposits. Menard (1967), however, believed that depressed mantle in small ocean basins like the gulf does not occur in simple response to sediment loading, but that most or all of the required high-density material is in the mantle. Moore (1972) pointed to the possibility that high-density mantle beneath the Gulf of Mexico may be attributable to the greater age of the basin (Jurassic or older), as compared with the average age for the western Atlantic (Cretaceous). Sclater *et al.* (1971) suggested that thermal contractton resulting from dissipation of initial heat after sea-floor spreading causes midocean rises to sag more than 3 km during the period of 10^8 yr required to approach thermal equilibrium. Moore suggested that such thermal contraction could produce a density change in the upper mantle sufficient to balance the crust of the Gulf of Mexico isostatically with younger and thinner crusts of larger ocean basins.

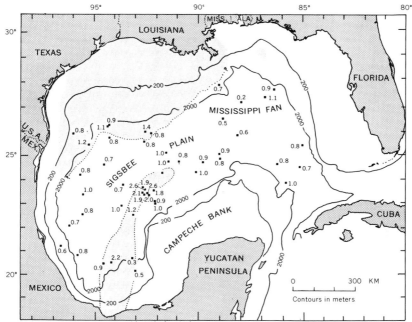

Fig. 7. Heat flow in microcalories per square centimeter per second in the Gulf of Mexico. Lightly shaded area describes extent of known salt-dome structures in the Gulf of Mexico. From Epp *et al.* (1970).

Caprock recovered from Challenger knoll by the Deep Sea Drilling Project (DSDP) confirmed that the Sigsbee Knolls are salt domes (Burk *et al.*, 1969). Gravity measurements over three of the Sigsbee Knolls by Talwani and Ewing (1966) indicated that the knolls are cored with low-density material which yields small negative anomalies of about −2 mgal. From the lack of associated magnetic anomalies, the presence of high heat-flow values, and the gravitational effect of the diapirs computed for various density contrasts, Talwani and Ewing (1966) were among the first investigators to draw the conclusion that the knolls were salt plugs.

Heat-flow determinations in the gulf (Fig. 7; Epp *et al.*, 1970) demonstrate that, with the exception of the Sigsbee–Campeche Knolls region, the abyssal plain has a uniformly low mean heat flow of 0.83 μcal/cm²-sec (HFU). In the drumstick-shaped area of diapirs (Figs. 1 and 7) outlined by Worzel *et al.* (1968), the mean heat flow is high (2.1 HFU), the highest values occurring not only on the tops of diapirs but in the sediments around them as well. Because a salt dome or a field of salt domes provides a low-resistance channel for thermal transfer equal to about six times that of oceanic sediment, Epp *et al.* (1970) interpreted the heat-flow values as additional evidence that salt forms the cores of the Sigsbee Knolls and underlies the field of high heat flow around them.

Fig. 8. Residual magnetic anomaly map of Gulf of Mexico. Area 1 compiled from Anderson *et al.*
Castillo (1974); and area 4, Heirtzler *et al.* (1966). Shaded area represents

1969) and Closuit *et al.* (1973); area 2, Ensminger and Matthews (1972); area 3, Moore and del egative magnetic residuals; *X* and *Y* are prominent negative anomalies.

Residual magnetic anomalies (Fig. 8) over the abyssal plain are generally subdued and for the most part range between ± 50 gammas. A long negative anomaly ($-300\,\gamma$, labeled X on Fig. 8), which Gough (1967) suggested was caused by a crustal downwarp, lies along the northeastern base of the Campeche Escarpment. The source of a prominent negative anomaly ($-300\,\gamma$, Y on Fig. 8) between the Sigsbee Knolls and Campeche Escarpment is unknown.

III. CONTINENTAL MARGINS

A. Northern Gulf of Mexico

The continental margin of the northern gulf from the De Soto Canyon to the Rio Grande is composed of a great thickness of clastic sediment deposited in offlapping wedges that have been deformed by salt intrusions during Cenozoic time. Sediment thickness reaches geosynclinal proportions along the Texas–Louisiana coast where more than 9 km of Tertiary clastic material is estimated to have accumulated (Fig. 9). Comprehensive recent papers on the regional structure of the northern gulf have been prepared by Lehner (1969), Dorman *et al.* (1972), Wilhelm and Ewing (1972), Emery and Uchupi (1972*a*), Garrison and Martin (1973), Worzel and Watkins (1973), and Uchupi (this volume).

Thick transgressive and regressive sections of Tertiary sediments overlie mainly carbonate beds of Cretaceous age in the Gulf Coastal Plain of Texas and Louisiana (Fig. 9). Redbeds and evaporite layers ranging in age from Late Pennsylvanian(?)–Permian(?) to Jurassic (Vernon, 1971) underlie the Cretaceous sequence, and on the northern margin of the Coastal Plain they rest unconformably upon the truncated rocks of the Ouachita fold belt (Flawn *et al.*, 1961), which rim the gulf basin from central Alabama to the Rio Grande (Fig. 1).

The structural complexity of the northern gulf margin has resulted from the mobility of Middle Jurassic(?) Louann Salt which has intruded the younger sedimentary section from De Soto Canyon to northern Mexico and from the inner Coastal Plain to the Sigsbee Escarpment (Figs. 1 and 9). Vertical and lateral migration of salt have produced faulting, slumping, and local thickening and thinning of stratigraphic units, all of which characterize the deformation pattern over and around individual diapirs. On a regional scale, the mobility of salt and of pressured shale masses (Bruce, 1973) has caused long-term subsidence of thick sedimentary sections along down-to-the-basin "growth fault" belts (Ocamb, 1961) which parallel the coastline from the Rio Grande to the Mississippi Delta (Fig. 1). The field of diapiric structures that dominates the deformational character of the northern gulf margin abruptly terminates along the foot of the Texas–Louisiana slope where a wall of salt occurs under the

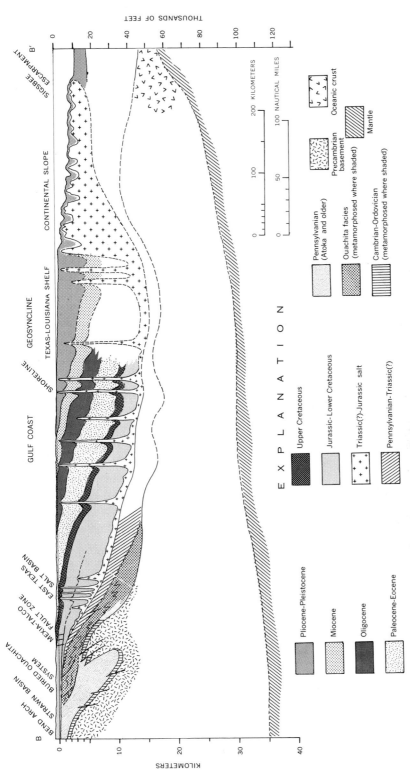

Fig. 9. Diagrammatic cross section of northern gulf margin southeastward from the Bend arch to the Sigsbee Escarpment. Adapted from Lehner (1969); Amery (1969); Vernon (1971); Emery and Uchupi (1972a); Powell and Woodbury (1971); Shinn (1971); Tipsword et al. (1971); Lofton and Adams (1971); Holcomb (1971); and Rainwater (1971). Location of section shown in Fig. 2.

Sigsbee Escarpment. Along much of the length of the escarpment, this salt wall apparently is in near-vertical contact with younger sedimentary layers of the Continental Rise and abyssal plain. de Jong (1968) and Amery (1969) have reported the apparent intrusive overflow of the salt mass into the Continental Rise along part of the escarpment south of the Louisiana coast.

A relatively smooth transition from near-oceanic crust beneath the Sigsbee Plain to typical continental crust in northeast Texas is suggested from refraction data (Fig. 3). The upper part of the sedimentary section, however, is considerably more complicated than the geometry shown in Figs. 3 and 9. Refraction data of Hales et al. (1970a) show that the Gulf Coast geosyncline is defined by low-velocity layers ($Vp = 1.8$–3.3 km/sec) which represent the thick accumulation of Tertiary (and older?) sediments. From the velocities shown on the crustal section, the total sedimentary thickness in the geosyncline may exceed 16 km.

Hales et al. (1970a) interpreted the shoreward offset (Fig. 3, point X) in the low-velocity layer ($Vp = 1.8$–2.2 km/sec) as a fault beneath the middle shelf, but they were unable to correlate onshore and nearshore velocity structure seaward of this offset, probably because of scatter from salt structures near and along their line of observations. At this same locality, the intermediate-velocity layer ($Vp = 5.2$–5.8 km/sec) either pinches out or abuts a layer having a P-wave velocity of 6.3–6.4 km/sec. Dorman et al. (1972), from projection of Cram's (1961) refraction data to the coast (Fig. 3), concluded that the intermediate-velocity layer may pinch out. Extrapolations from the stratigraphic section onshore (Fig. 9) indicate that the upper part of this intermediate layer is composed of Cretaceous and Jurassic rocks. Lower Cretaceous units in Texas consist mainly of carbonate beds (Rainwater, 1971), which can account for the higher velocities (5.7–5.8 km/sec) recorded along the onshore profile (Fig. 3; Hales et al., 1970a). The lower velocities (5.2–5.4 km/sec) recorded by Dorman et al. (1972) and Cram (1961) may also indicate limestone or even salt.

The higher velocities (6.3–6.4 km/sec) observed by Hales et al. (1970a) beneath the middle and outer shelf may indicate a lateral velocity change in the intermediate-velocity layer ($Vp = 5.2$–5.8 km/sec) representative of a southward facies change from less to more dense Cretaceous limestone. Alternatively, these data may indicate continental–oceanic crustal transition under the northern gulf shelf (Fig. 10; Worzel and Watkins, 1973).

A velocity layer correlative with Louann Salt (assumed $Vp - 4.6$–5.3 km/sec) is not indicated by any of the refraction data from the onshore and shelf areas, probably because of the velocity inversion that would be created by overlying higher-velocity carbonate rocks. Under the Continental Slope, however, refraction data reveal a velocity layer ($Vp = 5.3$ km/sec) at a depth of 5–7 km that could possibly be correlated with salt or with an intrusive

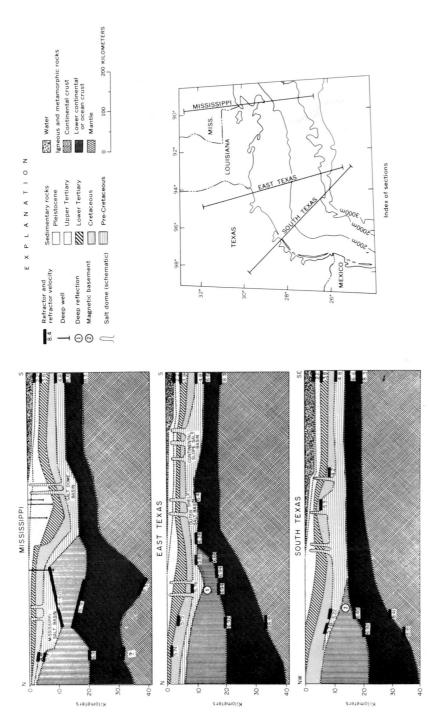

Fig. 10. Crustal sections across continental margin, northern Gulf of Mexico; velocities in km/sec. From Worzel and Watkins (1973).

ridge of basalt similar in composition to Layer 2 (Ewing *et al.*, 1960; Antoine and Ewing, 1963). Worzel and Watkins (1973) interpreted the 5.3 km/sec arrival as top of salt (Fig. 10), noting, however, that the velocity is anomalously high for salt in comparison with velocities from other nearby refraction data. Whether salt or igneous material, the structural position of this layer forms a ridge or horst of higher-velocity material that separates low-velocity sediments in the Gulf Coast geosyncline from those in the Sigsbee basin. A free-air gravity high (0-mgal closure) above the Sigsbee Escarpment (Fig. 5) also suggests a ridgelike mass of higher-density material under the Texas–Louisiana slope; negative gravity anomalies would normally be expected over the large volume of low-density salt that intrudes the terrigenous section (Worzel and Bryant, 1973).

Residual magnetic anomalies (Fig. 8) along the terrigenous margin of the northern gulf are broad and of low amplitude ($\pm 50\ \gamma$), thus reflecting the great depth of magnetic basement. The irregular anomaly pattern in this magnetically "quiet" area of the gulf may be related to drastic thickening and thinning of sediments, to changes in the composition of the basement, or to the variable magnetite content of shallow sand bodies. Over the Louisiana shelf, several east- to northeast-trending anomalies with amplitudes of about 50 γ were described by Heirtzler *et al.* (1966, Fig. 3), who suggested that they resulted from sources buried deeper than 5 km; Yungul (1971) has speculated that these apparent linear anomalies may be evidence of a spreading or rifting episode, but more recent magnetic data (Closuit *et al.*, 1973) indicate that "spreading" linear anomalies are not present.

Free-air and Bouguer gravity anomalies (Worzel and Bryant, 1973; H. L. Krivoy, unpublished data; Dehlinger and Jones, 1965) roughly parallel the northern Gulf Coast from Florida to south Texas (Figs. 5 and 6). An arcuate positive Bouguer anomaly parallels the coast from the Rio Grande to the Mississippi Delta, approximately along the axis of the Gulf Coast geosyncline. The anomaly, over a belt of very thick, low-density sediments, suggests that either the mantle rises to a relatively shallow depth under the geosyncline or that the crust and (or) upper mantle beneath the inner shelf have excess density. In either case, the positive gravity-anomaly field, which extends onshore into the delta areas of the Mississippi and Rio Grande Rivers, strongly suggests sedimentary outbuilding from continental onto transitional or oceanic crust.

B. Western Gulf of Mexico

The continental margin of the western Gulf of Mexico is considerably narrower than the other margins of the basin. The Coastal Plain is bordered on the west by the Sierra Madre Oriental, a Laramide fold belt transversely

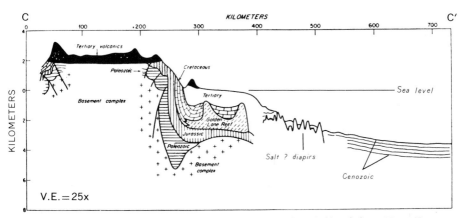

Fig. 11. Diagrammatic geologic section across eastern Mexico, shelf and slope. From Emery and Uchupi (1972a, Fig. 185). Location shown in Fig. 2.

cut by the Cenozoic–Quaternary Trans-Mexican volcanic belt (Fig. 1). The northern part of the western margin is underlain by the Tamaulipas carbonate platform, which was built at the same time as the Florida and Yucatan platforms during the Cretaceous (Emery and Uchupi, 1972a). The platform is bordered on the west by the El Abra reef and on the southeast by the Golden Lane atoll (Viniegra and Castillo-Tejero, 1970), which formed at the same time as other Lower Cretaceous reefs that nearly surround the gulf basin (Fig. 11). In the southwestern gulf, the continental margin widens to form a shelf and slope province closely akin structurally and physiographically to the northern gulf off Louisiana and Texas (Garrison and Martin, 1973).

The Continental Slope of the western Gulf of Mexico southward from Texas to the western edge of the Campeche Bank is composed of several clearly definable structural provinces (Fig. 1). In the region seaward of south Texas and northern Mexico, the structure of the shelf and slope is a continuation of the salt dome province of the northern gulf margin (Bryant et al., 1968; Garrison and Martin, 1973). The Continental Slope has been deformed principally by chains of salt massifs which lie on trend with the anticlinal folds in the Mexican ridge province to the south (Bryant et al., 1968; Garrison and Martin, 1973; Martin, 1973). The foot of the slope is marked by an escarpment formed by the front of a mobile salt mass. Unlike the near-vertical and overflow character of the Sigsbee Escarpment, the Rio Grande salt front (Garrison and Berryhill, 1970) has produced large folds in the thick abyssal units along the base of the slope (Garrison and Martin, 1973).

The central part of the Mexican slope is characterized by the Mexican ridges (Fig. 1), a system of symmetrically folded strata that extends southward almost 400 km along the slope from lat. 24° N to near junction with the Campeche knolls province. Folds are exposed on the lower slope but are buried in the upper slope and shelf. These ridges, first reported by Bryant *et al.* (1967), form a strikingly regular regional pattern with wave lengths of 10–12 km. Many individual ridges are continuous over great distances and have relief of as much as 500 m; some ridges appear to be cored with diapiric material (Bryant *et al.*, 1968; Garrison and Martin, 1973).

Near lat. 22° N, the fold system is broken by a gap several kilometers wide that marks a change in the trend of the fold system. Folds are present in the gap but have a much lower amplitude than elsewhere. The abrupt termination of high-amplitude folds at this gap led Bryant *et al.* (1968) to speculate that a major fault zone, possibly an extension of the Zacatecas fault in central Mexico, crosses the slope at this point. Massingill *et al.* (1973) attributed the gap to a change in basement trend rather than to faulting. Garrison and Martin (1973) suggested that the gap and the attendant change in fold trends have resulted from separate large-scale submarine landslides on decollements north and south of the gap. Conclusive evidence in support of any of these ideas is lacking, however.

Origin of the folds on the Mexican slope is still speculative. Smoothness of the magnetic field (Fig. 8) over the slope, except in the southernmost part of the region where anomalies are related to volcanic bodies, suggested to Emery and Uchupi (1972a) that basement has not been involved in the folding. In contrast, Massingill *et al.* (1973) interpreted Ensminger and Matthews' (1972) residual magnetic chart (Fig. 8) as indicative of an irregular basement surface whose deformation in Jurassic–Early Cretaceous times localized evaporite deposition and whose uplift later during Laramide time initiated salt flow. Jones *et al.* (1967), from analogy with mathematical models by Selig and Wermund (1966), suggested that the ridges were caused by the initial deformation of an evaporite layer which, under loading, flowed into a system of parallel ridges as a result of gravity sliding on a decollement surface. Gravity data over some of the ridges indicate a correlation between negative Bouguer anomalies (3–10 mgal) and the positions of ridge crests, which suggest that at least some of the folds have cores of low-density material, possibly salt or shale (Garrison and Martin, 1973; Moore and del Castillo, 1974). Although the evidence is indirect, the gravity data, the nearly constant wavelength of the ridges throughout the province, and the diapiric character of many of the folds together imply that the Mexican ridges have resulted from the tectonic mobility of salt.

Deep crustal structure beneath the continental margin of the western gulf is little known. One reversed profile on the lower slope southwest of the Golden Lane reef (Swolfs, 1967) and the refraction observations of Hales *et al.* (1970b) are the only such crustal investigations reported from the western gulf margin (Fig. 2). About 7 km of sedimentary material ($Vp = 2.0$–3.0 km/sec) overlying 5.5 km of crystalline basement ($Vp = 4.7$ km/sec) were found by Swolfs (Fig. 4). The lower crustal layer has a velocity of about 7.0 km/sec, similar to that of oceanic crust. Arrivals from the mantle were not observed by Swolfs, but he placed the M-discontinuity at a depth of about 22 km on the basis of minimum-depth calculations and on crustal thickness inferred from onshore data (Woollard, 1959). Swolfs' mantle depth is in good agreement with that estimated by Hales *et al.* (1970*b*).

The third distinctive province of the continental margin of eastern Mexico is in the Golfo de Campeche, where a province of diapirs, the Campeche knolls, strikingly resembles, both physiographically and structurally, the Texas–Louisiana slope (Fig. 1). Salt domes on the slope are essentially continuous southward beneath thick shelf sediments into the Isthmian salt basin of southern Mexico. Northward, the field of diapirs narrows and trends toward the Sigsbee Knolls, but rather than piercement structures, low-amplitude folds appear to be present between the two areas (Garrison and Martin, 1973). The Campeche knolls are separated on the east from the carbonate Campeche Bank by Campeche Canyon. To the west, the knolls are bordered by a narrow area of undisturbed abyssal sediments which separates the rugged sea floor of the Campeche knolls from the foot of the Mexican ridges. Seismic-refraction observations on the northwestern edge of the knolls province (Swolfs, 1967) reveal a crustal geometry similar to that beneath the Sigsbee Plain to the northeast; the mantle is at a depth of approximately 16 km.

Magnetic measurements made along the western gulf margin and in the Golfo de Campeche (Krivoy, 1970; Ensminger and Matthews, 1972; Moore and del Castillo, 1974) have disclosed (1) a strong linear positive anomaly along long. 93° W, asymmetrical to the Campeche knolls province (Fig. 8); (2) dipolar anomalies associated with buried volcanic centers in the shelf and upper slope off southeastern Mexico; and (3) low-amplitude, long wavelength anomalies having an easterly trend over the western part of the Golfo de Campeche. The anomaly along the 93° W meridian has an amplitude of 300 γ and a breadth of about 36 km. A chain of buried volcanoes, a basement horst, or a steeply dipping dike system were suggested by Krivoy (1970) as possible magnetic sources for the 93° W anomaly. Ensminger and Matthews (1972) favored the interpretation of a basement horst bounded by grabens, because such a basement structure in conjunction with salt deposition and subsequent deep burial could trigger salt diapirism. It should be pointed out, however, that the magnetic anomaly is markedly asymmetrical to the belt of salt structures.

Steep magnetic gradients over the shelf and upper slope off the Mexican coast near Veracruz (Fig. 8) characterize areas of volcanic rocks, which are believed to represent offshore extensions of the Trans-Mexican volcanic belt (Fig. 1; Moore and del Castillo, 1974). The anomalies are dipolar and have total amplitudes of as much as 450 γ. Positive anomalies are south of negative anomalies and thus are crudely aligned with the earth's present magnetic field. This and the shallow depth of burial of pluglike masses seen in reflection profiles drew Moore and del Castillo to the conclusion that submarine eruptions occurred in this part of the gulf during the last 700,000 yr.

East-oriented magnetic anomalies in the western Golfo de Campeche having amplitudes of about 75 γ have been interpreted by Moore and del Castillo (1974) as caused by deeply buried oceanic crust emplaced during an episode of sea-floor spreading. However, we feel that the anomalies are not comparable in length, breadth, or amplitude to typical spreading anomalies. More simply, the alignment of these anomalies with the Trans-Mexican volcanic belt suggests that they may result from basement structure or emplacement of volcanic rocks along fractures associated with rifting in the volcanic belt.

Gravity-anomaly maps (Figs. 5 and 6) suggest a narrower zone of transition from oceanic to continental crust in the western gulf than that along the northern margin. Along the central Mexican slope, where a negative free-air gravity field would be expected if salt underlies the Mexican ridges, three centers of positive gravity are shown. These anomalies may indicate buried masses of high-density material, possibly volcanic centers associated with Cenozoic volcanism in the Coastal Plain north of the Golden Lane. An elongate positive free-air anomaly over the shelf and upper slope parallels the coast south of lat. 21° N. It extends both north and south of the seaward terminus of the Trans-Mexican volcanic belt and includes the volcanic terrane of the Sierra de Tuxtla (Fig. 1). Moore and del Castillo (1974) have suggested that the anomaly reflects a southward bend of the volcanic belt toward volcanic centers in southern Mexico and Guatemala. On the Bouguer gravity map, positive anomalies are associated with the volcanic centers along the Mexican coast, but their trend is nearly perpendicular to the major negative Bouguer anomaly associated with the Trans-Mexican volcanic belt.

C. Carbonate Platforms

The eastern and south-central regions of the Gulf of Mexico have been sites of widespread carbonate accumulations since at least Early Cretaceous time. During Cretaceous time, a carbonate depositional environment prevailed around the periphery of the gulf, resulting in thick accumulations of limestones, chalks, and anhydrites from south Florida (Fig. 12) through Louisiana and Texas and on the Tamaulipas and Yucatan platforms in Mexico. Deep-water limestones and carbonate oozes recovered from drill holes on the Continental Slope off south Texas and in the approaches to the Straits of Florida (Lehner, 1969; Worzel and Bryant, 1973) suggest that carbonate deposition was perhaps gulf-wide during Late Mesozoic time. Very early in Tertiary time, however, a change in depositional regime, caused by Laramide mountain-building to the

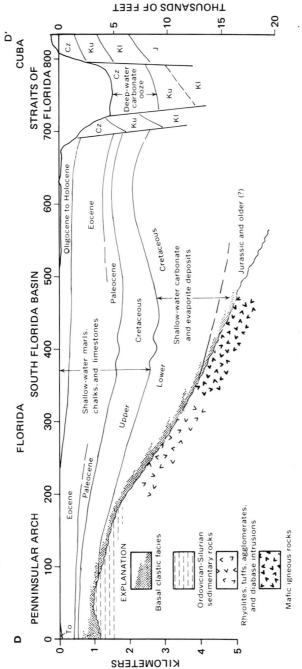

Fig. 12. Cross section through Florida and Cuba. Adapted from Emery and Uchupi (1972a), Rainwater (1971), and Brooks (1973). Cz, Cenozoic; To, Oligocene; Ku, Upper Cretaceous; Kl, Lower Cretaceous; J, Jurassic. Location shown in Fig. 2.

west and northwest, occurred in the western gulf, and since then, terrigenous deposition has prevailed. In the eastern and south-central gulf, terrigenous clastic materials were not available in similar quantities during the Tertiary, and carbonate sedimentation and reef growth continued in pace with subsidence to produce the massive Florida and Yucatan platforms.

In most respects, the west Florida continental margin is analogous to the Campeche Bank. The two areas, separated by a gap of less than 150 km, form a broad area of shallow-water and bathyal carbonate accumulations which is continuous eastward with the Bahama platform. Very little is known about the structure and stratigraphy of the Yucatan platform, of which the Campeche Bank is the submerged part. Present knowledge is based on a few wells on the peninsula and limited geophysical data across the shelf reported by Ewing et al. (1960), Antoine and Ewing (1963), and Paine and Meyerhoff (1970). Except for parts of the upper slope and the escarpments that front the platform in the gulf, seismic-reflection profiling across the Campeche Bank has been fruitless because of multiple echoes reflected from competent carbonate beds. Similarly, the interior framework of much of the west Florida platform is poorly known, except from the refraction work of Antoine and Ewing (1963) and the extrapolations from onshore and nearshore well data made by Oglesby (1965).

Cores and dredge samples recovered from several localities along the Florida Escarpment, and from the Campeche Escarpment in the Catoche Tongue area, support interpretations from many seismic-reflection profiles that an extensive reef system was present in the gulf during Early Cretaceous time (Fig. 1). Apparently the reef extended through what is now the Coastal Plain of Louisiana, Texas, and Mexico and perhaps was connected with reef systems of equivalent age in Cuba and along the periphery of the Bahama Banks (Antoine et al., 1967; Uchupi and Emery, 1968; Bryant et al., 1969; Emery and Uchupi, 1972a; Garrison and Martin, 1973). The characteristic acoustic pattern of the reef, expressed by nested hyperbolic reflections in the scarp faces, is observed in most reflection records across the Florida Escarpment north of lat. 27° N and south of lat. 24° N, but the absence of stacked anticlinelike reflections in the scarp face between these areas suggests a lack of continuity in the trend (Bryant et al., 1969), possibly because of post-Cretaceous erosion of the escarpment (Antoine and Pyle, 1970). Such acoustical patterns are seen only in reflection profiles crossing the northeastern face of the Campeche Escarpment, but not to the west, suggesting that the reef does not extend completely around the perimeter of the Campeche Bank (Garrison and Martin, 1973).

Seismic-refraction studies of Antoine and Ewing (1963) on the central part of the Campeche Bank and across the south Florida shelf (Fig. 2) indicate that both platforms are underlain by near-horizontal layers having similar seismic velocities, thicknesses, and depths (Figs. 3 and 13). Antoine and Ewing's correlations of seismic-velocity structure with stratigraphic units from wells on the Florida and Yucatan Peninsulas indicate equivalent ages for the velocity layers on both platforms. Cenozoic beds ($Vp = 1.8$–2.8 km/sec) on both banks show a small variation in layer thickness, which averages a little less than 1 km along the profiles. On the Campeche Bank, the top of the 5.1-km/sec layer appears to correlate reasonably well with a horizon of Upper Cretaceous(?) andesite (Fig. 3) penetrated in wells in northern Yucatan (Antoine and Ewing, 1963; Bass and Zartman, 1969).

The seismic-velocity structure of the northeastern part of the Campeche Bank (Fig. 13) appears dissimilar to that under the bank to the west (Fig. 3) and from that of the west Florida platform to the east. However, the 4.9 km/sec layers here probably correspond to Lower and Upper Cretaceous beds, and the 3.4 km/sec layer probably includes Tertiary carbonates more dense than elsewhere on the two banks. Lower Cretaceous and older rocks rest on high-velocity basement ($Vp = 6.4$ km/sec). Absence of an intermediate-velocity layer ($Vp = 5.5$–6.0 km/sec), as observed in other parts of the Yucatan and west Florida platforms, indicates a difference in basement composition under the eastern Campeche Bank (Ewing *et al.*, 1960). Displacement of the 4.9 km/sec layer downward beneath the abyssal straits strongly suggests that the flanking reefs formed along fault-line scarps. Free-air and Bouguer gravity anomalies (Figs. 5 and 6) over the northeastern Campeche Bank, the west Florida platform, and the abyssal plain between them indicate that the three regions are essentially in mass balance. In the absence of deep crustal and mantle refraction along section E–E' (Fig. 13), little can be said about the crustal properties of the region, particularly whether or not the crust between the escarpments is near-oceanic in composition and geometry.

The top of the 4.9 km/sec layer in the abyssal plain between the Florida and Campeche Escarpments corresponds with the deepest continuous reflector observed on reflection profiles recorded aboard the USNS *Kane* (Fig. 13; see also Fig. 3), which outlines the buried edges of the carbonate banks in the eastern and southern gulf. Dredge-haul and core samples at many points along these escarpments invariably have yielded material of Early Cretaceous age; thus, this refracting and reflecting horizon probably approximates the top of Lower Cretaceous beds. Some of the reefal material may be older than Early Cretaceous (Emery and Uchupi, 1972a). The Early Cretaceous reef, which underlies the carbonate escarpments at many places in the eastern gulf, is shown diagrammatically in Fig. 13. The true width and vertical continuity of these reefs is unknown. From many observations it is apparent that the reefs ceased to grow rather abruptly at the end of Early Cretaceous time; thus, their tops may correspond to Aptian–Albian shallow-water isobaths, which, except for broad structural warping along the Florida Escarpment (Pyle and

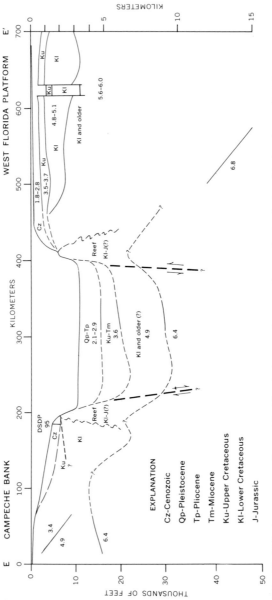

Fig. 13. Velocity structure of west Florida platform and eastern Campeche Bank. Modified and compiled from Emery and Uchupi (1972a), Antoine and Ewing (1963), Ewing et al. (1960), Oglesby (1965), and Worzel, Bryant, et al. (1973). Stratigraphy in abyssal plain from studies by R. G. Martin. Velocities given in km/sec; location of section shown in Fig. 2.

Antoine, 1973), suggest remarkably uniform subsidence of the carbonate platforms in the southeastern gulf during post-Early Cretaceous time.

Magnetic investigations in the eastern gulf have provided some clues to the deeper structure of the platforms and their escarpments but have yielded no conclusive evidence about the origin of their foundations. Local magnetic anomalies of as much as 400 γ in amplitude were found over the Florida and Campeche Escarpments by Miller and Ewing (1956), who suggested that the anomalies indicated buried volcanic bodies because they were comparable in width and amplitude with those over seamounts. Later work by Heirtzler et al. (1966) demonstrated that elongate anomalies extending across the Campeche Bank (Fig. 8) were longer than reported by Miller and Ewing (1956), and they suggested that these anomalies were probably caused by large crustal structures nearly 11 km deep rather than by buried volcanic centers; isolated high-amplitude anomalies over the central part of the west Florida platform were attributed either to volcanic rocks buried at shallow depths or to mafic igneous intrusions. Gough (1967) showed an extensive patch of anomalies, possibly indicative of volcanic rocks, stretching southwest across the west Florida platform and onto the northeast edge of the Campeche Bank (Fig. 14). Anomalies similar to those in Gough's volcanic province have been reported by Pyle et al. (1969) from Jordan, Catoche, and Pinar del Rio Knolls in the approaches to the Yucatan Channel and in the Straits of Florida (Fig. 1).

The remarkable linearity of the scarp-faces of both the Florida and Campeche Escarpments has suggested to many investigators that these steep slopes were either fault-line scarps or localized by major faults along which the basin subsided (Weaver, 1950; statements of Weaver et al. in Jordan, 1951; Dietz, 1952; Eardley, 1954; and more recently, Moody, 1973). Crustal faulting along the escarpments, however, has not been resolved by geophysical studies. Dehlinger and Jones (1965) and Henderson (1963) concluded that observed gravity anomalies could be accounted for by models either including or excluding crustal faulting along the escarpments. The lack of magnetic expression of the Florida Escarpment led Gough (1967) to rule out basement faulting along its trend; an elongate negative magnetic anomaly associated with the northeast face of the Campeche Escarpment (X on Figs. 8 and 14) was shown by Gough (1967) to fit a model of downwarped oceanic crust more closely than it would a graben. Magnetic expression, however, depends on the nature and depth of magnetic basement.

Over the northern part of the west Florida platform, Heirtzler et al. (1966) and Gough (1967) have shown elongate northeast-trending anomalies similar in amplitude and trend to those in northern peninsular Florida (King, 1959, Fig. 1, p. 2845). They suggested that a large positive anomaly on the shelf near lat. 29° N was related to a ridge of Paleozoic basement (Heirtzler et al., 1966) having an Appalachian structural trend and possibly correlative

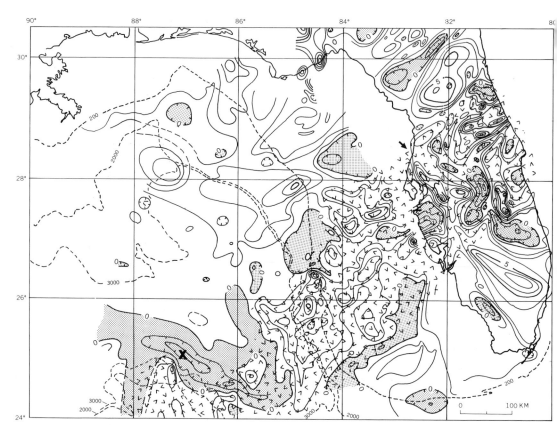

Fig. 14. Residual magnetic anomaly map of eastern Gulf of Mexico. Contour interval 1.0 mG. Redrawn from Gough (1967). Negative residual anomalies are shaded; *X*, a prominent negative anomaly, is discussed in the text. *V* pattern denotes area of volcanic-type magnetic anomalies; arrows indicate trend of magnetic gradient discussed in text. Bathymetry in meters.

with the East Coast anomalies described by Drake *et al.* (1963). Structure contours based on refraction data in the northeastern gulf by Antoine and Harding (1963, 1965) have shown that a major structural feature, the Middle Ground arch (Fig. 1; Winston, 1969), trends westward across the shelf to the edge of the west Florida platform, crudely coincident with the axis of the large positive magnetic anomaly described by Heirtzler *et al.* (1966). The residual magnetic-anomaly pattern in this region, however, has been shown by later work to be considerably more complicated (Fig. 8).

Regional trends of the Bouguer anomaly and residual magnetic fields over the west Florida platform and peninsular Florida are in close correspondence (Figs. 6 and 14). In general, all major trends in both fields strike northeast in northern Florida and northwest in southern peninsular Florida. Off the west margin of south Florida, both fields trend roughly northeast (Figs. 6 and 8). A steep northwest-trending gravity and magnetic gradient that transects southern Florida from near lat. 26°30' N to lat. 28° N may indicate a major basement fault or crustal boundary (Figs. 6 and 14). The positive Bouguer and magnetic anomalies to the southeast may represent thinned crust. The Bouguer gravity field northeast of this trend indicates a continental-type crustal geometry (Krivoy and Pyle, 1972). To the southwest, the Bouguer field is dotted by small areas of relative negative anomalies, suggesting that much of south Florida and the adjacent part of the west Florida margin may have a continental foundation like the Bahamas. Some of these negative anomalies may be caused by intrabasement silicic masses.

North of this area, a landward salient of positive Bouguer anomaly has been described as caused either by an area of transitional crust where reef progradation occurred over an oceanic embayment or an area where crust formed by rifting during an episode of rotation of Florida (Krivoy and Pyle, 1972). Farther north on the platform, a trend of gravity highs corresponds to the axial trace of the Middle Ground arch. The Bouguer anomaly field becomes progressively more negative northward from the arch, reflecting an increase in crustal and sediment thickness into the eastern end of the Gulf Coast geosyncline.

The Bouguer gravity field over the Yucatan Peninsula and the Campeche Bank suggests continental crustal thickness over much of the northern and western part of the region, but considerably thinner crust over the easternmost margin. A crude correlation between negative Bouguer gravity (Fig. 6) and positive magnetic anomalies (Fig. 8; Heirtzler et al., 1966) can be seen on the Campeche Bank, suggesting the presence of low-density but magnetic volcanic masses or thick sedimentary rocks overlying magnetic basement of average density.

IV. SEISMICITY OF GULF OF MEXICO

The Gulf of Mexico and adjacent Coastal Plain are essentially aseismic (Fig. 15). However, in the extreme southwestern corner of the gulf along the Trans-Mexican volcanic belt and in the region of the Sierra de Tuxtla and southeastward (Fig. 1), shallow to deep-focus earthquakes are known to occur with great frequency (Fig. 15). Epicenters define what appear to be two distinct zones of seismicity, which, for identification here, are referred to as (1) the Trans-Mexican seismic zone and (2) the Pacific seismic zone.

The Trans-Mexican seismic zone extends across Mexico almost due east along lat. 19° N, approximately the southern margin of the Late Cenozoic–Quaternary Trans-Mexican volcanic belt. Hypocenters along this lineament

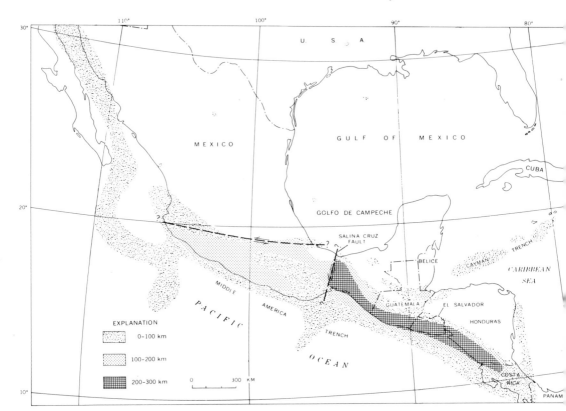

Fig. 15. Generalized seismicity map of Gulf of Mexico, Mexico, and Central America for the period 1961–1967, showing postulated transcurrent faulting in Mexico. Epicenter pattern and focal depths from Barazangi and Dorman (1969).

generally range from 0–200 km in depth, no foci occurring deeper than 200 km during the period 1961–1967 (Barazangi and Dorman, 1969).

Extending from approximately long. 95° W, southeastward along the Central American arc, the Pacific seismic zone is defined by a large concentration of earthquakes whose foci may be as deep as 300 km. An abrupt decrease in the frequency of earthquakes having hypocenters at depths of 100–200 km and an absence of foci deeper than 200 km (Barazangi and Dorman, 1969) west and northwest of long. 95° W, are almost coincident with the proposed Salina Cruz left-lateral fault (Viniegra, 1971; Fig. 9), which crosses the Isthmus of Tehuantepec from the Pacific into the Gulf.

Epicenters along the southern margin of the Trans-Mexican volcanic belt are believed to result from the transtensional rifting of Mexico approximately along lat. 19° N. Bends and apparent offsets in the Laramide foldbelt in this region of eastern Mexico suggest left-lateral motion along this zone. The extent of transtensional faulting into the Golfo de Campeche is undetermined.

The abrupt decrease in deep-seated seismicity west of long. 95° W may result from changes in direction, rate, and/or angle of underthrusting in the Mid-American Trench. Alignment of shallow-focus earthquakes along the Salina Cruz fault suggests that the fault is presently active.

V. ORIGIN OF THE GULF OF MEXICO

A. Continental Foundering or Oceanization

Foundering of a continental block has long been suggested as an origin for the Gulf of Mexico (Schuchert, 1935; Weaver, 1950; Rainwater, 1967; Viniegra, 1971, and others). This concept seems to be based on the idea that a southerly landmass supplied the quartzose, feldspathic, and other clastic detritus in the Paleozoic Ouachita system. Many years ago, Miser (1921) termed this southern landmass "Llanoria" and suggested that it occupied parts of south Arkansas, Louisiana, and south Texas. Other investigators (Eardley, 1954) presumed that the landmass extended farther south over the site of the present gulf basin. Recent stratigraphic studies, however, have indicated that multiple sources may have supplied clastic detritus to the Ouachita system: longitudinal filling from easterly, northeasterly, or south-easterly directions has been inferred by Cline (1960); Glick (1973), Haley and Stone (1973), and others, and this raises doubts about the extent of a southern landmass. Further objections to the postulate of continental foundering were encountered when it was established that an oceanic-type layer exists in the central gulf. Although favoring an alternate hypothesis of origin, Wilhelm and Ewing (1972) mentioned that an oceanization process (Belousov, 1968) for the Gulf of Mexico merits consideration.

B. Phase Changes

Several general types of polymorphic phase transitions have been proposed to account for large regional subsidence and uplift. A familiar example is conversion of gabbro to a denser phase, such as eclogite, which results in a net volume reduction and, hence, subsidence of the free surface (Lovering, 1958). Other transitions that may be geologically significant include serpentinization or deserpentinization reactions on a regional scale (Sheridan, 1969). Such phase transitions are controlled by temperature and pressure in the upper mantle and crust. Lowering of temperature may accompany convective motion of the mantle which could effectively convert a thick crustal block to a thinner one having physical properties of an oceanic block. In other words, the Gulf of Mexico could have formed over a "cold spot" in the mantle. It should be pointed out that the volume reduction from gabbro to eclogite is about 15%. In order to achieve 5 km of surficial subsidence, a phase change would have to be distributed through 60 km of lower crust and upper mantle. Hales (1973) suggested that a phase change at 57 km may be indicated by the seismic discontinuity found beneath the Gulf of Mexico by Hales et al. (1970b). Other than the seismic discontinuity and low heat-flow values, little evidence exists for or against a polymorphic phase-transition origin for the gulf basin.

C. Rifting and Spreading

The concepts of rifting and continental drift are not entirely new ideas for the origin of the gulf basin. Wegener's (1929) model for opening of the Gulf of Mexico during the breakup of Pangea differs little, in movements and time frame, from models being proposed today. Carey (1958) proposed that opening of the Gulf of Mexico began during the Late Jurassic when nuclear Central America—the Yucatan block on the east and the Honduran block on the west—was rifted away from the North American continent along the approximate axis of the Gulf Coast geosyncline and subsequently was rotated counterclockwise into the present positions of Guatemala and Honduras. Freeland and Dietz' (1971) model differs from that of Carey (1958) in that the predrift positions of the Central American blocks were transposed, the basin opened in Late Triassic, and Yucatan and Honduras were rotated in a clockwise direction to arrive in their approximate present position by the end of Early Cretaceous time. Other discussions of gulf rifting have been presented by Yarborough (1967), Wilhelm and Ewing (1972), and Walper and Rowett (1972).

Yungul (1971) proposed the possibility of a sea-floor-spreading episode in the Gulf of Mexico based on his interpretation of low-amplitude linear magnetic anomalies over the Louisiana shelf. Moore and del Castillo (1974) also proposed a sea-floor-spreading model for the gulf in which Yucatan and a nuclear Cuba were rifted away from North America along a zone spreading south of and peripheral to the buried Ouachita system (Fig. 1). Moore and del Castillo based their hypothesis on low-amplitude, east-trending magnetic anomalies in the western Golfo de Campeche, which they consider to result from deeply buried spreading strips.

In reviewing hypotheses of rifting or spreading, it is clear that these concepts are based on the following types of circumstantial evidence: (1) a crude geomorphic fit between scarps bounding the north, south, and eastern margins of the deep gulf and between margins of sialic crust identified from refraction studies; (2) a need to explain oceanic crust in the central gulf; (3) the lack of direct evidence in the present central gulf for a southern provenance for clastic material in the Ouachita system; thus the source may at present be farther south (nuclear Central America); (4) weak evidence of linear "spreading-type" magnetic anomalies; and (5) an apparent (and unproven) gap in the salt in the central gulf basin.

In the absence of direct evidence for a spreading center and major fracture zones, which, if present, are masked by Jurassic and younger rocks, it appears that rifting and spreading hypotheses are as conjectural as any of the others.

D. Ancient Ocean Basin

Several groups have proposed that the Gulf of Mexico has been in existence since early Paleozoic and possibly since Precambrian time. The evidence from lithofacies, orogenic belts, paleoclimatology, and assumed sedimentation rates has been summarized by Meyerhoff (1967) and Paine and Meyerhoff (1970).

Again, the age of the rocks beneath the salt, or the age of the acoustic basement in the central gulf, must be determined before the minimum age of the gulf can be established. It is certainly pre-Jurassic, but no lower age limit can be fixed on the basis of data presently available.

VI. SUMMARY

Crust having a thickness and compressional wave velocity comparable to typical oceanic crust underlies the Sigsbee Plain in the central part of the Gulf of Mexico, where as much as 9 km of Mesozoic–Cenozoic sediments has accumulated. The nature of the crust under the abyssal plain in the eastern gulf is poorly known. The quasi-oceanic layer in the central part of the basin and the M-discontinuity are depressed more than 6 km below normal oceanic levels, possibly owing to a combination of phase changes resulting from a mantle cold spot, heat loss after rifting or spreading, and sediment loading. Except for some local mass anomalies, gravity data indicate that the gulf basin is essentially in isostatic balance.

Refraction, gravity, and heat-flow data in the Sigsbee Plain all suggest that evaporite deposits are limited to the Sigsbee–Campeche Knolls region. Seismic-reflection profiles, gravity, and heat-flow measurements provide strong evidence that a single salt basin or chain of salt basins extends southward from the knolls through the eastern part of the Golfo de Campeche onshore into southern Mexico. The difference of 4 km in elevation between the salt that feeds the Sigsbee Knolls and that onshore in the Isthmus of Tehuantepec could have been accomplished by differential subsidence of the central part of the basin relative to the margins rather than through a process of deep-water salt deposition.

Seismic-refraction investigations across the northern gulf continental margin have outlined the gross geometry of the Gulf Coast geosyncline and indicate that thin, transitional-type crust extends beneath much of the shelf. Construction of a detailed crustal model in the northern gulf from refraction data has been limited by attenuation of seismic signals by lateral variations in sediment thickness, velocity inversions, and the highly complicated structural pattern of the Gulf Coast salt dome province. Gravity anomalies and refraction data suggest a possible basement ridge beneath the lower Continental Slope which may have had considerable influence on salt accumulation and the tectonic evolution of the margin.

The deeper structure of the continental margin of the western gulf is virtually unknown. Correlation of refraction data from profiles on the slope with distant profiles in the Sigsbee Plain and on the Mexican Plateau indicate a smooth westward thickening of crust into the margin; gravity data reveal a similar pattern.

In the southwestern gulf, near Veracruz, Mexico, detailed gravity and magnetic surveys define volcanic centers beneath the shelf and slope which are nearly perpendicular to the eastward extension of the late Cenozoic Trans-Mexican volcanic belt into the gulf. Correlation between ridge crests on the Mexican slope and Bouguer gravity lows has suggested that the Mexican ridges have cores of low-density material.

Refraction work and correlation of refracting horizons with stratigraphy in onshore wells has established basic similarities in depths, thicknesses, and ages of velocity layers in the west Florida and Campeche carbonate platforms. Magnetic anomalies over both platforms suggest structural relief of the basement surface. High-amplitude magnetic anomalies over peninsular Florida and its western platform suggest a pre-Early Cretaceous volcanic history for southern Florida. Similar anomalies associated with major knolls in the southeastern gulf may indicate contemporaneous volcanism in the Straits of Florida and Yucatan channel. Bouguer gravity anomalies over the west Florida and Campeche platforms indicate that for the most part these massive carbonate banks have been built on a continental foundation.

Nearly 20 years of geophysical and geological investigations in the Gulf of Mexico have provided considerable information about the structure, stratigraphy, and Mesozoic–Cenozoic history of the basin. Our knowledge, however, is still fragmental, and the origin, history, and detailed geology of the gulf are still far from known. Some of the key questions remaining to be answered about the gulf basin are the following: (1) provenance of the clastic facies of the Ouachita system; (2) composition and extent of oceanic crust beneath the deep gulf; (3) cause for abnormal depths to crust and mantle beneath the abyssal plain; (4) nature and age of the material overlying oceanic crust; (5) presence or absence of crustal faults in the basin; (6) extent and origin of evaporite deposits in the gulf region; (7) deep crustal structure of the continental margins; (8) the tectonic relationship of the southern gulf to the Caribbean region; and (9) the cause of the apparently rapid subsidence of the gulf basin since late Mesozoic time, as indicated by the large accumulation of Cenozoic sediments in the Gulf Coast geosyncline and the central gulf basin and the abrupt termination of reef growth along the escarpments in the eastern gulf at the end of Early Cretaceous time.

REFERENCES

American Geophysical Union, Special Committee for the Geophysical and Geological Study of the Continents, 1964, Bouguer gravity anomaly map of the United States (exclusive of Alaska and Hawaii): U.S. Geol. Surv., Washington, D.C., 2 sheets, scale 1:2,500,000.

Amery, G. B., 1969, Structure of Sigsbee scarp, Gulf of Mexico: *Am. Assoc. Petr. Geol. Bull.*, v. 53, p. 2480–2482.

Anderson, C. N., Vogt, P. R., Bracey, D. R., and Kontis, A. L., 1969, A magnetic survey in the eastern Gulf of Mexico and its relation to the east coast aeromagnetic survey: *Trans. Am. Geophys. Union*, v. 50, p. 207.

Antoine, J. W., Bryant, W., and Jones, B., 1967, Structural features of continental shelf, slope and scarp, northeastern Gulf of Mexico: *Am. Assoc. Petr. Geol. Bull.*, v. 51, p. 257–262.

Antoine, J. W., and Ewing, J., 1963, Seismic refraction measurements on the margins of the Gulf of Mexico: *J. Geophys. Res.*, v. 68, p. 1975–1996.

Antoine, J. W., and Harding, J. L., 1963, Structure of the continental shelf, northeastern Gulf of Mexico: Texas A. & M. Univ. Tech. Rep. 63-13T, 18 p.

Antoine, J. W., and Harding, J. L., 1965, Structure beneath continental shelf, northeastern Gulf of Mexico: *Am. Assoc. Petr. Geol. Bull.*, v. 49, p. 157–171.

Antoine, J. W., and Pyle, T. E., 1970, Crustal studies in the Gulf of Mexico: *Tectonophysics*, v. 10, p. 477–494.

Ballard, J. A., and Feden, R. H., 1970, Diapiric structures on the Campeche shelf and slope, western Gulf of Mexico: *Geol. Soc. Am. Bull.*, v. 81, p. 505–512.

Barazangi, M., and Dorman, J., 1969, World seismicity maps compiled from ESSA, Coast and Geodetic Survey, epicenter data, 1961–1967: *Seismol. Soc. Am. Bull.*, v. 59, p. 369–380.

Bass, M. N., and Zartman, R. E., 1969, The basement of Yucatan Peninsula [abstract]: EOS, v. 50, n. 4, p. 313.

Belousov, V. V., 1968, Some problems of development of the Earth's crust and upper mantle of oceans: Am. Geophys. Union Geophys. Mon. 12 (NAS-NRC Publ. 1687), p. 449–459.

Brooks, H. K., 1973, Geological oceanography, in: A Summary of Knowledge of the Eastern Gulf of Mexico, 1973, Jones, J. I., et al., eds.: Florida State Univ. System, Inst. Oceanography, p. IIE-1–IIE-50.

Bruce, C. H., 1973, Pressured shale and related sediment deformation: Mechanism for development of regional contemporaneous faults: *Am. Assoc. Petr. Geol. Bull.*, v. 57, p. 878–886.

Bryant, W. R., Antoine, J., and Ewing, M., 1967, Structure of the Mexican continental shelf and slope [abstract]: Geol. Soc. Am. Spec. Pap. 101, p. 28–29.

Bryant, W. R., Antoine, J., Ewing, M., and Jones, B., 1968, Structure of Mexican continental shelf and slope, Gulf of Mexico: *Am. Assoc. Petr. Geol. Bull.*, v. 52, p. 1204–1228.

Bryant, W. R., Meyerhoff, A. A., Brown, N. K., Jr., Furrer, M. A., Pyle, T. E., and Antoine, J. W., 1969, Escarpments, reef trends, and diapiric structures, eastern Gulf of Mexico: *Am. Assoc. Petr. Geol. Bull.*, v. 53, p. 2506–2542.

Burk, C. A., Ewing, M., Worzel, J. L., Beall, A. O., Jr., Berggren, W. A., Bukry, D., Fischer, A. G., Pessagno, E. A., Jr., 1969, Deep-sea drilling into the Challenger Knoll, central Gulf of Mexico: *Am. Assoc. Petr. Geol. Bull.*, v. 53, p. 1338–1347.

Carey, S. W., 1958, The tectonic approach to continental drift, in: *Continental Drift—A Symposium*: Univ. Tasmania, Geology Dept., Hobart, Australia, p. 177–355.

Cline, L. M., 1960, Stratigraphy of the late Paleozoic rocks of the Ouachita Mountains, Oklahoma: Oklahoma Geol. Surv. Bull. 85, 113 p.

Closuit, A. W., Wolaver, T. W., and Jack, H. C., 1973, Residual magnetic intensity contour map—Western Gulf of Mexico: U.S. Naval Oceanographic Office Manuscript Chart (unpublished).

Cram, I. H., Jr., 1961, A crustal structure refraction survey in south Texas: *Geophysics*, v. 26, p. 560–573.

de Jong, A., 1968, Stratigraphy of the Sigsbee scarp from a reflection survey [abstract]: Soc. Explor. Geol., Fort Worth Mtg. Program, p. 51.

Dehlinger, Peter, and Jones, B. R., 1965, Free-air gravity anomaly map of the Gulf of Mexico and its tectonic implications, 1963 edition: *Geophysics*, v. 30, p. 102–110.

Dietz, R. S., 1952, Geomorphic evolution of continental terrace (continental shelf and slope): *Am. Assoc. Petr. Geol. Bull.*, v. 36, p. 1802–1819.

Dorman, J., Worzel, J. L., Leyden, R., Crook, J. N., and Hatziemmanuel, M., 1972, Crustal section from seismic refraction measurements near Victoria, Texas: *Geophysics*, v. 37, p. 325–336.

Drake, C. L., Heirtzler, J. R., and Hirshman, J., 1963, Magnetic anomalies off eastern North America: *J. Geophys. Res.*, v. 68, p. 5259–5275.

Eardley, A. J., 1954, Tectonic relations of North and South America: *Am. Assoc. Petr. Geol. Bull.*, v. 38, p. 707–773.

Emery, K. O., and Uchupi, Elazar, 1972a, Western North Atlantic Ocean: Topography, rocks, structure, water, life and sediments: Am. Assoc. Petr. Geol. Mem. 17, 532 p.

Emery, K. O., and Uchupi, Elazar, 1972b, Caribe's oil potential is boundless: *Oil Gas J.*, v. 70, n. 50, p. 156, 158, 160, 162.

Ensminger, H. R., and Matthews, J. E., 1972, Origin of salt domes in Bay of Campeche, Gulf of Mexico: *Am. Assoc. Petr. Geol. Bull.*, v. 56, p. 802–807.

Epp, David, Grim, P. J., and Langseth, M. G., Jr., 1970, Heat flow in the Caribbean and Gulf of Mexico: *J. Geophys. Res.*, v. 75, p. 5655–5669.

Ewing, J., Antoine, J., and Ewing, M., 1960, Geophysical measurements in the western Caribbean Sea and in the Gulf of Mexico: *J. Geophys. Res.*, v. 65, p. 4087–4126.

Ewing, J., Edgar, J. T., and Antoine, J. W., 1970, Structures of the Gulf of Mexico and Caribbean Sea, in: *The Sea. Volume 4. Ideas and Observations on Progress in the Study of the Seas*: Wiley–Interscience, New York, Pts. 2–3, p. 321–358.

Ewing, J. I., Worzel, J. L., and Ewing, M., 1962, Sediments and oceanic structural history of the Gulf of Mexico: *J. Geophys. Res.*, v. 67, p. 2509–2527.

Ewing, M., Worzel, J. L., Ericson, D. B., and Heezen, B. C., 1955, Geophysical and geological investigations in the Gulf of Mexico, Part I: *Geophysics*, v. 20, p. 1–18.

Flawn, P. T., Goldstein, A., Jr., King, P. B., and Weaver, C. E., 1961, The Ouachita system: Texas Univ. Publ. 6120, 401 p.

Freeland, G. L., and Dietz, R. S., 1971, Plate tectonic evolution of Caribbean–Gulf of Mexico region: *Nature*, v. 232, n. 5305, p. 20–23.

Garrison, L. E., and Berryhill, H. L., Jr., 1970, Possible seaward extension of the San Marcos Arch [abstract]: Geol. Soc. Am. Abstr. with Programs, 2, p. 285–286.

Garrison, L. E., and Martin, R. G., 1973, Geologic structures in the Gulf of Mexico basin: U.S. Geol. Surv. Prof. Pap. 773, 85 p.

Garrison, L. E., Reimnitz, E., and Martin, R. G., 1972, Acoustic-reflection profiles, western continental margin, Gulf of Mexico, 1970 cruise 70-02 of R/V *Cadete Virgilio Uribe*: Nat. Tech. Inf. Serv. Rep. PB-207-593, 19 p., 15 figs.

Glick, E. E., 1973, Paleotectonic history of carboniferous rocks of Arkansas [abstract]: Geol. Soc. Am. Abstr. with Programs, 5, p. 258–259.

Gough, D. I., 1967, Magnetic anomalies and crustal structure in eastern Gulf of Mexico: *Am. Assoc. Petr. Geol. Bull.*, v. 51, p. 200–211.

Hales, A. L., 1973, The crust of the Gulf of Mexico: A discussion: *Tectonophysics*, v. 20, n. 1–4, p. 217–225.

Hales, A. L., Helsley, C. E., and Nation, J. B., 1970a, Crustal structure study on Gulf Coast of Texas: *Am. Assoc. Petr. Geol. Bull.*, v. 54: p. 2040–2057.

Hales, A. L., Helsley, C. E., and Nation, J. B., 1970b, P travel times for an oceanic path: *J. Geophys. Res.*, v. 75, p. 7362–7381.

Haley, B. R., and Stone, C. G., 1973, Paleozoic stratigraphy and depositional environments in the Ouachita Mountains, Arkansas [abstract]: Geol. Soc. Am., Abstr. with Programs, 5, p. 259–260.

Heirtzler, J. R., Burckle, L. H., and Peter, George, 1966, Magnetic anomalies in the Gulf of Mexico: *J. Geophys. Res.*, v. 71, p. 519–526.

Henderson, G. C., 1963, Preliminary study of the crustal structure across the Campeche Escarpment from gravity data: *Geophysics*, v. 28, p. 736–744.

Holcomb, C. W., 1971, Hydrocarbon potential of Gulf Series of western Gulf basin, in: *Future Petroleum Provinces of the United States—Their Geology and Potential*, Vol. 2, Cram, I. H., ed.: Am. Assoc. Petr. Geol. Mem. 15, p. 887–900.

Houtz, R. E., Ewing, J. I., and LePichon, X., 1968, Velocity of deep-sea sediments from sonobuoy data: *J. Geophys. Res.*, v. 73, p. 2615–2641.

Jones, B. R., Antoine, J. W., and Bryant, W. R., 1967, A hypothesis concerning the origin and development of salt structures in the Gulf of Mexico sedimentary basin: *Gulf Coast Assoc. Geol. Soc. Trans.*, v. 17, p. 211–216.

Jordan, G. F., 1951, Continental slope off Apalachicola River, Florida: *Am. Assoc. Petr. Geol. Bull.*, v. 35, p. 1978–1993.

King, E. R., 1959, Regional magnetic map of Florida: *Am. Assoc. Petr. Geol. Bull.*, v. 43, p. 2844–2854.

King, P. B., 1969, Tectonic map of North America: U.S. Geol. Surv., Washington, D. C., 2 sheets, scale 1:5,000,000.

Kirkland, D. W., and Gerhard, J. E., 1971, Jurassic salt, central Gulf of Mexico, and its temporal relation to circum-Gulf evaporites: *Am. Assoc. Petr. Geol. Bull.*, v. 55, p. 680–686.

Krivoy, H. L., 1970, A magnetic lineament in the Bay of Campeche [abstract]: Geol. Soc. Am. Abstr. with Programs, 2, p. 288.

Krivoy, H. L., and Pyle, T. E., 1972, Anomalous crust beneath West Florida Shelf: *Am. Assoc. Petr. Geol. Bull.*, v. 56, p. 107-113.

Krivoy, H. L., Pyle, T. E., and Eppert, H. C., Jr., in press, Bouguer gravity anomaly map of Gulf of Mexico: U.S. Geol. Surv.

Lehner, P., 1969, Salt tectonics and Pleistocene stratigraphy on continental slope of northern Gulf of Mexico: *Am. Assoc. Petr. Geol. Bull.*, v. 53, p. 2431–2479.

Lofton, C. L., and Adams, W. M., 1971, Possible future petroleum provinces of Eocene and Paleocene, western Gulf basin, in: *Future Petroleum Provinces of the United States— Their Geology and Potential*, Vol. 2, Cram, I. H., ed.: Am. Assoc. Petr. Geol. Mem. 15, p. 855–886.

Lovering, J. F., 1958, The nature of the Mohorovicic discontinuity: *Am. Geophys. Union Trans.*, v. 39, p. 947–955.

Malloy, R. J., and Hurley, R. J., 1970, Geomorphology and geologic structure: Straits of Florida: *Geol. Soc. Am. Bull.*, v. 81, p. 1947–1972.

Martin, R. G., 1972, Structural features of the continental margin, northeastern Gulf of Mexico: U.S. Geol. Surv. Prof. Pap. 800-B, p. B1-B8.

Martin, R. G., 1973, Salt structure and sediment thickness, Texas–Louisiana Continental Slope, northwestern Gulf of Mexico: U.S. Geol. Surv. Open-File Rep., 21 p.

Massingill, J. V., Bergantino, R. N., Fleming, H. S., and Feden, R. H., 1973, Geology and genesis of the Mexican Ridges: Western Gulf of Mexico: *J. Geophys. Res.*, v. 78, p. 2498–2507.

Menard, H. W., 1967, Transitional types of crust under small ocean basins: *J. Geophys. Res.*, v. 72, p. 3061–3073.

Meyerhoff, A. A., 1967, Future hydrocarbon provinces of Gulf of Mexico–Caribbean region: *Gulf Coast Assoc. Geol. Soc. Trans.*, v. 17, p. 217–260.

Miller, E. T., and Ewing, M., 1956, Geomagnetic measurements in the Gulf of Mexico and in the vicinity of Caryn Peak: *Geophysics*, v. 21, p. 406–432.

Miser, H. D., 1921, Llanoria, the Paleozoic land area in Louisiana and eastern Texas: *Am. J. Sci.*, v. 2, p. 61–89.

Moody, J. D., 1973, Petroleum exploration aspects of wrench-fault tectonics: *Am. Assoc. Petr. Geol. Bull.*, v. 57, p. 449–476.

Moore, G. W., 1972, Crust and mantle of the Gulf of Mexico: *Nature*, v. 238, n. 5365, p. 452–453.

Moore, G. W., and Castillo, Luis del, 1974, Tectonic evolution of the southern Gulf of Mexico: *Geol. Soc. Am. Bull.*, v. 85, p. 607–618.

Moore, G. W., and Staff, 1971, U.S.G.S.–I.D.O.E., Leg 1: Bahia de Campeche: *Geotimes*, v. 16, n. 11, p. 16–17.

Murray, G. E., 1961, Geology of the Atlantic and Gulf coastal province of North America: Harper & Row, New York, 692 p.

Nowlin, W. D., Jr., Harding, J. L., and Amstutz, D. E., 1965, A reconnaissance study of the Sigsbee Knolls of the Gulf of Mexico: *J. Geophys. Res.*, v. 70, p. 1339–1347.

Ocamb, R. D., 1961, Growth faults of south Louisiana: *Gulf Coast Assoc. Geol. Soc. Trans.*, v. 11, p. 139–175.

Oglesby, W. R., 1965, Folio of south Florida basin—A preliminary study: Florida Geol. Surv. Map Ser. 19, 3 p., 10 figs.

Paine, W. R., and Meyerhoff, A. A., 1970, Gulf of Mexico basin: interactions among tectonics, sedimentation, and hydrocarbon accumulation: *Gulf Coast Assoc. Geol. Soc. Trans.*, v. 20, p. 5–44.

Powell, L. C., and Woodbury, H. O., 1971, Possible future petroleum potential of Pleistocene, western Gulf basin, in: *Future Petroleum Provinces of the United States—Their Geology and Potential*, Vol. 2, Cram, I. H., ed.: Am. Assoc. Petr. Geol. Mem. 15, p. 813–823.

Pyle, T. E., and Antoine, J. W., 1973, Structure of the west Florida platform, Gulf of Mexico: Texas A. & M. Univ. Tech. Rep. 73-7-T, 168 p.

Pyle, T. E., Antoine, J. W., Fahlquist, D. A., and Bryant, W. R., 1969, Magnetic anomalies in Straits of Florida: *Am. Assoc. Petr. Geol. Bull.*, v. 53, p. 2501–2505.

Rainwater, E. H., 1967, Resumé of Jurassic to Recent sedimentation history of the Gulf of Mexico basin: *Gulf Coast Assoc. Geol. Soc. Trans.*, v. 17, p. 179–210.

Rainwater, E. H., 1971, Possible future petroleum potential of Lower Cretaceous, western Gulf basin, in: *Future Petroleum Provinces of the United States—Their Geology and Potential*, Vol. 2, Cram, I. H., ed.: Am. Assoc. Petr. Geol. Mem. 15, p. 901–926.

Raitt, R. W., 1963, The crustal rocks, in: *The sea*, Vol. 3, Hill, M. N., ed., Wiley–Interscience, New York, p. 85–102.

Schuchert, C., 1935, *Historical Geology of the Antillean–Caribbean Region*: John Wiley and Sons, New York, 811 p.

Sclater, J. G., Anderson, R. N., and Bell, M. L., 1971, Elevation of ridges and evolution of the central eastern Pacific: *J. Geophys. Res.*, v. 76, p. 7888–7915.

Selig, Franz, and Wermund, E. G., 1966, Families of salt domes in the Gulf coastal province: *Geophysics*, v. 31, p. 726–740.

Sheridan, R. E., 1969, Subsidence of continental margins: *Tectonophysics*, v. 7, p. 219–230.

Shinn, A. D., 1971, Possible future petroleum potential of Upper Miocene and Pliocene, western Gulf basin, in: *Future Petroleum Provinces of the United States—Their Geology and Potential*, Vol. 2, Cram, I. H., ed.: Am. Assoc. Petr. Geol. Mem. 15, p. 824–835.

Swolfs, H. S., 1967, Seismic refraction studies in the southwestern Gulf of Mexico: College Station, Texas, Texas A. & M. Univ., unpublished Master's thesis.

Talwani, M., and Ewing, M., 1966, A continuous gravity profile over the Sigsbee Knolls: *J. Geophys. Res.*, v. 71, p. 4434–4438.

Tipsword, H. L., Fowler, W. A., Jr., and Sorrell, B. J., 1971, Possible future petroleum potential of lower Miocene–Oligocene, western Gulf basin, in: *Future Petroleum Provinces of the United States—Their Geology and Potential*, Vol. 2, Cram, I. H., ed.: Am. Assoc. Petr. Geol. Mem. 15, p. 836–854.

Uchupi, Elazar, and Emery, K. O., 1968, Structure of continental margin off Gulf Coast of United States: *Am. Assoc. Petr. Geol. Bull.*, v. 52, p. 1162–1193.

Vedder, J. G., and Staff, 1971, U.S.G.S.–I.D.O.E.: Leg 2: preliminary report: *Geotimes*, v. 16, n. 12, p. 10–12.

Vernon, R. C., 1971, Possible future petroleum potential of pre-Jurassic, western Gulf basin, in: *Future Petroleum Provinces of the United States—Their Geology and Potential*, Vol. 2, Cram, I. H., ed.: Am. Assoc. Petr. Geol. Mem. 15, p. 954–979.

Viniegra O., Francisco, 1971, Age and evolution of salt basins of southeastern Mexico: *Am. Assoc. Petr. Geol. Bull.*, v. 55, p. 478–494.

Viniegra O., Francisco, and Castillo-Tejero, C., 1970, Golden Lane fields, Veracruz, Mexico, in: *Geology of Giant Petroleum Fields*, Halbouty, M. T., ed.: Am. Assoc. Petr. Geol. Mem. 14, p. 309–325.

Walper, J. L., and Rowett, C. L., 1972, Plate tectonics and the origin of the Caribbean Sea and the Gulf of Mexico: *Gulf Coast Assoc. Geol. Soc. Trans.*, v. 22, p. 105–116.

Weaver, Paul, 1950, Variations in history of continental shelves: *Am. Assoc. Petr. Geol. Bull.*, v. 34, p. 351–360.

Wegener, Alfred, 1929, *Die Entstehung der Kontinente und Ozeane*, 4th ed., rev.: Friedr. Vieweg & Sohn, Braunschweig (translated by John Biram, 1966, The origin of continents and oceans: Dover, New York, 246 p.).

Wilhelm, O., and Ewing, M., 1972, Geology and history of the Gulf of Mexico: *Geol. Soc. Am. Bull.*, v. 83, p. 575–600.

Winston, G. O., 1969, A deep glimpse of west Florida's platform: *Oil Gas J.*, v. 67, n. 48, p. 128–129, 132–133.

Woollard, G. P., 1959, Crustal structure from gravity and seismic measurements: *J. Geophys. Res.*, v. 64, p. 1521–1544.

Woollard, G. P., Machesky, L., and Monges Caldera, J., 1969, A regional gravity survey of northern Mexico and the relation of Bouguer anomalies to regional geology and elevation in Mexico: Hawaii Inst. Geophys. [Data Rep.] HIG-69-13, 52 p.

Worzel, J. L., 1965, *Pendulum Gravity Measurements at Sea*, 1936–1959: John Wiley & Sons, New York, 422 p.

Worzel, J. L., and Bryant, W. R., 1973, Regional aspects of deep sea drilling in the Gulf of Mexico, Leg 10, in: *Initial reports of the Deep Sea Drilling Projects*, Vol. 10, Worzel, J. L., Bryant, W., *et al.*, eds.: U.S. Gov. Printing Office, Washington, p. 737–748.

Worzel, J. L., Bryant, W. R., *et al.*, 1973, Initial Reports of the Deep Sea Drilling Projects, Vol. 10, Worzel, J. L., Bryant, W., *et al.*, eds.: U.S. Gov. Printing Office, Washington, 748 p.

Worzel, J. L., Leyden, R., and Ewing, M., 1968, Newly discovered diapirs in Gulf of Mexico: *Am. Assoc. Petr. Geol. Bull.*, v. 52, p. 1194–1203.

Worzel, J. L., and Watkins, J. S., 1973, Evolution of the northern Gulf Coast deduced from geophysical data: *Gulf Coast Assoc. Geol. Soc. Trans.*, v. 23, p. 84–91.

Yarborough, H., 1967, Geologic history of the Gulf basin [abstract]: *Gulf Coast Assoc. Geol. Soc. Trans.*, v. 17, p. 160.

Yungul, S. H., and Dehlinger, P., 1962, Preliminary free-air gravity anomaly map of the Gulf of Mexico from surface-ship measurements and its tectonic implications: *J. Geophys. Res.*, v. 67, p. 4721–4728.

Yungul, S. H., 1971, Magnetic anomalies and the possibilities of continental drifting in the Gulf of Mexico: *J. Geophys. Res.*, v. 76, p. 2639–2642.

Chapter 3

GEOPHYSICAL STUDIES IN THE CARIBBEAN SEA

J. E. Case

U.S. Geological Survey
345 Middlefield Road, Menlo Park, California 94025

I. INTRODUCTION

Since the pioneering pendulum gravity measurements of Vening-Meinesz and Wright (1930) and Hess (1933, 1938), the Caribbean Sea, with its multitude of geologic problems, has received intensive study as new geophysical techniques evolved. General features of crustal structure are now partly known through seismic-refraction and gravity measurements, and the locations of major sedimentary basins and many shallow structures have been defined using seismic-reflection profiles and gravity measurements. Correlation of reflection data with stratigraphic information obtained from deep drill holes, such as those of the Deep Sea Drilling Project (DSDP), has led to a rudimentary knowledge of time-stratigraphic and sedimentological features of the southern Caribbean basins. The general gravity field of the Caribbean region, both onshore and offshore, is rather well established. Enough heat-flow data are available in the Caribbean Sea to determine some regional patterns, but far more measurements are required for true definition of the heat-flow field. The magnetic field of the Caribbean, as of December 1973, is a troublesome problem. Essentially the entire Caribbean has been surveyed, but the published surveys tend to cover rather spotty areas. Some surveys were made by ships and others by aircraft; some results have been presented as total-intensity maps, and others as residual-anomaly maps. Thus, at this time, a clear picture of Caribbean-wide magnetic patterns cannot be obtained from the published literature.

Theories of origin of the Caribbean Sea and its marginal lands are highly varied, as can be seen in the reviews by Meyerhoff and Meyerhoff (1972) and Nagle (1971). Some investigators have proposed schemes involving tensional rifting and extreme rotation of various Caribbean blocks (Carey, 1958; Freeland and Dietz, 1971) as part of a continental drift process. Others regard the Caribbean, especially the southern part, as part of a Pacific plate that has moved eastward or remained relatively static as North and South America drifted westward at a faster rate (North, 1965; Malfait and Dinkelman, 1972; Molnar and Sykes, 1969). Still others have regarded the Caribbean as a foundered continental mass (Škvor, 1969), perhaps formed by a process of oceanization or basification. Funnell and Smith (1968), Ball and Harrison (1969), and Harrison and Ball (1973) have invoked north–south extension combined with left-lateral shear. Many investigators in recent years have assumed that the volcanic Lesser Antilles are a product of westward underthrusting of Atlantic oceanic crust beneath the island arc. Meyerhoff and Meyerhoff (1972) postulated that the Caribbean Basin has existed since Jurassic time and was bordered on its northern and southern sides by orthogeosynclines.

In this chapter, an attempt will be made to inventory the status of regional geophysical studies and to present some of the principal interpretive findings. More detailed descriptions of many structural features and regional tectonic syntheses will be found in other chapters of this volume. New gravity studies along the Venezuelan Borderland will be described at considerable length because the full interpretations have not been published elsewhere.

From the data in hand, it is clear that many of the onshore geologic features of the Caribbean region extend into the marine domain and that boundaries between "oceans" and "continents" are becoming increasingly blurred as more data accumulate.

II. GENERAL CRUSTAL STRUCTURE

A knowledge of the composition and structure of the Caribbean crust and upper mantle is crucial to solve the question of how the Caribbean evolved. It is well known that the Caribbean crust is heterogeneous and has seismic velocities that tend to be intermediate between typical oceanic and typical continental values. Does this mean that the Caribbean crust is oceanic but atypical? That the crust has been oceanized, basified, or serpentinized? Is the crust mostly continental but containing sufficient mafic igneous rocks to impart a quasi-oceanic seismic character? Particularly acute is the question of

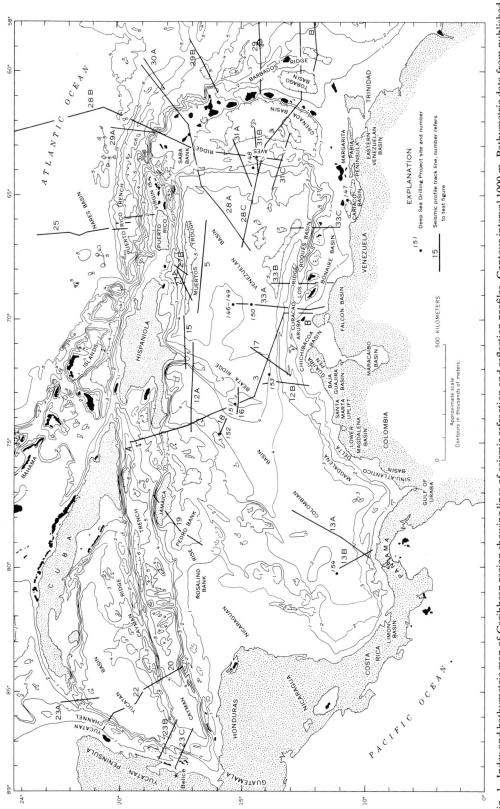

Fig. 1. Index and bathymetric map of Caribbean region showing lines of seismic refraction and reflection profiles. Contour interval 1000 m. Bathymetric data from published maps of U.S. Naval Oceanographic Office and unpublished compilations by R. R. Murchison, J. Rebman, and F. L. Marchant, U.S. Naval Oceanographic Office, Silver et al. (1975), R. G. Martin, Krause (1971), Weeks et al. (1971), and Fink (1972).

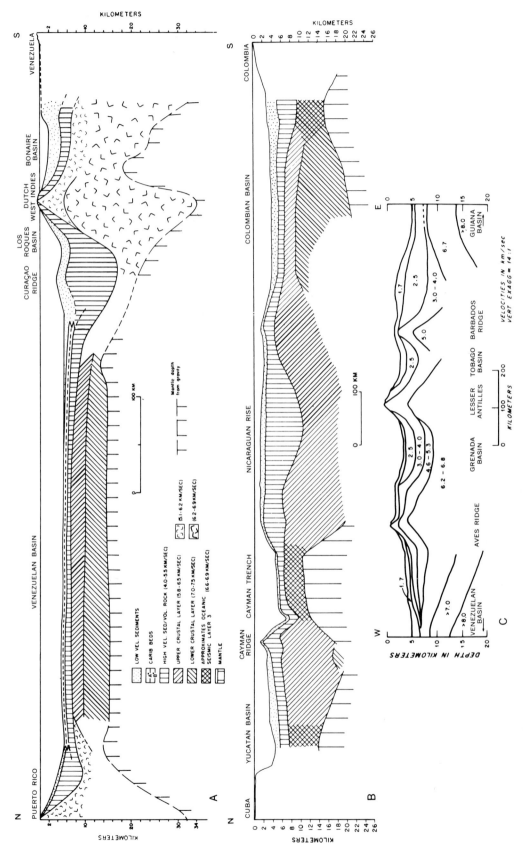

Fig. 2. Generalized crustal models across the Caribbean Sea. (A) Section from Puerto Rico to Venezuela; (B) section from Cuba to Colombia; (C) section from eastern Venezuelan Basin to the Atlantic Ocean basin. From Ewing et al. 1970, Fig. 12; Bunce et al. 1970, Fig. 2. Published with permission of the authors and John Wiley & Son, Inc.

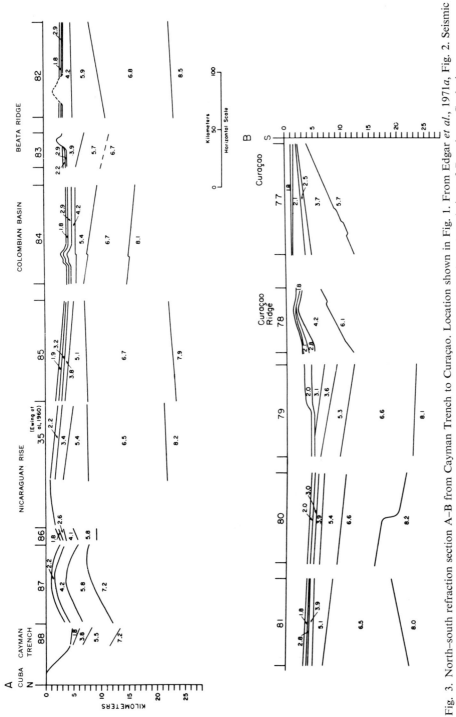

Fig. 3. North–south refraction section A–B from Cayman Trench to Curaçao. Location shown in Fig. 1. From Edgar *et al.*, 1971*a*, Fig. 2. Seismic velocities in km/sec. Published with permission of the authors and The American Association of Petroleum Geologists.

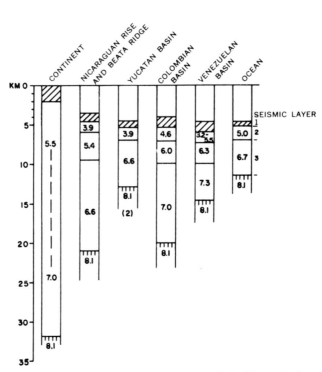

Fig. 4. Comparison of velocity structure of Caribbean basins
and ridges with those of continents and ocean basins. From
Edgar *et al.*, 1971*a*, Fig. 9. Seismic velocities in km/sec. Published
with permission of the authors and The American Association of
Petroleum Geologists.

crustal composition beneath the Cayman Ridge, Nicaraguan Rise, Beata
Ridge, and Aves Ridge.

Comprehensive summaries of crustal seismic-refraction data were presented during the early 1970's by Ewing *et al.* (1970) and Edgar *et al.* (1971*a*).
Locations of some refraction profiles are shown in Fig. 1, and representative
crustal models across the Caribbean are shown in Figs. 2 and 3. Ewing *et al.*
(1970) compared the velocity structure of the Caribbean region with that of
typical ocean basins and concluded that the Caribbean crust is generally thicker
than "normal" oceanic crust but its velocity structure is definitely not that of
a typical continent (Fig. 4).

III. SEISMIC REFLECTION

In the southern Caribbean Sea, several prominent subbottom seismic
reflectors are widespread and provide a basis for extended correlations. Most

prominent are seismic Horizons A'' and B'' (Fig. 5), which have now been identified as to age and general rock type by drilling. Horizon A'' is a cherty or siliceous–calcareous sediment sequence of about early Eocene age, and Horizon B'' is near the contact of a chert or siliceous limestone sequence with basalt, dolerite, or diabase (Edgar *et al.*, 1971*b*). The age of the oldest sediments overlying this igneous basement where penetrated by drill holes is now known to range from Late Turonian/Coniacian to Maestrichtian (Edgar, Saunders *et al.*, 1973).

Above and below Horizon A'' are acoustically transparent or less strongly reflecting beds, known as the Carib Beds (Ewing *et al.*, 1967, p. 92; 1968, p. 317–318). These have been identified as pelagic and hemipelagic sediments and sedimentary rocks (Edgar *et al.*, 1971*b*). In many parts of the southern Caribbean, particularly in marginal troughs, strong reflectors occur in the upper part of the sedimentary section, and these have generally been regarded as turbidites. Many of these beds, where dated, are Pleistocene or Holocene in age. Carbonate reefs, limestones, and limestone terrace deposits have been identified in some areas. Where these occur at shallow depth, they commonly cause seismic "ringing" or multiples that obscure deeper seismic reflectors. Where drilled or dredged, the "basement" includes a spectrum of igneous rocks from ultramafic through basaltic or gabbroic to granitic in composition; metamorphic rocks have been found at some localities. In many places, "basement," as recorded on single-channel, unprocessed seismic records, may be carbonate deposits.

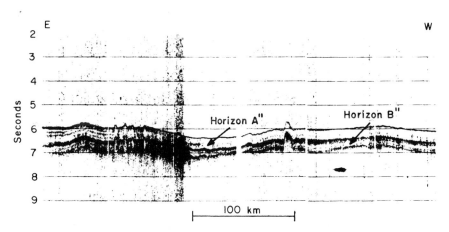

Fig. 5. Profiler record in the northern Venezuelan Basin showing two prominent subbottom reflectors A'' and B'' and the acoustically transparent Carib Beds above and below A''. See Fig. 1 for location of section. From Ewing *et al.*, 1970, Fig. 14. Published with permission of the authors and John Wiley & Sons, Inc.

IV. GRAVITY ANOMALY FIELD

Relatively complete free-air and Bouguer anomaly data for the Caribbean, based on pendulum measurements at sea, were presented by Worzel (1965). Data from shipborne gravimeters have accumulated rapidly during the interval 1959–1972, and free-air and Bouguer anomaly maps of the Caribbean region, utilizing much of the combined data, have been compiled by Bowin (in press) (Figs. 6 and 7) and by the Lamont–Doherty Geological Observatory (Talwani, oral communication, May 1973).

Conspicuous free-air anomalies include the negative anomaly (−200 to −300 mgal) that occurs over the Puerto Rico Trench* and extends east of the Lesser Antilles to Barbados Ridge. Another negative anomaly (−100 to −150 mgal) extends from Barbados Ridge, across southern Trinidad, and on land over the Eastern Venezuela sedimentary basin. A negative anomaly follows the southern margin of the Caribbean from west of Aves Ridge to the Magdalena Delta region. Still other negative anomalies occur off the north Panama coast and along the Cayman Trench system. Elsewhere in the Caribbean Sea, anomalies are of low amplitude, ranging from −50 to +50 mgal, and this suggests general isostatic equilibrium for most of the internal Caribbean region.

Bouguer anomalies range from +200 to +350 mgal in the interior Caribbean region, except where thick sedimentary sequences or thickened crust is suspected. Bouguer anomalies over the Antilles range from 0 to +150 mgal, in contrast to values in northern South America, most of Central America, and Yucatan, where values are generally below +50 mgal or are negative. Bouguer anomalies of the Santa Marta region of northern Colombia and eastern Panama, however, exceed +100 mgal.

V. SEISMICITY

Precise knowledge of the seismicity of the Caribbean has lagged somewhat because of the sparse distribution of sensitive recording sites prior to the 1960's. Earthquake data in the Caribbean area for 1950–1964 were presented by Sykes and Ewing (1965), and data for 1954–1962 for the Central America and southern Mexico region were described by Molnar and Sykes (1969). Figure 8 summarizes the earthquakes in the region for 1961–1968, from data compiled by the U.S. Coast and Geodetic Survey. In the Caribbean region, belts of shallow-focus earthquakes follow the Cayman Trench fault complex (Fig. 1). Hispaniola, the Puerto Rico Trench, and the Lesser Antillean arc

---

* For most place names and geologic features, see Fig. 1. Place names follow usage of Board of Geographic Names.

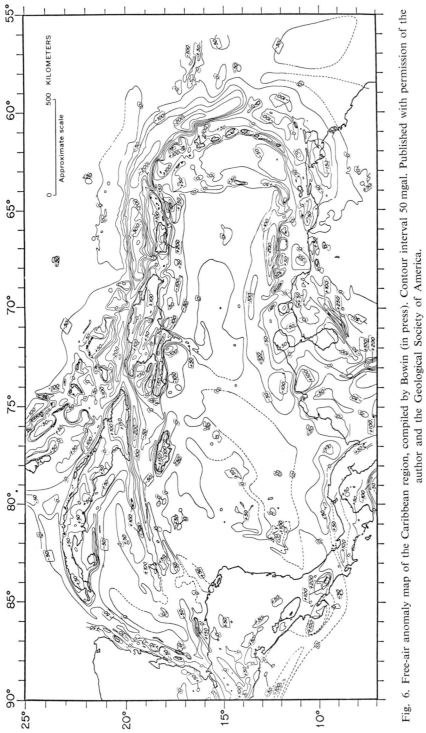

Fig. 6. Free-air anomaly map of the Caribbean region, compiled by Bowin (in press). Contour interval 50 mgal. Published with permission of the author and the Geological Society of America.

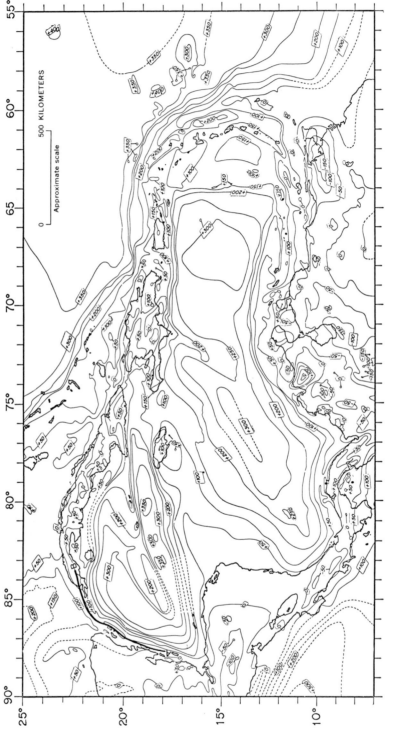

Fig. 7. Bouguer anomaly map of Caribbean region, compiled by Bowin (in press). A–B shows relative negative anomaly. Reduction density 2.67 g/cm³, except Cuba, where it is 2.3 g/cm³. Contour interval 50 mgal. Published with permission of the author and the Geological Society of America.

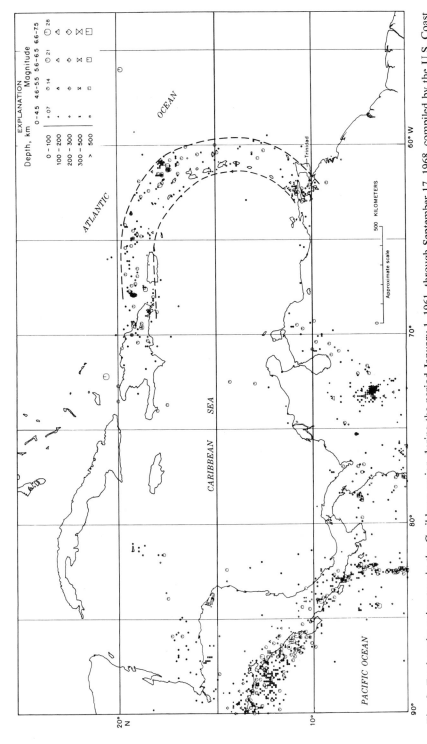

Fig. 8. Earthquake epicenters in the Caribbean region during the period January 1, 1961, through September 17, 1968, compiled by the U.S. Coast and Geodetic Survey. Dashed lines outline earthquake zone from Hispaniola through the Lesser Antillean arc. These data show less dispersion in area northeast of Trinidad than epicenters of Fig. 1 of Sykes and Ewing (1965) for period 1950 through 1964. From Bowin, 1972. Published with permission of the author and the Geological Society of America.

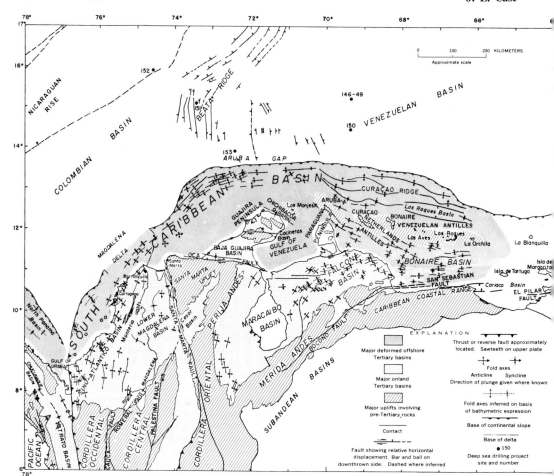

Fig. 9. Simplified tectonic map of the continental margin of part of northern South America. Principal sources of data: northern Colombia, Irving, 1971; Venezuela, King, 1969; Beata Ridge, Roemer, 1973; eastern Panama, Panama Direccion General de Recursos Minerales, 1972; fold trends off Colombia, Krause, 1971, Shepard, 1973, Emery and Uchupi, 1972a, Roemer, 1973. Fold trends in the vicinity of Bonaire Basin and Curaçao Ridge are modified from data of Silver et al. (1972, 1975) and Scientific Staff, 1972.

have abundant earthquakes. Earthquakes having focal depths greater than 100 km are largely confined to a band extending from eastern Hispaniola around the interior Antillean arc to northwest of Trinidad. Principal earthquake activity in northern South America occurs along the Bocono and Santa Marta fault systems of Venezuela and Colombia (Fig. 9), but a diffuse band of shallow-focus earthquakes extends across western Colombia and into eastern Panama (Case et al., 1971). In the interior Caribbean basins, a few earthquakes have occurred near Beata Ridge, and an interesting patch of shallow-focus earthquakes lies off Nicaragua–Costa Rica, on the Nicaraguan Rise.

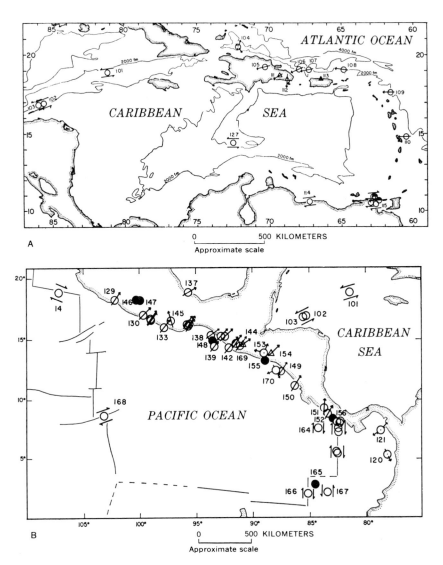

Fig. 10. Azimuths of slip vectors for earthquake mechanisms in the Caribbean region (A) and the middle America region (B). Modified from Molnar and Sykes, 1969, Figs. 12 and 14. The sense of motion for strike-slip faulting is given by a pair of parallel opposed arrows that surround an open symbol. For thrust faults, an arrow through an open symbol gives the direction of underthrusting. When the ambiguity of which nodal plane is the fault plane cannot be resolved, both possible slip vectors are given. Normal faults are given by closed symbols; neither possible slip vector is shown. Earthquakes deeper than 100 km are shown as triangles and those shallower than 100 km as circles. Hinge faulting which occurs near 10° N 62° W is indicated by half-closed circles; the fault planes in these cases strike eastward, and the north side moves down with respect to the south side along steeply dipping planes. Published with permission of the authors and the Geological Society of America.

Tomblin (1972) has studied the distribution of earthquakes in the eastern Caribbean region and shows, as did Sykes and Ewing (1965, Fig. 4), that distribution of hypocenters defines a Benioff zone that dips westward beneath the Antillean arc.

Sykes and Ewing (1965, p. 5073) noticed that hypocenters of the earthquakes through the Greater Antilles between 64° and 69° W revealed that the vertical distribution of activity is more complicated than that of a simple dipping focal surface. This point was reiterated by Khudoley and Meyerhoff (1971) who replotted hypocenters across Puerto Rico and Hispaniola separately. From their plot, one can infer that a major zone of earthquakes dips southward beneath the islands from the Puerto Rico Trench, that a secondary zone dips northward from the Muertos Trough, and that the data in no way preclude strike-slip faults in the Puerto Rico Trench or Muertos Trough.

Molnar and Sykes (1969) have determined fault-plane solutions (Fig. 10) for some earthquakes in the Caribbean region. Although complicated in detail, the major features are (1) underthrusting of the Pacific (Cocos) plate from the Middle American Trench beneath Central America; (2) left-lateral strike-slip motion along the Cayman Trench; (3) oblique underthrusting from the Puerto Rico Trench beneath northeastern Hispaniola; (4) underthrusting of Atlantic oceanic crust beneath the Lesser Antillean arc; (5) right-lateral strike-slip motion, combined with hinge faulting, near the Paria Peninsula of Venezuela, along the San Sebastian–El Pilar system of northern South America.

VI. HEAT FLOW

Regional patterns of heat flow (Fig. 11) in the Caribbean have been reviewed by Epp et al. (1970), and those in the Cayman Trench and Yucatan Basin by Erickson et al. (1972). Heat flow in much of the region seems to be "normal"; that is, most values are in the range of 1–2 μcal/cm^2-sec (HFU). The Cayman Trench has high mean heat flow in the western part, 2.07 HFU, and highest values tend to occur in the deepest part of the trench. From these relationships, Erickson et al. (1972) suggested an origin for the trough by extension normal to the axis of the trough and/or strike-slip faulting related to the eastward movement of the Caribbean lithospheric plate relative to the Atlantic plate. Epp et al. (1970) found that below-average values of heat flow occur east of the Antillean arc, and above-normal values to the west.

VII. GEOPHYSICAL DATA IN THE CARIBBEAN REGION

The discussion of geophysical data bearing on various structural-geomorphic segments of the region will deal first with the Colombian Basin and

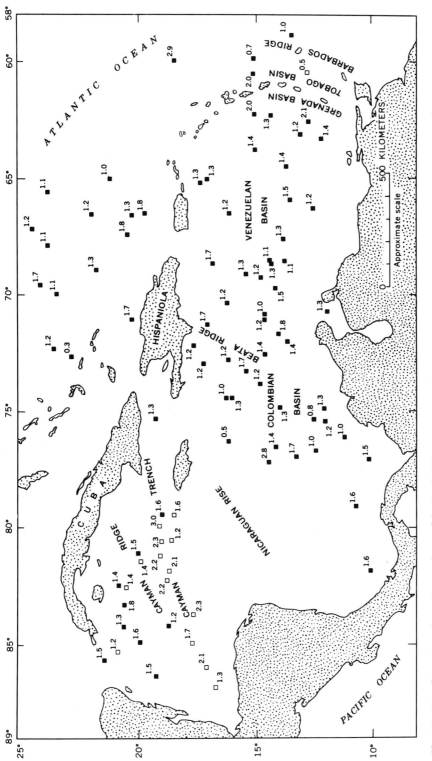

Fig. 11. Heat-flow values in μcal · cm⁻² · sec⁻¹ in the Caribbean Sea and adjacent Atlantic basins. Closed squares are data points from Epp *et al.*, 1970; open squares are data points from Erickson *et al.*, 1972.

Beata Ridge (Fig. 1), and then with the other regions in an approximate clockwise direction.

A. Colombian Basin

In the Colombian Basin, Ewing *et al.* (1960) and Edgar *et al.* (1971*a*) have shown that mantle material having velocities of about 7.8–8.2 km/sec is overlain by a lower crustal layer 6–12 km thick having velocities of 6.7–7.5 km/sec (Figs. 2 and 3). This lower crustal layer, in turn, is overlain by an upper crustal layer 2–3 km thick having velocities of about 5.8–6.5 km/sec. Above the upper crustal layer is a layer about 1.5–4 km thick, probably composed of relatively high-velocity (4.0–5.5 km/sec) sedimentary and igneous rocks. The upper part of this sequence may include or coincide with seismic Horizon B″, the prominent deep reflector that is widespread on seismic profiles in the Colombian and Venezuelan Basins (Figs. 5 and 12). Low-velocity sediments range in thickness from about 1.1 to 2.1 km.

Along the southern margin of the Colombian Basin, Krause (1971) found that sediments increase in thickness southward from about 0.6 to 1.5 secs* (Fig. 12B). The nearest DSDP drill hole (site 153, Fig. 1) bottomed in basalt beneath Coniacian sediments at a depth of about 760 m below the sea floor (Edgar *et al.*, 1971*b*). Krause's profile clearly shows a filled trench at the southern edge of the Colombian Basin; the upper part of the trench contains turbidites. In Aruba Gap (Fig. 9), Hopkins (1973) found more than 1 sec (± 1 km) of layered material, presumably a volcanic-sedimentary sequence below B″. He inferred that the basement is probably as old as Early Cretaceous or even Jurassic. The deeper and outer segment of the Continental Slope contains at least 1 sec of folded and faulted turbidites, presumably of Tertiary age. This zone of deformation extends from Curaçao Ridge, at the south margin of the Venezuelan Basin, southwestward (beneath Magdalena Delta sediments?), and extends onshore as the Sinu-Atlantico basin, in northwest Colombia. The deformed belt off the Colombian coast is also shown by Emery and Uchupi (1972*a*) and in the chapter by Uchupi in his Fig. 15, profiles B and C. Krause's profile shows a landward belt of prograded sediments extending up on the shelf off the Santa Marta–Guajira region.

Ewing *et al.* (1967) and Edgar *et al.* (1971*a*) showed that Horizons A″ and B″ can be traced partly across the Colombian Basin, in an east–west direction, from Beata Ridge onto the flanks of the Nicaraguan Rise (Fig. 12A). Southward across the basin, A″ and B″ can be traced beneath a filled trench

---

* Here and elsewhere in this report, reflection times are expressed in two-way travel times. Sediments in the region probably have a mean velocity of about 2 km/sec; thus 1 sec of reflection time approximately equals 1 km of sediment thickness.

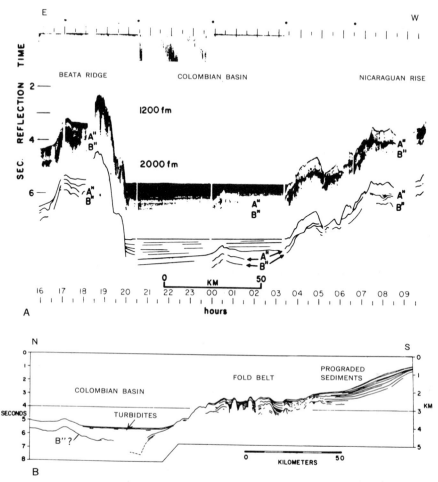

Fig. 12. Reflection profiles in the Colombian Basin. Locations shown in Fig. 1. (A) Profiler record across northern Colombian Basin showing presence of A″ and B″ on Beata Ridge, Nicaraguan Rise, and beneath flat stratified turbidites of abyssal plain (airgun, C-10). From Ewing *et al.* (1967) and Edgar *et al.* (1971a, Fig. 12). (B) Seismic profile across Colombian Basin north of Guajira Peninsula. Note progradation of sediment, fold belt, and turbidite-floored Colombian basin. Modified from Krause (1971). Published with permission of the authors. The American Association of Petroleum Geologists, and the Geological Society of America.

along the northern margin of Panama (Fig. 13A). The Carib beds bend down beneath the Panamanian margin and are overlain by horizontally bedded turbidites. Other profiles (Edgar *et al.*, 1971a) show that turbidites in the trench are deformed in the outer ridge off northern Panama (Fig. 13B). A similar relation is seen on profile A, Fig. 15, of Uchupi's chapter. This Tertiary fold belt north of Panama is similar to that off Colombia, and a major regional

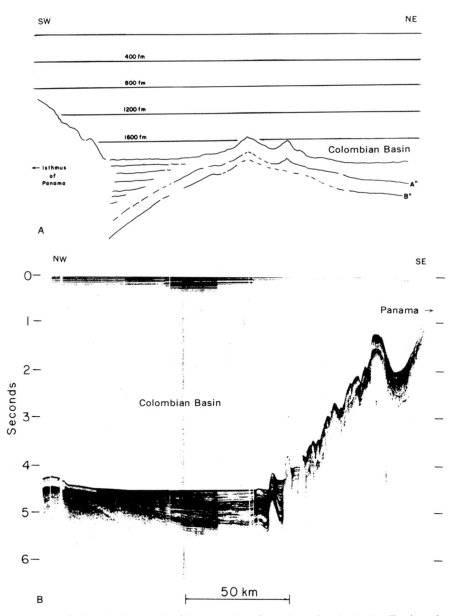

Fig. 13. Reflection profiles north of Panama. Locations shown in Fig. 1. (A) Tracing of reflection profiler record in the Colombian Basin northeast of the Isthmus of Panama. Vertical exaggeration is ×25. From Ewing *et al.*, 1967. Length of profile about 400 km. (B) Profiler record in the southern Colombian Basin showing turbidites restricted to the marginal trough and acoustically transparent sediments occurring seaward of the trough. Note deformation of turbidite beds at the base of the northern slope off Panama. From Edgar *et al.*, 1971a, Fig. 17. Published with permission of the authors, the University of Miami Press, and The American Association of Petroleum Geologists.

tectonic problem is whether the two fold belts are continuous and have a common origin or whether they formed independently. Fold axes in Colombia trend northeast, and those off Panama trend northwest to west (Fig. 9). As a major thrust fault may be present between the fold belts in the Gulf of Uraba (Irving, 1971), an independent origin for the fold belts is plausible.

Detailed studies of the Magdalena Delta (Shepard *et al.*, 1968; Shepard, 1973) reveal that prograded deltaic sediments have been intruded by mud diapirs. In places the sediments are complexly folded and faulted, and slump folds and valleys are common.

Heezen and Muñoz (1968) and Rezak *et al.* (1972) have discussed the origin of the thick sequence of turbidites in the central part of the Colombian Basin. They inferred the turbidites to have been derived principally from the Magdalena Delta region to the southeast. This sequence extends northward across the basin to the vicinity of Hispaniola. The sequence is not present in much of the central Venezuelan Basin to the east, and this is probably due to a damming effect of Beata Ridge, especially if the principal source was the Rio Magdalena (see also discussion by Bouma *et al.*, 1972).

Free-air anomalies (Fig. 6) in the Colombian Basin range from relative highs of about 0 to +35 mgal in the central part of the basin to −100 mgal along the northern coast of Panama and −125 mgal off the Colombian coast near the Santa Marta uplift and Guajira Peninsula (Worzel, 1965, Bowin, in press). Bouguer anomalies (Fig. 7) range from +306 mgal in the central part of the basin to about +50 mgal off the coast of eastern Panama and as low as −50 mgal off the Guajira Peninsula. The marginal free-air and Bouguer anomaly lows partly reflect the presence of thick low-density Tertiary sediments in deformed sedimentary basins along the continental margin (Bowin and Folinsbee, 1970; Talwani, 1966; Talwani, personal communication, 1968; Krause, 1971; Case and MacDonald, 1973). The outer basin and fold belt north of the Santa Marta–Guajira area is evidently a continuation of the thick accumulation of sediments that underlies Curaçao Ridge farther east (Edgar *et al.*, 1971a; Silver *et al.*, 1972). Other Tertiary basins, reflected by gravity minima, splay from or join with the more northerly basin (Fig. 9): the Chichibacoa basin, northeast of the Guajira Peninsula; the Baja Guajira basin, just north of the Santa Marta uplift; and the Magdalena basin west of the Santa Marta uplift (Case and MacDonald, 1973).

The negative gravity anomaly off the north coast of Panama coincides with the sedimentary basin detected by Ewing *et al.* (1967). The axis of the negative gravity anomaly north of Panama may extend onshore near the Panama–Costa Rican border, but the onland gravity anomalies are not well defined.

Gravity surveys in northwestern Colombia and eastern Panama have been interpreted by Case *et al.* (1971) and Case (1974) who found that prominent positive anomalies occur over such uplifts as the Serrania del Darien and Sautata arch which have a pre-middle Eocene core of basaltic to andesitic rocks, locally intruded by dioritic to granitic plutons. Prominent negative anomalies coincide with Tertiary sequences in the Chucunaque basin of eastern Panama and the Atrato–San Juan basin of northwestern Colombia. Farther west and south, oceanic crust of Campanian or older age near the Pacific margin of Panama and Colombia causes Bouguer anomalies in excess of +100 mgal (Fig. 7). The sequence of pillow basalt with associated cherts of eastern Panama and northwest Colombia is essentially an outcrop of Horizon B″.

Magnetic anomalies in the Colombian Basin are irregular and trend approximately east–west (Fig. 14; Krause, 1971; Christofferson, 1973). Christofferson (1973) has inferred that these anomalies are related to an east-trending spreading center which lay either north or south of the present physiographic boundaries of the basin. He favored the interpretation that the crust within

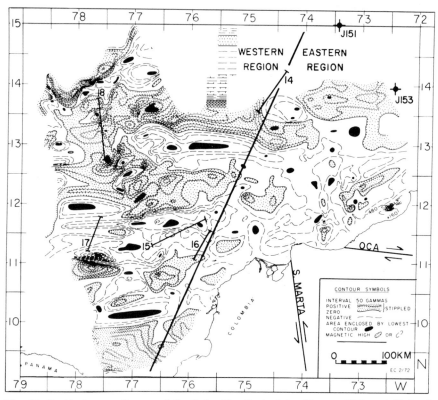

Fig. 14. Residual magnetic map of the Colombian Basin (after Krause, 1971, and Christofferson, 1973). The heavy line is an inferred northeast-trending fault having right-lateral displacement that divides the Colombian Basin into distinct magnetic provinces. Published with permission of the authors and the Geological Society of America.

the Colombian Basin survey area was generated from the south during the interval 67 to 76 m.y. ago. Krause (1971, p. 49) found that anomalies in the deep basin off Colombia trend firmly east, in contrast to marginal anomalies which trend roughly parallel to the coast.

Only a few shallow-focus earthquakes have been reported in the interior Colombian Basin near Panama and Costa Rica (Fig. 8). The heat flow ranges from 0.8 to 1.7 HFU (Fig. 11).

B. Beata Ridge

Refraction profiles across Beata Ridge (Fig. 3) show that the crust is about 23 km thick and that the mantle beneath the crust has a velocity as high as 8.5 km per sec.

The ridge is a complex faulted horst (Fig. 15) containing many outcrops of basement rocks, graben with ponded sediments, and a variable stratigraphic section of Late Cretaceous to Pleistocene age (Ewing et al., 1967; Fox et al., 1970, Edgar et al., 1971a; Rezak et al., 1972; Roemer et al., 1973). Fox and others dredged basalt and diabase from the basement. Shallow shelf sediments dredged by Fox and others indicate that Beata Ridge was higher in early Tertiary time and has later subsided, but Edgar, Saunders, et al. (1973) found no indication of neritic sedimentation in DSDP site 151. Horizons A″ and B″ can be traced across the fault blocks of the ridge in places (Figs. 16 and 17). At DSDP site 151 (Fig. 1; Edgar et al., 1971b; Edgar, Saunders, et al., 1973), the sediment down to 367 m is foraminiferal nannoplankton ooze, chalk, and marl of Tertiary age. The section cored contains a hiatus from the early Oligocene through the middle Eocene. Santonian beds are calcareous clays containing organic, glauconitic, and volcanic clay. Amygdaloidal basalt corresponds to a 0.4-sec reflector, probably B″ (Roemer et al., 1973).

Free-air anomalies over Beata Ridge range from about 0 to +50 mgal, and Bouguer anomalies from +200 to +250 mgal (Figs. 6 and 7). Bowin (in press) regarded the topography of the ridge as being compensated at depth; thus, the ridge cannot be a simple horstlike uplift of oceanic crust and mantle, as refraction data also indicate a thicker crust. The ridge seems to be aseismic, except at its southern end, and heat-flow values range from 1.0 to 1.7 HFU.

C. Nicaraguan Rise

Few mantle determinations have been published for the area beneath the main Nicaraguan Rise. A crustal layer having velocities of 5.8–6.7 km/sec apparently extends to at least 20 km below sea level (Figs. 2 and 3). Above this, several layers having velocities of 1.9–5.5 km/sec range in thickness from about 4 to 8 km. Arden (1969) assigned the upper part of this section to

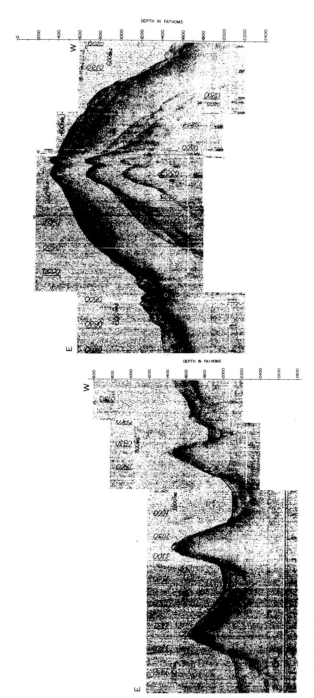

Fig. 15. Reflection profile of east flank and crest of Beata Ridge showing fault-block structure. After Rezak et al, 1972. Location shown in Fig. 1. Length of profile about 150 km. Published with permission of the authors and the Caribbean Geological Conference.

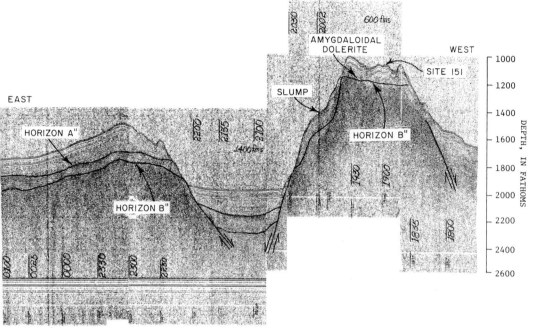

Fig. 16. Seismic reflection profile across central part of Beata Ridge near DSDP site 151 (after Roemer *et al.*, 1973). Length of profile about 90 km. Published with permission of the authors and Texas A. & M. University.

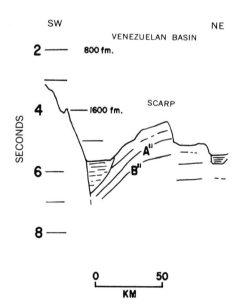

Fig. 17. Reflection profile across a southeast lobe of Beata Ridge, north of the Guajira Peninsula. From Edgar *et al.* (1971*a*, Fig. 16). Location shown in Fig. 1. Published with permission of the authors and The American Association of Petroleum Geologists.

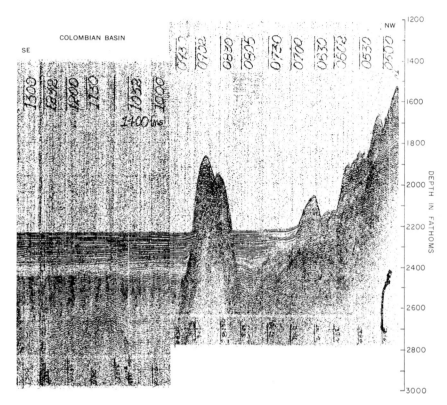

Fig. 18. Reflection profile from the Colombian Basin onto the south flank of the Nicaraguan Rise. From Rezak *et al.* (1972). Location shown in Fig. 1. Length of profile about 57 km. Published with permission of the authors and the Caribbean Geological Conference.

Tertiary and Cretaceous rocks and the lower part to lower Mesozoic limestone, volcanic, and metamorphic rocks.

The south flank of the Nicaraguan Rise is highly faulted, and the slope is devoid of sediment cover in many places (Rezak *et al.*, 1972; Fig. 18). The section shown in Fig. 18 is near DSDP site 152 (Fig. 1) where the Cretaceous–Tertiary boundary was reached at about 245 m. An upper acoustic reflector [not evident on the profile by Rezak *et al.* (1972)] is Eocene chert; the second is Campanian basalt with limestone inclusions (Edgar *et al.*, 1971*b*, p. 15). A major question concerns the nature of acoustic basement on the crest of the rise. Tertiary and Cretaceous limestones are abundant in some places on the crest (Arden, 1969; Cook, 1967), and they apparently serve as effective basement for single-channel unprocessed seismic-reflection records, especially in shallow water. The rise appears to be highly faulted, with the structural lows containing as much as 1 km of sediments (Emery and Uchupi, 1972*b*, Rezak *et al.*, 1972). As surficial deposits are displaced, some of the faulting is recent.

Power-spectra analysis shows that more than 1 sec (± 1 km) of Tertiary limestones underlie younger Tertiary bedded sediments (clastic materials?) near Pedro Bank, 100–125 miles southwest of Jamaica (Cook, 1967). According to Arden (1969), two prominent deep seismic horizons have been detected in the long section from Pedro Bank across Jamaica (Fig. 19). The sedimentary section exceeds 4 km in many places, on the basis of data from seismic reflection, drill holes, and onland Jamaican geology. A volcanic ridge may be present beneath the western part of Pedro Bank, and a deep basin containing more than 5 km of sediments, beneath the eastern part of Pedro Bank and Walton Bank.

Well data near the Nicaraguan coast indicate a sedimentary section that thickens seaward to more than 4 km. The sedimentary rocks are inferred to include Cretaceous limestone and calcareous shale overlain by Paleocene and Eocene marine shale and siltstone. Eocene limestones are about 2000 m thick. Miocene redbeds and Pliocene sands are more than 1000 m thick.

Near the coast of Honduras and Nicaragua, Pinet (1972, 1973) mapped a series of piercement structures, which he believed originated from an evaporite basin, perhaps one continuous with the Chiapas basin of northern Central America. Meyerhoff (1973), however, suggested that the piercement structures may be composed of serpentinite rather than evaporites.

Published gravity data across the Nicaraguan Rise are very scanty. Free-air anomalies apparently range from -50 to $+50$ mgal and Bouguer anomalies from $+50$ to $+100$ mgal, lower values being near Jamaica (Figs. 6 and 7). On Jamaica, free-air anomalies along the coast range from $+50$ mgal at the western end to more than $+100$ mgal along the east and north coast. Bouguer anomalies on Jamaica have a large range, from less than $+50$ mgal in the south to more than $+100$ mgal over much of the island. The Bouguer anomaly map by Bowin indicates that the anomalies over the Nicaraguan Rise are about 200–300 mgal more negative than over the Colombian Basin and Cayman Trench (Fig. 7). Part of this is the effect of differences in water depth. The Nicaraguan Rise is 3000–4000 m shallower than the adjoining basins, and the Bouguer effect of a 3000 m slab is 207 mgal (for a reduction density of 2.67 g/cm³). Thus, the apparent Bouguer low is probably a combination of shallow water, local blocks of thick sediments, and thicker low-density crust (20–25 km) under the rise than under the adjacent basins. Bouguer anomalies on land at the margin of Nicaragua and Honduras are weakly positive—about 0 to $+30$ or $+40$ mgal, so that continental crust, if it is present, must be rather thin. Bouguer anomalies on land do not become strongly negative until well inland, where values range from -50 to -100 mgal in the Honduras–Nicaraguan highlands.

Few magnetic data have been reported for the Nicaraguan Rise. Donnelly (1971) stated that a roughly circular area of conspicuously low total magnetic

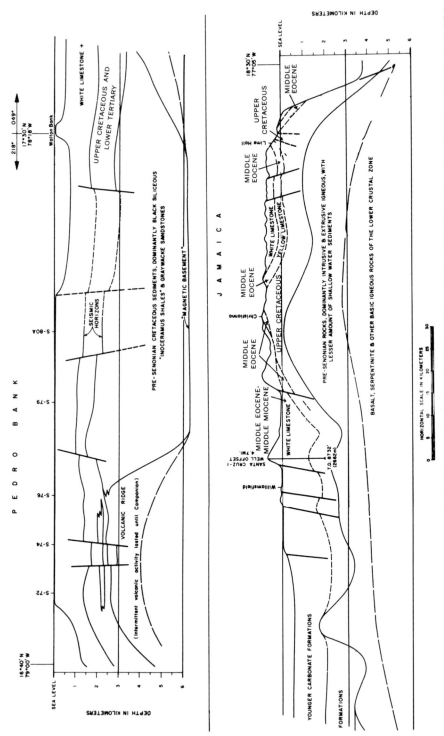

Fig. 19. Jamaica–Pedro Bank. Interpretation of seismic, magnetic, and geologic data. From Arden, 1969. Location of profile shown in Fig. 1. Published with permission of the author and the Gulf Coast Association of Geological Societies.

intensity occurs over the border between the Nicaraguan Rise and the Colombian Basin.

Several prominent topographic lineaments trend northeast across the rise. One of the most northerly, termed the Pedro Bank fracture zone by Krause and Jacobson (in press), is on strike with or en echelon with a group of onland lineaments defined on radar imagery in Nicaragua. The group of shallow-focus earthquakes off the Nicaraguan coast is partly within the area of the offshore lineaments. Surficial faults of the Managua earthquake in 1972 trended northeast and had left-lateral displacement (Brown *et al.*, 1973), and one of the possible fault-plane solutions for the Managua earthquake is a left-lateral strike-slip motion (Algermissen *et al.*, 1973).

I believe that this group of facts supports a tectonic setting similar to that along the Cayman–Motagua–Polochic left-lateral system to the north in Guatemala. Activity on the Nicaragua–Jamaica system has been less intense and may have involved much less "transtension" than along the Cayman system. The earthquake seems to have reflected left-lateral shear between nuclear Central America and a south Caribbean plate. In short, the Managua earthquake may have been more closely related to Caribbean motions than to underflow of the Pacific plate beneath nuclear Central America.

No heat-flow determinations have been reported for the main rise, and critical areas for heat-flow measurements are along the bathymetric lineaments. One heat-flow determination on a fault at the south boundary of the Nicaraguan Rise is 2.8 HFU, like values along the Cayman Trench (Fig. 11).

Arden (1969) has presented arguments that the Nicaraguan Rise is underlain by a basaltic and serpentine crust. But other investigators, Carey (1958), Freeland and Dietz (1971), for example, regard part of the Nicaraguan Rise as a continental segment that has rotated relatively away from Yucatan and South America. Resolution of the question of oceanic versus continental crust is critical to deriving a satisfactory model of the tectonic evolution of the region.

D. Cayman Trench–Cayman Ridge

Ewing *et al.* (1960) showed that the western Cayman Trench is underlain by thin crust and shallow mantle and has a velocity structure fairly similar to that of typical ocean basins (Figs. 2 and 3). In longitudinal refraction profiles in the eastern Cayman Trench, they also found that the mantle occurs at depths of 10–12 km and that the crustal velocity structure is similar to that determined from refraction profiles farther west. The crust thickens and deepens northeastward beneath the Oriente Deep and Cuban margin, however, where mantle was found at 20 km. From a combined analysis of refraction and gravity data across the trench, Bowin (1968) also inferred that mantle occurs at depths of about 9 km below sea level beneath the trench but at depths of 19 km immediately north of the trench under Cayman Ridge and at 22 km south of the trench in the Nicaraguan Rise. He concluded that the Cayman Trench was produced by extension and "upwelling" of mantle material. Serpentinized lherzolite and other ultramafic rocks were dredged from the Cayman Trench by Eggler *et al.* (1973), and the samples add weight to this interpretation, but Fox and Schreiber (1970) reported granodiorite from the eastern part of the trench. Complexity of the trench is clear from many dredge hauls on the north wall of Cayman Trench, where Heezen *et al.* (1973) reported the following crustal sequence:

> The deepest hauls recovered peridotite, serpentinite, gabbro, dolerite, basalt and their metamorphic equivalents. Somewhat higher on the escarpment, granodiorite, dacite, diorite, andesite, basalt, metasediments and ignimbrites were recovered. Conglomerate, subarkose, arenite, graywacke and argillite, including "red beds," overlie and apparently were derived from the calc-alkaline assemblage. Upper Eocene and Oligocene neritic limestone and Miocene and Pliocene pelagic ooze and micritic limestone form the uppermost strata.

Because the shallow seismic structure of the region is treated more fully in other chapters of this volume, only a single reflection profile reported by Dillon *et al.* (1972) will be described. Figure 20 shows the geologic structure from the Yucatan Basin on the northwest to near Swan Island on the southeast. The north flank of the Nicaraguan Rise off Swan Island is extremely steep, and a major fault appears near km 475 and a second near km 445. These scarps appear to be essentially free of unconsolidated sediments. The northwest flank of Cayman Trench is likewise a steep faulted boundary. Subsidiary faults occur in the trench between kms 360 and 390. Acoustic basement crops out as a low knoll near km 430, and another buried hill is near km 390. Turbiditelike sediments are of variable thickness in the floor of the trench, reaching a maximum of about 1 sec (± 1 km) in thickness. The topography of Cayman Ridge is irregular and has several basement spikes, and the northern slope of the ridge is apparently complexly faulted and possibly folded. Farther west, the ridge is somewhat narrower and is clearly a horstlike fault block with sediments preserved on the crest (Fahlquist and Davies, 1971). Malin and Dillon (1973) believed that Cayman Ridge structure extends onland into the Motagua–Polochic fault complex. Erickson *et al.* (1972) have suggested that the Cayman Trench is geologically youthful, having originated in the early Tertiary, that tectonic activity has been largely restricted to the margins, and that the central and eastern parts of the trench are isolated from sedimentary sources, whereas terrigeneous clastic materials were deposited in the western part of the trench (as much as 1.5 sec thick).

Free-air anomalies range from below -200 mgal south of Cuba to about $+50$ mgal along the margin of Cayman Trench. Over Cayman Ridge, they

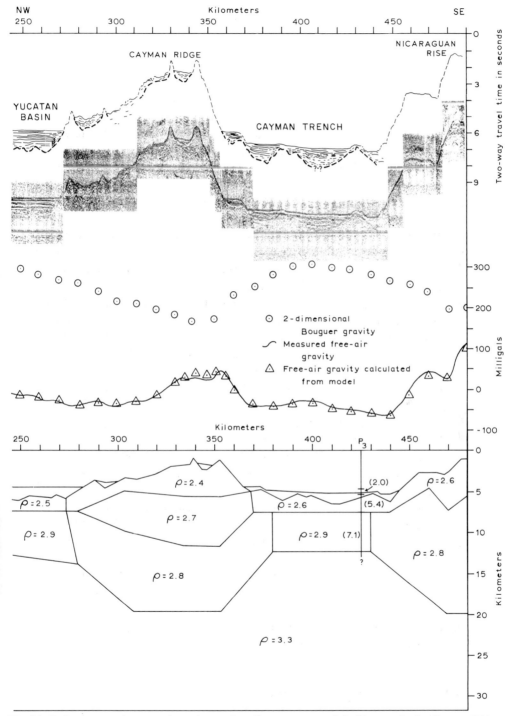

Fig. 20. Reflection, gravity anomaly, and crustal profiles across part of the Yucatan Basin, Cayman Ridge, Cayman Trench, to the Nicaraguan Rise. Location of profile shown in Fig. 1. From Dillon *et al.*, 1972, Fig. 1. Published with permission of the authors and Elsevier Scientific Publishing Company.

range from -50 to more than $+100$ mgal (Fig. 6). Bouguer anomalies are strongly positive, ranging from about $+150$ to $+200$ mgal south of Cuba to more than $+400$ mgal over deep parts of the trench. Bouguer anomalies are in the range $+150$ to $+200$ mgal over Cayman Ridge (Fig. 7). Bowin (1968) showed that the gravity data are consistent with refraction data of Ewing *et al.* (1960) in that mantle material is shallow beneath the trench. As previously noted, he interpreted the data as indicating that the trench was produced by crustal extension accompanied by upwelling of dense mantle material to shallow depths. Bowin concluded that some eastern Cayman Trench structures are on trend with tectonic features in Hispaniola rather than connecting with the Puerto Rico Trench system.

Banks and Richards (1969) and Pinet (1971) have reported on the magnetic field in the Gulf of Honduras and western Cayman Trench region. Pinet (1971) interpreted northeast-trending positive anomalies off the coast of Honduras as caused by blocks of ultramafic masses offset by northwest-trending right-lateral strike-slip faults. Banks and Richards found complex high-amplitude anomalies over the Cayman Ridge area (Fig. 21). Gough and Heirtzler (1969) found long linear magnetic anomalies on both north and south walls of Cayman Trench (Fig. 21); patterns become irregular west of $82°$ W on the north wall, and some topographic features have associated magnetic expression, but others do not. They concluded that the linear magnetic anomalies arise from juxtaposition of magnetic and nonmagnetic material in the walls of the trench caused by large strike-slip displacements with simultaneous openings in tension.

The Cayman Trench is one of the more seismic regions of the Caribbean. Comparison of maps compiled by Sykes and Ewing (1965), Molnar and Sykes (1969), and Barazangi and Dorman (1969), covering roughly the period 1950–1967, suggests that the western two-thirds of the trench is more seismic than the eastern third. Epicenters lie along a north-trending fracture zone in the central quarter of the trough (Holcombe *et al.*, 1973). The fault-plane solutions for earthquakes in the Cayman Trench indicate left-lateral strike-slip faulting, parallel to the trench (Molnar and Sykes, 1969, p. 1654).

Heat flow ranges from 1.3 to 2.3 HFU in the trench proper (Fig. 11); most values exceed 2.0 HFU. Values on Cayman Ridge range from 1.2 to 1.5 HFU. Erickson *et al.* (1972, p. 1257) concluded that the high heat-flow anomaly in Cayman Trench is in accord with both strike-slip movement and lateral extension and that both mechanisms were significant in formation of the trough.

Holcombe *et al.* (1973) have proposed that the north-trending topography in the central quarter of the Cayman Trench (Fig. 1) represents a small locus of sea-floor spreading. Whether this area of sea-floor spreading is an active cause of plate displacement or a result of passive or secondary transtensional opening between two plates appears to be an open question.

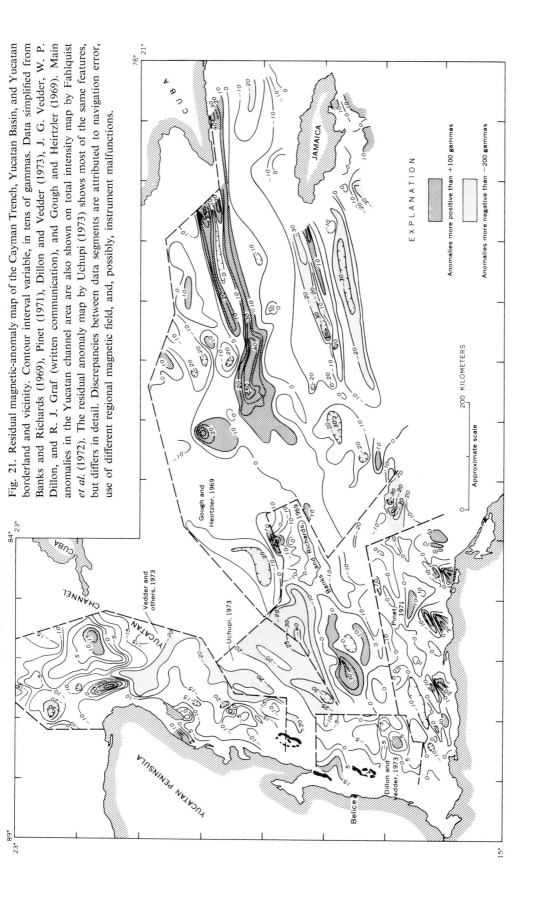

Fig. 21. Residual magnetic-anomaly map of the Cayman Trench, Yucatan Basin, and Yucatan borderland and vicinity. Contour interval variable, in tens of gammas. Data simplified from Banks and Richards (1969), Pinet (1971), Dillon and Vedder (1973), J. G. Vedder, W. P. Dillon, and R. J. Graf (written communication), and Gough and Heirtzler (1969). Main anomalies in the Yucatan channel area are also shown on total intensity map by Fahlquist *et al.* (1972). The residual anomaly map by Uchupi (1973) shows most of the same features, but differs in detail. Discrepancies between data segments are attributed to navigation error, use of different regional magnetic field, and, possibly, instrument malfunctions.

E. Yucatan Basin

Under the Yucatan Basin the crust is about 14 km thick, somewhat thinner than beneath Cayman Ridge, and the crustal velocity structure is crudely similar to that of the Colombian Basin (Ewing *et al.*, 1960). The central Yucatan Basin is remarkably flat and contains as much as 2 sec of sediments in some areas (Fig. 22). More than 1 sec of the upper sequence is composed of turbiditelike reflectors (Dillon *et al.*, 1972, Uchupi, 1973). Beneath is a sequence of variable thickness having the acoustic characteristics of pelagic sediments. The topography of the acoustic basement is variable, as shown in the figure. A prominent basement high, near km 195, appears to be a buried hill, but intrusive relations are not precluded by the seismic data. The northwest flank of the Yucatan Basin is a complex faulted margin, as discussed in the next section.

Free-air anomalies are in the range -50 to $+50$ mgal across the basin (Fig. 6). Bouguer anomalies range from $+250$ to more than $+300$ mgal, similar to values in the Colombian Basin (Fig. 7).

Few magnetic data have been published for the main basin, but surveys from peripheral areas that have overlapped the basin indicate that the anomalies tend to be of low amplitude and subdued (see, for example, Fig. 3 of Uchupi, 1973). The basin is aseismic, and heat-flow values range from 1.2 to 1.8 HFU (Fig. 11).

F. Yucatan Borderland–Yucatan Channel

Several investigations have demonstrated that the eastern Yucatan border-land is composed of a series of parallel, northeast-trending horsts and graben, with a sedimentary basin to the northwest, just off Cozumel (Fahlquist *et al.*, 1972; Pyle *et al.*, 1973; Dillon *et al.*, 1972; Vedder, Dillon, and Graf, written communication, 1973; Uchupi, 1973; Dillon and Vedder, 1973). Fahlquist *et al.* (1972) showed that two parallel ridges on the northwest side of the Yucatan abyssal plain trend northeast from the Yucatan borderland toward Cuba (Fig. 23). The main trough between these ridges contains more than 1 sec of sediments (gently deformed turbidites?). Farther southwest, the double ridge system disappears (Fig. 22, Dillon *et al.*, 1972), but the basin northwest of the ridge is deeper, having more than 2 sec of sediments. Structural complexity and the block-faulted nature of the borderland off Belice is clearly shown by the profiles of Uchupi (1973) and Dillon and Vedder (1973) (Fig. 23). Using combined data from various investigators, it has been possible to prepare an isopach map of minimum sediment thickness and generalized structure of the borderland (Fig. 24).

Vedder *et al.* (1973) dredged mafic volcanic rocks off Cabo San Antonio,

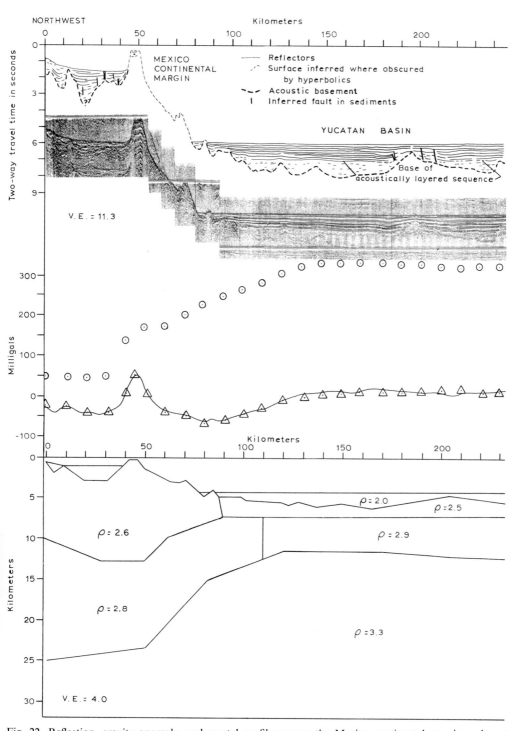

Fig. 22. Reflection, gravity anomaly, and crustal profiles across the Mexico continental margin and part of the Yucatan Basin, modified from Dillon *et al.* (1972). Location shown in Fig. 1. Published with permission of the authors and the Elsevier Scientific Publishing Company.

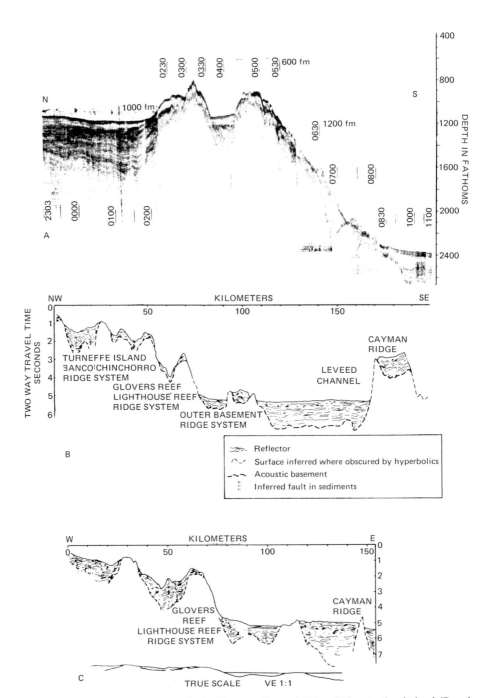

Fig. 23. Seismic-reflection profiles in the Yucatan Channel (A) and Yucatan borderland (B and C). From Fahlquist et al. (1972) and Dillon and Vedder (1973). Locations shown in Fig. 1. Length of profile A is approximately 190 km. Published with permission of the authors, the Caribbean Geological Conference, and the Geological Society of America.

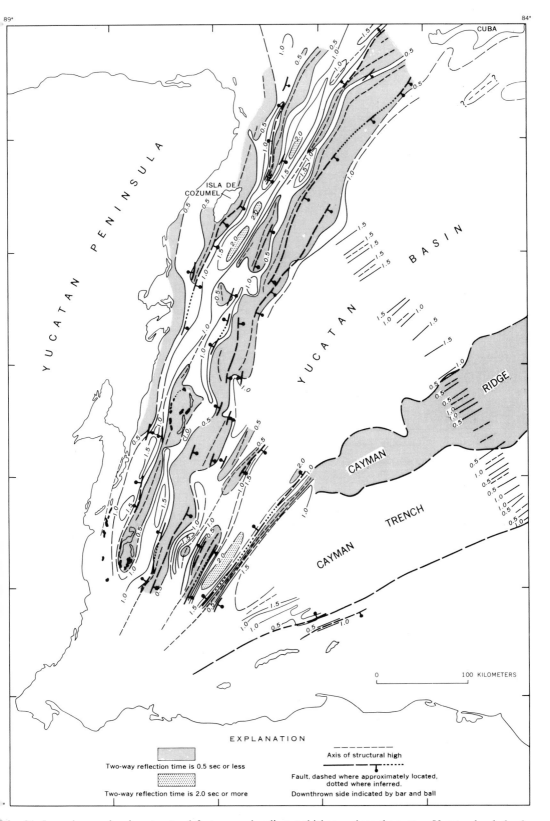

Fig. 24. Isopach map showing structural features and sediment thickness along the eastern Yucatan borderland. Modified from Uchupi (1973), Fahlquist *et al.* (1972), Pyle *et al.* (1973), Vedder *et al.* (1972), and Dillon and Vedder (1973). Minimum sediment thickness is expressed as two-way reflection time, in seconds.

westernmost Cuba, and low-grade metamorphic rocks (phyllite and marble) in the central part of the Yucatan channel and west of Isla de Cozumel. Radiometric ages of the metamorphic rocks suggest that the minimum age of the metamorphism was pre-Late Cretaceous (59.3 m.y. to 92.5 m.y.). They concluded that the phyllites were produced by older, perhaps middle Mesozoic, metamorphism, but probably an episode that correlates with a Late Permian to Late Triassic episode of orogeny identified on land.

Pyle *et al.* (1973) also dredged metamorphic rocks from the central part of Yucatan channel, and radiometric dates of 65 m.y. provide a minimum "Laramide" age for the latest thermal event affecting these rocks.

Detailed gravity surveys have been conducted along the eastern Yucatan margin by Vedder, Dillon, and Graf (written communication, 1973). Elongate ridges and valleys trending north–northeast are correlative with free-air positive and negative anomalies. Near Cozumel, for example, free-air anomalies are about +50 mgal; they decrease seaward to about −60 to −70 mgal over a basin and then increase to +50 or +60 mgal near 86° W over a horstlike uplift.

Gravity surveys indicate a prominent negative anomaly (A–B on Fig. 7) over the sedimentary basin that parallels the Yucatan borderland, whose axis extends southward between Turneffe and the mainland and northeastward off Cozumel. This negative anomaly extends across the Yucatan channel and may continue as the negative anomaly that dominates the axial region of western Cuba. These relations strongly suggest that no major post-Tertiary horizontal offsets occur between Cuba and the Yucatan margin (see also Baie, 1970). The Tertiary basin in Cuba evidently has been covered by the southward-moving thrusts of the Pinar del Rio province (Khudoley and Meyerhoff, 1971).

Onland in the Yucatan Peninsula, the Bouguer anomalies are strongly positive, as much as +100 mgal along the northeast coast. Westward, however, mean Bouguer anomalies are about zero. Thus, it may be postulated that oceanic crust underlies the northeastern Yucatan margin and that continental crust is dominant farther west (Krivoy, personal communication, 1972). This suggests that a block of "Cuban"-type crust extends beneath the Yucatan channel to at least 87° W.

Dillon and Vedder (1973) and Vedder, Dillon, and Graf (written communication, 1973) have presented generalized residual magnetic-anomaly maps off Belice and Yucatan (Fig. 21). Fahlquist *et al.* (1972) found a closed gentle magnetic high in the central part of Yucatan channel and a strong positive anomaly of short wave length west of the channel, which is related to a possible igneous intrusion in the shallow subsurface.

The Yucatan borderland is aseismic. No heat-flow determinations have been made.

Although differing in some important details, most recent investigators have proposed a variant of sphenochasmic rifting to explain the Cayman Trench–Yucatan Basin–Yucatan borderland structure. The models range from combinations of sphenochasmic rifting followed by extreme strike-slip faulting (Pinet, 1972), through sphenochasmic rifting followed by intermediate strike-slip faulting (Dillon and Vedder, 1973), to dominant sphenochasmic rifting followed by only modest strike-slip displacement (Uchupi, 1973). Meyerhoff (1962) believed that the Cayman (Bartlett) Trench began during the middle Cretaceous orogeny and that sinistral offset might be as much as 200 km, but probably is much less.

G. Cuba and Vicinity

Much of the Bouguer anomaly field in Cuba is positive (Fig. 7), reflecting the abundant mafic to ultramafic rocks that constitute the basement framework of the island. Lower values of gravity crudely coincide with Tertiary basins and with areas of metamorphic complexes such as those on the Isle of Pines and Sierra de Trinidad (see Khudoley and Meyerhoff, 1971). Data of Soloviev et al. (1964a) and Ipatenko and Sashina (1971) show that, in detail, some of the serpentine masses coincide with positive anomalies, and others do not. This may reflect not only the degree of serpentinization (and hence densities) of original peridotitic masses, but also the existence of rather thin thrust sheets. Where strong negative anomalies coincide with serpentinite areas, as in the Pinar del Rio region, thin surficial thrust sheets and lower-density (Tertiary?) sedimentary masses beneath the thrust plates may be suspected. Free-air and Bouguer anomalies in southeastern Cuba suggest continuity of the Cayman Ridge structural block with the Oriente block, as inferred by King (1969).

Magnetic surveys of Cuba have been described by Soloviev et al. (1964b), and those for the Bahama region by Emery and Uchupi (1972b), Uchupi et al. (1971), and Drake et al. (1963), and will not be reviewed here. Reflection-seismic data over Bahama–Florida Straits have been summarized by Uchupi in another chapter of this volume.

H. Hispaniola

Analysis of the gravity field in Hispaniola is presented in the chapter by Bowin. It can be pointed out here, for completeness, that Bouguer anomalies range from about -25 to $+125$ mgal (Reblin, 1973, Bowin, in press) and that negative anomalies correlate very closely with Tertiary sedimentary basins and with parts of the metamorphic-granitic belt in the central part of the island. Gravitational expression of possible landward extensions of the Cayman

Trench structure is somewhat masked by the negative Bouguer anomalies of the sedimentary basins.

Magnetic maps around the periphery of Hispaniola seem to be nonexistent, although a few profiles have been published. Virtually all Hispaniola is seismic (Molnar and Sykes, 1969, 1971). Deeper focus earthquakes seem to be concentrated on the southeast corner of the island (Bracey and Vogt, 1970). A heat-flow value off the northern coast of the island is 1.7 HFU (Epp et al., 1970).

I. Puerto Rico Trench

Many refraction and gravity lines have been obtained from the Nares Abyssal Plain southward across the Puerto Rico Trench. Excellent summaries of the results have been presented by Talwani et al. (1959), Bunce and Fahlquist (1962), and Bowin (1972). Mantle (velocities, 8.1–8.2 km/sec) occurs at depths of about 10–12 km north of the Puerto Rico Trench (Fig. 25) and extends to depths of 18–20 km beneath the trench, apparently shallows somewhat beneath the Continental Slope or inner wall of the trench, and then is inferred to be at a depth of about 30 km beneath the island of Puerto Rico. Talwani et al. (1959) and Bunce and Fahlquist (1962) have presented slightly different interpretations of the crustal velocity structure beneath the trench, but both groups derived a grabenlike structure extending from the surface through the crust.

Origin of the trench has long been controversial. Is it a pure tensional, pure strike-slip, underthrust, or combined feature? Evidence is conflicting, but a combination of strike-slip faulting with underthrusting seems most consistent with available evidence. The grabenlike nature is characteristic of transtensional faulting.

Bunce and Hersey (1966), Bunce (1966), and Bunce et al. (1974), have discussed the seismic-reflection results of the region north of the Puerto Rico Trench and some results in the deep trench. One profile presented by Bunce (1966) and Bunce et al. (1974) shows about 0.75 to 1.2 sec of sediments (including some turbiditelike reflectors) in the floor of the deep trench. In some places the basement below the abyssal plain has a rough topography.

Several varieties of serpentinized peridotite, Cenomanian limestone, Cretaceous siliceous rocks, and Eocene sedimentary rocks were dredged from the north wall of the trench by Bowin et al. (1966). Chase and Hersey (1968) also dredged tholeiitic basalt, probably of Cenomanian age, and upper Cretaceous and lower Tertiary sediments—chert, claystone, and minor limestone—from the northern wall. Schneidermann et al. (1972) dredged shallow-water middle Cretaceous to Eocene limestones from the south wall of the Puerto Rico Trench.

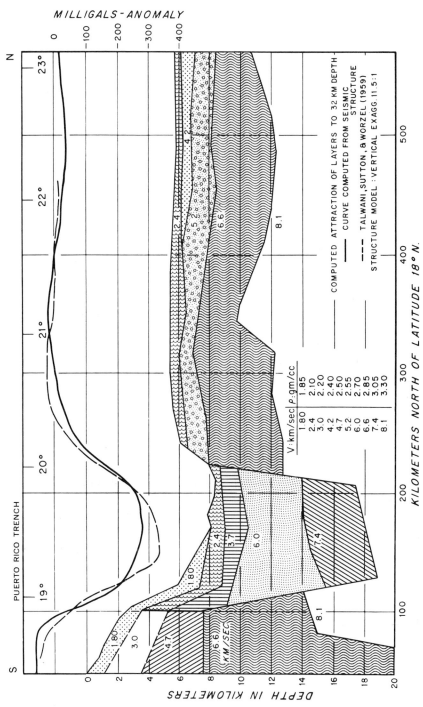

Fig. 25. Representative crustal model across the Puerto Rico Trench. Modified from Bunce and Fahlquist (1962). Location of profile shown in Fig. 1, Published with permission of the authors and the American Geophysical Union: Bunce, E. T., and Fahlquist, D. A., 1962, *J. Geophys. Res.*, v. 67, n. 10, Fig. 3, p. 3955–3972, copyright by the American Geophysical Union.

The gravity-anomaly field in Puerto Rico is positive, in keeping with the abundance of mafic volcanic rocks in the lower part of the section. Bouguer anomalies range from about +90 to +165 mgal over the island (Griscom, 1971; Shurbet and Ewing, 1956, Bowin, in press). Residual lows are present over some granodioritic bodies in eastern Puerto Rico and beneath southern Vieques Sound. Low values also coincide with areas of Tertiary sedimentary basins. A residual negative anomaly occurs in Mona Passage between Hispaniola and Puerto Rico, and this anomaly may reflect a fairly large Tertiary sedimentary basin, perhaps an offshore continuation of the north coast basin of Puerto Rico.

The magnetic field north of the Bahamas–Antilles Islands has been reviewed by Bracey (1968). He found two sets of linear magnetic-anomaly trends that are nearly perpendicular. One set trends northeast at the margin of the deep Atlantic off the Bahamas (presumably these are "spreading" anomalies from the mid-Atlantic ridge); the other set trends northwest and is especially prominent north of Hispaniola (Fig. 26). Bracey related these features to a major east–west couple in the regional stress pattern. Bunce et al. (1974) suggested that the Antilles Outer Ridge is the boundary between two magnetic provinces and near the junction of two ancient oceanic plates which accreted as Africa separated from North America and South America.

Griscom and Geddes (1966) speculated from analysis of aeromagnetic data that the rocks north of the Puerto Rico Trench may be similar in age, rock type, and structure to the basement rocks of Puerto Rico. The magnetic patterns are consistent with an origin by block faulting for the Anegada Passage, just east of the Virgin Island platform, and for the platform itself. They inferred a possible chain of submarine volcanoes south of and parallel to the Puerto Rico–Virgin Islands platform. They concluded that the aeromagnetic anomaly pattern is consistent with the hypothesis that the Puerto Rico Trench originated by downwarping on the northern side of a thrust fault which dips south beneath the islands.

East of the northern Lesser Antilles, a few linear anomalies are prominent, some trending east and others north–northeast (Bunce et al., 1974); some coincide with topographic features, but others do not, and hence probably originate at deeper levels. The Puerto Rico Trench and its extensions are magnetically quiet for the most part.

J. Muertos Trough

As Ewing et al. (1967) and Edgar et al. (1971a) showed, sediments of the Venezuelan Basin thicken slightly northward across the basin. Reflection profiles indicate that Horizon A″ dips beneath Muertos Trough and the deformed southern margin of Puerto Rico and the Dominican Republic (Fig. 27).

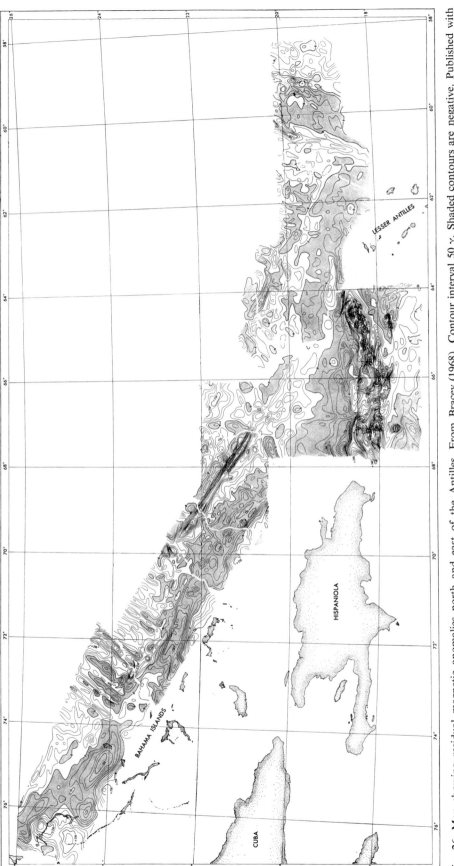

Fig. 26. Map showing residual magnetic anomalies north and east of the Antilles. From Bracey (1968). Contour interval 50 γ. Shaded contours are negative. Published with permission of the author and the Society of Exploration Geophysicists.

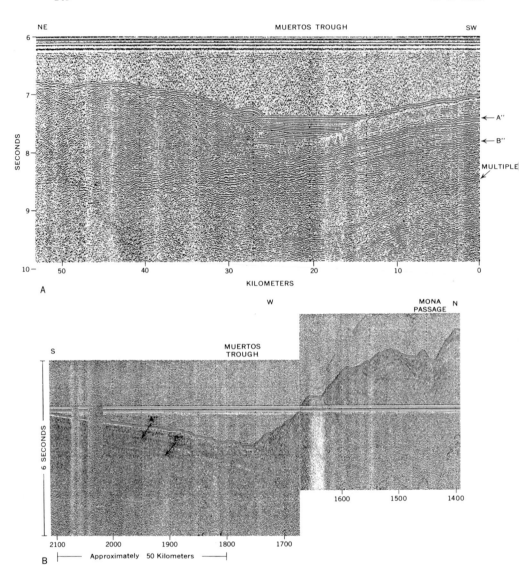

Fig. 27. Reflection profiles across Muertos Trough south of Puerto Rico. From Garrison *et al.* (1972*a*) and Silver *et al.* (1972). Location shown in Fig. 1. Profile A is diagonal to the trough. Published with permission of the authors and the American Geological Institute.

Moreover, Horizon B″ can be traced for 15 km northward from the trench axis far beneath the deformed margin of southern Hispaniola and Puerto Rico (Fig. 27). Garrison *et al.* (1972*a*) have summarized the geologic relations as follows:

"In the region south of Hispaniola and Puerto Rico, the floor of the Venezuela Basin joins the south flank of the Greater Antilles ridge along a linear east–west depression called the Muertos Trough. Although this feature lacks the strongly negative gravity anomaly of the Puerto Rico Trench ... structural relations in the Muertos Trough resemble those of the eastern Puerto Rico Trench. The landward wall of Muertos Trough is formed by a ramp of acoustically transparent material similar to that east of the Lesser Antilles, and its seaward wall is formed by downward-bent oceanic crust that is traceable for several kilometers northward beneath the ramp deposits [Figure 27]. Horizons A″ and B″, identified in JOIDES drillings in the Venezuela Basin as Eocene and Late Cretaceous (Coniacian) age, respectively, appear to be offset by normal faulting under the trough fill. Turbidite layers on the north side of Muertos Trough have been folded, uplifted, and incorporated into the ramp itself, suggesting a southward migration of the trough axis. It is thus possible to interpret the Muertos Trough as a zone of underthrusting similar to that in the eastern Puerto Rico Trench. The lack of seismic activity may imply a rate of relative plate motion much slower than that in the trench, but folding of the young turbidite layers suggests that movement has occurred in relatively recent times."

The Muertos Trough structural belt may extend across southern Hispaniola as the Enriquillo Cul-de-Sac graben system. Thus a zone of underthrusting may pass laterally in the zone of transtensional faulting represented by the Enriquillo Cul-de-Sac system.

K. Anegada Passage and Vicinity

This structurally complex area is at the intersection of the Virgin Islands platform, northern end of Aves Ridge and Saba Bank, northwestern end of the Lesser Antilles, and the Puerto Rico Trench (Fig. 1). Donnelly (1965) has proposed sets of northeast, northwest, and near east-trending faults on the platform between Vieques and St. Thomas. A major left-lateral fault through Anegada Passage has long been inferred (Hess, 1938, 1966; King, 1969, Garrison et al., 1972c), and the fault projects into the zone of apparent underthrusting of the Muertos Trough fault complex.

The gravity-anomaly field in the junction area of the Virgin Islands, northern end of Aves Ridge, and western Lesser Antilles is extremely varied. Free-air and Bouguer anomalies have wide ranges over short distances, reflecting not only the rugged topography but also the interplay between rock masses of varying density (Figs. 6 and 7). A regional Bouguer low that apparently trends across the Virgin Islands is interrupted and perhaps offset near Anegada Passage. A broader regional low occurs over Saba Bank and between Aves Ridge and St. Christofer and Nevis. A large part of this Bouguer low may be simply an effect of sea-floor topography, but onland values of these islands (+70 mgal) are much lower than values on Antigua (+160 mgal), so a thick sedimentary sequence must be suspected beneath Saba Bank.

L. Lesser Antillean Arc

Northeast and east of the Antillean arc, Ewing *et al.* (1957) and Officer *et al.* (1959) showed that the mantle, having velocities of 7.9–8.4 km/sec, is at depths of 12–15 km below sea level and that the lower crustal layer, having velocities of 6.6–6.9 km/sec, is 4–10 km thick (Fig. 28A,B). An upper crustal layer, having velocities of 4.5–6.0 km/sec, is only a few km thick in the Atlantic Basin but increases in thickness beneath the Puerto Rico Trench. The structure derived beneath the Virgin Islands segment of the Antilles is complex, but mantle appears to be at depths greater than 20 km. The M-discontinuity clearly descends beneath the Antillean islands from both the Atlantic and Caribbean sides accompanied by crustal thickening.

Farther south around the Antillean arc, Officer *et al.* (1959) showed that the mantle occurs at depths of 13 km in the Venezuelan Basin (Fig. 28A,C), where 8.1 km/sec material is overlain by about 7 km of crust having velocities of 6.0–7.2 km/sec, and this, in turn, is overlain by 1–2 km of low-velocity sediments. Mantle was not detected beneath Aves Ridge, but it apparently is present at depths of at least 18 km beneath the western flank of the ridge. The top of the 6.2–6.7 km/sec crustal layer is shallow beneath Aves Ridge, deep beneath Grenada Basin, shallow beneath the lesser Antilles, and deepens toward Tobago Trough. Seismic velocities as low as 4.9–5.3 were recorded at depths of 4–10 km beneath Barbados Ridge (Fig. 28C).

From reflection profiles, the crust of the Atlantic plate underthrusts (at least geometrically) the eastern margin of the Antilles (Chase and Bunce, 1969). This concept was further documented by Bunce *et al.* (1970). Garrison *et al.* (1972a, 1972b; Fig. 29, this report) and Marlow *et al.* (1974) showed that underthrusting is present near the Puerto Rico Trench at lat. 16° N, but that at lat. 19° N underthrusting is not so apparent, and they concluded that strike-slip faulting becomes dominant between profiles A and B shown in Fig. 29. More than 1.5 sec of pelagic sediments and possible pelagic scrapings are preserved above Horizon B on the landward wall of the trench near 16° N. From the data of Bunce *et al.* (1970), it would appear that underthrusting extends northward to at least 18°30' N (Fig. 30A).

Bunce *et al.* (1970) have shown that underthrusting extends south to about 14° N (Fig. 29C), but becomes less clearly defined farther south near 12° N (Fig. 30B). The Barbados Ridge, however, is highly crumpled and deformed at 12° N, and deep underthrusting may not be resolvable seismically. As much as 2 sec of sediments is preserved east of Barbados Ridge.

The most recent discussions of gravity-anomaly belts in the Lesser Antillean arc have been presented by Bunce *et al.* (1970) and Bowin (1972; in press). A free-air minimum, the classic "negative strip," extends as an arc south and east of the main Puerto Rico Trench and its extensions (Fig. 6).

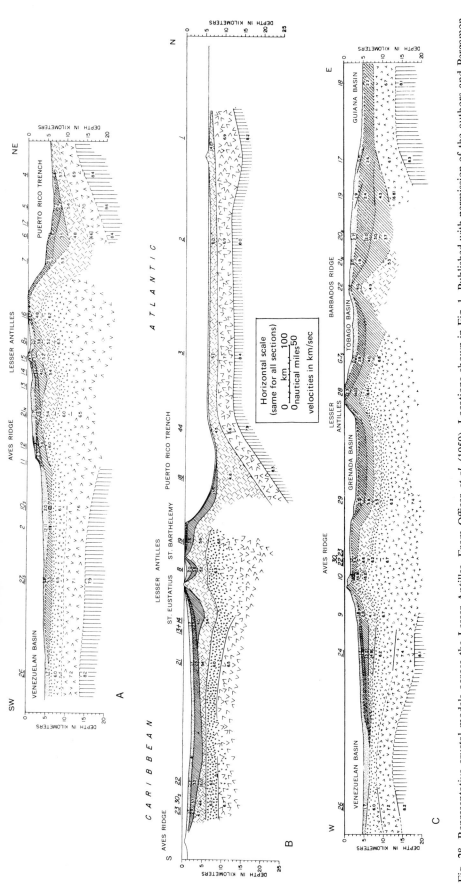

Fig. 28. Representative crustal models across the Lesser Antilles. From Officer *et al.* (1959). Locations shown in Fig. 1. Published with permission of the authors and Pergamon Press, Ltd.

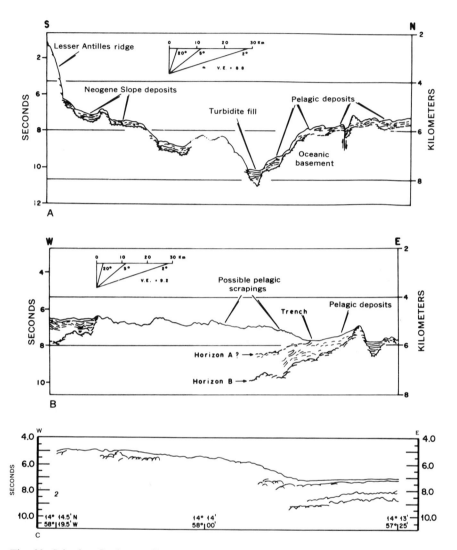

Fig. 29. Seismic-reflection profiles across extension of Puerto Rico Trench. Profiles A and B from Garrison *et al.*, 1972*a*; Profile C from Bunce *et al.*, 1970. Locations shown in Fig. 1. Length of profile C about 100 km. Published with permission of the authors, the American Geological Institute, and John Wiley & Sons, Inc.

The minimum passes through Barbados. A Bouguer-anomaly minimum extends from about 19° N, 62° W, southward past Trinidad, nearly coincident with the free-air minimum (Fig. 7). To the south, the main negative Bouguer anomaly extends onshore across the eastern Venezuela sedimentary basin. It seems clear that this anomaly coincides with a region of thick low-density sediments, and possibly also with thickened or low-density crust.

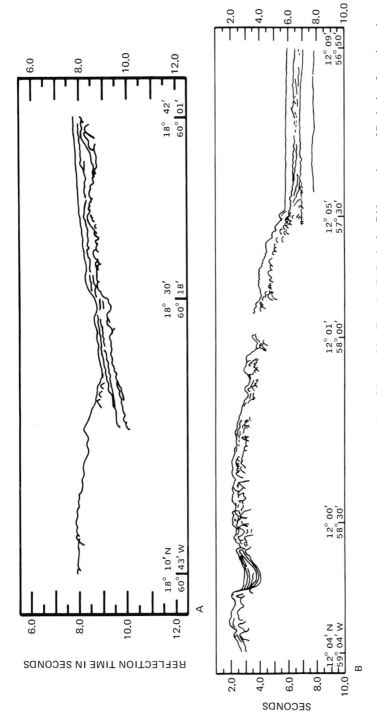

Fig. 30. Continuous seismic profiles. (A) Southeastern extension of Puerto Rico Trench; (B) Barbados Ridge southeast of Barbados. Locations shown in Fig. 1. From Bunce et al., 1970. Length of profile A about 95 km and profile B about 220 km. Published with permission of the authors and John Wiley & Sons, Inc.

A free-air and Bouguer-anomaly maximum extends through the islands of the main Antillean arc (Andrew *et al.*, 1970). Data presented here (Fig. 7), however, show that this is a double maximum: one maximum coincides exactly with the islands from Martinique southward, but a second maximum parallels it offshore to the east, generally about $\frac{1}{2}°$ from the islands.

Farther west, a small positive free-air anomaly overlies Aves Ridge. The Grenada Basin has Bouguer anomalies that are somewhat lower than those over the southern Lesser Antilles; they are probably related to the thick sequence of sediments.

According to Bunce *et al.* (1970), Barbados Ridge has a smooth magnetic field, and anomalies over the Aves Ridge and Antillean arc are characterized by high-amplitude and short wave length in comparison with the relatively smooth field over the Grenada and Tobago Basins. The anomaly field over the southern parts of both Aves Ridge and Antilles arc is smoother than that over the northern parts.

M. Grenada Basin–Aves Ridge

Sediments in the Grenada Basin appear to become progressively thicker from north to south: 1–2 sec ($\pm$1–2 km) in the north and 2 secs or more in the south (Bunce *et al.*, 1970). Turbidites (as much as 1 sec) and hemipelagic or pelagic sediments at depth are both preserved at the extreme southern end of the basin (Silver *et al.*, 1972). Basement topography beneath the sediments is rougher, and deformation of the sediments is greater in the north than in the south (see Bunce *et al.*, 1970, plate III).

Seismic-reflection profiles clearly indicate that the Aves Ridge is an anticlinal or horstlike uplift (Fig. 31). In many places basement blocks composed of mafic, intermediate volcanic, and granitic rocks are exposed (Fox *et al.*, 1971; Nagle, 1972). Horizon B'' can be traced from the Venezuelan Basin up the western flank of the ridge, but it is lost on the eastern flank. As much as 2 sec ($\pm$2 km) of pelagic sediments and turbidites is preserved across the broad crest of the ridge.

Fox *et al.* (1971) proposed that Aves Ridge is underlain by granitic crust and that volcanic activity started in the early Tertiary. This led Meyerhoff and Meyerhoff (1972) and others to the conclusion that Aves Ridge is probably a volcanic arc precursor to the Lesser Antilles arc. If these postulates are correct, it appears that a double volcanic arc existed in early Tertiary time, because local volcanism in the Lesser Antilles dates from Eocene time.

N. Southeast Corner of Caribbean

Lattimore *et al.* (1971) and Bassinger *et al.* (1971) have described the region between Venezuela–Trinidad and the southern Lesser Antilles. Their

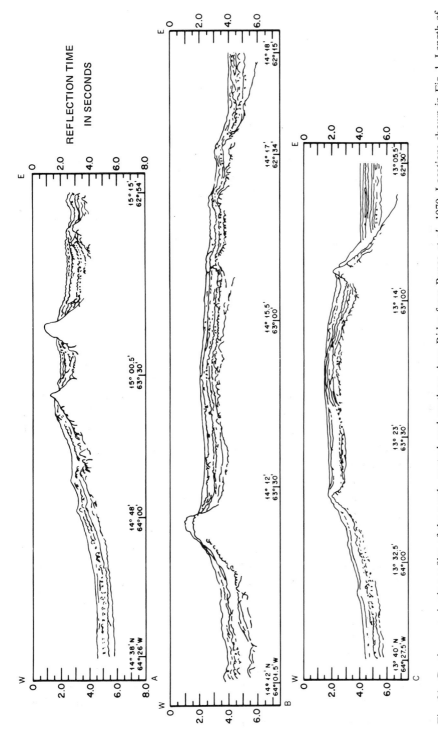

Fig. 31. Continuous seismic profiles of the north, central, and southern Aves Ridge from Bunce *et al.*, 1970. Locations shown in Fig. 1. Length of profile A is about 200 km, profile B about 200 km, and C about 215 km. Published with permission of the authors and John Wiley & Sons, Inc.

results will not be reviewed in detail, but principal structures are shown in Fig. 32. Briefly, Lattimore *et al.* infer an "igneous ridge" between Los Testigos and Grenada and a basin south of Grenada that is a prong of the Tobago Basin; they also identified several faults just north of the Paria Peninsula and Trinidad. Bassinger *et al.* (1971) also identified a series of anticlines or ridges separated by synclines north of the extension of the El Pilar fault system. More complete structural and stratigraphic data for the region based on detailed seismic surveys and several drill sites have been presented by Feo-Codecido (1973). The basin between Tobago and Grenada contains at least 6 km of Tertiary sediments, according to Feo-Codecido.

O. Venezuelan Basin

The crust of the Venezuelan Basin is thinner than that of the Colombian Basin, but the velocity structure is generally similar (Figs. 2–4).

In the Venezuelan Basin, the sedimentary sequence above Horizon B″ ranges from about 250 m at DSDP site 150 to more than 1.3 sec ($\pm$1.3 km) at the south margin of the basin off Curaçao. Eastward, about 1.5 km of sediments border the west flank of Aves Ridge; northward, about 1.5 km of sediments descend into Muertos Trough on the northern side of the basin (Fig. 27). The southern edge of the Venezuelan Basin is a filled trench. Figure 33 shows the seismic structure from north to south across the southern part of the Venezuelan Basin. Most of the sediments in the basin have a pelagic character, but at the south margin of the basin a wedge of turbidites is preserved just north of Curaçao Ridge (Fig. 33A). Eastward, the total sedimentary section thickens north of Los Roques where almost 4 sec of material is present above a deep reflector, possibly B″ (Fig. 33C). The upper 2–3 sec of material includes turbiditelike reflectors, and the lower part has a pelagic reflective character. In this and other profiles, the turbidites of the Venezuelan Basin are deformed as they pass southward into Curaçao Ridge.

Holcombe and Matthews (1973) reported that structure contours of Horizon B″ in the Venezuelan Basin reveal an orthogonal structural fabric: one set of lineaments trends northeast and appears to predate B″ and the other trends northwest, predating B″ and partly postdating A″. Isopach and structure contour maps of the Venezuelan Basin have been presented by Edgar, Holcombe, *et al.* (1973).

As far as can be determined from available gravity coverage, the Venezuelan Basin has a remarkably constant gravity field (Figs. 6 and 7). Bouguer anomalies range from +250 to +325 mgal; free-air anomalies, away from the marginal deeps, are about −25 to +25 mgal, indicative of isostatic equilibrium. The magnetic-anomaly field of the Venezuelan Basin is rather subdued, having amplitudes of 100–200 γ for the most part (Ewing *et al.*, 1960, Fig. 5). From

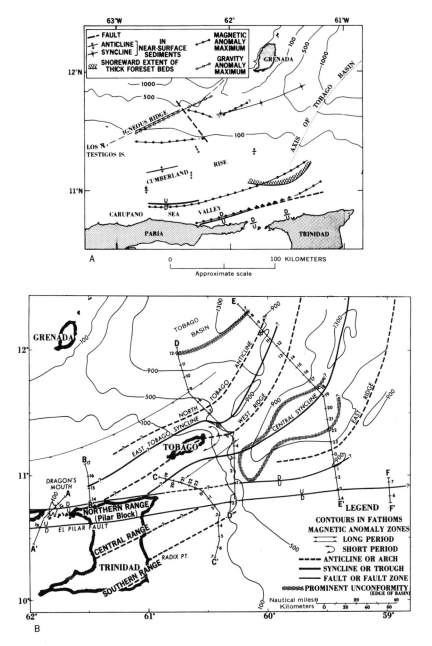

Fig. 32. Generalized structural maps of the Grenada–Trinidad region. (A) Trends and structural patterns of Paria shelf and parts of Tobago Basin and Grenada platform (contours are in fathoms). From Lattimore et al., 1971. (B) Major structural trends and location of representative traverses around Tobago. Barbados anticlinorium is complex of structures lying east of Tobago Basin through East ridge. From Bassinger et al., 1971. Published with permission of the authors and The American Association of Petroleum Geologists.

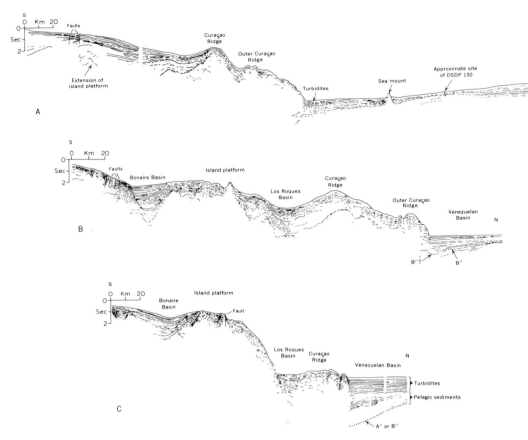

Fig. 33. Seismic-reflection profiles across the Venezuelan Basin and northern Venezuelan borderland. From Silver *et al.*, 1972, and Silver *et al.*, 1975. Locations shown in Figure 1. (A) Profile between Aruba and Curaçao; (B) profile between Curaçao and Bonaire; (C) profile across eastern end of Curaçao Ridge.

aeromagnetic profiles and a few ship-borne profiles, Donnelly (1973) found that magnetic anomalies in the Venezuelan Basin trend northeast, and he concluded that the magnetic "basement" lies considerably beneath the igneous rocks which terminated drilling at five DSDP sites. The interior Venezuelan Basin is virtually aseismic, except for a few epicenters in the vicinity of Muertos Trough (Fig. 8). Heat-flow values range from 1.1 to 1.7 HFU (Fig. 11).

P. Venezuelan Borderland

South of the Venezuelan Basin, the main geomorphic features of the Venezuelan borderland include Curaçao Ridge, Los Roques Basin, an island platform containing the Netherlands and Venezuelan Antilles, and Bonaire

Basin (Fig. 9). This area has recently been intensively surveyed by Silver *et al.* (1972, 1975).

A fundamental regional tectonic problem is the location of the northern margin of the Paleozoic and Precambrian craton of northern South America. Is there an abrupt or transitional boundary between continental "granitic" crust and oceanic "basaltic" Caribbean crust? Seismically, a transition zone evidently lies beneath the Venezuelan Borderland (Edgar *et al.*, 1971a, p. 838, Fig. 22). Seismic data on crustal properties are relatively complete in the Caribbean but are nearly nonexistent on the Venezuelan mainland.

In their comprehensive study of the gravity field of the Venezuelan Andes and adjacent basins, Hospers and Van Wijnen (1959) assumed a crustal thickness of 30 km both north and south of the Mérida Andes, southwest of the area described in this report, where Bouguer anomalies range between 0 and -150 mgal—"continental" values. They assumed a mean crustal density of 2.67 g/cm^3 and an upper mantle density of 3.27 g/cm^3. The crust was inferred to thicken southward across the Maracaibo basin to 40 km beneath the northwestern Mérida Andes and then to thin abruptly to 30 km beneath the southern Mérida Andes. Farther south, the crust thickens again to about 34 km beneath the Barinas–Apure basin.

Lagaay's (1969, Fig. 14) model across the Netherlands Antilles showed a gradual crustal thickening, from a depth of about 17 km in the Venezuelan Basin to 24 km under Curaçao Ridge, a slight upward bulge under Curaçao,. and a thickening to 30 km at Venezuela. In his analysis of the isostatic-anomaly profile, a mean crustal density of 2.67 g/cm^3 was assumed. In view of the seismic evidence of Edgar *et al.* (1971a) that the crust is dense and heterogeneous, having seismic velocities of 6.3 to 7.6 km/sec (densities 2.8 to 3.3 g/cm^3), it appears that his model is oversimplified.

Hambleton (cited by Worzel, 1965) constructed a crustal model across the Venezuelan borderland along longitude 68° W based on free-air anomalies and seismic information (Fig. 34). In this model, the crust extends to depths of 16 km beneath the Venezuelan Basin, thins to 12 km near the base of the continental slope, thickens to about 16 km beneath Curaçao Ridge, thickens to about 34 km between Bonaire and Las Aves, thins to 16 km beneath Bonaire Basin, and thickens to 25 km at the Venezuelan margin. About 14 km of low-density (2.1–2.4 g/cm^3) material underlies Curaçao Ridge, and shallow dense rocks (2.85 g/cm^3) are within 2 km of the surface between Bonaire and Aves Passage. A mantle density of 3.4 g/cm^3 and a lower crustal density of 3.0 g/cm^3 were assumed. Perhaps the most conspicuous thing about Hambleton's model is that relatively dense rocks closely approach the surface near Las Aves, and the total crust is much thicker beneath Las Aves than to the north beneath Los Roques Basin and to the south near Bonaire Basin. Silver *et al.* (1975) derived a crustal model that is similar to Hambleton's.

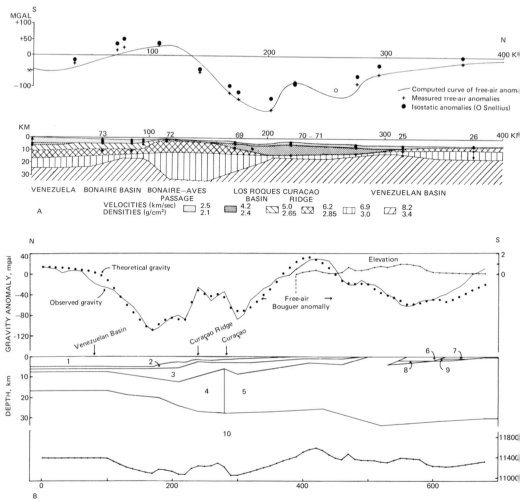

Fig. 34. Crustal models across Venezuelan margin. (A) Section at 68° W, constructed by Hambleton (Worzel, 1965) from free-air anomalies and seismic information. The black dots in the crustal model indicate the position of the observed seismic discontinuities. From Fig. 21, Lagaay, 1969. (B) Crustal section across Venezuelan borderland near Curaçao, from Folinsbee, 1972. Numbers refer to type of material and assumed densities: (1) water, 1.03 g/cm³; (2) sediments, 2.2 g/cm³; (3) sediments on land and probably mixed basalts and sediments beneath ocean, 2.4 g/cm³; (4) oceanic basement, 2.8 g/cm³; (5) continental basement, 2.7 g/cm³; (6) Cretaceous sediments, 2.5 g/cm³; (7) sediments, 2.3 g/cm³; (8) Cretaceous sediments, 2.6 g/cm³; (9) sediments, 2.4 g/cm³; (10) mantle, 3.3 g/cm³. Published with permission of the authors and Elsevier Scientific Publishing Company.

A somewhat different crustal model from the Venezuelan Basin through Curaçao and across the Venezuelan Andes has been presented by Folinsbee (1972; Fig. 34, this paper). In this model, the continental crust is assumed to extend out to Curaçao. The onshore crustal structure is gravitationally derived, and thus it is subject to multiple interpretations of density–depth relationships. The main point is that a crust of about 30 km beneath the Venezuelan mainland, 28 km beneath Curaçao, and 15–18 km beneath the Venezuelan Basin is generally consistent with gravity data.

Among other possibilities, the thickened and relatively dense crust beneath the Netherlands–Venezuelan Antilles could be associated with formation of a constructional island arc on a basement of older crust, or by multiple stacking of typical oceanic crust in a zone of convergence between an early Caribbean plate and nuclear South America (Silver *et al.*, 1975).

Curaçao Ridge (Fig. 33) is a deformed wedge of thick low-velocity sediments, presumably of Coniacian and younger age. Some of the turbidites and pelagic materials in the Venezuelan Basin have been deformed in the ridge (Fig. 33C). Los Roques Basin, a trenchlike feature south of Curaçao Ridge, is locally filled with turbidites at its eastern end. At its western end it passes into a broad basin north of Aruba (Figs. 9 and 33B). The island platform (Beets, 1972) is cored with a complex basement assemblage of basalts and diabase, largely of Late Cretaceous and older(?) age, intruded by Upper Cretaceous granodioritic plutons. Siliceous sediments and sandstones of Late Cretaceous age overlie the basement. The basement surface is essentially the site of outcrop of Horizon B″, so the structural relief of Coniacian or older rocks from the south margin of the Venezuelan Basin to Bonaire is as much as 9 km.

The sedimentary basins of Curaçao Ridge and the Bonaire Basin are connected by smaller basins between Curaçao and the Paraguana Peninsula and between Curaçao and Bonaire (Fig. 9; Scientific Staff, Leg 4, 1971 Cruise, UNITEDGEO I, 1972). Although some sediments are present near the eastern end of Bonaire Basin (Peter, 1972), they are much thinner than to the west. The continuity of Bonaire Basin with Grenada Basin is interrupted by cross faults trending northwest, such as those forming the graben between La Orchila and La Blanquilla. The shallow structure of the Venezuelan margin as described by Ball *et al.* (1971) consists of many normal faults dipping seaward. Cariaco Basin, a transtension feature, is bordered by faults that are projections of the El Pilar system on the east and the San Sebastian system on the west (Fig. 9), but the trench is offset by a presumed extension of the northwest-trending Urica fault (Peter, 1972, p. 292).

One of the most significant regional gravity anomalies along the Caribbean margin is the pronounced gravity high that extends from the volcanic windward Antilles (Andrew *et al.*, 1970; Bush and Bush, 1969) through Los Testigos

(Lattimore *et al.*, 1971) to Margarita (Fig. 35). This positive anomaly, shown on free-air, Bouguer, and isostatic-anomaly maps, extends westward through La Tortuga and has been traced to about 10°45′ N, 67° W (Scientific Staff, Leg 4, 1971 Cruise, UNITEDGEO I, 1972). The anomaly overlies metamorphosed Cretaceous and older(?) rocks on Margarita and seems to be caused by a raised complex ridge of basement rocks. Its continuity from the Tertiary volcanic Antilles to the Mesozoic basement complex of the Venezuelan borderland is doubly significant; first, the western part is related to a raised block of pre-Tertiary rocks. This suggests that the *basement* beneath the outer Antillean volcanic arc also may be composed of dense pre-Tertiary crystalline rocks. Fink (1972) has reported Jurassic basement rocks on Désirade, where trondhjemites intrude spilitic basalts that have interbedded cherts. Rocks analogous to the Tortuga–Margarita sequence probably serve as the deep foundation for the volcanic Antilles. Second, the near continuity of this anomaly, even though the axis is offset somewhat between Margarita and Tortuga, precludes a "South Caribbean fault" having substantial post-Mesozoic strike-slip displacement between Margarita and Grenada, as pointed out by Peter and Lattimore (1971), Meyerhoff and Meyerhoff (1972, p. 51–53), and Weeks *et al.* (1971, p. 1750–1751).

Gravity anomalies have an extremely large variation along the Venezuelan borderland, in keeping with the great variations in water depth and the complex geology. Free-air anomalies range from more than −200 mgal over Los Roques Basin to more than +180 mgal on Bonaire (Fig. 35; Lagaay, 1969, Table 3). Simple Bouguer anomalies range from +340 mgal over the Venezuelan Basin to −50 mgal near the western end of Cariaco Basin (Scientific Staff, Leg 4, 1971 Cruise, UNITEDGEO I, 1972, Fig. 7).

Southward over the Venezuelan Basin, free-air anomalies range from 0 to −100 mgal, and Bouguer anomalies from +340 to about +240 mgal, a typical range of values across margins of some ocean basins (Worzel, 1965). This regional southward decrease in values is partly the effect of a thickening wedge of sediments just north of the base of the continental slope. The decrease in Bouguer-anomaly values north of the base of the slope, however, cannot be attributed solely to changes in sediment thickness. Neither can it be caused by changes in water depth (and hence in Bouguer corrections), as the depth range is less than 1000 m in the abyssal plain, and depth changes are in the opposite direction from the changes in Bouguer-anomaly values; that is, Bouguer anomalies are increasingly negative toward the extreme south edge of the Venezuelan Basin, despite the fact that water depths are greater than in the basin to the north. Thus, part of the gradient must reflect a southward decrease in crust or upper mantle density or a southward increase in crustal thickness (Fig. 34).

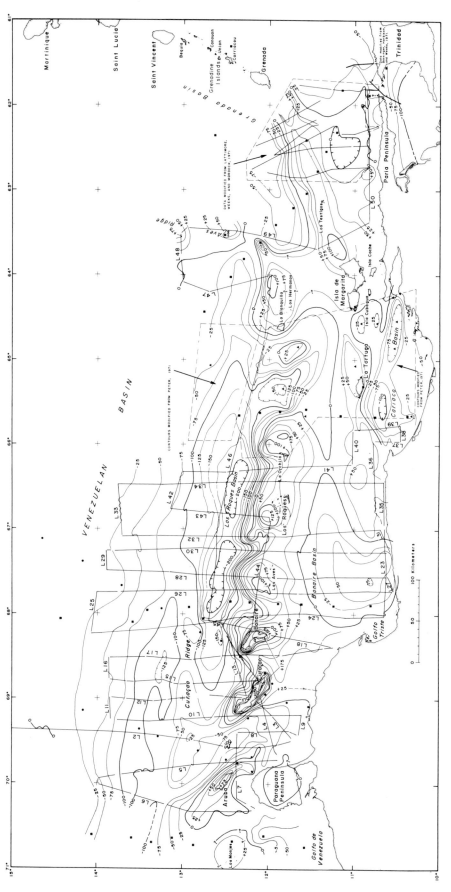

Fig. 35. Free-air anomaly map, Venezuelan borderland. From Scientific Staff, Leg 4, 1971 Cruise, UNITEDGEO I, 1972. Contour interval 25 mgal. Contours on Aruba, Curaçao, and Bonaire are simple Bouguer anomalies. Squares indicate pendulum stations reported by Bush and Bush (1969) and Worzel (1965); closed circles indicate pendulum stations reported by Lagaay (1969); triangles indicate stations reported by Peter (1971) and Worzel (1965).

Outer Curaçao Ridge and the main Curaçao Ridge have associated *relative* positive free-air anomalies, although values are still negative. These local highs are almost certainly due in part to topographic effects, as is the local Bouguer anomaly high along lines 10, 12, and 15 (near lat. 13°10′ N) which occurs over a topographic low (Scientific Staff, Leg 4, 1971 cruise, UNITED-GEO I, 1972).

One of the most significant regional anomalies is the great low over Los Roques Basin where free-air anomalies are more negative than −200 mgal (Fig. 35). Part of this low iş a topographic effect; water depths range from 4600–5000 m in Los Roques Basin—similar to depths in the Venezuelan Basin—but the anomalies over the Los Roques Basin are at least 100 mgal more negative than those over the Venezuelan Basin. A regional negative Bouguer anomaly, where values range from 60 mgal to 120 mgal, runs near the Los Roques Basin. A residual negative anomaly also overlies Curaçao Ridge north of Curaçao and Bonaire. These regional gravimetric relations indicate that the thick wedge of deformed sediments extends from Curaçao Ridge beneath Los Roques Basin. Refraction data of Edgar *et al.* (1971a) further indicate that at least 5 km of low-velocity material underlies Los Roques Basin and that as much as 14 km of low-velocity material underlies Curaçao Ridge.

The regional southward decrease in free-air and Bouguer-anomaly values is interrupted by a series of strongly positive values over the islands from Aruba to La Blanquilla. The coincidence of free-air and Bouguer anomalies as great as +75 to +180 mgal over Aruba, Curaçao, and Bonaire with a suite of basement rocks having oceanic affinities (pillow basalt, chert, and other marine sediments) indicates that these islands are raised blocks of dense, probably oceanic, crust. Similar anomalies over Las Aves, Los Roques, La Orchila, La Blanquilla–Los Hermanos, and the gravity high detected by Peter (1971, 1972) southwest of La Blanquilla further indicate that the basement framework comprises raised blocks of dense crust. According to Peter (1971, p. 152–153):

> The positive free-air anomaly belt of the Aves Ridge extends into the Blanquilla platform where it joins the positive values associated with the Tortuga–Margarita Bank (Talwani, 1966). West of the Blanquilla platform the large negative free-air anomaly values of the Los Roques Canyon interrupt this belt, but west of the canyon the same belt appears to follow the islands of the Aruba–Orchila chain (Talwani, 1966; Lagaay, 1969).

This zone is also one of positive isostatic anomalies that extends from Aves Ridge to the Guajira Peninsula and beyond (Bush and Bush, 1969; Case and MacDonald, 1973).

The group of positive anomalies over the islands is broadly aligned and extends from Los Monjes on the west to Aves Ridge on the east, but both free-air and Bouguer-anomaly contours indicate that many of these high-standing basement blocks are not directly connected and are probably separated by a series of northwest-trending, en echelon basement faults.

Major free-air and Bouguer-anomaly minima between Aruba and Curaçao indicate that this area is filled with a great thickness of sedimentary rocks, connecting the sedimentary wedge of Bonaire Basin with that of Los Roques Basin–Curaçao Ridge. Steepened gravity gradients and seismic-reflection data indicate a major fault zone along the southwest margin of Curaçao, close to the island shore. Similarly, steep gradients along the northeast coast of Aruba suggest a fault margin. Thus, the sedimentary basin may occur in a grabenlike structural setting between the islands. These faults trend northwest and are approximately in the position inferred by MacDonald *et al.* (1971, Fig. 1) between Aruba and Curaçao to account for the contrast in metamorphic grade between the islands.

Other zones of steepened northwest-trending free-air gravity gradients appear along the southwest flank of Aruba, southwest flank of Bonaire, between Las Aves and Los Roques, on both flanks of the graben between La Ochila and La Blanquilla, and southwest of La Blanquilla. These steepened gradients coincide with northwest-trending steep topographic gradients and in part simply reflect the topographic grain. However, the steepened northwest-trending gradients remain on the Bouguer-anomaly map along the flanks of Aruba and Bonaire, along the northeast flank of the graben east of La Ochila and the area southwest of La Blanquilla, so that real lithologic contrasts or structural zones are indicated. Peter (1971, Fig. 38, p. 121–122, p. 133–136), for example, has presented seismic evidence for northwest-trending grabenlike fault troughs in between La Ochila and Blanquilla which may be underlain by a considerable thickness of sedimentary rocks judging from the nègative free-air and residual negative Bouguer anomalies. George Peter (written communication, 1972) suggested, further, that the gravity low indicates a graben that cuts the entire crust.

The sense of displacement along these transverse fault zones is difficult to determine. Most of the faults clearly have a large vertical component of displacement from both seismic and gravitational evidence. If one attempts to align the islands and their submarine extensions by restoration along strike-slip faults, a series of left-lateral displacements would be required on a system of northwest-trending faults. Galavis and Louder (1970, Fig. 1A) inferred left-lateral displacement on the fault along the southwest margin of Aruba. In an analysis of the regional isostatic-anomaly field, Lagaay (1969, Fig. 13, p. 44, 75) noted offsets in the field in the vicinity of the Netherland Antilles. These offsets could be fitted by a system of north-to-northwest-trending right-lateral faults (between Aruba and Curaçao) or by a system of north-to-north-east-trending left-lateral faults (between Curaçao and Bonaire). Despite such hints of strike-slip displacements, the general pattern seems to be one of an elongate block of uplifted oceanic crust that was pulled apart by east–west extension creating rifted zones between the blocks.

A broad regional free-air gravity low is present over Bonaire Basin (Fig. 35). Although the general configuration of the low parallels the topography, in detail the axis of the low is offset south of the main basin axis. Bouguer-anomaly contours across the basin show a general southward decrease. Hence, it seems clear that deeper crustal density changes are partly masking effects of the topographic trough and sedimentary basin. Bouguer anomalies are strongly negative south of the mainland border, near Caracas, indicating typical continental crust at depth (Folinsbee, 1972, Fig. 9).

A general gravity low or area of flattened gravity gradient extends eastward from the main Bonaire Basin to Grenada Basin, a sedimentary basin, between Aves Ridge and the Windward Antilles. In a broad regional sense these sedimentary basins are on trend, although apparently interrupted by a cross structural high, structural saddle, or cross faults between Margarita–La Tortuga and La Orchila–Blanquilla, according to Peter's (1971, 1972) gravity and seismic data. A regional negative isostatic anomaly extends from Grenada Basin across Bonaire Basin to the Falcon basin (Bush and Bush, 1969, Fig. 6).

Most of the Venezuelan borderland is magnetically rather quiet (Fig. 36) except over the uplifted blocks of basement rocks, where residual anomalies of more than 200 γ are associated with the island platform (Silver et al., 1975). Peter (1972, p. 291) found a similar relationship between La Orchila and Blanquilla and in the area north of the Cariaco Basin where the total-intensity contour map (Fig. 37) essentially outlines areas of igneous basement domes. Peter (1972, p. 291) postulated that his magnetic data, when combined with that of Ball et al. (1971), corroborate the proposed northward extension of the Urica fault, which appears to terminate magnetic anomalies north of the eastern deep of the Cariaco Basin.

Seismicity of the Venezuelan borderland has been reviewed by Molnar and Sykes (1969) and by Tomblin (1972). Seismic activity is concentrated along the San Sebastian–El Pilar fault zone. Right-lateral strike-slip displacement seems to be dominant, except toward the east where hinge-faulting has been determined by Molnar and Sykes (1969). As far as known, no sea-floor heat-flow determinations have been made south of the Venezuelan Basin.

Q. Geophysical Studies in Northern South America

The gravity field on land in northern South America has been the subject of considerable study in recent years. Hospers and Van Wijnen (1959), Folinsbee (1972), Graterol (1973), and Bonini et al. (1973) have discussed the gravity field in parts of Venezuela. Gravity lows coincide with all the major sedimentary basins. Crustal thickness apparently is about 25–40 km. Bonini et al. (1973)

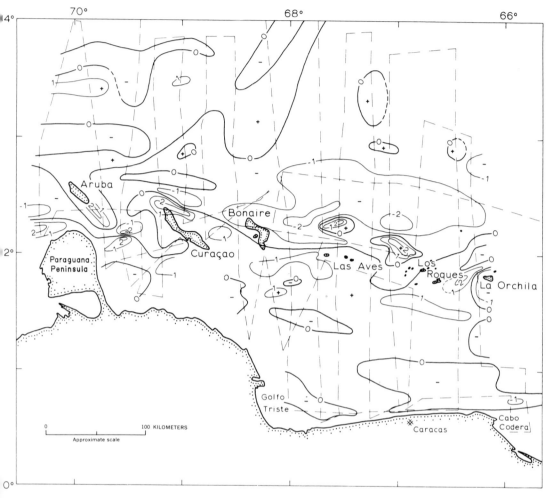

Fig. 36. Residual magnetic anomaly map of the Venezuelan borderland. Silver *et al.*, 1975. Dashed lines indicate ship tracks. Contour interval 100 γ. Published with permission of the authors and the Geological Society of America.

found that the Villa de Cura Belt, in the southern part of the Caribbean coastal ranges, is not thick, a conclusion compatible with the hypothesis that the Villa de Cura is an allochthonous body. In northern Colombia, the crust seems to be substantially thinner under the Guajira Peninsula and Santa Marta uplift (Fig. 9), both areas being characterized by strongly positive Bouguer and isostatic anomalies (Case and MacDonald, 1973). Negative anomalies coincide with major basins such as the Lower Magdalena and Sinu–Uraba. Negative anomalies likewise coincide with the Antioquian batholith of the Andean Cordillera Central near 6° N, 74°30′ W. Case *et al.* (1971) have

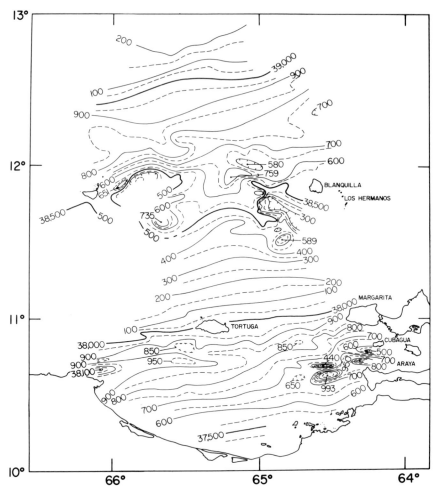

Fig. 37. Magnetic total intensity for the year 1969 near the Cariaco Basin and area to the north. From Peter, 1972. Contour interval 50 γ. Published with permission of the author and the Caribbean Geological Conference.

inferred that the boundary between dominant oceanic crust to the west and continental crust to the east coincides with the Cauca–Romeral fault system between the Colombian Cordillera Occidental and Cordillera Central (Fig. 9), but northward, the position of the boundary is questionable in the Santa Marta region and may occur beneath the uplift. It evidently lies between the Netherlands Antilles and the Venezuelan mainland off Venezuela. Bonini *et al.* (1973) concluded that the steep gradient associated with the Cordillera de la Costa may be explained by a deep crust-mantle fault with a displacement of 12 km, downthrown to the south.

No onland seismic determinations of crustal thickness and structure had been made in Colombia until early 1973. Case *et al.* (1971, Fig. 8) derived a continental crust having a thickness of about 30–35 km beneath the Cordillera Central of the Andean system at about lat. 6° N. Gravitational evidence suggests that the crust thickens southward along the Andes toward Ecuador (Case *et al.*, 1973).

VIII. SUMMARY

Of the many hypotheses on the origin of the Caribbean Sea and its borderlands, some of which are described in detail in other chapters of this volume, none can be proved or disproved from the available geophysical data. The problem is that most geophysical data provide information on the *present* distribution of rocks, motions, and potential fields, but the data may not represent past distributions with respect to some fixed coordinate system. For example, from seismology—epicentral, hypocentral, and first-motion studies—we obtain an idea of current relative motions of large segments of the lithosphere. These data, however, tell us nothing directly about motions that may have occurred in the Late Cretaceous. We must rely on the rocks themselves for "final" solutions about past events.

From seismic-reflection data, it appears that as much as 2 km of sediments and sedimentary rocks overlies a "basement" in the Yucatan, Colombian, and Venezuelan Basins. In these basins a pelagic or hemipelagic sequence ranges from a few hundred to 1000 m in thickness, and, in much of the Yucatan and Colombian Basins, this sequence is overlain by several hundred to 1000 m of turbidites.

The thick turbidite sequences in the Colombian Basin and Yucatan Basin constitute a significant major problem in the evolution of the Caribbean. What is the composition and age of the sequence? How do the turbidites correlate with episodes of uplift and/or volcanism in the surrounding land areas? If the turbidites in the Colombian Basin near Hispaniola turn out to have a substantial quartz fraction and if they are products of erosion of northern South America, deposited completely across the basin from the Rio Magdalena, then we must reexamine classical concepts of high quartz content and proximity of "continental" sources.

In many places in the Colombian and Venezuelan Basins, discontinuous seismic reflectors have been identified below Horizon B″ (of Coniacian–Maestrichtian Age). Whether these reflectors represent sills or whether they represent sequences of pre-Coniacian sedimentary rocks is unknown. Assuming that seismic horizon B″ is not greatly time transgressive, it can be shown that the true crust of the southern Caribbean—from the south flank of the Nicaraguan Rise, across the Colombian Basin, Beata Ridge, and the Venezuelan Basin—formed in Coniacian or *older* time.

Away from the relatively undisturbed sediments of the deep basins, highly deformed, even chaotic, sediments occur between the east flank of the Lesser Antilles and the zones of apparent underthrusting along the negative gravity strip farther east. Tertiary fold belts have now been identified along many continental slopes of the interior Caribbean, as along Curaçao ridge, along the Colombian margin, north of Panama, and off the south continental slope of Puerto Rico. In such places, seismic Horizon B″ appears to dive beneath the continental slope—geometrically a zone of underthrusting. Are these fold-thrust occurrences tectonic in origin, related to compressive stresses across interior Caribbean margins, or are they simply a product of gravitative downslope sliding of poorly consolidated sediments? Regardless of the ultimate mechanism of origin, turbidites and pelagic sediments along the southern margin of the Venezuelan Basin are progressively deformed in the Curaçao Ridge fold complex. Tracing of reflector units from DSDP sites indicates that these sediments in the filled trench at the south margin of the Caribbean are of Coniacian and younger age. Deformed sediments in the Curaçao Ridge, then, are almost certainly of Late Cretaceous to Tertiary or Holocene age. Moreover, this fold belt extends north of the Santa Marta–Guajira region, across the Magdalena delta, and passes onshore to form the Sinu–Atlantico basin of northern Colombia (Fig. 9). Duque Caro (1972) has demonstrated that depositional depths of Upper Cretaceous–Paleocene deposits of the northern Colombian coast exceeded 2000 m. Deposits of early Eocene time are not represented. Deposits of middle and late Eocene time were locally laid down in shallow water, but sediments of Oligocene–early Miocene time were again deposited in water greater than 2,000 m. Only in Miocene and later time did water depths shoal again.

The Tertiary fold belt off the northern coast of Panama probably extends onshore as the Limon basin near the Panama–Costa Rican border, although it may be locally offset by faults.

In contrast to the fold belts along some margins of the interior Caribbean, many areas are characterized by block faulting, especially the eastern Yucatan borderland, Cayman Ridge, Cayman Trench, the Nicaraguan Rise, and Beata Ridge. Other areas seem to contain both fold and fault structures, for example, Aves Ridge and the Venezuelan borderland south of Curaçao Ridge. On both Aves Ridge and Beata Ridge, rock evidence indicates deep-water facies in Late Cretaceous, shallow-water facies (at least locally) during middle Eocene, and deep-water facies during Oligocene–Holocene. Is this because shallow atolls bordered volcanoes which were adjacent to deep ocean basins in Eocene time or because there have been several episodes of extreme vertical uplift and depression of the southern Caribbean Basin?

Free-air, Bouguer, and isostatic-anomaly maps indicate that a major negative anomaly parallels the Lesser Antilles from about 18° N, 61° W to southern Trinidad, although it is locally interrupted near Barbados Ridge. The anomaly continues onland in the eastern Venezuelan sedimentary basin. This anomaly partly correlates with a thick sequence of low-density sedimentary rocks. Along much of its seaward extent it likewise correlates with the zone of apparent underthrusting where sediments have apparently been scraped off, stacked, and compressed into a wedge of great thickness. It is most significant that *west* of 18° N, 62° W, the negative Bouguer anomaly is not present over the Puerto Rico Trench, even though the negative free-air and isostatic anomalies persist. This indicates that no great thickness of sediments is present in the main Puerto Rico Trench, as is clear from reflection profiles that show 1 km or less of sediments above acoustic basement. From the combined evidence of seismic reflection and gravity, underthrusting is dominant outside the Lesser Antillean arc, and strike-slip (or transtensional) faulting constitutes an important component of dislocations along the Puerto Rico Trench.

The isostatic-anomaly map by Bush (1969) is very misleading in the vicinity of the Lesser Antilles, because their map was based on only a few control points. On their map, positive isostatic anomalies range from 25 to 46 mgal, but, in fact, many anomalies computed by Andrew *et al.* (1970) exceed +100 mgal, so that the Lesser Antilles have a positive anomaly that is nearly as great as the negative anomaly to the east. Thus, the Lesser Antilles are not in isostatic equilibrium, and considerable excess mass is present beneath the islands. Comparatively high seismic velocities at rather shallow depths are indicated by the refraction data of Officer *et al.* (1959). It may be postulated that the rather thick and dense crust beneath the Lesser Antilles results from multiple stacking or obduction of oceanic crust in early stages of convergence between an ancestral Caribbean plate and an ancestral Atlantic plate, perhaps during Jurassic–Early Cretaceous time (see chapter in this volume by Fox and Heezen).

Over the Virgin Islands, Puerto Rico, Hispaniola, Jamaica, and parts of Cuba, Bouguer anomalies are strongly positive, especially away from the Tertiary basins and graben. Such positive Bouguer anomalies, especially at higher topographic elevations, likewise indicate isostatic imbalance. Land-refraction surveys are needed to determinate if the crust is thickened and of high velocity, as seems to be the case in the Lesser Antilles.

Even though the Yucatan, Colombian, and Venezuelan Basins seem to have a crustal structure that is thicker (13–19 km) and of atypical velocity as compared with standard oceanic crust, gravity data suggest that these basins are nearly in isostatic equilibrium. Moreover, when seismic-refraction velocities are converted to densities and the mass per unit area is computed (Edgar *et al.*, 1971a), it is clear that the main basins have only small mass deficiencies

$(88.2–92.3 \times 10^5 \text{ g/cm}^2)$ as compared with "standard" oceanic mass $(93.1 \times 10^5$ g/cm²). Similarly, the thicker crust beneath Beata Ridge (22 km) and Aves Ridge $(>15 \text{ km})$ is nearly in isostatic balance, judged from the low values of free-air and isostatic anomalies and the values of mass per unit area (88.5–92.8 g/cm²)—nearly identical with the mass per unit area of the main basins. One may say that virtually all of the interior Caribbean is in internal isostatic equilibrium, even though it may have slightly less mass per unit area than a typical ocean basin.

Many gravity anomalies along the south Caribbean margin indicate isostatic imbalance. The negative anomaly over the thick sequence of sediments along the south Caribbean margin extends from Curaçao Ridge and Los Roques Basin westward to the Magdalena delta region. Relative negative free-air and Bouguer anomalies occur over the sedimentary basin off the north coast of Panama.

Strongly positive free-air, Bouguer, and isostatic anomalies occur over the Netherlands and Venezuelan Antilles, Paraguana Peninsula (Graterol, 1973), Alta Guajira Peninsula, and Santa Marta uplift. These areas are clearly out of isostatic equilibrium at present and indicate that continental crust is thin or absent. Only south of the El Pilar–Bocono fault systems, the current belts of high seismicity, do normal or continental values of gravity prevail. Yet the crust beneath the Netherlands Antilles is apparently thicker than that to the north in the Venezuelan Basin. Do these positive gravity anomalies reflect zones of obduction or multiple stacking of crustal layers in older zones of convergence?

Such belts of positive gravity anomalies must be related to one of the most crucial regional geological problems: what do such "orthogeosynclinal" belts as those of the Greater Antilles and Venezuelan coastal ranges represent? According to current concepts of plate tectonics, they are zones of plate convergence, either in a subduction or obduction mode. According to alternative contraction hypotheses, they represent volcanism and deformation in very ancient zones of weakness. The near continuity of the positive-anomaly belt, although locally offset, from the Greater Antilles, through the Lesser Antilles, to the Venezuelan borderland suggests that the "orthogeosyncline" is a nearly continuous feature.

From the combined geological and geophysical evidence along the Cariaco Basin–El Pilar fault zone, central and western Puerto Trench, and Cayman Trench, it is clear that these structures are transtension faults; that is, the dominant displacements are horizontal, but substantial components of vertical displacements are present. In the interior parts of the fault zones, pull-apart, tensional features are essential parts of the zone of distributive faulting. These fault zones show a spectrum of relative motions—from nearly pure thrusting east of the Antillean arc to pure strike-slip faulting along segments of the

Cayman and El Pilar systems. These fault zones are *not* single faults, but rather are complex zones of multiple subparallel traces.

I will close with the comment that geophysical studies in the Caribbean seem to be at a plateau. In the next phase of Caribbean studies, emphasis should be placed on multichannel seismic-reflection surveys so that more definitive information can be obtained in areas covered by shallow water, such as the Nicaraguan Rise, and so that sub-B″ reflectors can be mapped. Seismic-refraction programs should be organized to shoot and record both onshore and offshore to clarify the structure of the crust and upper mantle across the boundary between oceans and continents. An attempt should be made to compile all existing magnetic data on a common reference datum. Far more heat-flow measurements are required in such areas as the Nicaraguan Rise. Finally, the huge volume of geophysical work that has been accomplished during the 1960's and 1970's requires careful synthesis with onland geology.

ACKNOWLEDGMENTS

I wish to thank numerous colleagues for providing access to data and manuscripts that were unpublished at the time this summary was prepared. Helpful review at various stages of manuscript preparation was provided by H. R. Berryhill, Jr., T. W. Donnelly, P. J. Fox, L. E. Garrison, R. G. Martin, A. E. M. Nairn, F. G. Stehli, and Elazar Uchupi.

REFERENCES

Algermissen, S. T., Dewey, J. W., Langer, C., and Dillinger, W., 1973, The Managua, Nicaragua, earthquake of December 23, 1972; Location, focal mechanism, and intensity distribution [abstract]: Seismol. Soc. Am., Ann. Mtg., 1973, Golden, Colorado, Program, p. 11.

Andrew, E. M., Masson Smith, D., and Robson, G. R., 1970, Gravity anomalies in the Lesser Antilles: Great Britain, Inst. Geol. Sci., Geophys. Pap. No. 5, 21 p.

Arden, D. D., Jr., 1969, Geologic history of the Nicaraguan Rise: *Trans. Gulf Coast Assoc. Geol. Soc.*, v. 19, p. 295–309.

Baie, L. F., 1970, Possible structural link between Yucatan and Cuba: *Am. Assoc. Petr. Geol. Bull.*, v. 54, n. 11, p. 2204–2207.

Ball, M. M., and Harrison, C. G. A., 1969, Origin of the Gulf and Caribbean and implications regarding ocean ridge extension, migration, and shear: *Trans. Gulf Coast Assoc. Geol. Soc.*, v. 19, p. 287–294.

Ball, M. M., Harrison, C. G. A., Supko, P. R., Bock, W., and Maloney, N. J., 1971, Marine geophysical measurements on the southern boundary of the Caribbean Sea, in: *Caribbean Geophysical, Tectonic, and Petrologic Studies*, Donnelly, T. W., ed.: Geol. Soc. Am. Mem. 130, p. 1–33.

Banks, N. G., and Richards, M. L., 1969, Structure and bathymetry of western end of Bartlett Trough, Caribbean Sea, in: *Tectonic Relations of Northern Central America and the Western Caribbean—the Bonacca Expedition*, McBirney, A. R., ed.: Am. Assoc. Petr. Geol. Mem. 11, p. 221–228.

Barazangi, Muawia, and Dorman, James, 1969, World seismicity maps compiled from ESSA, Coast and Geodetic Survey, epicenter data, 1961–1967: *Seismol. Soc. Am. Bull.*, v. 59, p. 369–380.

Bassinger, B. G., Harbison, R. N., and Weeks, L. A., 1971, Marine geophysical study northeast of Trinidad–Tobago: *Am. Assoc. Petr. Geol. Bull.*, v. 55, n. 10, p. 1730–1740.

Beets, D. J., 1972, Lithology and stratigraphy of the Cretaceous and Danian succession of Curaçao: *Uitgaven "Natuurwetenschappelijke studjekring voor Suriname en de Nederlandse Antillen,"* Utrecht, No. 70, 165 p.

Bonini, W. E., Acker, Christopher, and Buzan, George, 1973, Gravity studies across the western Caribbean Mountains, Venezuela [Abstract]: *II Congreso Latinoamericano de Geologia, 11–16 Nov., 1973, Caracas, Venezuela*: Resumenes, p. 52–53.

Bouma, A. H., Rezak, R., Antoine, J. W., Bryant, W., and Fahlquist, D. A., 1972, Deep-sea sedimentation and correlation of strata off Magdalena River and in Beata Strait: *Trans. Sixth Caribbean Geological Conference, Margarita, Venezuela, 1971*, p. 430–438.

Bowin, C. O., 1968, Geophysical study of the Cayman Trough: *J. Geophys. Res.*, v. 73, n. 16, p. 5159–5173.

Bowin, C. O., 1972, Puerto Rico Trench negative gravity anomaly belt, in: *Studies in Earth and Space Sciences*, Shagam, R., *et al.*, eds.: Geol. Soc. Am. Mem. 132, p. 339–350.

Bowin, C. O., in press, The Caribbean: Gravity field and plate tectonics: *Geol. Soc. Am. Bull.*

Bowin, C. O., and Folinsbee, Allin, 1970, Gravity anomalies north of Panama and Colombia [abstract]: EOS (*Trans. Am. Geophys. Union*), v. 51, n. 4, p. 317.

Bowin, C. O., Nalwalk, A. J., and Hersey, J. B., 1966, Serpentinized peridotite from the north wall of the Puerto Rico Trench: *Geol. Soc. Am. Bull.*, v. 77, n. 3, p. 257–269.

Bracey, D. R., 1968, Structural implications of magnetic anomalies north of the Bahama–Antilles Islands: *Geophysics*, v. 33, n. 6, p. 950–961.

Bracey, D. R., and Vogt, P. R., 1970, Plate tectonics in the Hispaniola area: *Geol. Soc. Am. Bull.*, v. 81, n. 9, p. 2855–2859.

Brown, R. D., Jr., Ward, P. L., and Plafker, George, 1973, Geologic and seismologic aspects of the Managua, Nicaragua, earthquakes of December 23, 1972: U.S. Geol. Surv. Prof. Pap. 838, 34 p.

Bunce, E. T., 1966, The Puerto Rico Trench, in: *Continental Margins and Island Arcs*, Poole, W. H., ed.: Can. Geol. Surv. Pap. 66-15, p. 165–175.

Bunce, E. T., and Fahlquist, D. A., 1962, Geophysical investigation of the Puerto Rico Trench and outer ridge: *J. Geophys. Res.*, v. 67, n. 10, p. 3955–3972.

Bunce, E. T., and Hersey, J. B., 1966, Continuous seismic profiles of the outer ridge and Nares Basin north of Puerto Rico: *Geol. Soc. Am. Bull.*, v. 77, n. 8, p. 803–811.

Bunce, E. T., Phillips, J. D., Chase, R. L., and Bowin, C. O., 1970, The Lesser Antilles arc and the eastern margin of the Caribbean Sea, in: *The Sea*, Vol. 4, Pts. 2–3, Maxwell, A. E., ed.: Wiley–Interscience, New York, p. 359–385.

Bunce, E. T., Phillips, J. D., and Chase, R. L., 1974, Geophysical study of Antilles Outer Ridge, Puerto Rico Trench, and northeast margin of Caribbean Sea: *Am. Assoc. Petr. Geol. Bull.*, v. 58, n. 1, p. 106–123.

Bush, S. A., and Bush, P. A., 1969, Isostatic gravity map of the eastern Caribbean region: *Trans. Gulf Coast Assoc. Geol. Soc.*, v. 19, p. 281–285.

Carey, S. W., 1958, The tectonic approach to continental drift, in: Continental Drift, a Symposium, Carey, S. W., ed.: Tasmania Univ., Hobart, Geol. Dep., p. 177–355.

Case, J. E., in press, Oceanic crust forms basement of eastern Panama: *Geol. Soc. Am. Bull.*, v. 85, n. 4, p. 645–652.

Case, J. E., and MacDonald, W. D., 1973, Regional gravity anomalies and crustal structure in northern Colombia: *Geol. Soc. Am. Bull.*, v. 84, n. 9, p. 2905–2916.

Case, J. E., Duran S., L. G., Lopez R., Alfonso, and Moore, W. R., 1971, Tectonic investigations in western Colombia and eastern Panama: *Geol. Soc. Am. Bull.*, v. 82, n. 10, p. 2685–2712.

Case, J. E., Barnes, Jerry, Paris Q., Gabriel, Gonzales I., Humberto, and Viña, Alvaro, 1973, Trans-Andean geophysical profile, southern Colombia: *Geol. Soc. Am. Bull.*, v. 84, n. 9, p. 2895–2904.

Chase, R. L., and Bunce, E. T., 1969, Underthrusting of the eastern margin of the Antilles by the floor of the western North Atlantic Ocean, and origin of the Barbados ridge: *J. Geophys. Res.*, v. 74, n. 6, p. 1413–1420.

Chase, R. L., and Hersey, J. B., 1968, Geology of the North Slope of the Puerto Rico Trench: *Deep-Sea Res.*, v. 15, n. 3, p. 297–317.

Christofferson, Eric, 1973, Linear magnetic anomalies in the Colombia basin, central Caribbean Sea: *Geol. Soc. Am. Bull.*, v. 84, n. 10, p. 3217–3230.

Cook, E. E., 1967, Geophysical reconnaissance in the north-western Caribbean: Soc. Explor. Geophys., 1967, Ann. Mtg., Preprint, 17 p.

Dillon, W. P., and Vedder, J. G., 1973, Structure and development of the continental margin of British Honduras: *Geol. Soc. Am. Bull.*, v. 84, n. 8, p. 2713–2732.

Dillon, W. P., Vedder, J. G., and Graf, R. J., 1972, Structural profile of the northwestern Caribbean: *Earth Planet. Sci. Lett.*, v. 17, p. 175–180.

Donnelly, T. W., 1965, Sea-bottom morphology suggestive of post-Pleistocene tectonic activity of the eastern Greater Antilles: *Geol. Soc. Am. Bull.*, v. 76, n. 11, p. 1291–1294.

Donnelly, T. W., 1971, A preliminary analysis of aeromagnetic profiles in the Gulf of Mexico and Caribbean areas [abstract]: *Trans. Fifth Caribbean Geol. Conf., St. Thomas, Virgin Islands, 1968*, p. 9.

Donnelly, T. W., 1973, Magnetic observations in the eastern Caribbean Sea, in: 1973, *Initial Reports of the Deep Sea Drilling Project*, Vol. 15, Edgar, N. T., Saunders, J. B., *et al.*, eds.: U.S. Government Printing Office, Washington, p. 1023–1029.

Duque Caro, Hermann, 1972, Ciclos tectonicos y sedimentarios en el norte de Colombia y sus relaciones con la paleocologia: Inst. Nac. de Inv. Geol.-Min., Bogota, Colombia, Bol. Geologico, v. 19, n. 3, p. 1–23.

Drake, C. L., Heirtzler, J. R., and Hirshman, J., 1963, Magnetic anomalies off eastern North America: *J. Geophys. Res.*, v. 68, n. 18, p. 5259–5275.

Edgar, N. T., Ewing, J. I., and Hennion, John, 1971a, Seismic refraction and reflection in Caribbean Sea: *Am. Assoc. Petr. Geol. Bull.*, v. 55, n. 6, p. 833–870.

Edgar, N. T., Holcombe, Troy, Ewing, John, and Johnson, William, 1973, Sedimentary hiatuses in the Venezuelan Basin, in: *Initial Reports of the Deep Sea Drilling Project*, Vol. 15, Edgar, N. T., Saunders, J. B., *et al.*, eds.: U.S. Government Printing Office, Washington, p. 1051–1062.

Edgar, N. T., Saunders, J. B., *et al.*, 1971b, Deep Sea drilling project—Leg 15: *Geotimes*, v. 16, n. 4, p. 12–16.

Edgar, N. T., Saunders, J. B., *et al.*, eds., 1973, *Initial Reports of the Deep Sea Drilling Project*, Vol. 15: U.S. Government Printing Office, Washington, 1137 p.

Eggler, D. H., Fahlquist, D. A., Pequegnat, W. E., and Herndon, J. M., 1973, Ultrabasic rocks from the Cayman Trough, Caribbean Sea: *Geol. Soc. Am. Bull.*, v. 84, n. 6, p. 2133–2138.

Emery, K. O., and Uchupi, Elazar, 1972a, Caribe's oil potential is boundless: *Oil Gas J.*, v. 70, n. 50, p. 156–162.

Emery, K. O., and Uchupi, Elazar, 1972b, Western North Atlantic Ocean: Topography, rocks, structure, water, life, and sediments: Am. Assoc. Petr. Geol. Mem. 17, 532 p.

Epp, David, Grim, P. J., and Langseth, M. G., Jr., 1970, Heat flow in the Caribbean and Gulf of Mexico: J. Geophys. Res., v. 75, n. 29, p. 5155–5669.

Erickson, A. J., Helsley, C. E., and Simmons, Gene, 1972, Heat flow and continuous seismic profiles in the Cayman Trough and Yucatan Basin: Geol. Soc. Am. Bull., v. 83, n. 5, p. 1241–1260.

Ewing, John, and Ewing, Maurice, 1970, Seismic reflection, in: The Sea, Vol. 4, Pt. 1, Maxwell, A. E., ed.: Wiley–Interscience, New York, p. 1–51.

Ewing, J. I., Officer, C. B., Johnson, H. R., and Edwards, R. S., 1957, Geophysical investigations in the eastern Caribbean: Trinidad Shelf, Tobago Trough, Barbados Ridge, Atlantic Ocean: Geol. Soc. Am. Bull., v. 68, n. 7, p. 897–912.

Ewing, John, Antoine, John, and Ewing, Maurice, 1960, Geophysical measurements in the western Caribbean Sea and in the Gulf of Mexico: J. Geophys. Res., v. 65, n. 12, p. 4087–4126.

Ewing, John, Talwani, Manik, Ewing, Maurice, and Edgar, Terence, 1967, Sediments of the Caribbean: Miami Univ. Stud. Trop. Oceanogr., v. 5, p. 88–102.

Ewing, John, Talwani, Manik, and Ewing, Maurice, 1968, Sediment distribution in the Caribbean Sea: Trans. Fourth Caribbean Geol. Conf., Port-of-Spain, Trinidad and Tobago, 1965, p. 317–323.

Ewing, J. I., Edgar, N. T., and Antoine, J. W., 1970, Structure of the Gulf of Mexico and Caribbean Sea, in: The Sea, Vol. 4, Pts. 2–3, Maxwell, A. E., ed.: Wiley–Interscience, New York, p. 321–358.

Fahlquist, D. A., and Davies, D. K., 1971, Fault-block origin of the western Cayman Ridge, Caribbean Sea: Deep-Sea Res., v. 18, n. 2, p. 243–253.

Fahlquist, D. A., Antoine, J. W., Bryant, W. R., Bouma, Arnold, and Pyle, Thomas, 1972, Seismic reflection profiles in the Yucatan Channel: Trans. Sixth Caribbean Geol. Conf., Margarita, Venezuela, 1971, p. 367–371.

Feo-Codecido, Gustavo, 1973, Un esbozo geologico de la plataforma continental Margarita–Tobago (preprint): II Congreso Latinoamericano de Geologia, 11–16 Nov., 1973, Caracas, Venezuela, 35 p.

Fink, L. K., Jr., 1972, Bathymetric and geologic studies of the Guadeloupe region, Lesser Antilles Island Arc: Mar. Geol., v. 12, n. 4, p. 267–288.

Folinsbee, R. A., 1972, The gravity field and plate boundaries in Venezuela: Mass. Inst. Technol., Ph.D. Dissert., 159 p.

Fox, P. J., and Schreiber, Edward, 1970, Granodiorites from the Cayman Trench [abstract]: Geol. Soc. Am., Abstr. with Programs, v. 2, n. 7, p. 553.

Fox, P. J., Ruddiman, W. F., Ryan, W. B. F., and Heezen, B. C., 1970, The geology of the Caribbean crust, I: Beata Ridge: Tectonophysics, v. 10, n. 5–6, p. 495–513.

Fox, P. J., Schreiber, Edward, and Heezen, B. C., 1971, The geology of the Caribbean crust: Tertiary sediments, granitic and basic rocks from the Aves ridge: Tectonophysics, v. 12, n. 2, p. 89–109.

Freeland, G. L., and Dietz, R. S., 1971, Plate tectonic evolution of Caribbean–Gulf of Mexico region: Nature, v. 232, p. 20–23.

Funnell, B. M., and Smith, A. G., 1968, Opening of the Atlantic Ocean: Nature, v. 219, p. 1328–1333.

Galavis S., J. A., and Louder, L. W., 1970, Preliminary studies on geomorphology, geology and geophysics on the continental shelf and slope of northern South America (preprint): Eight World Petr. Cong., Caracas, Venezuela, Sept. 1970, 26 p.

Garrison, L. E., et al., 1972a, USGS-IDOE Leg. 3: Geotimes, v. 17, n. 3, p. 14–15.

Garrison, L. E., *et al.*, 1972*b*, Acoustic reflection profiles—Eastern Greater Antilles: U.S. Dep. Commer. Nat. Tech. Inf. Serv. PB2-07596.

Garrison, L. E., Martin, R. G., Jr., Berryhill, H. L., Jr., Buell, M. W., Jr., Ensminger, H. R., and Perry, R. K., 1972*c*, Preliminary tectonic map of the eastern Greater Antilles region: U.S. Geol. Surv. Misc. Geol. Inv. Map I-732.

Gough, D. I., and Heirtzler, J. R., 1969, Magnetic anomalies and tectonics of the Cayman Trough: *Roy. Astron. Soc. Geophys. J.*, v. 18, p. 33–49.

Graterol G., V. R., 1973, Anomalia de Bouguer de la Peninsula de Paraguana [abstract]: *II Congreso Latinoamericano de Geologia, 11–16 Nov., 1973, Caracas, Venezuela*: Resumenes, p. 189.

Griscom, Andrew, 1971, Tectonic implications of gravity and aeromagnetic surveys on the north and south coasts of Puerto Rico [abstract]: *Trans. Fifth Caribbean Geol. Conf., St. Thomas, Virgin Islands, 1968*, p. 23–24.

Griscom, Andrew, and Geddes, W. H., 1966, Island-arc structure interpreted from aeromagnetic data near Puerto Rico and the Virgin Islands: *Geol. Soc. Am. Bull.*, v. 77, n. 2, p. 153–162.

Harrison, C. G. A., and Ball, M. M., 1973, The role of fracture zones in sea floor spreading: *J. Geophys. Res.*, v. 78, n. 12, p. 7776–7785.

Heezen, B. C., and Muñoz J., N. G., 1968, Magdalena turbidites in deep-sea sediments [abstract]: *Trans. Fourth Caribbean Geol. Conf., Port-of-Spain, Trinidad and Tobago, 1965*, p. 342.

Heezen, B. C., Perfit, M. R., Dreyfus, M., and Catalano, R., 1973, The Cayman Ridge [abstract]: Geol. Soc. Am., Abstr. with Programs, v. 5, n. 7, p. 663.

Hess, H. H., 1933, Interpretation of geological and geophysical observations, in: *U.S. Hydrographic Office Navy-Princeton gravity expedition to the West Indies in 1932*: U.S. Hydrographic Office, Washington, D.C., p. 27–54.

Hess, H. H., 1938, Gravity anomalies and island arc structure with particular reference to the West Indies: *Proc. Am. Philos. Soc.*, v. 79, n. 1, p. 71–96.

Hess, H. H., 1966, Caribbean research project, 1965, and bathymetric chart, in: *Caribbean Geological Investigations*, Hess, H. H., ed.: Geol. Soc. Am. Mem. 98, p. 1–10.

Holcombe, T. L., and Matthews, J. E., 1973, Structural fabric of the Venezuela Basin, Caribbean Sea [abstract]: Geol. Soc. Am., Abstr. with Programs, v. 5, n. 7, p. 671–672.

Holcombe, T. L., Vogt, P. R., Matthews, J. E., and Murchison, R. R., 1973, Evidence for sea-floor spreading in the Cayman Trough: *Earth Planet. Sci. Lett.*, v. 20, p. 357–371.

Hopkins, H. R., 1973, Geology of the Aruba Gap abyssal plain near DSDP site 153, in: *Initial Reports of the Deep Sea Drilling Project*, Vol. 15, Edgar, N. T., Saunders, J. B., *et al.*, eds.: U.S. Government Printing Office, Washington, p. 1039–1050.

Hospers, J., and Van Wijnen, J. C., 1959, The gravity field of the Venezuelan Andes and adjacent basins: Koninkl. Nederlandse Akad. Wetens., Afd. Natuurk., Verh., Eerste Reeks, v. 23, n. 1, 95 p.

Ipatenko, S., and Sashina, N., 1971, Sobre el levantamiento gravimetrico en Cuba [Gravity survey in Cuba]: Havana, Cuba, Ministerio de Minas, 14 p.

Irving, E. M., 1971, La evolucion estructural de los Andes mas septentrionales de Colombia: Colombia, Inst. Nac. de Inv. Geol.-Min., Bol. Geol., v. 19, n. 2, 89 p.

Khudoley, K. M., and Meyerhoff, A. A., 1971, Paleogeography and geological history of Greater Antilles: Geol. Soc. Am. Mem. 129, 199 p.

King, P. B. (compiler), 1969, Tectonic map of North America: U.S. Geol. Surv., Washington, D.C., 2 sheets, scale 1:5,000,000.

Krause, D. C., 1971, Bathymetry, geomagnetism, and tectonics of the Caribbean Sea north of Colombia, in: *Caribbean Geophysical, Tectonic, and Petrologic Studies*, Donnelly, T. W., ed.: Geol. Soc. Am. Mem. 130, p. 35–54.

Krause, D. C., and Jacobson, A. L., in press, Pedro Bank fracture zone, western Caribbean: *Geol. Soc. Am. Bull.*

Lagaay, R. A., 1969, Geophysical investigations of the Netherlands Leeward Antilles: Koninkl. Nederlandse Akad. Wetens., Afd. Natuurk., Verh., Eerste Reeks, v. 25, n. 2, p. 1–86.

Lattimore, R. K., Weeks, L. A., and Mordock, L. W., 1971, Marine geophysical reconnaissance of continental margin north of Paria Peninsula, Venezuela: *Am. Assoc. Petr. Geol. Bull.*, v. 55, n. 10, p. 1719–1729.

MacDonald, W. D., Doolan, B. L., and Cordani, U. G., 1971, Cretaceous–Early Tertiary metamorphic K–Ar age values from the south Caribbean: *Geol. Soc. Am. Bull.*, v. 82, n. 5, p. 1381–1388.

Malfait, B. T., and Dinkelman, M. G., 1972, Circum-Caribbean tectonic and igneous activity and the evolution of the Caribbean plate: *Geol. Soc. Am. Bull.*, v. 83, n. 2, p. 251–271.

Malin, P. E., and Dillon, W. P., 1973, Geophysical reconnaissance of the western Cayman Ridge: *J. Geophys. Res.*, v. 78, n. 32, p. 7769–7775.

Marlow, M. S., Garrison, L. E., Martin, R. G., Trumbull, J. V. A., and Cooper, A. K., 1974, Tectonic transition in the northeastern Caribbean: *J. Res. U.S. Geol. Surv.* v. 2, n. 3, p. 289–302.

Meyerhoff, A. A., 1962, Bartlett fault system: Age and offset: *Trans. Third Caribbean Geol. Conf., Kingston, Jamaica, 1962*, p. 1–9.

Meyerhoff, A. A., 1973, Diapirlike features offshore Honduras: Implications regarding tectonic evolution of Cayman Trough and Central America: Discussion: *Geol. Soc. Am. Bull.*, v. 84, n. 6, p. 2147–2152.

Meyerhoff, A. A., and Meyerhoff, H. A., 1972, Continental drift, IV: The Caribbean "plate": *J. Geol.*, v. 80, n. 1, p. 34–60.

Molnar, Peter, and Sykes, L. R., 1969, Tectonics of the Caribbean and Middle America regions from focal mechanisms and seismicity: *Geol. Soc. Am. Bull.*, v. 80, n. 9, p. 1639–1684.

Molnar, P., and Sykes, L. R., 1971, Plate tectonics in the Hispaniola area: Discussion: *Geol. Soc. Am. Bull.*, v. 82, n. 4, p. 1123–1126.

Nagle, Fred, 1971, Caribbean geology, 1970: *Bull. Mar. Sci.*, v. 21, n. 2, p. 376–439.

Nagle, Frederick, 1972, Rocks from seamounts and escarpments on the Aves Ridge: *Trans. Sixth Caribbean Geol. Conf., Margarita, Venezuela, 1971*, p. 409–413.

North, F. K., 1965, The curvature of the Antilles: *Geol. Mijnbouw*, v. 44, n. 3, p. 73–86.

Officer, C. B., Ewing, J. I., Edwards, R. S., and Johnson, H. R., 1957, Geophysical investigations in the eastern Caribbean: Venezuelan basin, Antilles island arc, and Puerto Rico Trench: *Geol. Soc. Am. Bull.*, v. 68, n. 3, p. 359–378.

Officer, C. B., Ewing, J. I., Hennion, J. F., Harkrider, D. G., and Miller, D. E., 1959, Geophysical investigations in the eastern Caribbean: Summary of 1955 and 1956 cruises, in: *Physics and Chemistry of the Earth*, Vol. 3, Ahrens, L. H., et al., eds.: Pergamon Press, London, p. 17–109.

Panamá, Dirección General de Recursos Minerales, 1972, Mapa geologico preliminar, Republica de Panamá: Panamá Direccion General Recursos Minerales, open-file map, scale 1:1,000,000.

Peter, George, 1971, Geology and geophysics of the Venezuelan continental margin between Blanquilla and Orchil[l]a Islands: Univ. Miami, unpublished Ph.D. dissert., 206 p.

Peter, George, 1972, Geologic structure offshore north-central Venezuela: *Trans. Sixth Caribbean Geol. Conf., Margarita, Venezuela, 1971*, p. 283–294.

Peter, George, and Lattimore, R. K., 1971, Geophysical investigations of the Venezuelan continental margin [abstract]: EOS (*Trans. Am. Geophys. Union*), v. 52, n. 4, p. 251.

Pinet, P. R., 1971, Structural configuration of the northwestern Caribbean plate boundary: *Geol. Soc. Am. Bull.*, v. 82, n. 7, p. 2027–2032.

Pinet, P. R., 1972, Diapirlike features offshore Honduras: Implications regarding tectonic evolution of Cayman Trough and Central America: *Geol. Soc. Am. Bull.*, v. 83, n. 7, p. 1911–1921.

Pinet, P. R., 1973, Diapirlike features offshore Honduras: Implications regarding tectonic evolution of Cayman Trough and Central America: Reply: *Geol. Soc. Am. Bull.*, v. 84, n. 6, p. 2153–2158.

Pyle, T. E., Meyerhoff, A. A., Fahlquist, D. A., Antoine, J. W., McCrevey, J. A., and Jones, P. C., 1973, Metamorphic rocks from northwestern Caribbean Sea: *Earth Planet. Sci. Lett.*, v. 18, p. 339–344.

Reblin, M. T., 1973, Regional gravity survey of the Dominican Republic: Univ. Utah, unpub. M.S. thesis, 122 p.

Rezak, Richard, Antoine, J. W., Bryant, W. R., Fahlquist, D. A., and Bouma, A. H., 1972, Preliminary results of Cruise 71-A-4 of the R/V ALAMINOS in the Caribbean: *Trans. Sixth Caribbean Geol. Conf., Margarita, Venezuela, 1971*, p. 441–449.

Roemer, L. B., 1973, Geology and geophysics of the Beata Ridge-Caribbean: College Station, Texas, Texas A. & M. Univ., unpub. M.S. thesis, 93 p.

Roemer, L. B., Bryant, W. R., and Fahlquist, D. A., 1973, Geology and geophysics of the Beata Ridge-Caribbean: Texas A. & M. Univ. Tech. Rep. 73-14-T, 92 p.

Schneidermann, Nahum, Beckmann, J. P., and Heezen, B. C., 1972, Shallow water carbonates from the Puerto Rico trench region: *Trans. Sixth Caribbean Geol. Conf., Margarita, Venezuela, 1971*, p. 423–425.

Scientific Staff, Leg 4, 1971 Cruise, UNITEDGEO I, International Decade of Ocean Exploration, 1972, Regional gravity anomalies, Venezuela continental borderland: U.S. Geol. Surv., open-file report, 24 p.

Shepard, F. P., 1973, Sea floor off Magdalena delta and Santa Marta area, Colombia: *Geol. Soc. Am. Bull.*, v. 84, n. 6, p. 1955–1972.

Shepard, F. P., Dill, R. F., and Heezen, B. C., 1968, Diapiric intrusions in foreset slope sediments off Magdalena delta, Colombia: *Am. Assoc. Petr. Geol. Bull.*, v. 52, n. 11, p. 2197–2207.

Shurbet, G. L., and Ewing, Maurice, 1956, Gravity reconnaissance survey of Puerto Rico: *Geol. Soc. Am. Bull.*, v. 67, n. 4, p. 511–534.

Silver, E. A., et al., 1972, Acoustic reflection profiles—Venezuela continental borderland: U.S. Dep. Commer., Nat. Tech. Inf. Serv. PB2-07597.

Silver, E. A., Case, J. E., and MacGillavry, H. J., 1975, Geophysical study of the Venezuelan Borderland: *Geol. Soc. Am. Bull.*, v. 86, n. 2, p. 213–226.

Škvor, Vladimír, 1969, The Caribbean area: A case of destruction and regeneration of continent: *Geol. Soc. Am. Bull.*, v. 80, n. 6, p. 961–968.

Soloviev, O. N., Skidan, S. A., Skidan, I. K., Pankratov, A. P., and Judoley, C. M., 1964a, Comentarios sobre el mapa gravimetrico de la isla de Cuba: *Tecnologica*, v. 2, n. 2, p. 8–19.

Soloviev, O. N., Skidan, S. A., Pankratov, A. P., and Skidan, I. K., 1964b, Commentarios sobre el mapa magnetometrico de Cuba: *Tecnologica*, v. 2, n. 4, p. 5–23.

Sykes, L. R., and Ewing, Maurice, 1965, The seismicity of the Caribbean region: *J. Geophys. Res.*, v. 70, n. 20, p. 5065–5074.

Talwani, Manik, 1966, Gravity anomaly belts in the Caribbean [abstract], in: *Continental Margins and Island Arcs*, Poole, W. H., ed.: Can. Geol. Surv. Pap. 66-15, p. 177.

Talwani, Manik, Sutton, G. H., and Worzel, J. L., 1959, A crustal section across the Puerto Rico Trench: *J. Geophys. Res.*, v. 64, n. 10, p. 1545–1555.

Tomblin, J. F., 1972, Seismicity and plate tectonics of the eastern Caribbean: *Trans. Sixth Caribbean Geol Conf., Margarita, Venezuela, 1971*, p. 277–282.

Uchupi, Elazar, Milliman, J. D., Luyendyk, B. P., Bowin, C. O., and Emery, K. O., 1971, Structure and origin of southeastern Bahamas: *Am. Assoc. Petr. Geol. Bull.*, v. 55, n. 5, p. 687–704.

Uchupi, Elazar, 1973, Eastern Yucatan continental margin and western Caribbean tectonics: *Am. Assoc. Petr. Geol. Bull.*, v. 57, n. 6, p. 1075–1085.

Vedder, J. G., *et al.*, 1971, USGS-IDOE Leg 2: *Geotimes*, v. 16, n. 12, p. 10–12.

Vedder, J. G., *et al.*, 1972, Acoustic reflection profiles—East Margin Yucatan Peninsula: U.S. Dep. Commer., Nat. Tech. Inf. Serv. PB 2-07595.

Vedder, J. G., MacLeod, N. S., Lanphere, M. A., and Dillon, W. P., 1973, Age and tectonic implications of some low-grade metamorphic rocks from the Yucatan Channel: *J. Res. U.S. Geol. Surv.*, v. 1, n. 2, p. 157–164.

Vening-Meinesz, F. A., and Wright, F. E., 1930, The gravity measuring cruise of the U.S. Submarine S-21: U.S. Nav. Obs., Publ. 13, Appendix I, 94 p.

Weeks, L. A., Lattimore, R. K., Harbison, R. N., Bassinger, B. G., and Merrill, G. F., 1971, Structural relations along Lesser Antilles, Venezuela, and Trinidad–Tobago: *Am. Assoc. Petr. Geol. Bull.*, v. 55, n. 10, p. 1741–1752.

Worzel, J. L., 1965, *Pendulum Gravity Measurements at Sea, 1936–1959*: John Wiley & Sons, New York, 422 p.

Chapter 4

BASEMENT ROCKS BORDERING THE GULF OF MEXICO AND THE CARIBBEAN SEA*

Philip O. Banks

Department of Earth Sciences
Case Western Reserve University
Cleveland, Ohio

I. INTRODUCTION

This paper reviews briefly what is presently known about the lithologies, distribution, and ages of basement rocks surrounding the Gulf of Mexico and Caribbean Sea. The term basement is interpreted to mean pre-Mesozoic rocks, that is, rocks whose times of formation predate the initiation of the Gulf and Caribbean basins as recognized today. A major question is whether or not pre-Mesozoic rocks underlie the floors proper of these basins. No paleontologic evidence for any has yet been found, and geophysical evidence, by itself, does not determine rock ages. Therefore, it is of considerable interest that Pequegnat *et al.* (1971) report a K–Ar age of 319 ± 5 m.y. for the 2–0.2 μ fraction of an unusual violet-colored siltstone, dredged from one of the Sigsbee Knolls in the central Gulf of Mexico. This age suggests that Upper Paleozoic

* Contribution No. 108, Department of Earth Sciences, Case Western Reserve University, Cleveland, Ohio.

rocks occur beneath the Gulf, having been brought to an exposed position at the Knolls by diapiric movements. Its significance, though, depends critically on whether authigenic glauconite is the only potassium-bearing mineral in the analyzed material, as the authors claim, or whether detrital illite or similar mineral is also present, for in that case the measured age could be older than the deposition age of the siltstone. Consequently this one result does not fully resolve the problem, and a challenge remains for further work.

At present, the only certain knowledge of pre-Mesozoic rocks is from surrounding land areas, principally the continents. Because not all the continental pre-Mesozoic is relevant to the topic of this book, some sort of criterion is needed to limit the coverage. The view is adopted here that the pre-Mesozoic basement rocks can be regarded as passive components of a framework that was progressively disturbed by tectonism (and degraded by erosion) during the evolution of the Gulf and Caribbean basins, and that their relevance stems from the clues they may ultimately provide about the evolving configurations of that framework. From this point of view, it suffices to restrict attention chiefly to a rather narrow bordering zone, generally not exceeding 100–200 km in width, in which pre-Mesozoic rocks reach their closest known proximity to the Gulf and Caribbean shores.

Well-founded comprehensive regional interpretations of the basement geology are not yet possible over much of the area of interest. Mesozoic and younger sediments associated with the developing basins have blanketed and obscured the earlier rocks for variable distances inland; much mapping remains to be done; and stratigraphic, structural, and temporal relations are difficult to unravel because of one or more overprints of deformation and metamorphism. Drillhole penetration has provided invaluable information but unfortunately is restricted to areas of economic potential. This review, therefore, is organized mainly according to the patchwork of local circumstances that comprise knowledge of the basement geology, avoiding all but the most elementary regional speculations.

The approach will be geographical, following a counterclockwise traverse from Florida around to Venezuela and back through the Antilles. In discussing local areas along this traverse, references will be made only to the more recent literature. Space does not permit reviewing the entire history of geologic investigations and thoughts concerning each area; such information can be obtained from the references or from other papers in this volume. A valuable general reference is that of Murray (1961), who presents an admirably thorough discussion of the geologic setting of the Gulf of Mexico. No recent summary of comparable scope is available for the Caribbean.

II. SOUTHEASTERN UNITED STATES

The entire northern margin of the Gulf of Mexico is blanketed with Mesozoic to Recent sedimentary rocks of the Gulf Coastal Plain, Mississippi Embayment, and Florida Platform. A considerable amount of information about the basement rocks beneath this cover is available from deep drilling. Over much of the area the buried basement can be correlated more or less directly with rocks of Precambrian to Paleozoic age exposed at the surface farther inland, although a notable exception is the basement of Florida where lithologies occur that seem to have no counterparts in the nearby exposed Piedmont. The greatest difficulties are in clarifying the nature of transitions from one geologic province to another, the only data available being the discontinuous scraps of information provided by borehole data.

A useful summary map of the basement rocks of the United States, including the region dealt with here, has been compiled by Bayley and Muehlberger (1968).

A. Florida and Georgia

The distribution of basement rocks beneath Florida and southernmost Georgia was largely worked out by Applin (1951). Bass (1969) and Milton and Grasty (1969) added new petrographic observations and measured radiometric ages for several samples. Briefly, the situation deduced by these authors is as follows. Under a small area in east-central peninsular Florida the basement consists of interlayered amphibolite, schist, and gneiss and veined, sheared, and altered granite. These rocks are bordered on the north, west, and south by a terrain of relatively unaltered rhyolitic tuffs, agglomerates, and ignimbrites. Northward the rhyolitic terrain gives way to overlying mildly deformed Lower Ordovician to Lower Devonian clastic sedimentary rocks that extend beneath much of northern Florida and southernmost Georgia. North of the Paleozoic rocks rhyolite appears again in a small area under southeastern Georgia, but beyond a relatively narrow strip along the border with Florida the coastal plain of Georgia is underlain by schists, gneisses, and amphibolites of greenschist or higher facies that are similar to rocks of the exposed Piedmont Province.

Basalts and diabases have been encountered in several wells in the region and probably for the most part are dikes or sills of various ages. However, basalt is dominant in the two southernmost wells in Florida, and Bass (1969) suggests that there it may represent the margin of a distinctly basaltic southern province.

Extensive rhyolitic lithologies and weak or absent deformation and metamorphism of Paleozoic sedimentary rocks are not known elsewhere along the easterly side of the southern Appalachian orogen. The principal hypotheses are that the Florida basement is "in place" and simply represents a rapid change in orogenic facies that is more difficult to recognize elsewhere on the nearby Atlantic seaboard or continental shelf; that the Florida basement has been offset by transcurrent faulting from possible correlative sequences farther west; or that the Florida basement is an "exotic" block whose present position is a consequence of the vagaries of plate tectonic motions and is unrelated to the Appalachians proper.

Radiometric ages have not yet demonstrated the presence of unequivocal Precambrian rocks beneath Florida. Bass (1969) obtained Rb–Sr isochron ages of about 530 m.y. from mineral fractions of the granites and metamorphic rocks of east-central Florida. This may be a minimal estimate because of alteration. K–Ar ages of various other samples range widely from a low of 89 m.y. for basalt overlying the granite-metamorphic terrain (Milton and Grasty, 1969) to a high of 520 m.y. for biotite from hornfels underlying a diorite sill (Bass, 1969). The sill itself gave a whole-rock K–Ar age of 480 m.y. (Muehlberger *et al.*, 1966). Geologic and radiometric evidence are thus in agreement that the oldest rocks are at least Cambrian in age, but chronologic details are obscured by uncertainties inherent in the radiometric measurements.

B. Alabama and Mississippi

Except for southernmost Georgia and the Florida Peninsula, discussed above, the dominant element of the basement geology as far west as central Mississippi is the Appalachian orogen. The exposed southernmost Appalachians are subdivided into several more-or-less parallel NE–SW trending provinces or belts, which differ in their lithologies and styles of deformation, and which generally can be recognized continuing southwest and west beneath the Coastal Plain cover.

On the southeastern side of the orogen is the Piedmont Province, consisting of variably metamorphosed schists and gneisses derived chiefly from volcanic and clastic sedimentary materials and intruded by granitic bodies. Fullagar (1971), working mainly in North and South Carolina, has obtained whole-rock Rb–Sr isochron ages for several Piedmont granites that cluster into three groups: 595–520 m.y., 415–385 m.y., and about 300 m.y. These results, and numerous others, indicate that some of the intruded rocks are Cambrian or older and that several thermal disturbances have affected the Province.

The northwestern side of the orogen is the Valley and Ridge Province, characterized by folded and thrusted but essentially unmetamorphosed Paleozoic shelf-type carbonates and clastics. The intensity of deformation decreases rapidly west and northwest into the broad arches and basins of the continental interior. In the subsurface to the south the structural trends of this province swing toward the west, possibly indicating continuity with the Ouachita System (see below).

The core of the southern and central Appalachians is the Blue Ridge Province, along which crystalline basement rocks and overlying volcanic and clastic sedimentary sequences of Upper Precambrian age are exposed. The basal gneisses and granites are mostly about 1000 m.y. old (Tilton *et al.*, 1960), although ages as high as 1300 m.y. have been measured locally (Davis *et al.*, 1962). This complex extends southwest into northern Georgia; beyond that point the central portion of the orogen appears to consist of uppermost Precambrian clastic rocks and Paleozoic slates, cherts, and sandstones. The latter are best known in the Talladega Belt, adjacent to the Valley and Ridge Province.

In western Mississippi the Coastal Plain deposits conceal Paleozoic strata rather distinct from the lithologies typical of the Appalachian Paleozoic. These strata are part of the Ouachita System discussed in the next section. The transition between Appalachian and Ouachita regimes, which is entirely concealed, has been a topic of considerable debate. Most suggestions involve some combination of two limiting cases: the transition is an abrupt change in facies and strike of an otherwise continuous belt, or one belt truncates or over-rides the other by some system of faulting. An unequivocal solution to this problem will require more subsurface information.

III. THE OUACHITA SYSTEM

From western Mississippi to western Texas and adjacent parts of Mexico, basement rocks closest to the Gulf of Mexico belong to the Paleozoic Ouachita System. This great belt is concealed beneath younger rocks for much of its length, being exposed principally in the Ouachita Mountains of Oklahoma–Arkansas and Marathon Uplift of west Texas, and in a few smaller areas near the Mexico–Texas border. The known portions of the Ouachita System are restricted to its northerly and westerly extremities where cover is either absent or sufficiently thin to allow frequent penetration. These portions follow a narrow sigmoidal trend from the Ouachita Mountains southward across central Texas, thence westward around the Llano Uplift to the Marathon region.

Flawn *et al.* (1961) and King (this volume) have summarized the geology of the Ouachita System. It is composed of variously folded, faulted, and locally metamorphosed sedimentary rocks ranging in age from late Cambrian to Permo-Carboniferous. The Lower and Middle Paleozoic is characterized by dark graptolitic and siliceous shales, limestones, and cherts, while the Upper Paleozoic is characterized by flysch deposits of lighter-colored shales, siltstones, and sandstones. In most places metamorphism, if present, is incipient to weak, but in south-central Texas phyllite, schist, slate, metaquartzite, and marble have been encountered in the subsurface. Similar rocks also occur locally to the west. This suite probably represents an interior zone of the Ouachita orogen. The main phase of orogenic deformation appears to have occurred in mid-Pennsylvanian time in the eastern portion of the belt and to have been somewhat later to the west. Late Paleozoic volcanic rocks and granitic intrusives are known from isolated exposures in northern Mexico, but elsewhere igneous activity seems to have been relatively unimportant until postorogenic relaxation, when diabase dikes and small bodies of alkaline character invaded the deformed Paleozoic rocks.

On its continental side the Ouachita System laps onto and is thrust over Precambrian crystalline rocks of the continental interior. The geochronology and generalized geology of the Precambrian craton have been summarized by Muehlberger *et al.* (1966) and Muehlberger, Denison, and Lidiak (1967). Radiometric ages are distributed in a crudely zonal fashion, with younger rocks to the south and older ones to the north. Immediately adjacent to the Ouachita belt, ages of 1000–1200 m.y. are characteristic, as in the Llano Uplift (Zartman, 1964) where gneisses and schists were metamorphosed and intruded by granite at about that time. Known occurrences of such rocks are everywhere a considerable distance inland from the Gulf of Mexico, and their Gulfward extent beneath overlying rocks is problematical.

IV. MEXICO

Knowledge of the distribution of pre-Mesozoic rocks in Mexico is much less detailed than in the southern United States because of thick Mesozoic and Cenozoic cover, disturbance by Laramide tectonic activity, and a relative scarcity of deep wells. At scattered localities in Coahuila, Chihuahua, and Nuevo Leon, Middle and Upper Paleozoic clastic sedimentary strata, minor carbonates, and rocks of volcanic derivation are exposed in local uplifts. King (this volume) has reviewed these occurrences. In places the rocks are metamorphosed to low-grade schistose and phyllitic assemblages and are intruded by granitic stocks of probable Permian or Triassic age. Denison *et al.* (1969) obtained radiometric ages on metasedimentary rocks from Sierra del Carmen and Acatita Valley of 263 m.y. (K–Ar, muscovite), 275 m.y. (Rb–Sr, whole rock/muscovite pair), and 204 m.y. (K–Ar, whole rock). Granodiorites from Potrero de la Mula and Acatita Valley yielded ages of 206 m.y. (K–Ar, hornblende) and 203 m.y. (K–Ar, biotite). Thermal activity thus appears to have been widespread in the region in Permian–Triassic time.

The principal objective of Denison *et al.* (1969) was to investigate possible source areas for boulders in the Haymond Formation (Pennsylvanian) of the Marathon Basin. Several samples of granitic gneiss and schist from the boulder assemblage gave a combined whole rock/muscovite Rb–Sr isochron age of 396 ± 9 m.y., considerably older than any of the investigated exposures and suggesting the possibility of an undiscovered or buried Silurian–Devonian metamorphic terrain in the region.

In southern Tamaulipas, west of Ciudad Victoria, Paleozoic beds including Middle Silurian limestone are exposed in fault contact with gneisses and schists in the core of the Huizachal–Peregrina Anticlinorium. Detrital zircons from the gneisses yielded Pb-alpha ages of 924–1350 m.y. (Fries et al., 1971), while two K–Ar mica ages measured on the metamorphics gave 740 and 315 m.y. (Fries et al., 1962a). Although the structural relations between the schists and gneisses are not fully established, it appears certain that Upper Precambrian rocks occur at this locality, perhaps metamorphosed about 1000 m.y. ago.

Southward through Mexico, fragmentary outcrop and borehole evidence indicates that a depositional belt subparallel to the eastern coast probably existed during much of the Middle and Upper Paleozoic (López–Ramos, 1969). Definite Lower Paleozoic (Cambro-Ordovician) limestone, shale, and sandstone are recognized only at the southern end of this belt in the Nochistlán area of Oaxaca (Pantoja and Robison, 1967).

Known and possible occurrences of Precambrian rocks in Mexico have been reviewed by de Cserna (1971). Continuing southward from Tamaulipas, pre-Carboniferous muscovite, biotite, cordierite, and garnet gneisses are exposed in the core of the Huyacocotla Anticlinorium in the state of Hidalgo. Pb-alpha ages of detrital zircons from these gneisses range from 1210 to 800 m.y. (Fries and Rincón-Orta, 1965; Fries et al., 1971), similar to the gneisses near Ciudad Victoria. At Taxco, south of the Mexican Volcanic Belt, a Pb-alpha age of 1020 m.y. was obtained from zircons of a pre-Middle Cretaceous metarhyolite (de Cserna et al., 1971). In contrast to the detrital zircons mentioned above, this result can be interpreted as a direct approximation to the primary age of the rhyolite.

In the Sierra Madre del Sur of Oaxaca, Paleozoic strata overlie a complex of gneisses, schists, marbles, and various intrusive bodies including pegmatites. The mean of several Rb–Sr, K–Ar, and Pb-alpha age determinations from the pegmatites is 884 m.y. (de Cserna, 1971). Anderson and Silver (1971) have confirmed more precisely the ages of these rocks through isotopic analyses of zircons. They obtained nearly concordant results of 1080 m.y. for a biotite gneiss and a syntectonic pegmatite, and of 975 m.y. for a post-tectonic pegmatite. Fries et al. (1962b) proposed the name Oaxacan Orogeny for the deformation, metamorphism, and intrusion that took place in this region about 1000 m.y. ago, and suggested that the Oaxacan orogenic belt may have been temporally equivalent to and laterally continuous with the 1000–1200 m.y. Grenville orogenic belt of the United States and Canada.

V. NUCLEAR CENTRAL AMERICA

Nuclear Central America includes that portion of Mexico south and east of the Isthmus of Tehuantepec, the countries of Guatemala, British Honduras, and Honduras, and the northern part of Nicaragua. Pre-Mesozoic rocks are exposed along a mountainous arcuate belt, convex to the south, extending roughly E–W across the region. Recent summaries of regional relationships are given by Dengo and Bohnenberger (1969), Kesler (1971), and Dengo (this volume). Detailed mapping of pre-Mesozoic rocks has been performed in central and western Guatemala, but elsewhere most of the information is of reconnaissance character.

A. Chiapas

The mountains of Chiapas are underlain chiefly by granitic intrusive rocks with subordinate amounts of banded gneiss, augen gneiss, and rare marble. Pantoja *et al.* (1971) report a 700 m.y. Pb-alpha age from a pink granite that intrudes schist and gneiss in the Arriaga area. Whether this reflects a continuation of the Oaxacan terrain or is a distinctly younger sequence cannot be ascertained until more work is done in the region.

B. Guatemala

The principal exposures of basement rocks in Guatemala occur in a belt bounded on the north by the Cuilco–Chixoy–Polochíc fault zone and on the south by the Motagua fault zone. Large bodies of serpentinite and peridotite occur adjacent and subparallel to these fault zones and were postulated by McBirney (1963) to represent the ultimate basement of the area. Excepting the serpentinites, the oldest rocks are gneisses, biotite–garnet schists, marbles, and amphibolites of the Chuacús Group (McBirney, 1963) and Western Chuacús Group (Kesler *et al.*, 1970). In central Guatemala the Chuacús Group is intruded by the Rabinal and Matanzas granites. Throughout the region these crystalline rocks are overlain by shales, sandstones, and limestones of the Pennsylvanian–Permian Santa Rosa Group.

Gomberg *et al.* (1968) analyzed zircons from a gneiss of the Chuacús Group and from the Rabinal granite. Their results formed a discordant array that was interpreted to signify an age of 1075 m.y. for the zircons in the gneiss and an age of 345 m.y. for the time of crystallization of the Rabinal granite. It was suggested that the 1075 m.y. age might be inherited from a source terrain and hence is an older limit to the time of original deposition of the Chuacús Group, a view that has been adopted by McBirney and Bass (1969a). Thus, while the Chuacús Group is clearly pre-Pennsylvanian, it may be no older than Middle or Early Paleozoic.

C. British Honduras

Pre-Mesozoic rocks are exposed in the upfaulted Maya Mountains of British Honduras. The major rock types are weakly metamorphosed argillites, sandstones, and limestones, at least partly of Pennsylvanian–Permian age. Several granitic stocks intrude the metasediments and have locally produced higher grades of metamorphism. Along the southwestern edge of the outcrop area rhyolitic material is abundant.

Dixon (1956) believed that there are two series of sediments separated by a regional unconformity: an older Maya Series and a younger Macal Series, the latter being correlative with the Santa Rosa Group of Guatemala. The rhyolitic rocks, termed the Bladen Porphyry, were interpreted to be post-Macal shallow intrusives. Work subsequent to Dixon's is reviewed by Bateson (1972) and Hall and Bateson (1972). They find no evidence for a regional unconformity and propose that the entire sedimentary sequence be renamed Santa Rosa Group. In their interpretation the rhyolites are flows and pyroclastics interbedded with the sediments (Bladen Volcanic Member). Three samples of the Bladen rhyolites have given a whole-rock Rb–Sr isochron age of about 300 m.y. (Hurley *et al.*, 1968*a*), tending to support the interpretation of Bateson and Hall. Bateson (1972) lists limited radiometric data for some of the granites. K–Ar ages from two of the bodies are 206–213 m.y., also in agreement with stratigraphic interpretations. However, two samples of a third body, the Mountain Pine granite, gave Rb–Sr ages of 280–300 and 390 m.y., respectively. The latter age in particular suggests that temporal relations in this area are still not well understood and that rocks of at least Middle Paleozoic age might be present.

D. Yucatan

Wells drilled in the northern and central Yucatan Peninsula have encountered andesitic to rhyolitic flows and breccias beneath Cretaceous limestones and shales. On the basis of preliminary Rb–Sr data Bass and Zartman (1969) state that the volcanics are possibly Middle Paleozoic in age. López-Ramos (this volume) quotes a Rb–Sr age of about 410 m.y. for rhyolites from Yucatan No. 1. Apparently, therefore, these rocks are older than the Bladen rhyolites of the Maya Mountains. No other surface exposures of rocks that might be correlative with the volcanics of Yucatan are known in the region.

In the southern portion of the Peninsula and in adjacent parts of Guatemala and British Honduras, drilling reveals the presence of schistose and granitic rocks which probably represent subsurface extensions of the basement complexes of the Guatemalan Cordillera and the Maya Mountains.

Fragments of sericitic schist and marble have been dredged from the Yucatan Channel (Vedder *et al.*, 1971), which resemble some of the lithologies on the mainland and also in Cuba. Vedder *et al.* (1973) obtained K–Ar ages on impure micas from these rocks ranging from 59 to 93 m.y., but point out that the ages are probably minimal due to chloritization. Consequently, it is uncertain whether these fragments represent an extension of pre-Mesozoic basement to the edge of the continental shelf or are associated with younger tectonothermal events.

E. Honduras and Nicaragua

McBirney and Bass (1969*a*) describe a eugeosynclinal assemblage of metamorphosed greywackes, greenstones, slates, and cherts south of the Motagua fault which they named the El Tambor Formation. To the east and south, this grades into an extensive terrain dominated by phyllite with subordinate carbonate. Thick limestones overlying phyllite are believed to correlate with the Permian Chochal Formation of Guatemala. Regional considerations suggest that the El Tambor Formation underlies the phyllites and may be equivalent in age to the Chuacús Group north of the Motagua fault.

Few details are known of the phyllitic terrain of Honduras and northern Nicaragua. Local variations are illustrated in the stratigraphic sections of Mills *et al.* (1967). Pushkar (1968) published whole rock Rb–Sr data for a single sample of phyllite from Honduras, from which McBirney and Bass (1969*a*) calculated an age of 412 m.y. This age is not particularly sensitive to the choice of initial Sr ratio but its significance depends on a number of assumptions about the behavior of the Rb–Sr system in such rocks. Consequently, it is no more than suggestive of a Middle Paleozoic age for at least some of the material.

F. Bay Islands

The Bay Islands lie in the Gulf of Honduras on the south side of the Bartlett Trough. Their geology has been summarized by McBirney and Bass (1969*b*). Metasedimentary rocks consisting of greywacke, shale, chert, and thin limestone are present but vary in proportions from one island to another. Also present are serpentinites, amphibolites, pyroxene, hornblendite, and hornblende–biotite melagabbro. On Guanaja the metasediments are recrystallized progressively eastward to quartz–mica schist and gneiss, and grade into sodic granite. It seems likely, in spite of differences in detail, that these assemblages correlate with the El Tambor Formation and related rocks south of the Motagua Fault on the mainland, although at present no proof of such correlation exists.

VI. SOUTHERN CENTRAL AMERICA

From northern Nicaragua to Colombia no pre-Mesozoic rocks are known to exist. The foundation of this region consists of basalts, diabases, pillow lavas, diorite intrusives, and associated rocks of Cretaceous and possibly Jurassic age.

VII. COLOMBIA

Inland from the coast, Precambrian and Paleozoic rocks crop out in the Central and Eastern Cordilleras of Colombia (Pinson et al., 1962; Jacobs et al., 1963). The Central Cordillera proper terminates northward at the Magdalena River, but similar lithologies reappear near the northern coast in the Sierra Nevada de Santa Marta. MacDonald and Hurley (1969) have outlined briefly the geology of the Santa Marta massif. The oldest rocks are high-grade biotitic and amphibolitic banded gneisses, intruded by dioritic plutons. At scattered localities they are overlain by Devonian sandstone and Carboniferous limestone. MacDonald and Hurley obtained a whole-rock Rb–Sr isochron age of about 1400 m.y. from the gneiss near Dibulla, and they suggest that the Sierra Nevada de Santa Marta and the Central Cordillera lie at or near the western margin of the Precambrian Guyana Shield of the continental interior.

The Santa Marta massif is bounded on the north by the E–W trending Oca Fault. North of the fault, metamorphic and igneous rocks are again exposed in the uplands of the Guajira Peninsula, where they consist of granitic gneisses, mica schists, amphibolites, diorites, and granodiorites (MacDonald, 1972). MacDonald (1964) reported a K–Ar age of 195 m.y. for a pegmatite intruding the schists. On the northern coast of the peninsula, lower-grade metamorphics and intrusives of Mesozoic and Tertiary age are known (MacDonald et al., 1971); consequently the 195 m.y. age is probably minimal due to the possibility of Ar loss during younger events, and the schists may be late Paleozoic or older. The presence of Precambrian rocks on the peninsula is indicated by a preliminary zircon isotopic age of 1250 m.y. for the Jojoncito gneissic granite, obtained by the writer.

The Eastern Cordillera of Colombia bifurcates north of Bogotá to form the Sierra de Perijá, along the Colombia–Venezuela border west of Lake Maracaibo, and the Mérida Andes of Venezuela, south and southeast of Lake Maracaibo. Bowen (1972) has described in some detail the pre-Cretaceous stratigraphy of the northern part of the Sierra de Perijá. The oldest fossiliferous material is Lower Devonian, unconformably overlying a complex of quartzite, schist, phyllite, slate, granitic gneiss, amphibolite, and intrusive granite.

These rocks are considered to be in part Lower Paleozoic and in part Precambrian in age. Martin-Bellizzia (1968) quotes a radiometric age of 800 m.y. for orthoclase from the El Palmar granite, but no details are given. She also lists other ages from granites of this region ranging from 200 to 370 m.y.

The Devonian section consists mostly of sandstones, shales, and marls, and is overlain by Permo-Carboniferous limestones and redbeds presumed to be correlative with similar lithologies in the Mérida Andes.

VIII. VENEZUELA

A. Venezuelan Andes

The geology of the central portion of the Mérida Andes of Venezuela has been summarized by Shagam (1972a, b and this volume). The basal unit is the Iglesias Complex, which is subdivided into two principal facies. To the north, quartzofeldspathic gneisses and mica schists of amphibolite facies metamorphism, locally reaching sillimanite grade, comprise the Sierra Nevada Facies. On the southern flank of the range, quartz–muscovite–chlorite schists of greenschist facies metamorphism comprise the Bella Vista Facies. Intermediate to these two in position and grade is a third facies, the Tostós Facies, which possibly correlates with the Pennsylvanian Mucuchachi Formation instead of the basal complex. A probable Precambrian age for some or all of the Iglesias Complex is indicated by a K–Ar age of 660 m.y. for granite intruding the Bella Vista schists (Compañia Shell and Creole, 1964; details not given).

Graptolitic shales, sandstones, and fine-grained limestones of Middle Ordovician to Upper Silurian age overlie the Bella Vista Facies in the southern part of the range. In the central part, dark slates, variegated shales, and redbeds of probable Pennsylvanian age overlie the Iglesias Complex and are overlain in turn by Permian limestones and shales. Shagam (1972a) postulates several periods in the Paleozoic during which granitic plutons were intruded into the section. Martin-Bellizzia (1968) lists available radiometric ages from the Andes, most of which are K–Ar and Rb–Sr mineral ages and probably have been biased by subsequent tectonothermal events. Nevertheless, at the very least, Middle and Late Paleozoic intrusive activity is suggested by the radiometric data.

B. Caribbean Mountain System

The Caribbean Mountain System is a complex terrain of folded, faulted, and metamorphosed sedimentary and volcanic strata underlying most of the coastal region of Venezuela. Extensive exposures occur in the Cordillera de la Costa and its adjacent foothills to the south, along the Peninsula of Araya–Paria, in the Cordillera Norte of Trinidad, and in several of the Venezuela Caribbean islands (Margarita, La Orchila, etc.). Bellizzia (1972*a,b*) has summarized the general features of the geology of this region.

Scattered paleontologic and radiometric data indicate that most of the rocks making up the Caribbean Mountain System are Mesozoic or younger. Possible pre-Mesozoic basement may, however, be represented by the Sebastopol and El Tinaco Complexes. The former underlies the Late Jurassic–Cretaceous Caracas Group of the Cordillera de la Costa; the latter crops out just south of the Cordillera and is overlain by post-Caracas Group metasediments and metavolcanics. The complexes consist of hornblende, biotite, and quartzofeldspathic gneisses, mica schists, and migmatites, metamorphosed to the almandine–amphibolite facies and intruded by trondhjemitic and granitic bodies. K–Ar ages have been unreliable indicators of true age for these rocks because of strong overprints from Mesozoic orogenic activity. A possible Middle Paleozoic or earlier age is suggested by a single whole-rock Rb–Sr determination of 425 m.y. reported by Hurley *et al.* (1968*b*) for a gneiss from the Sebastopol Complex. In addition to this scant evidence, the existence of pre-Mesozoic basement in the coastal region of Venezuela is suggested by U–Pb ages of 262–265 m.y. for sphene from a granite of the Mesa de Cocodite, Peninsula of Paraguaná (Martin-Bellizzia, 1968; details not given). It is conceivable that pre-Mesozoic rocks extend under much of the coastal lands and into the offshore area, but at present there is no firm evidence of this.

C. Guyana Shield

The deformed and metamorphosed rocks of the coast ranges of Venezuela give way southward to less deformed and unmetamorphosed Mesozoic and Cenozoic deposits, and, south of the Rio Orinoco, Precambrian crystalline rocks of the Guyana Shield appear at the surface. They extend for an undetermined distance northward beneath the overlying cover. Of particular interest is that some of the northernmost exposed rocks are also some of the oldest in South America. These are the granulitic gneisses and migmatites of the Imataca Complex, which have yielded a whole-rock Rb–Sr isochron age of 3100–3400 m.y. (Hurley *et al.*, 1970). Other aspects of the geology of the Guyana Shield and a summary of radiometric data are given by Martin-Bellizzia (1968). North of the limit of outcrop the distribution and extent of Precambrian lithologies in the subsurface is not well known, and no radiometric data are available.

IX. ANTILLES

Pre-Mesozoic rocks are not found on the islands of the Lesser Antilles, but in the Greater Antilles there are scattered localities where metamorphosed rocks of unknown age underlie unmetamorphosed to weakly metamorphosed Middle to Late Jurassic or Cretaceous strata. The antiquity of these metamorphosed complexes has been a subject of considerable debate. Khudoley and Meyerhoff (1971) have summarized the occurrences of such rocks and the arguments concerning their age.

In Cuba, possible pre-Jurassic rocks are exposed in the Sierra de Trinidad of southern Las Villas Province, in eastern Oriente Province, and on the Isle of Pines (Pardo, this volume). The Sierra de Trinidad exposures consist of the Trinidad Formation and the San Juan Marble. The Trinidad Formation is composed of epidote–amphibolite facies rocks including micaceous and graphitic schists, quartzites, coarsely crystalline gneiss, chlorite, actinolite, and serpentine schists, and amphibolite and metagabbro. The San Juan Marble is a thick sequence of metamorphosed limestone whose stratigraphic position relative to the schists is uncertain. Similar lithologies occur on the Isle of Pines, although the presence there of staurolite and sillimanite schists indicates a somewhat higher grade of metamorphism. Serpentinized peridotite, metagabbro, gabbro, and granodiorite intrude the metamorphics of the Isle of Pines. Stratigraphically the Trinidad Formation and San Juan Marble appear to underlie weakly metamorphosed shales, siltstones, and sandstones that have been correlated with the Middle to Late Jurassic San Cayetano Formation.

North of the Sierra de Trinidad and separated from the above units by a major fault, a belt of amphibolites is intruded by the Sancti Spiritus batholith. A whole-rock K–Ar age of 180 m.y. from a granodiorite of this batholith is the oldest radiometric age so far obtained in Cuba (see Meyerhoff et al., 1969). If it is correct, it suggests an Early Jurassic age for the batholith and therefore by inference a pre-Jurassic age for the metamorphic rocks. K–Ar mineral ages from the metamorphic rocks are Cretaceous or younger, probably reflecting the influence of later events.

In Jamaica, altered volcanic and volcaniclastic rocks of probable Lower Cretaceous age are the oldest units (Roobol, 1972). They are intruded by the Above Rocks granodiorite which has yielded a radiometric age of about 65 m.y., confirmed by K–Ar, Rb–Sr, and U–Pb methods (Chubb and Burke, 1963).

Metamorphic rocks occur at several localities on Hispaniola. Mica schists, amphibolite, serpentinite, and talcose and chloritic metavolcanics are found in the Plaine du Norde area of Haiti. Banded schistose marble crops out on Tortuga Island and on the Samana Peninsula of the Dominican Republic. Metamorphosed keratophyric and basic igneous rocks, mostly of greenschist

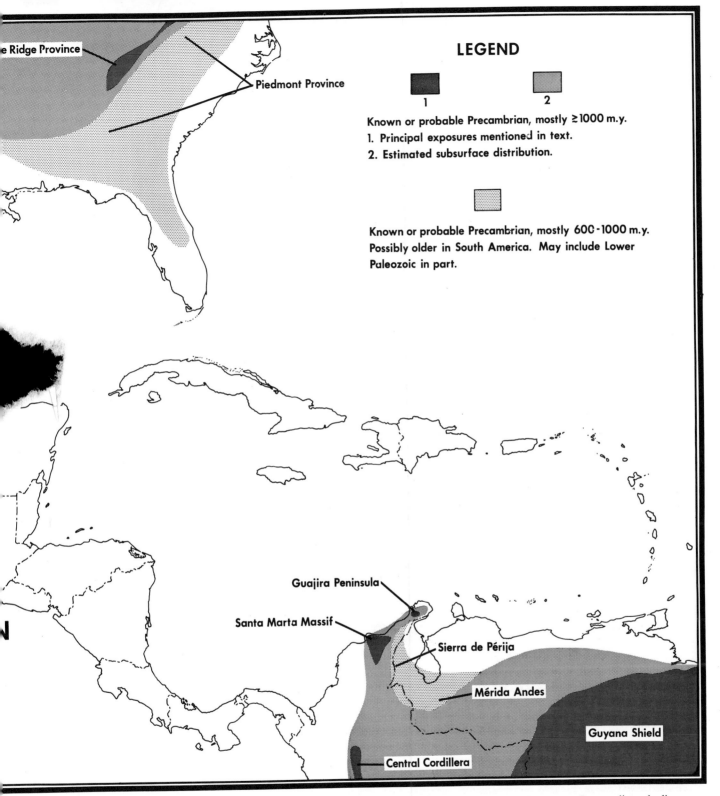

LEGEND

1   2

Known or probable Precambrian, mostly ≥1000 m.y.
1. Principal exposures mentioned in text.
2. Estimated subsurface distribution.

Known or probable Precambrian, mostly 600-1000 m.y.
Possibly older in South America. May include Lower
Paleozoic in part.

Ridge Province

Piedmont Province

Guajira Peninsula

Santa Marta Massif

Sierra de Périja

Mérida Andes

Guyana Shield

Central Cordillera

gion. Darkest shading represents principal areas of exposure, which consist predominantly of rocks 1000 m.y. old or more. Intermediate shading repre-
quity. Lightest shading represents areas in which rocks of inferred pre-Paleozoic age occur, either exposed or in the subsurface, but which have not

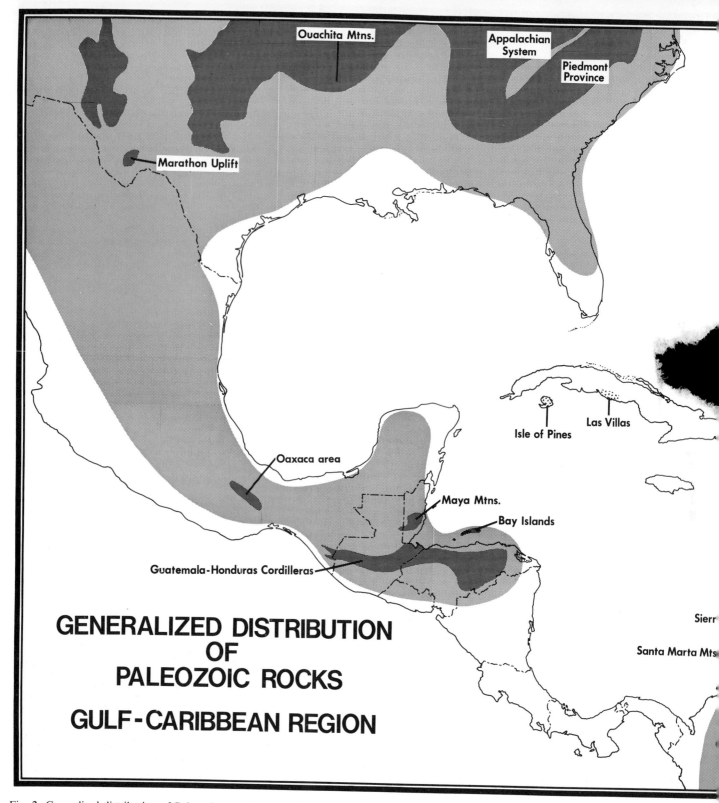

Fig. 2. Generalized distribution of Paleozoic rocks in the Gulf–Caribbean region. Darkest shading represents principal areas of exposure, intermediate s[...] and the lighter shadings represent successively less well established occurrences. Note that the Paleozoic is temporally undifferentiated and,

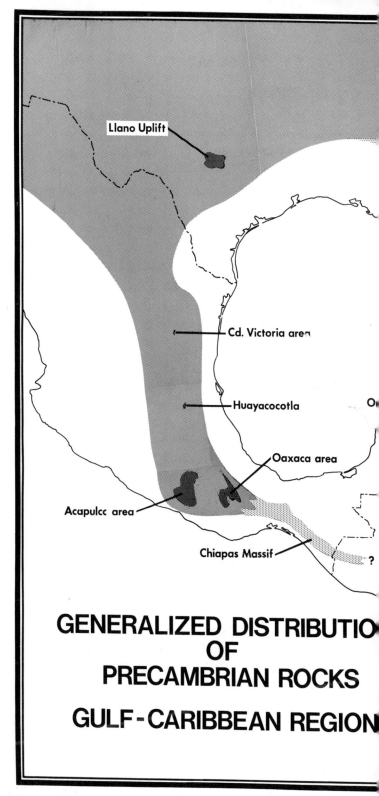

Fig. 1. Generalized distribution of Precambrian rocks in the Gulf–Caribbean r
sents schematically the inferred subsurface distribution of rocks of similar ant
been shown to be as old as 1000 m.y.

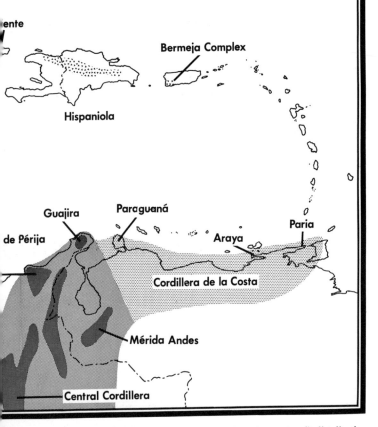

ente

Bermeja Complex

Hispaniola

Guajira

Paraguaná

Paria

de Périja

Araya

Cordillera de la Costa

Mérida Andes

Central Cordillera

ading represents schematically the inferred subsurface (or restored) distribution,
consequently, that the map is not a paleogeographic reconstruction.

facies, occur in the central part of the Dominican Republic. The latter area has yielded the oldest radiometric age from Hispaniola, a K–Ar age of 127 m.y. from a hornblendite phase of a tonalitic intrusion in the metavolcanics (Bowin, 1966; Khudoley and Meyerhoff, 1971).

The Bermeja Complex of southwestern Puerto Rico consists of gneiss, amphibolite, and serpentinite. Stratigraphically the complex is pre-Campanian. A K–Ar age of 110 m.y. has been reported for hornblende from the gneiss (Mattson, 1966).

It seems clear that if pre-Mesozoic rocks are present in the Greater Antilles, considerably more effort will be needed to find samples that have retained a radiometric memory of their original age.

X. SUMMARY

This review has attempted to outline the surface and subsurface occurrences of basement rocks—defined here to be rocks of Paleozoic or Precambrian age—in the regions bordering the Gulf of Mexico and Caribbean Sea. Such occurrences can be categorized according to the confidence with which the age assignment is made: known, probable, contentious, or speculative. Known occurrences speak for themselves. Probable occurrences are those for which direct evidence of age is lacking, but for which the postulated age assignment is reasonable from other considerations. Contentious occurrences are wholly uncertain; their age could be either pre-Mesozoic or younger. Speculative occurrences, as a class, have not been treated in this review. They include regions where the only sampled rocks are demonstrably Mesozoic or younger but may, at depth, be underlain by older material, as on the continental shelves. Also included in the speculative class are inferences based on geophysical information. For instance, the thick crust beneath the Antilles (e.g., Talwani et al., 1959) could be interpreted to represent accumulation of materials over a time span longer than the Mesozoic and Cenozoic. There are, however, plausible alternate interpretations, such as tectonic thickening, that can account for the same data, and that place such topics beyond the scope of this review.

Figures 1 and 2 summarize the distribution of pre-Mesozoic rocks in the Gulf–Caribbean area. Because of the limitations of regional generalizations and the broad age groupings of rocks considered in this review, these maps depict only the grossest features of the areal distributions. Subsurface boundaries are, of course, schematic and subject to differences of opinion. Many outstanding problems must be solved before the regional picture is satisfactorily established, and some of the more obvious are reflected in the following questions. What is the true subsurface extent of the Precambrian shields of North and South America? Is the Oaxacan Belt of Mexico continuous with the Precambrian of the United States and Canada? Are the oldest rocks of Guatemala and Honduras Precambrian or Paleozoic? How extensive are the presumed Paleozoic complexes underlying the Caribbean Mountain System? Do rocks of identifiable pre-Mesozoic age underlie any part of the Antilles or the deeper floors of the Gulf or Caribbean? Until questions such as these are answered with some definitiveness, unique interpretations of the crustal evolution of the Gulf–Caribbean area are not likely to be achieved.

REFERENCES

Anderson, T. H., and Silver, L. T., 1971, Age of granulite metamorphism during the Oaxacan Orogeny, Mexico (abstract): Geol. Soc. Am. Abstr. with Programs, v. 3, p. 492.

Applin, P. L., 1951, Preliminary report on buried pre-Mesozoic rocks in Florida and adjacent states: U.S. Geol. Survey Circ. 91, 28 p.

Bass, M. N., 1969, Petrography and ages of crystalline basement rocks of Florida—Some extrapolations: Am. Assoc. Petr. Geol. Mem. 11, p. 283–310.

Bass, M. N., and Zartman, R. E., 1969, The basement of Yucatan Peninsula (abstract): EOS (*Trans. Am. Geophys. Union*), v. 50, n. 4, p. 313.

Bateson, J. H., 1972, New interpretation of the geology of Maya Mountains, British Honduras: *Am. Assoc. Petr. Geol. Bull.*, v. 56, p. 956–963.

Bayley, R. W., and Muehlberger, W. R., 1968, Basement Rock Map of the United States: U.S. Geol. Surv., Washington D.C.

Bellizzia, G. A., 1972a, Sistema Montañoso del Caribe, borde sur de la Placa Caribe. Es una cordillera Aloctona?, in: *VI Caribbean Geol. Conf.*, Petzall, C., ed.: Caracas, p. 247–258.

Bellizzia, G. A., 1972b, Is the entire Caribbean Mountain Belt of northern Venezuela allochthonous?: Geol. Soc. Am. Mem. 132, p. 363–368.

Bowen, J. M., 1972, Estratigrafia del Precretaceo en la parte norte de la Sierra de Perijá: *Cuarto Cong. Geol. Venezuela*: Bol. Geol., Pub. Esp. No. 5, Caracas, p. 729–761.

Bowin, C. O., 1966, Geology of central Dominican Republic (a case history of part of an island arc): Geol. Soc. Amer. Mem. 98, p. 11–84.

Chubb, L. J., and Burke, K. C., 1963, Age of the Jamaican granodiorite: *Geol. Mag.*, v. 100, p. 524–532.

Compañia Shell de Venezuela and Creole Petroleum Corp., 1964, Paleozoic rocks of Merida Andes, Venezuela: *Am. Assoc. Petr. Geol. Bull.*, v. 48, p. 70–84.

Davis, G. L., Tilton, G. R., and Wetherill, G. W., 1962, Mineral ages from the Appalachian Province in North Carolina and Tennessee: *J. Geophys. Res.*, v. 67, p. 1987–1996.

De Cserna, Z., 1971, Precambrian sedimentation, tectonics, and magmatism in Mexico: *Geol. Rundschau*, Band 60, p. 1488–1513.

De Cserna, Z., Fries, C., Jr., Rincón-Orta, C., Solorio-Munguía, J., and Schmitter-Villada, E., 1971, Edad precámbrica tardía del Esquisto Taxco, Guerrero: *Univ. Nac. Autón. Méx., Inst. Geol., Bol.* 100, p. 27–36.

Dengo, G., and Bohnenberger, O., 1969, Structural development of northern Central America: Am. Assoc. Petrol. Geol. Mem. 11, p. 203–220.

Denison, R. E., Kenny, G. S., Burke, W. H., Jr., and Hetherington, E. A., Jr., 1969, Isotopic ages of igneous and metamorphic boulders from the Haymond Formation (Pennsylvanian), Marathon Basin, Texas, and their significance: *Geol. Soc. Am. Bull.*, v. 80, p. 245–256.

Dixon, C. G., 1956, Geology of southern British Honduras with notes on adjacent areas: Belize Govt. Printer, 85 p.

Flawn, P. T., Goldstein, A., Jr., King, P. B., and Weaver, C. E., 1961, The Ouachita System: Univ. Texas Publ. No. 6120, 401 p.

Fries, C., Jr., and Rincón-Orta, C., 1965, Nuevas aportaciones geocronológicas y técnicas empleadas en el Laboratoria de Geocronometría: *Univ. Nac. Autón. Méx., Inst. Geol., Bol.* 73, p. 57–133.

Fries, C., Jr., Schmitter, E., Damon, P. E., Livingston, D. E., and Erickson, R., 1962a, Edad de las rocas metamórficas en los cañones de La Peregrina y de Caballeros, parte centro-occidental de Tamaulipas: *Univ. Nac. Autón. Méx., Inst. Geol., Bol.* 64, p. 55–69.

Fries, C., Jr., Schmitter, E., Damon, P. E., and Livingston, D. E., 1962b, Rocas precámbricas de edad Grenvilliana de la parte central de Oaxaca en el sur de México: *Univ. Nac. Autón. Méx., Inst. Geol., Bol.* 64, p. 45–53.

Fries, C., Jr., Rincon-Orta, C., Silver, L. T., Solorio-Munguía, J., Schmitter-Villada, E., and de Cserna, Z., 1971, Nuevas aportaciones a la geocronología de la Faja Tectónica Oaxaqueña: *Univ. Nac. Autón. Méx., Inst. Geol., Bol.* 100, p. 11–26.

Fullagar, P. D., 1971, Age and origin of plutonic intrusions in the Piedmont of the southeastern Appalachians: *Geol. Soc. Am. Bull.*, v. 82, p. 2845–2862.

Gomberg, D. N., Banks, P. O., and McBirney, A. R., 1968, Guatemala: Preliminary zircon ages from Central Cordillera: *Science*, v. 162, p. 121–122.

Hall, I. H. S., and Bateson, J. H., 1972, Late Paleozoic lavas in Maya Mountains, British Honduras, and their possible regional significance: *Am. Assoc. Petr. Geol. Bull.*, v. 56, p. 950–956.

Hurley, P. M., *et al.*, 1968a, Age of the Bladen Porphyry, British Honduras, in: *Variations in Isotopic Abundances of Strontium, Calcium, and Argon and Related Topics*, MIT-1381-16, Sixteenth Annual Progress Report for 1968: Mass. Inst. of Tech., Cambridge, p. 79–80.

Hurley, P. M., *et al.*, 1968b, Basement gneiss, Cordillera de La Costa, Venezuela, in: *Variations in Isotopic Abundances of Strontium, Calcium, and Argon and Related Topics*, MIT-1381-16, Sixteenth Annual Progress Report for 1968: Mass. Inst. of Tech., Cambridge, p. 81.

Hurley, P. M., Kalliokowski, J., Fairbairn, H. W., and Pinson, W. H., Jr., 1970, Further work on the Katarchean Imataca Series, Venezuela, in: *Variations in Isotopic Abundances of Strontium, Calcium, and Argon and Related Topics*, MIT-1381-18, Eighteenth Annual Progress Report for 1970: Mass. Inst. of Tech., Cambridge, p. 5–8.

Jacobs, C., Bürgl, H., and Conley, D. L., 1963, Backbone of Colombia: Am. Assoc. Petr. Geol. Mem. 2, p. 62–72.

Kesler, S. E., 1971, Nature of ancestral orogenic zone in Nuclear Central America: *Am. Assoc. Petr. Geol. Bull.*, v. 55, p. 2116–2129.

Kesler, S. E., Josey, W. L., and Collins, E. M., 1970, Basement rocks of western Nuclear Central America: The Western Chuacús Group, Guatemala: *Geol. Soc. Am. Bull.*, v. 81, p. 3307–3322.

Khudoley, K. M., and Meyerhoff, A. A., 1971, Peleogeography and geological history of Greater Antilles: Geol. Soc. Amer. Mem. 129, 199 p.

López-Ramos, E., 1969, Marine Paleozoic rocks of Mexico: *Am. Assoc. Petr. Geol. Bull.*, v. 53, p. 2399–2417.

MacDonald, W. D., 1964, Geology of the Serrania de Macuira area, Goajira Peninsula: Ph. D. Thesis, Princeton Univ., Princeton, New Jersey.

MacDonald, W. D., 1972, Caracteristicas estructurales principales de La Peninsula de La Guajira (Colombia–Venezuela) y el Caribe sur-central, *Cuarto Cong. Geol. Venezuela*: Bol. Geol., Pub. Esp. No. 5, Caracas, p. 2463–2476.

MacDonald, W. D., Doolan, B. L., and Cordani, U. G., 1971, Cretaceous–Early Tertiary metamorphic K–Ar age values from the South Caribbean: *Geol. Soc. Am. Bull.*, v. 82, p. 1381–1388.

MacDonald, W. D., and Hurley, P. M., 1969, Precambrian gneisses from northern Colombia, South America: *Geol. Soc. Am. Bull.*, v. 80, p. 1867–1872.

Martin-Bellizzia, C., 1968, Edades isotópicas de rocas Venezolanas: *Bol. Geol. Caracas*, v. 10, n. 19, p. 356–380.

Mattson, P. H., 1966, Geological characteristics of Puerto Rico, in: *Continental Margins and Island Arcs*: Geol. Surv. Can. Pap. 66-15, p. 124–138.

McBirney, A. R., 1963, Geology of a part of the Central Guatemalan Cordillera: *Univ. Calif. Publ. Geol. Sci.*, v. 38, p. 177–242.

McBirney, A. R., and Bass, M. N., 1969a, Structural relations of pre-Mesozoic rocks of northern Central America: Am. Assoc. Petr. Geol. Mem. 11, p. 269–280.

McBirney, A. R., and Bass, M. N., 1969b, Geology of Bay Islands, Gulf of Honduras: Am. Assoc. Petr. Geol. Mem. 11, p. 229–243.

Meyerhoff, A. A., Khudoley, K. M., and Hatten, C. W., 1969, Geological significance of radiometric dates from Cuba: *Am. Assoc. Petr. Bull.*, v. 53, p. 2494–2500.

Mills, R. A., Hugh, K. E., Feray, D. E., and Swolfs, H. C., 1967, Mesozoic stratigraphy of Honduras: *Am. Assoc. Petr. Geol. Bull.*, v. 51, p. 1711–1786.

Milton, C., and Grasty, R., 1969, "Basement" rocks of Florida and Georgia: *Am. Assoc. Petr. Geol. Bull.*, v. 53, p. 2483–2493.

Muehlberger, W. R., Denison, R. E., and Lidiak, E. G., 1967, Basement rocks in continental interior of United States: *Am. Assoc. Petr. Geol. Bull.*, v. 51, p. 2351–2380.

Muehlberger, W. R., Hedge, C. E., Denison, R. E., and Marvin, R. F., 1966, Geochronology of the midcontinent region, United States. 3. Southern area: *J. Geophys. Res.*, v. 71, p. 5409–5426.

Murray, G. E., 1961, *Geology of the Atlantic and Gulf Coastal Province of North America*: Harper and Row, New York, 692 p.

Pantoja-A., J., and Robison, R. A., 1967, Paleozoic sedimentary rocks in Oaxaca, Mexico: *Science*, v. 157, p. 1033–1035.

Pantoja-A., J., Rincón-Orta, C., Fries, C., Jr., Silver, L. T., and Solorio-Munguía, J., 1971, Contribución a la geocronología del Estado de Chiapas: *Univ. Nac. Autón. México, Inst. Geol., Bol.* 100, p. 47–58.

Pequegnat, W. E., Bryant, W. R., and Harris, J. E., 1971, Carboniferous sediments from Sigsbee Knolls, Gulf of Mexico: *Am. Assoc. Petr. Geol. Bull.*, v. 55, p. 116–123.

Pinson, W. H., Jr., Hurley, P. M., Mencher, E., and Fairbairn, H. W., 1962, K–Ar and Rb–Sr ages of biotites from Colombia, South America: *Geol. Soc. Am. Bull.*, v. 73, p. 907–910.

Pushkar, P., 1968, Strontium isotope ratios in volcanic rocks of three island arc areas: *J. Geophys. Res.*, v. 73, p. 2701–2714.

Roobol, M. J., 1972, The volcanic geology of Jamaica, in: *VI Caribbean Geol. Conf.*, Petzall, C., ed.: Caracas, p. 100–107.

Shagam, R., 1972a, Evolucion tectónica de Los Andes Venezolanos, *Cuarto Cong. Geol. Venezuela*: Bol. Geol., Pub. Esp. No. 5, Caracas, p. 1210–1261.

Shagam, R., 1972b, Andrean research project, Venezuela: Principal data and tectonic implications: Geol. Soc. Amer. Mem. 132, p. 449–463.

Talwani, M., Sutton, G. H., and Worzel, J. L., 1959, A crustal section across the Puerto Rico trench: *J. Geophys. Res.*, v. 64, p. 1545–1555.

Tilton, G. R., Wetherill, G. W., Davis, G. L., and Bass, M. N., 1960, 1000-million-year-old minerals from the eastern United States and Canada: *J. Geophys. Res.*, v. 65, p. 4173–4180.

Vedder, J. G., *et al.*, 1971, USGS IDOE Leg 2: *Geotimes*, v. 16, n. 12, p. 10–12.

Vedder, J. G., MacLeod, N. S., Lanphere, M. A., and Dillon, W. P., 1973, Age and tectonic implications of some low-grade metamorphic rocks from the Yucatan Channel: *J. Res. U.S. Geol. Surv.*, v. 1, p. 157–164.

Zartman, R. E., 1964, A geochronologic study of the Lone Grove Pluton from the Llano Uplift, Texas: *J. Petrol.*, v. 5, p. 359–408.

Chapter 5

THE OUACHITA AND APPALACHIAN OROGENIC BELTS

Philip B. King

U.S. Geological Survey
Menlo Park, California

I. INTRODUCTION

Orogenic belts of Paleozoic age border the southeastern and southern sides of the North American continent in Canada, the United States, and Mexico. The Appalachian orogenic belt extends 3700 km along the southeastern side, from Newfoundland to Alabama, plunging under the continental shelf on the northeast and under the deposits of the Gulf Coastal Plain on the southwest. The Ouachita orogenic belt begins near the southwestern terminus of the Appalachian belt and extends 1800 km across the southern United States and probably into Mexico (Fig. 1). Unlike the widely exposed Appalachian belt, the Ouachita belt is only exposed for 450 km of this distance—in the Ouachita Mountains of Arkansas and Oklahoma, and in the Marathon region and several smaller areas in western Texas. The remainder is covered by Mesozoic and Cenozoic deposits of the Gulf Coastal Plain, but much is known about the covered part from the records of drilling.

This account will deal with the whole known extent of the Ouachita belt, but will include only the southwestern segment of the Appalachian belt, in Georgia and Alabama (cf. Fig. 2).

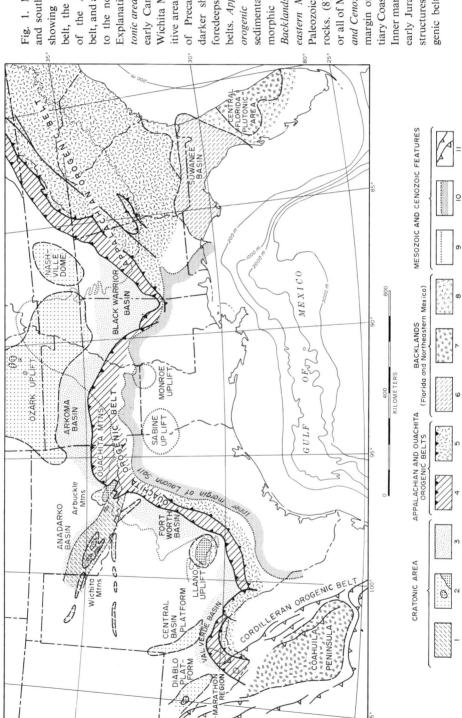

Fig. 1. Map of south-central and southeastern United States, showing the Ouachita orogenic belt, the southwestern segment of the Appalachian orogenic belt, and adjoining tectonic units to the north, south, and west. Explanation of patterns: *Cratonic area:* (1) Late Proterozoic-early Cambrian aulocogene, in Wichita Mountains belt. (2) Positive areas, with small outcrops of Precambrian basement in darker shading. (3) Basins or foredeeps bordering the orogenic belts. *Appalachian and Ouachita orogenic belts:* (4) Deformed sedimentary rocks. (5) Metamorphic and plutonic rocks. *Backlands in Florida and northeastern Mexico:* (6) Flat-lying Paleozoic strata. (7) Plutonic rocks. (8) Volcanic rocks, part or all of Mesozoic age. *Mesozoic and Cenozoic features:* (9) Inner margin of Cretaceous and Tertiary Coastal Plain deposits. (10) Inner margin of Louann Salt, of early Jurassic age. (11) Frontal structures of Cordilleran orogenic belt.

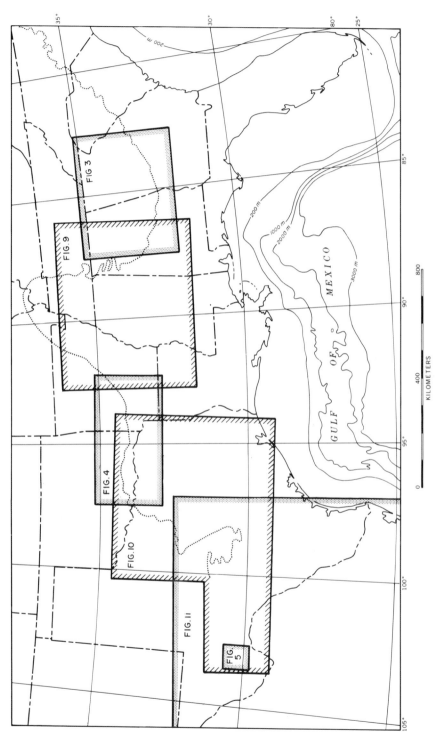

Fig. 2. Map of south-central and southeastern United States, showing areas covered by figures 3, 4, 5, 9, 10, and 11.

II. CRATON NORTH OF THE OROGENIC BELTS

The two orogenic belts face northwestward and northward toward the craton of the continental interior, where variable thicknesses of little disturbed Paleozoic rocks lie on a Precambrian basement. Near the orogenic belts, as elsewhere, the craton is diversified by positive and negative areas (Fig. 1).

A. Negative Areas

Especially prominent negative areas are the foredeeps in front of the orogenic belts, which grew during orogenic phases that provided much of their sedimentary filling. The foredeeps include: on the east, the Black Warrior basin of Alabama and Mississippi, in the angle between the Appalachian and Ouachita belts; the Arkoma basin of Arkansas and Oklahoma, north of the Ouachita Mountains; the Fort Worth basin of north-central Texas, west of the buried extension of the Ouachita belt; and the Val Verde basin of southwestern Texas, north of the Marathon segment of the Ouachita belt. The history and structure of these basins will be mentioned in the discussions of the particular segments of the orogenic belts which they adjoin.

B. Positive Areas

The positive areas include the following: the Nashville dome and Ozark uplift of Tennessee, Missouri, and adjacent states, north of the Black Warrior and Arkoma basins; the Arbuckle and Wichita Mountain structures of southwestern Oklahoma; the Llano uplift of central Texas; and the Central Basin platform and Diablo platform of western Texas, north of the Val Verde basin. All of them are at least partly masked by younger rocks, but the exposed parts of some bring Precambrian basement rocks to the surface. Only the Arbuckle and Wichita structures and the Llano uplift impinge directly on the orogenic belt, the others lying at a distance from it.

The positive areas vary in size, history, and degree of tectonic mobility. The Nashville dome, Ozark uplift, and Llano uplift are passive features that grew secularly during Paleozoic time. The Central Basin platform and Diablo platform were mildly deformed during the later Paleozoic. Contrasted with these, the Arbuckle and Wichita structures are truly orogenic features, formed by germanotype deformation of a linear belt within the craton.

C. Arbuckle–Wichita Belt

The Arbuckle–Wichita belt originated in late Proterozoic–early Cambrian time as a fault-bordered aulocogene, which was filled by more than 6500 m of graywackes, volcaniclastics, and lavas, into which floored felsic and mafic plutons were injected along the southern side (Ham *et al.*, 1964, p. 35). Over these, 10,000–13,000 m of Upper Cambrian to Permian sediments were deposited, the lower part miogeosynclinal and including the great mass of carbonates of the Arbuckle Group, the upper part synorogenic or postorogenic clastics resulting from deformational pulses within the belt. In late Mississipian time the Wichita Mountains and their buried extensions were raised as a chain of horsts in the southern part of the former aulocogene; in the late Pennsylvanian the Arbuckle Mountains were raised at the north edge of the aulocogene (Ham *et al.*, 1964, p. 161–162). Between the two uplifts, and westward along the north flank of the Wichita Mountains, the Ardmore and Anadarko basins subsided deeply and were filled by synorogenic Pennsylvanian sediments, to the accompaniment of steep compressional folding (Tomlinson and McBee, 1959). Postorogenic Permian redbeds are the final deposits in the Anadarko basin and also spread over the truncated edges of much of the rest of the deformed complex.

The Arbuckle–Wichita orogenic belt trends east–southeast and impinges on the front of the Ouachita orogenic belt immediately south of its outcrops in the Ouachita Mountains in a complex manner that will be discussed later (Fig. 4). The rock sequence in the Arbuckle–Wichita belt resembles somewhat that of the Southern Appalachians farther east, although its structural style differs, but if any connection exists between them, it is now concealed beneath the allochthonous rocks of the Ouachita orogenic belt.

III. APPALACHIAN OROGENIC BELT

The Appalachian orogenic belt, in its southwestern or Georgia–Alabama segment, has an exposed breadth of 320 km, fronting northwestward on flat-lying Pennsylvanian rocks of the craton, and passing southeastward beneath Cretaceous and Tertiary Coastal Plain deposits. It is divisible into longitudinal provinces of lithically and structurally like rocks, whose erosion has produced the geomorphic Valley and Ridge province, Blue Ridge province, and Piedmont province (Fig. 3).

A. Valley and Ridge Province

The northwestern, or Valley and Ridge province, 65 or 80 km wide, is formed of miogeosynclinal Cambrian to Pennsylvanian sediments about 10,000 m thick, strongly folded and thrust-faulted, but not metamorphosed (Butts, 1926, p. 61–230; Butts and Gildersleeve, 1948). Basal Lower Cambrian clastics are followed by a carbonate sequence 3000 m thick of Cambrian to Middle

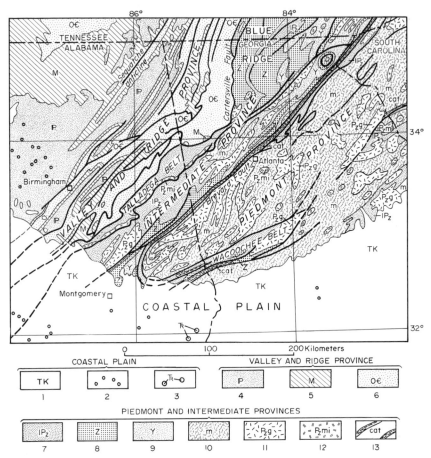

Fig. 3. Geologic map of southwestern segment of Appalachian orogenic belt, in Georgia and Alabama. Generalized from Geologic Map of the United States (1974). Explanation of patterns: *Coastal Plain*: (1) Cretaceous and Tertiary. (2) Wells drilled into Paleozoic rocks. (3) Wells drilled into Triassic rocks. *Valley and Ridge province*: (4) Pennsylvanian. (5) Mississippian. (6) Cambrian and Ordovician (with thin Silurian and Devonian at top in places). *Piedmont and Intermediate provinces*: (7) Lower Paleozoic. (8) Upper Proterozoic supracrustal rocks. (9) Middle Proterozoic metamorphic basement. (10) Metamorphic complex (paragneiss, orthogneiss, and migmatite). (11) Paleozoic granite. (12) Paleozoic mafic and ultramafic intrusives. (13) Cataclastic rocks.

Ordovician age. Subsurface data indicate that half or more of this sequence wedges out and overlaps on the craton, where Upper Cambrian is commonly the basal deposit. Presumably the carbonates were laid down on a shelf along the edge of a proto-Atlantic Ocean, and there is no indication of lands or sediment sources to the southeast. The middle Paleozoic (Middle Ordovician to Devonian) is no more than 300 m thick in this segment, and the clastic wedges that are prominent in the Central Appalachians are missing; an interesting component is the Silurian red iron ore that is exploited commercially near Birmingham, Alabama. The higher Paleozoic (Mississippian and Pennsylvanian) is more clastic and 4000 m thick, marine below and passing up into coal measures. The upper Mississippian (Chesterian) Floyd Shale oversteps the older parts of the sequence southeastward down to the Lower Ordovician, but without structural discordance. Following it is the coarser Parkwood Formation at the Mississippian–Pennsylvanian boundary, and the great mass of coal-bearing lower Pennsylvanian Pottsville Group. The Pennsylvanian is preserved in synclinal downfolds within the Valley and Ridge province to a much greater extent than farther northeast. The upper Paleozoic clastic rocks were derived principally from the Ouachita belt to the southwest, rather than from the interior of the Appalachian belt.

The structures of the miogeosynclinal belt include long, parallel, thrust-faulted folds, vergent northwestward. Many or all of the folds and faults formed over surfaces of decollement in weak strata low in the sequence. The northwestern anticlines are long and narrow (including the Sequatchie anticline in front of the main orogen), and are separated by broader, flat-keeled synclines. Farther southeast, folding is steeper throughout, but many of the faults have sinuous traces, implying a low angle of dip.

B. Intermediate Province

Between the Valley and Ridge and Piedmont provinces is an intermediate belt 40–120 km wide, which includes the morphological Blue Ridge province of the Central Appalachians and its lithic and structural continuation in the Piedmont terrain farther southwest (Fig. 3). Its rocks are generally older, more uplifted, and more metamorphosed than those of the Valley and Ridge province, and it is separated from them by a fundamental, deep-seated, low-angle thrust—the Great Smoky thrust and its extension into Georgia as the Cartersville thrust (the separation is less plain in Alabama).

In Tennessee and northern Georgia, much of the province is formed by the mass of nonvolcanic clastic rocks of the upper Proterozoic Ocoee Supergroup (King et al., 1958), capped by some outliers and infolds of lower Paleozoic and resting along its southeastern side on middle Proterozoic paragneisses and orthogneisses, with billion-year-old ("Grenville") radiometric dates. Southwest of a constriction at Cartersville, Georgia, younger rocks, probably lower Paleozoic, are more extensive, although the middle and upper Proterozoic apparently persist along the southeastern side, where they are complexly intermingled in the metamorphic terrane (Bentley and Neathery, 1970).

From Cartersville southwestward, stratigraphic relations are plainest in the less metamorphosed northwestern strip of the Intermediate province—the Talladega belt. An outlying plate west of the belt near the edge of the Coastal

Plain is of proved Mississippian age (Carrington, 1973), but the main body is a homoclinal sequence 10,000 m thick, with sericitic slate below, followed by the ridge-making Cheaha Sandstone and by the Jemison Chert, Erin Slate, and other undefined units. The Jemison has yielded abundant invertebrates of Devonian age, and the Erin a few plant remains ascribed a Carboniferous age (Butts, 1926). Ages of the remaining units are less certainly proved, but it seems clear that the Talladega belt contains a lengthy Paleozoic sequence of a different, more internal facies from that in the Valley and Ridge province.

The Intermediate province is bordered on the southeast by the Brevard fault zone, which pursues a remarkably straight course from southern Virginia to the edge of the Coastal Plain in Alabama. Throughout its course it is followed by a narrow zone of mylonitization and cataclasis and marks an absolute difference between the rocks and structures on either side. The age and origin of the Brevard zone has been much debated and variously interpreted; probably it has had a long history and a composite origin. During its early phases it may have been the root zone for thrust sheets farther northwest (Burchfiel and Livingston, 1967) and (or) the front of thrust sheets farther southeast (Hatcher, 1971), but during its later phases it clearly had a large but undetermined component of strike-slip displacement (Reed and Bryant, 1964).

C. Piedmont Province

The Piedmont province narrows southwestward at the surface as a result of Coastal Plain overlap, although sparse well data indicate that it continues southwestward in full strength beneath the cover (Fig. 3). The southwestern part, in Alabama and adjacent Georgia, has recently been mapped and classified in detail by the Alabama Geological Survey (Bentley and Neathery, 1970); the remainder of Georgia is less understood in regional terms.

The country rock in the segment is a metamorphic complex of felsic and mafic paragneisses, migmatites, and orthogneisses, containing embedded granitic plutons and a few mafic plutons of various Paleozoic ages. The paragneisses originated as supracrustal sediments and volcanics; their original age is enigmatic, but they are probably partly of late Proterozoic and partly of early Paleozoic age. The metamorphic climax is attained in the northwestern part of the Piedmont (Morgan, 1972), where the rocks are of amphibolite facies, with sillimanite and andalusite; regional metamorphism increases progressively toward it across the Intermediate province (except for retrograding in the Brevard zone) and then decreases southeastward.

In North and South Carolina the granitic plutons fall into early, middle, and late Paleozoic groups, dated radiometrically at 595–520 m.y., 415–385 m.y., and near 300 m.y. (Fullagar, 1971, p. 2859). Data are less complete in the Georgia–Alabama segment, but those available suggest that most of the plutons belong to the early and middle Paleozoic groups. Regional metamorphism was probably nearly contemporaneous with the plutonic activity (Butler, 1972).

Complications develop toward the southwestern end of the exposed Piedmont. Near the edge of the Coastal Plain is the Wacoochee belt, 25 km broad, which extends 160 km west–southwest from Warm Springs, Georgia, into Alabama (Hewett and Crickmay, 1937, p. 25–30, pl. 2). It contains an openly folded metasedimentary sequence, unfossiliferous but almost certainly lower Paleozoic, with the basal Hollis Quartzite, overlain by the Manchester Schist and (in Alabama) the Chewacla Marble. The belt is bordered north and south by faults that dip away from it, which are followed by zones of shearing and mylonitization many kilometers wide. Between the Wacoochee belt and the Brevard zone on the northwest the Piedmont metamorphic complex lies in a synform 50 km wide and plunging northeast, bordered up the plunge to the southwest by faults that splay off the Brevard zone to connect, near the edge of the Coastal Plain, with the fault on the north side of the Wacoochee belt. Possibly the whole synform is allochthonous, with its rocks derived from south of the Wacoochee belt (Bentley and Neathery, 1970, Fig. 17, p. 73).

D. Buried Extensions of Appalachian Belt

Paleozoic miogeosynclinal rocks like those of the Valley and Ridge province have been penetrated by drilling along its southwestward extension across Alabama into Mississippi (Fig. 9). Available well control suggests that they form at least two south-dipping panels of lower Paleozoic carbonates and overlying Mississippian and Pennsylvanian clastics, with probable thrusts on the north sides, facing the Black Warrior basin. In Mississippi, their strike bends from southwest to west–northwest (Thomas, 1972, p. 82–85; 1973).

South of the miogeosynclinal rocks, a few wells penetrate slate, phyllite, and marble, which may be on a southwestward extension of the Talladega belt. Farther east, beneath the Coastal Plain in southern Alabama and Georgia, wells penetrate schist, gneiss, and granite like those in the Piedmont province, but they are too widely spaced to indicate a structural pattern; neither the Brevard fault zone nor the synform southeast of it have been identified. In addition, some wells have entered red shale and sandstone with associated diabase intrusives, probably correlative with the Upper Triassic Newark Group exposed farther north in the Piedmont and, like the Newark, postorogenic to the Paleozoic structures.

E. Orogenic History

The times of deformation in the Southern Appalachians cannot be determined in detail, especially in its southwestern segment in Georgia and

Alabama. The terminal orogeny is certainly later than the lower Pennsylvanian Pottsville Group of the miogeosynclinal sequence on the northwest, which is deformed with the rocks beneath, and is older than the Newark Group whose equivalents underlie the Coastal Plain to the south.

Some regional unconformities occur lower down in the miogeosynclinal sequence (such as that beneath the Chesterian Floyd Shale), but they appear to result largely from epeirogenic movements. The only clastic wedge in the southwestern segment which would indicate tectonic activity elsewhere in the orogenic belt is that in the upper Mississippian and lower Pennsylvanian, which seems to have been derived from the Ouachita belt to the southwest rather than from the Appalachian belt to the southeast.

Middle Paleozoic clastic wedges indicative of tectonic activity in the interior of the Appalachian belt, which are prominent in the Central Appalachians, are lacking in the southwestern segment. Nevertheless, such activity is suggested here also by granitic plutons and regional metamorphism of early to middle Paleozoic age in the Piedmont province like those farther northeast.

IV. OUACHITA OROGENIC BELT

The Ouachita orogenic belt, exposed and concealed, has a pattern of salients and recesses like those in the Appalachian belt, with similar lengths but with much greater convexity (Fig. 1). The exposed parts in the Ouachita Mountains and Marathon Region each lie at the apices of the salients, between and east of which are deep recesses now concealed by younger deposits; a deep recess lies between the Appalachian and Ouachita belts in Alabama and Mississippi, and another between the Ouachita Mountains and Marathon Region southeast of the Llano uplift in Texas.

The known parts of the Ouachita orogenic belt are on the updip margin of the Gulf Coastal Plain, whose Jurassic, Cretaceous, and Cenozoic deposits form coastward thickening wedges; hence, the rocks of the orogenic belt become progressively deeper coastward, and at a distance of 80–100 km from the orogenic front they pass beyond the reach of ordinary commercial drilling (3000–5000 m)—especially beyond the boundary where the early Jurassic Louann Salt wedges in at the base of the Coastal Plain sequence. Thus, knowledge of the Ouachita orogenic belt is confined to its margins, and drill penetrations are lacking in coastward areas, where its internal parts would be expected.

Early field investigations by Purdue and Miser in Arkansas, by Taff and Honess in Oklahoma, and by Udden and Baker in western Texas demonstrated the existence of deformation in the Ouachita belt fully as powerful as that in the Appalachians (Purdue and Miser, 1923; Miser and Purdue, 1929; Taff, 1902; Honess, 1923; Udden, 1917; Baker and Bowman, 1917), but its sig-

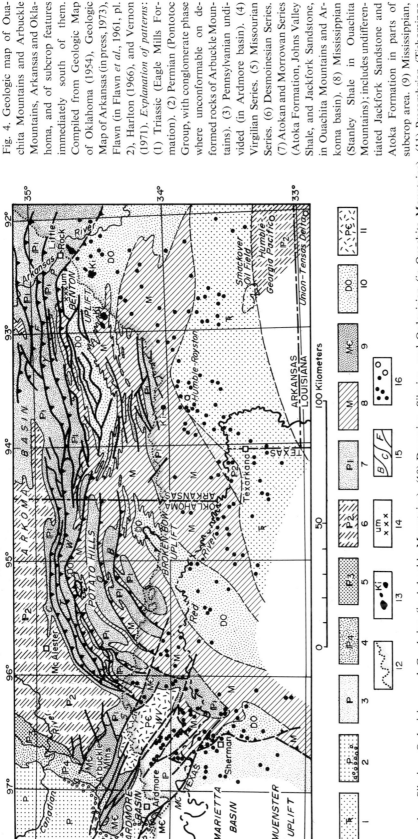

Fig. 4. Geologic map of Oua-chita Mountains and Arbuckle Mountains, Arkansas and Okla-homa, and of subcrop features immediately south of them. Compiled from Geologic Map of Oklahoma (1954), Geologic Map of Arkansas (in press, 1973), Flawn (in Flawn et al., 1961, pl. 2), Harlton (1966), and Vernon (1971). Explanation of patterns: (1) Triassic (Eagle Mills For-mation). (2) Permian (Pontotoc Group, with conglomerate phase where unconformable on de-formed rocks of Arbuckle Moun-tains). (3) Pennsylvanian undi-vided (in Ardmore basin). (4) Virgilian Series. (5) Missourian Series. (6) Desmoinesian Series. (7) Atokan and Morrowan Series (Atoka Formation, Johns Valley Shale, and Jackfork Sandstone, in Ouachita Mountains and Ar-koma basin). (8) Mississippian (Stanley Shale in Ouachita Mountains); includes undiferen-tiated Jackfork Sandstone and Atoka Formation in parts of subcrop area. (9) Mississippian,

Devonian, Silurian, Ordovician, and Cambrian (in Arbuckle Mountains). (10) Devonian, Silurian, and Ordovician (in Ouachita Mountains). (11) Precambrian (Tishomingo Granite and other plutonic units, in Arbuckle Mountains). (12) Edge of Cretaceous and Tertiary Coastal Plain deposits. (13) Cretaceous intrusives. (14) Soapstone and serpen-tinite (eastern Ouachita Mountains). (15) Initials, to indicate names of faults, as follows: B - Boktukola, C - Choctaw, F - Fourche la Fave, MC - Meers-Crimer, O - Octavia, P - Panther Creek, S - Sulfur, T - Ti Valley, W - Windingstair, WV - Washita Valley, Y - Y City. (16) Wells drilled into Paleozoic and Triassic rocks; open circles indicate wells specifically mentioned in text.

nificance was belittled by synthesizers of the time. The Ouachita belt and the Arbuckle–Wichita belt were commonly grouped together because of their apparent continuity of trends in southern Oklahoma, despite the alpinotype rocks and structures of the one and the germanotype rocks and structures of the other; the rocks of both were supposed to have originated in a "Ouachita Embayment" (Schuchert, 1923, p. 181–184, Fig. 3) and to have been deformed along a "cross-fold" emanating from one of the salients in the Appalachians (Keith, 1923, p. 329–330).

The relations were clarified by the brilliant tectonic analysis of van der Gracht (1931a,b) based on European analogies; he demonstrated that the Ouachita belt was an orogenic feature fully the equal of the Appalachian belt in continental significance; his insights have been justified by the vastly greater fund of information that has accumulated since his day.

Nevertheless, some geologists continued to maintain that the structures in the Ouachita Mountains are a rather minor wrinkle in the craton, without regional importance, which pass southward as well as northward into less deformed rocks (Misch and Oles, 1957; Tomlinson, 1959). Although this has been refuted by later data, a modern variant is that the apices of the two salients, where the rocks are exposed in the Ouachita Mountains and Marathon Region, were the centers of maximum sedimentation and deformation, and that the belt is less developed in the intervening areas (where subsurface data are sometimes equivocal) (Flawn in Flawn *et al.*, 1961, p. 168; Folk, 1973, p. 723); this question is still unresolved, although a likely alternative is that parts of the Ouachita structure in the intervening areas have been telescoped or eliminated by large-scale thrusting.

A. Exposed Parts of the Ouachita Belt

In both the Ouachita Mountains and Marathon Region the belt consists of rocks of early Paleozoic to Pennsylvanian age, strongly deformed after the fashion of the Paleozoic rocks in the Valley and Ridge province of the Appalachians, and like them overfolded and thrust toward the craton (Figs. 4 and 5). However, the rock sequence differs from that of the Appalachian miogeosyncline; it consists of a condensed leptogeosynclinal sequence in the lower Paleozoic and an enormously overthickened mass of flysch in the Carboniferous above. The rocks and structures of the exposures in the Ouachita Mountains and Marathon Region are remarkably alike, suggesting an original continuity, and hence may be treated together here.

1. *Lower Paleozoic Leptogeosynclinal Sequence*

The exposed lower Paleozoic sequence in the Ouachita Mountains is 1300–2300 m thick (Miser and Purdue, 1929, p. 25–59), and in the Marathon

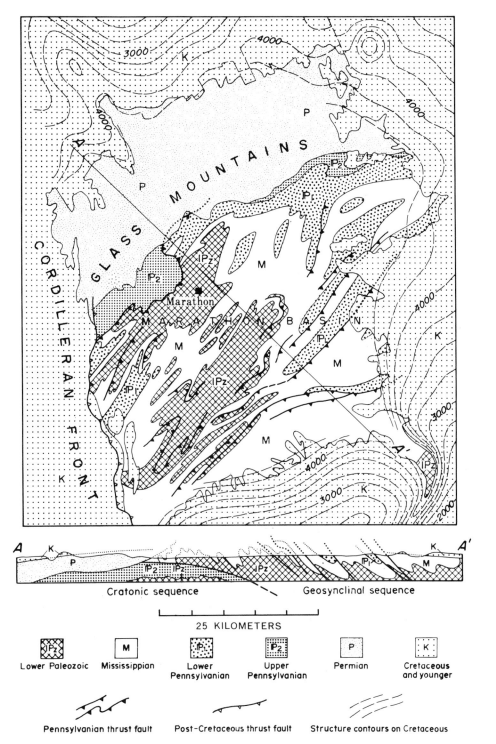

Fig. 5. Geologic Map of Marathon Region, west Texas. Generalized from King (1937, pl. 23).

Region a little more than 1000 m (King, 1937, p. 22–55); these thicknesses are less than those of equivalent strata in the nearby craton and very much less than those in the Arbuckle–Wichita trough, where they attain 3000 m.

The lowest exposed rocks are Upper Cambrian in the Marathon Region (Wilson, 1954a, p. 254) and slightly younger (Tremadocian) in the Ouachita Mountains (Repetski and Ethington, 1973); the base of the sequence is unknown (Table I). Deep wells in the core areas of both regions have drilled into great thicknesses of highly deformed shaly rocks, at least part of which are older than those at the surface; large negative gravity anomalies in both regions (more than −100 mgal in the Ouachita Mountains) indicate that the basement lies at great depth (Fig. 12).

TABLE I

Approximate Correlation of Pre-Carboniferous Rocks in Ouachita Mountains with Those in the Marathon Region

System and series	Ouachita Mountains	Marathon Region
Lower Mississippian	Arkansas Novaculite	Caballos Novaculite
Devonian		
Silurian	Missouri Mountain Shale	Hiatus
	Blaylock Sandstone	
Upper Ordovician	Polk Creek Shale	Maravillas Chert*
	Bigfork Chert	
Middle Ordovician	Womble Shale	Woods Hollow Shale*
	Blakeley Sandstone*	Fort Peña Formation
Lower Ordovician	Mazarn Shale	Alsate Shale
	Crystal Mountain Sandstone	Marathon Limestone*
	Collier Shale	
Upper Cambrian	Not exposed	Dagger Flat Sandstone

* Occurrence of exotic bouldery debris.

The lower Paleozoic rocks are a distinctive sequence of argillaceous and cherty rocks and of minor sandstones, somewhat more calcareous at Marathon than in the Ouachitas, with graptolites as the dominant fossils in the Ordovician and Silurian, and conodonts higher up. The sandstone units, such as the Blaylock, express obscure early tectonic pulses within the orogenic belt. The Blaylock Sandstone wedges out rapidly northward across the Ouachita Mountains; the Silurian is very thin in the northern belts of the mountains and is missing entirely in the Marathon Region, where there seems to have been a long failure of deposition. Chert units occur in the Upper Ordovician (Bigfork and Maravillas), which have cherty limestone equivalents in the nearby cratonic sequences. However, the most distinctive siliceous units are the overlying novaculites (Arkansas and Caballos), which stand in prominent white ridges. Their few conodonts and other fossils indicate that these rather thin units (65–295 m thick) span the whole of Devonian and early Mississippian time (Hass, 1951). The source and origin of the widespread cherty and siliceous formations of the Ouachita sequence have been much debated and are still enigmatic (Goldstein and Hendricks, 1953; McBride and Thomson, 1970, p. 74–92; Folk, 1973, p. 722–723).

During Ordovician time, at least, the cratonic border of the Ouachita belt must have been a steep shelf break, which was a source of exotic bouldery debris in the geosynclinal deposits. The debris in the Ordovician formations of the Marathon Region is largely shallow-water carbonate with shelly fossils of various ages (Wilson, 1954a, p. 258–263; 1954b, p. 2469; Young, 1970, p. 2306–2307); that in the Ouachita Mountains includes crystalline rocks (Stone and Sterling, 1962, p. 388–389). The existence of bouldery debris derived from a shelf break suggests that the geosynclinal sequence accumulated in waters of considerable depth. Many other features of the lower Paleozoic geosynclinal sediments also suggest deep-water deposition, although there is not agreement among all sedimentologists.

2. Upper Mississippian–Lower Pennsylvanian Flysch Sequence

Succeeding the novaculites of the Ouachita belt is a great mass of clastic rocks of late Mississippian and early Pennsylvanian age (Meramecan to Atokan). Because of their thickness, they are the dominant surface rocks in both the Ouachita Mountains and Marathon Region, where their more argillaceous parts form hilly lowlands, and their sandier parts linear ridges and mountains.

The clastics are a sequence of flysch or turbidite, which has all the characters of the classical flysch of the orogenic belts of middle Europe (Cline, 1960, p. 87–99; McBride, 1970). The flysch succeeds the novaculite abruptly and conformably (although possibly with a small hiatus), and marks a drastic change in the sedimentary and tectonic regime of the geosyncline; slow siliceous sedimentation in a leptogeosyncline was terminated by tectonic mobility and by deep subsidence of troughs in the orogenic belt, where the clastic debris rapidly accumulated.

The flysch sequence is represented in the Ouachita Mountains by the Stanley Shale, Jackfork Sandstone, Johns Valley Shale, and Atoka Formation; and in the Marathon Region, by the Tesnus Formation, Dimple Limestone, and Haymond Formation (Table II). The lower formations (Stanley, Jackfork, and Tesnus) attain their greatest thickness toward the southeast; in the southeastern part of the Ouachita Mountains they are more than 6500 m thick (Goldstein and Hendricks, 1962, p. 389; Cline, 1960, p. 12), and in the southeastern part of the Marathon Region more than 2300 m (King, 1937, p. 55). They thin dramatically toward the craton on the northwest, where they are virtually

TABLE II

Approximate Correlation of Carboniferous Strata in Ouachita Mountains with Those in the Marathon Region

Series	Ouachita Mountains	Marathon Region
Wolfcampian	Absent	Neal Ranch Formation
Virgilian Missourian Desmoinesian	Hartshorne Sandstone and higher formations in Arkoma Basin	Gaptank Formation
Atokan	Atoka Formation	Haymond Formation*
Morrowan	Johns Valley Shale,* Wapanucka Limestone, and related units	Dimple Limestone
Morrowan	Jackfork Sandstone (or Group)	Tesnus Formation
Chesterian Meramecian	Stanley Shale (or Group)	Tesnus Formation
Osageian	Hiatus?	Hiatus?
Kinderhookian — Devonian	Arkansas Novaculite	Caballos Novaculite

* Occurrence of exotic bouldery debris.

absent. The thinning has been exaggerated by the crustal shortening produced by folding and thrusting, but it nevertheless expresses a steep northwestern face of the geosynclinal trough (Goldstein and Hendricks, 1962, p. 402). The upper flysch units (Atoka and Haymond) are preserved in remnants across much of the breadth of exposure in both regions, but the Atoka at least seems to have its depositional maximum in the frontal belt of the Ouachita Mountains and the adjacent part of the Arkoma basin, where it attains a thickness of as much as 6200 m (Reinemund and Danilchek, 1957). Northward across the Arkoma basin, the Atoka changes rapidly into shallow-water deposits with beds of coal and thins to a feather edge on the flank of the Ozark uplift 160 km to the north. The southward thickening and facies change of the Atoka is related to contemporaneous growth faults, downthrown to the south, many of which do not reach the surface and are proved by subsurface data.

The Stanley, and much of the Tesnus, consist of interbedded shales and graywackes, with some persistent layers of siliceous shale (spiculite), and with a few thin beds of volcanic ash in the southern Ouachita Mountains. The sandstone in the Jackfork is thicker-bedded and more cleanly washed. Large parts of the Atoka and Haymond are thinly interbedded sandstone and shale. The Dimple Limestone of the Marathon Region is a seeming anomaly, but much of it is resedimented and is a flysch or turbidite like the rest (Thompson and Thomasson, 1969, p. 67–70).

Notable features of the upper part of the sequence in both regions are the boulder-beds or wildflysch, which interrupt the usual deposition of the finer-grained, evenly bedded flysch (Cline, 1960, p. 60–85; McBride, 1966, p. 24–29, 48–52). Those in the Ouachita Mountains are Morrowan (in the Johns Valley Shale); those in the Marathon Region are Atokan (in the Haymond Formation). The Johns Valley boulder-beds occur for more than 160 km along the frontal belts of the Ouachita Mountains; the Haymond boulder-beds are more scantily preserved. Both deposits consist of exotic clasts of all sizes which lie helter-skelter in a muddy matrix, the most spectacular components being giant blocks and slabs of carbonate rocks 30–100 m across. The boulders and slabs in the Ouachita occurrences are sedimentary rocks derived from the craton, mostly Cambrian to Devonian but a few from younger formations (such as hard layers from the Chesterian Caney Shale). The large slabs in the Marathon occurrences are likewise from the craton, but they are mingled with smaller boulders from the earlier geosynclinal formations and roundstone cobbles of granite and rhyolite with radiometric dates of 370–410 m.y. (middle Paleozoic) that were apparently derived from the backlands (Denison et al., 1969, p. 247–249).

Whence came this great mass of flysch, and why? Its sudden appearance in the sequence in mid-Mississippian time implies a radical change in the erosional regime of the source area, probably related to tectonic events in the Ouachita and Appalachian belts. Traditionally, the source was ascribed to Llanoria, a mythical backland of the Ouachita belt (Miser, 1921, p. 64–74), but complications are indicated by modern data—of paleocurrents measured on turbidite structures in the sandstones, and of the mineralogy of the sandstones.

In the Stanley of the interior Ouachitas, and in the Tesnus and Haymond at Marathon, dominant paleocurrents were northwestward or westward across the trough (Johnson, 1962, 1966; McBride, 1966, p. 54–56) (Figs. 6 and 7). In the Jackfork and Atoka of the northern and western Ouachitas, paleocurrents were dominantly westward and southwestward, longitudinal to the trough (Reinemund and Danilchek, 1957; Briggs and Cline, 1967, p. 991–994; Morris, 1971, p. 398–399) (Fig. 6). With the change in direction of paleocurrents, the sandstones change from immature sediments derived from older sedimentary, metamorphic, and crystalline rocks to more mature sediments derived from distant sources. In the Atoka of the Arkoma basin to the north, ripple marks and cross-beds in the shallow-water sediments indicate paleocurrents southward toward the trough, which turn westward where the deepwater facies begins (Agterberg and Briggs, 1963).

The older flysch thus was derived from relatively nearby sources to the southeast, probably from tectonic ridges and welts in the backland—a modern variant of the old Llanoria concept (Johnson, 1966, p. 156–157). The younger flysch came from one or more sources farther east—from the eastern parts of the Ouachita belt, from the Appalachian belt, or from the craton (Briggs and Cline, 1967, p. 996–997; Morris, 1971, p. 398–402). The volume of the contribution from the craton is not clear; it may have been much or little. Paleocurrent studies of the upper Mississippian–lower Pennsylvanian clastics of the northeastern interior region indicate southward transport of clastics which could have entered the Ouachita trough east of the Ozark uplift; transport into the trough from the craton farther west is less likely, as this part was widely masked by lower Paleozoic carbonate rocks.

The wildflysch, or boulder-beds, express an exaggerated phase of the subsidence of the flysch trough, during which clasts of many sizes slumped or slid from its margins onto the floor. The Johns Valley boulders may have broken from the scarps of growth faults along the edge of the craton, like the growth faults known in the Arkoma basin a little farther north. Similar growth faults along the edge of the craton in the Marathon Region may have triggered the sliding of the large blocks into the Haymond boulder-bed, but these were mingled with clasts derived from orogenic sources in the geosyncline and its backland. Unlike the longitudinal transport of the sandstones in the flysch, transport of the bouldery material and its muddy matrix was lateral from the nearby sides of the trough.

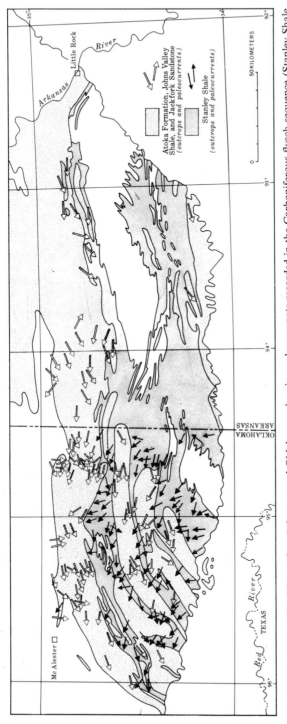

Fig. 6. Map of Ouachita Mountains, Arkansas and Oklahoma, showing paleocurrents recorded in the Carboniferous flysch sequence (Stanley Shale, Jackfork Sandstone, Johns Valley Shale, and Atoka Formation). Compiled from Reinemund and Danilchek (1957). Johnson (1966), Briggs and Cline (1967), and Morris (1971).

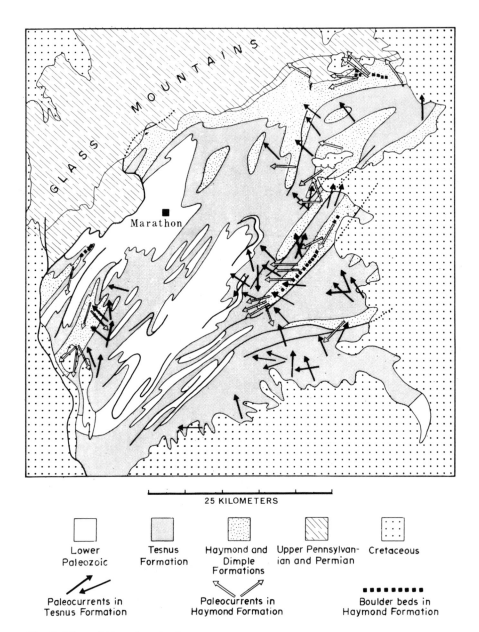

Fig. 7. Map of Marathon Region, west Texas, showing paleocurrents recorded in flysch of Tesnus and Haymond Formations. Compiled from Johnson (1962) and McBride (1966).

3. *Upper Pennsylvanian Orogenic and Postorogenic Sequence*

Upper Pennsylvanian strata (Desmoinesian and younger) do not occur in the internal parts of the Ouachita belt, either in the Ouachita Mountains or Marathon Region, and they were probably never deposited there. They are exposed in the craton to the north and northwest, in the western part of the Arkoma basin, and along the northern edge of the Marathon Region next to the Permian rocks in the Glass Mountains (Figs. 4 and 5); those in the latter area continue northward in subsurface into the Val Verde basin. Most of the upper Pennsylvanian is a shallow-water deposit, quite different from the earlier flysch of the orogenic belt, marine in the Marathon Region but including coal measures in the Arkoma basin. (The deposits of the Val Verde basin are an exception to this generalization, as we shall see later.)

The later Pennsylvanian was an orogenic time in the Ouachita belt, whose several pulsations affected sedimentation in the adjoining craton. The first pulsation was during Desmoinesian time, although beginning earlier and ending later from place to place; the sediments of that age are mostly preserved well away from the centers of deformation, where actual unconformities related to the pulsation are unimpressive. Beginning of deformation in the Ouachita Mountains is suggested by the shift of maximum accumulation of the Atoka Formation northward to the orogenic front, but the first Ouachita-derived clasts appear in conglomerates in the lower part of the Desmoinesian formations (Hartshorne Sandstone of Arkoma basin, Deese Formation of Ardmore basin, and Gaptank Formation at Marathon). Moreover, beneath the Coastal Plain south of the Ouachita Mountains, wells drilled through the Mesozoic within 25 km of the exposures of deformed Ouachita rocks encounter flat-lying shelf carbonates and shales with Desmoinesian fusulinids (Vernon, 1971, p. 967–972). One well (Humble Oil Co., No. 1 Royston, Hempstead County, Arkansas) penetrated 490 m of Desmoinesian strata beneath the Triassic Eagle Mills Formation and entered steeply dipping, cleaved sandstones and shales of Ouachita type beneath. Clearly, in this segment, deformation of the internal part of the Ouachita belt was essentially completed by Desmoinesian time. It is tempting to extrapolate this record (but with less proof) to infer that most of the complex structures of the belt, both here and at Marathon, were produced during this mid-Pennsylvanian pulsation.

Nevertheless, orogeny continued through the remaining Pennsylvanian and into the early Permian (Goldstein and Hendricks, 1962, p. 427). The Atoka and younger Pennsylvanian formations of the Arkoma basin are thrown into broad folds resembling those in the Ouachita Mountains but diminishing northward, and the Ouachita belt interacted complexly with the Arbuckle–Wichita belt on the west which was also in process of deformation during Pennsylvanian time. Details of the later Pennsylvanian orogeny in this segment of the Ouachita belt cannot be worked out in detail; the nearest postorogenic deposits are the basal Permian redbeds (Pontotoc Group) that overlap the western end of the Arbuckle uplift.

The later Pennsylvanian record is clearer in the Marathon segment of the Ouachita belt farther southwest. The upper Pennsylvanian of the northwestern part of the Marathon Region differs materially from the Gaptank Formation of the normal sequence, exposed in the eastern part of the region. It is thicker, more clastic, more like flysch, and more intensely deformed. It underlies the great frontal thrust of the Marathon Region (Dugout Creek fault), and drilling indicates that it lies, not on the geosynclinal rocks of the Ouachita belt, but on cratonic lower Carboniferous and lower Paleozoic (Flawn *et al.*, 1961, p. 237–238) (section A–A', Fig. 5). Infolded with the upper Pennsylvanian strata are others with early Permian (early Wolfcampian) fusulinids. The deformed complex is overlain on the north with right-angled unconformity by coarse upper Wolfcampian conglomerates (Lenox Hills Formation) at the base of the Glass Mountains Permian sequence (King, 1937, p. 82).

These relations epitomize a second and terminal pulsation of the Pennsylvanian orogeny in the Ouachita belt—of late Virgilian–early Wolfcampian age—and one that seems to have been of greater magnitude in the Marathon segment of the belt than farther east.

That the pulsation is a regional feature is indicated by subsurface relations to the north, in the Val Verde basin (Flawn *et al.*, 1961, p. 136–138). This basin, or foredeep, extends in subcrop more than 320 km along the northern border of the segment, from Brewster County to Val Verde County, southwest Texas, and contains 4300 m or more of upper Pennsylvanian–lower Wolfcampian fine-grained clastics or flysch—a new flysch trough like the earlier Carboniferous troughs, but displaced north of the orogenic belt onto the edge of the craton. The strata below and above the flysch are of normal thickness, indicating that it marks a unique sedimentary and tectonic event triggered by the late Virgilian–early Wolfcampian pulsation.

In the Marathon segment, Permian marine rocks of Wolfcampian to Guadalupian age border the Ouachita belt on the north, in the Glass Mountains, and are tilted gently northward toward the West Texas Permian Basin, but are not otherwise deformed (King, 1931, p. 117–120) (section A–A', Fig. 5). They overlap unconformably on the edge of the deformed Ouachita complex; erosional debris from the complex is common for 300 m or more above the case of the sequence, indicating that the Ouachita belt was being vigorously eroded through a large fraction of the Permian period.

4. *Tectonics*

Structures in the exposed parts of the Ouachita orogenic belt are much like those in the Appalachian Valley and Ridge province—lengthy folds and thrusts vergent toward the craton, with major décollement thrusts along parts of the orogenic front. Folds in the Ouachita Mountains have about the same dimensions as those in the Appalachians, but those in the Marathon region are smaller and narrower.

In most of the Ouachita Mountains the structures trend east and west, but they curve strongly southwestward at the western end in Oklahoma and slightly southeastward at their eastern end in Arkansas (Fig. 4), reflecting the position of the mountains on the apex of a salient in the orogenic belt. The Arkansas part of the Ouachitas is dominated by a central anticlinorium, or Benton uplift, which brings up the older parts of the sequence, with flysch formations north and south of it; but in Oklahoma, the anticlinoria of older rocks are smaller and lower, the largest being the Broken Bow uplift in the southeast, and the flysch formations are the dominant surface rocks. Formerly the Oklahoma part was believed to be more allochthonous and more overthrust on the craton, and the Arkansas part more autochthonous and more simply folded; recent work demonstrates that complexity of the Arkansas part is as great or greater than that of the Oklahoma part and that both are equally allochthonous.

The frontal belt of the Ouachita Mountains is dominated by great thrusts directed toward the Arkoma basin and the craton, which have originated in zones of décollement along weak layers low in the flysch sequence (Hendricks et al., 1947) (Fig. 8). These break the rocks into a succession of slices in a belt 25 km wide, in which the transition between the geosynclinal and cratonic sequences in telescoped (Goldstein and Hendricks, 1962, p. 402). In Oklahoma the frontal thrust is the Choctaw fault; it extends only about 40 km into Arkansas, but others like it continue eastward through the Atoka Formation to the edge of the Mississippi Embayment (Stone, 1966, p. 213). Minor décollement thrusts extend northward from the orogenic front beneath the Arkoma basin, where they have been proved by drilling (they are not to be confused with the earlier growth faults which also occur beneath the basin) (Berry and Trumbley, 1968, p. 91–93). The thrusts at the Ouachita front are less fundamental than others farther back—the Ti Valley fault of Oklahoma and the analogous Y City fault in Arkansas—which form the northern border of the true Ouachita geosynclinal sequence.

South of the Ti Valley fault in Oklahoma are more widely spaced thrusts —the Windingstair, Octavia, and Boktukola faults—which enclose blocks of broadly folded flysch formations. A notable variant is the Potato Hills uplift a little south of the Windingstair fault, which brings up the novaculite and older formations and encloses a window of an upper branch of the Windingstair fault (Miser, 1929, p. 18–20; Arbenz, 1968).

The anticlinoria farther back in the Ouachita Mountains—the Benton uplift of Arkansas and the Broken Bow uplift of Oklahoma—are both highly

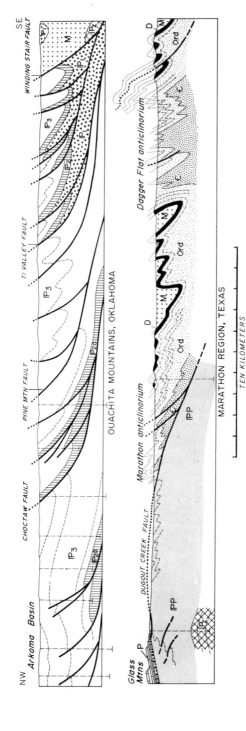

Fig. 8. Structure sections comparing the frontal thrusts of the Ouachita Mountains and Marathon segments of the Ouachita orogenic belt. *Above*: Frontal thrusts between Ouachita Mountains and Arkoma basin, eastern Oklahoma. Based on Berry and Trumbly (1968, section AA'). *Below*: Dugout Creek thrust of Marathon Region. Based on King (1937, section EE', pl. 21). Well control in both sections is indicated by vertical dashed lines. Letter symbols as follows: *Above*: IPz, lower Paleozoic of interior Ouachitas; M, Stanley Shale; IP1, Jackfork Sandstone; IP2, Johns Valley Shale; IP2a, equivalent strata of frontal Ouachitas, including Wapanucka Limestone; IP3, Atoka Formation. *Below*: €, Upper Cambrian; Ord, Ordovician; D, Caballos Novaculite; IPz, lower Paleozoic of cratonic sequence; M, Tesnus Formation; IPP, upper Pennsylvanian and lower Wolfcampian foredeep deposits; P, upper Wolfcampian (Lenox Hills Formation).

complex but differ in plan, the first following the structural grain, and the second crossing it transversely and formed by a culmination on the lesser folds. The older rocks within the anticlinoria are tightly folded, extensively inverted, and metamorphosed to low-grade slates and phyllites (mostly green-schist facies, rarely amphibolite facies), and are permeated by hydrothermal quartz veins (Miser, 1943, p. 102–106). In the northern part of the Benton uplift 25 km west of Little Rock, Arkansas, are half a dozen small pods of soapstone or serpentinite, evidently highly altered shreds of upthrust mantle material.

The rocks of the two anticlinoria have a compound structural fabric, with successive superposed folds and cleavages too complex for easy generalization (Viele, 1966, p. 256–265; 1973). Earlier and greater structures with north-ward vergence have been obscured and partly obliterated by structures with southward vergence. For example, in the Broken Bow uplift, cleavage and fold axes dip north, and at its northern edge early north-verging thrusts have been rotated to a northward dip and have been mistaken for normal faults (Honess, 1923, p. 214–240, 260–261, pl. 1). Southward-verging folds and thrusts dominate all the southern flank of the Benton uplift (Purdue and Miser, 1923; Miser and Purdue, 1929) and occur entirely across it at its easternmost exposures, yet here and there relics of earlier north-verging structures are preserved.

The Ouachita Mountains structures change profoundly at their western end, just under the edge of the Coastal Plain deposits, as a result of their impingement on the structures of the Arbuckle–Wichita belt (Flawn et al., 1961, p. 171–172; Harlton, 1966). Here, the cover is only a few hundreds of meters thick, and well penetrations of the Paleozoic rocks are closely spaced, furnishing unusually accurate data on the subcrop relations (Fig. 4). The Ouachita frontal structures turn southwest and south and are faulted and recessed around the southeast-trending Arbuckle–Wichita structures. The Arbuckle uplift, with its core of Tishomingo Granite (1050–1350 m.y. old), extends 55 km southeastward beyond its outcrop and creates a pronounced reentrant in the Ouachita front, bordered north and south by major high-angle faults, probably with components of both dip-slip and strike-slip displacement. The frontal Ouachita thrusts (Choctaw, Ti Valley, etc.) are truncated southward by the high-angle faults and cannot be correlated specifically with subcrop Ouachita structures south of the Arbuckle–Wichita belt.

Clearly, there has been a complex interaction between the Ouachita structures and the Arbuckle–Wichita structures, as both were in process of deformation during the Pennsylvanian, but details of the chronology are not entirely obvious. The Ouachita allochthon was emplaced over the intracratonic Arbuckle–Wichita autochthon during the Desmoinesian, as indicated by Ouachita-derived detritus in the Deese Formation of the Ardmore Basin (Tomlinson and McBee, 1959, p. 39–40). Later in the Pennsylvanian the Arbuckle horst was raised, and the subcrop trace of the Ouachita front was offset and faulted, with components of vertical and lateral displacement.

Ouachita structures in the Marathon Region resemble those in the Ouachita Mountains but are on a smaller scale (King, 1937, p. 119–130). Two anticlinoria (miniature replicas of the Broken Bow and Benton uplifts) bring up formations as old as the Cambrian, ringed by novaculite ridges and surrounded by areas of the flysch formations (Fig. 5). A notable feature of the structure is the alternation of competent and incompetent units, resulting in disharmonic folding and faulting. The Ordovician and Cambrian stand isoclinal folds; the novaculite has been split into numerous thrust slices, then folded to create a remarkably complex pattern; the flysch formations lie in more open anticlines and synclines. Although the deformation is severe, the rocks show only the weakest symptoms of incipient metamorphism.

The frontal thrust on the northwest, the Dugout Creek fault, differs from the décollement thrusts along the orogenic front in the Ouachita Mountains (Fig. 8). It was formed during the Virgilian–early Wolfcampian orogenic pulse and shears off geosynclinal rocks that had already been thoroughly deformed during the Desmoinesian pulse, displacing them many kilometers over their foredeep, the Val Verde basin. Although it is exposed only in the northwestern part of the Marathon region, deep wells indicate that it extends entirely across the region, and probably beyond.

B. Concealed Parts of the Ouachita Orogenic Belt

In the Gulf Coastal Plain, the buried connections between the different exposures of the Ouachita orogenic belt and its connection with the Appalachian belt are indicated by drill data. These data are available only for 80–100 km behind the orogenic front, and even where available they vary in abundance and quality. Where the belt is at shallow depth and where economic interest is great, well penetrations are close together; in other places, penetrations are half a dozen or less to a county, and in some counties none at all. Many of the penetrations of the belt are recorded only by logs, but cores are available from some of the wells, allowing a more definitive geological study.

From the drill data, subcrop areal geological maps can be made (for example, Flawn et al., 1961, pls. 2 and 3). In closely drilled areas these can be made with confidence. Where there are fewer data the mapper must rely more on his own inference and predilection; in complex areas several different interpretations are possible from the same sets of data.

1. *Segment East of Ouachita Mountains*

The junction between the Ouachita and Appalachian orogenic belts is beneath the eastern part of the Gulf Coastal Plain, in the 550-km gap between their areas of exposure (Fig. 9).

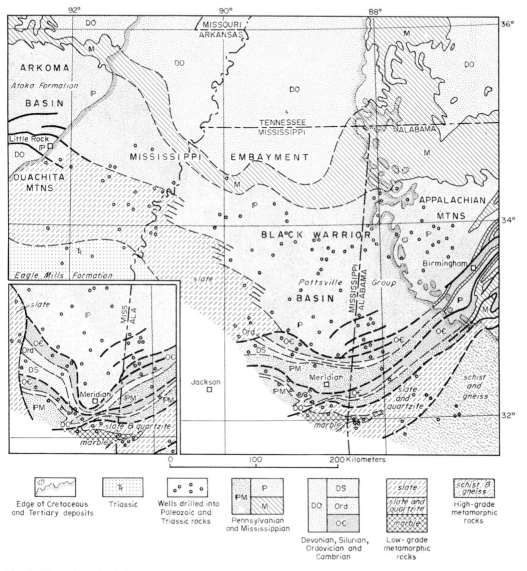

Fig. 9. Map of the buried segment of the orogenic belt between Appalachian and Ouachita Mountains, Alabama, Mississippi, and Arkansas, showing inferred subcrop geology, and the intersection of the Appalachian and Ouachita belts. Inset shows one of several possible alternative interpretations of the intersection. Based on Thomas (1972, 1973), supplemented by King (in Flawn *et al.*, 1961, pl. 3) and other sources.

Well data indicate that the Appalachian belt extends southwestward across Alabama into Mississippi, and the Ouachita belt southeastward across Arkansas into Mississippi, with the Black Warrior Basin, or foredeep, in the angle between them (Mellen, 1947), filled by more than 3000 m of Carboniferous clastics—equivalents of the Floyd, Parkwood, and Pottsville Formations of Alabama.

As indicated earlier, rocks and structures like those of the Valley and Ridge Province and the Talladega Belt can be traced by well data from the Appalachians in Alabama southwestward into eastern Mississippi, where their trend changes to west–northwest. An extension of the Ouachita belt southeastward across Arkansas into western Mississippi is indicated by more scanty well penetrations into slaty and cherty low-grade metamorphic rocks of uncertain age but probably equivalent to those in the core of the Ouachita Mountains. In addition, some wells farther northeast have also been called "Ouachita wells," as they penetrate disturbed and sheared clastic rocks, but these rocks are probably part of the Carboniferous sequence of the Black Warrior Basin close to the orogenic front (Flawn et al., 1961, p. 88).

The exact nature of the Appalachian–Ouachita junction in central Mississippi is still problematical, and several interpretations are possible from the available well data (compare Fig. 9 with insert). In the Black Warrior Basin to the north, Ouachita-type elements interfinger from the west into lower Paleozoic shelf deposits—black shale in the Ordovician and Silurian and novaculite in the Devonian—suggesting an approach to a shelf break into deeper water (Thomas, 1972, p. 92–96). A shelf break between the Black Warrior and Arkoma basins must have existed during Carboniferous time, between the shallow-water coal measures of the Pottsville in one, and the contemporaneous deep-water flysch of the Atoka in the other. Extending this idea, Thomas (1972, 1973) proposes that the change from the Appalachian into the Ouachita orogenic belt is primarily a change in sedimentary facies rather than a structural feature. Nevertheless, the rocks of the internal Ouachita Mountains are everywhere thrust over their foredeep in the Arkoma basin, and such a tectonic contact must continue a smaller or greater distance farther east.

2. Segment in North-Central Texas

South of its junction with the Arbuckle–Wichita belt in southern Oklahoma, the concealed Ouachita belt trends south–southwest around the southeastern side of the Llano uplift, in a course more or less coincident with the Balcones and Luling–Mexia fault zones in the Mesozoic cover rocks (Fig. 10).

Its western orogenic front, facing the Forth Worth basin, or foredeep, is clearly defined by well data. In many places, penetrations of Ouachita rocks are closely juxtaposed to penetrations of cratonic rocks; some wells for a few

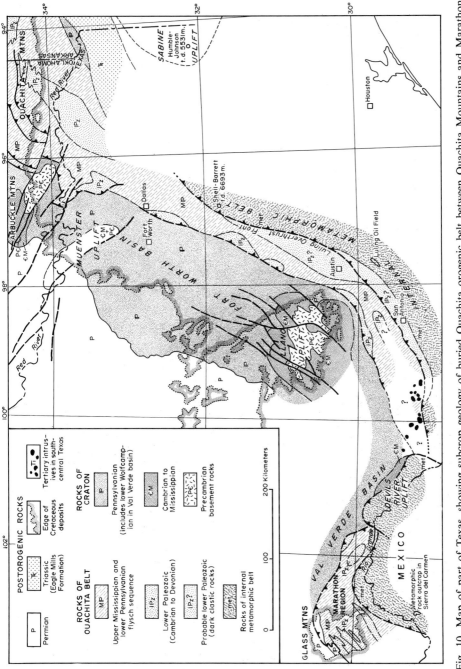

Fig. 10. Map of part of Texas, showing subcrop geology of buried Ouachita orogenic belt between Ouachita Mountains and Marathon Region. Based on Flawn (in Flawn *et al.*, 1961, pl. 2), with additions from later data.

kilometers east of the front pass through Ouachita rocks into cratonic rocks, indicating a low-angle thrust contact Flawn *et al.*, 1961, p. 167). An even greater allochthony of this segment is demonstrated by the 6693 m Shell Oil Co., no. 1 Barrett well, Hill Country, 40 km east of the Ouachita front (Fig. 10) (Rozendal and Erskine, 1971). Beneath the Mesozoic the well passed through 3178 m of phyllite and quartzite of the internal metamorphic belt (described below), into 2495 m of carbonate rocks with basal quartzite, evidently cratonic lower Paleozoic, and ended in a basement of quartz diorite. All the rocks below the Mesozoic have been metamorphosed, including the carbonate and the quartz diorite, and have yielded middle to late Paleozoic radiometric dates of 257–280 m.y., probably expressing metamorphic and orogenic events.

In the frontal belt of the Ouachita structure, 15–40 km wide, the rocks are recognizably like the lower and upper Paleozoic formations of the Ouachita Mountains. Lower Paleozoic rocks occur in places along the orogenic front and along the southwestward prolongation of the Broken Bow uplift of Oklahoma, but sandstone and shale like the Stanley of the Ouachita Mountains are more extensive. Southeast of the frontal belt in central Texas is another belt 15–50 km wide of weakly metamorphosed dark clastic rocks of undetermined age but possibly lower Paleozoic of a more internal facies than on the outcrop (Flawn *et al.*, 1961, p. 77–78).

The innermost known belt of the Ouachita structure, southeast of the one just described, is formed of phyllite, slate, and quartzite, which have been subjected to low- to middle-grade metamorphism with a strong shearing component (greenschist to amphibolite facies) (Flawn *et al.*, 1961, p. 79–81). The belt has been penetrated for several hundred kilometers east and west of the Llano uplift, and includes the long-known sub-Cretaceous metamorphic basement of the Luling oil field southeast of Austin. Its southeastward extent is undetermined, as within 50 km downdip it passes beyond reach of drilling. The metamorphic contrast between the rocks of the belt and those immediately northwest suggests a major structural discontinuity, the "Luling overthrust front" (Flawn *et al.*, 1961, p. 79, 169). This internal metamorphic belt may be tectonically analogous to the Piedmont province of the Appalachian orogenic belt.

3. *Segment in Southwestern Texas*

West of the Llano uplift the continuity of the Ouachita belt is lost for 225 km in a complex area in Uvalde, Kinney, and Val Verde Counties, and only reappears in the salient around the Marathon Region (Fig. 10). Much of the segment is bordered on the north by the Val Verde basin, or foredeep, described earlier.

In the eastern part of the complex area well control is sparse, and the relations are obscured by profuse Tertiary intrusions. In the western part of the area is the Devils River uplift (Galley, 1958, p. 397; Flawn et al., 1961, p. 144–145), a high-standing subcrop feature trending northwest, along the edge of the Val Verde basin. The rather plentiful wells on the uplift penetrate many types of rock, but none certainly attributable to the frontal Ouachita belt. In the southwestern part of the uplift are low-grade, highly sheared marble, phyllite, and quartzite, probably correlative with those of the internal metamorphic belt of farther east. These are juxtaposed against, and probably thrust over, carbonate rocks of the lower Paleozoic cratonic sequence and over metavolcanics, probably Precambrian basement. If the frontal part of the Ouachita belt was originally continuous across the Devils River uplift, it has been obliterated by thrusting of the internal belt over the foreland.

Farther west, the frontal belt reappears, and many wells penetrate lower and upper Paleozoic rocks correlative with formations of the outcrop area in the Marathon Region. Farther southeast, near the Rio Grande, several wells penetrate metamorphic rocks of the internal belt, and these come to the surface at the base of the Sierra del Carmen, south of the Rio Grande in Mexico (Fig. 10; locality 1, Fig. 11) (Flawn and Maxwell, 1958)—the only known outcrop of the internal belt. Radiometric dates on rocks of this outcrop are 263–275 m.y., or late Paleozoic (Denison et al., 1969, p. 251), probably expressing metamorphic events much later than the age of the rocks themselves.

4. Postorogenic Deposits

Deposits of various ages that are postorogenic to the Ouachita deformation are penetrated in many places beneath the Jurassic in the Gulf Coastal Plain and especially in the segment south of the Ouachita Mountains, where they conceal the internal part of the orogenic belt.

The best known and probably most extensive deposits are red continental clastics with profuse diabase intrusives, which are lithically like the Upper Triassic Newark Group exposed in the Appalachian Piedmont Province farther northeast. They are extensive in a broad area in southern Georgia and Alabama south of the Appalachian Piedmont, where they are as much as 1600 m thick (McKee et al., 1959, pls. 5, 9). They are encountered again south of the Ouachita Mountains in southern Arkansas and adjacent states, where they are termed the Eagle Mills Formation.

The Eagle Mills is typified by the record of the Humble Oil Co., No. 1 Royston well, Hempstead County, Arkansas (Fig. 4), where it is 2286 m thick (the maximum known for the unit), between Lower Cretaceous and the Desmoinesian shelf carbonates and shales referred to earlier (Scott et al., 1961, p. 7–10; Vernon, 1971, p. 968–974). At several levels near the middle

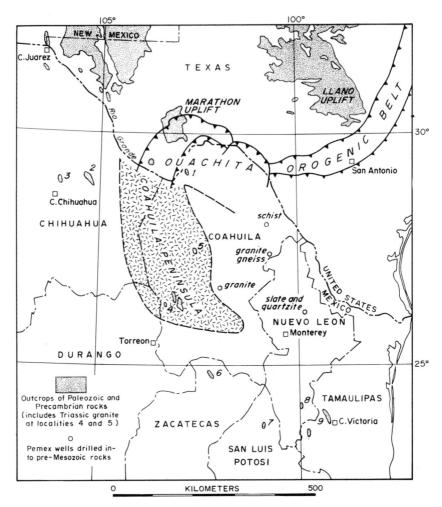

Fig. 11. Map of northeastern Mexico, showing outcrops of Paleozoic and Precambrian rocks; the locations of Pemex wells that were drilled into the pre-Mesozoic basement; and outline of Coahuila Peninsula as inferred by Kellum *et al.* (1936). Numbered localities, described in text, are as follows: (1) Sierra del Carmen, (2) Mina Plomosas-Placer de Guadalupe area; (3) Sierra del Cuervo; (4) Las Delicias area; (5) Potrero de la Mula; (6) Caopus-Rodeo; (7) Catorce; (8) Aramberri; (9) Ciudad Victoria area, including Peregrina Canyon.

it contains the leaves of the cycad *Macrotaenopteris*, which also occurs in the Chinle Formation farther west and the Newark Group farther east. The Eagle Mills is known to extend for more than 150 km farther east and west, but it has an abrupt boundary on the north, probably faulted; very likely it occurs in a fault trough like the fault troughs of the Newark Group in the Piedmont Province.

Other deposits of gray and red continental deposits occur elsewhere beneath the Coastal Plain which contain spores of Permian age, but their extent and continuity is less known.

The oldest postorogenic deposits are the shelf carbonates and shale with Desmoinesian fusulinids which fringe the northern margin of the Eagle Mills trough, next to the deformed earlier Paleozoic of the Ouachita orogenic belt. Postorogenic Pennsylvanian rocks were also penetrated in two wells on the Monroe uplift south of the Eagle Mills trough—Union Producing Co., No. 1A Tensas Delta, Morehouse Parish, Louisiana, and the Humble Oil Co., No. 1 Georgia-Pacific Corp., Ashley County, Arkansas. The beds in the first well, termed Morehouse Formation, were penetrated to a thickness of 390 m, and those in the second to 2381 m, without reaching their base (Flawn et al., 1961, p. 350; Vernon, 1971, p. 968, 970).

Far out in the Coastal Plain on the Sabine uplift, 240 km south of the Paleozoic outcrops in the Ouachita Mountains, several wells have reached pre-Jurassic rocks, among them the 5513 m Humble Oil Co., No. 1 G. R. Johnson well, Panola County, Texas. Beneath red and gray spore-bearing Permian, it entered dark clastic "flysch" containing spores of Pennsylvanian or Mississippian age (Vernon, 1971, p. 966, 972, 977). I hazard a guess that these are postorogenic Pennsylvanian rocks rather than deformed Ouachita rocks.

V. PALEOZOIC OF NORTHEASTERN MEXICO

Northeastern Mexico is a domain of Mesozoic rocks, a large part of which were involved in the Cordilleran orogenies of Mesozoic and Cenozoic time. Paleozoic and Precambrian rocks emerge only in widely spaced inliers (Fig. 11), where they have been brought to the surface in the cores of anticlines or along the edges of uplifted fault blocks. These inliers are insufficient to establish a detailed pattern of pre-Mesozoic tectonics.

Pertinent outcrops are as follows (numbers are keyed to those in Fig. 11):

(1) Sierra del Carmen, northern Coahuila, at west base of escarpment, 12 km east of Boquillas on Rio Grande. Phyllite, marble, and quartzite of low-grade (greenschist) metamorphism; with radiometric dates of 263–275 m.y. Overlain by Lower Cretaceous (Flawn and Maxwell, 1958; Denison et al., 1969, p. 251).

(2) Mina Plomosas–Placer de Guadalupe area, Chihuahua, 80 km northeast of Chihuahua City. Ordovician, Silurian, and Devonian shaly limestone, dolomite, and chert, 540 m; Mississippian and Pennsylvanian shales and carbonates, 360 m; Permian shales with reef limestones, 1150 m. Overlain by Jurassic without obvious structural discordance (Bridges, 1964, 1965).

(3) Sierra del Cuervo, Chihuahua, 30 km north of Ciudad Chihuahua. Steeply folded flysch like that of the Haymond Formation at Marathon, but with Wolfcampian fusulinids at top, 1000 m. Overlain unconformably by Lower Cretaceous (de Cserna and Diaz, 1959).

(4) Las Delicias, Coahuila, 100 km northeast of Torreon. Permian (mainly Leonardian and Guadalupian) volcaniclastic shale and graywacke, with fossiliferous reefy carbonate lenses and blocks, 3000 m. Steeply folded along northeast axes and intruded by middle Triassic granite; overlain with structural discordance by Lower Cretaceous (King *et al.*, 1944, p. 7–23; Newell, 1957; Denison *et al.*, 1969, p. 251).

(5) Potrero de la Mula, west-central Coahuila, about 160 km north of Las Delicias. Granodiorite with a middle Triassic radiometric date (206 m.y.), as at Las Delicias. Overlain unconformably by Lower Cretaceous (Kellum *et al.*, 1936, p. 976–977; Denison *et al.*, 1969, p. 251).

(6) Caopus-Rodeo, northeastern Zacatecas.

(7) Catorce, northern San Luis Potosi.

(8) Aramberri, southern Nuevo Leon. Small widely separated areas in the cores of mountain uplifts of phyllite, schist, and metavolcanics of undetermined ages; may include both Precambrian and Paleozoic. Equivocal radiometric dates on the metavolcanics range from 141–200 m.y. (Flawn *et al.*, 1961, p. 101–102; Denison *et al.*, 1969, p. 253).

(9) Perigrina Canyon and other localities, in Sierra Madre Oriental 20 km west of Ciudad Victoria, Tamaulipas. Separate fault blocks or thrust slices of (a) gneiss with 924–1350 m.y. radiometric dates; (b) low-grade chlorite and sericite schist; (c) Silurian and Devonian black graptolite shale and novaculite, 230 m; and (d) Mississippian to Permian (Leonardian) flysch, 1450 m. Overlain unconformably by Jurassic (Flawn and Diaz, 1959; Fries, 1962, p. 55–69; de Cserna, 1970, p. 261–263; 1971, p. 546–547).

In addition to these outcrops, a few Pemex wells in eastern Coahuila and Nuevo Leon have bottomed in pre-Mesozoic schist, metaquartzite, granite, and granite gneiss; the granite gneiss has yielded a radiometric date of 358 m.y. (Flawn *et al.*, 1961, p. 103; Denison *et al.*, 1969, p. 252).

Rather surprisingly, none of these exposures of pre-Mesozoic rocks and structures are convincingly like those in the well-characterized Ouachita orogenic belt in the United States—even those in eastern Chihuahua, only 130 km across the Rio Grande from the last exposure of the belt in The Solitario. Does the belt change abruptly in character along the strike in Mexico, or has the belt been eliminated or displaced here by some form of pre-Mesozoic structure?

A major feature in northeastern Mexico is the Coahuila peninsula (Kellum *et al.*, 1936, p. 972–973) which occupies most of the western half of the State

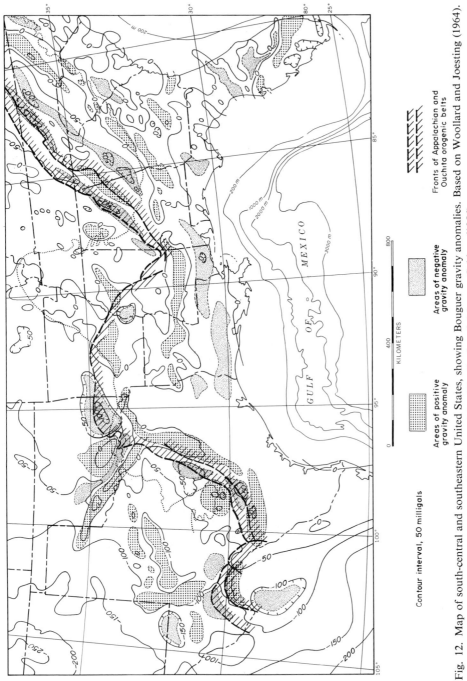

Contour interval, 50 milligals

Areas of positive gravity anomaly

Areas of negative gravity anomaly

Fronts of Appalachian and Ouchita orogenic belts

Fig. 12. Map of south-central and southeastern United States, showing Bouguer gravity anomalies. Based on Woollard and Joesting (1964). Supplemented by Woollard and Monges Caldera (1956).

of Coahuila, south of the Marathon Region and The Solitario. It had a strong influence on Cordilleran tectonics, as Mesozoic rocks are strongly deformed west, south, and east of it, whereas they have only been broadly folded or domed over it; it also coincides with prominent Bouguer gravity anomalies (Fig. 12) (Woolard and Monges Caldera, 1956, Figs. 6 and 10). Probably it originated as a positive massif as a result of orogeny and consolidation in the backland of the Ouachita belt, but this consolidated Paleozoic basement is not exposed; the only basement exposed is the Triassic granodiorite at Potrero de la Mula (item 5 above). However, the rocks in the Pemex wells farther east are very likely parts of the Ouachita backland.

The areas of outcrop of Paleozoic rocks listed above are west and south of the Coahuila peninsula, and their distribution is controlled by the south-eastward and southward trends of the uplifts in the post-Paleozoic Cordilleran orogenic belt, especially in the Sierra Madre Oriental. From their distribution, it has commonly been assumed that the trends of the Paleozoic rocks and structures in eastern Mexico were the same direction, from Chihuahua to Guatemala. A Paleozoic geosyncline in this position has been postulated (Lopez-Ramos, 1969, p. 2402–2403), whose rocks were deformed into a Huastecan orogenic belt in the latter part of Paleozoic time (de Cserna, 1960, p. 598–601). While these speculations are appealing, they are based on very fragmentary evidence.

VI. THE FLORIDA ELEMENT

In Georgia and Florida the base of the Coastal Plain deposits is nearer the surface than in the Gulf Coastal Plain farther west (within less than 1000 m in northern Florida, deepening southward), so that the pre-Mesozoic basement has been penetrated in many wells. Crystalline rocks like those in the Piedmont province of the Appalachians have been penetrated as far as southern Georgia, beyond which are subcrop tectonic units not duplicated elsewhere in the southern United States (Applin, 1951).

In northwestern Florida and extreme southern Georgia is the Suwanee basin of lower Paleozoic rocks (Fig. 1)—called a "basin" out of deference to the surrounding crystalline rocks, although its strata are essentially horizontal. Fossiliferous marine shales and sandstones of Ordovician, Silurian, and Devonian age are topped by terrestrial middle Devonian deposits; known thickness of the sequence is about 1000 m, but the base has not been penetrated (Flawn *et al.*, 1961, p. 90–91). The faunas are unlike any in the Appalachian or Ouachita belts and have their greatest affinities with those in parts of Europe and Africa.

In east-central Florida is an area of plutonic rocks (granites, granodiorites, and diorites) (Fig. 1) which have yielded an assortment of late Proterozoic and early Paleozoic radiometric ages (480–630 m.y.) (Milton and Grasty, 1969). Ages with this time range are unusual in the Appalachian crystalline area, except in the "Avalonian" belt close to the coast, and are much like those in western Africa.

Northeast of the Suwanee basin, between it and the plutonic area, and southwest of the latter are belts of volcanic rocks—rhyolitic lavas and tuffs, and amygdaloidal basalts. A Paleozoic age has been assumed for them, but those on the southwest have Jurassic radiometric dates (143–183 m.y.), and there is no compelling reason why the remainder should not be Mesozoic also.

These subcrop units are followed or bordered broadly by belts of magnetic and Bouguer gravity anomalies (Fig. 12) (King, 1959), suggesting that they are fundamental crustal units. Moreover, the east-coast magnetic anomaly which lies near the edge of the continental shelf northeastward to New England and beyond bends westward near the 31st parallel and extends inland across southern Georgia immediately north of the units of the Florida element (Taylor *et al.*, 1968, p. 756–765). This anomaly seems to mark the edge of the original North American continental plate, suggesting that the units of the Florida element have been accreted to it at a later time (Taylor *et al.*, p. 777).

One thus receives the impression that the older units of the Florida element (the Suwanee basin and the east-central Florida plutonic area) are separate plates, unrelated to each other or to the Piedmont crystallines of the Appalachian orogenic belt north of them, accidentally assembled, and welded together along belts of Jurassic volcanics. There is much merit in the proposal (Wilson, 1966, p. 680) that these plates are fragments of transoceanic continental material joined to the North American continent during a late Paleozoic closing of the Atlantic and rifted from their parents by subsequent reopening of the Atlantic during Mesozoic and later time.

REFERENCES

Agterberg, E. P., and Briggs, Garrett, 1963, Statistical analysis of ripple marks in Atokan and Desmoinesian rocks in the Arkoma basin of east-central Oklahoma: *J. Sediment. Petrol.*, v. 33, n. 2, p. 393–410.

Applin, P. L., 1951, Preliminary report on buried pre-Mesozoic rocks in Florida and adjacent states: *U.S. Geol. Surv. Circ.* 91, 28 p.

Arbenz, J. K., 1968, Structural geology of the Potato Hills, Ouachita Mountains, Oklahoma, in: *A Guidebook to the Geology of the Western Arkoma Basin and Ouachita Mountains, Southeastern Oklahoma*, Cline, L. M., ed.: Okla. City Geol. Soc. Guidebook, p. 109–121.

Baker, C. L., and Bowman, W. F., 1917, Geologic exploration of the southeastern front range of trans-Pecos Texas: Texas Univ. (Bur. Econ. Geol.) Bull. 1753, p. 63–172.

Bentley, R. D., and Neathery, T. L., 1970, Geology of the Brevard zone and related rocks of the inner Piedmont of Alabama: Al. Geol. Soc., 8th Ann. Field Trip Guidebook, 119 p.

Berry, R. M., and Trumbley, W. D., 1968, Wilburton gas field, Arkoma basin, Oklahoma, in: *A guidebook to the Geology of the Western Arkoma Basin and Ouachita Mountains, Southeastern Oklahoma*, Cline, L. M., ed.: Okla. City Geol. Soc. Guidebook, p. 86–103.

Bridges, L. W., II, 1964, Stratigraphy of the Mina Plomosas–Placer de Guadalupe area, in: Geology of the Mina Plomosas–Placer de Guadalupe area, Chihuahua, Mexico: West Texas Geol. Soc. Publ. 64-50 (Field Trip Guidebook), p. 50–61.

Bridges, L. W., II, 1965, Geologia del area de Plomosas, Chihuahua, in: Estudios geologicos del Estado de Chihuahua, Parte 1: *Univ. Nac. Auton. Mex., Inst. Geol., Bol.* 74, p. 1–134.

Briggs, Garrett, and Cline, L. M., 1967, Paleocurrents and source areas of late Paleozoic sediments of the Ouachita Mountains, southeastern Oklahoma: *J. Sediment. Petrol.*, v. 37, n. 4, p. 985–1000.

Burchfiel, B. C., and Livingston, J. L., 1967, Brevard zone compared to Alpine root zones: *Am. J. Sci.*, v. 265, n. 4, p. 241–256.

Butler, J. R., 1972, Age of Paleozoic regional metamorphism in the Carolinas, Georgia, and Tennessee southern Appalachians: *Am. J. Sci.*, v. 272, n. 4, p. 319–333.

Butts, Charles, 1926, The Paleozoic rocks, in: The geology of Alabama: Ala. Geol. Surv. Spec. Rep. 14, p. 41–230.

Butts, Charles, and Gildersleeve, Benjamin, 1948, Geology and mineral resources of the Paleozoic area in northwest Georgia: *Ga. Geol. Surv. Bull.* 54, 176 p.

Carrington, T. L., 1973, Paleozoic age and structure of formations within the metasedimentary Talladega Group, Chilton County, Alabama [abstract]: Geol. Soc. Am. Abstr. with Programs, v. 5, n. 3 (South-Central Sec., 7th Ann. Mtg.), p. 250.

Cline, L. M., 1960, Late Paleozoic rocks of the Ouachita Mountains: *Okla. Geol. Surv. Bull.* 85, 113 p.

de Cserna, Zoltan, 1960, Orogenesis in time and space in Mexico: *Geol. Rundschau*, Band 50, p. 595–605.

de Cserna, Zoltan, 1970, The Precambrian of Mexico, in: *The Precambrian*, Vol. 4, Rankama, Kalervo, ed.: Wiley–Interscience, New York, p. 253–270.

de Cserna, Zoltan, 1971, Taconian (early Caledonian) deformation in the Huastecan structural belt of eastern Mexico: *Am. J. Sci.*, v. 271, n. 5, p. 544–550.

de Cserna, Zoltan, and Diaz G., Teodoro, 1956, Estratigrafia Mesozoica y tectonica de la Sierra Chihuahua; Permico de Placer de Guadalupe, Chih.; geohidrologia de la Region Lagunera; estratigrafia Mesozoica y tectonica de la Sierra Madre Oriental entre Mapimi, Dgo. y Monterrey, N. L.: Excursion A-13, Congreso Geologico Intern. XX, 120 p.

Denison, R. E., Kenney, G. S., Burke, W. H., Jr., and Hetherington, E. A., Jr., 1969, Isotopic ages of igneous and metamorphic boulders from the Haymond Formation, Marathon Basin, Texas, and their significance: *Geol. Soc. Am. Bull.*, v. 80, n. 2, p. 245–256.

Flawn, P. T., and Maxwell, R. A., 1958, Metamorphic rocks of the Sierra del Carmen, Coahuila, Mexico: *Am. Assoc. Petr. Geol. Bull.*, v. 42, n. 9, p. 2245–2249.

Flawn, P. T., and Diaz G., Teodoro, 1959, Problems of Paleozoic tectonics in north-central and northern Mexico: *Am. Assoc. Petr. Geol. Bull.*, v. 43, n. 1, p. 224–230.

Flawn, P. T., Goldstein, August, Jr., King, P. B., and Weaver, C. E., 1961, The Ouachita system: Texas Univ. (Bur. Econ. Geol.) Publ. 6120, 401 p.

Folk, R. L., 1973, Evidence for peritidal deposition of Devonian Caballos Novaculite, Marathon Basin, Texas: *Am. Assoc. Petr. Geol. Bull.*, v. 57, n. 4, p. 702–725.

Fries, Carl, Jr., 1962, Estudios geocronologicos de rocas Mexicanas: *Univ. Nac. Auton. Mex., Inst. Geol., Bol.* 64, 151 p.

Fullagar, P. D., 1971, Age and origin of the plutonic intrusions in the southern Appalachians: *Geol. Soc. Am. Bull.*, v. 82, n. 11, p. 2845–2862.

Galley, J. E., 1958, Oil and geology in the Permian basin of Texas and New Mexico, in: *Habitat of Oil*: Am. Assoc. Petr. Geol., p. 395–446.

Goldstein, August, Jr., and Hendricks, T. A., 1953, Siliceous sediments of Ouachita facies in Oklahoma: *Geol. Soc. Am. Bull.*, v. 64, n. 4, p. 421–441.

Goldstein, August, Jr., and Hendricks, T. A., 1962, Late Mississippian and Pennsylvanian sediments of Ouachita facies, Oklahoma, Texas, and Arkansas, in: *Pennsylvanian System in the United States*, Branson, C. C., ed.: Am. Assoc. Petr. Geol., p. 385–430.

Ham, W. E., Denison, R. E., and Merritt, C. A., 1964, Basement rocks and structural evolution of southern Oklahoma: *Okla. Geol. Surv. Bull.* 95, 302 p.

Harlton, B. H., 1966, Relation of buried Tishomingo uplift to Ardmore basin and Ouachita Mountains, southeastern Oklahoma: *Am. Assoc. Petr. Geol. Bull.*, v. 50, n. 7, p. 1365–1374.

Hass, W. H., 1951, Age of Arkansas Novaculite: *Am. Assoc. Petr. Geol. Bull.*, v. 35, n. 12, p. 2526–2541.

Hatcher, R. D., Jr., 1971, Stratigraphic, petrologic, and structural evidence favoring a thrust solution of the Brevard problem. *Am. J. Sci.*, v. 270, n. 3, p. 177–202.

Hendricks, T. A., Gardner, L. S., Knechtel, M. M., and Averitt, Paul, 1947, Geology of the western part of the Ouachita Mountains of Oklahoma: U.S. Geol. Surv. Oil and Gas Invest. Prelim. Map 66.

Hewett, D. F., and Crickmay, G. W., 1937, The Warm Springs of Georgia; their geologic relations and origin, a summary report: U.S. Geol. Surv. Water-Supply Pap. 819, 40 p.

Honess, C. W., 1923, Geology of the southern Ouachita Mountains of Oklahoma, Part 1: Stratigraphy, structure, and physiographic history: *Okla. Geol. Surv. Bull.* 32, 278 p.

Johnson, K. E., 1962, Paleocurrent study of Tesnus Formation, Marathon Basin, Texas: *J. Sediment. Petrol.*, v. 32, n. 4, p. 781–792.

Johnson, K. E., 1966, A depositional interpretation of the Stanley Group of the Ouachita Mountains, Oklahoma, in: *Field Conference of Flysch Facies and Structure of the Ouachita Mountains*: Kansas Geol. Soc., 29th Field Conf. Guidebook, p. 140–163.

Keith, Arthur, 1923, Outlines of Appalachian structure: *Geol. Soc. Am. Bull.*, v. 34, n. 2, p. 309–380.

Kellum, L. B., Imlay, R. W., and Kane, W. G., 1936, Evolution of the Coahuila peninsula, Mexico. Part 1: Relation of structure, stratigraphy, and igneous activity to an early continental margin: *Geol. Soc. Am. Bull.*, v. 47, n. 7, p. 969–1008.

King, E. R., 1959, Regional magnetic map of Florida: *Am. Assoc. Petr. Geol. Bull.*, v. 43, n. 12, p. 2844–2854.

King, P. B., 1930, Geology of the Glass Mountains, Part 1, Descriptive geology: Texas Univ. (B. Econ. Geol.) Bull. 3038, 167 p.

King, P. B., 1937, Geology of the Marathon region, Texas: U.S. Geol. Surv. Prof. Pap. 187, 148 p.

King, P. B., Hadley, J. B., Neuman, R. B., and Hamilton, Warren, 1958, Stratigraphy of the Ocoee Series, Great Smoky Mountains, Tennessee and North Carolina: *Geol. Soc. Am. Bull.*, v. 69, n. 8, p. 947–966.

King, R. E., Dunbar, C. O., Cloud, P. E., Jr., and Miller, A. K., 1944, Geology and paleontology of the Permian area northwest of Las Delicias, Coahuila, Mexico: Geol. Soc. Am. Spec. Pap. 52, 172 p.

Lopez-Ramos, Ernesto, 1969, Marine Paleozoic rocks of Mexico: *Am. Assoc. Petr. Geol. Bull.*, v. 53, n. 12, p. 2399–2417.

McBride, E. F., 1966, Sedimentary petrology and history of the Haymond Formation (Pennsylvanian), trans-Pecos Texas: Texas Univ. (Bur. Econ. Geol.) Rep. Invest. 57, 101 p.

McBride, E. F., 1970, Flysch sedimentation in the Marathon Region, Texas, in: *Flysch Sedimentology in North America*, Lajoie, J., ed.: Geol. Assoc. Can. Spec. Pap. 7, p. 67–83.

McBride, E. F., and Thomson, A. F., 1970, The Caballos Novaculite, Marathon Region, Texas: Geol. Soc. Am. Spec. Pap. 122, 129 p.

McKee, E. D., Oriel, S. S., Ketner, K. B., MacLachlan, M. E., Goldsmith, J. W., Mac-Lachlan, J. C., and Mudge, M. R., 1959, Paleotectonic maps of the Triassic System: U.S. Geol. Surv. Misc. Geol. Invest., Map I-300, 33 p.

Mellen, F. F., 1947, Black Warrior basin, Alabama and Mississippi: Am. Assoc. Petr. Geol. Bull., v. 31, n. 10, p. 1808–1816.

Milton, Charles, and Grasty, Robert, 1969, "Basement" rocks of Florida and Georgia: Am. Assoc. Petr. Geol. Bull., v. 53, n. 12, p. 2483–2493.

Misch, Peter, and Oles, K. F., 1957, Interpretation of Ouachita Mountains of Oklahoma as autochthonous folded belt; preliminary report: Am. Assoc. Petr. Geol. Bull., v. 41, n. 8, p. 1899–1905.

Miser, H. D., 1921, Llanoria, the Paleozoic land area in Louisiana and eastern Texas: Am. J. Sci., 5th ser., v. 2, n. 8, p. 61–89.

Miser, H. D., 1929, Structure of the Ouachita Mountains of Oklahoma and Arkansas: Okla. Geol. Surv. Bull. 50, 30 p.

Miser, H. D., 1943, Quartz veins in the Ouachita Mountains of Arkansas and Oklahoma; their relations to structure, metamorphism, and metalliferous deposits: Econ. Geol., v. 38, n. 2, p. 91–118.

Miser, H. D., and Purdue, A. H., 1929, Geology of the DeQueen and Caddo Gap quadrangles, Arkansas: U.S. Geol. Surv. Bull. 808, 195 p.

Morgan, B. A., 1972, Metamorphic map of the Appalachians, scale 1:2,500,000: U.S. Geol. Surv. Misc. Geol. Invest., Map I-724.

Morris, R. C., 1971, Stratigraphy and sedimentology of the Jackfork Group, Arkansas: Am. Assoc. Petr. Geol. Bull., v. 55, n. 3, p. 387–402.

Newell, N. D., 1957, Supposed Permian tillites in northern Mexico are submarine slide deposits: Geol. Soc. Am. Bull., v. 68, n. 11, p. 1569–1576.

Purdue, A. H., and Miser, H. D., 1923, Description of the Hot Springs district, Arkansas: U.S. Geol. Surv. Geol. Atlas, Folio 215.

Reed, J. C., Jr., and Bryant, Bruce, 1964, Evidence for strike-slip faulting along the Brevard zone in North Carolina: Geol. Soc. Am. Bull., v. 75, n. 12, p. 1177–1196.

Reinemund, J. A., and Danilchek, Walter, 1957, Preliminary geologic map of the Waldron quadrangle and adjacent areas, Scott County, Arkansas: U.S. Geol. Surv. Oil and Gas Invest. Map OM-192.

Repetski, J. E., and Ethington, R. L., 1973, Conodonts from graptolite facies in the Ouachita Mountains, Arkansas and Oklahoma [abstract]: Geol. Soc. Am. Abstr. with Programs, v. 5, n. 3 (South-Central Section 4th Ann. Mtg.), p. 277.

Rozendal, R. A., and Erskine, W. S., 1971, Deep test in Ouachita structural belt, central Texas: Am. Assoc. Petr. Geol. Bull., v. 55, n. 11, p. 2008–2017.

Schuchert, Charles, 1923, Site and nature of the North American geosynclines: Geol. Soc. Am. Bull., v. 34, n. 2, p. 151–229.

Scott, K. R., Hayes, W. E., and Fietz, R. P., 1961, Geology of the Eagle Mills Formation: Gulf Coast Assoc. Geol. Soc. Trans., v. 11, p. 1–14.

Stone, C. G., 1966, General geology of the eastern frontal Ouachita Mountains and southeastern Arkansas valley, Arkansas, in: Field Conference on Flysch Facies and Structure of the Ouachita Mountains: Kansas Geol. Soc. 29th Field Conf. Guidebook, p. 195–221.

Stone, C. G., and Sterling, P. J., 1962, New lithologic markers in Ordovician rocks, eastern Ouachita Mountains of Arkansas: Am. Assoc. Petr. Geol. Bull., v. 46, n. 3, p. 387–390.

Taff, J. A., 1902, Description of the Atoka quadrangle, Indian Territory: U.S. Geol. Surv. Geol. Atlas, Folio 79.

Taylor, P. T., Zietz, Isidore, and Dennis, L. S., 1968, Geologic implications of aeromagnetic data for the eastern continental margin of the United States: *Geophysics*, v. 33, n. 5, p. 755–780.

Thomas, W. A., 1972, Regional Paleozoic stratigraphy in Mississippi between the Ouachita and Appalachian Mountains: *Am. Assoc. Petr. Geol. Bull.*, v. 56, n. 1, p. 81–106.

Thomas, W. A., 1973, Southwestern Appalachian structural system beneath the Gulf Coastal Plain: *Am. J. Sci.*, v. 273, Byron N. Cooper Memorial Number, p. 372–390.

Thomson, A. F., and Thomasson, M. R., 1969, Shallow to deep water facies development in the Dimple Limestone (lower Pennsylvanian), Marathon Region, Texas: Soc. Econ. Paleontol. Mineral. Spec. Pap. 14, p. 57–78.

Tomlinson, C. W., 1959, Ouachita problems, in: *The Geology of the Ouachita Mountains; a Symposium*, Cline, L. M., *et al.*, ed.: Dallas Geol. Soc. and Ardmore Geol. Soc., p. 1–19.

Tomlinson, C. W., and McBee, William, Jr., 1959, Pennsylvanian sediments and orogenies of Ardmore district, Oklahoma, in: *Petroleum Geology of Southern Oklahoma*, vol. 2: Am. Assoc. Petr. Geol. and Ardmore Geol. Soc., p. 3–52.

Udden, J. A., 1917, Notes on the geology of the Glass Mountains: Texas Univ. (Bur. Econ. Geol.) Bull. 1753, p. 3–59.

Van der Gracht, W. A. J. M. van Waterschoot, 1931a, The Permo-Carboniferous orogeny in south-central United States: *Am. Assoc. Petr. Geol. Bull.*, v. 15, n. 8, p. 991–1057.

Van der Gracht, W. A. J. M. van Waterschoot, 1931b, The Permo-Carboniferous orogeny in the south-central United States: Verhandelingen der Koninklijke Akademie van Wetenschappen te Amsterdam, Afdeeling Natuur Kunde (Tweede Sectie), Deel 27, n. 3, 170 p.

Vernon, R. C., 1971, Possible future petroleum potential of pre-Jurassic, west Gulf basin, in: *Future Petroleum Provinces of the United States; Their Geology and Potential*, Cram, I. A., ed.: *Am. Assoc. Petr. Geol. Bull.*, Mem. 15, v. 2, p. 954–979.

Viele, G. W., 1966, The regional structure of the Ouachita Mountains of Arkansas; a hypothesis, in: *Field Conference on Flysch Facies and Structure of the Ouachita Mountains*: Kansas Geol. Soc., 29th Field Conf. Guidebook, p. 245–278.

Viele, G. W., 1973, Tectonic history of eastern Ouachita Mountains, Arkansas [abstract]: Geol. Soc. Am. Abstr. with Programs, v. 5, n. 3 (South-Central Sec., 7th Ann. Mtg.), p. 284–285.

Wilson, J. L., 1954a, Late Cambrian and early Ordovician trilobites from the Marathon uplift, Texas: *J. Paleontol.*, v. 28, n. 3, p. 249–285.

Wilson, J. L., 1954b, Ordovician stratigraphy of the Marathon folded belt, Texas: *Am. Assoc. Petr. Geol. Bull.*, v. 38, n. 12, p. 2455–2475.

Wilson, J. T., 1966, Did the Atlantic close and then reopen?: *Nature*, v. 211, n. 5050, p. 676–681.

Woollard, G. P., and Monges Caldera, Julio, 1956, Gravidad, geologia regional y estructura cortical en Mexico: *Univ. Nac. Auton. Mex., Inst. Geofis. An.*, v. 2, p. 60–112.

Woollard, G. P., and Joesting, H. R., 1964, Bouguer gravity anomaly map of the United States, by the American Geophysical Union's Special Committee for the Geophysical and Geological Study of the Continents, and the U.S. Geological Survey: U.S. Geol. Surv., scale 1 : 2,500,000.

Young, L. M., 1970, Early Ordovician sedimentary history of the Marathon geosyncline, trans-Pecos Texas: *Am. Assoc. Petr. Geol. Bull.*, v. 54, n. 12, p. 2303–2316.

Chapter 6

TECTONIC EVOLUTION OF THE SOUTHERN CONTINENTAL MARGIN OF NORTH AMERICA FROM A PALEOZOIC PERSPECTIVE*

James Helwig

Department of Geology
Case Western Reserve University
Cleveland, Ohio

I. INTRODUCTION

The objective of this paper is to discuss tectonic hypotheses for the Paleozoic evolution of the southern continental margin of North America building on the geological synthesis of P. B. King (this volume) to which the reader is directed for many references. At the outset, it must be acknowledged that our present grasp of Paleozoic paleogeography and tectonics is generally insufficient to permit unambiguous tectonic interpretation, plate or otherwise, of any Paleozoic mountain system. This is especially true for the southern Appalachian, Ouachita, and Huastecan systems which often are poorly exposed. It is nevertheless worthwhile to review major aspects of paleogeographic and tectonic history, however poorly known, in order to formulate simple models and actualistic plate-tectonic analogues that can be of value in understanding the geologic history of the region.

* Contribution No. 106, Department of Geology, Case Western Reserve University, Cleveland, Ohio.

II. PALEOGEOGRAPHIC EVOLUTION

The Paleozoic regional stratigraphy and structure of the northern margin of the Gulf of Mexico establish a tectonic and paleogeographic framework (King, this volume). The fundamental elements comprise the North American craton, an adjacent Palaeozoic orogenic belt with miogeosyncline and eugeosyncline (including the southern Appalachian, Ouachita, and Huastecan segments) and, finally, a hypothetical ancestral land or sea occupying the present Gulf of Mexico.

During retreat and advance of Paleozoic seas, the craton served alternately as a sediment source and a site of shelf sedimentation. The craton was a minor sediment source for the Appalachian miogeosyncline but evidently constituted a more significant source of sediment for the Ouachita system. Although much of the detritus in the Carboniferous clastic wedge of the Ouachita Mountains was derived from a southern source, part probably came from the north.

Prior to the Taconic orogeny, the Appalachian miogeosyncline was a stable continental edge accumulating thick carbonates and a thick basal cratonic-derived clastic sequence to form a continental terrace sedimentary prism. Subsequently, it became an elongate basin and sediment trap for detritus eroded from the internal zones of the fold belt, accumulating thick clastic wedges during episodes of deformation. In the southernmost Appalachians, the only thick clastic wedge is Pennsylvanian, suggesting a transition into the Ouachita belt. By late Pennsylvanian time the entire Appalachian miogeosyncline was an alluvial plain destined to remain above sea level.

The paleogeography of the eugeosynclinal part of the southernmost Appalachians is poorly known. This region includes the Intermediate and Piedmont provinces of King (this volume) where the age and structural relations of metamorphic rock units remain controversial. It certainly included complex tectonic ridges and furrows, with superposed magmatic arcs and regional high T/P metamorphic belts, and was subject to complex deformation including the formation of nappes. The lithologies and structures indicate that it was the site of an active continental margin bordered by a subduction zone and at least a small ocean basin, particularly in earlier Paleozoic time. However, the belt was entirely emergent by Permian time, as plutonism, metamorphism, and deformation drew to a close. An abrupt southern limit of tectonism is proven by flat-lying Ordovician, Silurian, and Devonian shelf sediments in the Florida element, which could be a transposed fragment of Africa welded to North America (King, this volume).

The frontal belt of the Ouachita system is an allochthonous pericratonic trough which accumulated thin sediments in pre-Carboniferous time and very thick flysch during the Carboniferous. The presence of turbidites and siliceous sediments showing features of deep ocean-floor deposition through all but the youngest parts of this Paleozoic sequence suggests deposition beyond the edge of the Paleozoic continent. Therefore, the southern margin of North America seems characterized by a longer period of stability than the Appalachian margin and apparently behaved as a passive continental margin through early and medial Paleozoic time (Thomas, 1972). However, both paleocurrent/petrographic data from the flysch and subsurface occurrences of igneous and metamorphic rocks in the buried internal zone of the Ouachita system demonstrate that a southern sialic tectonic welt existed. The flysch and its late Carboniferous to Permian deformation attest to the activization of the continental margin at that time, presumably by northward thrusting of this enigmatic internal tectonic welt toward the craton. Subsurface data (Flawn *et al.*, 1961) are limited to a narrow belt, but geophysical data (Shurbet, 1968; Hales *et al.*, 1970) suggest that this sialic, mostly metasedimentary internal zone may be several hundred kilometers wide, comparable in width to the Appalachian eugeosynclinal belt. Absolute-age studies show that rocks now in the northern part of the zone, accessible to the drill, underwent Silurian through Permian igneous and metamorphic activity (Denison *et al.*, 1969; Rozendal and Erskine, 1971).

Sporadic inliers south of the Marathon region, west Texas, suggest a continuity of the Appalachian–Ouachita mobile belt into Mexico. The stratigraphy of these occurrences shows similarities with both the Appalachian and Ouachita segments (King, this volume) and exhibits Permian volcanics and younger deformation than the more northerly segments.

Paleozoic paleogeography in the region south of available surface and subsurface data, including the Gulf of Mexico, is entirely speculative. Traditionally the Gulf has been considered the site of a sunken land mass, Llanoria (Miser, 1921). Although paleomagnetism indicates Africa collided with North America during Appalachian orogeny, no data exist to indicate similar collision of a continental block with the Ouachita segment, which may explain the poor development of molasse facies.* Marine geological and geophysical studies in the Gulf of Mexico (Ewing *et al.*, 1970; Martin and Case, this volume) show that the present structure is essentially oceanic crust overlain by about 6 to 12 km of sediment, about one-half of which must be pre-Jurassic, suggesting a Paleozoic age for the oceanic crust (Wilhelm and Ewing, 1972). This inference is compatible with the geology on land which shows termination of orogenic activity in the Ouachita system as early as pre-Morehouse Formation (late Pennsylvanian) time. Siltstones dredged from near diapirs in the Sigsbee Knolls in the central Gulf have yielded glauconite that is K–Ar dated as Pennsylvanian (Pequegnat *et al.*, 1971), but it is questionable whether the minerals dated are authigenic and even whether the published x-ray patterns are diagnostic of glauconite. In summary, the constraints of land and marine geology are insufficient to demonstrate existence of either continental or oceanic crust in the Gulf of Mexico region in late Paleozoic time.

---

* See note added in proof, p. 255.

III. TECTONIC EVOLUTION

The Paleozoic tectonic evolution of the southern margin of North America occurred in the context of a pre-drift continental arrangement. The most probable type of pre-drift arrangement shows good fit of Atlantic continental margins and geology (Bullard *et al.*, 1965) but problematical overlap of North and South America. This overlap is a major geotectonic problem with consequences for the late Paleozoic tectonic (as well as subsequent) history, a perspective perhaps not sufficiently considered. Two types of hypotheses seem to provide reasonable solutions of this problem (Fig. 1).

One hypothesis places the southern tip of North America in the Gulf of Mexico by rotation of the pre-Mesozoic basement blocks of Yucatan and nuclear Central America (Carey, 1958; Freeland and Dietz, 1971) to produce a late Paleozoic continental mass in the Gulf region. The second hypothesis invokes an opposite rotation of the embarrassing overlapped region of the Bullard fit to place the southern tip of North America off the west coast of South America (Hamilton, 1966; Walper and Rowett, 1972). Following this latter hypothesis, one may alternatively consider the Gulf of Mexico to be occupied by the northwest shoulder of South America (Walper and Rowett, 1972) or to be a small ocean basin (Fig. 2) in late Paleozoic time.

To evaluate these hypotheses, we must go back even farther in geologic time. Continental displacements during all of Paleozoic time, inferred from paleomagnetism (Smith *et al.*, 1973), must also be related to the tectonic evolution of the southern continental margin of North America. It is perhaps feasible and certainly desirable to consider how the tectonic evolution as a whole may be interpreted within the framework of plate tectonic theory and analogous Meso-Cenozoic models.

Using the foregoing guidelines and the paleogeographic constraints previously discussed, I present one simple possible plate arrangement for Carboniferous and Permian time (Fig. 2). This reconstruction provides one plausible framework for understanding the tectonic and paleogeographic evolution of the region. It must be emphasized that this reconstruction is speculation and not a hypothesis because constraints on interpretation are few.

The relationship of the Appalachian and Ouachita belts is a question of major interest. The Appalachian belt comprises many tectonic elements assembled at an active continental margin. The early Paleozoic seems dominated by magmatic arcs indicative of oceanic underthrusting, whereas both wrench faulting ("California type" orogenesis) and continental collision ("Himalayan type" orogenesis) characterize the later Paleozoic. Geologic evidence, particularly in the northern Appalachians (e.g., Webb, 1969), shows dominant dextral strike-slip faulting parallel to the strike of the mountain belt; likewise, paleomagnetic data are interpreted to show dextral sliding of North America

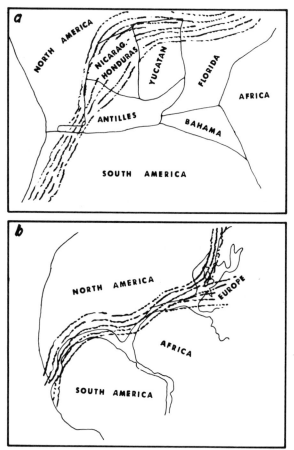

Fig. 1. Two major hypotheses for the Late Paleozoic configuration of the Americas showing trend of Appalachian–Ouachita–Huastecan orogenic belts. (a) Reconstruction of Carey, 1958 (simplified), involves 180° to 270° counterclockwise rotations of Yucatan and Nicaragua–Honduras blocks to fill Gulf of Mexico. Problematical aspects of this reconstruction and similar ones (Freeland and Dietz, 1971) include severing of apparently continuous belts in Mexico and nuclear Central America and the lack of similarity of the geology of blocks juxtaposed with the Ouachita system. (b) Reconstruction of Walper and Rowett, 1972, involves clockwise rotation of southern tip of North America to fit opposite northwest Andes. Problematical aspects of this reconstruction include destruction of fit of the Atlantic continents (Le Pichon and Fox, 1971; Le Pichon and Hayes, 1971), continental overlap of Florida, and geological misfit of northeast Venezuela with the Ouachita system.

past northwest South America and northwest Africa during Devonian and Carboniferous time (Smith *et al.*, 1973).

The Brevard Zone of the southern Appalachians could be such a major dextral fault (King, this volume) and has been interpreted as such (Reed and Bryant, 1964). However, the Brevard has also been interpreted as the root zone for nappes and thrusts (Burchfiel and Livingston, 1967; Hatcher, 1971) or as a left lateral strike-slip fault (Bryant and Reed, 1971). There is no conclusive evidence to support a unique interpretation of the Brevard Zone, and recent mapping in western Georgia shows at least one metamorphic rock unit crossing the zone without apparent offset (Hurst, 1973).

The tectonics of the Appalachians, as a whole, and the Brevard Zone, in particular, are of concern here because if both the Ouachita and Appalachian systems evolved in response to motion of a single adjacent plate, then the different strike of the Ouachita system necessitates significantly different but related movements contemporaneous with Appalachian movements. Specifically, due to the geometry of plates on a sphere (McKenzie and Parker, 1967), plate motions producing crustal compression and consumption in the Appalachians would require dextral faulting and only minor convergent motion on Ouachita trends. Compression and consumption in the Ouachita system would necessitate sinistral faulting on Appalachian trends (Fig. 3).

Since the Appalachians apparently record mostly convergent or (questionable) dextral motions, the reconstruction (Fig. 2) interprets the Ouachita

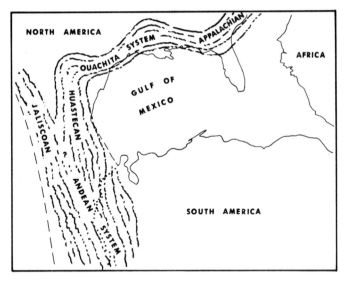

Fig. 2. Proposed Late Paleozoic reconstruction of the Americas. This reconstruction retains a modified Bullard fit which tends to align the Huastecan belt with the northwest Andes as implied by Le Pichon and Fox (1971). The Ouachita reentrant served to preserve a remnant of Paleozoic ocean crust as the Gulf of Mexico. Yucatan and nuclear Central America must be placed within the northwest Andes in this reconstruction, following Walper and Rowett (see Fig. 1b). It is assumed that the northernmost blocks of the Andes have been displaced into the southern Caribbean Meso-Cenozoic orogenic belt (Shagam, this volume). One problematical aspect of this reconstruction is the evidence of widespread Triassic rifting in the circum-Gulf region.

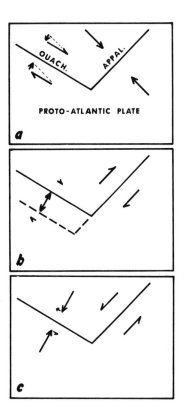

Fig. 3. Diagram illustrating interdependent nature of movements in Appalachian and Ouachita belts. It is assumed that the orogenic belts are responding to the motions of a single adjacent proto-Atlantic plate situated south and east of the North American craton. (a) Convergence on Appalachian trend causes dominantly right-lateral transcurrent motion on Ouachita trend; (b) right-lateral motion on Appalachian trend causes predominant extension on Ouachita trend with minor right-lateral movement; (c) left-lateral motion on Appalachian trend causes convergence on Ouachita trend and minor left-lateral motion. This model would have application to the salient-reentrant structure of the entire Ouachita system (King, this volume). However, it is noteworthy that unequivocal large transcurrent faults remain to be demonstrated in either mountain belt (Ham *et al.*, 1964; Hurst, 1973).

system and its southern border as a plate boundary with dominant dextral transform faulting and subordinate tensional or compressional components reflecting dextral and convergent motions, respectively, on the Appalachian system. This reconstruction is the simplest plate model compatible with (although not required by!) the limited data and provides plausible explanations of contrasts between the Appalachian and Ouachita systems and other major geological features as follows:

1. *The paucity of volcanism and plutonism in the Ouachita system as compared to the Appalachians* (*and possibly the Huastecan belt also*) *due to a very limited amount of subduction.* In this respect, the Ouachita system may be analogous to the Andaman arc as described by Rodolfo (1969; Fig. 4).* However, note that the lack of igneous rocks may also be explained as due to lack of exposure or fragmentation of the interior of the Ouachita system.

2. *The historically persistent sharp break between shelf and deep water on the southern continental margin of North America due to absence of persistent compressive tectonics.* Abundant boulder beds (Tables 1 and 2 of King, this

* See note added in proof, p. 255.

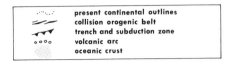

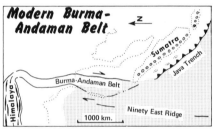

Fig. 4. A comparison of the hypothetical reconstructed Ouachita system (Fig. 2) with the active Burma–Andaman system (after Rodolfo, 1969). If the Appalachian "Himalayan" type collision belt were transposed with the Huastecan volcanic arc (Las Delicias volcanics, King, this volume), the geometric analogy would be complete.

volume) attest to a persistent faulted continental slope. Such normal faulting is compatible with pre-medial Carboniferous dextral faulting in the Appalachians (Fig. 3).

3. *The abrupt medial Carboniferous compression and emergence of the internal zone of the Ouachita system due to a change in plate motions which terminated large dextral motions along Appalachian trends and joined Africa to North America.* The Ouachita geosyncline in its pre-Stanley "starved basin" phase may be interpreted as a marginal basin of extensional origin, analogous to the modern Andaman basin (Fig. 4). The subsequent deposition of the flysch and its emplacement as nappes on the craton edge could hence reflect closing of this hypothetical marginal basin in medial through late Carboniferous time. Contemporaneous sinistral and convergent movements along Appalachian trends would be anticipated. (However, only convergence producing crustal uplift is clearly evident from consideration of the great Pennsylvanian clastic wedges of the Southern Appalachians.)

4. *The problems of flysch provenance and its paleotectonic setting (Morris, 1971; 1973) attributed to deposition in a marginal basin.* Northward transport as nappes and subsequent folding with both north and south overturning may

be a consequence of squeezing-out of an original oceanic basement beneath the flysch and older sediments (basement perhaps represented by isolated altered serpentinites in the Ouachita Mountains), as in certain parts of the Alps. This setting is in contrast to that of Appalachian flysch sequences which were deposited over the craton edge.

5. *A Paleozoic age for the Gulf of Mexico due to preservation of Proto-Atlantic ocean crust in the continental reentrant bounded by the Ouachita system* (Fig. 2). Facts which tend to support this explanation include the thick Jurassic and older (?) sediments in the Gulf which show thickening to the north (Ewing, Edgar, and Antoine, 1970; Wilhelm and Ewing, 1972; Martin and Case, this volume); the exceptionally thick quasi-oceanic crust of the Gulf, which may be the oldest oceanic crust on the earth; and the presence of relatively undeformed late Pennsylvanian strata in the subsurface south of the Ouachita Mountains (King, this volume). If the Gulf crust is Paleozoic, widespread Jurassic evaporites could be interpreted as either deep basinal deposits or shallow water deposits depending on their geographic position, possible drying up of the Gulf, and possible elevation of old oceanic crust by thermal expansion during renewed Mesozoic sea-floor spreading (Ewing, Edgar, and Antoine, 1970).

6. *The geological parallels of the Northern Andes and the Appalachians* (*Shagam, this volume*) *due to original proximity and continuity.*

IV. DISCUSSION

At this point it is appropriate to pull the reins on speculation and emphasize that the reconstruction of Fig. 2 and the Andaman Arc analogue (Fig. 4) have consequences that are perhaps less compatible with certain data than other reconstructions.

The age of the crust in the Gulf of Mexico is hypothesized here (Fig. 2) to be pre-Mesozoic. The reconstructions (Fig. 1) of Freeland and Dietz (1971) or Walper and Rowett (1972) require a Triassic–Jurassic age. There is much geologic data compatible with these latter views and involving an early Mesozoic rifting origin for the Gulf of Mexico. There is abundant evidence of Triassic graben formation and basaltic effusions along the entire Paleozoic tectonic belt from Newfoundland to Honduras and including the circum-Gulf region (Durham and Murray, 1967). All drift models and sea-floor reconstructions (Ladd, 1973) require North and South America to move apart during the Mesozoic, producing an extensional tectonic regime. The sedimentation and distribution of Jurassic salt is perhaps easier to explain in terms of Triassic–Jurassic rifting and sea-floor generation with consequent shallow restricted basins which could later subside with the cooling ocean crust. The strikingly similar Mesozoic–Cenozoic history of the Florida and Yucatan–Campeche (Emery and Uchupi, 1972) platforms could readily be explained by a rift origin.

Large post-Permian counterclockwise rotation of the southern tip of North America is required by both the writer's reconstruction and that of Walper and Rowett, but clockwise rotation is required by the Carey or Freeland and Dietz reconstructions. Paleomagnetic studies of Mesozoic and older rocks in the proposed rotated blocks are urgently needed to test these speculations. The late Paleozoic configuration of orogenic belts is substantially different if the two schemes of rotation are compared. The pattern of Paleozoic orogenic belts makes little sense in the clockwise rotation scheme of Carey or of Freeland and Dietz. The counterclockwise rotation scheme, on the other hand, considers a continuity and identity of the Appalachian, Ouachita, Huastecan, Central American, and Andean Paleozoic mountain belts as a fundamental basis of reconstruction (Walper and Rowett, 1972). Continuity of the belts implies that the Jaliscoan and Huastecan belts of Mexico converge southward into the Andean belt. It is possible to provide for this tectonic continuity and retain either continental or oceanic structure in the Gulf of Mexico region in late Paleozoic time (Figs. 1 and 2).

V. CONCLUSIONS

The tectonic evolution of the Paleozoic mountain chains surrounding the Gulf of Mexico–Caribbean region is far more complex than our perception of it today or in the forseeable future. There is simply too much geology hidden from observation.

This paper has presented three geologic hypotheses (speculations, really) that offer reasonable explanations for the tectonic evolution of the southern continental margin of North America (Figs. 1 and 2). The first hypothesis (Carey, 1958; Freeland and Dietz, 1971) invokes the rifting and clockwise rotation of the continental crustal blocks of Yucatan and nuclear Central America away from the southern margin of North America in early Mesozoic time to open up the Gulf of Mexico. Since these blocks apparently comprise mostly Paleozoic orogenic belts, the reconstruction requires the existence of a problematical huge orogenic plexus south of the Ouachita belt in late Paleozoic time. The second hypothesis (Walper and Rowett, 1972) explains the opening of the Gulf of Mexico by separation of northwestern South America from the Ouachita belt in Triassic time and requires an intimate geologic correspondence of late Paleozoic history in the southern Appalachian, Ouachita, Huastecan, Central American, and Andean belts. A third hypothesis proposes a Paleozoic age for the Gulf of Mexico (Fig. 2) but in many respects is similar to the Walper and Rowett reconstruction. Important differences are the interpretation of the Gulf as a remnant of the Proto-Atlantic Ocean and interpretation of the Ouachita system as the site of a Paleozoic stable continental margin flanked by a marginal sea and arc but no continental mass.

None of these hypotheses can be tested at present. However, they are considered more likely than hypotheses involving oceanization and/or static positioning of the various blocks through time because of the contrary evidence of relative movements of North and South America (Smith *et al.*, 1973; Ladd, 1973).

From a Paleozoic perspective, and considering the overlap problem of the Bullard fit, it appears that a much closer connection of Paleozoic mountain belts and events from the Appalachians to the Andes is reasonable. Future research should concentrate on the Paleozoic geology of the northwest Andes and of the region between the Americas. The reconstructions already discussed may then be tested and improved, and the concepts of plate tectonics may be more critically applied to reconstructions. Reconstructions based upon late Paleozoic geology can constitute independent evidence of the validity of tectonic reconstructions showing the Meso-Cenozoic evolution of the Caribbean Sea and island arcs.

Sharpening our models of late Paleozoic geography and tectonics by further work in the exposed Ouachita system is unlikely. Careful geochronological and paleomagnetic study of pre-Mesozoic rocks in Central America and the northwest Andes will prove more fruitful. Preliminary paleomagnetic evidence of significant post-Jurassic tectonic rotations in the Caribbean region is already available (MacDonald and Opdyke, 1972). Refinement of paleogeographic and tectonic history through Mesozoic–Cenozoic time is also now liable to control provided by knowledge of the relative positions of the two Americas established through study of Atlantic sea-floor magnetic anomalies (Ladd, 1973).

Two particularly refractory problems need to be solved. One is the nature of the internal zone of the Ouachita system.* Is it a continental fragment, or an island arc, or perhaps originally part of South America? The second is the age and origin of the Gulf of Mexico, a topic that has been speculatively discussed here and in other papers in this volume. Unambiguous solutions to these problems can only be provided by deep holes drilled through the floor and margins of the Gulf of Mexico.

ACKNOWLEDGMENTS

I am grateful to Philip B. King for correspondence and discussions, and I have drawn freely from data and references presented by him in this volume.

* See note added in proof, p. 255.

The interpretations and speculations presented are mostly taken from the
literature, and I hope that I have adequately acknowledged their sources.
New interpretations discussed in this paper are my responsibility but have
evolved through continuing stimulating discussion with P. B. King and Stephen
Franks.

REFERENCES

Bryant, B., and Reed, J. C., Jr., 1971, Geology of the Grandfather Mountain Window and
    Vicinity, North Carolina and Tennessee: U.S. Geol. Surv. Prof. Pap. 615, 190 p.
Bullard, E. C., Everett, J. E., and Smith, A. G., 1965, The fit of the continents around the
    Atlantic, in: A symposium on continental drift: *Philos. Trans. Roy. Soc. London, Ser. A*,
    v. 258, p. 41–51.
Burchfiel, B. C., and Livingston, J. L., 1967, Brevard zone compared to Alpine root zones:
    Am. J. Sci., v. 265, p. 241–256.
Carey, S. W., 1958, The tectonic approach to continental drift, in: *Continental Drift—A
    symposium*, Carey, S. W., convener: Univ. Tasmania, Hobart, p. 177–355.
Denison, R. E., Kenny, G. S., Burke, W. H., Jr., and Hetherington, E. A., Jr., 1969,
    Isotopic ages of igneous and metamorphic boulders from the Haymond Formation
    (Pennsylvanian), Marathon Basin, Texas, and their significance: *Geol. Soc. Am. Bull.*,
    v. 80, p. 245–256.
Durham, C. O., Jr., and Murray, G. E., 1967, Tectonism of the Atlantic and Gulf Coastal
    Plains: *Am. J. Sci.*, v. 265, p. 428–441.
Emery, K. O., and Uchupi, Elazar, 1972, Western North Atlantic Ocean: Am. Assoc. Petr.
    Geol. Mem. 17, 532 p.
Ewing, J., Edgar, N. T., and Antoine, J. W., 1970, Structure of the Gulf of Mexico and
    Caribbean Sea, in: *The Sea*, Vol. 4, Pt. 2, A. E. Maxwell, ed.: Wiley–Interscience, New
    York, p. 321–358.
Flawn, P. T., Goldstein, A., Jr., King, P. B., and Weaver, C. E., 1961, The Ouachita system:
    Univ. Texas (Bur. Econ. Geol.) Publ. 6120, 401 p.
Freeland, G. L., and Dietz, R. S., 1971, Plate tectonic evolution of Caribbean–Gulf of Mexico
    region: *Nature*, v. 232, p. 20–23.
Hales, A. L., Helsley, C. E., and Nation, J. B., 1970, Crustal structure study on Gulf Coast
    of Texas: *Am. Assoc. Petr. Geol. Bull.*, v. 54, p. 2040–2057.
Ham, W. E., Denison, R. E., and Merritt, C. A., 1964, Basement rocks and structural evolu-
    tion of southern Oklahoma: *Okla. Geol. Surv., Bull.* 95, 302 p.
Hamilton, W., 1966, Formation of the Scotia and Caribbean Arcs: Geol. Surv. Can. Pap.
    66-15, p. 178–187.
Hatcher, R. D., Jr., 1971, Structural, petrologic, and stratigraphic evidence favoring a thrust
    solution to the Brevard problem: *Am. J. Sci.*, v. 273, p. 643–670.
King, P. B., 1969, The tectonics of North America—A discussion to accompany the tectonic
    map of North America: U.S. Geol. Surv. Prof. Pap. 628, scale 1:5,000,000, 94 p.
Ladd, J. W., 1973, Relative motion between North and South America and the evolution of
    the Caribbean: Geol. Soc. Am., Abstr. with Programs, v. 5, n. 7, p. 705.
Le Pichon, X., and Fox, P. J., 1971, Marginal offsets, fracture zones, and the early opening
    of the North Atlantic: *J. Geophys. Res.*, v. 76, p. 6294–6308.
Le Pichon, X., and Hayes, D. E., 1971, Marginal offsets, fracture zones, and the early opening
    of the South Atlantic: *J. Geophys. Res.*, v. 76, p. 6283–6293.

MacDonald, W. D., and Opdyke, N. D., 1972, Tectonic rotations suggested by paleomagnetic results from northern Colombia, South America: *J. Geophys. Res.*, v. 77, p. 5720–5730.

McKenzie, D. P., and Parker, R. L., 1967, The North Pacific: an example of tectonics on a sphere: *Nature*, v. 216, n. 1267, p. 1276–1280.

Miser, H. D., 1921, Llanoria, the Paleozoic land area in Louisiana and eastern Texas: *Am. J. Sci.*, 5th Ser., v. 2, p. 61–89.

Morris, R. C., 1971, Stratigraphy and sedimentology of the Jackfork Group, Arkansas: *Am. Assoc. Petr. Geol. Bull.*, v. 55, p. 387–402.

Morris, R. C., 1973, Petrology of Stanley–Jackfork Sandstones, Ouachita Mountains, Arkansas: Geol. Soc. Am., Abstr. with Programs, v. 5, n. 3, p. 274.

Pequegnant, W. E., Bryant, W. R., and Harris, J. E., 1971, Carboniferous sediments from Sigsbee Knolls, Gulf of Mexico: *Am. Assoc. Petr. Geol. Bull.*, v. 55, p. 116–123.

Reed, J. C., Jr., and Bryant, B., 1964, Evidence of strike-slip faulting along the Brevard zone in North Carolina: *Geol. Soc. Am. Bull.*, v. 75, p. 1177–1196.

Rodolfo, K. S., 1969, Bathymetry and marine geology of the Andaman Basin, and tectonic implications for southeast Asia: *Geol. Soc. Am. Bull.*, v. 80, p. 1203–1230.

Rozendal, R. A., and Erskine, W. S., 1971, Deep test in Ouachita structural belt of central Texas: *Am. Assoc. Petr. Geol. Bull.*, v. 55, p. 2008–2017.

Shurbet, D. H., 1968, Upper mantle structure beneath the margin of the Gulf of Mexico: *Geol. Soc. Am. Bull.*, v. 79, p. 1647–1650.

Smith, A. G., Briden, J. C., and Drewry, G. E., 1973, Phanerozoic world maps, in: Organisms and continents through time: Special Papers in Palaeontology, n. 12, Palaeont. Assoc., London, p. 1–42.

Thomas, W. A., 1972, Regional Paleozoic stratigraphy in Mississippi between Ouachita and Appalachian Mountains: *Am. Assoc. Petr. Geol. Bull.*, v. 56, p. 81–106.

Walper, J. L., and Rowett, C. L., 1972, Plate tectonics and the origin of the Caribbean Sea and the Gulf of Mexico: *Trans. Gulf Coast Assoc. Geol. Soc.*, v. 22, p. 105–116.

Webb, G. W., 1969, Paleozoic wrench faults in Canadian Appalachians, in: North Atlantic—geology and continental drift: Am. Assoc. Petr. Geol. Mem. 12, p. 754–786.

Wilhelm, O., and Ewing, M., 1972, Geology and history of the Gulf of Mexico: *Geol. Soc. Am. Bull.*, v. 83, p. 575–600.

NOTE ADDED IN PROOF

Since the completion of this paper, there have been several important new contributions to late Paleozoic reconstructions:

Graham, S. A., Dickinson, W. R., and Ingersoll, R. V., 1975, Himalayan–Bengal model for flysch dispersal in the Appalachian–Ouachita system: *Geol. Soc. Am. Bull.*, v. 86, p. 273–286.

Morris, R. C., 1974, Sedimentary and tectonic history of Ouachita Mountains, in: Dickinson, W. R., ed., *Tectonics and Sedimentation*: Soc. Econ. Paleontologists and Mineralogists Spec. Publ. 22, p. 120–142.

Van der Voo, R., and French, R. B., 1974, Apparent polar wandering for the Atlantic-bordering continents: late Carboniferous to Eocene: *Earth Sci. Rev.*, v. 10, p. 99–119.

Woods, R. D., and Addington, J. W., 1973, Pre-Jurassic geologic framework, northern Gulf Basin: *Trans. Gulf Coast Assoc. Geol. Soc.*, v. 23, p. 92–108.

Chapter 7

GEOLOGICAL SUMMARY OF THE YUCATAN PENINSULA*

E. Lopez Ramos

Petróleos Mexicanos
Mexico

I. INTRODUCTION

This paper presents an integrated, though generalized, study of the Geology of the Yucatán Peninsula based upon what the writer believes to be the most adequate available information. The study covers the Yucatan Shelf from 16°–22° north and from 86°–91° west, including the eastern parts of the states of Yucatan and Campeche, the territory of Quintana Roo, the northern part of Guatemala, and the northeastern part of British Honduras (Belice) (Fig. 1). This is an area of approximately 300,000 km². On the north and northeast the continental slope at the edge of Campeche bank forms a natural boundary for the region. To the east the continental slope also bounds the area, but there descends into the Caribbean Sea. The southern margin is formed by the Arco de la Libertad (Fig. 2), a positive feature including the Guatemalan Petén and the Maya Mountains of British Honduras. The western boundary is formed by the coastal plains of the Gulf of Mexico in southwestern Campeche and eastern Tabasco.

---

* A fuller version of this paper appears in Spanish in the *Boletin Asociación Mexicana de Geólogos Petroleros*, v. XXV, no. 1–3, 1973.

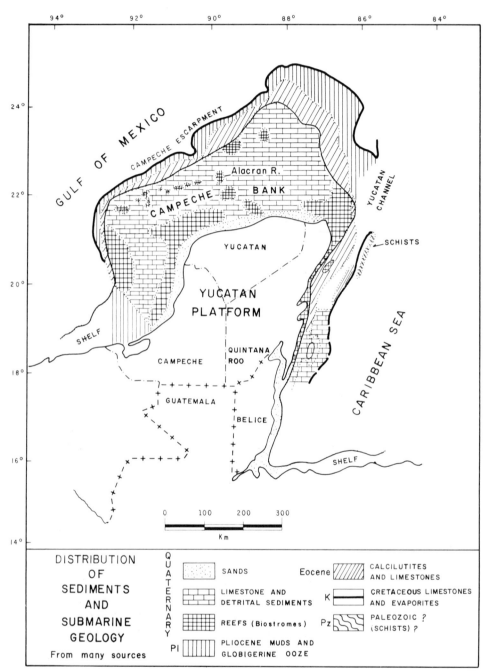

Fig. 1. General geographic relationships of the Yucatan Peninsula and general surface geology of the submerged portion of the Yucatan Platform.

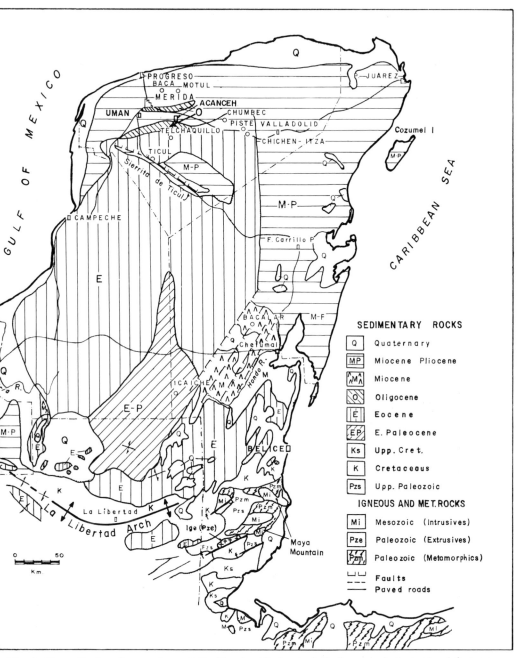

Fig. 2. Generalized surface geology of the Yucatan Peninsula including principal structural features.

Throughout the Yucatan region the older portion of the geological column is known only from subsurface work, and even the younger rocks are imperfectly known from geological work on the surface. However, surface studies began early with the work of Sapper (1896) but proceeded slowly and have been less revealing than one might hope. This is because of the heavy vegetative cover, poor access, lack of surface drainage and, therefore, exposures in valley walls and water courses, the generally flat-lying attitude of

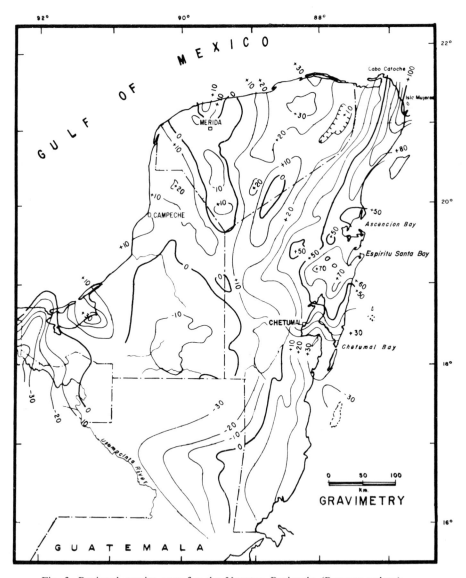

Fig. 3. Regional gravity map for the Yucatan Peninsula (Bouguer values).

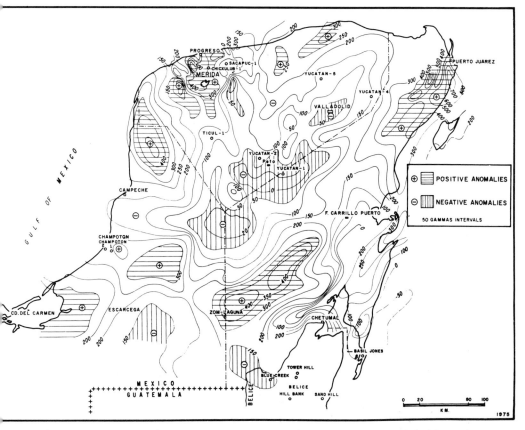

Fig. 4. Magnetic map of the Yucatan Peninsula showing principal regions of positive and negative anomalies.

the beds in an area of relatively low relief, and a heavy, obscuring cover of caliche over the dominantly carbonate exposures.

Tercier (1932) compiled the then existing information into a useful summary which has provided a foundation for subsequent work. A Pemex geologist M. Alvarez (1954) carried out a reconnaissance study of the part of Quintana Roo lying west of Rio Hondo and recognized fossiliferous marine Miocene of shallow water aspect which he designated the Rio Dulce formation. A geological survey of the Sierrita de Ticul (Fig. 2) was carried out under the direction of R. Gutierrez Gill for Petróleos Mexicanos in 1953. In 1955 Pemex geologists, together with R. Robles Ramos, recognized that caliche-covered Eocene limestones existed between Valladolid and Mérida. In the course of an exploration program for phosphates, I. Hernandez discovered the existence of two parallel bands of marine Oligocene rocks lying 5 and 25 km south of Mérida (Fig. 2). The most complete general study of the geology of the Yucatan

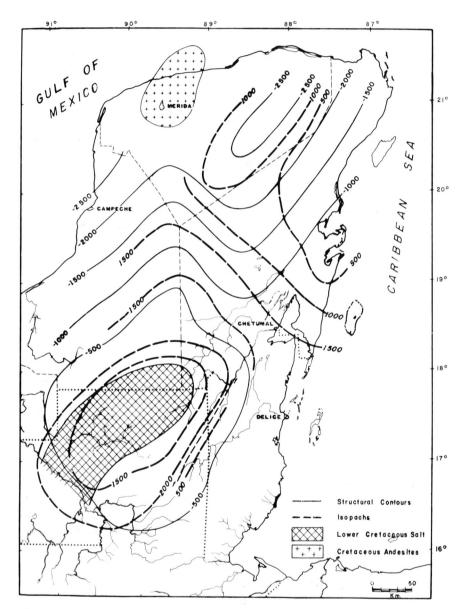

Fig. 5. Structural contours on top of the Yucatan Evaporites and, when known, isopachs of this unit and regions thought to contain Lower Cretaceous salt. Also shown is the region of Cretaceous andesites near Mérida.

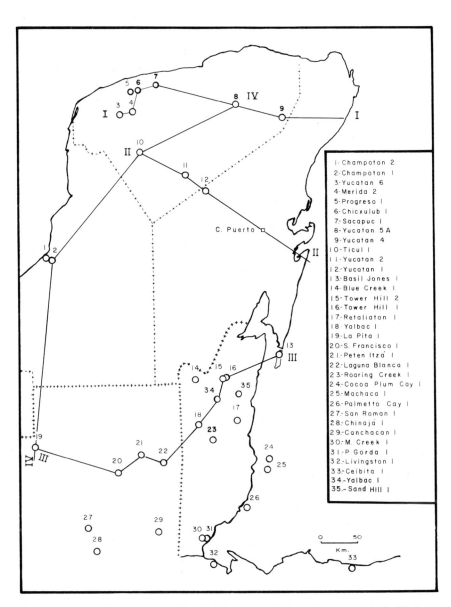

1- Champoton 2
2- Champoton 1
3- Yucatan 6
4- Merida 2
5- Progreso 1
6- Chicxulub 1
7- Sacapuc 1
8- Yucatan 5 A
9- Yucatan 4
10- Ticul 1
11- Yucatan 2
12- Yucatan 1
13- Basil Jones 1
14- Blue Creek 1
15- Tower Hill 2
16- Tower Hill 1
17- Retaliaton 1
18- Yalbac 1
19- La Pita 1
20- S. Francisco 1
21- Peten Itza 1
22- Laguna Blanca 1
23- Roaring Creek 1
24- Cocoa Plum Cay 1
25- Machaca 1
26- Palmetto Cay 1
27- San Roman 1
28- Chinaja 1
29- Canchacan 1
30- M. Creek 1
31- P. Gorda 1
32- Livingston 1
33- Ceibita 1
34- Yalbac 1
35.- Sand Hill 1

Fig. 6. Names and locations of wells which have provided much of the geological information available for the Yucatan Peninsula. Note the positions of cross sections I–IV which
are seen in Figs. 7–10.

and parts of British Honduras has been that of the New Orleans Geological Society (1962).

Serious geophysical studies began with an intensive survey by Petróleos Mexicanos in 1947. Gravimetric, magnetic, and seismic surveys were included in the study. In general, because of the carbonate shelf facies of the rocks and various external factors, the geophysical surveys were difficult to interpret. Many local anomalies were discovered, but at least some, such as the magnetic and gravity high at Mérida Progresso, turned out to be superficial expressions of buried extrusive igneous rocks and did not appear encouraging in terms of petroleum exploration (Figs. 3 and 4).

It was not until Pemex began a drilling program in 1952 with the Chicxulub No. 1 near Mérida (Fig. 5) that large quantities of significant data became available. To the present, 10 wells have been drilled in this program. These wells together with others in British Honduras and Guatemala (Fig. 6) provide the bulk of the reliable geologic information on the Yucatan.

II. SURFACE GEOLOGY AND GEOMORPHOLOGY

In a general way the Yucatan Peninsula can be divided into three physiographic regions. The northern part of the peninsula is a broad coastal plain characterized by low relief and an exceedingly low gradient, so that in its entire extent from the Sierrita de Ticul to the Gulf of Mexico, a distance of about 150 km, it rises only a few meters above sea level. Some 60 km south of Mérida the Sierrita de Ticul forms a second province rising 50 to 100 m above this plain. The Sierrita trends northwest–southeast for a distance of 110 km and, to the northeast, terminates in short scarps while ending in a gentle slope and an undulating terrane to the southwest. The third province lies beyond the Sierrita de Ticul and is composed of the "central plains" of the peninsula. This large province extends south into Guatemala and runs from the Gulf of Mexico to the Caribbean Sea. Throughout the central and northern part of the Yucatan there is essentially no surface drainage. The limestone surface is marked by abundant *cenotes* or sinkholes which are doubtless integrated into an underground drainage system and do provide some exposure. Surface drainage and more normal exposures exist in the south and southeast as the Arco de la Libertad and the Maya mountains are approached.

A. Geological Column

South of the Yucatan Peninsula in Guatemala and in British Honduras, basement is overlain by marine Upper Paleozoic sediments (Bonis *et al.*, 1970). Above the Paleozoic and more ancient rocks lies an unconformable sequence

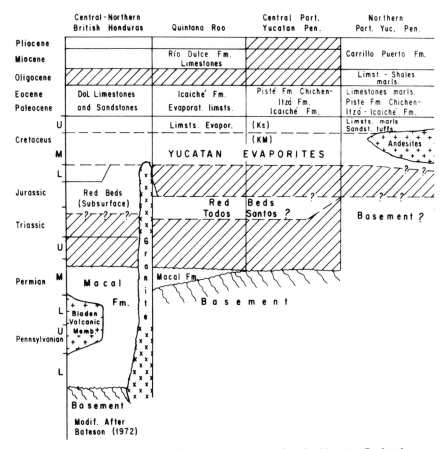

Fig. 7. Generalized stratigraphic correlation chart for the Yucatan Peninsula.

of red beds belonging to the Todos Santos formation (Fig. 7). These red beds are in turn overlain by very extensive evaporites which occur throughout the Peninsula and for which the writer hereby formally proposes the name Yucatan Evaporites (Fig. 5). A general idea of the geology may be obtained by examining Table I and Fig. 2.

B. Upper Cretaceous Rocks

Sapper (1937) reported fragments of rudistids and the presence of *Dicyclina sp.* in the region north of the Maya mountains. Flores (1952) carried out a geological survey near the Maya Mountains in British Honduras where he found a section of Upper Cretaceous limestones and dolomites apparently of a lagoonal or backreef facies. Though Upper Cretaceous evaporites exist in the entire region north of the Maya Mountains, they are not known to outcrop anywhere. The known fauna of the Upper Cretaceous rocks lacks such deep-water fossils as *Globotruncana* but includes many shallow water forms such as *Dicyclina, Lockhartia, Nummoloculina, Valvulina*, miliolids, rotalids, and rudistid fragments.

C. Lower Paleocene–Eocene

There is scattered information suggesting the presence of limestones and dolomites with secondary black cherts in northern British Honduras. Typical fossils found here are *Miscellanea* cf *M. Bermudezi* and *Hexagonocyclina caudri*, which suggest a Paleocene age.

D. Middle and Lower Eocene

The rocks of the Chichén Itzá Formation consists largely of microcrystalline massively bedded limestones. Most outcrops have provided a microfauna suggestive of Middle Eocene age and including a variety of foraminifera and calcareous algae (*Clypeina*) indicative of shallow water.

E. Upper Eocene

Along the Mérida–Chichén Itzá railroad and near the towns of Libre Unión, Chumbec, and San Isidro are massive white saccharoidal limestones containing an Upper Eocene microfauna including many miliolids and apparently of shallow-water origin.

F. Oligocene

South of Mérida, Petróleos Mexicanos carried out a survey in the search for phosphorites and, thanks to outcrops provided by dynamite blasts used in the survey, Lower Oligocene rocks were recognized. These rocks consist of thick-bedded (1 to 1.5 m), poorly stratified calcarenite containing a shallow water foraminiferal fauna. The Upper Oligocene is a cream to white limestone, calcarenitic in part and containing enormous numbers of more or less well preserved molluscan shells in addition to algae and shallow-water foraminifera.

G. Miocene

The Carrillo Puerto formation in the northern part of the Peninsula is clearly Miocene, but some doubt exist as to whether it is Middle or Upper Miocene because of the absence of planktonic foraminifera which might provide a concise definition of age. The formation has yielded shallow-water foraminifera, algae, and ostracods. Marine Miocene rocks have also been found in the southern part of Quintana Roo by M. Alverez, Jr., a Pemex geologist who surveyed this region. These rocks are believed to be Lower Miocene and are grayish white to pink limestone coquinas with mollusks. They are correlated with the Rio Dulce formation of Guatemala.

H. Miocene–Pliocene

Rocks of this age are widely distributed in the Peninsula, often lying with slight discordance on older rocks. Their lithology is quite variable, ranging from crystalline white limestone to coquina and conglomerate. Though bedding is poor, there seems to be no evidence of attitudes other than the horizontal.

I. Pleistocene–Holocene

Rocks of this age occur in a marginal position around parts of the Peninsula, where they consist of calcarenites and calcilutites. It may also be assumed that much of the surface caliche which covers a large portion of the Peninsula is of Recent or sub-Recent age.

III. SUBSURFACE GEOLOGY

Because of the very poor nature of surface outcrops, much of the reliable information on the Yucatan Peninsula comes from wells drilled by Pemex. The writer is grateful to a variety of geologists who were involved in the drilling of these wells and have kindly provided information. Acknowledgment is made to R. Acosta (1951–1955), A. Flores (1964–1966), and D. Godoy R. (1965). Mrs. J. C. de Sansores provided the correlations between wells and additional subsurface information which appeared in an unpublished geological report by I. Hernandez (1966). Table I lists the wells drilled in the Yucatan Peninsula and gives the depths at which various horizons were encountered. Figure 6 provides a location map for wells drilled in the Yucatan Peninsula and also in the British Honduras and adjacent Guatemala.

A. Basement

The Yucatan No. 1 drilled in the central Yucatan Peninsula encountered rhyolite at a depth of 3,200 m. Geochronological work on this rhyolite (Rb–Sr) suggests an age of about 410 m.y. (Silurian) for the rocks and also the probable existence of a metamorphic event at about 300 m.y. (Mississipian). The Bladen volcanics of the Maya Mountains in British Honduras have yielded a some-

TABLE I

Wells Drilled in the Yucatan Peninsula[a]

	Yucatan					Ticul	Sacapuc	Chicxulub	Champotón	
	1	2	4	5-A	6	1	1	1	1	2
Pliocene–Miocene (F. Carrillo Puerto)	OC[b]	OC	OC	OC	OC	OC	OC	OC	OC	OC
Oligocene	—	—	—	—	76	—	375	298	—	—
Eocene (Upper–Middle) (M. Piste, F. Chichén Itzá)	20	1	13	53	426	—	619	483	OC	OC
Eocene (Middle–Lower) Paleocene (Icaiché)	195	208	193	214	736	195	714	666	304	435
Upper Cretaceous										
Maestrich (ks)	276	322	249	292	986	525	932	901	564	—
Without a formation name Turonian	1476	1381	1176	1809	—	1745	1240(?)	—	1684	1650
Middle Cretaceous (km)	2258	2153	1891	2587	—	1900	—	—	1879	1805
Cenomanian–Albian	2914									
Triassic–Jurassic	3058	3298	2349							
Todos Santos (red beds)	3173									
Basement (Breccia and andesites)			2390		1245		1415	1258		
Igneous rocks (extrusive)										
Total depth	3202	3474	2398	2983	1631	3145	1516	1569	2413	2146

[a] Depths in meters below sea level.
[b] OC = outcrop; F = formation; M = member.

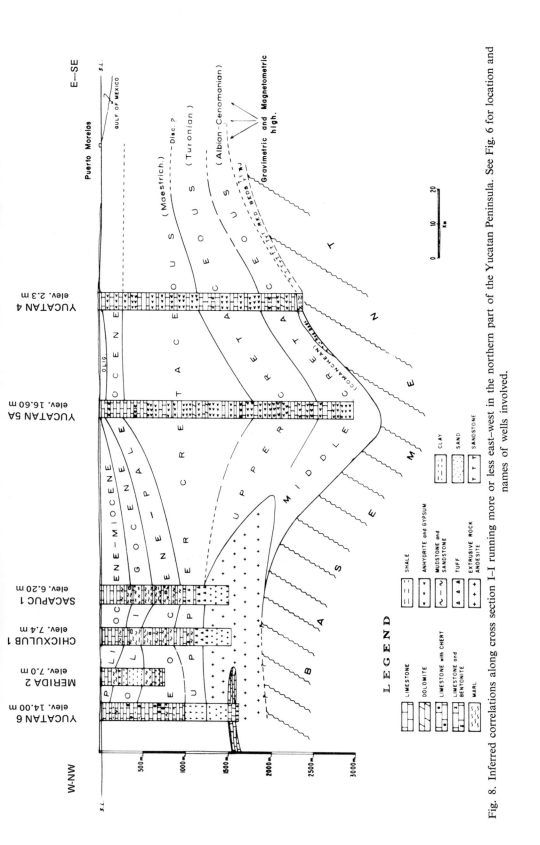

Fig. 8. Inferred correlations along cross section I–I running more or less east-west in the northern part of the Yucatan Peninsula. See Fig. 6 for location and names of wells involved.

what similar (300 m.y.) date in that region. Apparently the rhyolite porphyries of the Yucatan No. 1 occur in a quartz chlorite schist which was found in core No. 48 at a depth of 3,219 m.

Considerably further to the northeast, the Yucatan No. 4 after drilling to 3,290 m below sea level encountered basement below the red beds of the Todos Santos formation. The drilled basement rocks consisted of 8 m of light-gray, very hard, yellowish brown, weathering, slightly metamorphosed quartzite which showed occasional slickensides. This quartzite has been somewhat weathered and suggests a significant period of exposure to the atmosphere before the deposition of the Todos Santos red beds.

In the southern portion of the Peninsula several wells have reached basement with the following results:

British Honduras	Depth	Petrology
Basil Jones 1	2190 m	schist
Tower Hill 1	2140	granite
Guatemala		
San Francisco No. 1-A	1940 m (?)	—

As may be seen, the information on the basement rocks of the Peninsula is rather poor, although we know there are schists, granites, quartzites, and volcanic rocks of considerable antiquity. It is not yet possible to provide any general distribution maps.

In addition to those wells which struck basement, a number encountered andesite flows. The Sacapuc No. 1 struck andesite at 1415 m, and Chicxulub No. 1 encountered andesite at a depth of 1258 m. Both of these wells are near Mérida, where it is believed that a considerable area contains what are probably Cretaceous andesite flows.

B. Red Beds

In a number of places in the Yucatan Peninsula, red beds appear between the Paleozoic basement and Cretaceous rocks. There is no direct evidence of the age of these rocks, and in the geological map of Bonis et al. (1970) as well as in the work of Viniegra (1971), they are considered as Cretaceous–Jurassic continental rocks. The writer believes that these rocks may be of Triassic–Jurassic age and so considers them here, though the evidence is not compelling.

In Guatemala and British Honduras, almost all wells have encountered red beds overlying Paleozoic sediments or schist. In the northern portion of the Yucatan Peninsula, these red beds have been encountered in the Yucatan

No. 1 and Yucatan No. 4, in both of which they underlie Cretaceous evaporites and overlie basement rocks (Figs. 8 and 9). Where encountered, the red beds are made up of siltstones and sandstones and occasionally contain some quartz gravels. Colors range through white, yellowish white, greenish white, and reddish brown, and include interlayered beds of green and emerald-green bentonite. In the Yucatan No. 1 interlayered within the sand body is 36 m (3176 to 3140 m) of ligh gray cryptocrystalline dolomitic limestone. This facies is of interest because it suggests a relatively small transgression during the Triassic–Jurassic. (?)

Figure 6 shows that only 35 deep wells have been drilled in the 300,000 km² considered in this study. The spacing is obviously not very close, and it is quite possible that rocks exist of which we have at present no knowledge. Available information suggests, however, that no wells in the Yucatan Peninsula contain definite marine Jurassic or pre-Albian Cretaceous rocks. The La

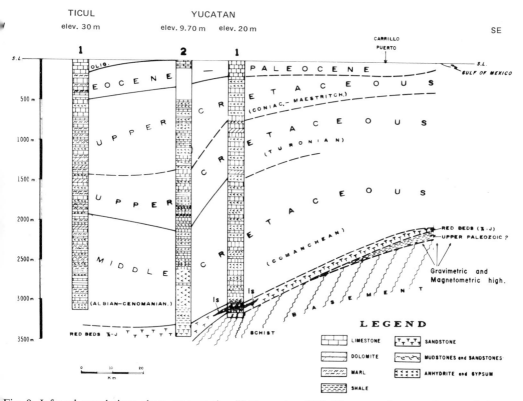

Fig. 9. Inferred correlations along cross section II–II running NW–SE across the central part of the Yucatan Peninsula. See Fig. 6 for location and names of wells involved.

Pita No. 1 in Guatemala did, however, encounter, without wholly penetrating, rocks interlayered with salt which contained a microfauna of Aptain–Barremian age. It appears probable, therefore, that Lower Cretaceous and possibly older marine rocks do occur in the area to the northeast of the La Pita well (Fig. 10). In the center of the Peninsula north of the Arco de la Libertad where salt domes have been detected, it is also possible that extensive pre-Albian Cretaceous and older rocks exist (Fig. 5). Further to the north, the oldest definitely dated rocks are Albian–Cenomanian and Upper Cretaceous.

C. Cretaceous

In the northern portion of the Yucatan Peninsula only a few wells have gone through the whole Cretaceous sequence. One of the better studied of these is the Yucatan No. 1, which is used here to give an idea of the general stratigraphy. Correlation was done on electrical logs, and units have been designated Y-I, Y-M, and Y-S. In the following treatment, approximate ages have been attached to these electric log units.

1. *Comanchean Cretaceous. Unit Y-I*

The petrological and paleontological characteristics of unit Y-I allow its consistent recognition, though there is sufficient lithological variation to allow subdivision if necessary. The lower portion of the interval from 3089 to 2840 m consists of anhydrite, bentonite, pelitic tuff, and limestones. The lower portion of this sequence contains important limestones and numerous pyroclastics; the middle portion consists principally of bentonites and pyroclastics; the upper portion contains a predominance of white, cream, and light-brown anhydrite. From 2840 to 2265 m the rock is composed largely of light-colored anhydrite with thin intercalations of limestone and dolomite. Bentonitic layers occur throughout the interval but are concentrated in the lower part.

Correlation. The dating of unit Y-I is dependent on the fauna of thin limestones at the top which contain Nummoloculines and, in general, the coastal or subreef fauna of Albian–Cenomanian time. The deeper portion of the interval with its frequent anhydrites has not yet yielded any fossils, and its age is, therefore, somewhat indeterminate. It is possible that the base of the section may be Neocomian or older, as appears to be the case in parts of Guatemala and in the La Pita No. 1 well already mentioned.

Wells in northern British Honduras and in the north-central part of Guatemala as well as throughout the Yucatan Peninsula itself show the presence of Cretaceous evaporitic rocks (Figs. 6–12). It, therefore, appears that the entire region was a shallow, wide platform which must have constituted a

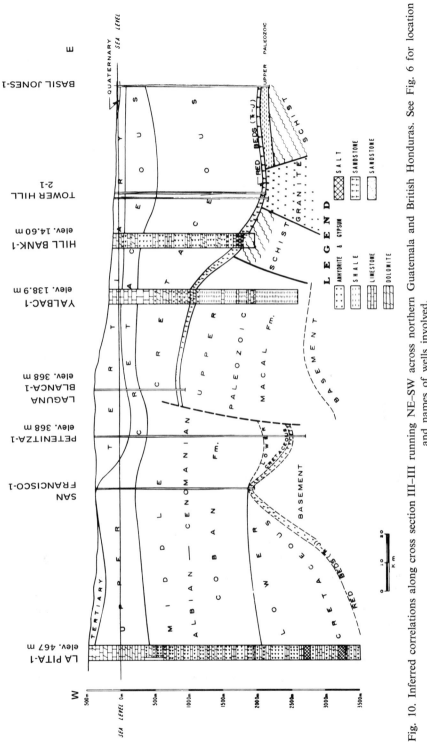

Fig. 10. Inferred correlations along cross section III–III running NE–SW across northern Guatemala and British Honduras. See Fig. 6 for location and names of wells involved.

very effective evaporating pan. These extensive evaporites may be related to those to the north, well known from drilling around the Gulf of Mexico.

2. *Coniacian–Santonian. Unit Y-M*

As seen in the Yucatan No. 1 (Fig. 9), the strata assigned to Y-M are as follows: From 2265 to 2040 m, the rocks consist largely of limestone with a few interlayers of anhydrite. In general, the limestones are crystalline, fine to medium grained, light brown to creamy and, in some places, especially near the top, cryptocrystalline or chalky. From 2040 to 1829 m the unit consists mostly of anhydrite which becomes darker upward and tends to include more gypsum. Thin layers of dolomitic limestone and crystalline dolomite occur within this unit. Between 1829 and 1760 m the section consists largely of limestone and dolomitic limestone containing microfossils. From 1760 to 1581 m, anhydrite with thin intercalations (1–4 m) of dolomitic limestone and crystalline limestone occurs. Between 1581 and 1487 m the section consists of dolomitic limestones and limestones which have produced microfossils. The bottom of the interval contains oolites and pseudo-oolites.

Correlation. The stratigraphic unit Y-M shows certain characteristics that differ from those of overlying and underlying units and allow its physical correlation. The most important of these characteristics are the relative abundance of rudistids, the presence of oolitic and pseudo-oolitic limestone, and the presence of cryptocrystalline limestone. Though ages are not well known, it appears that this unit could be equivalent to the Campur Formation of Guatemala (Vinson, 1962) and, if so, would be of Coniacian–Santonian age (Figs. 8–11). The Campur, however, does not seem to include the abundance of anhydrites found in the sections described.

3. *Late Cretaceous. Unit Y-S*

Unit Y-S, as seen in the Yucatan No. 1, consists largely of anhydrite with thin and sporadically developed limestones and dolomites. From 1487 to 1317 m it consists principally of dolomitic limestones, cryptocrystalline limestones, calcarenites, and limestones with microfossils, oolites, and pseudo-oolites. From 1317 to 1175 m, white to creamy-white and light-brown anhydrite constitutes the bulk of the unit. From 1171 to 1015 m, anhydrite becomes uncommon, and limestone predominates. Between 1015 and 657 m, thin limestone layers are infrequent, and the formation consists largely of anhydrite. From 657 to 440 m, anhydrite is again predominant and broken only by thin limestone layers. From 440 to 265 m there is again anhydrite broken by thick bodies of limestone breccia and cryptocrystalline limestone, and the sequence contains, near the bottom, thin layers of green bentonite.

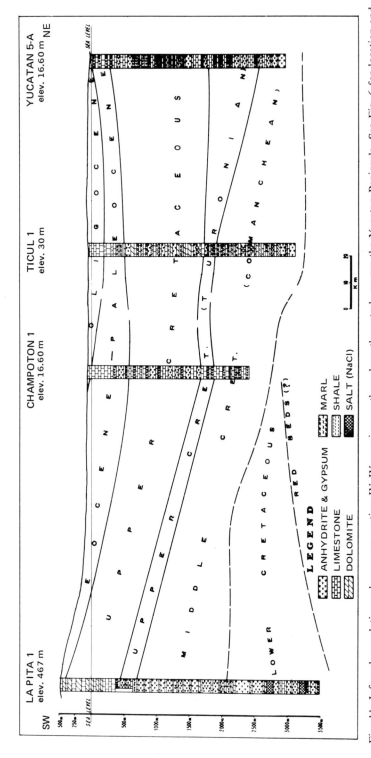

Fig. 11. Inferred correlations along cross section IV–IV running north and north easterly across the Yucatan Peninsula. See Fig. 6 for location and names of wells involved.

Correlation. This unit is readily correlated by virtue of its petrological and faunal composition and its electrical log characteristics. There is evidently a considerable facies change as one moves to the north. In the Yucatan No. 1 in west-central Quintana Roo, the unit is predominantly anhydrite with subordinate amounts of limestone. The Ticul No. 1 in western Yucatan shows a great increase in the abundance of limestone. Still further to the north the Sacupuc No. 1 near the coast contains no evaporites. The fauna which occurs in the fossiliferous portion of the section seems definitely to indicate a late Cretaceous age.

D. Tertiary

1. *Paleocene*

The Paleocene is less well known in the Yucatan No. 1 than the Chicxulub No. 1 well which, therefore, is used here as a basis for its description. In this well the Paleocene is considered to extend from 920 to 810 m on the basis of microfaunal information and on the basis of rather clear boundaries with the overlying Eocene and underlying Cretaceous. The sediments are largely light-colored, fine-grained limestones interlayered with marl and occasional gray and greenish-gray lutites.

2. *Eocene*

Again the section in the Chicxulub No. 1 is better known than that in the Yucatan No. 1 and is used as a basis for the following comments: Microfossils are quite abundant and indicate that the Lower Eocene extends from 810 to 685 m; the Middle Eocene, from 685 to 595 m; and the Upper Eocene, from 595 to 525 m. Lithologically speaking, the Eocene section is made up of white and light-gray limestones interlayered with plastic marls and a little dark-gray limestone (Fig. 8).

3. *Oligocene*

In the northern part of the Peninsula the Yucatan No. 6, Sacupuc No. 1, and Chicxulub No. 1 have gone through the Oligocene. In the Chicxulub well the unit is formed of white, gray, and cream-colored marls with intercalations of clay of the same colors and with some compacted cream-gray limestones. Micropaleontological study has allowed the division of the Oligocene into Upper, Middle, and Lower units as follows: Lower Oligocene, 525 to 455 m; Middle Oligocene, 455 to 415 m; and Upper Oligocene, 415 to 370 m. According to the unpublished work of Mrs. Sansores (1966), this fauna is of a bathyal type in contrast with the neritic Oligocene fauna found at outcrops south of the Mérida (Fig. 2) and may indicate a basin deepening toward the north of the Peninsula and possible absence of Oligocene deposits in the South.

4. *Miocene–Pliocene*

The Miocene–Pliocene Carrillo Puerto formation is well developed in the Chicxulub No. 1. Generally speaking, the formation consists of white to cream-colored, soft plastic marls with intercalations of clay and a few limestones. The rocks contain an abundant and well preserved microfauna, suggestive of moderately deep water.

5. *Pliocene–Pleistocene*

Near the site of the Chicxulub No. 1 are outcrops of cream to cream-white, partly oolitic limestones. The age of these limestones has not been determined, but it seems most probable that they are of Pliocene age. Due to circulation loss, the recovery of samples from the Chicxulub well was poor, but observation of drilling speeds suggests that these probable Pliocene rocks extend to a depth of 205 m.

IV. SUBMARINE GEOLOGY ADJOINING THE PENINSULA

Figure 1 shows, in a general way, the submarine geology of the Yucatan Peninsula and its north and northwestern extension into the Campeche Bank. According to information provided by Lynch in 1954 and published by the New Orleans Geological Society (1962), as well as surveys carried out by Bonet (1956), the sequence of marine outcrops is highly irregular and is principally formed of autochthonous limestones including some reefs, such as Alacrán Island. Detailed work by Logan *et al.* (1969) suggests that to a depth of about 172 m the rocks consist mostly of Pleistocene and Pliocene carbonates. According to Wilhelm and Ewing (1972), these rocks discordantly overlie a strip of Eocene sediments which in turn rest on Cretaceous rocks that appear on the bottom of the ocean around the Campeche Bank to the north and north-west, and also in the deepest portion of the Yucatan Channel, some 60 km to the west of Cozumel Island.

Recent information (Geotimes USGS-1971, Leg. 2, p. 10–12) on submarine sampling in the Yucatan Channel and near 100 km from the northeast point of the Peninsula indicates the presence of schist and marble exhibiting a north–south trend (Fig. 1). If the presence of schist and marble is confirmed, it is conceivable that these rocks might correlate with those found in the Basil Jones No. 1 in northeastern British Honduras.

V. METAMORPHIC AND IGNEOUS ROCKS

It is noted in Table I that only the Chicxulub No. 1, Sacapuc No. 1, and Yucatan No. 6 wells encountered extrusive andesite rocks with interlayered tuffs in the marine Upper Cretaceous rocks of the Yucatan Peninsula. These wells were drilled on the gravimetric and magnetic highs caused by the igneous rocks. Only the Yucatan No. 6 penetrated the andesitic flows to encounter Cretaceous anhydrites at the bottom of the hole.

Far to the south in the Maya Mountains of British Honduras, Mesozoic granites appear to make up the core of the range. Figure 2 indicates that the extrusive rocks nearest to the Maya Mountains occur considerably to the southwest and are rhyolitic lavas interbedded in marine rocks of Upper Paleozoic age. It should be recalled that the Yucatan No. 1 in west-central Quintana Roo encountered a rhyolite porphyry of probable Paleozoic age which might be correlated with rocks to the South.

Metamorphic rocks have been found only in the Yucatan wells Nos. 1 and 4 and the Basil Jones No. 1 (Figs. 9 and 10). The age of the schist encountered in these wells is probably Paleozoic.

VI. TECTONICS

The geological map of the Yucatan Peninsula (Fig. 2), while doubtless greatly limited by the difficulty of surface work in this area, suggests little in the way of tectonism expressed in the pattern of rock distribution. On a larger scale, one can see that the Paleozoic foreland of Yucatan was submerged during the Mesozoic but remained as a subsiding platform where relatively thin continental sedimentary cover was accumulated (Figs. 8–11). The Coban basin between the Yucatan platform and the hinge was divided into two deep portions separated by a Pre-Cretaceous structural high called Arco de la Libertad (Vinson and Brineman, 1963). According to the suggestions of Lloyd and Dengo (1960), the Arco de la Libertad is a part of a great fault block which includes the Maya Mountains.

The presence of volcanic rocks (andesite and tuffs) in the Upper Cretaceous of the northern part of the Peninsula indicates that uplift may have continued in this region to the end of the Mesozoic. Red beds underlying Mesozoic rocks appear largely absent in the north, and there is in fact the suggestion that from southwest to northeast the basement is overlain by successively younger rocks. For example, Lower Cretaceous rocks occur in the region of the La Pita No. 1, Middle Cretaceous evaporitic rocks in the Yucatan No. 2, and Upper Cretaceous marls and tuffs in the Chicxulub No. 1 well.

The thick and extensive sequence of evaporitic rocks in the Yucatan Peninsula and Campeche Bank indicates that this region remained as a rather stable, slowly subsiding platform during much of the Mesozoic. Cessation of evaporitic accumulation in the Tertiary could be due to a slight northward tilt opening the evaporite basin or to an increase in eustatic sea level, but it does not suggest major tectonic activity.

It would be interesting to speculate about plate tectonics with respect to the Yucatan Peninsula, but unfortunately no information which might allow constrained speculation is available. However, the change in trend of folding between the Sierra Madre in Chiapas (NW–SE) and that of the fold and fault system in northern Guatemala and British Honduras (NE–SW) does allow speculation that there could have been a rotation of the Peninsula from the east to its present position.

VII. CONCLUSIONS

Figure 12 suggests a paleogeographic history for the Yucatan Peninsula based on the wells that have been drilled and such additional information as

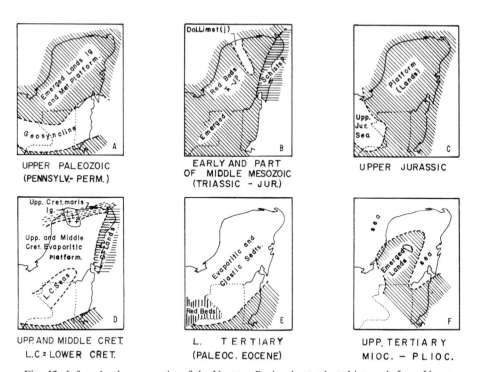

Fig. 12. Inferred paleogeography of the Yucatan Peninsula at selected intervals from Upper Paleozoic to Upper Tertiary.

can be pieced together. Doubtless, a large part and, perhaps, all of the Peninsula, had a cratonic character in the Paleozoic. Figure 12 indicates that it was largely emergent until the Triassic–Jurassic (?), as is confirmed by the presence of largely continental red beds present in various wells and assumed to be of this age. A minor transgression of the sea from the north is indicated by the marine tongue in the red-bed sequence in this region (Fig. 12B). An Upper Jurassic sea is known to have slightly entered the region from the west and to have extended into portions of Guatemala (Fig. 12C). Lower Cretaceous (Fig. 12D) seas developed in the south, and perhaps more widely and with them came the evaporitic regime which characterized the Middle and Upper Cretaceous throughout the Peninsula.

During the Upper Cretaceous (Fig. 12D) and part of the Tertiary (Fig. 12E,F), the northern portion of the Peninsula received marly limestones which may suggest the deepening of the seas in this direction or the presence of some kind of relatively nearby terrigenous source for argillaceous materials. South of parallel 20°30′, marine Oligocene sediments disappear probably by erosion beneath the Carillo Puerto formation (Pliocene–Miocene) which lies unconformably on Eocene and Oligocene rocks in big areas of the Peninsula. The present shape of the Peninsula was attained near the end of the Pliocene and continued into the Quaternary despite the fact that large reef trends are still developing north of the Campeche Bank. Sedimentation in this region is mainly calcareous because the lack of surface drainage on the Peninsula limits the input of terrigenous material.

ACKNOWLEDGMENT

The author acknowledges the help of Francisco Viniegra, Petróleos Mexicanos Exploration Manager, for providing the facilities to complete this paper.

REFERENCES

Alvarez, M., 1954, Exploración geologica preliminar del Rio Hondo, Quintana Roo: *Assoc. Mex. Geol. Petr. Bol.*, v. 6, p. 207–213.

Bateson, J. H., 1972, New interpretation of geology of Maya Mountains, British Honduras: *Am. Assoc. Petr. Geol. Bull.*, v. 56, p. 956–963.

Bryant, W. R., Antoine, J., Ewing, M., and Jones, B., 1968, Structure of the Mexican continental shelf and slope, Gulf of Mexico: *Am. Assoc. Petr. Geol. Bull.*, v. 52, p. 1204–1228.

Bonet, F., 1956, Nota Preliminar Sobre la Constitución de los Arrecifes Coralinos de la Sonda de Campeche, México, Resumen XX, Congreso Geol. International. México. See also: Bonet, F., 1967, Biogeología Subsuperficial del Arrecife Alacranes, Yucatán, Inst. de Geol. de la U.N.A.M., Bol. 80, 192 p.

Bonis, S., Bohnenberger, H., and Dengo, G., 1970, *Mapa Geológico de la Republica de Guatemala*, 1a Edición: Instituto Geográfico Nacional.

Butterlin, J., and Bonet, F., 1960, Microfauna del Eoceno Inferior de la Peninsula de Yucatán: Paleontologia Mexicana, No. 7, Inst. de Geol. de la U.N.A.M., 18 p.

Dengo, G., and Bohnenberger, H., 1970, Structural development of northern Central America: Am. Assoc. Petr. Geol. Mem., p. 203–220.

Dustano, G. R., 1965, Bosquejo Geológico de la Península de Yucatán: Tesis Professional ESIA ipn México, 55 p.

Edwards, C., 1957, Quintana Roo; Mexico's Empty Quarter (Report of Field Work Carried out under Office of Naval Research, Contract 222): Dept. of Geography, Univ. Cal., Berkeley (mimeographed).

Ericson, D. B., and Heezen, B. C., 1955, Geophysical and geological investigation in the Gulf of Mexico—Part 1: *Geophysics*, v. 20, p. 1–18.

Ewing, M., and Worzel, J. L., 1970, Deep Sea Drilling Project—Leg 10: *Geotimes*, v. 15, n. 6, p. 11–13.

Flores, G., 1952, Geology of Northern British Honduras: *Am. Assoc. Petr. Geol. Bull.*, v. 36, p. 404–408.

Halbuty, M. T., 1967, Salt Domes, Gulf Region, United States and Mexico: Gulf Publ. Co. 425 p.

Kirkland, D. W., and Gerhard, J. E., 1971, Jurassic salt, Central Gulf of Mexico and its temporal relation to circum-Gulf evaporites: *Am. Assoc. Petr. Geol. Bull.*, v. 55, p. 680–686.

Lloyd, J. J., and Dengo, G., 1960, Posibilidades petrolíferas de la cuenca de Petén, Guatemala: Fac. de Ingeniería, Univ. de San Carlos, Bol. 9, p. 1–11.

Logan, B. W., et al., 1969, Carbonate sediments and reefs, Yucatan Shelf, Mexico: Am. Assoc. Petr. Geol. Mem. 11, p. 1–198.

López Ramos, E., 1969, Geología del SE de México y Norte de Guatemala, en: *Trabajos Presentados en la II Reunión de Geólogos de América Central*: ICAITI, p. 57–67.

López Ramos, E., 1969, Marine paleozoic rocks of México; *Am. Assoc. Petr. Geol. Bull.*, v. 53, p. 2399–2417.

López Ramos, E., 1972, *Geología General*, Edicíon Escolar, 262 p., Mexico, D. F.

Meyerhoff, A. A., 1967, Future hydrocarbon provinces of Gulf of Mexico. Caribbean region: *Gulf Coast Assoc. Geol. Soc. Trans.*, v. 17, p. 217–260.

Moore, G. W., et al., 1971, U.S.G.S.—I.D.O.E. Leg 1: Bahía de Campeche: *Geotimes*, v. 16, n. 11, p. 16–17.

Murray, G., 1961, *Geology of the Atlantic and Gulf Coastal Province of North America*: Harper and Row, New York, 696 p.

New Orleans Geological Society, 1962, *Guide Book, Field Trip to Peninsula de Yucatán*, 120 p.

Paine, W. R., and Meyerhoff, A. A., 1970, Interactions among tectonics, sedimentation and hydrocarbon accumulation: *Gulf Coast Assoc. Geol. Soc. Trans.*, v. 20, p. 5–44.

Petróleos Mexicanos, 1966–69, Informes geológicos inéditos.

Sapper, K., 1896, Sobre la geografía física y la geologia de la Peninsula de Yucatán: Inst. Geol. de Mexico, Bol. 3, Mexico, D. F.

Sapper, K., 1937, *Mittelamerika*; *Handbuch der Regionalen Geologia*, v. 8, pt. 4, Hft. 29: Carl Winter, Heidelberg, 160 p.

Schuchert, C., 1935, *Historical Geology of the Antillean and Caribbean Region*: John Wiley and Sons, New York.

Tercier, J., 1932, Geological investigation in Peninsula de Yucatán. (Informe inédito).

Vedder, J. G., et al., 1971, U.S.G.S.—I.D.O.E. Leg 2: *Geotimes*, v. 16, n. 12, p. 10–12.

Viniegra, F., 1971. Age and evolution of salt basins of southeastern Mexico, *Am. Assoc. Petr. Geol. Bull.*, v. 55, p. 478–494.

Vinson, G. L., 1962, Upper Cretaceous and Tertiary Stratigraphy of Guatemala: *Am. Assoc. Petr. Geol. Bull.*, v. 46, p. 425–456.

Vinson, G. L., and Brineman, J. H., 1963, Nuclear Central America, hub of the Antillean transverse belt, in: Backbone of the Americas: Am. Assoc. Petr. Geol. Mem. 2, p. 101–112.

Wilhelm, O., and Ewing, M., 1972, Geology and history of the Gulf of Mexico: *Geol. Soc. Am. Bull.*, v. 88, p. 575–600.

Worzel, J. L., Leyden, R., and Ewing, R., 1968, Newly discovered diapirs in Gulf of Mexico: *Am. Assoc. Petr. Geol. Bull.*, v. 52, p. 1194–1203.

Chapter 8

PALEOZOIC AND MESOZOIC TECTONIC BELTS IN MEXICO AND CENTRAL AMERICA

Gabriel Dengo

Instituto Centroamericano de Investigación y Tecnología Industrial (ICAITI)
Guatemala City, Guatemala

I. INTRODUCTION

The purpose of this synthesis is to present the structural framework which developed during the Paleozoic and Mesozoic on the western side of the Gulf of Mexico and the Caribbean. As defined by King (1969, p. 84), a foldbelt is "a linear belt that has been subjected to folding and other deformation during the orogenic phase of a tectonic cycle. Each foldbelt evolved during a time span different from that of any others, and details of their histories differ, although they can be grouped broadly, according to worldwide time of orogeny." King cites as synonyms *mobile belt, orogenic belt, mountain belt* (less properly) and the French term *region de plissément.* The area under discussion falls mainly within King's Cordilleran foldbelt and Antillean foldbelt (King, 1969, and Tectonic Map of North America).

The Cordilleran foldbelt, as described by King (1969, p. 64), extends 8,000 km from Alaska to northern Central America. The Texas Lineament separates it into two mayor parts with different topography, tectonic history, and style of deformation. The southern portion includes most of Mexico, with the exception of Baja California and the Gulf of Mexico, and extends to

Central America as far south as northern Nicaragua; it is, therefore, the part discussed here.

The major features of the Cordilleran foldbelt are the result of Mesozoic orogenies. However, it has a history of Paleozoic sedimentation and deformation; but the early deformational features are largely masked by Mesozoic ones, and the patterns are not always clear.

Southern Central America (Nicaragua to Panama) is classified as a part of the Antillean foldbelt, which differs from other foldbelts of North and South America in being "an oceanic rather than a continental feature" (King, 1969, p. 77).

The geological history of the foldbelts, particularly of the Cordilleran one, vary from place to place, and it is difficult to make broad generalizations. In order to present this discussion, the author has chosen to describe the region in terms of "tectonic belts" which are more restricted both in time and in areal extent, but which are considered as units of a major foldbelt. A tectonic belt, in the sense used here, is the area of major deformation within a region of large accumulation of sedimentary rocks, be it a geosyncline or a basin, and includes igneous events.

This approach has made necessary the use of terms which are apparently incongruous, such as "metamorphic tectonic belt" and "sedimentary tectonic belt," which are applied to Paleozoic features at present characterized by different degrees of deformation, and which are useful in distinguishing major tectonic characteristics of the region.

In preparing this synthesis, the author has faced several problems. One is presenting a coherent summary of the Paleozoic metamorphic belt. This is due to lack of data in many places and to the uncertain age of the meta-sedimentary rocks throughout the region. The author's incomplete direct knowledge of some areas, particularly in northern Mexico, has been another shortcoming, and, as a result it has been necessary to rely mostly on the literature. Finally, the use of appropriate terminology for regional tectonic features at a time when some of the basic principles of global tectonics are being challenged, or debated, poses a most difficult problem.

The description of each of the tectonic belts deals mostly with factual data and, in some cases with existing problems that require further work, specifically more detailed mapping and dating. Broad geographical and geological coverage of the subject matter made it necessary to present a synthesis of the state of knowledge which basically represents only a guide to the most pertinent literature on the subject. The following works deal with the geology of the region, or of large parts of it, and provide other references which are not mentioned in this paper: Schuchert (1935), Garfias and Chapin (1949), Guzmán and de Cserna (1963), Weyl (1961), Dengo (1968), and King (1969).

Formational names, with a few necessary exceptions, have been purposely avoided; their inclusion would have required a much longer discussion. Although the author does not intend to discuss the area in relation to the tectonics of the Gulf of Mexico and the Caribbean Sea, which is the responsibility of other authors, the last part of this paper presents some problems of a regional nature as well as suggesting some areas for research.

Parts of this paper were critically reviewed by Eng. Ernesto López Ramos, Dr. Thomas W. Donnelly, Dr. Alan Nairn, and Dr. Francis G. Stehli. The author is indebted to them for their help. Special acknowledgment is due to Dr. Samuel Bonis who kindly helped in the revision of the text and whose cooperation has been very valuable in this occasion as well as in the preparation of several of the author's previous articles. Miss María del Carmen Almengor and Miss Rosángela Escobar typed the manuscript and drafted the illustrations.

II. REGIONAL SETTING AND TECTONIC UNITS

In order to provide the reader with a background on the geography, including political divisions, and the geology of the region, two illustrations (Figs. 1 and 2) are presented, together with a short description of the morphotectonic units as previously discussed by Alvarez (1961), Guzmán and de Cserna (1963), Dengo and Bohnenberger (1969), and Dengo (1968). These units are established on the basis of their relief and their internal constitution. They represent the present-day distribution of geologic and physiographic features and, therefore, provide a good basis for discussion of regional geology.

A. Mexico

1. *Sierra Madre Oriental*

This is the most extensive unit of the region, extending from the United States–Mexico border to the southeast, along Mexico and northern Central America, where it forms the northern ranges of the Sierras of Northern Central America. Between parallels 19° and 20°30′ N, it is interrupted by the Trans-Mexico Volcanic Belt. The largest part of the Sierra Madre Oriental consists of folded Mesozoic sedimentary rocks, predominantly limestones, and Paleocene marine clastic sediments. Metamorphic rocks crop out in the States of Tamaulipas and Hidalgo, and highly folded Paleozoic sedimentary rocks occur in Chihuahua and Tamaulipas.

Within intermountain valleys there are younger Tertiary volcanic rocks and terrigneous sediments. Guzmán and de Cserna (1963) include the Mesa

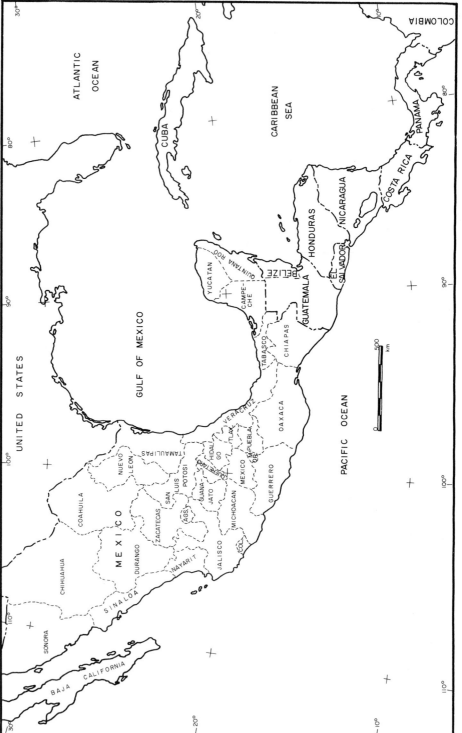

Fig. 1. Political divisions: Mexico and Central America

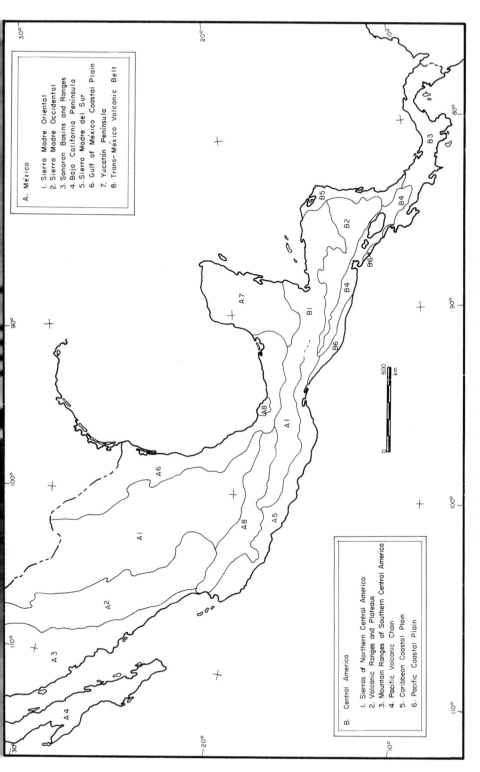

A. México

1. Sierra Madre Oriental
2. Sierra Madre Occidental
3. Sonoran Basins and Ranges
4. Baja California Peninsula
5. Sierra Madre del Sur
6. Gulf of México Coastal Plain
7. Yucatán Peninsula
8. Trans-México Volcanic Belt

B. Central America

1. Sierras of Northern Central America
2. Volcanic Ranges and Plateaus
3. Mountain Ranges of Southern Central America
4. Pacific Volcanic Chain
5. Caribbean Coastal Plain
6. Pacific Coastal Plain

Fig. 2. Morphotectonic units: Mexico and Central America.

Central, which other authors consider a separate unit, within the Sierra Madre
Oriental. The sedimentary rocks are intruded by acid to intermediate igneous
bodies, mainly of granodioritic to dioritic composition, which localize some
of the important mineral deposits in Mexico. Structurally the Sierra Madre
Oriental is complex and highly deformed in parallel folds cut by thrust and
normal faults.

2. *Sierra Madre Occidental*

This division parallels the Sierra Madre Oriental on its western side from
the United States–Mexico border southward to the Trans-Mexico Volcanic
Belt. It is formed mainly of andesitic to basaltic volcanic rocks at the base
and covered by dacitic and rhyolitic ignimbrites and tuffs of great thickness.
These rocks are usually horizontal or slightly tilted. They were considered to
be of Tertiary age until recently, when it became possible to date part of them
radiometrically in Sinaloa and Sonora where they were found to be of Upper
Cretaceous age (oral communication by S. Clabaugh, University of Texas,
Austin, 1972).

3. *Sonoran Basins and Ranges*

This unit is located between the Sierra Madre Occidental and the Gulf of
California. According to Guzmán and de Cserna (1963),

> ... the region consists of highly dissected block-mountains which rise above
> the land of a semidesert plain.... . Geologically, it contains Precambrian
> granites, metamorphics, and sediments, which are overlain by Cambrian fine
> clastics and carbonates. The Paleozoic sedimentary record is completed by
> the presence of Permo-Carboniferous carbonates—with the apparent lack of
> middle Paleozoic sediments. The Paleozoic sediments are overlain uncon-
> formably by an Upper Triassic–Lower Jurassic (?) partly marine, coal-bearing
> clastic sequence, which in turn is covered by Lower Cretaceous clastics and
> carbonates and by Upper Cretaceous clastics and volcanics. Tertiary con-
> tinental clastics and volcanics complete the sedimentary sequence. Cretaceous
> granitic stocks and Tertiary dioritic stocks and hypabyssal bodies are the
> principal representatives of the igneous intrusive activity which affected the
> region.

4. *Baja California Peninsula*

This peninsula is formed by a mountain range 1,650 km long with a
maximum width of 130 km. It is characterized by a group of extensive batholiths
flanked on the west by metamorphic rocks of possible Mesozoic age, overlain
unconformably by Upper Cretaceous clastic sediments, Tertiary marine sediments, and volcanic rocks. The main structural features of the peninsula are

due to block faulting. The graben structure of the Gulf of California is also related to large-scale faulting.

5. *Sierra Madre del Sur*

These mountains extend from the Trans-Mexico Volcanic Belt, along the Pacific side of Mexico, to the Guatemalan border and continue across Central America as the central ranges of the Sierras of Northern Central America. Geologically, the Sierra is formed to a large extent by Precambrian and Paleozoic metamorphic and intrusive rocks, overlain locally by Mesozoic sedimentary rocks. The rocks are cut locally by younger hypabyssal intrusive bodies.

6. *Gulf of Mexico Coastal Plain*

The coastal plain, located between the Sierra Madre Oriental and the Gulf of Mexico, extends from the Rio Bravo, or Rio Grande, to the Yucatan Peninsula. It is characterized by Quaternary sediments on the surface which cover Tertiary and Cretaceous marine sedimentary rocks. Several Tertiary basins are well defined, particularly in the south, but do not fall within the scope of this paper.

7. *Yucatan Peninsula*

Although a lowland area, it is considered a separate unit because it has a different geological history. On the surface it is characterized mainly by Tertiary marine carbonate rocks. Another article in this volume deals specifically with the Yucatan Peninsula.

8. *Trans-Mexico Volcanic Belt*

This is a young unit, formed by upper Tertiary and Quaternary volcanic rocks with many small and several large cones, which extends across Mexico in a N 80° W direction and transects the Sierra Madre Oriental and the Gulf of Mexico Coastal Plain between parallels 19° and 20°30′ N.

B. Central America

1. *Sierras of Northern Central America*

According to Dengo and Bohnenberger (1969), this morphotectonic unit is formed by a series of subparallel mountain ranges that extend from the

Mexican border to the east through Guatemala, Belize, Honduras, and northern Nicaragua, ending in the Bay Islands off the Honduras coast. The western part of the ranges trends generally northwest, the central part east–west, and the eastern part northeast. Two submarine topographic features in the Caribbean, the Cayman Ridge, and the Nicaragua Rise seem to be a structural continuation of the Sierras.

The northern ranges are formed by tightly folded Permian and Cretaceous rocks, mainly carbonates, thrust-faulted to the north and cut by younger normal faults. This part is a continuation of the Sierra Madre Oriental of Chiapas. The middle group of ranges consists mostly of metamorphic rocks, granitic batholiths, and ultramafic bodies, while Paleozoic and Mesozoic sedimentary rocks are present only locally. This group is the continuation of the Sierra Madre del Sur of Chiapas. The southern group lies within the political limits of Central America and does not seem to have a counterpart in Mexico. It consists mostly of Paleozoic, low-grade metamorphic rocks and Mesozoic sedimentary rocks, intruded by Mesozoic and Tertiary stocks and batholiths. Tectonic deformation of these ranges is less intense than in the central and northern ones.

2. Volcanic Ranges and Plateaus

This unit is characterized by Tertiary volcanic rocks (ranging from basalt to rhyolite) and extensive areas of ignimbrites and associated terrigenous volcanic sediments. It covers most of the southern part of northern Central America. In Guatemala, El Salvador, and western Honduras the Tertiary volcanic rocks form several small, block-faulted ranges, whereas in southern Honduras and Nicaragua they form extensive plateaus. Both petrologically and structurally, the ranges and plateaus are similar to the Sierra Madre Occidental in Mexico, although they are younger.

3. Mountain Ranges of Southern Central America

This is a geologically and physiographically complex unit that extends from southern Nicaragua through Costa Rica and Panama and continues to the south as the coastal ranges and valleys of western Colombia. As a tectonic unit, it has been designated as the Southern Central American Orogen (Dengo, 1962a, Lloyd, 1963). It is characterized by a basement of Mesozoic basic volcanic rocks and is intruded by different types of Late Cretaceous and Tertiary stocks and batholiths. Part of the unit is also covered by Quaternary volcanic deposits. It was tectonically deformed during Late Cretaceous–Eocene time and also at the end of the Miocene. Much of the deformation was controlled by vertical movements, some as young as Pliocene.

4. *Pacific Volcanic Chain*

A chain of Quaternary volcanoes, many of which are active, parallels the Pacific coast from the Mexico–Guatemala border to central Costa Rica. The chain continues to the southeast in Panama as isolated volcanic cones.

The volcanic system trends NW–SE, and the individual cones are related to faults parallel with this trend, particularly with a large graben—the Nicaragua Depression—which extends from El Salvador across Nicaragua and ends in northern Costa Rica. Some of the volcanic centers are localized along smaller north–south faults (Dengo, Bohnenberger, and Bonis, 1970). The chain is genetically related to the Middle America Trench which parallels the Pacific coast of southern Mexico and Central America. It should be noted that this chain is in a different tectonic setting than the Trans-Mexico Volcanic Belt.

5. *Caribbean Coastal Plain*

The Caribbean coastal plain is more a physiographic feature than a morphotectonic unit. It extends from western Honduras south along the coast of Nicaragua in an area where both the Sierras of Northern Central America and the volcanic ranges and plateaus have been eroded nearly to sea level.

6. *Pacific Coastal Plain*

A narrow coastal plain extends along the Pacific from the Isthmus of Tehuantepec to El Salvador and continues, with interruptions, southward through Nicaragua. The boundary between the coastal plain and the Pacific Volcanic Chain is probably a series of major faults burried under aluvial and volcanic-slope deposits. Its morphology is the result of regional faulting and rapid accumulation of fluviatile sediments from the rising Quaternary cones.

III. PALEOZOIC TECTONIC BELTS OF MEXICO AND NORTHERN CENTRAL AMERICA

A. Paleozoic Metamorphic Rocks of Mexico and Northern Central America

1. *State of Knowledge*

Much of the Precambrian and early Paleozoic history of Mexico and Central America is poorly known because of incomplete mapping and because, in parts of the region, the outcrops are few and far apart. In some areas, like Chiapas and Guatemala, it is still difficult to separate the Paleozoic from possible Precambrian. Another chapter of this volume deals with the known ages of the basement rocks of the region.

As a background to the general interpretation and areal distribution of these rocks, reference is made here to several geological and tectonic maps, considered to represent the "state of knowledge" up to the time of their compilation. Reference to other works is made in subsequent sections. In the case of Mexico, the "Tectonic Map of Mexico," scale 1:2,500,000 (de Cserna, 1961), is very useful as a starting point for this discussion. This map is based on the principle of "structural belts," defined as "elongated regions that became consolidated during a geotectonic cycle." The geotectonic cycle as defined on the map to a large extent coincides with the definition of the similar term used by King (1969, p. 43). The structural belts proposed for Mexico have also been discussed at greater length by Guzmán and de Cserna (1963). The Paleozoic metamorphic rocks are included by de Cserna in the Jaliscoan (equivalent in part to Cordilleran foldbelt of King, 1969) and Huastecan (equivalent to Ouachita foldbelt of King, 1969) structural belts. de Cserna includes in both of these belts Paleozoic metamorphic rocks as well as younger sedimentary rocks of Paleozoic and, in some cases, Mesozoic age. In this discussion, and on the basis of his field experience in Guatemala and Chiapas, the author has chosen to treat the Paleozoic metamorphic rocks separately from the sedimentary or only slightly metamorphosed younger Paleozoic rocks. This is a more simplistic approach, but it is also more factual than interpretative.

Another important map is the "Carta Geológica de la República Mexicana," scale 1:2,000,000 (Hernández Sánchez-Mejorada and López Ramos, 1968). This map presents the geologic units according to the more traditional system of grouping them by ages. It shows the distribution of Paleozoic metamorphic rocks as one unit and presents more up-to-date information on the differentiation of this unit and the Precambrian metamorphic rocks. It also shows the metamorphic rocks in the State of Sinaloa as different from those in southern Mexico and as belonging to the Mesozoic.

In the case of Central America, reference is made here to two maps: "Geologic Map of Central America," scale 1:1,000,000 (Roberts and Irving, 1957) and "Metallogenic Map of Central America," scale 1:2,000,000 (Dengo et al., 1969). The first shows the metamorphic rocks of Guatemala, Honduras, and Nicaragua under three units: (1) Precambrian metamorphic rocks, (2) Rocks of Paleozoic age, and (3) Pre-Cretaceous metamorphic rocks.

Because of the uncertainty about Precambrian rocks in Central America, in the compilation of the metallogenic map all the units were grouped under the single heading of Paleozoic metamorphic rocks (Pre-Pennsylvanian).

Another recent compilation, the "Tectonic Map of North America," scale 1:5,000,000 (King, 1969), presents the Paleozoic metamorphic rocks of southern Mexico and Central America as a unit under the designation "metamorphic complex of the Cordilleran Foldbelt," indicating that it may also include Precambrian rocks, which are probably derived from early geosynclinal sediments. The metamorphic rocks in the State of Sinaloa and the northern part of Baja California are described as Old-eugeosynclinal deposits in part deformed by one or more Paleozoic orogenies, but also part of the Cordilleran Foldbelt.

The areas covered by "Paleozoic" (?) metamorphic rocks and the localities mentioned here are shown in Fig. 3. The following notes refer to specific localities where these rocks are considered to be of Paleozoic age.

2. Areal Distribution

Mexico. In northern Mexico, that is, in the large region to the north of the Trans-Mexico Volcanic Belt, the outcrops of metamorphic rocks of possible Paleozoic age are restricted to relatively small areas.

In the State of Sonora, north of Hermosillo, towards the United States–Mexico boundary, several areas of metamorphic rocks are known, but unfortunately there is very little information regarding their composition, origin, and age (see "Carta Geológica de la República Mexicana"). de Cserna, in his map, includes them as part of the Huastecan structural belt.

In the northeastern part of the country the outcrops of metamorphic rocks are fewer and restricted to smaller areas. In the State of Tamaulipas, near Ciudad Victoria, gneiss and schists described by Carrillo-Bravo (1961) and Fries and Rincón-Orta (1965) are interpreted by de Cserna (1971a) as being of post Precambrian and pre-Middle Silurian age. Although the field relations are complex, the interpretation presented by de Cserna seems convincing. Green shists and gneisses near Aramberri, Nuevo León, are also considered as lower Paleozoic and may be of the same age as those near Ciudad Victoria.

In the geological map of the Apizolaya quadrangle, scale 1:100,000, in the State of Zacatecas, Córdoba (1965) shows an area of Precambrian metarhyolites overlain by chlorite schist and phyllite of possible lower Paleozoic age which in turn is overlain unconformably by unmetamorphosed red beds of Triassic age.

The metamorphic rocks in the State of Sinaloa, previously mentioned, at present are considered to be of Mesozoic age in the "Carta Geológica de la República Mexicana," and probably belong to a different tectonic belt. However, a recent map of this state (Carta Geológica del Estado de Sinaloa, 1:500,000, Instituto de Geología, 1970) indicates the existence of Precambrian, Paleozoic, and Mesozoic metamorphic rocks.

In southern Mexico, south of the Trans-Mexico Volcanic Belt, and particularly on the Pacific side of the country, metamorphic and igneous rocks

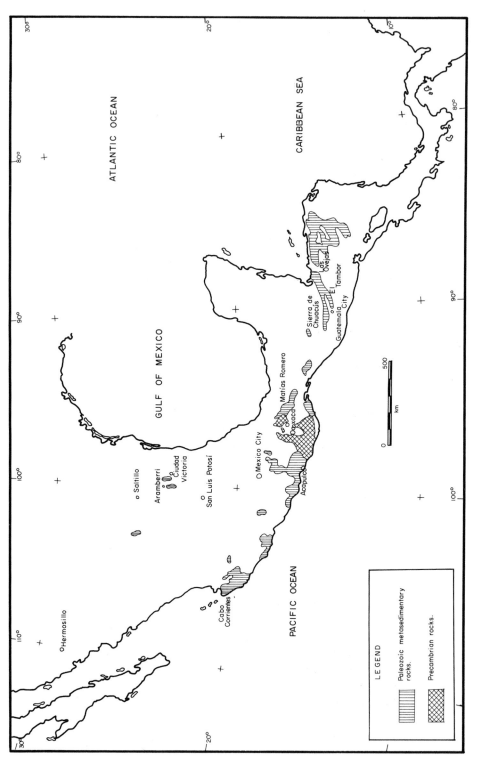

Fig. 3. Paleozoic metamorphic rocks: areas of outcrop and localities mentioned in the text.

cover an extensive area from Cabo Corrientes to the Mexico–Guatemala border. In the State of Oaxaca an extensive area of Precambrian metamorphic and igneous rocks is known. Except for this area, the other metamorphic rocks are considered to be Paleozoic, even though detailed information is available on only a few localities.

A thick sequence of metasedimentary rocks has been studied by de Cserna (1965) between El Ocotito and Acapulco, State of Guerrero, which he named the Xolapa complex. The rocks are mainly biotite schists and gneisses of the amphibolite metamorphic facies. These rocks are considered to be of early Paleozoic age (Cambrian to Devonian), although the evidence is not conclusive.

In the Isthmus of Tehuantepec, outcrops of phyllite (35 km S of Matías Romero, Oaxaca) and argillite at the Benito Juárez Dam have also been considered to be of early Paleozoic age (Webber and Ojeda, 1957; López Ramos, 1969a, 1969b). According to Pantoja Alor (oral communication to López Ramos, 1973) there are also Precambrian rocks in this area.

On the basis of structural trends in the Precambrian of Oaxaca and those of undated metamorphic rocks around it, Kesler (1973) finds support for an early Paleozoic age of the latter group. Kesler (1973) considers the rocks in the Sierra Madre del Sur, in Chiapas, to be mainly early Paleozoic and of metaigneous (quartz–dioritic) nature.

Guatemala. The metamorphic rocks that form the core of the Central Cordillera of Guatemala have been mapped in more detail than any other similar rocks in the region. In spite of this, knowledge is still restricted to certain areas, and there are many unresolved problems in regard to age and correlation, partly because some of the ranges are separated by faults of possible large lateral displacement (i.e., Polochic and Motagua faults of central Guatemala).

The original detailed study of these rocks in Guatemala was undertaken by McBirney (1963) who, under the name of the Chuacús series, described metasedimentary rocks of great thickness, consisting of garnet–biotite and staurolite–sillimanite schists and gneisses interlayered with marbles and amphibolites. A German geologic mission, headed by Hans Nicolaus, worked in Guatemala between 1967 and 1970 and mapped a more extensive area, including that studied earlier by McBirney. To date only one map has been published as a result of this work (van den Boom *et al.*, 1971). Radiometric age determinations of these rocks (Gomberg, Banks, and McBirney, 1968) indicate that some of them are probably of Devonian age and that some contain zircons derived from a Precambrian terrain. Kesler, Josey, and Collins (1970) mapped the western extension of the Chuacús series, close to the Guatemala–Mexico border, but were not able to trace them into Mexico.

South of the Motagua river (and the Motagua fault zone) both meta-sedimentary rocks and metavolcanic rocks crop out extensively and have been named the El Tambor Formation (Williams, McBirney, and Dengo, 1964; McBirney and Bass, 1969).

Schwarts (1972) has found a group of metamorphic rocks (Las Ovejas Group) south of the Motagua fault zone which are older than El Tambor. They consist of siliceous marbles interbanded with pelitic schists and amphibolites, migmatites, and minor metaquartzites. Whether these rocks are correlative with the Chuacús series or whether they are of a different age is unknown. The fact that they are overlain by Mesozoic sedimentary rocks gives some support for their possible Paleozoic age.

Honduras–Nicaragua. In northern Honduras, medium grade metamorphic rocks occur along the ranges that parallel the Guatemalan border in its eastern part and may be equivalent to the Chuacús series. In the central and eastern part of the country extensive outcrops of graphitic phyllites are known. This type of lithology is also found in southeastern Guatemala (Departamento de Chiquimula) and in the northern part of Nicaragua.

In central Honduras, Fakundiny (1970) has found older metamorphic rocks consisting of boulder conglomerates and graphite–quartz schist unconformably underlying phyllites and marble. Again, the correlation of the phyllites with other known sequences presents a problem. They may represent correlatives of the Chuacús in lower metamorphic grade or they may be younger. Some authors (i.e., Figge, 1966) have considered the possibility of a Mesozoic age, while others (Clemons, 1966) have correlated them with the unmetamorphosed Pennsylvanian of central Guatemala. This type of meta-sedimentary rock is not known south of northern Nicaragua.

Western Caribbean. In the western Caribbean, near the Central American coast, evidence of a metamorphic basement has been found. Phyllite and marble were dredged from the lower part of the continental slope between the Yucatan Peninsula and Cuba (Vedder, MacLeod, Lamphere, and Dillon, 1973; Dillon and Vedder, 1973). This find supports the interpretation that the subparallel ridges along the Belize coast represent a basement of continental crust. In the Nicaragua Rise, as far east as Rosalind Bank, wells have bottomed in meta-sedimentary rocks, indicating a submarine extension of the metamorphic basement exposed in Honduras and Nicaragua.

3. Correlation and Chronological Problems

This summary of the geographic distribution and age assignment of metamorphic rocks leads us to consider several important questions: Are all the rocks under discussion of Paleozoic age, or does an extensive Precambrian basement exist? In the event that a better separation becomes possible between the Precambrian and Paleozoic metamorphic rocks, do the Paleozoic rocks belong to one or more tectonic belts? What are the ages of the Paleozoic metamorphic rocks and of the episodes of metamorphism?

B. Paleozoic Sedimentary Rocks of Mexico and Northern Central America

1. *Areal Distribution*

Knowledge of Paleozoic sedimentary rocks in Mexico and Central America is incomplete. Excellent summaries have recently been presented by López Ramos (1969a, 1969b) which show that it is not yet possible to determine if the Paleozoic sediments were deposited in a single geosynclinal area or in several basins. Despite lack of data which makes it difficult to compile paleogeographic maps, López Ramos has presented a series of useful interpretations. The following notes are largely based on his papers and additional selected references.

In general terms, if one considers that the present geographic configuration has not changed, it is possible to outline a major geosynclinal area that extends along the central part of Mexico to the SE, bends to the east across Guatemala, and ends in British Honduras (Belize). This is probably a continuation of the Ouachita Geosyncline of the United States, although, as might be expected, the stratigraphy varies considerably from one part to another, as a result of differences in sediment provenance. However, the fact that there are discontinuities from one area to another lends itself to other possible interpretations, such as that of crustal blocks previously located in a different geographic position. At present there is not sufficient information to support these alternatives.

Whether the Paleozoic metasedimentary rocks and the unmetamorphosed Paleozoic sedimentary rocks formed part of a single geosyncline, deformed at various times with different intensities, is a problem that needs much study. From the following descriptions it is evident that in some areas of Mexico there are unmetamorphosed sedimentary rocks as old as Cambrian. On the other hand, there are also metasediments of Ordovician and Silurian age in Mexico (de Cserna, 1971b) and of possible Devonian age in Guatemala (McBirney and Bass, 1969).

According to López Ramos (1969a, 1969b), a main geosynclinal trough extended from western Coahuila to Tamaulipas, through Zacatecas, San Luis Potosí, Querétaro, and northeastern Hidalgo (Huayacocotla anticlinorium).

Four main areas of Paleozoic sedimentary rocks are known as the Sonora, Chihuahua, Tlaxiaco, and Chiapas–Guatemala basins. With the exception of the Sonora basin which lies to the west of the main geosyncline, the others are aligned along a common axis which probably represents the central part of the trough. These areas and the localities mentioned are shown in Fig. 4. The sedimentary history of each basin is different, and there is doubt whether the Chiapas–Guatemala basin was continuous at any time with the others. One major problem in unraveling Paleozoic sedimentation is the small number of pre-Permian occurrences.

Sonora Basin. This is a basin of carbonate sedimentation in which only the Silurian system seems to be absent. The Cambrian is well represented by a 1,400-m thick sequence of limestone, sandstone, quartzite, and shale near Caborca which extends eastward into the Chihuahua basin. The early Ordovician, consisting of approximately 100 m of limestones, crops out about 75 km E of Hermosillo. The presence of Silurian rocks is questionable, although rocks of this age are known in the adjoining Chihuahua basin. The Devonian consists of 279 m of limestone and dolomite, as described by Alvarez (1962) in the Murciélagos Mountains. South of Agua Prieta, in the Cabullona area (or Cabullona basin, according to Viveros, 1965), 100 m of dark-gray Devonian limestone unconformably overlies sandstones of Cambrian age. Whether this unconformity represents a local disturbance or a major tectonic event is a problem in the regional interpretation of Paleozoic history that merits study.

Thick Lower Mississippian limestones and dolomites are known 150 km ESE of Cananea, where they may reach as much as 1,500 m. Even if the entire Mississippian system is not represented in the Sonora basin, a complete section is known to the north, in the Big Hatchet Mountains, New Mexico. In Sonora the Lower Pennsylvanian has not been identified. Limestones of Late Pennsylvanian age, which reach as much as 700 m, are known W and ESE of Cananea and also NW of Nacozari. Rocks of Early to Middle Permian age (Wolfcampian–Leonardian), varying in thickness between 150 and 800 m, are known in several localities, indicating their widespread distribution. They are mainly limestones, some with reefal facies. The Upper Permian, as in the other basins, is absent, suggesting a regional episode of emergence or of orogeny.

Chihuahua Basin. Although López Ramos restricted this name to the northern part of the geosyncline, the area discussed in this summary will be extended to cover the known Paleozoic localities in the central Mexican, states of Zacatecas, San Luis Potosí, and Querétaro. The total estimated thickness of sediments is of the order of 3,000 m in the Villa Aldama area. The lower and middle Paleozoic rocks, including the Pennsylvanian, are predominantly carbonates, while the upper Paleozoic is made up of flysch type rocks (Villa Aldama), probably equivalent to those of western Coahuila, Tamaulipas, Zacatecas, San Luis Potosí, and Querétaro. The basin reached its maximum geographic extention during the Permian.

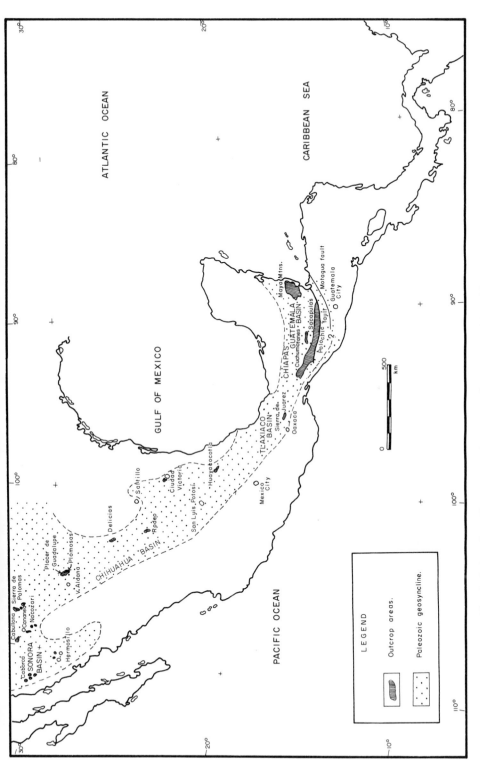

Fig. 4. Paleozoic sedimentary rocks: areas of outcrop and localities mentioned in the text.

The Cambrian, near Ciudad Victoria, Tamaulipas, consists of a thin conglomerate (40 m) and seems to be restricted to the northern part of the basin. Cambrian rocks may also have been deposited in the area of Placer de Guadalupe and in the Palomas region, in Chihuahua. Early Ordovician shales, 275 m thick, crop out in Chihuahua to the NW of Placer de Guadalupe; and Late Ordovician–Silurian to Middle Devonian limestones and dolomites, 40 m thick, have been identified S of Mina Plomosas. In the State of Tamaulipas, it is possible that limestones which unconformably underlie well dated Silurian rocks, in the area of Cañón de la Peregrina, west of Ciudad Victoria, are also of Ordovician age.

The Silurian system, as in other areas, is not well represented in this basin. Besides the above-mentioned occurrence, it has been identified in the Cañón de Caballeros, to the west of Ciudad Victoria, where it consists of 90 m of dark-gray limestones interbedded with black shales (Carrillo-Bravo, 1959, 1961). Other known localities are in the area of Mina Plomosas and Placer de Guadalupe. Devonian limestones, usually cherty, were mentioned by Alvarez (1962) as occurring in the areas of Mina Plomosas and Placer de Guadalupe. Middle and Late Devonian shales, novaculite and sandstone, of the order of 100 m thick, have been identified in the Cañón de la Peregrina, NW of Ciudad Victoria, in the Huizachal–Peregrina anticlinorium (Carrillo-Bravo, 1961).

The Mississippian System is extensively represented in northern Mexico and extends to the south, within the same basin, to Durango and San Luis Potosí. It may even have extended as a continuous depositional area to the Tlaxiaco basin in Oaxaca. In Chihuahua, NW of Sierra Palomas, it consists of 500 m of limestone (Late Mississippian), while to the east, in Tamaulipas, NW of Ciudad Victoria, it is characterized by approximately 200 m of sandstones and shales (Early Mississippian). Pennsylvanian rocks are also widespread, and probably represent the conformable continuation of sedimentation above Mississippian rocks, in spite of the fact that in some areas the section is incomplete. Well data in the Tampico area points out the possibility that the Pennsylvanian sea opened toward the Gulf of Mexico. Some of the better known localities are in Chihuahua, in the Sierra de las Palomas, where over 1,000 m of limestone crop out, and NW of Villa Aldama, where 2,500 m of Pennsylvanian shales have been measured. In Tamaulipas, NW of Ciudad Victoria, limestones, sandstones, and shales reach a thickness of 200 m.

The Permian is the most widespread Paleozoic system in the region under discussion. In the Chihuahua basin proper Permian rocks are known in many localities. In the extension of the basin into central Mexico, in Zacatecas, Querétaro, and Hidalgo, the Permian is also well represented. In broad terms it consists mainly of carbonates and, to a lesser degree, of shales. Some of the important localities are: in the Sierra de las Palomas, where limestones and dolomites of Wolfcampian age reach as much as 1,036 m; and in Placer de Guadalupe, where Wolfcampian–Leonardian reef limestones and shales are 1,400 m thick. Upper Permian rocks have not been identified in the Chihuahua basin. For other localities and details, the reader is referred to López Ramos (1969*b*, p. 2413).

Tlaxiaco Basin. López Ramos (1969*b*) considers this area, in the State of Oaxaca, as a subbasin of the geosyncline, perhaps because it is more restricted. In this area Paleozoic rocks are known also in the Sierra de Juárez and in Nochistlán, approximately 70 km NW of Oaxaca City. They have been described by Pantoja and Robison (1967). Rocks of Cambrian, Ordovician, and Carboniferous ages have been identified, while the Silurian and Devonian are missing. This leads to the conclusion that it was an isolated basin during the early Paleozoic and became integrated into the main geosyncline during the late Paleozoic (Pennsylvanian or Mississippian).

The Cambrian–Ordovician, which crops out 15 km NNE of Nochistlán, consists of 30 m of thin-bedded limestone and shale lying on Precambrian metamorphic rocks. Northeast of Nochistlán, sandstones, shales, and conglomerates 192 m thick have been assigned to the Mississippian, while siltstones and sandstone S of Ixtáltepec are considered to be of Pennsylvanian–Permian age. The stratigraphic sections between this basin and the Chiapas–Guatemala basin to the east are very different, and, therefore, it is possible that they never formed a continuous depositional area.

Chiapas–Guatemala Basin. The basin is known from west central Chiapas, in Mexico, across Guatemala, to the Maya Mountains of British Honduras (Belize). Its possible continuation into the Caribbean is unknown, although some speculations have been made (Meyerhoff, 1966; Dengo and Bohnenberger, 1969). No pre-Pennsylvanian sedimentary rocks have been dated on the basis of fossils in this area, and, if any of the older Paleozoic systems are represented here, it is by metasediments.

The lower part of the sedimentary sequence consists of slightly metamorphosed coarse conglomerates, well exposed near Sacapulas, in north-central Guatemala, and conglomerates and coarse sandstones in the Maya Mountains of British Honduras, where they had been originally considered as part of the "metamorphic complex" (Dixon, 1956) but are now known to be part of the upper Paleozoic sedimentary sequence (Bateson and Hall, 1971). These rocks are conformably overlain by a thick sequence of black shales and argillites, the upper part of which carries Late Pennsylvannian–Early Permian fossils. Whether the lower part of the sequence is of Mississippian age, as has been proposed by some authors (Eardley, 1954, p. 715; Maldonado-Köerdell, 1954, p. 124), is still undefined and could only be established by radiometric dating. The total thickness of this clastic sequence is probably greater than 3,500 m.

The Permian, in Chiapas and Guatemala, is characterized by alternations of limestone and shale, conformable over the lower coarse clastic unit, and a thick limestone unit above, reaching 2,400 m in Chiapas (Gutiérrez-Gil and Thompson, 1956) and 1,500 m in the Cuchumatanes Mountains of Guatemala (Clemons and Burkart, 1971). In the Maya Mountains of British Honduras the Permian consists almost entirely of black shales, with only thin intercalations of limestone. The known Permian rocks in this basin are of Wolfcampian and Leonardian age. The Upper Permian has not been identified.

Unmetamorphosed Paleozoic sedimentary rocks are not known in Central America, south of central Guatemala.

C. Paleozoic Deformational and Igneous Episodes

Deformational trends are really the features that define tectonic belts as used in this paper. As is known from other areas, deformational trends usually follow the earlier trends of geosynclinal deposition. In the case of the Paleozoic rocks of Mexico and Central America, correlation problems make it difficult to identify either deformational episodes or trends. An attempt will be made here to arrive at some conclusions based on the data previously summarized.

The early Paleozoic tectonic history is, of course, the most difficult to understand. From the scanty information of the Ciudad Victoria area one can infer a deformational episode that caused overthrusting of pre-Middle Silurian schists above Precambrian rocks (de Cserna 1971a). In southern Mexico, near Acatlán (Puebla), de Cserna (1971a) has described an upper Ordovician metavolcanic and metasedimentary sequence with an adjacent thrust slice. In the Sonora Basin an unconformity seems to exist between the Cambrian and the Devonian, while in the Tlaxiaco Basin the Late Ordovician is missing and the Cambro-Ordovician sedimentary rocks are overlain by Mississippian and younger rocks. These scattered facts, some taken from López Ramos (1969b) and others from de Cserna (1971a), have led the latter author to postulate a Taconic orogeny in Mexico, extending at least as far south as Oaxaca. Evidence for a similar orogeny has not been found thus far in Chiapas or Central America. As has been pointed out by Rodgers (1971), the Taconic orogeny of eastern North America was not a single event at the end of the Ordovician Period but a series of orogenic episodes spread over part of that period.

If the age of the Chuacús group in Guatemala is Devonian and older, and if these rocks were metamorphosed at the end of the Devonian, as radiometric dates of 374 m.y., suggest (McBirney and Bass, 1969; Dengo, 1968, p. 13), a strong orogenic episode must have taken place at that time. This was suggested several years ago by Eardley (1954, p. 718), who considered it as equivalent to the Acadian orogeny of North America. The Paleozoic metamorphic basement of northern Central America forms an arcuate trend from Chiapas to the Caribbean. Analyses of the structural trends in different areas has led Kesler (1971) to believe that this is an ancestral pattern which controlled later folding in the area. It is not yet possible to assign an age to the deformational episodes that affected the metamorphic rocks of Honduras (Fakundiny, 1970).

As was pointed out in the description of the Paleozoic sedimentary rocks, Late Permian sediments are not known anywhere in the region. Several authors following the summary of Central American geology by Sapper (1937), have considered that a strong and widespread orogeny took place during Late Permian to Early Jurassic time. This idea was challanged by McBirney (1963), who considered it mainly an episode of uplift and land emergence. Field evidence found by Bohnenberger north of Nebaj, in northern Guatemala (Dengo and Bohnenberger, 1969, p. 211), where sedimentary Paleozoic rocks overturned to the north are unconformably overlain by Late Jurassic–Early Cretaceous sediments, gives support to the original idea. Uplift and folding during Late Permian is also supported by field evidence in the Cuchumatanes Mountains of Western Guatemala (Burkart and Clemons, 1972). Whether this orogeny extended along the Paleozoic geosyncline in Mexico is still an unresolved problem. However, for northern Mexico, Alvarez (1962) has postulated a Late Permian orogenic episode.

Little can be said in regard to the igneous episodes and to their relation to regional deformational events. Near Acatlán, Puebla, a radiometrically dated granitic pluton of Late Ordovician age (440 ± 50 m.y. by lead-alpha method, according to de Cserna, 1971b) may be correlated with the so-called Taconic orogenic disturbances. The so-called Chiapas batholith in the southeastern part of the State of Chiapas is classified as Paleozoic in the "Carta Geológica de la República Mexicana" and is considered as middle Paleozoic by Guzmán and de Cserna (1963, p. 115). This is not a single igneous body but a complex of metaigneous rocks (Kesler, 1973) and intrusives, probably of different ages—some as young as Cretaceous—since other intrusions of this age are common along the Pacific coast of Mexico and in central Guatemala.

The age of the intrusive rocks associated with the metamorphics in central Guatemala has been a debatable subject. McBirney and Bass, 1969, have established a Devonian radiometric age for the Rabinal granite which is associated with the Chuacús series; Moody (in Dengo, 1969, p. 13) gives an age of 235 ± 35 m.y. for a granite pluton in the Maya Mountains of Belize. These intrusives can be correlated, respectively, with the Late Devonian and the Late Permian orogenic events. Other localities of Permian igneous rocks are known in northern Mexico and in Belize, practically at the two extremes of the region of Paleozoic rocks. de Cserna et al. (1970) have described a Lower Permian rhyolite flow in the Placer de Guadalupe area in northeast Chihuahua.

In Belize, Bateson and Hall (1971) have determined that a rhyolite pyroclastic unit, in the Maya Mountains, is interlayered in the Lower Permian sedimentary rocks. These volcanic rocks had earlier been interpreted by Dixon (1956) as an intrusive body.

D. Tectonic Belts

De Cserna, in his Tectonic Map of Mexico (1961), and Guzmán and de Cserna (1963) have included the Paleozoic rocks as part of two structural belts, defined as follows on the map:

> The *Jaliscoan Structural Belt* extends from northwestern Sonora to southeastern Chiapas paralleling the present Pacific coast of Mexico. It became consolidated during the *Jaliscoan geotectonic cycle*, which lasted from the Cambrian to the end of the Paleozoic (?).
>
> The *Huastecan Structural Belt* extends from northern Coahuila to Zacatecas and Tamaulipas, continuing southwestward into Chiapas, parallel to the present coast of the Gulf of Mexico; it became consolidated during the *Huastecan geotectonic cycle*, which lasted from the Cambrian until the end of the Middle Jurassic. The Huastecan structural belt may be a southward continuation of the Ouachita structural belt.

Dengo and Bohnenberger (1969) and Dengo *et al.* in the Metallogenic Map of Central America (1969), following the same lines of thought, included the metamorphic rocks of northern Central America as part of the Jaliscoan structural belt. King (1969) presented a different concept, and includes most of the Paleozoic rocks as part of the more extensive (in time and space) Cordilleran foldbelt. Newer data, and the paleogeographic interpretations of the sedimentary Paleozoic rocks by López Ramos (1969b), will probably make it necessary to revise and change the above-mentioned classifications and to establish more restricted tectonic belts as part of a larger foldbelt in the sense of King's definition (1969).

Guzmán and de Cserna (1963, Fig. 5, p. 120) proposed some broad paleogeographic interpretations for the Paleozoic, with a Precambrian cratonic area in the central part of Mexico, extending from the United States border as far south as Oaxaca, with geosynclinal sedimentation to the east (Huastecan cycle) and to the west (Jaliscoan cycle).

For Central America, Dengo (1968) and Dengo and Bohnenberger (1969) proposed a geosynclinal area covering all of Guatemala, El Salvador, Honduras, and northern Nicaragua, south of a cratonic area in the present position of the Yucatan Peninsula and an open sea to the south. López Ramos (1969b) on the other hand, proposes for the sedimentary Paleozoic rocks a different situation with a geosynclinal trough along central and eastern Mexico, at times connected with the Gulf of Mexico, and with land masses on both sides of it. An interpretation of the known Precambrian localities in Mexico by Kesler and Heath (1970) presents a structural belt of Grenville age which lends support to the ideas of Guzmán and de Cserna (1963) for the pre-Paleozoic conditions, but not for the distribution of land masses in middle and late Paleozoic time.

Although it is difficult to reconcile all these different ideas, the author is inclined to believe that the Paleozoic metamorphic rocks belong to two different tectonic belts, one on each side of the central cratonic backbone of Mexico, and added to it by accretion. Downwarping of the axis of the craton, initiated as local basins and finally forming a single geosyncline, formed the locus of sedimentation for the Paleozoic sedimentary rocks. Therefore, this feature should be considered as a separate tectonic belt.

So far, the author has purposely avoided discussing the possibility of plate boundaries within the region during the Paleozoic, mainly for lack of enough supporting evidence. However, since he favored the idea of independent deformation in adjoining crustal blocks (Dengo, 1967), even before the concept of plate tectonics was fully established as a model, this possibility should be taken into consideration in trying to understand the existence of metamorphosed and unmetamorphosed rocks of similar age within the same region.

IV. MESOZOIC TECTONIC BELTS OF MEXICO AND CENTRAL AMERICA

A. Mexico and Northern Central America

1. Regional Distribution of Mesozoic Sedimentary Rocks

Major Sedimentary Areas. The state of knowledge of the Mesozoic sedimentary rocks of Mexico and northern Central America presents quite a different situation from that for the Paleozoic. Extensive areas are covered by these rocks in sequences that permit accurate age determination on the basis of fossil content. The occurrence of petroleum, particularly in northeastern and eastern Mexico, has contributed to the detailed knowledge of the subsurface stratigraphy.

In northern and eastern Mexico, Mesozoic rocks form the bulk of the Sierra Madre Oriental, which extends from the Big Bend of the Rio Bravo, or Rio Grande, to the southeast, through the States of Coahuila, Nuevo León, Tamaulipas, and San Luis Potosí for a distance of over 1,000 km, and has a width between 100 and 150 km (Garfias and Chapin, 1949, p. 55).

The Sierra Madre Oriental, according to other authors (Alvarez, 1961; Guzmán and de Cserna, 1963), continues along the States of Oaxaca and

Chiapas in southeastern Mexico, south of the Trans-Mexican Volcanic Belt. From Chiapas to the east, across Guatemala, the sedimentary Mesozoic rocks crop out extensively along the frontal ranges of the Sierras of Northern Central America (Dengo and Bohnenberger, 1969) and extend to the north over the Yucatan platform.

The major depositional area of Mesozoic sedimentary rocks, from the United States–Mexico boundary to Guatemala, has been referred to as the Mexican Geosyncline since the work of Schuchert (1935). Whether this geosyncline continued eastward over the present-day Greater Antilles is still a debatable subject. Other areas of Mesozoic sedimentary rocks are known throughout the region, and some probably represent embayments or smaller basins of the Mesozoic seas. Most prominent among these are the Balsas basin in the state of Guerrero, Mexico, and the Ulúa and Mosquitia basins in Honduras and northern Nicaragua. Mesozoic features and localities are shown on Fig. 5.

Sierra Madre Oriental. The Sierra Madre Oriental proper, that is, between the United States–Mexico boundary and the Trans-Mexico Volcanic Belt, is one of the most prominent topographic features in the region. Its present configuration is controlled to some extent by pre-Mesozoic structural conditions which resulted from the Permian–Triassic orogeny (Appalachian). The main features were an extensive landmass along the central and western part of Mexico, now concealed to a large extent by the volcanic rocks of the Sierra Madre Occidental, a long and narrow area of downwarping which was the locus of the Mesozoic or Mexican Geosyncline, and a shelf area to the east, along the present-day Gulf of Mexico coastal plain. A large uplifted area controlled Mesozoic deposition on the north. This area is known as the Coahuila Peninsula (for detailed references, see Murray, 1961, p. 133–134).

Recent information on the stratigraphy of parts of the Sierra Madre Oriental and references to earlier works are found in the study by de Cserna (1956) of the area between Torreón and Monterrey and in the works of Weidie and Wolleben (1969) and Bishop (1970). Summaries have been presented by Garfias and Chapin (1949), Benavides (1956), and Murray (1961, p. 297–320). The earlier extensive reconnaisance of the area, done in 1920–1921, is that of Baker (1971). Lack of space allows only a brief description here.

The Triassic system has not been definitely identified in the northern part of the Sierra Madre Oriental but may be represented by red conglomerates and sandstones in the Huizachal valley, SW of Ciudad Victoria (Garfias and Chapin, 1949). The Huizachal Group, from this area, has been discussed by Mixon, Murray, and Díaz (1959), and it is mainly of Jurassic age.

The Lower Jurassic consists of terrigenous red shales, siltstones, and sandstones, up to 800 m thick, in the Sierras del Toro and San Julián, which,

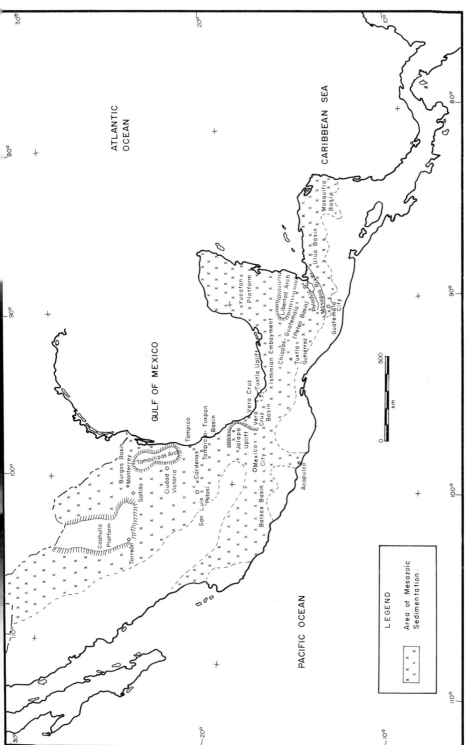

Fig. 5. Mesozoic sedimentary rocks: areas of deposition and localities mentioned in the text.

according to de Cserna (1956) are a molasse deposit previous to the Late Jurassic marine transgression. Gypsum deposits, up to 100 m near Monterrey, are also of possible Early Jurassic age (Weidie and Wolleben, 1969).

The Upper Jurassic (Oxfordian to Portlandian) consists of a sequence, over 1,500 m thick, of nearshore shaly limestones and intercalated sandstones and shales, for which detailed descriptions and localities are given by Imlay (1938) and de Cserna (1956), considered as a litoral facies of the early stages of the Mexican Geosyncline.

The entire Cretaceous system is represented in the Sierra Madre Oriental. The older series crop out in the central and northern part of the Sierra, where they conformably overlie Late Jurassic rocks. The cumulative thickness of the Cretaceous, in the Torreón–Monterrey area, according to the figures given by de Cserna (1956, p. 10), may exceed 3,000 m for the Lower Cretaceous and 7,000 m for the Upper Cretaceous formations.

In other areas of the Sierra, such as southern Nuevo León and southwest Tamaulipas, the complete Cretaceous system is also known. Sections that range in age from Turonian to Maestrichtian are known in San Luis Potosí, in the area of Cárdenas and Tamasopo. Calcareous strata predominate in the Cretaceous sequence, with the exception of a lower clastic sublitoral facies, and, in the uppermost formations of Coniacian to Maestrichtian age, a flysch facies of fine clastics.

The frontal ranges of the central part of the Sierra Madre Oriental are made up mainly of litoral reefal facies of Aptian to Cenomanian age. These rocks are very extensive in the Gulf coastal plain and offshore in the Gulf of Mexico, where they form important petroleum reservoirs. Details of their stratigraphy are given, among others, by Mena Rojas (1955), Flores-Revueltas (1955), Benavides (1956), Barnetche and Illing (1956), Pessagno, Jr. (1969), and Coogan, Debout, and Maggio (1972). Three major embayments, separated by local uplifts, are known along the Coastal Plain area. They are the Burgos basin in northeastern Mexico, separated by the Tamaulipas arch to the south from the Tampico–Tuxpan basin, which in turn is separated on the south by the Jalapa uplift from the Veracruz basin. These basins also formed the locus for thick Tertiary sediments (see Fig. 5).

Balsas Basin. To the south and west of the depositional axis of the Mexican Geosyncline, there are several areas of Mesozoic sedimentation, of which the most extensive is the Balsas basin in the State of Guerrero. The oldest Mesozoic formation, considered to be of Late Triassic–Early Jurassic age, consists of 800 m of clastic volcanic rocks, conglomerates, and sandstones, which are of continental origin (Bolívar, 1963; de Cserna, 1965). Unconformably over these rocks rest Albian to Cenomanian argillaceous limestones up to 500 m thick. These are separated by a nonconformity from Turonian calcilutites and calca-

renites 30 m thick. An upper clastic sequence, up to 1,000 m thick, and of Senonian age, consists of calcarenites, calcareous shales, siltstones, and sandstones. The Mesozoic sedimentary conditions and history of this basin were
obviously quite different from those of the geosyncline.

Southern Mexico–Guatemala. Mesozoic rocks occur over a large area in
Chiapas, northern Guatemala, and Belize in the Sierra Madre Oriental and
the Sierras of northern Central America. Isolated outcrops and subsurface
data indicate times of depositional continuity with northeastern Mexico,
although some of the features are very different, in particular, the salt basins
of southeastern Mexico.

The stratigraphic sequence starts with a red-bed molasse deposit (Todos
Santos Formation) which had been previously assigned to the Triassic or
Jurassic. Recent information (Richards, 1963; Anderson et al., 1973) places
it as Late Jurassic–Early Cretaceous. In eastern Chiapas, however, some
outcrops are still considered as possibly Triassic (Mapa Geológico de Chiapas,
López Ramos, 1972).

The age and stratigraphic relations of the salt deposits have also been
debatable. Recently Viniegra (1971), on the basis of subsurface data in the
Isthmus of Tehuantepec, has shown that there are salt deposits of two different
ages. The oldest is dated as pre-Kimeridgian, but not older than Oxfordian.
It extends from Chinameca east to the Guatemalan border, and it probably
underlies the red-bed sequence. The second group of salt beds lies between
Upper Jurassic strata below and Lower Cretaceous beds above and may
extend from Chiapas east across Guatemala to Belize. It is possible, then,
that part of the salt deposition was coeval with continental red-bed deposits.

Except for a thin litoral sandstone, which marks the beginning of a marine
transgression from the north, the Cretaceous sequence consists mainly of
carbonate and evaporitic rocks which attain a great thickness (over 3,500 m)
in the frontal ranges of the Sierra Madre in Chiapas and in the Sierras of
northern Central America and the Petén basin of northern Guatemala–Belize.
Limestone and dolomites along the frontal ranges grade northward into
anhydrites. Lack of fossils, particularly in the anhydrite, prevents exact age
determinations. The lower unit (Sierra Madre Formation in Chiapas, or Cobán
Formation and equivalents in Guatemala) has been considered by Vinson
(1962) to range from Neocomian to Turonian, with only one part well dated
as Aptian–Albian. The upper limestone over it (Campur Formation) is more
fossiliferous and is of Coniacian–Santonian age.

It is difficult to establish regional unconformities in the sequence. Local
disturbances along the frontal ranges mark depositional breaks, but the
subsurface information in the deeper part of the original basin points to pos-

sible continuous sedimentation from Neocomian to Santonian time. An important post-Santonian stratigraphic break is marked by the deposition of a flysch sequence, of Maestrichtian to early Eocene age, about 1,000 m thick (Chubb, 1959; Bermúdez, 1963) along a narrow trough parallel to the frontal ranges of the Sierras.

Southern Guatemala–Honduras. From central Guatemala (south of the Motagua river) to northern Nicaragua, including a large part of Honduras and a small area in El Salvador, Mesozoic sedimentary rocks crop out and show some differences from those of Chiapas and northern Guatemala.

According to Mills *et al.* (1967), there were two depositional basins, namely, the Ulúa in north central Honduras and southern Guatemala and the Mosquitia in southeastern Honduras–northern Nicaragua, which they considered to be of intracontinental nature because of their position to the south of the major depositional area of the Mexican Geosyncline. There is the possibility that the depositional area was more extensive, even continuous with northern Guatemala, but that as a result of erosion and extensive cover of Tertiary volcanic rocks the original extent has been obscured.

The only sedimentary rocks of Late Triassic–Early Jurassic age known today in Central America occur in central Honduras and consist of plant-bearing dark-gray sandstones and siltstones of the El Plan Formation (Carpenter, 1954; Mills *et al.*, 1967), which attain a thickness of 300 m. These rocks are highly contorted and are unconformably overlain by a thin sandstone unit probably equivalent to the Late Jurassic–Early Cretaceous continental red-bed sequence (Todos Santos Formation). Mills *et al.* (1967, Fig. 30) have suggested that the El Plan rocks were deposited along a NE-trending trough that extended from central Honduras toward the Caribbean.

Red-beds of varying thickness crop out in several areas and are probably equivalents of the similar sequence in Chiapas and northern Guatemala. However, their age has not been determined. They occur under Cretaceous carbonates whose lower age limit is yet unknown but which are at least of Albian age.

A thick sequence of limestones, including a variety of shallow-water depositional facies not yet well understood, attains as much as 1,000 m in thickness in central Honduras; it is known as the Yojoa Group of Aptian to possible Cenomanian (?) age and has been divided into several formations (Mills *et al.*, 1967; Burkart, Clemons, and Crane, 1973). It is overlain by widely distributed red clastic continental rocks up to 900 m thick. These are locally interbeded with thin limestones of Cenomanian age, which indicate a different depositional history than that described for the Chiapas–northern Guatemala area.

2. *Mesozoic Deformational and Igneous Episodes*

It is difficult to present a complete picture of the deformational and igneous events that took place over such a large area during the Mesozoic era. Obviously only the major orogenies affected the complete region, and those are the only ones considered in the following summary. Tectonic disturbances that are only of local importance are not within the scope of this paper.

After the late Paleozoic–early Mesozoic orogeny, most of the area was emergent, and Triassic sedimentary rocks were deposited only locally, as in the Huizachal Valley in northeastern Mexico, the Balsas Basin, and in a very restricted area in central Honduras. Jurassic sedimentation was extensive in northeastern Mexico, and, according to de Cserna (1970, Fig. 4) there was an Early Jurassic deformation which produced the Zacatecas–Guanajuato thrust front. This information, and the fact that the Late Triassic–Early Jurassic rocks in central Honduras were also deformed before subsequent Mesozoic deposition in that area, points to an intra-Jurassic deformational episode probably equivalent to phases of the Nevadan orogeny. There is not enough information to determine if there was a large tectonic belt throughout the region resulting from that orogeny.

The red-bed molasse deposits of Late Jurassic–Early Cretaceous age in southern Mexico and northern Central America point to taphrogenic conditions previous to the extensive marine sedimentation which began with the Early Cretaceous transgression from the Gulf of Mexico. Although most of the Cretaceous carbonate deposition was characterized by shallow-water and shelf conditions, a deep and elongated area subsided more rapidly, allowing a greater thickness of sediments to accumulate, and it is on this basis that the term Mexican Geosyncline has been used (see Fig. 5). Cretaceous deposition varied along the entire area, and, as indicated in the previous sections, local unconformities or discontinuities mark intra-Cretaceous tectonic disturbances, some of which are only of local importance. In northern Mexico a break between the Lower Cretaceous (Comanchean) and the Upper Cretaceous (Gulfian) is well established, and it is possible that it also occurred in Chiapas and northern Central America, although the evidence is poor, except in southern Guatemala and Honduras.

The most widespread regional deformation occurred during Late Cretaceous to early Eocene time and corresponds to the Laramide Orogeny. This deformation produced strong folding and thrusting and was accompanied by concomitant flysch deposition, largely of Paleocene age, along elongated troughs in front (towards the Gulf of Mexico) of the deformed belt. In different areas along the strongly deformed belt, intense folding has been interpreted as a result of *decóllement* of carbonate rocks sliding over evaporitic deposits or as result of evaporite diapirism. This seems to be the case in eastern Chihuahua (Gries and Haenggi, 1970; Weidie and Martínez, 1970), Chiapas (Viniegra, 1971), and Alta Verapaz in Guatemala (Dengo and Bohnenberger, 1969). The main Mesozoic tectonic belt, which controls much of the present-day topographic configuration, is indicated in Fig. 6.

Guzmán and de Cserna (1963) have indicated the possibility of "eugeosynclinal" Mesozoic rocks to the west of the Mexican Geosyncline. Although the information is not conclusive, there is the possibility that they are concealed by Tertiary volcanic rocks of the Sierra Madre Occidental, and that the Mesozoic metamorphic rocks of the State of Sinaloa are part of that sequence. However, on account of their geographic position, the writer is inclined to consider these rocks as related more to Mesozoic interaction between the Pacific Ocean and the continental mass of western Mexico, and as part of a different tectonic belt, also of Mesozoic age.

In an unpublished paper Wilson (1974) presents evidence for possible "eugeosynclinal" Cretaceous rocks in central Guatemala and for the complexities of tectonic deformation in that area and Honduras during Cretaceous time. It is still difficult to determine if this new information can be correlated to other areas in the region. However, it does not alter the generalizations presented here.

Several Mesozoic "granitic" (mostly dioritic to granitic) bodies are known in the region, and several of them have been dated radiometrically as middle to Late Cretaceous. A series of batholiths and stocks is known along the Pacific coast of Mexico from Bahía Banderas to the Isthmus of Tehuantepec, and its possible continuation may be traced to the east across Guatemala, Honduras, and northern Nicaragua (Fig. 6). Most of them seem to be related to activity and deformation during Aptian–Albian time or at the end of the Cretaceous and are, therefore, related to the above-mentioned tectonic disturbances. One interesting fact is that in Mexico the line of intrusives and the belt of greatest deformation of the Mesozoic sedimentary rocks are widely separated, while they seem to converge in northern Central America.

Serpentinite and serpentinized peridotite are widespread in central Guatemala and northern Honduras where they occur parallel to several major fault zones (Polochic, Motagua, Chamelecón; see Fig. 6). These rocks are now believed to have been emplaced during Cretaceous time (probably middle to Late Cretaceous) and should, therefore, be related to the Laramide deformational events (Bonis, 1967; Dengo and Bohnenberger, 1969; Dengo, 1972). The significance of their location in only one part of the Mesozoic tectonic belt is discussed later.

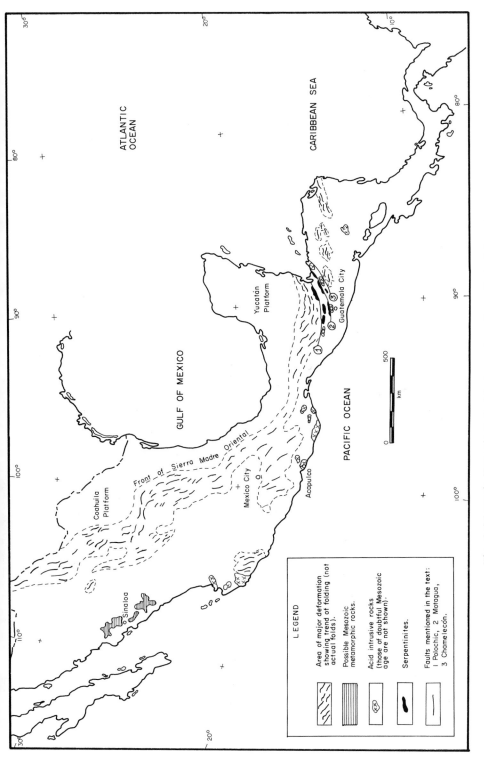

Fig. 6. Mesozoic Tectonic Belt: Mexico and northern Central America.

B. Southern Central America

1. *The Basement of Southern Central America*

The morphotectonic unit defined as the Mountain Ranges of southern Central America is geologically very different from the northern part of Central America. This unit, which extends from southern Nicaragua to the southeast and east, along Costa Rica and Panama into western Colombia, has a Mesozoic basement formed largely by basic volcanic rocks.

The major tectonic events, which gave the unit its present characteristics, happened during Tertiary and Quaternary time. Therefore, it is more properly a Cenozoic than a Mesozoic tectonic belt. If it is considered, however, that important tectonic events during the late Mesozoic gave rise to its basement and to its oldest known sedimentary rocks, it seems appropriate to describe here those events and the involved rock types, even though they are not yet well known or understood. The distribution of basement and sedimentary Cretaceous rocks is shown in Fig. 7.

In Costa Rica this basement is known as the Nicoya complex and consist of basalts, including pillow lavas and agglomerates, diabase and gabbro intrusives, associated with dense dark graywackes, cherts, and siliceous aphanitic limestones. More acid intrusive rocks are restricted to very small areas. Part of the basement represents a typical ophiolite suite (Dengo, 1962*a* and 1962*b*). In Panama the basement crops out more extensively in the eastern mountain ranges, as well as in some of the Pacific peninsulas (del Giudice, 1973). In this country, however, there are more acidic intrusives, for instance, in the Azuero Peninsula and in the San Blás Mountains. Associated with the basement are ultramafic rocks which crop out in the Santa Elena Peninsula of northwest Costa Rica and in Azuero and the eastern Darién Mountains of Panama. As part of the basement, or overlying it, are cherts of lower Campanian age both in Costa Rica and Panama (Dengo 1962*b*; Henningsen and Weyl, 1967; Bandy and Casey, 1973; Case, 1973). On this basis, and with additional information from a few radiometric dates on basalts from the Nicoya Peninsula (Barr and Escalante, 1969) and diorites from Azuero (del Giudice, 1973), it is possible to assign the basement to early Campanian; it may be even older.

The similarity of these rocks with parts of eastern Ecuador was pointed out by Dengo (1962*b*). Recent studies by Case *et al.* (1971) and by Goosens and Rose, Jr. (1973) support the extension of a belt of Upper Cretaceous (and perhaps older) basic volcanic rocks and associated sediments along the Pacific Coast of northwestern South America to the Nicoya Peninsula in Costa Rica.

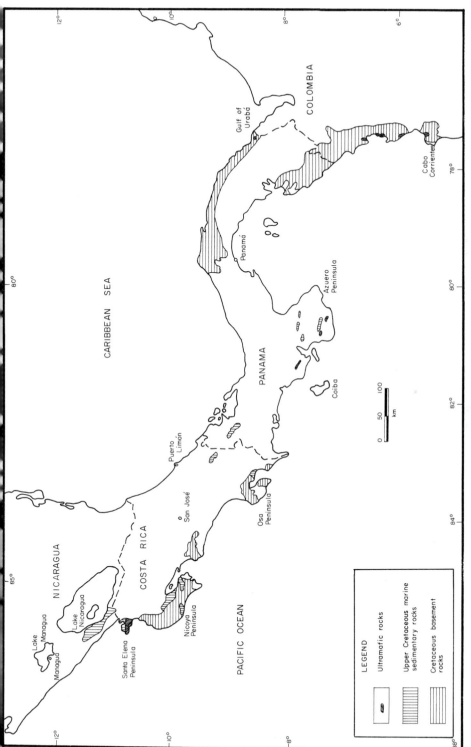

Fig. 7. Mesozoic basement and Cretaceous sedimentary localities of southern Central America.

2. *Mesozoic Sedimentary Rocks*

Besides the lower Campanian cherts, Upper Cretaceous marine sedimentary rocks crop out at different places in the area. They are best exposed in southwestern Nicaragua (Zoppis-Bracci, and del Giudice, 1958) and the Nicoya Peninsula of Costa Rica (Dengo, 1962*b*), where they consist of as much as 2,370 m of sandstone and dense siltstone, largely tuffaceous, with thin interbedded limestones of Campanian to Maestrichtian age. Isolated outcrops of marine sedimentary rocks of the same age are known in eastern Costa Rica and western Panama (Fisher and Pessagno, 1965), in Azuero (del Giudice, 1973) and in northwestern Colombia (Case *et al.*, 1971). From this fragmentary information, it is difficult to establish if the Campanian–Maestrichtian marine sequence was deposited along a continuous area or if it was restricted to very small basins separated from each other by earlier uplifts, probably islands, as suggested by Lloyd (1963).

3. *Deformational Episodes*

The basement of southern Central America is regarded by several authors as elevated oceanic crust (Dengo, 1962*a*; Henningsen and Weyl, 1967; Case *et al.*, 1971; Bandy and Casey, 1973). The early Campanian cherts, in particular those of the Golfo de San Miguel in Panama, which are associated with extensive pillow basalts, have been correlated by Bandy and Casey (1973) and by Case (1974) with the lower reflector horizon of the ocean basins, specifically with reflector "B" identified in the Caribbean by Ewing *et al.* (1965).

The fact that there are Campanian–Maestrichtian clastic sedimentary rocks in the area indicates that the basement had emerged during the Campanian. In southwestern Nicaragua and adjacent parts of Costa Rica, north of the Santa Elena Peninsula, marine deposition was continuous from Maestrichtian to early Eocene. On the other hand, in the central and southern part of the Nicoya Peninsula, only a short distance from the Santa Elena Peninsula, there is evidence of tectonic deformation during Paleocene and probably Eocene (Dengo, 1962*b*).

The type of deformation of this area at the end of the Cretaceous and beginning of the Tertiary is difficult to decipher but should be related to interaction between the Caribbean and the Pacific crusts. The correlations established before point to a tectonic belt in southern Central America and western South America which began to develop in Late Cretaceous and became well defined during the Tertiary (Dengo, 1968; Butterlin, 1972; Case *et al.*, 1971). The characteristics of this tectonic belt are different from those of other belts of Mexico and northern Central America.

V. PALEOZOIC AND MESOZOIC TECTONIC BELTS OF MEXICO AND CENTRAL AMERICA AND GLOBAL TECTONICS

The approach used here to define tectonic belts is more the traditional one of geosynclinal areas and deformation along elongated zones than the modern one of tectonic plates. An effort has been made to present facts rather than interpretations. Although the author favors a plate tectonic interpretation, in his opinion practically all the explanations for the origin of the Gulf of Mexico and the Caribbean, following plate models, have been based largely on geometric rather than geological considerations. Not all the geological data available have been taken into consideration for such reconstructions as those proposed by Carey (1958), Freeland and Dietz (1971), and Malfait and Dinkelman (1972), to mention only some of the most elaborate ones.

In trying to make a contribution to the understanding of the region, some of the important problems which are evident from the previous description will be emphasized in this section even though they may appear quite obvious. These are problems which need further research before a satisfactory interpretation of the regional tectonics can be presented. The main ones, in the author's opinion, are the following, summarized in chronological order:

1. The age definition and correlation of the Paleozoic metamorphosed sedimentary rocks. Some of the proposed reconstructions of the region call for small crustal blocks originally in quite different positions than they are today, which moved and became joined. For this to be the case, major breaks should exist at least along the Isthmus of Tehuantepec and along the Motagua or the Polochic fault zones. It is essential to know the ages of the Paleozoic metamorphic rocks in each one of the blocks to either support or disregard said interpretations, or to propose other possible structural relations.

2. A similar problem exists with the correlation of the Paleozoic sedimentary rocks. The main problem here is to establish if the present distribution of these rocks represents the original configuration of a geosyncline, or if there were isolated basins with different geological histories, originally located in different relative positions. As was indicated before, the lower Paleozoic rocks are the more difficult to trace from one area of outcrop to another. In particular, the different pre-Carboniferous stratigraphy between the Tlaxiaco and the Chiapas–Guatemala basins indicates some interruption or break in the Isthmus of Tehuantepec.

The late Paleozoic stratigraphy gives enough support to the single geosyncline interpretation, although there are enough differences again between the Chiapas–Guatemala basin and those to the north to leave the possibilities open for a major depositional break in the Tehuantepec area. Since no Permian sedimentary rocks are known south of the Motagua fault zone, it is difficult to determine the tectonic relations at that time between the two sides of the fault zone. Was the Permian south of the Motagua totally eroded? Is it represented by undated low-grade metasediments? Or were the blocks north and south of the Motagua in very different relative positions during Permian time?

3. The Permo-Triassic puzzle. The absence of Late Permian sediments throughout the region was stressed before in support of the proposed orogeny and uplift at that time, continuing into the Early Triassic. Direct stratigraphic information for the time span between Late Permian and Triassic is scanty, in particular in southern Mexico and Central America. Better age definitions as well as correlations of the "known" Triassic and subsequent Jurassic red beds and evaporites, and correlations with similar sequences in the United States and northern South America, will be necessary to support or reject some of the proposed theories for the origin of the Gulf of Mexico and the Caribbean, which involve crustal separations and motions.

4. Laramide orogeny. Most authors agree that from Late Jurassic through Cretaceous, Mexico and northern Central America were essentially as they are today, with a large area along the Gulf margin submerged as a major marine depositional trough or geosyncline. The entire area was strongly deformed during the Laramide orogeny with thrusting and overturning toward the Gulf of Mexico. The Laramide orogeny was also strong around the Caribbean, with thrusting and overturning away from it, that is, to the north in Cuba and to the south in northern Venezuela.

Whether the Gulf of Mexico is considered to have developed a simatic crust in late Paleozoic (Wilhelm and Ewing, 1972), or whether it was the result of plate movements, the fact is that during the Laramide orogeny it must have behaved as a stable area. Practically the same is true for the Caribbean, which has behaved as a plate at least since late Mesozoic (Ewing *et al.*, 1965).

Some of the Laramide features are better explained by plate tectonic models. For instance, the large faults of northern Central America (Polochic, Motagua, Chamelecón) seem to have been well defined during the Laramide orogeny, although some authors (Anderson, 1968; Bonis, 1968) have presented evidence for an older age. These faults mark the northern boundery of the Caribbean plate. The serpentinite belts of northern and southern Central America are also interpreted as features of the Caribbean plate boundaries (Dengo, 1972). Another feature of the area which is interpreted as due to plate interactions is the basement of southern Central America which could also be considered as the result of the Laramide orogeny, using this term in a broad sense.

5. Recent studies of seismic focal mechanisms (Molnar and Sykes, 1969) leave no doubt as to the present existence and direction of movements of the Caribbean and Cocos plates in relation to the Americas plate. Some of the

previous discussion points to the existence of these plates at least since late Mesozoic. The Laramide tectonic belts can be easily interpreted in terms of plate movements. Whether similar criteria are applicable to the Paleozoic tectonic belts still remains one of the major geological problems in the Mexico–Central America region.

REFERENCES

Alvarez, Jr., M., 1961, Provincias fisiográficas de la República Mexicana: *Soc. Geol. Mex. Bol.*, v. 24, p. 5–20.

Alvarez, Jr., M., 1962, Orogenias Pre-Terciarias en México: *Asoc. Mex. Geol. Petr. Bol.*, v. 14, p. 23–25.

Anderson, T. H., 1968, Pre-Pennsylvanian and later displacements along Chixoy–Polochic fault trace, northwestern Guatemala: Geol. Soc. Am. Abstr. 1968, Spec. Pap. 121, p. 6–7.

Anderson, T. H., Burkart, B., Clemons, R. E., Bohnenberger, O. H., and Blount, D. N., 1973, Geology of the western Altos Cuchumatanes, northwestern Guatemala: *Geol. Soc. Am. Bull.*, v. 84, p. 805–826.

Baker, C. L., 1971, Geologic reconnaissance in the eastern Cordillera of Mexico: Geol. Soc. Am., Spec. Pap. 131, p. 83.

Bandy, O. L., and Casey, R. E., 1973, Reflector horizons and bathymetric cycles, eastern Panama: *Geol. Soc. Am. Bull.*, v. 84, p. 3081–3086.

Barnetche, A., and Illing, L. V., 1956, The Tamabra limestone of the Poza Rica Oilfield (Veracruz, Mexico): XX Congr. Geol. Intern., México, 38 p.

Barr, K. W., and Escalante, G., 1969, Contribución al esclarecimiento del problema del Complejo de Nicoya, Costa Rica: *Publ. Geol. ICAITI*, v. 2, p. 43–47.

Bateson, J. H., and Hall, I. H. S., 1971, Revised Geologic nomenclature of Pre-Cretaceous rocks of British Honduras: *Am. Assoc. Petr. Geol. Bull.*, v. 55, p. 529–530.

Benavides, L., 1956, Notas sobre la geología petrolera de México: XX Congr. Geol. Intern., México, Symposium Yacimientos Petróleo y Gas, v. 3, p. 351–362.

Bermúdez, P. J., 1963, Foraminíferos del Paleoceno del Departamento de El Petén, Guatemala: *Soc. Geol. Mex. Bol.*, v. 26, p. 1–56.

Bishop, B. A., 1970, Stratigraphy of Sierra de Picachos and vicinity, Nuevo León, Mexico: *Am. Assoc. Petr. Geol. Bull.*, v. 45, p. 1245–1270.

Bolívar, J. M., 1963, Geología del área delimitada por el Tomatal, Huitzuco y Mayanalán, Estado de Guerrero: *Univ. Nac. Auton. Mex., Inst. Geol.*, v. 69, p. 1–35.

Bonis, S., 1967, Age of Guatemalan serpentinite (abstract): Geol. Soc. Am. Program, New Orleans Meet., p. 18.

Bonis, S., 1968, Evidence for a Paleozoic Cayman Trough: Geol. Soc. Am. Abstr. 1968, Spec. Pap., 121, p. 32.

Burkart, B., and Clemons, R. E., 1972, Late Paleozoic orogeny in northwestern Guatemala: VI Caribbean Geol. Conf. Proc., Venezuela, p. 210–213.

Burkart, B., Clemons, R. E., and Crane, D. C., 1973, Mesozoic and Cenozoic stratigraphy of southeastern Guatemala: *Am. Assoc. Petr. Geo . Bull.*, v. 57, p. 63–73.

Butterlin, J., 1972, La posición estructural de los Andes de Colombia: IV Congr. Geol. Venezolano Mem., v. 2, p. 1185–1200.

Carey, W. C., 1958, The tectonic approach to continental drift: Continental Drift Symp., Univ. of Tasmania, p. 177–355.

Carpenter, R. H., 1954, Geology and ore deposits of the Rosario mining district and the San Juancito Mountains, Honduras, Central America: *Geol. Soc. Am. Bull.*, v. 65, p. 23–28.

Carrillo-Bravo, J., 1959, Notas sobre el Paleozoico de la región de Ciudad Victoria, Tamps.: *Asoc. Mex. Geol. Petr. Bol.*, v. 11, p. 671–680.

Carrillo-Bravo, J., 1961, Geología del anticlinorio de Huizachal–Peregrina, al N. W. de Ciudad Victoria, Tamps.: *Asoc. Mex. Geol. Petr. Bol.*, v. 13, p. 1–98.

Case, J. E., 1974, Oceanic crust forms basement of eastern Panama: *Geol. Soc. Am. Bull.*, v. 85, p. 645–652.

Case, J. E., Durán, L. G., López, A., and Moore, W. R., 1971, Tectonic investigations in western Colombia and eastern Panama: *Geol. Soc. Am. Bull.*, v. 82, p. 2685–2712.

Chubb, L. J., 1959, Upper Cretaceous of central Chiapas, Mexico: *Am. Assoc. Petr. Geol. Bull.*, v. 43, p. 725–756.

Clemons, R. E., 1966, Geology of the Chiquimula quadrangle, Guatemala, Central America: Ph.D. Thesis, Univ. of Texas at Austin (mimeographed), 124 p.

Clemons, R. E., and Burkart, B., 1971, Stratigraphy of northwestern Guatemala: *Soc. Geol. Mex. Bol.*, v. 32, p. 143–158.

Coogan, A. H., Bebout, D. G., and Maggio, C., 1972, Deposition environments and geologic history of Golden Lane and Poza Rica trend, Mexico, an alternative view: *Am. Assoc. Petr. Geol. Bull.*, v. 56, p. 1419–1447.

de Cserna, Z., 1956, Tectónica de la Sierra Madre Oriental de México, entre Torreón y Monterrey: XX Congr. Geol. Intern., México, 87 p.

de Cserna, Z., 1965, Reconocimiento geológico en la Sierra Madre del sur de México, entre Chilpancingo y Acapulco, Estado de Guerrero: *Univ. Nac. Auton. Mex., Inst. Geol.*, v. 62, p. 1–77.

de Cserna, Z., 1970, Mesozoic sedimentation magmatic activity and deformation in northern Mexico, in: *The Geologic Framework of the Chihuahua Tectonic Belt, a Symposium in Honor of Ronald K. Deford*: West Texas Geol. Soc., p. 99–116.

de Cserna, Z., 1971a, Taconian (Early Caledonian) deformation in the Huasteca structural belt of eastern Mexico: *Am. J. Sci.*, v. 271, p. 544–550.

de Cserna, Z., 1971b, Precambrian sedimentation, tectonics, and magmatism in Mexico: *Geol. Rundsch.*, v. 60, p. 1488–1512.

de Cserna, Z., Rincón-Orta, C., Solorio-Munguía, J., and Schmitter-Villada, E., 1970, Una edad radiométrica temprana de la región de Placer de Guadalupe, noroeste de Chihuahua: *Soc. Geol. Mex. Bol.*, v. 31, p. 65–73.

del Giudice, D., 1973, Características geológicas de la República de Panamá (unpublished), presented at the Science and Man in the Americas Conference, Mexico, July 4, 1973.

Dengo, G., 1962a, Tectonic-igneous sequence in Costa Rica, in: *Petrologic Studies, Buddington Volume*: Geol. Soc. Am., p. 133–161.

Dengo, G., 1962b, Estudio geológico de la región de Guanacaste, Costa Rica: Inst. Geogr., San José, Costa Rica, 112 p.

Dengo, G., 1967, Geologic structure of Central America: *Univ. Miami Stud. Trop. Oceanogr.*, v. 5, p. 56–73.

Dengo, G., 1968, Estructura geológica, historia tectónica y morfología de América Central, Centro Regional de Ayuda Técnica, México, 50 p.

Dengo, G., 1969, Problems of tectonic relations between Central America and the Caribbean: *Gulf Coast Assoc. Geol. Soc. Trans.*, v. 19, p. 311–320.

Dengo, G., 1972, Review of Caribbean serpentinites and their tectonic implications; Geol. Soc. Am. Mem. 132, p. 303–312.

Dengo, G., and Bohnenberger, O. H., 1969, Structural development of northern Central America: Am. Assoc. Petr. Geol. Mem. 11, p. 203–220.

Dengo, G., Bohnenberger, O. H., and Bonis, S., 1970, Tectonics and volcanism along the Pacific marginal zone of Central America: Geol. Rundsch., v. 59, p. 1215–1232.

Dillon, W. P., and Vedder, J. G., 1973, Structure and development of the continental margin of British Honduras: Geol. Soc. Am. Bull., v. 84, p. 2713–2737.

Dixon, C. G., 1956, Geology of southern British Honduras, with notes on adjacent areas: Govt. Printer, Belize, 85 p.

Eardley, A. J., 1954, Tectonic relations of north and south America: Am. Assoc. Petr. Geol. Bull., v. 38, p. 707–773.

Ewing, J., Talwani, M., Ewing, M., and Edgar, T., 1965, Sediments of the Caribbean: Univ. Miami Stud. Trop. Oceanogr., v. 5, p. 88–102.

Fakundiny, R. H., 1970, Geology of the El Rosario quadrangle, Honduras, Central America: Ph.D. Thesis, Univ. of Texas at Austin (mimeographed), 234 p.

Figge, K., 1966, Die stratigraphische Stellung der metamorphen Gesteine NW-Nicaraguas: Neues Jahrb. Geol. Paläeontol., v. 4, p. 234–247.

Fisher, S. P., and Pessagno, E. A., Jr., 1965, Upper Cretaceous strata in northwestern Panama: Am. Assoc. Petr. Geol. Bull., v. 49, p. 433–444.

Flores-Revueltas, J., 1955, Los arrecifes de la cuenca de Tampico–Tuxpán, México: Asoc. Mex. Geol. Petr. Bol., v. 7, p. 397–500.

Freeland, G. L., and Dietz, R. S., 1971, Plate tectonic evolution of the Caribbean–Gulf of Mexico region: Nature, v. 232, p. 20–23.

Fries, C., Jr., and Rincón-Orta, C., 1965, Nuevas aportaciones geocronológicas y técnicas empleadas en el laboratorio de geocronometría: Univ. Nac. Auton. Mex. Inst. Geol. Bol., v. 73, p. 57–133.

Garfias, R. V., and Chapin, T. C., 1949, Geología de México, Edit. Jus, p. 109–115.

Gomberg, D. N., Banks, P. O., and McBirney, A. R., 1968, Guatemala—preliminary zircon ages from Central Cordillera: Science, v. 162, p. 121–122.

Goosens, P. J., and Rose, W., Jr., 1973, Chemical composition and age determination of tholeiitic rocks in the basic igneous complex, Ecuador: Geol. Soc. Am. Bull., v. 84, p. 1043–1052.

Gries, J. C., and Haenggi, W. T., 1970, Structural evolution of the eastern Chihuahua tectonic belt, in: The Geologic Framework of the Chihuahua Tectonic Belt, a Symposium in Honor of Ronald K. Deford: West Texas Geol. Soc., p. 119–137.

Gutiérrez-Gil, R., and Thompson, M. L., 1956, Geología del Mesozoico y estratigrafía pérmica del Estado de Chiapas: XX Congr. Geol. Intern., México, p. 1–82.

Guzmán, E. J., and de Cserna, Z., 1963, Tectonic history of México: Backbone of the Americas: Am. Assoc. Petr. Geol. Mem. 2, p. 113–129.

Henningsen, D., and Weyl, R., 1967, Ozeanische Kruste im Nicoya-Complex van Costa Rica (Mittelamerika): Geol. Rundsch., v. 57, p. 33–47.

Imlay, R. W., 1938, Studies of the Mexican geosyncline: Geol. Soc. Am. Bull., v. 49, p. 1651–1671.

Kesler, S. E., 1971, Nature of ancestral orogenic zone in nuclear Central America: Am. Assoc. Petr. Geol. Bull., v. 55, p. 2116–2129.

Kesler, S. E., 1973, Basement rock structural trends in southern Mexico: Geol. Soc. Am. Bull., v. 84, p. 1059–1064.

Kesler, S. E., and Heath, S. A., 1970, Structural trends in the southernmost North American Precambrian, Oaxaca, Mexico: Geol. Soc. Am. Bull., v. 81, p. 2471–2476.

Kesler, S. E., Josey, W. L., and Collins, E. M., 1970, Basement rocks of western nuclear

Central America: The western Chuacús group, Guatemala: *Geol. Soc. Am. Bull.*, v. 81, p. 3307–3322.

King, P. B., 1969, The tectonics of north America. A discussion to accompany the tectonic map of north America: U.S. Geol. Surv. Prof. Pap. 628, scale 1:5,000,000, p. 1–94.

Lloyd, J. J., 1963, Tectonic history of the south Central American orogen: Backbone of the Americas: Am. Assoc. Petr. Geol. Mem. 2, p. 88–100.

López Ramos, E., 1969a, Geología del sureste de México y norte de Guatemala: *Publ. Geol. ICAITI*, v. 2, p. 57–67.

López Ramos, E., 1969b, Marine Paleozoic rocks of Mexico: *Am. Assoc. Petr. Geol. Bull.*, v. 53, p. 2399–2417.

Maldonado-Köerdell, M., 1954, Nomenclatura, bibliografía y correlación de las formaciones Arqueozoicas y Paleozoicas de México: *Asoc. Mex. Geol. Petr. Bol.*, v. 6, p. 113–138.

Malfait, B. T., and Dinkelman, M. G., 1972, Circum-Caribbean tectonic and igneous activity and the evolution of the Caribbean plate: *Geol. Soc. Am. Bull.*, v. 83, p. 251–272.

McBirney, A. R., 1963, Geology of a part of the central Guatemalan Cordillera: *Univ. Calif. Publ. Geol. Sci.*, v. 38, p. 177–242.

McBirney, A. R., and Bass, M. N., 1969, Structural relations of pre-Mesozoic rocks of northern Central America, Am. Assoc. Petr. Geol. Mem. 11, p. 269–280.

Mena Rojas, E., 1955, Estudio geológico-económico del Cretácico superior y medio al este de la Faja de Oro: *Asoc. Mex. Geol. Petr. Bol.*, v. 7, p. 327–366.

Meyerhoff, A. A., 1966, Bartlett fault system: age and offset: *III Caribbean Geol. Conf. Trans.*, Jamaica, p. 1–9.

Mills, R. A., Hugh, K. E., Feray, D. E., and Swolfs, H. C., 1967, Mesozoic stratigraphy of Honduras: *Am. Assoc. Petr. Geol. Bull.*, v. 51, p. 1711–1786.

Mixon, R. B., Murray, G. E., and Díaz, T., 1959, Age and correlation of the Huizachal Group (Mesozoic), State of Tamaulipas, Mexico: *Am. Assoc. Petr. Geol. Bull.*, v. 43, p. 757–771.

Molnar, P., and Sykes, L. R., 1969, Tectonics of the Caribbean and middle America regions from focal mechanisms and seismicity: *Geol. Soc. Am. Bull.*, v. 80, p. 1639–1694.

Murray, G. E., 1961, Geology of the Atlantic and Gulf coastal province of North America: Harper and Row, New York, 692 p.

Pantoja, A. J., and Robinson, R. A., 1967, Paleozoic sedimentary rocks in Oaxaca, México: *Science*, v. 157, p. 1033–1035.

Pessagno, E. A., Jr., 1969, Upper Cretaceous stratigraphy of western Gulf coast area of México, Texas, and Arkansas: Geol. Soc. Am. Mem. 111, p. 1–139.

Richards, H. G., 1963, Stratigraphy of earliest Mesozoic sediments in southeastern Mexico and western Guatemala: *Am. Assoc. Petr. Geol. Bull.*, v. 47, p. 1861–1870.

Rodgers, J., 1971, The taconic orogeny: *Geol. Soc. Am. Bull.*, v. 82, p. 1141–1178.

Sapper, K., 1937, *Mittelamerika, Handbuch der Regionalen Geologie*: Steinman und Wilckens, Heidelberg, 160 p.

Schuchert, C., 1935, *Historical Geology of the Antillean–Caribbean Region*: John Wiley & Son, New York, p. 332–343.

Schwartz, D. P., 1972, Petrology and structural geology along the Motagua fault zone (abstr.), VI Caribbean Geol. Conf. Proc., Venezuela, p. 299.

Vedder, J. G., MacLeod, N. S., Lanphere, M. A., and Dillon, W. P., 1973, Age and tectonic implications of some low-grade metamorphic rocks from the Yucatán Channel: *J. Res. U.S. Geol. Surv.*, v. 1, p. 157–164.

Viniegra, O. F., 1971, Age and evolution of salt basins of southeastern México: *Am. Assoc. Petr. Geol. Bull.*, v. 48, p. 70–84.

Vinson, G. L., 1962, Upper Cretaceous and Tertiary stratigraphy of Guatemala: *Am. Assoc. Petr. Geol. Bull.*, v. 46, p. 425–456.

Viveros, M. A., 1965, Estudio geológico de la Sierra de Cabullona, Municipio de Agua Prieta, Estado de Sonora: Univ. Nac. Autón. Méx. (unpublished thesis), 82 p.

Webber, B. N., and Ojeda, J., 1957, Investigación sobre lateritas fósiles en las regiones sureste de Oaxaca y sur de Chiapas: *Inst. Nac. Invest. Recur. Nat., Mex.*, v. 37, p. 1–67.

Widie, A. E., and Martínez, J. D., 1970, Evidence for evaporite diaprism in northeastern Mexico: *Am. Assoc. Petr. Geol. Bull.*, v. 54, p. 655–661.

Weidie, A. E., and Wolleben, J. A., 1969, Upper Jurassic stratigraphic relations near Monterrey, Nuevo León, Mexico: *Am. Assoc. Petr. Geol. Bull.*, v. 53, p. 2418–2420.

Weyl, R., 1961, *Die Geologie Mittelamerikas*: Geb. Borutraeger, Berlin, 266 p.

Wilhelm, O., and Ewing, M., 1972, Geology and history of the Gulf of Mexico: *Geol. Soc. Am. Bull.*, v. 83, p. 575–600.

Williams, H., McBirney, A. R., and Dengo, G., 1964, Geologic reconnaissance of southeastern Guatemala, *Univ. Calif. Publ. Geol. Sci.*, v. 50, p. 1–56.

Wilson, H. H., 1974, Cretaceous sedimentation and orogeny in nuclear Central America: *Am. Assoc. Petr. Geol. Bull.*, v. 58, p. 1348–1396.

Zoppis-Bracci B., L., and del Giudice, D., 1958, Geología de la costa del Pacífico de Nicaragua: *Serv. Geol. Nicaragua Bol.*, v. 2, p. 19-68.

Maps

Córdoba, D. A., 1965, Hoja Apizolaya 13 r-1(9), 1:100,000: Instituto de Geología, México.

de Cserna, Z., 1961, Tectonic map of México, 1:2,500,000: Geol. Soc. America.

Dengo, G., Levy, E., Bohnenberger, O. H., and Caballeros, R., 1969, Mapa metalogenético de América Central, 1:2,000,000: Publ. Geol. ICAITI 3.

Hernández-Sánchez-Mejorada, S., and López Ramos, E., 1968, Carta geológica de la República mexicana, 1:2,000,000: Comité de la Carta Geológica de México.

Instituto de Geología, México, 1970, Carta geológica del Estado, de Sinaloa, 1:500,000.

King, P. B., 1969, Tectonic map of North America, 1:5,000,000: U.S. Geol. Sur.

López Ramos, E., 1972, Carta geológica del Estado de Chiapas, 1:500,000: Instituto de Geología, México (ozalid reproduction).

van den Boom, G., Müller, A., Nicolaus, H. J., Paulsen, S., and Reyes, J., 1971, Mapa geológico general, 1:125,000, Baja Verapaz y parte sur de la Alta Verapaz (Guatemala): Bundesanstalt für Bodenforschung, Hannover.

Roberts, R. J., and Irving, E. M., 1957, Mineral deposits of Central America, 1:1,000,000: U.S. Geol. Surv. Bull. 1034.

Chapter 9

THE NORTHERN TERMINATION OF THE ANDES

Reginald Shagam

Department of Geology
Ben Gurion University of the Negev
Beer Sheva, Israel

I. INTRODUCTION

In terms of topography, the northernmost extensions of the Andes branch into long fingerlike ranges separated by low-lying sediment-filled basins. Geologic definition of the northern Andes is a far more complex problem. For many, the term "Andes" has the connotation of Tertiary block uplifts or Germano-type orogeny, in contrast to the orogenic model associated with island arcs. It is common practice, therefore, to regard the Southern Caribbean Mountains, i.e., the coastal ranges of northern Venezuela and Trinidad, as having affinities with the Caribbean Island Arc, whereas the Mérida ranges are classed as Andean. There are others such as Gansser (1973) who include all ranges on the continental mainland as Andean and consider the Caribbean Island Arc to begin at the southernmost of the Lesser Antilles. On the gross scale of this report, the two kinds of orogenic belt are closely associated in space: With allowance for major lateral offsets, one may claim geologic continuity from the lesser Antilles into the northern ranges of Trinidad–Venezuela and thence into the Guajira Peninsula and the northwestern apex of the Santa Marta block continuing into the Cordillera Central (cf. Alvarez, 1971). It has been an arbitrary decision on my part to exclude from discussion

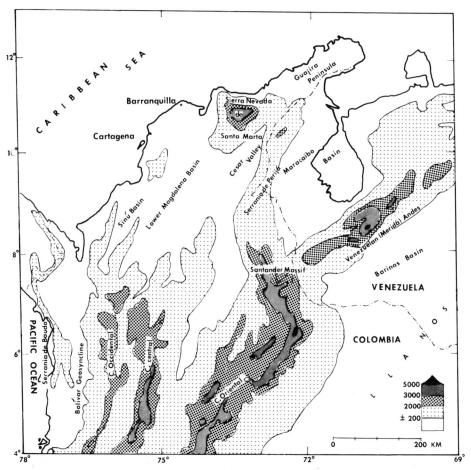

Fig. 1. Principal elements of the physical geography of the northern termination of the Andes, and the nomenclature employed in this review.

here the northern ranges of Trinidad–Venezuela and the Guajira Peninsula. For convenience, the southern limit of the region under discussion has been set at the latitude of Bogotá (4°40′ N). The geographic area and the nomenclature of its component parts are shown in Fig. 1.

II. OUTLINE OF PREVIOUS WORK AND SCOPE OF THE PAPER

In a valuable introduction to his last work, published posthumously, Burgl (1973) outlined the development of geological research in Colombia. Additional data are provided by Irving (1971). Schubert (1969) and the writer (1972a) have summarized previous work in the Venezuelan Andes, and Bowen

(1972) has done the same for the Perijá Mountains. In broadest outline, geologic researches in Colombia and western Venezuela have progressed through four stages. The late 18th to the middle of the 19th century was the period of the great naturalist explorers such as Humboldt. From the late 19th century up to World War I, the explorers were succeeded by another class of intrepid investigators with specific geologic-paleontologic training who gave the first regional reports with sketch maps. Among this class were Sievers, Stille, and Scheibe. A third stage stemmed from the search for petroleum brought into focus by world conflict, and this extended from the early 1920's until approximately 1960. In this interval, the dictates of efficiency led to the establishment of adequate cartographic and laboratory facilities to back up the work of hundreds of field geologists. These studies were largely focused on sedimentary formations of Cretaceous–Tertiary age. For a variety of reasons, petroleum geologists did engage in some reconnaissance studies of older sedimentary and crystalline terrains. An outstanding example is the map and reports of Kundig (1938) and Kehrer (1938) reflecting the work of some eight geologists over a 2-month time span in the Venezuelan Andes. A more detailed study was subsequently published by Cía. Shell and Creole (1964), but for the most part studies in pre-Cretaceous terrains were of reconnaissance, single-traverse type reflecting not indifference but the limited time available to one normally occupied elsewhere. Growth of the petroleum industry stimulated the establishment of national geological surveys and departments of geology at the principal universities. Thus was ushered in the fourth stage of development in which there has been a growth in the native schools of geology and increasing emphasis on quantitative studies accomplished by coordination of research efforts with workers from foreign universities and governmental agencies.

With the above outline in mind, one may appreciate the almost exponential growth in Andean research with time. Yet coverage is uneven. Large areas of the Venezuelan Andes, Cordillera Oriental, Santa Marta block, and Cordillera Central have been mapped at scales sufficiently large to bring out considerable detail [see, for example, Garcia and Campos (1972), Ramírez and Campos (1972), Kovisars (1971), Useche and Fierro (1972), Ward *et al.* (1969, 1970), McLaughlin *et al.* (1969), Feininger *et al.* (1970), and Tschanz *et al.* (1969)]. The mapping in the Perijá Mountains remains rudimentary [see Miller (1962) and Bowen (1972)]. Detailed areal mapping by the Dirección de Geología of the Ministry of Mines and Hydrocarbons, Venezuela, began, however, in 1973 under the supervision of A. Espejo. Field data for the western ranges of Colombia are even more sparse, and only recently has the National Geological Service begun intensified regional mapping programs there.

There is neither space nor point to catalog here all the work that has been done in the region since the early writings of Humboldt. I have attempted to avoid lengthy accounts of the evolution of geologic thought pertaining to the northern Andes by using a relatively small number of recent papers, mainly of the broad summary type, as a sort of datum for this report even though my conclusions diverge radically in many respects. Among papers repeatedly quoted are Irving (1971), Burgl (1973), Cediel (1972), and Tschanz *et al.* (1969, 1974) for the Colombian Andes, Miller (1962) and Bowen (1972) for the Perijá Ranges, and Shagam (1972*a,b*) and Grauch (1971, 1972) for the Venezuelan Andes. Interested researchers may readily find complete bibliographies through these works should they wish to follow the development of tectonic ideas. Though it is convenient to refer to relatively few papers, all papers noted in the bibliography have been consulted in the original.

It has been my aim to provide a summary of the principal field data (Sec. IV) and of other studies of tectonic significance (Sec. V) and then to integrate these into a broad geotectonic framework (Sec. VI). I have attempted to give food for thought to interested investigators by questioning some of the basic data currently in the literature and by attempting to set tectonic limits rather than to fabricate new syntheses.

III. TERMINOLOGY

The terms "internal" and "external" are used in the sense of direction of thrusting, of tectonic transport; in the region of the northern Andes, "external" means toward the Caribbean and Pacific, which is *away* from the Craton.

IV. GEOLOGIC HISTORY

A. The Venezuelan Andes

In the simplest terms, the Geology of the Venezuelan Andes may be described in terms of a crystalline (Precambrian?) core corresponding to the central and highest portion of the ranges, overlain toward both the NE and SW by successive skirts of Upper Paleozoic, Mesozoic, and Tertiary formations. An element of asymmetry is provided by the occurrence, on the southeast flank of the ranges, of a narrow belt of Precambrian schists in fault contact to the northwest with a parallel narrow belt of early Paleozoic sedimentary rocks.

Stratigraphic relationships in these Andes are summarized in Fig. 2. Principal unconformities appear to occur in the Precambrian–Mid-Ordovician interval, during the Devonian–Mississippian, at or toward the close of the Permian, and approximately at the Eocene–Oligocene boundary. Additional unconformities indicative of tectonic events are identified (some tentatively) in Fig. 2. On the basis of lithological characteristics and thickness, it may be reasonably concluded that thick piles of sediment were accumulated in unstable subsiding basins closely related to major tectonic activity during the Precambrian, early Paleozoic (especially the late Silurian), Pennsylvanian–Permian, Triassic–Jurassic, Cretaceous, and middle to late Tertiary. Of these accumulations, a portion of the Pennsylvanian–Permian section, the Triassic–Jurassic section, and most of the Tertiary section is of continental type and has been described by many workers, probably unwisely, as "molasse." In contrast, the early Paleozoic section includes graptolitic black shales, fossiliferous siliceous limestones, and a thick sequence of interbedded coarse marine conglomerates and fossiliferous shales. A portion of the Pennsylvanian–Permian section closely resembles Alpine "flysch" and is overlain by fossiliferous shallow-water marine shales and limestones. The Cretaceous–early Tertiary section reflects an extensive marine transgression over the northern margin of the continent (see Zambrano et al., 1970). On a gross scale, this transgression may be viewed in terms of epeirogenic tectonism, but locally there were deep unstable basins in which great thicknesses of sediment accumulated. One of these, the Uribante basin, was located over the southwestern portion of these Andes; another, the Barquisimeto basin, overlapped onto the northeasternmost portion of the ranges.

Granitic plutonism has been the dominant igneous activity throughout the geologic history of these Andes. Largely on the basis of field relationships with some supporting radiometric data, the writer (1972b) concluded that the ages of granitic events become progressively younger as one crosses these Andes from southeast to northwest. Intrusions in the Precambrian, Bella Vista schists on the southeast flanks of the Andes are believed to be of pre-mid-Ordovician age; Cía. Shell de Venezuela and Creole Petroleum Corporation (1964) report a Rb/Sr age of 660 ± 30 m.y. for one such mass. Hence, these oldest granitic intrusions appear to be of late Precambrian–early Cambrian age. The youngest plutons occur in the northwesternmost portion of the ranges in a broad region extending from the vicinity of Valera southwestward to the region of La Grita. These plutons appear to be in sedimentary contact with the Triassic–Jurassic La Quinta Formation and are believed to have been emplaced at the close of the Permian or in earliest Triassic time.

What may have been a major volcanic event is now represented by dykes and sills of olivine basalt (lamprophyric in some outcrops) which occur in the Bella Vista schists and which appear to postdate the oldest granites but predate the mid-Ordovician–Silurian section. A petrographically distinct basalt occurs in small volumes in association with the Lower Paleozoic section but is probably coeval with basalt flows and dykes which occur in minor volume in the Sabaneta Formation of Pennsylvanian–early Permian age. In the vicinity of La Grita, rhyolitic to dacitic lavas and tuffs occur at the base of the La Quinta Formation and are considered to represent the volcanic equivalents of the youngest plutonic event referred to above.

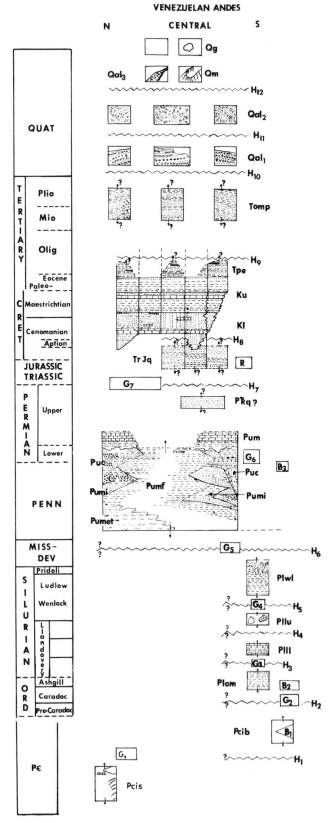

Fig. 2.

Fig. 2. Stratigraphic column of the Venezuelan Andes. Qg, modern glaciers. Qal$_3$, youngest alluvial gravels as thin sheets related to modern drainage, and small fans and cones. Qm, well-preserved-moraines (late Pleistocene) and related minor alluvial fans. Qal$_2$, poorly sorted, poorly stratified gravels as valley and channel fills in Qal$_1$ or as fans on Qal$_1$; may occupy some glaciated valleys; deposited largely as mud flows. Qal$_1$, roughly sorted, well-stratified gravels as terraces and fans; oldest deposits; may pre-date oldest glaciation in part or whole. Tomp, Oligo-Mio-Pliocene "Molasse"; thick friable gravels, ss., and mudstones deposited along both flanks of ranges and as thinner deposits in central graben following Boconó fault; grades from continental to lacustrine and marine in N; continental only elsewhere. Tpe, Lower Tertiary section (Paleocene–Eocene); massive ss. with interbedded carbonaceous shales, minor coals; represents final basin-fills; lacustrine to continental at top, grading down into marine Cretaceous deposits. Ku, Upper Cretaceous section. Dominant marine sh. and ls. with some massive qz. ss.; middle of section is La Luna Formation composed of dark gray fetid sh., concretionary ls., laminated cherts, and phosphatic fish beds. Kl, Lower Cretaceous section. Dominant qz. ss. with lesser interbedded sh. and ls.; prominent marker unit is Apón Formation of dark gray fossiliferous ls. (Apt.–Alb.); section may grade into red bed ss. and sh. of La Quinta Formation. TrJq, Trias–Jura, La Quinta Formation; massive red bed ss., siltst. with some cg. and sh.; continental to lacustrine deposits; scarce ostracods, fish scales, teeth; rhyolitic tuffs at base. PTrq?, red beds not distinguishable from TrJq with which they may prove to be coeval; distinguished here on basis of unconformable overlap by Kl section; may correlate with Jordan Formation in C. Oriental, Colombia. *Upper Paleozoic section*: Pum, fossiliferous lower marine sh. upper ls. (Palmarito Formation); early Permian. Puc, deltaic continental red-bed facies (Mérida, N; Sabaneta, S). Sabaneta facies is dominantly tan lenticular cgtic. ss.; red micaceous sandy siltst. are dominant towards top; minor altered vesicular basalt lavas and aphanitic to nodular algal ls. occur in upper part; transitional upwards into Pum and believed to interfinger laterally through Pumi into Pumf. Pumi, mixed (continental-marine) facies; red to tan cg., ss., breccias interbedded with Mucuchachí-type slates. Pumf, marine flysch (Mucuchachí facies); laminated, graded slates, siltst., wacke ss., minor cg.; rare cryst. ls.; rare fossils: some brachs. mainly crinoid fragments; Mid-Pennsylvanian age in part. Pumet, greenschist to amphibolite grade (and./ky/sil-gar-staur; assemblages); metamorphic rocks (= Tostós facies) believed portion of Pumf. *Lower Paleozoic section*: Plwl, Wenlock–Ludlow section; fossiliferous blue-gray sh. interbedded with gray-green ss., coarse cg. Pllu, Upper Llandovery; siliceous, aphanitic to fine-grained ls. with thin shell beds; only as few blocks in Plwl. Plll, Lower Llandovery; blue-green, fine-grained siliceous ls. with shell beds interbedded with green sh.; laminated cherts and minor ss. Plom, Mid-Ordovician; dark gray, graptolitic sh. with minor fossiliferous siltst., ss., sandy ls. *Precambrian Iglesias Complex*: Pcib, Bella Vista Formation; Greenschist facies, chl-musc-qz-ab schist, laminated cryst. cherts, and other granoblastic siliceous rocks. Pcis, Sierra Nevada facies; Amphibolite facies, gar-bi-musc-sil schists, gneisses; amphibolites; portion may be highly metamorphosed part of Upper Paleozoic section; migmatitic near intrusive granites. G$_1$–G$_7$, Granitic plutons; G$_1$ represented only by rootless pegmatitic pods in Sierra Nevada facies; G$_2$–G$_7$ are granitic plutons representing progressive decrease in age from S. ($\pm$660 m.y.) to N. (end-Permian). (G$_7$ intruded approx. where G$_1$ occurs). B$_1$, olivine basalt, lamprophyric (bi-) in some outcrops, intruding Bella Vista schists and postdating G$_2$ granites; B$_2$, altered basaltic (nonolivinic) dykes (?) in Lower Paleozoic section; probably related to B$_3$. B$_3$, altered basaltic (nonolivinic) lavas in Upper Paleozoic section. R, Rhyolitic to dacitic lavas and tuffs at base of La Quinta Formation at type locality; believed surface equivalents at G$_7$ granitic plutons. H$_1$–H$_{12}$, principal hiatuses recognized.

Grauch (1971, 1972) and the writer (1972b) have summarized the available evidence concerning events of regional metamorphism. Regional metamorphism of the Bella Vista Formation to the greenschist facies is known to have occurred in pre-mid-Ordovician time, and whether exactly coeval or not, it is considered orogenically related to the earliest granitic event (G_2) in that region. In the north-central region, Grauch (1972) has outlined the case for a major event of regional metamorphism at the end of the Permian, which is presumably related tectonically to the youngest granites (G_7). This metamorphic event is believed to have been imprinted on the previously metamorphosed crystalline rocks of the Precambrian (?) basement. Whether the earlier metamorphism was coeval with that of the Bella Vista schists is not known. Tectonic considerations suggest that it was a Precambrian event, but areal geological relationships permit only the inference that it occurred in pre-Pennsylvanian time.

The principal structural features of these Andes have been summarized by the writer (1972b), and the most significant structural features from the geotectonic viewpoint of this report may be listed as follows:

a. Major deformation appears to have occurred at times corresponding to hiatuses H_2, H_6, H_7, and H_9. Hypothetical H_1 may actually have been a hiatus coeval with H_2 but geographically separate. Furthermore, in view of the apparent areal restriction of rocks of early Paleozoic age to the southern piedmont of the ranges, it cannot be shown that hiatus H_6 extended over the north-central part of these Andes. In order to maintain a model approximately consistent in time perspective, all unconformities from the end of the Cretaceous to the present are viewed as subsidiary stages to one major geotectonic hiatus.

b. There is a marked contrast in structural style across the Paleozoic–Mesozoic boundary. Whereas pre-Mesozoic deformation (up to H_7) involved intense lateral compression reflected in tight folds, thrust faults, and related features, post-Paleozoic deformation was characterized by uplift and tilting of fault blocks with zones of opposite tilt separated by a keystone-type graben (cf. Shagam, 1972b, Fig. 5, p. 458). The most important graben (see Fig. 3) corresponds currently to a profound topographic depression running the length of the ranges and associated with the Boconó fault.

c. At least part of the deformation produced in H_7 took place by décollement over the crystalline basement.

d. Locally, for example, in the region south and southwest of Mérida (Shagam, 1972b, Fig. 1, and 1972a, Fig. 4), there are angular discordances in the trends of structural features produced in the different deformations, but in the overall view the trends (including those produced during Tertiary uplift) are roughly parallel.

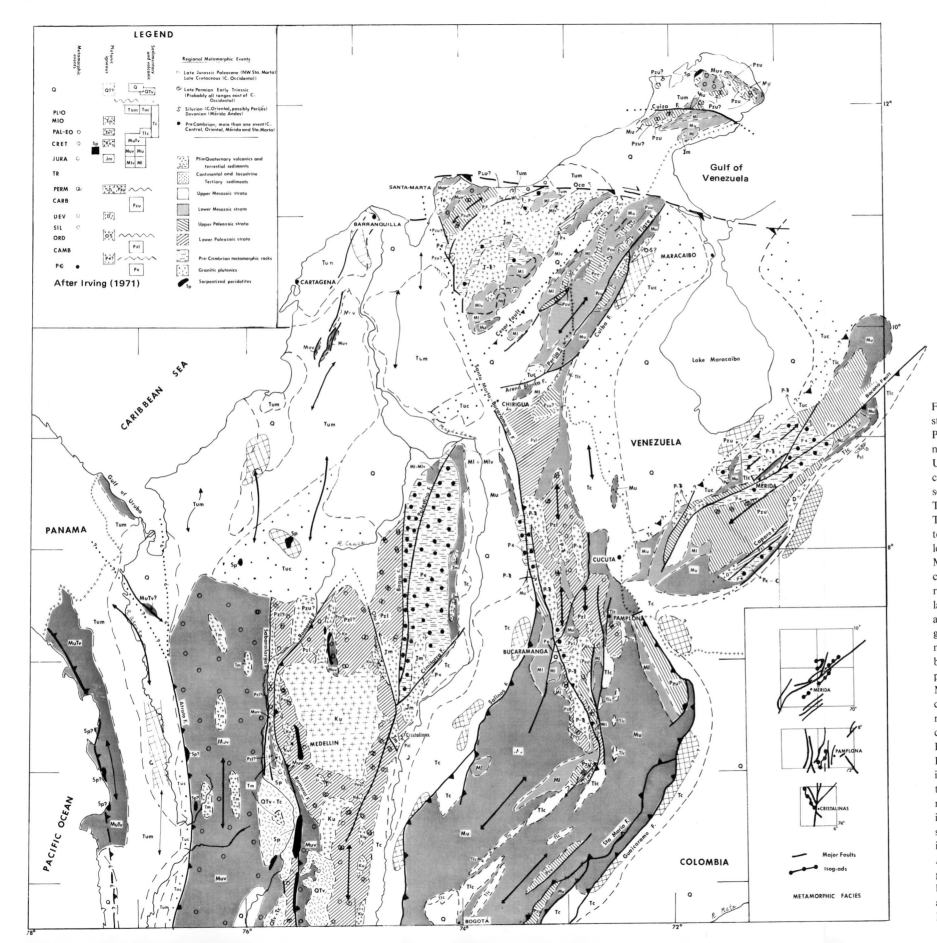

Fig. 3. The following are the symbols for stratified rocks: Q, Quaternary alluvium. QTv, Plio-Quaternary Volcanics and terrestrial sediments in the Cauca Valley. Tum, marine Upper Tertiary section; mainly clastic. Tuc, continental and lacustrine Upper Tertiary section. Tlc, continental and lacustrine Lower Tertiary section. Tc, continental and lacustrine Tertiary section comprising deposits of early to late Tertiary age, e.g., mid-Magdalena valley, or of unknown specific Tertiary age. MuTv, late Cretaceous to Eocene basic volcanic and marine (deep-water) sedimentary rocks. Muv, volcanic Upper Mesozoic (Jur.–late Cretaceous) rocks of the C. Occidental and N. W. Santa Marta; metamorphosed to greenschist facies. Mu, Upper Mesozoic sedimentary section (nonvolcanic) composed of basal clastics grading upward to dominantly pelitic–carbonate section. Mlv, volcanic Lower Mesozoic (Trias–Jura) rocks; marine and continental. Ml, Lower Mesozoic (Trias–Jura) rocks; mainly continental red beds, but includes some marine deposits. Pzu, Upper Paleozoic section, mainly marine; Devonian–Permian age; basal clastics passing upwards into Permo-Carb carbonate/shale section; latter grades upwards and laterally into deltaic red beds (Devonian deposits not recognized in Venezuelan Andes); Pzl, Lower Paleozoic section; mainly (?) marine; Cambro–Ord. age in C. Oriental; Ord–Sil. age in Venezuelan Andes. Pe, Precambrian metamorphic rocks; granulites in C. Central and Santa Marta block; amphibolite grade in Santander Massif and Venezuelan Andes; in Perijás may be Paleozoic in part or whole.

It may be said in summary of the tectonic evolution of the Venezuelan Andes that orogenic cycles involving accumulation of thick sedimentary piles and culminating in deformation, regional metamorphism, and/or granitic plutonism occurred at intervals corresponding to hiatuses H_2 (late Precambrian), H_6 (Devonian–Mississippian), and H_7 (late Permian–early Triassic). An additional cycle corresponding to H_1 remains hypothetical. The axes of the three orogenic cycles are parallel but appear to have moved progressively northwestward with time. The last orogenic cycle is that of hiatus H_9 and its related substages continuing to the present. This cycle is essentially one of block uplift differing from previous cycles in structural style and the absence of thermal events. As illustrated in Fig. 15, the uplift involved a component externally to the northwest. The zone of maximum uplift corresponds approximately with the axes of the H_7 orogeny.

Discussion

The writer (1972*a,b*) has outlined the uncertainties which remain to be investigated and tested in the tectonic outline presented above. Some of these uncertainties merit further discussion.

Grauch (oral discussion, summer 1973) has noted that aluminosilicate isograds (see Grauch, 1972) appear in many areas of the northern Andes, approximately coincident with what has been mapped as the contact between the crystalline (Precambrian?) basement and rocks of the overlying Upper Paleozoic section. One possible interpretation of this metamorphic geometry is that the unconformity between the two rock units became a favored site for metamorphic reactions because of the concentration there of volatile components. More likely, according to Grauch, is the possibility that the mapped "contact" is in reality a metamorphic feature; i.e., the appearance of massive, amphibolite-grade gneisses is merely the final stage in the metamorphic transition from slates through staurolitic schists. If so, the obvious question is whether the entire "Precambrian" core of the Venezuelan Andes might not be a highly metamorphosed facies of the Upper Paleozoic section. This would, *inter alia*, provide a simple explanation for the regional parallelism of structures which the writer has ascribed above to Precambrian and late Permian deformations, respectively. The so-called Precambrian trends would quite simply be reinterpreted as part and parcel of the same late Permian deformation. One possible way to test for the existence of an older (i.e., pre-late Permian) metamorphic basement is illustrated in Fig. 4. The situation corresponding roughly to alternative B is exemplified in the Chacantá–Lagunillas area, southwest of Mérida (see Fig. 5). There is an additional complication in the real situation in that there has been décollement in this area (Fig. 5). It may hence be argued that the current setting of the amphibolite facies "basement" enclosed by low- or subgreenschist facies slates is the result of postmetamorphic tec-

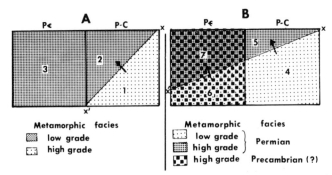

Fig. 4. Diagrammatic representation of principal metamorphic problems. Areas under Pε and P–C represent Precambrian and Permian–Carboniferous terrains, respectively. Diagonals x–x′ represent isograds, and arrows show prograde direction. (A) Grauch (see p. 333) suggests that isograd has been mapped as a formational contact and that area of high-grade schist (2) mapped as Precambrian is in reality part of the Permian–Carboniferous section; (B) Preferred relationships in order to test whether metamorphic events preceded the late Permian event. If the Precambrian was already a high-grade metamorphic terrain prior to the late Paleozoic metamorphism, then assemblages encountered in the prograde direction going from 6 to 7 would be quite different from those encountered in going from 4 to 5. These relationships may obtain in the region south of Mérida (see Fig. 5).

tonism which brought the contrasting metamorphic facies into juxtaposition. The writer believes that there may have been lateral displacement of metamorphic isograds, but whether the décollement preceded the metamorphism or not remains to be proven. If the late Permian metamorphism was indeed overprinted on a previously existing metamorphic terrain, it does not prove that the latter was Precambrian. The Iglesias Group basement of the north-central Andes could conceivably represent the as-yet-unidentified Devonian–Mississippian section or be coeval with the Lower Paleozoic section exposed on the southern flanks of these Andes, or even be of Cambrian age. None of these possibilities can be absolutely disproved, but on the basis of regional distribution and characteristics of lithology and sedimentary facies none are considered likely. There is some radiometric evidence (see below) which favors a Precambrian age for the Iglesias Group.

Radiometric age determinations on granitic plutons in the Venezuelan Andes are currently being measured by L. Burkley at Case Western Reserve University. Initial results suggest that some modification of the writer's simple outline of progressive plutonism northwestward with time will be required. The interim results do not contradict the proposed shift of orogenic axes with time, outlined above, but do suggest that in some areas there may have been additional earlier events. Regional metamorphic overprints are associated with the Precambrian plutons in the Bella Vista belt and with the late Permian–early Triassic plutons in the northern portion of the ranges. Corresponding metamorphic rocks have not been reported for the Ordovician (?) to Devonian

Fig. 5. Metamorphic relationships in the Mérida Andes region. Moving in the prograde direction, A–A′ (NE corner of map) parallels the direction 1–2 in Fig. 4A. By inference, the amphibolite facies schists of the Sierra Nevada Formation currently mapped as Precambrian may be (partly?) coeval with the Upper Paleozoic section. By contrast, the relationships in moving from B–B′ (west central on map) parallel those predicted in moving from 6–7 in Fig. 4B. By inference, the Sierra Nevada Schists in the area north of Chacantá may already have been of amphibolite grade prior to regional metamorphism at the close of the Paleozoic. The fact that sillimanite is yet to be reported from the schists of the Chacantá region whereas it is abundant in the Sierra Nevada near A–A′ supports the suggestion (i.e., all sillimanite is of late Paleozoic age). On the other hand, the greenschist facies, Tostos schists, may be an allochthon (see regional structural

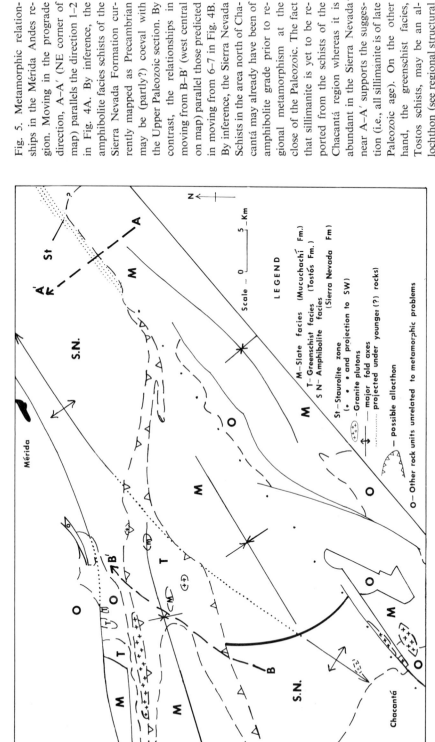

LEGEND

M–Slate facies (Mucuchachí Fm.)
T–Greenschist facies (Tostós Fm.)
S N–Amphibolite facies
        (Sierra Nevada Fm)

M

St–Staurolite zone
(• • • and projection to SW)

— — major fold axes

.... projected under younger (?) rocks)

+++ –Granite plutons

— possible allochthon

O– Other rock units unrelated to metamorphic problems

Scale – 0   5   Km

N

Mérida   S.N.

S.N.   Chacantá

features) whose root zone may have been located along the projected former possible extension of the staurolite zone. If metamorphism occurred before lateral tectonic movement, then the stipulation of an earlier metamorphic event to explain the schists of the Chacantá region does not necessarily follow.

plutons tentatively identified in the outcrop belt of the Lower Paleozoic section. In the early stages of field mapping, the writer assigned to the Bella Vista schists in the headwaters of the Rio Caparo some rocks which might well be metamorphosed portions of the Lower Paleozoic section.

A structural interpretation of importance in this report concerns the Boconó fault, a major lineament trending northeast along the length of these Andes and extending to the coast some 140 km west of Caracas. This fault constitutes the more southern of the boundary faults defining the keystone graben referred to in the outline. A detailed study of Pleistocene glacial and geomorphologic features by Giegengack and Grauch (1972, 1973a,b) and Giegengack et al. (1973), and of structural, petrologic, and regional data by the writer and A. Bellizzia (in preparation), suggests that throughout its history, including all of the Tertiary, movement on this fault was predominately of dip-slip type. A maximum of 250 m of strike-slip motion can be shown for post-late-Pleistocene time; even then, a component of dip-slip motion as great or greater than the strike-slip component can be shown to have occurred in some areas. If this fault is to be considered a plate boundary in the context of "plate tectonics," it can only be reasonably viewed in such a light since the latest Pleistocene. This matter is further discussed in Sec. V of this review.

B. Cordillera Oriental of Colombia and the Perijá Mountains

It is convenient here to outline separately the geology of the Cordillera Oriental and of the Perijá Mountains, the transition from one range to the other occurring approximately at latitude 9° N.

The Cordillera Oriental (South of latitude 9° N) does not display in outcrop pattern the elements of simple symmetry that characterize the Venezuelan Andes, but there are many notable similarities in the geology of the two ranges. The stratigraphic column is summarized in Fig. 6. A Precambrian crystalline basement, the Bucaramanga Gneiss, outcrops in the Santander massif and is unconformably overlain by a thick metamorphosed marine section (Quétame and Silgara Series). The last named are unconformably overlain in turn by a Devonian–Permian section which appears to represent a transgressive–regressive cycle: Basal conglomerates and sandstones pass transitionally up into shales (Devonian) and in turn into an increasingly carbonate-rich section (Pennsylvanian–Permian) terminating in shaley red beds.

As in the Venezuelan Andes, the Paleozoic section is unconformably overlain by a thick red-bed section which passes transitionally upward into marine deposits of the Cretaceous transgression. That portion of the Cordillera trending northeast through Bogotá was the approximate locus of a particularly deep Cretaceous basin in which some 12 km of marine sediment accumulated. The post-Paleocene Tertiary section consists of continental deposits reflecting erosion of the newly emergent cordillera following the Cretaceous–Paleocene transgression.

Field evidence supported by radiometric ages (Rb/Sr and K/Ar) indicate that there were plutonic granitic events during the two intervals: late Ordovician to late Silurian and late Triassic to early Cretaceous. The radiometric data (Goldsmith *et al.*, 1971) suggest to the writer the probability of an early Silurian event during the former interval, but there may have been three distinct plutonic events during the latter interval. The majority of plutons yield middle to late Triassic K/Ar ages. One pluton, the Rio Negro batholith, yields two near-identical earliest Jurassic ages, and a rhyolite porphyry believed related to the plutonic events gives a K/Ar age of early Cretaceous. Volcanic activity is represented by small volumes of siliceous eruptives in the Permian Jordan Formation. A few rhyolite porphyry dykes and eruptives believed related to the plutonic events gave early Cretaceous K/Ar ages.

Regional metamorphism occurred during the late Precambrian, again during the late Ordovician–late Silurian interval, and possibly in the Mississippian and Permian. The rocks of the Santander massif were converted to amphibolite-grade metamorphics by the late Precambrian event. The Silgara Formation was metamorphosed to "low to medium grade" (Ward *et al.*, 1970, legend to sheet H-13, Pamplona). Only subgreenschist facies products are recognized from the other postulated metamorphic events.

Geostructural features worthy of note may be summarized as follows:

a. The marked contrast in structural style across the Permian–Triassic boundary, which is such a striking feature in the Venezuelan Andes, occurs to a far lesser degree in the Cordillera Oriental. Uplift and tilting of blocks of post-Paleozoic rock is evident, but, in contrast, there has been significant folding of the section within each block.

b. The parallelism in the structural trends produced in deformations widely separated in time is, as in the Venezuelan Andes, a characteristic feature. One additional aspect of this parallelism should be noted: In the region of the Santander massif, Irving (1971, p. 29–30) notes that the current axis of the massif (trending north to north–northwest) coincides with the orogenic axes of the late Ordovician, late Devonian, and late Triassic to Jurassic deformations. On the other hand, in that portion of the Cordillera which trends northeast through Bogotá, structural trends in the Cretaceous–Tertiary section are also oriented northeast, although late Ordovician and late Devonian trends are northerly as in the Santander massif.

It appears at first sight that there were orogenic cycles culminating in deformation regional metamorphism and/or granitic plutonism in the late Precambrian, late Ordovician to late Silurian, late Devonian to early Mississippian, and in the Triassic to early Jurassic. The latest cycle began at the end of the Paleo-

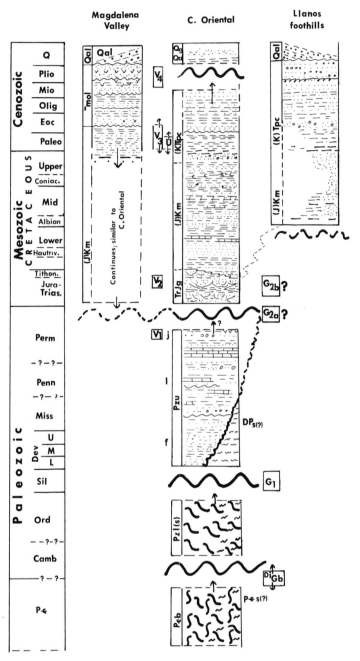

Fig. 6. Stratigraphic column of Cordillera Oriental. Qal, Qg, Quaternary fluvial, glacio-fluvial deposits; mainly fans and terraces flanking the Cordillera Central. Tmol, Tertiary molasse shed from west (Cordillera Central) in four major cycles: (1) Lower and Middle Gualanday; (2) Upper Gualanday, La Cira; (3) Honda; (4) Mesa; cycles have lower coarse, upper finer facies; significant volcanic tuffs occur in 3 and 4. (K)Tpc, Paralic-continental late Cretaceous–late Tertiary deposits; initial late Cretaceous paralic coals, clays, and sandstones

cene, was particularly pronounced during the Miocene, and continues to the present. This cycle was essentially one of vertical uplift devoid of thermal events. In the Bucaramanga area the overturning to the west of the sedimentary pile against the boundary fault of the cordillera suggests that here, as in the Mérida ranges, uplift involved a component outward to the west. In the northeast-trending portion of the ranges northeast of Bogotá, uplift appears to have been more symmetrical, with outward components to both the northwest and the southeast.

The available stratigraphic data for the Perijá ranges are summarized in Fig. 7. The amphibolite-grade metamorphic rocks of the Perijá Formation constitute a basement considered by most workers to be of early Paleozoic age (i.e., rudely correlative with the Quétame–Silgara sequences of the Cordillera Oriental rather than with the Bucaramanga Gneiss). Unconformably overlying the metamorphic basement is a thick (6–7 km) Upper Paleozoic section described by Bowen (1972) in terms of a transgressive Devonian portion (see Fig. 7) overlain discordantly by marine and continental (red bed) deposits

---

grade up into continental shales, silts, sandstones, and conglomerates of extensive internal alluvial flats; initial Maestrichtian–Paleocene cycle followed by Eocene–Miocene and Pliocene cycles; generally finer-grained facies in Cordillera Oriental coarsens to the west (grades into Tmol) and southeast (Guayana). (J)Km, marine (Jura)–Cretaceous facies best exposed in Cordillera Oriental; three major transgressive cycles in the Tithonian, Hauterivian, and Albian; final regressive phase beginning in the Coniacian; transgressive cycles have basal conglomerates, arenites, and concluding shale and limestone facies; final regressive cycle includes chert–shale section (= Luna) but is notably more sandy. TrJg, Trias–Jura Giron Formation; continental red beds, mainly sandstones and shales; molasse facies; see also TrJq in Fig. 2. Pzu, Devonian–Permian, Upper Paleozoic section; overall transgressive–regressive cycle; basal marine conglomerates, sandstones, and shale (Floresta, f) grading up to more shaly and carbonate-rich facies (Labateca, l) overlain by mainly carbonate facies (Bocas, b) and capped by a regressive red silt, sandstone, and volcanic facies (Jordan, j); portion may have suffered metamorphism in Perm ($\sim$) and may have been included with Silgara Fm [DPs(?)]; cf. Pzu in Figs. 2 and 7; Pzl(s), Lower Paleozoic (Cambro?–Ord) Silgara Fm; mainly low-grade phyllitic (chl–mica) schists, quartzites, and minor marbles; contains sillimanite–staurolite assemblages in some areas. Pcb, Precambrian, Bucaramanga Gneiss; mainly amphibolite-grade gneisses, schists, (bi–hbl) amphibolites, and some orthogneisses; retrograde portions may have been included with Silgara Fm [Pcs(?)]. *Metamorphism*: affected by late Permian metamorphism ($\sim$); includes Pzu, Pzl(s), and Pcb; affected by late Ordovician–late Silurian metamorphism; includes Pzl(s) and Pcb; affected by Precambrian metamorphism (?); includes Pcb. *Igneous rocks*: Volcanic: V4, andesitic–rhyolitic tuffs in Miocene and Pliocene of Magdalena valley molasse; V3, minor volcanics including basalts, andesites, dacites, and rhyolites in the Cretaceous of the western side of the Cordillera Oriental; believed of Tertiary age. V2, minor andesites and spilite (/), basalt found locally in Giron Fm; V1, acid-welded tuffs, altered intermediate to basic flows in Jordan Fm. *Plutonic rocks*: U, minor small ultramafic bodies (Muzo area); age and origin unknown; G2b, principal batholitic event based on K/Ar ages (cf. Goldsmith *et al.*, 1971). G2a, principal batholitic event proposed in this review based on geotectonic criteria. G1, diorite, tonalite, granodiorite orthogneisses in Santander massif. Di, Gb, dioritic, gabbroic orthogneisses in Pcb.

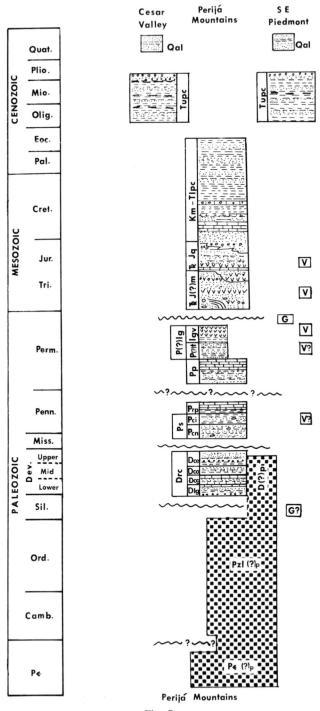

Fig. 7.

of Pennsylvanian age which are approximately correlative with the Sabaneta Formation of the Mérida Andes. In turn, a significant hiatus separates the Pennsylvanian rocks from the Permian section consisting of fossiliferous shales and limestones (Palmarito Formation) conformably overlain by fresh-water marls, shales, and volcanics (Tinacoa Formation and La Gé volcanics). The Upper Paleozoic section is discordantly overlain by Mesozoic–Cenozoic sedimentary formations almost indistinguishable from those of the Mérida Andes to the extent that much the same nomenclature is employed (see Fig. 7); indeed, some of the type localities of the Cretaceous section are located in the Perijá Ranges.

Fig. 7. Stratigraphic column of Perijá Mountains. Qal, Pleistocene sands, gravels, and clays (El Milagro, Necesidad); prominent alluvial terraces, fans in Cesar Valley. Tupc, Upper Tertiary section; paralic to continental deposits: mainly interbedded shales and siltstones; minor sandstone and lignite (Peroc, Macoa, Cuiba), overlain by sandstone and minor conglomerate and siltstone (Los Ranchos, La Villa). Km–Tlpc, Marine Cretaceous–paralic/ continental Lower Tertiary section; Lower Cretaceous basal sandstone (Rio Negro) overlain by fossiliferous sandy limestone (Apón, Maraca) with some calcareous glauconitic sandstone and shale (Lisure); Upper Cretaceous fetid calcareous shale, concretionary limestone, and laminated chert (Luna) overlain by shale and glauconitic sandstone (Mito Juán); Lower Tertiary (Orocué Gp.): alternating carbonaceous shale mudstone and some sandstone with coals (Catatumbo, Barco, Los Cuervos) terminating in sandstone (Mirador). TrJq, Triassic– Jurassic La Quinta Fm; lower part: red conglomerate, sandstone, and some marls, overlain by vesicular basalts and tuffs (gray, mauve, red); upper part: red siltstone and sandstone, terminating in coarse conglomerate (Seco Cg). TrJ(?)m, Triassic–Jurassic (?) Macoita Fm; basal carbonaceous tuffaceous sandstone and pyroclastics (trachytic?); upper graywacke sandstone and some conglomerate. P(?)lg, Permian (?) La Gé Group. lgv: La Gé volcanics; Dacitic to rhyolitic tuffs and flows. P(?)t, Tinacoa Formation; monotonous; dark gray, splintery tuffaceous shale, fine-grained graywacke sandstone, and black shaly limestone. Pp, Palmarito Fm; lower sandstone; upper massive fossiliferous limestone members. Prp, Pennsylvanian; Rio Palmar Fm; dominantly shelly to oolitic limestone; marls and calcareous mudstones towards base; some laminated cherts in middle of section. Ps, Pennsylvanian; Sabaneta Group. Pci, Caño Indio Fm; purple mudstone overlain by tan-purple massive sandstone; red silty shale and siltstone towards top; may include some trachytic(?) tuffs and andesitic breccias. Pcn, Caño del Noroeste Fm; conglomeratic arkosic sandstone lenses, green silty shale, and mudstone interbeds; some crinoidal limestones. Drc, Devonian Rio Cachirí Group. Dcc, Campo Chico Fm; interbedded carbonaceous silty shale, sandstone, and conglomerates, overlain by arkosic sandstone. Dco, Caño del Oeste Fm; siliceous shale and mudstone overlain by fossiliferous siltstone and sandstone. Dcg, Caño Grande Fm; uniform micaceous shales and mudstones and interbedded sandstones; fossiliferous silty marls occur near bottom and top of section. Dlg, Los Guineos Fm; quartzose breccias, sandstones, and interbedded silty shales, overlain by finer-grained sandstones and shales; fossiliferous black limestone and marls at top. Pzl(?)p, Lower Paleozoic (?) Perijá Fm; interbedded quartzites, sillimanitic schists; some amphibolites; lower grade towards Cesar valley; may include Precambrian rocks [Pc(?)p] and/or Devonian rocks [D(?)p]. *Igneous rocks*: v, volcanic rocks; see formational descriptions. G, plutonic rocks; granites and related varieties; see text for discussion of estimated ages.

Bowen (1972, p. 736–37) suggests that most of the granitic plutons in the Perijás are of pre-Devonian age without specifying a more precise age. Some may be of late Permian age, as indicated by K/Ar measurements listed by Mencher (1963, p. 75) and Martin-Bellizzia (1968, p. 369). The oldest volcanic rocks appear to be those of La Gé. No adequate description of the volcanic rocks is available; from descriptions of tuffaceous rocks by Hea and Whitman (1960) and by Bowen (1972), one may infer, however, that the volcanics are dacitic to rhyolitic in composition. Within the Jurassic La Quinta Formation (age assigned by Bowen, 1972, Fig. 2, p. 733) are the El Totumo volcanics which include quartz latites, trachytes, dacites, andesites, and possibly some basaltic types.

Pre-Devonian metamorphic rocks in the Colombian Perijás are described in terms of slates, phyllites, and other low-grade rocks by Trumpy (1943), Forero (1967), and others. In the headwaters of the Rio Cachirí (i.e., Caño Grande) in the Venezuelan Perijás, the pre-Devonian metamorphic rocks are in amphibolite facies. The writer observed abundant development of sillimanite in that section. Associated with the Palmar granite close to the eastern limit of the mountain belt are high-grade amphibolitic gneisses (Bowen, 1972, p. 735). One may interpret the sparse data in terms of two metamorphic terrains: younger (yet pre-Devonian) and of lower grade to the northwest, and older (Precambrian?) and higher grade to the southeast. Alternatively, it can be considered in terms of one pre-Devonian terrain which shows progressive increase in grade from northwest to southeast.

The principal structural elements show many similarities to those of the Mérida Andes. There is again the marked contrast in structural style about a temporal boundary near the end of the Paleozoic. Pre-Mesozoic rocks show intense folding and other evidence of having undergone intense compressive stresses, whereas deformation of post-Paleozoic rocks takes the form dominantly of uplift and tilting of faulted blocks associated with far more gentle folding. The present northeast trend of the Perijás is paralleled not only by the structural trends imparted in Tertiary uplift, but also by pre-Mesozoic structures including those of the metamorphic rocks of the Perijá Formation, so far as one can judge from the available data. Though not as clearly developed as in the Mérida Andes, the Perijás are also characterized by a central graben, following the Perijá–El Tigre fault system, separating two broad anticlinorial arches (Rio Negro–Totumo–Inciarte to the southeast, Becerril–Serranía del Valledupar to the northwest). Miller (1962, p. 1579) notes that the two arches are asymmetric with broad eastern and narrow western flanks. Both of these broad gentle upwarps plunge to northeast and southeast, so that deeper stratigraphic levels are progressively exposed as one moves from the terminations of the Perijás towards the center. The map pattern of the Arena Blanca and Cuiba Faults suggests that the entire Perijá range may constitute a thrust plate that has undergone eastward movement. It is more likely, however, that the Arena Blanca is a normal fault developed in response to volume adjustments in the uplift of the ranges. Miller (1962, p. 1587) regards the Cuiba fault as of high-angle reverse type.

In conclusion, one may specify a major orogeny associated with the metamorphism of the Perijá Formation and probably accompanied by in-trusion of the granitic plutons. This orogeny may have been a Precambrian event but is more likely to have occurred in late Ordovician–late Silurian time. Another orogenic phase appears to have occurred at or toward the close of the Permian. The Palmar granite and La Gé volcanics may be related to this event, but no known regional metamorphism has been identified. A third orogenic event, probably post-Cretaceous, took place in stages throughout the Tertiary much as in the Mérida Andes in the form of uplift and tilting unaccompanied by thermal events.

Discussion

Many of the uncertainties outlined for the Venezuelan Andes hold also for the Cordillera Oriental–Perijá chain. Goldsmith *et al.* (1971, p. D47) note that the contact between the Silgara Formation (Cambro-Ordovician?) and the Bucaramanga Gneiss (Precambrian?) is not clear. Although in general the two rock units are at significantly different metamorphic grade, portions of the Silgara Formation do reach staurolite–sillimanite assemblages. The quadrangle maps H-12 (Bucaramanga) and H-13 (Pamplona), both mapped by Ward *et al.* (1970), show isograds in the Silgara Formation in the regions 20 km north of Bucaramanga and 15 km west of Pamplona, trending sub-parallel to the mapped contact between the two rock units and suggesting that the problem outlined for the Precambrian–Pennsylvanian contact in the Venezuelan Andes may be present. The question again arises as to whether some or all of the Bucaramanga Gneiss might not be highly metamorphosed Silgara equivalent. Here too, as in the Venezuelan Andes, the structural grain of the two rock units is parallel. The radiometric determinations of Goldsmith *et al.* (1971), particularly a K/Ar age of 945 ± 40 m.y. on hornblende from a hornblende gneiss, clearly indicate that at least part of the Bucaramanga Gneiss is indeed Precambrian.

Doubt also remains about the postulated Cambro-Ordovician age of the Silgara Formation. On the one hand, the Devonian Floresta Formation appears to rest with angular unconformity on the Silgara metamorphics; on the other hand, Ward *et al.* (1970) show (quadrangle H-13, Pamplona) the isograd corresponding to the first appearance of garnet as coincident with the contact between the Silgara and Devonian sections in the region 6 km east of Santa Barbara (approximately 30 km ESE of Bucaramanga). This suggests again the possibility of uncertainty in distinguishing between lithostratigraphic contact and metamorphic isograd and raises the possibility that part of the Silgara Formation may be of Devonian age. Two whole-rock K/Ar ages reported by Goldsmith *et al.* (1971) on Silgara phyllite appear to reflect a late Permian–early Triassic thermal event and do not bear on the age of the formation.

Concerning the age of the Perijá Formation, similar problems occur in the Perijás. Bowen (1972, p. 735) prefers a pre-Devonian age for the former partly on the grounds that the Devonian section shows an almost total absence of metamorphism and intrusion by granites. Bowen (1972) rejects Hea and Whitman's (1960) suggestion that the Perijá Formation represents a metamorphosed portion of the Devonian Cachirí Group on the grounds that the latter does not have massive or thick-bedded sandstones (i.e., to match the quartzite layers interbedded with the schists of the Perijá Formation). The writer formed a different impression on viewing the section exposed in the headwaters of the Rio Cachirí in the company of A. Espejo. Thick, tough massive quartzites are interbedded with the calcareous shales in the basal portion of the Devonian section. Hea and Whitman's (1960) correlations seem a reasonable possibility to the writer, particularly after experience in the Venezuelan Andes. The possibility that the Perijá Formation may represent metamorphosed Cachirí Group is shown in the stratigraphic column of Fig. 7. Clearly, the problem of distinguishing between lithostratigraphic contact and metamorphic isograd has become the kernel of significant difference of opinion on the tectonic history of the Paleozoic. In the case of the late Paleozoic, the matter is further complicated by problems of fundamental stratigraphic philosophy. In part because of absence of Mississippian faunas, and in part on the basis of regional distribution of rock units, most writers consider the Upper Paleozoic section in terms of discontinuous stratigraphy widely interrupted in space and time by unconformities, discontinuities, etc. (cf. Irving, 1971, p. 16, 18; Burgl, 1973, p. 57, Bowen, 1972, p. 744, 747, 750; and others). In contrast, Cediel (1972) has integrated the work of Stibane (1968), Forero (in Cediel, 1972), and Stibane and Forero (1969) on the Devonian–Permian section in eastern Colombia in terms of a megasedimentary kindred defined above and below by major angular unconformities traceable throughout these Andes. In this view, the observed unconformities in the Upper Paleozoic section are seen in more reasonable perspective in terms of the progressive buildup of tectonism during the transgressive–regressive depositional cycle.

There is also divergence of opinion concerning metamorphism and plutonic events in the late Paleozoic. Irving (1971, p. 16) refers to metamorphism of Devonian rocks in the Santander massif but considers it a local perturbation and notes that no igneous activity is known. Burgl (1967, p. 437; 1973, p. 83) refers to mild folding and metamorphism corresponding to "Variscan" deformation. Cediel (1972) dismisses reference to metamorphosed Devonian strata as either determinative error or extremely local effects of dynamic or contact type. At the scale of this overview of tectonic relationships the difference of opinion between Cediel, Irving, and Burgl are relatively subtle and contrast sharply with some views put forth by others. Thus, Radelli (1967, p. 141–144), largely on the basis of stratigraphic criteria, concludes that there were plutonic granitic events in the early and late Devonian and late Mississippian to early Pennsylvanian in addition to the major Triassic plutonic events. Botero (1950) postulated a possible granitic event approximately at the end of the Devonian in the Floresta region, Boyaca Department. Most workers do not appear to accept the Devonian–Carboniferous plutonic events proposed by Radelli and by Botero. Stated by Cediel (1972) and largely implied by others is the view that the late Permian deformation, although a major event, may not have involved the entire spectrum of metamorphic and magmatic events normally associated with orogeny. The writer concedes that the orogeny at the close of the Paleozoic may not have been characterized by all the features commonly associated with classic or modern views of orogeny, but believes that major thermal events including regional metamorphism and granitic plutonism may have occurred in the Cordillera Oriental–Perijás as they did in the Venezuelan Andes. The matter of the coincidence of the Devonian boundary with a metamorphic isograd in the Santander massif (see Pamplona sheet of Ward et al., 1970) and the claim of Hea and Whitman (1960) that the schists of the Perijá Series may represent metamorphosed Devonian strata require careful restudy. In the writer's opinion, the apparent parallelism of structures of believed widely different ages also requires investigation.

The biggest uncertainty in the post-Paleozoic portion of the tectonic history of these ranges concerns the proposed dominant Triassic–Jurassic age of the granitic plutons. For reasons stated previously, the writer believes that K/Ar measurements need the backing of Rb/Sr and U/Pb ages for more thorough interpretation of the intrusive history of these plutons. The matter of the granitic plutons in the Santander massif and elsewhere is further considered in Sec. V in the discussion of the thermal events.

C. Cordillera Central

Above the mouth of the Rio Cauca the Rio Magdalena loops around a geologic terrain significantly different from the rest of the Cordillera Central (see Fig. 3). On the basis of the fragmentary information available (Fig. 8), the pre-Mesozoic history of this portion of the eastern flank of this range may be described as an extensive Precambrian (?), granulite facies, metamorphic terrain unconformably overlain in the Las Cristalinas area by slightly meta-

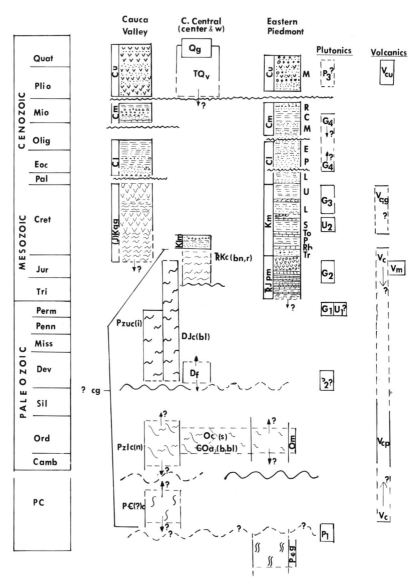

Fig. 8. Stratigraphic column of Cordillera Central. Qg, glacial and fluvio-glacial deposits.
TQv, late Cenozoic volcanics; basaltic to rhyolitic, mainly porphyritic andesites; Cu,
Pliocene molasse; conglomerates, sandstones, shales, and tuffs; Mesa Fm(M) in the East.
Cm, mid-Cenozoic molasse; basal shale and sandstone (Mugrosa, M) overlain by red shale
and sandstone (Colorado, C); overlain by sandstone, conglomerate, and shale (Real, R)
along the eastern Piedmont (cf. Fig. 5); probable equivalent in the Cauca Valley not clearly
defined. Cl, Lower Cenozoic section; basal alluvial basin fills (Lisama, L) overlain by molasse;
sandstone (La Paz, P) followed by sandstone, siltstone, and shale (Esmeraldas, E) in the
Eastern piedmont (Morales *et al.*, 1958); similar deposits with coals found in Cauca Valley,
but detailed stratigraphy and age (mainly Oligocene?) not known. Km, marine Cretaceous;

morphosed graptolitic shales of early Ordovician age. Elsewhere the bulk of the central and western portions of these Andes consists of greenschists and some amphibolite-grade metamorphics of both para and ortho type collectively referred to as the Cajamarca Group (Nelson, 1957). Similar rocks in

---

Lower Cretaceous basal sandstone (Tambor, Tr) overlain by limestone and marls (Rosa Blanca, RB), black shale (Paja, P), limestone (Tablazo, To), and shale, sandstone, and limestone (Simiti, Salto, S); Upper Cretaceous limestone, calcareous black shales, and cherts (Luna, L) overlain by a thick shale section (Umir, U). Klm, marine Lower Cretaceous (Aptian–Albian); sandstones, conglomerates, and shales in the Cordillera Central. (J)Kqg, Jurassic–Cretaceous Quebrada Grande Fm; lower and upper tuffaceous, quartzose, and carbonaceous phyllites, with minor limestone and conglomerate; intercalated with lower and upper altered (chlorite and epidote) greenstones, mainly porphyritic andesites, some basalts, spilites, and agglomerates. TrJpm, Triassic Payande Fm; basal arenites ("Pre-Payandé") overlain by sandy limestones, shales, and cherts (Payandé); Jura: Morrocoyal Fm.; black shale with fossiliferous limestones grading into red beds with acid volcanics. ?cg, Cajamarca Gp; greenschist–amphibolite facies metamorphism, actinolitic schists, graphitic schists, quartzose phyllites, metabasalts (albite–chorite), minor marble, and hornblende amphibolite; here considered to include Ayura–Montebello Gp (cf. Fig. 13). *Suggested age ranges*: TrKc(bn,r): Triassic–Cretaceous (Butterlin, 1972; Radelli, 1967); DJc(bl): Devonian–Jurassic (Burgl, 1973); Pzuc(i): Upper Paleozoic (Devonian–Permian) (Irving, 1972); Pzlc(n): Lower Paleozoic (Cambrian–Ordovician) (Nelson, 1957; Cediel, 1972); Pc(?)c: Precambrian; possible age, but not favored by any modern worker. Df, fossiliferous Devonian rocks; reported by Forero (Forero, in Cediel, 1972). Om; metamorphosed Ordovician rocks; low-grade metamorphics: slaty shales (some graptolitic), siltstones, orthoquartzites, and limestones. Oc(s), transitional facies from Om to Cajamarca Gp (Stibane, 1967). COa(b,bl), transition from Om to Ayura–Montebello Gp (Botero A., 1963; Burgl, 1973); cf. Fig. 12. Pcg, Precambrian granulite terrain; feldspathic gneiss, amphibolite, and marbles. *Metamorphism* (assigned ages depend on age range adopted for Cajamarca Gp): Permian (Irving, 1971); Jurassic (Burgl, 1973); early Cretaceous (Butterlin, 1972). Late Ordovician–late Silurian; may have undergone subsequent metamorphic event ($\sim$). Late(?) Precambrian; possible event not favored by modern workers. Precambrian; required event to explain granulites of Pcg. (Note: Structure and mineral assemblages of Kqg suggest possible late Cretaceous metamorphism in the Cauca Valley region.) *Igneous rocks*: Volcanics: Vcu, Upper Cenozoic volcanics; cf. TQv. Vqg, Quebrada Grande Volcanics; cf. Kqg. Vc, Cajamarca Gp volcanics; cf. Vcp: preferred age in this report. Vm, Morrocoyal volcanics; cf. TrJm. *Plutonics*: G4, granitic plutons in Cauca Valley; believed Eocene, may extend into Miocene. G3, quartz diorite batholiths (Antioquian); mainly end of the Cretaceous, early phase may have begun in mid-Cretaceous; include diorite and Gabbro facies. U2?, serpentine, serpentine dunite, minor wehrlite, harzburgite, lherzolite, and pyroxenite; believed mid-Cretaceous minimum age (Barrero *et al.*, 1969). G2: diorite and quartz diorite; early to mid-Jurassic in eastern Piedmont; some associated gabbro. G1, late Permian granitic plutons mainly in western Piedmont and the Cauca Valley. U1?, metamorphosed ultramafics (serpentine, talc, talc–serpentine); in the eastern part of the Cordillera Central; P1,2,3?, plutonic phases not identified but postulated on geotectonic basis: P1?, believed granitic complement to Pcg; probably removed by erosion; P2?, assumes orogeny and regional metamorphism of Cajamarca gp in late Ordovician–late Silurian; P3?, postulated as yet unexposed plutons equivalent to Vcu.

the area of Medellín and further north are known as the Ayura–Montebello Formation (Botero, 1963) and the Valdivia Group (Hall, in Irving, 1971, p. 21) and may possibly be correlatives of the Cajamarca Group as may the amphibolite-grade rocks of the Puqui Complex in the same region.

In contrast to the eastern Andean ranges the Triassic and Lower Jurassic appear in a fossiliferous marine facies (Payandé and Morrocoyal formations) on the eastern flank of this cordillera but grade up into red beds and volcanics which correlate with the Giron–La Quinta formations to the east. During the Cretaceous the Cordillera Central was largely a positive region; only the central portion at the latitude of the middle Magdalena valley was involved in the extensive marine transgression of the late Mesozoic and then only until the Albian. The western flank of the uplift was the site of basic marine volcanism and the deposition of a flyschlike section of wackes derived from the uplifted portions of these Andes at the time that marine clastics and carbonates were being deposited in the Cretaceous geosyncline along the eastern flank.

The early Tertiary section is paralic to continental and represents basin fill along both flanks of the Cordillera. These deposits overlie the Cretaceous section with relatively minor discontinuity. By contrast, the late Tertiary section is a thick wedge-shaped pile of sediment representing deposits stripped off the rising cordillera during "Andean" uplift which began in earliest Miocene time.

A major unconformity may reasonably be stipulated for some time in the Precambrian–early Ordovician interval. Possibly the same(?) hiatus separates the Cajamarca Group and correlatives from the Devonian exposures near Rovira. Irving (1971) postulates a major unconformity at the close of the Permian related to a major orogenic event. Along the central axis of the Cordillera, another erosional hiatus extends from Albian to Oligocene time. Several late Tertiary hiatuses are recognized in flanking deposits to the east and west of the axial zone and are here grouped as representing substages of so-called "Andean" orogeny.

Migmatitic terrain in the Precambrian basement probably represents a Precambrian granitic event, but no specific pluton of that age has been identified. Irving (1971, p. 44–48) refers to K/Ar ages which may reflect granitic events during the late Permian–early Triassic, Jurassic, mid-Cretaceous, late Cretaceous, and Eocene. Of special interest are the occurrences of metamorphosed serpentinites and related ultramafics associated with talc schists on the eastern flank of the ranges and fault-bound bodies of serpentinite associated with the Romeral fault on the west side of the Cordillera. The former may possibly have been emplaced in the Pennsylvanian–Permian; the latter are tentatively considered to be of mid-Cretaceous (pre-Antioquia batholith) age.

The earliest recognizable volcanic rocks are tuffs and basic lavas of the Cajamarca Group, metamorphosed to greenschists and amphibolites, which occur in the central and western portions of the Cordillera and are of pre-Devonian age according to Cediel (1972). Extensive basic volcanism next occurred in the mid-Jurassic along extensive portions of the eastern piedmont of the Cordillera. Probably beginning in the late Jurassic and continuing through the early Cretaceous, basic volcanism reappeared along the western flank. Middle to late Tertiary volcanic plugs occur on the western flank, in the Rio Cauca valley, and may be correlative in part with the voluminous intermediate to siliceous volcanism which occurred along the crestline of the Cordillera in its southernmost extensions (outside map area) from Pliocene into historic time.

At least one event of regional metamorphism may be ascribed to the oldest hiatus in the stratigraphic column of Fig. 8. Whether the regional metamorphism of the Cajamarca group occurred at the same time in relation to the same orogenic event is not known; it could have occurred at some time in the mid-Ordovician to Devonian interval. On the basis of K/Ar ages in the range 214–239 (both ±7 m.y.), Irving (1971) has suggested that the regional metamorphism of the Cajamarca Group occurred at the end of the Paleozoic. Plutons of Triassic to Tertiary age are considered to have given rise only to local contact metamorphism.

Published structural data are minimal for the Cordillera Central. Available maps (see, e.g., Feininger et al., 1970) suggest that structural grain in the Precambrian terrain subparallels that in the metamorphosed Cajamarca Group further west and that these trends (north) are in turn roughly parallel to the present trend of the Cordillera (north–north northeast). Folds in the metamorphic rocks appear to be asymmetric or overturned to the west with gentle plunges (0–20°) to the north or south and appear to have folded the metamorphic isograds. Post-Paleozoic (post-Jurassic?) structures for the most part take the form of outward tilt-blocks involving some flexural slip folding. On the west side of the ranges a major structural feature is the Romeral fault which is considered to pass from high-angle reverse type in the south to right-lateral transcurrent in the north (Irving, 1971, p. 62). Three other major faults, the Otu, Bagre, and Palestina, occur further east (see Fig. 3). The Bagre corresponds roughly with the contact between the Precambrian metamorphic terrain and the Cajamarca Group and may have seen considerable vertical displacement. The Otu and Palestina are considered to be of transcurrent type, as discussed by Feininger (1970). Horizontal displacement on the Palestina is nearly 28 km.

In summary, one may postulate at least one major orogenic event for some time in the late (?) Precambrian–early Ordovician. Irving (1971), on the basis of several late Permian–early Triassic K/Ar ages on granite, postulates another major orogeny corresponding thereto. Because the Precambrian granulites are only exposed in the eastern piedmont and because the Permian–Triassic

K/Ar ages come from the western flank, it is tempting to suggest that the orogenic axes moved from east to west, but the idea has no other evidence to commend it. In the Jurassic, the site of basic volcanism and granitic plutonism moved to crudely linear belts along the eastern flank of the Cordillera, but corresponding phases of orogenic sedimentation, regional metamorphism, and deformation have not been identified. Basic volcanism and granitic plutonism reappeared in the Jurassic–Cretaceous mainly in the region west of the present crestline. The intense folding of the volcanics in the Cauca valley and the appearance of metamorphic minerals such as pumpellyite, chlorite, epidote, and actinolite all point to a major orogenic event at the close of the Cretaceous with its principal axis located in a zone extending westward from the western piedmont into the Cordillera Occidental.

As in the cordilleras to the east, Tertiary orogeny took the form principally of uplift, and the nature and periodicity of the events has to be gauged by the characteristics of the sedimentary piles in the flanking basins of the Rios Magdalena and Cauca. In addition to uplift following the late Cretaceous orogeny, there appear to have been further events in the mid-Eocene, at the close of the Oligocene and the Miocene, and finally during an interval spanning the late Pliocene and most of the Pleistocene. Tertiary orogeny in the Cordillera Central differs notably from that in the ranges to the east in the occurrence of granitic plutonism during the Eocene and the continuation of intermediate to siliceous volcanism until very recent time.

Discussion

Almost the entire pre-Cretaceous history of this range is in question. But for the lone Ordovician and Devonian fossiliferous localities, there are neither paleontologic nor adequate isotopic data to support tentative ages assigned by various workers. The wheel has gone full circle since Nelson (1957) correlated the Cajamarca Group with the Cambro-Ordovician Quetame Formation of the Cordillera Oriental on the basis of regional stratigraphic relationships and petrographic similarities. Subsequently, a wide range of ages as postulated, including Trias–early Cretaceous [Butterlin (1972), based in part on believed gradation from metamorphics to Cretaceous sedimentary rocks], Devono-Jurassic [Burgl (1973), mainly on the basis of classical geosynclinal theory], and late Paleozoic [Irving (1971), inferred from K/Ar ages]. On the premise that the Cajamarca metamorphics are likely to be older than the unmetamorphosed Devonian finds at Rovira, Cediel (1972) favors the Cambro–Ordovician age suggested by Nelson. An early Paleozoic age for the original volcanism seems to be the likeliest possibility. However, the solution to the problem of ages in the Cordillera Central may prove to be highly complicated with virtually all writers having a share, smaller or larger, of the truth: i.e., the Cajamarca may prove to be a wastebasket for metamorphic rocks produced in a wide range of thermal events extending into Tertiary time. An example of possible complexities is provided by Burgl (1973, p. 36), who correlated high-grade metamorphics of the Ayura–Montebello Group with the lightly meta-morphosed Ordovician rocks at Las Cristalinas and viewed them in a distinct pre-Cajamarca eugeosynclinal setting. As discussed in Sec. VB, these views are partly challenged and it is concluded that Ayura–Montebello is best cor-related with the Cajamarca Group.

In view of the relationships noted in the Mérida Andes (see Figs. 4 and 5 and related discussion), where gradation from high- to low-grade metamor-phic facies may occur in short lateral distances, Cediel's premise that the Cajamarca Group is likely to pre-date the Devonian at Rovira need not be true. Verifying the ages of original volcanism and subsequent metamorphism of the Cajamarca Group will be crucial to geotectonic syntheses of the northern Andes. A late Paleozoic age for the volcanism could be interpreted in terms of the start of a volcanic province which spread continuously westwards reaching the Pacific coast in Tertiary time. A late Paleozoic regional metamorphism might suggest genetic relationship to the Permian metamorphism of the ranges further east. An early Paleozoic age for the volcanism (as preferred here) and/or metamorphism would require quite a different tectonic framework.

Concerning the sections of known Paleozoic age, the data are yet so sparse that it is not known whether there exists here the same bipartite stratigraphy (Lower vs. Upper Paleozoic sections separated by a major hiatus) that occurs in the ranges to the east. As previously mentioned, K/Ar measurements, however reliable and significant as data, cannot be relied on to give *original* ages of thermal events in terrain with a history of intense, long-continued orogenic activity.

Concerning the post-Paleozoic portion of the tectonic history, it is dif-ficult to accept unconditionally isolated events of volcanism and granitic plutonism without (or with minimal) accompanying effects such as regional metamorphism and deformation. That the emplacement of such enormous batholiths as the Antioquian (± 7000 km^2) in the late Cretaceous was ac-complished without a corresponding regional metamorphic overprint does not appear to the writer to be a reasonable hypothesis. Indeed, it is worth remarking that although Burgl (1973) considers the metamorphism of the Ayura–Montebello Group to have been an early Paleozoic (Caledonian?) event, his Fig. 7 (1973, p. 26) shows the Ayura–Montebello Group to occur as though it constituted a metamorphic aureole about the Antioquia batholith. This matter is further considered in Sec. V (see Fig. 13 and related discussion).

Much of the Jurassic, Cretaceous, and Eocene plutonism (and the Plio-Holocene volcanism) appears to have taken place in regions which were positive

or received only a very thin sedimentary veneer. It is difficult in such circumstances to find an adequate framework to explain such thermal events. Irving
(1971, p. 49) notes that the late Tertiary volcanics are associated with a major
fracture zone, suggesting that the volcanicity may simply have been a consequence of sudden release of pressure at depth in the crust. This does not
appear to be a suitable mechanism to explain extensive plutonic activity. The
apparent relatively abrupt change in loci of principal orogenic axes must also
be considered highly tentative pending more extensive radiometric and other
studies.

D. Sierra Nevada de Santa Marta

The Sierra Nevada de Santa Marta is an uplifted block shaped like an
equilateral triangle 180 km on a side and consisting dominantly of high-grade
metamorphics extensively intruded by granitic plutons. These crystalline rocks
are flanked on the northern and southeastern sides by volcanics and sediments
mainly of Mesozoic age. Regional interpretation is complicated by at least
three events of regional metamorphism (see Fig. 9).

Southeast of the Sevilla lineament the oldest rocks are Precambrian granulites (perthitic Qz–F rocks and Hbl–Px granulites) which may correlate with
those of the Cordillera Central. The Precambrian ages are confirmed by radiometric measurements of 1300 $\pm$ 100 m.y., 940 $\pm$ 47 m.y., and 752 $\pm$ 70 m.y.
(see Irving, 1971, p. 8–9, and Tschanz et al., 1974, p. 274–5 for sources). The
oldest rocks overlying the granulites are low-grade meta(?)-sedimentary rocks
(see Fig. 9) of the Chundua Group (Gansser, 1955) from the upper levels of
which deformed fossils of probable Pennsylvanian age have been recovered.
The Chundua rocks are known from relatively restricted localities in the
central to southeastern region of the block. To the northeast and southwest
the granulites are overlain by the late Permian to early Triassic Los Indios and
Corual Formations which consist of marine sedimentary rocks overlain by
basic and siliceous volcanics including spilitic tuffs. These rocks in turn are
overlain unconformably by a thick pile ($\pm$4 km) of spilitic or keratophyric
tuffs of the Guatapuri Formation, the deposition of which was followed
during the late (?) Jurassic and early Cretaceous by further extensive volcanic deposits, mainly of dacitic to rhyolitic composition and largely spilite-
free.

Jurassic–Cretaceous volcanism gives way on the southeastern side of the
massif to marine deposition of the Cretaceous transgression. The section grades
from basal clastics up through increasingly carbonate-rich and shaley facies,
culminating in Paleocene–Eocene paralic basin fills with coals. The resulting
stratigraphy is similar to that already noted for the same time span in the
Perijás and Venezuelan Andes. No further deposition occurred on the massif,

most of which stood high during post-Eocene time and shed sediment radially into the adjacent lows of the Cesar and Magdalena valleys and the Caribbean Sea.

Northwest of the Sevilla lineament the geologic history was complicated by additional events of regional metamorphism. Precambrian granulites within the lineament are in contact to the northwest with lower-grade gneisses and schists (almandine amphibolite) of the Sevilla Metamorphic complex. These rocks are of unknown age but are tentatively considered by Tschanz *et al.* (1974) to have undergone metamorphism in late Permian to Triassic time and again in the Jurassic. Northwest of the Sevilla metamorphic complex is the Santa Marta metamorphic belt (Gaira, San Lorenzo, and Taganga Formations) which ranges in grade from amphibolite facies on the southeast through greenschist to phyllite at the northwestern apex of the massif. The amphibolite and greenschist facies are separated by the Santa Marta batholith of Eocene age. Tschanz *et al.* (1974, p. 280) conclude tentatively that the amphibolite-grade schists of the Santa Marta belt (i.e., those between the Sevilla complex and the Santa Marta batholith) were metamorphosed in the late Jurassic and that the greenschists and phyllites northwest of the Santa Marta batholith were metamorphosed in the late Cretaceous.

The principal plutonic thermal events are summarized in Fig. 9. In the granulite terrain southeast of the Sevilla lineament there are extensive parallel batholithic belts which appear to become progressively younger, more shallowly emplaced, and more potassic towards the southeast (Tschanz *et al.*, 1974). The plutonism appears to have occurred from mid-Triassic to late Cretaceous time but to have attained a principal climax in the middle Jurassic (± 165–175 m.y.). Tschanz *et al.* (1974, p. 275) refer to a 250-m.y. K/Ar date of uncertain reliability on hornblende from a dioritic block collected in the Sevilla lineament. This age is the oldest measured to date, and no older Paleozoic or Precambrian granitic plutons have been identified. Northwest of the Sevilla lineament there were some initial intrusions in the early and late Cretaceous (Socorro and Latal plutons, respectively), but the climax came in the Eocene with the intrusion of the Santa Marta and Buritaca masses. Other plutonic rocks of potential tectonic significance include anorthosites of Precambrian age in the Sevilla lineament and small serpentinized harzburgites of Mesozoic (?) age in the Sevilla and Santa Marta belts.

Concerning volcanic thermal events, some basic metavolcanic rocks of late Permian to Triassic age are associated with the metamorphic rocks of the Sevilla lineament, but the great bulk of the volcanics are in the Corual and Guatapuri Formations of late Permian–early Triassic to early Jurassic age. These volcanics are spilitic and keratophyric and are overlain by the extensive rhyolitic volcanics of middle to late Jurassic–early Cretaceous age. The younger volcanic sequence probably represents surface equivalents of the major mid-

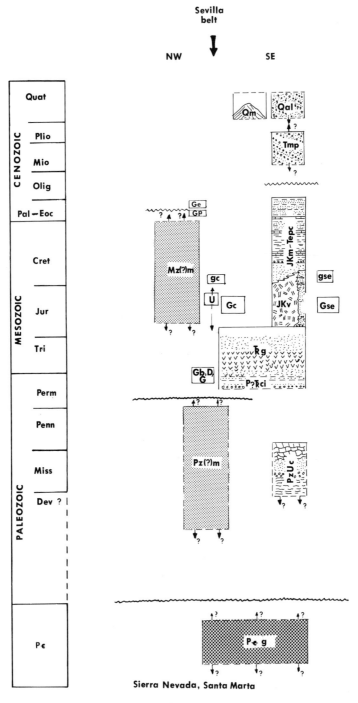

Fig. 9.

Jurassic plutonism. Indeed, some may be vesiculated portions of the roofs of granitic batholiths.

Among significant structural features is the pronounced northeasterly grain of the Santa Marta block, a trend which roughly parallels that of the structures in the Guajira Peninsula and in the Perijá Mountains. The metamorphic rocks appear to be tightly folded and mostly asymmetric or overturned to the southeast. Tschanz *et al.* (1974) suggest that intrusion of the vast Jurassic batholiths occurred in a dilational environment which involved major lateral movements within the massif of pre-Jurassic rocks. Among other examples of the latter, one may cite the possible allochthonous character of Precambrian granulites in the Sevilla lineament. In the field of structural geology there has been considerable discussion concerning the nature of the

---

Fig. 9. Stratigraphic sections for Sierra Nevada de Santa Marta. Qal, Quaternary alluvium, terrace and fan gravels; may extend into Pliocene. Qm, moraines, fluvial-glacio deposits. Tmp, Miocene–Pliocene marine, paralic, and continental deposits in basins flanking Santa Marta block; commonly arkosic clastics. Jkm–Tepc, marine Jurassic–Cretaceous to paralic and continental Lower Tertiary deposits; Lower Cretaceous sandstone and conglomerate (Rio Negro) overlain by sandy limestone, shale, and glauconitic sandstone (Aguas Blancas, Lagunitas); Upper Cretaceous shale, concretionary limestone, and cherts (Manaure, Laja), marine shale (Molino) terminating in shale, siltstones, and limestone (Hato Nuevo); L. Tertiary: interbedded sandstone, siltstone, shale, and minor limestone with coals, lignites, clays (Manantial, Cerrejon, Tabaco, Aguas Nuevas). JKv, ignimbritic breccias, minor flows, vitrophyres of quartz keratophyre, rhyolite, ryodacite, and dacite. JKs, mainly fine-grained clastic red beds ($\approx$ La Quinta, Giron Fms.); Guatapuri Fm.: spilitic and keratophyric volcanics passing up into tuffaceous red siltstones and sandstones; overlain by Triassic spilitic lavas. PT(?)ci, Permian–Triassic Corual and Los Indios Fms; graywacke siltstone, sandstones, and spilitic and basaltic volcanics (Corual); basal conglomerate overlain by shale, quartzites, and limestone terminating in shale, quartzites, and porphyries (Los Indios). PzUc, Mississippian(?)–Pennsylvanian, Chundua Group. Graphitic slate, quartzite, conglomerate and sandstone, terminating in crystalline limestone. *Metamorphic rock units*: Mz(?)m, Santa Marta metamorphic belt; greenschist to amphibolite facies metasedimentary rocks in northwestern Santa Marta block; unknown age, metamorphosed in late Jurassic–Paleocene; chlorite-sericite phyllite, greenstones (Taganga); mica–amphibolite schists (San Lorenzo); mica–amphibolite schists and marbles (Gaira). Pz(?)m, Sevilla metamorphic belt; amphibolite facies; hornblende gneisses, schists, and cataclasites (Muchachito); polymetamorphic, late Permian–Eocene. Pcg, Precambrian granulites; includes granitic gneisses, pyroxene amphibolites, marbles, and ultramafic gneisses (Los Mangos); also garnet and magnetite anorthositic gneisses; probably polymetamorphic. *Igneous rocks*: Volcanic rocks are included in description of rock formations above. Gp, Ge, batholiths of Paleocene (p) and Eocene (e) age in northwestern region. gc, gse, post-batholithic granitic stocks in central (c) and southeastern (se) belts. Gc, Gse, principal granitic batholithic phases in central (c) and southeastern (se) belts. Gb, D, G, early plutonic phase; includes granitic batholiths of western border zone and Rio Piedras complex; also gabbro and metadiorite along Sevilla lineament. U, ultramafic rocks; includes intrusive serpentinites and metamorphic talc-tremolite schists; age not known.

movement of the boundary faults of the Santa Marta block (see Tschanz
et al., 1974). The Oca fault along the north side is believed to show right-
lateral transcurrent movement of 65 km; some 110 km of left-lateral movement
is ascribed to the Santa Marta–Bucaramanga fault defining the southwestern
side. Alluvium obscures what is probably one or a set of normal faults trending
along the Cesar valley. Strike-slip motion on the Oca and Santa Marta faults
probably occurred in earliest Tertiary time. Dip-slip movement on the Oca
and Santa Marta faults, leading to some 12 km of structural relief between
the massif and the adjacent basins, is believed to have occurred beginning in
earliest Miocene time.

In summarizing the tectonic evolution of the Santa Marta block, one
may interpret the granulite terrain in terms of at least one Precambrian orogenic
event. No Cambrian–Silurian section has been identified, but the late Paleozoic
[Pennsylvanian(?) to Permian] section and the igneous-metamorphic events
in the area of the Sevilla lineament may be interpreted very tenuously in
terms of an orogenic cycle. Tschanz *et al.* (1974) consider the mid-Jurassic
batholiths southeast of the Sevilla lineament as anorogenic. Northwest of
that lineament there were metamorphic events in Permian, Triassic, Jurassic,
and Cretaceous time followed by plutonism in the Eocene. These several
thermal events may relate to one or more orogenic events. Orogeny involving simple uplift and tilting began in earliest Miocene time and probably involved more than one event of uplift, as in the Perijás and Mérida
Andes.

Discussion

As in the cases of the other ranges, uncertainties in tectonic interpretation
stem in part from lack of an adequate stratigraphic record. Interpretation of
Mesozoic events is complicated by metamorphism on the northern side, but
not on the southeastern, and also by differing sedimentary facies. For the
writer, the most puzzling feature concerns the spatial relationships (or rather
the lack of them) between major thermo-tectonic events. Whereas the young
terrain embracing metamorphic events of Permian–Triassic to Paleocene(?) age
is located northwest of the Sevilla lineament, the great bulk of Mesozoic
plutonism and volcanism occurred in the extensive region southeast of the
lineament. In that region, one now finds the anomalous situation of enormous
granitic batholiths in granulite-facies terrain. Furthermore, though some
writers [cf. Gansser (1955)] describe the Paleozoic formations in terms of
"graphitic" schists or "slates," the metamorphic effects are remarkably small
when one considers the enormous volume and proximity of the Mesozoic
batholiths. This raises anew the question of true intrusive ages and is particularly pertinent in the case of the region northwest of the Sevilla lineament

where at least three and possibly four overlapping metamorphic events occurred.

The principal thermal events and their spatial relationships are interpreted
by Tschanz *et al.* (1974) in terms of successive subduction zones which moved
northwest from the Sevilla lineament beginning in the late Permian or early
Triassic and continuing into the Paleocene. Metamorphism northwest of the
lineament is envisaged in a compressive environment, whereas batholithic
emplacement is viewed in a dilational environment. Fracturing and lateral
displacement of microcontinental cratonic fragments is suggested by Tschanz
et al. (1974, p. 283) to explain the dilational environment, but the absence of
high-pressure metamorphic facies northwest of the Sevilla lineament and a
suitable source for the vast batholiths in the granulite terrain to the southeast
remain as major unanswered questions.

The nature of the boundary faults remains a thorny problem. Ascribing
the strike-slip motion to latest Cretaceous–Paleocene time instead of post-
Miocene age claimed by Campbell (1968) does not solve a major geometric
and volumetric problem when strike-slip movement of several tens of kilometers
occurs along two sides of the triangular block. The change from lateral to
vertical movement along these fractures beginning in early Miocene time
(Tschanz *et al.*, 1974, p. 274) raises additional problems. Such structural
problems are reconsidered in Sec. V of this study.

Regional structural grain and similarity in the history of thermal events
and their products suggest geotectonic continuity of the Santa Marta block
with the Guajira Peninsula to the northeast and with the Cordillera Central
to the southwest, as noted by Alvarez (1971) and other workers. If so, the
Santa Marta–Bucaramanga fault system represents a pronounced discordance
superimposed on the older regional grain. The possibility that the Santa Marta
block may have been an independent geotectonic entity in Precambrian and
early Paleozoic time has only the uniqueness of the anorthosites and absence
of an identified early Paleozoic section to commend it, but in the light of plate
theory, this possibility requires further examination.

E. Cordillera Occidental and Pacific Coast Range

The relatively simple outline of the stratigraphy in these western ranges
is summarized in Fig. 10. No crystalline basement has been recognized, and
the section may be conveniently considered in terms of thick volcano-sedi-
mentary piles overlain by a largely nonvolcanic sedimentary section shed from
the rising Cordillera Central. The temporal boundary between these "super-
facies" corresponds approximately to the close of the Mesozoic in the region
of the Cauca valley but appears to migrate into the late Oligocene as one
approaches the Pacific coast.

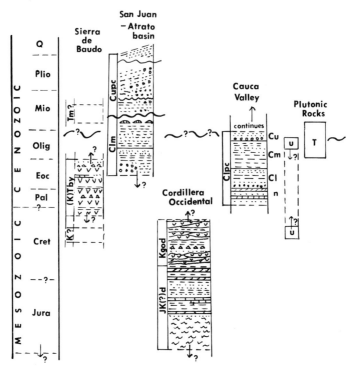

Fig. 10. Stratigraphic sections of the western ranges and associated basins. *Pacific Coast ranges* (*S. de Baudo*): (K)Tbv, Cretaceous(?)–Tertiary Baudo volcanics; basalts (including gabbro, spilite, and diabase) intercalated with agglomerate and tuffs; minor cherts, rhyolites, silicic shale, and mid-Eocene algal limestone. (K?, Tm?), late Cretaceous–Miocene sedimentary rocks occur in fault troughs in (K)Tbv (cf. Case *et al.*, 1971). *San Juan–Atrato basin*: Cupc, paralic–continental Upper Cenozoic section; carbonaceous conglomerate, sandstone, siltstone, and shale; molasselike facies mainly derived from east (cf. Nygren, 1950). Clm, marine Lower Cenozoic section; calcareous conglomerate, sandstone, and shale interbedded with marls and limestone; two gross, fining-upward cycles terminating with limestone; clastics mainly from the West through late Oligocene (cf. Nygren, 1950). *Cordillera Occidental*: Kcgd, Cretaceous Cañas Gordas, Diabase Gps; thick basic volcanic pile of basalt, diabase, and spilite with interbedded tuff, agglomerate, laminated cherts, and siliceous shale; associated ultramafic rocks. JK?d, Jurassic–Cretaceous(?) Dagua Gp; lower graphitic phyllitic slates; middle slates, quartzose sandstone, minor limestone; upper black cherts and siliceous slates; believed metamorphosed (subgreenschist?–greenschist?) in late Cretaceous to early Cenozoic. *Cauca Valley*: Clpc, paralic to continental Lower Cenozoic section; sandstone and black chert (Nogales) overlain by conglomerate, shale with coals, and sandstone (L. Cauca); shales with coals overlain by sandy shale (mid-Cauca) followed by quartzose conglomerate, shale with coals, and sandstone. (U. Cauca). *Igneous Rocks*: U, ultramafic rocks (including partly serpentinized gabbro?) associated with basic volcanics of Kcgd, (K)Tbv. T, tonalites in Cordillera Occidental.

In the Cordillera Occidental, the basal portion of the section is composed of basic lavas interbedded with slaty to phyllitic shales, siltstones, and sandy limestones terminating in cherts and flyschlike accumulations of slates with metamorphic minerals derived from the Cordillera Central. Overlying this section is an enormous pile (possibly 9 km thick) of basic submarine volcanic rocks interbedded with graywacke shales and siltstones. From scattered faunas in the intercalated sediments, it appears that the upper part (Diabasic Group) of the section is of Cretaceous age. The lower part (Dagua Group) certainly extends into the Jurassic, but its maximum age is not known. Burgl (1973, p. 92) concludes that the Dagua may range in age from late Paleozoic to Jurassic. In contrast, the post-Cretaceous section consists of continental to paralic gravels and sands which pass gradually westward into shallow-water epicontinental marine sediments.

In the Serrania de Baudo along the Pacific coast, the basic volcanics and associated tuffs are interbedded with fossiliferous Eocene and siliceous Oligocene sediments suggesting a progressive younging of volcanic activity westward with time. Gansser (1973) notes that basalt lavas enclose middle Eocene reef limestones.

Plutonic intrusions occur along the length of the Cordillera Occidental and are described as tonalites by Nelson (1957) and granodiorites by Gansser (1973). Irving (1971, p. 38 and 48) reports K/Ar ages of 24 and 36 m.y. on specimens from two batholiths. Some of the basic rocks in the Diabase Group have the characteristics of gabbro; in addition, serpentinite and other ultramafics are known from both the east and west slopes of the Cordillera; the latter probably related to the Atrato fault. In the Baudo range, the fragmentary data (see Gansser, 1973) point to the presence of small gabbroic and ultramafic bodies, but no plutons of intermediate to siliceous composition have been reported.

The Cordillera Occidental and the Serrania Baudo appear to be two anticlinoria with axes trending approximately north–south and composed of moderate to tight folds asymmetric or overturned to the west. The Atrato fault is a major east-dipping thrust forming the western limit of the Cordillera Occidental. Imbricate thrusting of the same orientation is common in that region.

The Mesozoic volcanic pile may be correlated with an orogenic event which extended over the northwest corner of the Santa Marta block to the Guajira Peninsula and possibly further east. Accumulation of the volcanic pile was followed by regional metamorphism at the end of the Cretaceous (in the Cordillera Occidental) and finally the intrusion of tonalitic plutons in early Oligocene to early Miocene time. During the Miocene and subsequently, the ranges were uplifted, a tectonism unaccompanied by thermal events.

Discussion

The most important uncertainty in the geology of these ranges concerns what underlies the exposed section. The apparent absence of Precambrian and Paleozoic "underlay" raises the possibility that the volcanic pile was deposited directly on relatively young (Triassic in the east, Jura-Cretaceous along the Pacific Coast?) oceanic crust. The corollary inference in this case is that continental crust represented in the Cordillera Central by Precambrian granulites must terminate (transitionally or abruptly) along some line east of the Cordillera Occidental (cf. Irving, 1971, plate 1). In the same framework a knowledge of the composition of the plutons in the Cordillera Occidental is important; one would expect to find representatives of a sodic or calc-sodic K-poor differentiation or partial-melting series.

Nelson (1957) refers to imbricate thrust faults which have resulted in repetition of the section in the Cordillera Occidental to the extent that stratigraphic reconstructions and estimates of thickness are highly tentative.*

F. Intermontane Areas

Flanking the several Andean ranges are sedimentary basins which record in their stratigraphic accumulations tectonic events in the adjacent ranges. Ideally, the sedimentary record in these intermontane basins may provide information concerning both initial and final phases of orogeny, as illustrated in Fig. 11. In reality, evidence of the oldest events is commonly hidden under a thick cover of molasselike deposits related to the youngest stages of orogenic uplift. In summary, the following comments may be made:

1. Sedimentary rocks, in places lightly metamorphosed, of known and probable Paleozoic age underlie the Mesozoic–Tertiary section in the region between the Precambrian Guayana shield and the innermost Andean cordillera. This basin is one of a tessellated pattern of basins which runs the length of the continent between the Andes and the shield massifs. Fossiliferous Cambrian–Ordovician rocks are known in outcrop from El Baul and La Macarena (Fig. 12A) and from Heliera No. 1 borehole located midway between the two outcrop areas. Other holes in Colombia and Venezuela have yielded slaty shales which, though unfossiliferous, must certainly be of Paleozoic age. From relationships observed in Peru–Bolivia–Argentina, the early Paleozoic section overlaps and thins in the direction of the Brazilian shield. Graptolitic black shales suggesting deep-water accumulations of fine sediment are known from the Caparo area of the Venezuelan Andes and Las Cristalinas in the Cordillera Central. Fragmentary as the information is, it is not unreasonable to propose that the relationships depicted in Fig. 11A prevailed, viz., that the

---

* See note added in proof, p. 415.

Fig. 11. Idealized sequence of sedimentary environments in Andean orogenies. (A) Marine transgression
accompanied by unstable subsiding basins as in the
Cretaceous (cf. Fig. 12E). (B) Simple overlap. (C)
Initial uplift of ranges inferred where sedimentary
basin may be laterally extensive but units show profound thickness variations; thicknesses increase from
margins toward uplift, then sharply decline over
uplift, as in early Tertiary history of Cordillera
Oriental and Mérida Andes and their adjacent basins.
(D) Uplifted range becomes positive and hence a
region of sedimentary provenance; former basin
divided into separate portions, and thick piedmont

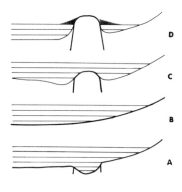

or "molasse" deposits skirt the new range; corresponds to late Tertiary and Quaternary
history of the Andes. Stages A–D may be identified in chronologic sequence from early
Cretaceous to the present, and with a little less certainty in the Devonian–Jurassic interval
(cf. Fig. 12C and D).

sedimentary regime obtaining in the Cambrian–Ordovician basin was interrupted, probably in the late Ordovician, by tectonism in the region of what
is now Andean terrain. Deposits corresponding to those in Fig. 11C have not
been identified for this early Paleozoic orogeny in the region of the Llanos
basins.

2. Conditions approximating those of Fig. 11 next appear in western
Venezuela from early Cretaceous to the present. The transgressing Cretaceous
sea extended across the Maracaibo basin, Perijás, and Andes into the northwestern portion of what is now the Barinas basin. The writer (1972*b*) has
detailed evidence reflecting equivalents to stages A, B, and C in Fig. 11. Stage
C appears to have begun at the close of the Cretaceous and to have proceeded
discontinuously with pronounced tectonic uplifts at the close of the Eocene,
in late Miocene or Pliocene time, and in the late(?) Pleistocene.

3. Van Houten and Travis (1968), Wellman (1970), and Anderson (1972)
have described nonmarine deposits in the Magdalena Valley corresponding to
stage D of Fig. 11. The deposits total some 12 km representing five tectonic
disturbances, initially in the Cordillera Central alone and subsequently in both
that range and the newly rising Cordillera Oriental, beginning in the late
Cretaceous and extending to Pliocene–Pleistocene time.

V. TECTONIC INDICATORS

Whereas Sec. IV was intended to convey an outline of the geologic history,
range by range, on the basis of the available field data it is proposed here to
attempt to discern the tectonic significance in the same data and to add thereto
information gleaned from additional sources such as geophysical studies.

A. Stratigraphic and Sedimentary Evidence

A most significant contribution to Andean stratigraphy has been made by Cediel (1972) who viewed the Upper Paleozoic section (Devonian–Permian) in the Cordillera Oriental and Mérida Andes in terms of a single transgressive–regressive sedimentary cycle enclosed above and below by major regional unconformities. Each unconformity represents major orogeny, and the younger event is further identified by the molasselike deposits of the Girón–La Quinta Formations. In contrast to Cediel's integrated concept of sedimentation in space and time, many other workers have noted and described an infinite variety of unconformities which, however real, are probably of local significance and which have served mainly to distract attention from unified concepts of tectonism.

A similar cycle appears to have been reinstituted following the deposition of the Girón–La Quinta molasse. Gradual but extensive marine transgression in the late Jurassic and Cretaceous was followed by regression in the early Tertiary accompanied by orogenic uplift and the development of another event of molasselike deposition which continues to the present.

Cambrian–Silurian sedimentation may have constituted another such cycle. No corresponding molasse facies has been identified, though the basal portion of the Devonian section may perhaps be considered in that light. Cediel (1972) describes the late Paleozoic cycle in terms of basal coarse-grained clastics passing upward to progressively more shaly and carbonate-rich sections and terminating with red beds, coals, and related facies indicative of the filling of the sedimentary basins. This fining-upward sequence characterizes the other cycles mentioned and may also be crudely discerned in the high-grade meta-morphic rocks constituting the basement in the Mérida Andes, but, as previously remarked, these rocks may yet prove to be metamorphosed equivalents of the Upper Paleozoic section and not of Precambrian age.

Excluding the Precambrian, there were thus at least two and possibly three sedimentary cycles in this region, and it is pertinent to inquire whether or not there is evidence in the sedimentary record to show either conformity with or independence of present Andean trends. The writer (1972b, p. 450) has noted the repeated development of unstable basins in the Venezuelan Andes in which thick sedimentary piles accumulated. The example of the Pennsylvanian–Permian section in the Mérida Andes is instructive. The writer has described the section in terms of a thick (± 5 km) marine flyschlike deposit which grades laterally to both north and south into deltaic tan and red clastics which denote shorelines within the present limits of the Venezuelan Andes. Both flysch and deltaic deposits grade upward into shallow-water marine pelites and carbonates. The problem is how to relate the environment of deposition of this section to the Devonian–Permian transgression of Cediel. As shown in

Fig. 12, the writer views the paleogeography in terms of progressive transgression from the west or northwest leading to the development of the thumblike Venezuelan embayment which was at the same time an unstable basin superimposed on the overall transgression. Stibane (1968 Figs. 36–38) suggests a simpler paleogeography in which virtually the entire western part of Venezuela was submerged in Devonian–Permian time; i.e., the southeast side of the "thumb" is continued roughly. north–northeast to the present coastline.

Progressive transgression from the west is supported by the fact that all faunas thus far collected from the late Paleozoic section of the Mérida Andes are of mid(?)-Pennsylvanian to Permian age. Possible Devonian faunas tentatively mentioned by Arnold (1966) are now discounted. The suggested thumblike embayment is supported by regional stratigraphic data. High-grade metamorphic (Precambrian?) rocks of the Maracaibo platform are currently overlain by sections of Mesozoic–Tertiary age. Evidently, whatever Devonian–Permian section may once have existed in that region has long since been removed by erosion. The 5–6 km of Pennsylvanian–Permian section in the Venezuelan Andes is matched by similar thicknesses of Devonian–Permian section in the Perijás and Cordillera Oriental (Santander massif). If one proposes that even half this thickness once extended over the Maracaibo platform in the region subtended by the three ranges, then a vast volume of sediment is unaccounted for inasmuch as the Triassic–Jurassic molasse facies (Girón–La Quinta) clearly comes from the immediate environs of its present occurrence. An acceptable alternative postulate would be that deep unstable troughs existed in the areas of the present ranges and that only a very thin platform facies (some tens of meters?) extended over the Maracaibo platform. This would correspond approximately to stage A of Fig. 11. In such a case, the paleogeographic map might, superficially, more closely resemble Stibane's. The obvious question is whether the Perijás might not have constituted another thumblike embayment, i.e., that there were two roughly parallel shorelines, one following the southeastern border of the Perijás and the other running northeast–southwest across the Santa Marta block. The available data are insufficient to permit serious hypothesis.

A far stronger case can be made to show that what the writer has proposed for the late Paleozoic obtained again during the Cretaceous transgression. Superimposed on the overall transgression from north and west is the establishment of unstable basins over portions of the Venezuelan Andes, Perijás, and Cordillera Oriental (see Fig. 12). The locations of the main Cretaceous basins do not precisely correspond (so far as one can ascertain) with those of Devonian–Permian time, but the general correspondence to the region of the present ranges is good, at least for the three ranges mentioned. Data for Cambrian–Silurian time is far too fragmentary to permit reasonable speculation. The writer prefers for the moment the hypothesis that Cambrian–Ordo-

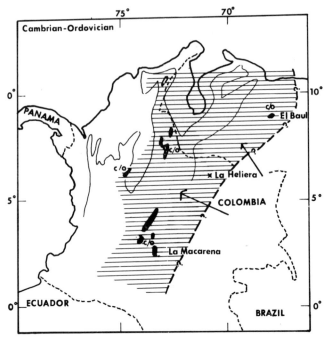

Fig. 12A.

Fig. 12B.

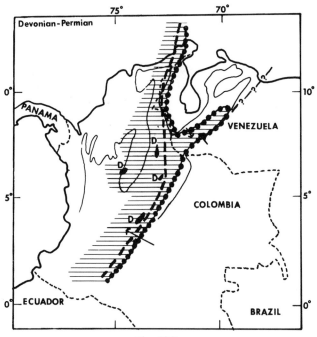

Fig. 12C.

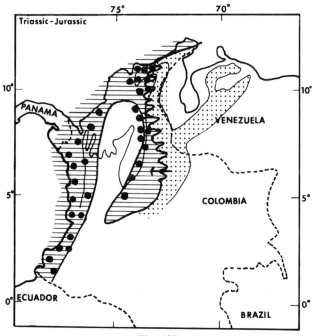

Fig. 12D.

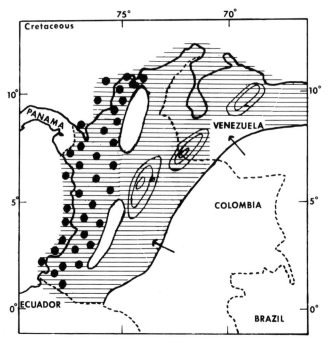

Fig. 12E.

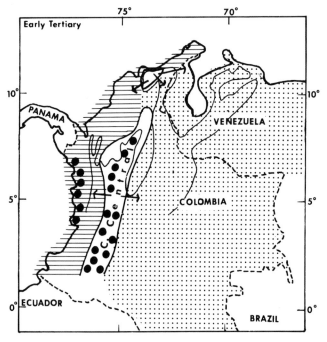

Fig. 12F.

vician sedimentation was independent of "Andean" control. An Andean effect may have begun with Silurian deposition in the Venezuelan Andes (Shagam, 1972*a* and *b*) and may have been maintained thereafter.

This effect of the superimposition of unstable sedimentary basins cannot be documented for the Cordillera Central and the ranges to the west. The Cordillera Central has been largely a positive region from the end of the Paleozoic to the present, so that at least to that extent one may speak of an ancestral Cordillera Central, as Irving (1971, p. 23, Fig. 5) has done. Except for the Devonian fossils near Rovira and the Ordovician material from Las Cristalinas, the rest of the pre-Mesozoic history is debatable, and paleogeography cannot be stipulated. As previously indicated, the pre-Jurassic history west of the Cordillera Central is not known.

B. Thermal Events and the Orogenic Concept

Since the pioneering work of Hall and of Dana in the mid-19th century, the concept of geosynclines and the processes of orogeny have developed to the degree that it is now clear that no two mountain belts have had identical histories. Considerable variation may occur within individual ranges (e.g., northern vs. southern Appalachians). Nonetheless, certain broad basic rela-

---

Fig. 12. Tentative paleogeographic outlines from Cambrian to early Tertiary time. Horizontal lines represent marine facies; stipple pattern represents continental facies; heavy dots represent volcanics. Small areas marked c/o (in A) and D (in C) are Cambro-Ordovician and Devonian outcrop areas, respectively. Arrows show directions of sediment transport. Faint line showing outlines of cordilleras indicates ±200 m contour. (A, B) First Phanerozoic transgression and sedimentary cycle. Silurian was probably the start of regressive phase, but tentative basin shown SE of the Venezuelan Andes (B) may have persisted from the Ordovician. Rare Silurian fossils in Perijás and Cordillera Oriental [see Bowen (1972) and Burgl (1973) for sources] indicate that Silurian seas may have been far more extensive than shown here. (C) Second Phanerozoic cycle showing proposed eastward migration of shoreline from Devonian (dashed line) to Permo-Carboniferous time (dotted) with development of thumblike embayment over Venezuelan Andes. Alternatively shoreline may have continued northeastward (dashed line with interrogations) as shown by Stibane (1968), in which case the Mérida Andes and Perijás probably constituted unstable deep basins superimposed on overall transgression. The writer prefers the former (thumblike embayment) alternative. (D, E) Third Phanerozoic cycle. Zigzag line is diagrammatic representation of interfingering of marine and continental facies. Light contours in the East show unstable subsiding basins superimposed on overall Cretaceous transgression (cf. comments in C). (F) Regressive phase of third Phanerozoic cycle but prior to uplift of the ranges east of the Cordillera Central. Available data interpreted to show repeated superimposition of depositional, erosional, and volcanic environments. Note, for example, deposition in Venezuelan Andes as in B, C, and E, uplift of Cordillera Central as in D, E, and F, and volcanism west of Cordillera Central as in D, E, and F.

tionships appear as a continuing thread through orogeny in space and time. The process climaxes with deformation and thermal events such as regional metamorphism and plutonic activity. The precise order varies from range to range, but most workers accept the orogenic concept to the extent, for example, that Precambrian orogenic belts are identified largely by the age of their granitic plutons which are commonly interpreted to represent the closing event.

Failure to understand this elemental concept was largely responsible for the long delay in recognizing late (?) Permian regional metamorphism in the Venezuelan Andes (Shagam, 1972b, p. 455–457). A glance at Irving's (1971, plate 1) map immediately raises similar queries. In the Santander massif of the Cordillera Oriental, an extensive metamorphic terrain is associated with several granitic batholiths. The metamorphism is considered to be of late Ordovician age, but the batholiths are dated as Triassic–Jurassic (principally late Triassic to early Jurassic). More detailed maps of portions of the massif (H-12, Bucaramanga; H-13, Pamplona) mapped by Ward et al. (1969, 1970) do show large areas of pre-Devonian orthogneiss of dioritic to granodioritic composition. Garnet/staurolite and staurolite/sillimanite isograds in the Silgara schists of presumed Cambrian–Ordovician age show a spatial distribution suggesting that they are related to the pre-Devonian orthogneisses. One is tempted to infer that the apparent discordance in the ages of the metamorphic and plutonic rocks in Irving's map is merely the consequence of cartographic oversimplification. The explanation may be partly correct but does not appear to be adequate to explain other relationships.

If the pre-Devonian batholiths were a product of a thermal event to which the regional metamorphism was also related, one might ask why the far larger Mesozoic batholiths do not appear to show a comparable associated regional metamorphism. One may conjecture that the country rocks had already been made over to metamorphic grades so high that the younger plutons caused no obvious recrystallization. Alternatively one may claim that the depth of erosion of the younger plutons is not sufficient to expose the principal aureolar affects. These are not fully credible explanations. In the discussion of the stratigraphic relationships of the Cordillera Oriental in Sec. IV, it was suggested that workers may have mapped as stratigraphic contacts what are in reality metamorphic contacts. It was noted, inter alia, that a garnet/biotite isograd coincides with the mapped contact between the Silgara schists of supposed Cambrian–Ordovician age and the Floresta Formation of Devonian age in a region adjacent to a large Triassic–Jurassic batholith. Also, analogy was drawn with the relationships in the Venezuelan Andes where it was shown that a major regional metamorphic event affected the upper Paleozoic section in the northern part of the Andes and that elsewhere the section appeared unmetamorphosed or only slightly metamorphosed. It was concluded that the Upper

Paleozoic section in the Santander massif was probably affected by a significant regional metamorphism related to the younger plutonic intrusions. On the flanks of the Santander massif, erosion has yet to expose the metamorphic aureole which is already exposed in the central portions of the massif, closer to the batholiths.

If one tentatively accepts major plutonic intrusion and accompanying metamorphism affecting the Upper Paleozoic section (and younger rocks?), it is pertinent to examine more closely the available evidence concerning the age limits of this major thermal event. Because regionally the Mesozoic overlies the Upper Paleozoic section with angular unconformity and inasmuch as granitic pebbles are abundant in the red-bed formations (Girón–La Quinta) which constitute the largely pre-Cretaceous basal portion of the Mesozoic section, it was long common practice to refer to ("Variscan") orogeny at the close of the Paleozoic accompanied by intrusion of the granitic plutons and followed by the development of a red-bed molasse spanning most of Triassic–Jurassic time. Lately there has been increasing dependence on radiometric measurements (mainly K/Ar) as the principal criterion for deciding intrusive ages of the plutons. Thus, Irving (1971, p. 27) noted that granitic pebbles recovered from the base of the molasse facies are considered by Goldsmith (see Irving, 1971) probably to have been derived from a granite dated radiometrically as early Jurassic. He infers in turn that the molasse is probably of middle to late Jurassic age possibly extending into the Cretaceous. Some workers have attempted to integrate the earlier tectonic approach with the radiometric measurements and have suggested that granitic intrusion and deposition of the molasse may have overlapped in time but have occurred in geographically separate areas. The writer is influenced by the relationships between the La Quinta red beds and the igneous rocks in the Venezuelan Andes. At the type locality (see Hargraves and Shagam, 1969) the red beds overlie with apparent conformity rhyolitic tuffs which are correlated with the last plutonic granitic event in those Andes. Initial U/Pb measurements on zircons from the tuffs (work currently in progress by L. Burkley and P. O. Banks at Case Western Reserve University) suggest ages closely approximating the Permian–Triassic boundary. The tuffs and corresponding plutonic event are considered to have related to the late Paleozoic orogeny which metamorphosed the Upper Paleozoic section including the early Permian Palmarito Formation. Here a late Permian orogeny with an end-Permian granitic event followed by deposition of Triassic–Jurassic molasse constitutes a reasonable reconstruction of tectonic events. Cediel's (1968, p. 54–60) view of the corresponding relationships in the region of the Santander massif are similar, though he does suggest that the intrusive ages may have partly overlapped the time (but not the place) of molasse deposition. The writer suggests that the K/Ar ages are younger than the true intrusive ages, perhaps because of Ar loss related to high ambient temperatures until such a time as uplift and erosion finally led to a closed system. Age measurements on minerals from the region affected by the Permian orogeny in the Venezuelan Andes are summarized by Olmeta (1968) and Martin-Bellizzia (1968). In some cases (e.g., for the Piñango pluton), Rb/Sr and K/Ar ages are discordant, the latter method giving younger values. Burkley's U/Pb ages are some 10–20% older than the K/Ar ages of 200, 181, and 196 (all $\pm$10 m.y.) listed by Olmeta which resemble the Triassic ages obtained by Goldsmith *et al.* (1971) on the granites of the Santander massif.

In summary, the writer suggests that the plutons of the Santander massif may be closer to late Permian than late Triassic–early Jurassic age and that their emplacement was accompanied by a regional metamorphic event. Proceeding on the above supposition, Cediel (1972) may be mistaken in claiming that reference to metamorphosed Devonian means either determinative error or extremely local effects of dynamic or contact type. By the same token, Irving (1971, p. 16) is mistaken in accepting the fact of metamorphism of Devonian rocks but considering it a local perturbation and not connected with igneous activity. On the other hand, Ward *et al.* (1970) who mapped the Pamplona sheet were *not* mistaken in showing a metamorphic isograd coincident with the contact between Silgara and Floresta formations. Whole-rock K/Ar ages of 221 and 198, both $\pm$8 m.y., on phyllites of the Silgara Formation reported by Goldsmith *et al.* (1971, Table 2 and discussion, p. D48) are credible results in the light of the above discussion. Similarly Rb/Sr ages of 230 and 242 m.y. on biotite from Precambrian(?) schists of the Mérida Andes reported by Bass and Shagam (1960) also fit the tectonic framework here outlined.

In the Perijás, a similar problem involving discordance between apparent ages of metamorphism and plutonism has already been raised in the pertinent discussion in Sec. IV. In the light of the strong evidence of orogeny and associated thermal events at the close of the Permian in the Mérida Andes, and the strong possibility of a similar situation in the Santander massif as discussed above, it appears to the writer reasonable to hypothesize that the same situation may have obtained in the Perijás.

The proposed thermal events at the close of the Paleozoic to earliest Triassic are considered to have been overprinted on the metamorphic rocks produced in the orogeny of late Ordovician to late Silurian time; it is not intended to imply that the initial metamorphism of the Quétame–Silgara formations occurred in the Permian. Angular unconformities showing a distinct hiatus in metamorphic grade between overlying Devonian and underlying Quétame–Silgara formations have been documented without doubt (e.g., in the Farallones de Medina of the Cordillera Oriental). On the other hand, the matter of metamorphic grade cannot be used in a reverse sense as a criterion of age or stratigraphic position. As has been emphasized above and in the related discussion in Sec. IV, some rocks assigned to the Precambrian or Lower Paleozoic sections on the basis of metamorphic grade may in reality prove to be younger inasmuch as there is now good evidence for major metamorphism at the close of the Paleozoic.

Prominent discordance between the late Permian (?) age of the regional metamorphism and the late Cretaceous age of the gigantic Antioquia batholith occurs in the Cordillera Central. There is far less stratigraphic control than for the ranges to the east, so that a reinterpretation here would have little meaning. Still, there are tantalizing scraps of evidence suggesting that major revisions in stratigraphic relationships and the distribution of thermal events in space and time are overdue. Some of the stratigraphic problems in this cordillera were discussed in Sec. IV where, *inter alia*, Burgl's (1973) views concerning the Ayura–Montebello and Cajamarca sections were summarized. The former is regarded as a eugeosynclinal equivalent of the miogeosynclinal Quétame association to the east, and the latter as correlative with the Devonian–Jurassic section in the Cordillera Oriental. Burgl's Fig. 7 (1973, p. 26) on which Fig. 13 is based shows the enormous region of outcrop of Ayura–Montebello metamorphics in the middle of which occurs the gigantic Antioquia batholith. It seems to the writer a distinct possibility that Ayura–Montebello is a giant contact aureole facies around the granite and that it is in reality a correlative of the Cajamarca group further to the south. The Ordovician rocks near Las Cristalinas are referred to as "barely metamorphosed" (Burgl, 1973, p. 27) but they may be correlative with the high-grade Ayura–Montebello schists and gneisses much as marked changes in metamorphic grade in short lateral distances have been noted in the Upper Paleozoic section of the Mérida Andes. Against the above background, the Permian ages reported by Irving (1971) as late Permian may be interpreted in terms of mid-Paleozoic ages reset by the late Cretaceous Antioquian event. The details of Forero's Devonian finds

Fig. 13. Problem of correlating between metamorphic sections. Ayurá–Montebello metamorphics are commonly at higher grade than those of the Cajamarca group in the Ibagué–Armenia area. Burgl (1973) correlates the former with the Cambro-Ordovician and the latter with Devono-Jurassic sections in the C. Oriental. However, Ayurà–Montebello rocks may be viewed as contact aureole on the grand scale, superimposed on the Cajamarca Group by the Cretaceous Antioquia batholith. Modified after Burgl, Fig. 7, 1973.

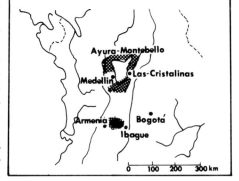

(referred to by Cediel, 1972) are not currently available to the writer but are clearly pertinent to the overall problem. Much as K/Ar dates in the Santander plutons are considered to yield ages younger than true intrusive ages, so the late Cretaceous K/Ar dating of the Antioquia batholith may also prove to be significantly younger than the true intrusive age.

Yet another discordant outcrop pattern occurs in the Santa Marta block. Southeast of the Sevilla lineament enormous Triassic–Jurassic batholiths are associated with Precambrian metamorphics. There is additional discordance in that the latter are dominantly granulites; this poses a serious problem in explaining the genesis of the granitic magmas (the problem also occurs on a smaller scale in the Cordillera Central). Tschanz et al. (1974) discuss the tectonic evolution of this block (and the origin of the granitic batholiths) in terms of a southeast-dipping subduction zone and note that the granites become progressively more K-rich and more shallowly emplaced toward the southeast. Theirs is a stimulating presentation in which clearly much careful thought has gone into the mechanics of batholithic emplacement (by extentional disruption related to a major transform movement synchronous with intrusion), but one is hard-pressed to find a reasonable source for the granitic liquids. Whereas it may be chemically reasonable to postulate dioritic to quartz–dioritic melts by anatectic melting of subducted oceanic crust, it is another matter to produce large volumes of K-rich, granitic to granodioritic melts by the same mechanism at depths up to almost 300 km. Solutions to the problem of magma generation by extensive lateral migration of granitic liquids or allochthonous blocks of granulite facies are not considered reasonable possibilities. There is a possible path to solution of the problem if one accepts the suggestions of some workers (e.g., Irving, 1971) that the Sevilla metamorphic belt is the continuation of the probably Permian metamorphic belt of the Cordillera Central. The latter is commonly three to five times wider in outcrop width than the Sevilla belt, and one may infer the possibility that, following late Permian–early Triassic orogeny, the Sevilla metamorphic belt was progressively subducted under the main Sierra Nevada tectonic belt (i.e., the region southeast of the Sevilla lineament). The mechanics of subduction may have involved either underfolding or underthrusting relationships. In either case, partial melting of Sevilla metamorphics would have resulted in granitic liquids which then would have had to migrate upwards *through* the anhydrous granulite facies to their final depth of emplacement (thereby explaining the occurrence of granulite xenoliths in the granites). Presumably the granulite facies rocks are the dehydrated refractory remains of some Precambrian orogeny in which a granitic fraction was removed and intruded at shallower depths. These oldest granites and their associated host rocks have apparently been completely removed by erosion. By analogy, the occurrence of the Precambrian granulite terrain in the Cordillera Central and of the

Mesozoic plutons which intrude it in the angles between the Palestina and Cimitarra faults would be explained in terms of an east-dipping subduction zone, the fossil trace of which may have been the Bagre fault. This view invites reconsideration of the Antioquia event in the Cordillera Central in terms of eastward subduction of the Cordillera Occidental under the Cordillera Central approximately at the Romeral fault.

The above discussion is intended to bring out the following implications of geotectonic significance:

a. In many ranges, discordance in age between plutons and metamorphic terrains is more apparent than real. There is reason to suspect that a major metamorphic overprint on older metamorphic terrains occurred in the Cordillera Oriental, Mérida Andes, and Perijás at the close of the Permian to earliest Triassic.

b. In the Santander massif and the Mérida and Perijá Andes, the tectonic setting of the thermal events is classical in the sense that one finds in spatial association marine geosynclinal piles, regional metamorphic and plutonic events, evidence of intense compressional deformations, molasselike end products, etc.— almost all the characteristics of the classic view of geosynclinal theory barring the occurrence of ultramafic rocks.

c. In the Santa Marta block, the Cordillera Central, and the ranges west to the Pacific coast, the relationships are best viewed within the framework of subduction, and many of the plutonic and metamorphic events appear to be spatially separate even if related. Tschanz *et al.* (1974) describe the Jurassic batholiths as anorogenic, in apparent reference to the lack of associated tectonic events that may be implied in the term orogeny.

d. In both the classic and the subduction frameworks, it is likely that age evaluations based solely on the K/Ar method are likely to be significantly younger than the true intrusive ages.

C. Petrologic Criteria

Very few detailed and modern mineralogic–petrologic studies have been made in the region under discussion. Most of the available data have been presented in previous sections of this study. In summary, one may list the following petrologic data as of geotectonic significance:

a. Mineral assemblages in the metamorphic terrains of the Santander massif and the eastern part of the Cordillera Central [see maps by Ward *et al.* (1969, 1970) and by Feininger *et al.* (1970)] suggest that they correspond approximately to Miyashiro's (1961) andalusite–sillimanite type (i.e., low-pressure type). Grauch (1971) regards the assemblages in the Mérida Andes as being of the low-pressure intermediate type (i.e., intermediate between Miyashiro's andalusite–sillimanite and kyanite–sillimanite types). The metamorphics of the Perijá ranges appear (on the most rudimentary field evidence) to be of the low-pressure type too. Miyashiro (1961) calls attention to the paired character of younger orogenic belts in the circum-Pacific region, and the mineral assemblages noted above correspond to the inner (continentward) metamorphic belt of Miyashiro. However, the outer (oceanward) metamorphic belt of the pair, characterized by the high-pressure jadeite–glaucophane type, has yet to be identified. Irving (1971, p. 32) notes a personal communication from P. Black concerning the appearance (coexisting?) of albite, epidote, chlorite, actinolite, sericite, and pumpellyite in specimens of the Cañasgordas Group (Cretaceous) of the Cordillera Occidental. Contrary to Irving's conclusions, the above assemblage could well represent a high-pressure or high-pressure intermediate group of metamorphics of Miyashiro's "outer" type. Perhaps the more typical minerals of the high-pressure group (glaucophane, jadeite, lawsonite) occur under the Cordillera Central in an as-yet-unexposed fossil subduction zone. Much the same explanation may be made of the absence of the high-pressure assemblages from the northwest corner of the Santa Marta block noted by Tschanz et al. (1974, p. 283); i.e., such high-pressure assemblages may exist at depth on fossil and present subduction zones. The late Cretaceous to early Tertiary volcanics along the Pacific Coast are not reported as metamorphosed. Late Tertiary metamorphism of the high-pressure type, if it occurred, must be represented on a subduction zone at a depth corresponding in vertical surface projection to the occurrence of the Cordillera Occidental. In such a case, the problem would be to define a corresponding inner low-pressure to low-pressure intermediate metamorphic belt.

b. The occurrence of ultramafic rocks in the region is summarized by Dengo (1972). A few small masses occur in the Sevilla and Santa Marta metamorphic belts northwest of the Sevilla lineament (Tschanz et al., 1969), but the great majority of such rocks occur in the Cordilleras Central and Occidental and in the coastal ranges. The sketchy petrographic data suggest that most of these masses were originally harzburgitic. Most are now described in terms of antigorite–bastite assemblages, suggesting that they have undergone metamorphism; but whether they are of regional or local hydrothermal character is not always clear. The ages of these rocks appear to be progressively younger as one moves from the vicinity of the Palestina fault in the Cordillera Central to the Pacific coast. Largely on the basis of the belief (not necessarily true) that metamorphism of the ultramafic rocks occurred at the same time as that of the host rocks in which they now occur, it has been suggested that these masses may range in age from Pennsylvanian–Permian in the Cordillera Central to early or mid-Tertiary along the coast.

D. Structural Geology

The most striking features of the northern Andes are their branching character, the kneelike bends in the Cordillera Oriental, the unique equidimensional character of the Santa Marta block, and the alternation of mountain ranges with low-lying sedimentary basins. On close inspection, it is seen that virtually all the boundaries of these component elements of the northern Andes are defined by major faults. Each major fault-enclosed block constitutes a structural unit. Interblock differences in structural characteristics are in part related to variation in stress environment with distribution in space and in part to variation in constituent materials. For the most part, intrablock differences in structural characteristics vary with the age of the deformed material. These factors are now considered.

E. Nature and Mechanics of Faulting

Inasmuch as major faults define the boundaries of the principal structural elements in the northern Andes, a knowledge of the nature and mechanics of faulting in the region is of first-order importance. Unhappily, even where the faults are well known in outcrop, considerable doubt remains as to their overall mechanical setting and age. Not unreasonably, many workers tend to postulate fault mechanics that fit their overall tectonic concept. Much the same course must be followed here, but the writer proposes first to lay at least the outline of factual foundation to support the subsequent argument.

The Boconó fault of the Venezuelan Andes was brought to international attention largely through the work of Rod (1956) and subsequently has repeatedly drawn attention in the literature where it is well established as a right-lateral strike-slip fault [e.g., Bell (1972); Schubert and Sifontes (1970)]. Careful studies of the displacement of Pleistocene moraines and related geomorphologic features by Giegengack et al. (1973) indicate beyond doubt right-lateral strike-slip movement of some hundreds of meters. In the same studies, Giegengack suggests that in some areas the motion is oblique and that the dip-slip movement probably far exceeds the strike-slip component. A wide variety of criteria for estimating the nature and amount of movement on the fault is currently being selected and tested by R. F. Giegengack, A. Bellizzia, R. I. Grauch, and the writer. To date, all criteria tested (ranging from displacement of late Permian aluminosilicate isograds to Pleistocene features) tentatively suggest dominance of dip-slip over strike-slip movement. Far more weighty than these criteria is a far larger-scale one of geophysical basis, viz., the relationship of the Venezuelan Andes to the two deep sedimentary basins (Maracaibo and Barinas) which flank the mountain belt. The pertinent data are summarized in the Figs. 3 and 14. Quite clearly uplift of the Andes has been counterbalanced

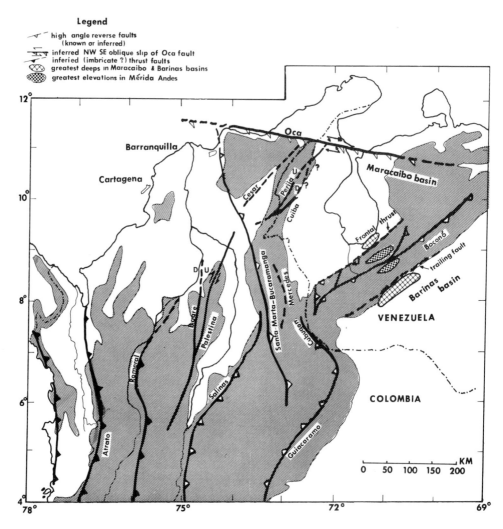

Fig. 14. Major fraction pattern. Pairs of similar high-angle reverse faults in the Mérida and Perijá Andes (trailing and Cuiba; Boconó and Perijá; frontal and Cesar) are viewed as related to anticlinal lithospheric folds (cf. Fig. 15). Note that the highest elevations in the Mérida Andes and the deepest portions of the adjacent basins fall on a symmetry plane normal to ranges. This strongly refutes idea of major dextral slip on the Boconó fault. Oblique slip on the Oca fault conforms mechanically with the sense of movement of the above faults. The strike-slip component on the Oca fault is believed to be <2× the dip-slip component. Santa Marta–Bucaramanga, Mercedes, and Cubugon faults in the Cordillera Oriental may be comparable to the frontal, Boconó, and trailing faults, respectively, of Mérida Andes. The more pronounced character of the frontal (i.e., Santa Marta–Bucaramanga) fault here may be a function of its more youthful age. In Cordillera Oriental (Bogotá leg) the Salinas and Guaicaramo reverse faults may be of high- or low-angle type (see Fig. 16). Romeral, Atrato, and other western faults are viewed as imbricate thrusts concave upward. Many faults are associated with ultramafic bodies. In the Cordillera Central the Palestina (28 km of dextral strike-slip) and Bagre (normal?) faults may be reactivated pre-Tertiary faults not genetically related to others shown.

by subsidence in the adjoining basins. The cross-sectional data on the basins implies start of uplift at the close of the Eocene. Elsewhere the writer (1972*b*; 1972*a*, Fig. 6 and caption) has suggested that there is good evidence to indicate that the initial phases of uplift began at the close of the Cretaceous, as indicated by the restriction of early Tertiary section to structural troughs located between blocks of opposite tilt. No early Tertiary section occurs in the tilt blocks, but one is present in conformable sequence on the Cretaceous section in the troughs. Depending on the geologic and/or ophthalmic bias of the viewer, one may stipulate barely detectable right- or left-lateral strike-slip movement on the basis of the geographic relationships of the basins to the ranges. However, it should surely be axiomatic that if the basin range relationship outlined above is to be related to plate tectonics and major horizontal movement along the Boconó fault, one should find far different topographic relationships from those currently obtaining (e.g., shallower, smeared-out, asymmetrically distributed basins with far thinner sedimentary accumulations). (On the other hand, in the mountains one might expect either two asymmetrically distributed highs or a single central one as at the present time but of significantly lower elevation.)

The situation depicted in the cross section points almost unequivocally to vertical structural relief between basins and mountain range of approximately 10 km. Based on the displacement of a staurolite isograd, there may have been as little as 1 km, but more likely about 5 km, of vertical displacement on the Boconó fault. Furthermore, stages of vertical movement are recorded in gravel and other deposits preserved in the graben of which the Boconó constitutes the southern limit. Significant deposits of Miocene–Pliocene(?) and Pleistocene age, separated by an angular unconformity, are known in addition to the thick Paleocene–Eocene section. The Boconó lineament was also a major sedimentary trap during the Triassic–Jurassic (La Quinta Formation) and Pennsylvanian–Permian (Sabaneta–Palmarito Formations), as outlined by the writer (1972*a*, Fig. 14, sections E-H), but the operative stress pattern need not have been identical with that obtaining during post-Cretaceous time. The current pattern of faults in the Mérida Andes closely resembles that found over diapiric structures such as salt domes. The distribution of subsidiary faults leading northwards off the Boconó is interpreted to be the result of the fact that the Venezuelan Andes did not rise vertically but rather were canted towards the northwest. The writer (1972*a*, p. 1234–1238) has summarized the evidence suggesting that the uplift of the Venezuelan Andes, beginning at the end of the Cretaceous and continuing to the late Pleistocene, had a crudely pyramidal form, with the major uplift occurring in the central portion of the mountain belt and accompanied by corresponding subsidence in the adjoining basins. Uplift was progressively less toward both the northeastern and southwestern terminations of the Venezuelan Andes.

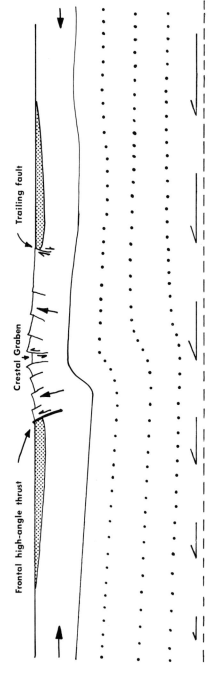

Fig. 15. Frontal and trailing high-angle reverse faults and crestal grabens of Andean ranges viewed as shallow brittle deformations related to lithospheric anticlinal flexures which pass into décollements at depth. Duplicated in series the profile would be a diagrammatic representation of the relationships of the Mérida and Perijá Andes to their respective adjacent basins. Flexuring leading to range uplifts is accompanied by subsidence of adjacent basins (stippled). Boconó fault corresponds to trailing fault of graben and as pictured here cannot be a plate boundary. Cf. Figures 14 and 16.

In such a setting, the lateral displacement (60–200 m) of late Pleistocene moraines, documented beyond doubt by Giegengack and Grauch (1972) and by other workers (e.g., Schubert and Sifontes, 1970), must be explained either in terms of local adjustments to available space by crustal blocks uncoupled at very shallow depths, or by a change in regimen of mechanics of faulting from high-angle reverse (or normal?) to wrench fault. If so, then the change has occurred in the late Pleistocene (plus or minus the last 20,000 years) to recent time, and the sum total of strike-slip motion on the Boconó may be only some few hundreds of meters.

In the Perijá Mountains, the Perijá–Tigre fault appears to play a similar mechanical role to the Boconó. Similarly, the postulated Cesár fault on the northwestern side and the Cuiba fault on the southeastern side would correspond to the boundary faults in the Venezuelan Andes. Miller (1962) notes many tectonic features in the Perijá ranges which parallel those outlined here for the Mérida Andes. Though not as clear as in the Mérida Andes, there is evidence for a similar pyramidal uplift of the ranges, the effects of uplift declining to both the northeast and the southwest from a central region approximately west of La Villa del Rosario. Miller (1962, p. 1579) writes of two main uplifts "separated by a longitudinal structural sag along the Perijá fault line." For the most part, the Cuiba appears to be a high-angle reverse fault dipping away from the Maracaibo platform and hence crudely a mirror image of the frontal Venezuelan Andean thrust. The Perijás were not uplifted to the same degree as the Mérida Andes, and, accordingly, the structural pattern and its relief is more restrained. Much as the uplift of the Mérida Andes was counterbalanced by subsidence in the Maracaibo and Barinas basins, so the Perijá uplift was counterbalanced by subsidence in the Cesár Valley and the northwestern Maracaibo basin. Also, much as Andean uplift took place in at least four distinct episodes, so there is reason to suspect (see Miller, 1962, p. 1594–1595) that similar discontinuous events of uplift occurred in the Perijás, though they may not have occurred at precisely the same times. The writer's views on faulting in the two ranges are summarized diagrammatically in Fig. 15.

In the tectonic framework outlined above, the writer prefers to view the Santa Marta–Bucaramanga fault as another frontal fault of high-angle reverse character. Campbell (1968) favors a wrench-fault interpretation for this fault and suggests that some 110 km of left-lateral displacement occurred in post-late Miocene time. Among other criteria, Campbell relies strongly on correlation of Cretaceous and Tertiary sections between the Middle Magdalena and Cesár basins, but Polson and Henao (1968) argue that these criteria are indecisive. Tschanz et al. (1974) concluded that up to 110 km of left-lateral displacement occurred in the early Tertiary (beginning in the late Cretaceous?) followed by large vertical displacement in post-early Miocene time. The left-lateral displacement is based on displacement of lithologic and geochronologic features of metamorphic basement, but the writer believes that a large portion of the displacement may be explained by large-scale folding related to dip- or oblique-slip movement on the fault.

The writer believes that the most striking feature of post-Cretaceous geology in the vast region from the Cordillera Central eastward across the Mérida Andes and the Maracaibo platform was the uplift of ranges and con-comitant subsidence of adjacent basins, as has been outlined above for the Mérida Andes–Perijá structures. The implications of the study on the Cenozoic deposits of the upper Magdalena valley by Van Houten and Travis (1968) are that similar structural conditions existed in the region of the Cordilleras Oriental and Central. Whereas initial uplift in the Venezuelan Andes occurred at the close of the Cretaceous and the major event came at the close of the Eocene, it appears that the Cordillera Oriental (or at least the Bogotá leg of it) did not "break surface" as a major mountain range until late Pliocene or even Pleistocene time. The difference is one of degree only; as summarized by Irving (1971, p. 39–41), there is reason to suspect that uplift of the Cordillera Oriental had already begun in Paleocene–Oligocene time even though the region was still negative. Sedimentation in the region of the Cordillera Oriental terminated at the close of the Oligocene but continued in the adjacent basins, indicating initial "emergence" of the Cordillera Oriental during the Miocene. Hence the fact of pronounced uplift at the close of the Cenozoic does not contradict the structural framework outlined by the writer. Within this framework, the writer prefers to view the Santa Marta–Bucaramanga fault as another frontal fault of high-angle reverse character related to the uplift of the Santander massif and the Santa Marta block. According to Campbell (1968, p. 259) there is about 12 km of relief across the fault measured on the "top of base-ment." In the Santander massif, there does not appear to be a structural feature directly comparable to the Perijá and Boconó crestal grabens, but the structural trench extending south through Pamplona (see Ward *et al.*, 1970) may play a comparable role. Cediel's (1968) description of the Girón Group as a molasse facies almost 5 km thick derived by the uplift and erosion of the Santander massif in Triassic–Jurassic time indicates that the Tertiary tectonism was not the first major event of uplift in the region. The presence of fault-bounded basins of Girón red beds in the Cordillera Oriental recalls similar La Quinta structural troughs in the Mérida Andes. The stratigraphic and structural similarities between the Pamplona and Boconó troughs are such that to infer a parallel mechanism of origin is surely reasonable.

Evidence concerning the nature of the faults external to the Cordillera Central and Occidental and Serrania Baudo is extremely limited and open to a wide range of interpretations. They may be of high-angle reverse type too, but the fact of more pervasive development of folding in these western ranges along with the apparently narrower character of the intervening basins suggests to the writer that they may be viewed as imbricate thrusts (possibly underthrust subduction zones?) on the grand scale. The markedly curved trace of the Romeral fault, the presence of abundant associated ultramafic bodies, and the orientation of adjacent fold axes conform to such an interpretation. The graben–horstlike relationships between some of the basins and ranges as suggested by Case et al. (1971) may represent subsequent isostatic or structural adjustments superimposed on the postulated thrust relationship.

It is convenient now to turn to consideration of the Oca fault, the prominent east–west fracture which forms the northern boundary of the Santa Marta block. Tschanz et al. (1974) refer to 65 km of right-lateral displacement in the early (?) Tertiary. MacDonald et al. (1971) suggested a few tens of kilometers of post-Paleocene right-lateral offset. Féo-Codecido (1972) considers the Oca to have been predominantly of high-angle reverse type until post-Eocene time when 15–20 km of right-lateral displacement occurred. The fact that the fault does not obviously appear to affect Eocene and younger deposits in the Falcón basin suggests that at least since that time the fault has not been a plate boundary in the context of sea-floor spreading. Much the same conclusion is reached by Stainforth (1969, especially Fig. 6). On the other hand, in a stimulating paper Vásquez and Dickey (1972) suggest that strike-slip motion in the Falcón area during the Tertiary may have taken the form of oblique fold wrinkles rather than actual shear fractures found further to the west. They conclude that dextral motion of possibly 195 km may have occurred between the start of the Tertiary and when the fault became inactive in the Pliocene. The writer believes it is simpler to accept the structural measurements made or estimated–perhaps 10 km of structural relief across the fault (Case and MacDonald, 1973, p. 2907), perhaps 15–20 km of right lateral displacement (MacDonald et al., 1971)—which imply oblique-slip movement in which lateral displacement is only about 2 times the dip-slip displacement.

In conclusion, an important philosophical lesson may be learned from the major Palestina and related fractures in the Cordillera Central. Following Feininger's (1970) study, the Palestina must surely be included in the small group of sacrosanct wrench faults. Horizontal right-lateral displacement of almost 28 km is proven beyond reasonable doubt by displacement of 10 unique lithologic, metamorphic, and structural features. The age of movement is not precisely known, but a post-Aptian–Albian age seems certain, and a minimum late Cretaceous age is a possibility on the basis of the fact that some dykes possibly related to the Antioquia batholith have not been displaced by the fault. If so, then all strike-slip movement would be pre-Tertiary and, therefore, unrelated to Tertiary plate movements. The orientation of the fault and the regional geology suggest to the writer that the Palestina may be a shallow feature (there is no geophysical data to show that it affects the

Moho) and that the strike-slip displacement reflects not plate-boundary
tectonics but rather adjustment of shallow crustal blocks in response to a
space problem induced by regional compressional deformation. Feininger
(1970, p. 1213–1215) points out the folly of attempting to interrelate all major
wrench faults in northern South America to a single regional stress field
operating through a specific restricted time span.

There are indications that the Boconó fault was active as far back as the
Pennsylvanian. This may be true of many of the major fractures mentioned
above. Elsewhere, the writer (1972a,b) has shown that a marked contrast in
structural styles occurs across a time zone corresponding approximately to
the Permian–Triassic boundary. Whereas pre-Mesozoic structural style commonly involves intense compression revealed in tight folds, strong development
of cleavage, thrust faults, and related features, post-Paleozoic structures are
commonly related to vertical movement and tilting. As noted by the writer
(1972b, Fig. 5, p. 458), post-Paleozoic deformation appears to have been related
to a tensional environment until the late Cretaceous followed by a short
span of relative structural stability prior to the start of a compressional environment at the close of the Cretaceous. The tensional environment was
mainly one of subsidence and sedimentation in fault-bound troughs, whereas
the compressional environment saw the gradual uplift of the ranges which
continued through the Pleistocene. Most of the deformation involved in
uplift was taken up by movement on the major boundary fractures and crestal
grabens. The latter formed in each block along zones located between belts
of opposite tilting. As outlined here, the major boundary fractures, tilt blocks,
and crestal grabens are viewed within the framework of brittle, near-surface
deformation, but in reality the writer believes that these are relatively minor
structural features related to gigantic lithospheric folds, as depicted in Fig. 15.
In this view, the width of the ranges is roughly a function of the thickness of
the folded lithosphere. Following classical structural doctrine, the fold will disappear in a décollement surface at some depth which is also a function of the
thickness of the folded layer but which cannot be precisely predicted because
the lithospheric fold is likely to combine some flowage along with concentric
flexure. Still, a rough approximation (see Fig. 15) suggests the depth of décollement occurs at about 150 km. Apparently, horizontal compression led to a
stripping-off at that depth and the upward propagation of the lithospheric anticlines into the only space available.

Presumably, the long direction of each block would be a function of
lithospheric strength, which would be decided in turn by the width and depth
of each range-fold. The equidimensional Santa Marta block would be related
solely to the mechanical accident (?) of the intersecting Oca and Santa Marta
faults. The various northerly to northeasterly trends of the ranges may
reflect in part inheritance of structural grain imparted in earlier orogenies and

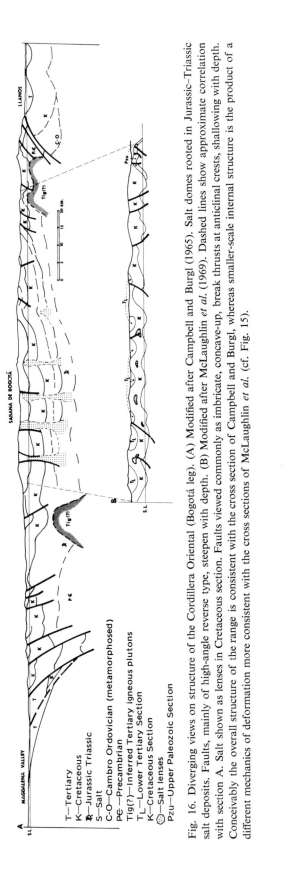

Fig. 16. Diverging views on structure of the Cordillera Oriental (Bogotá leg). (A) Modified after Campbell and Burgl (1965). Salt domes rooted in Jurassic–Triassic salt deposits. Faults, mainly of high-angle reverse type, steepen with depth. (B) Modified after McLaughlin *et al.* (1969). Dashed lines show approximate correlation with section A. Salt shown as lenses in Cretaceous section. Faults viewed commonly as imbricate, concave-up, break thrusts at anticlinal crests, shallowing with depth. Conceivably the overall structure of the range is consistent with the cross section of Campbell and Burgl, whereas smaller-scale internal structure is the product of a different mechanics of deformation more consistent with the cross sections of McLaughlin *et al.* (cf. Fig. 15).

in part local resolution of stresses produced in the mechanical interaction between the northwestern portion of the continent and the bounding oceanic regions.

Two aspects of post-Paleozoic deformation diverge from the outline presented above and deserve further discussion. The Pacific Coast ranges and the Cordillera Occidental (and possibly the western portion of the Cordillera Central) were far more profoundly affected by folding than the block uplifts to the east. Most modern workers view the tectonic environment of these western ranges in terms of eastward subduction of the Pacific under the continent. Whether or not the concept is correct, it brings out the view that these ranges may be considered as gigantic imbricate thrust slices rather than the more simple uplifts of the eastern ranges. The western ranges were thus more subject to penetrative internal compressional stresses. The second structural irregularity concerns the northeast-trending leg of the Cordillera Oriental through Bogotá. The internal structure of that region has been differently interpreted by Campbell and Burgl (1965) among others, on the one hand, and by McLaughlin et al. (1969) on the other. The basic difference is whether the dips of marginal thrust faults steepen or flatten with depth (see Fig. 16). Clearly, the latter view requires far more horizontal movement and is commonly associated with more intense folding. Whereas Campbell and Burgl (1965, plate 1) view salt plugs as rooted in Triassic–Jurassic evaporites, the map and sections of McLaughlin et al. suggest that they are dominantly related to the late Cretaceous Chipaque Formation and that they have played a significant role in the mechanics of both faulting and folding. The tight folding of the Cretaceous section is in marked contrast to the tilt block structures of the Cretaceous in the other ranges, and the interpretation of McLaughlin et al. seems reasonable to the writer.

It is tempting to view the deformation of the Bogotá leg of the Cordillera Oriental in terms of underthrusting from the north related to strike-slip movement on the Santa Marta–Bucaramanga fault. The explanation is unsatisfactory because Eocene formations are affected by the folding and faulting which probably continued through Pliocene–Pleistocene uplift, whereas whatever strike-slip motion there may have been on the Santa Marta–Bucaramanga fault must have taken place principally in Paleocene time or earlier. Moreover, the faults and folds appear to show symmetrical distribution and orientation across the Cordillera Oriental rather than a one-sided asymmetry one would expect of underthrusting from one side (north). Quite possibly then, the differences in internal structure between this range and the others to the north and east is related simply to the presence of the evaporite layers which served to enhance internal deformation, although the overall mechanics of uplift and fracture, through lithospheric flexure, was basically the same as in the other ranges.

Some of the structural features produced in pre-Mesozoic deformations which are of overall tectonic significance are now briefly reviewed. Structures of that age cannot be identified west of the Cordillera Central. In the remaining ranges, there may locally be strong angular discordance between the orientation of pre-Mesozoic and post-Paleozoic structures, but at the scale of each range there is marked parallelism, suggesting that at least in part the younger structures follow the structural grain imparted in earlier deformations. Parallelism of structures produced in pre-Mesozoic deformations, e.g., that of fold axes and some S-surfaces of Devonian rocks in the Santander massif, with those of the Cambrian–Ordovician and the Precambrian of the same region (cf. Irving, 1971, p. 16) may possibly reflect additional instances of younger deformation following previously established structural grain (see also Shagam, 1972b). A simpler explanation may lie in faulty dating of structures. In the case mentioned, as outlined in the discussion on the stratigraphy of the Cordillera Oriental in Sec. IV and in the section on thermal events in Sec. V, stratigraphic contact and metamorphic isograd may have become confused, so that structures mapped as Cambrian–Ordovician may yet prove to be of late Paleozoic age. Similarly, so-called Precambrian basement in the Mérida Andes may in reality be highly metamorphosed facies of Pennsylvanian slates, so that apparent parallelism of Precambrian and Permian structural features may be quite spurious. A further aspect of pre-Mesozoic structures makes inheritance of earlier structural grain unlikely. The earlier deformations were accompanied by major events of regional metamorphism and plutonic activity. In contrast to the simple upfolding of post-Paleozoic deformation (east of the Cordillera Central), it appears that preceding deformations involved downfolding and consequent rise of the geoisotherms. Whether one explains such deformation by mantle convection or not, it is almost certainly a process seated in the mantle, and therefore structural grain at shallow crustal levels is hardly likely to influence the orientation of subsequent deformations.

The writer (1972b, Table 1) summarized the principal structural data for the Mérida Andes and noted the presence of southeast-dipping imbricate thrust slices affecting the Paleozoic and Precambrian sections on the southeast flanks of those ranges, whereas those sections in the central to northern portion of the ranges are characterized by intense folds with axial planes (and cleavage) varying from near vertical to overturned to the northwest. Moreover, the late Paleozoic slates appear to have suffered décollement over massive amphibolitic gneisses and schists of the Precambrian (?). The axial plane and cleavage data are consistent with intense deformation within a crustal downbuckle. The writer has tentatively postulated that the décollement affecting the Pennsylvanian slates may also have truncated late Permian granites. If so, the décollement could reflect near-surface stripping-off rather than the product of intense lateral compression at deep crustal levels.

In addition to the décollement mentioned, evidence of extensive lateral transport occurs in the thrusting of the Precambrian Bella Vista Formation over the Lower Paleozoic section also referred to above. The most fruitful possibilities for finding large-scale nappelike structures appear to be located in the Cordillera Central. As suggested by Alvarez (1971) and Irving (1971), it appears reasonable to suggest continuity of the Cordillera Central into the Santa Marta massif and thence into the Guajira Peninsula. Allowing for some major offsets, one may project continuance of the belt into the Coastal Ranges (Southern Caribbean Mountains) of northern Venezuela, which is now viewed by many workers in terms of vast allochthonous masses which may have moved on décollement surfaces now characterized by serpentinites. Serpentinites are prominent in the Cordillera Central, and one may inquire whether the scattered metamorphosed bodies associated with the Otu and Palestina faults in Fig. 3 are not structural equivalents of the corresponding ultramafics further west. The question cannot be solved by debate, and at the present time available structural data are inadequate. Yet it seems reasonable to expect extensive lateral tectonic transport in this complex region.

F. Geomorphologic and Pleistocene Criteria

Certain easily recognizable and widely occurring geomorphologic features set limiting conditions of great importance on the mechanics of the late stages of tectonism in the Northern Andes. For example, the magnitude and age of the "Andean" uplift have been gauged by elevations and relationships to Pleistocene features of high-level erosion surfaces such as the "Altiplano" of Peru–Bolivia. The reliability of such criteria in the northern Andes may be judged by consideration of a widespread fossil surface on which the marine Cretaceous was deposited. The surface is approximately of late Jurassic age and is readily located throughout the region east of the Cordillera Central. The surface commonly occurs moderately to steeply inclined in tilted fault blocks, up to and in excess of elevations of 4,000 m, as shown in Fig. 17. Inasmuch as the surface is overlain by *marine* Cretaceous sedimentary rocks, it is clear that it must have been uplifted to its present position through sea level. Clearly, Andean uplift involved a significant element of rotation that began approximately at the close of the Cretaceous and became progressively greater with each "pulse" of uplift. By inference, subhorizontal high-level surfaces in the Andes have probably not been uplifted from former levels at or close to sea level but have been formed during or following uplift. As such, they are not reliable criteria for either the amount or age of uplift. On the other hand, the geometry of the late Jurassic surface indicates that uplift and tilting have resulted in crustal shortening ($=\cos$ dip angle), which for an average dip of about $40°$ amounts to about 20%. Necessarily, therefore,

Fig. 17. Tilted erosion surface in the Venezuelan Andes northwest of Mérida. Flatiron (background left) has basal fossiliferous marine limestone (Apon Formation of Aptian–Albian age) which rests on late Jurassic peneplain cut in Precambrian (?) schists. Stratigraphic relationships indicate that uplift through sea level and tilting must have occurred during "Andean" (Tertiary) uplift, and cannot be explained by vertical uplift of previously tilted block. Hence the uplift involved significant rotation which implies horizontal compression (cf. Fig. 15). The present high-level horizontal erosion surfaces, used by some writers as evidence of nonrotational vertical uplift of the Andes, must have been cut *following* uplift and tilting of fault blocks.

the boundary faults of the uplifted Andean blocks must bear a reverse fault relationship to the adjoining basins which requires a markedly compressional stress field in each case.

It is common practice to use deviation of drainage pattern along linear fractures as a test of strike-slip displacement. Campbell (1968) and Rod (1956) have done so for the Santa Marta and Boconó faults, respectively. The writer believes this criterion to be unreliable. Along the Boconó fault, drainage deviation is at least in part controlled by a downslope vector of gravity along the valley, so that for the south side of the fault, streams show right-hand drag in drainage flowing to the northeast, and left-hand drag where the drainage flows to the southwest.

The mapping of Pleistocene geological features is still in its infancy in this region, but enough has been done to indicate that significant tectonic data will emerge as detailed studies proceed. Polson and Henao (1968) comment that Raasveldt (1957), among others, found only two glacial stages represented on the Santa Marta uplift, which were interpreted by Burgl (1961) as corresponding to Riss and Wurm. Van der Hammen has begun the task of gathering detailed data on Pleistocene climates in Colombia (see, e.g., van der Hammen and González, 1963), and similar work is being done in Venezuela by Giegengack and Schubert; but available climatic, stratigraphic, and age data are currently insufficient to permit dependable interpretations. Giegengack and Grauch (1973a,b) conclude that sediments of at least one, probably two, and possibly three glacial periods have been preserved in the Venezuelan Andes, but the temporal relationships to the classic terrains of Europe and North America have not been established. Also it is not clear whether climatic changes are directly related to tectonics or not.

Near Bailadores in the southwestern Venezuelan Andes, in a region corresponding structurally to the trench associated with the Boconó fault, Giegengack et al. (1974, in press) report possible absolute subsidence of 800 m for glaciated terrain, which implies significant vertical displacement for late Pleistocene time.

G. Geophysical Criteria

Although far from complete, sufficient gravity, magnetic, and seismic studies have been done in the region of the northern Andes to allow for a considerable narrowing of the physical limits inside which tectonic reconstructions must fit. Figure 18 presents in simplified summary the simple Bouguer anomaly distribution compiled from the studies of Hospers and Van Wijnen (1959) for western Venezuela, Case et al. (1971) for western Colombia, and Case and MacDonald (1973) for northern Colombia. For the great bulk of the region, negative anomalies are associated with the ranges and their adjacent basins. The ranges do not appear to be compensated for by light roots, and the anomalies appear to be related to crustal downwarps associated with light sediment fill under the basins and complementary and contemporary uplift of the ranges. This conforms with the structural data previously discussed. The Mérida Andes and the Cordillera Central both show the same sense of asymmetry in mass distribution. Greatest mass deficiency occurs at or close to the external boundary of these two ranges and implies that there has been relative movement of crust from the internal towards the external regions. Although the mass anomaly of the basinal downwarp is roughly balanced by the range upwarp, both anomalies are negative; i.e., uplift and downwarp appear to be superimposed on an overall negative anomaly, suggesting that the region is currently under compressive deformation. Gravity coverage for the Cordillera Oriental is not available, but the possible trend of anomalies based on the information provided by Case et al. (1971) for the Magdalena Valley and by Hospers and Van Wijnen (1959) for the area south of the Tachira gap suggests that the same type and asymmetry of anomaly may

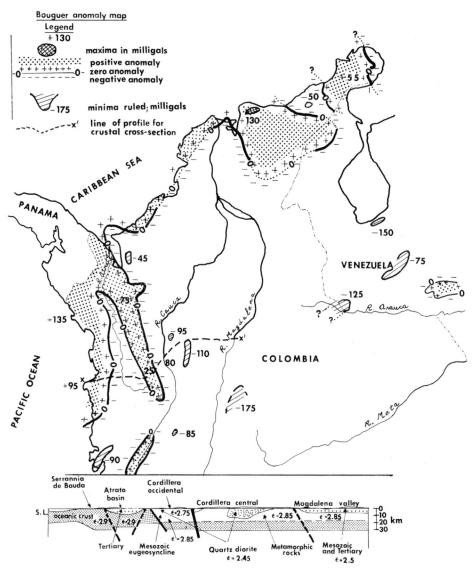

Fig. 18. Bouguer anomaly map. Positive anomalies in western Colombia and over Santa Marta block are principal exceptions to the overall negative-anomaly pattern. Although range uplifts are in approximate balance with the subsidence of adjacent basins, Bouguer anomalies are negative over both ranges and basins, indicating that a compressional environment continues to obtain. Compiled from Hospers and Van Wijnen (1959), Case *et al.* (1971), and Case and MacDonald (1973).

occur that is found associated with the Cordillera Central and the Mérida Andes.

The transition to positive Bouguer anomalies occurs close to the present coastlines and would show rough correlation therewith but for the remarkable positive anomalies associated with the Santa Marta block and the so-called West Colombia gravity high which extends from Panama to and beyond the Ecuador border. Case *et al.* (1971) interpret the regional anomaly characteristics to indicate that granitic crust occurs east of the West Colombia gravity high but is likely to be thin or absent under and west of the high, as shown in Fig. 18. The boundary between the change in crustal character corresponds crudely with the position of the Romeral–Cauca fault system. Mafic (oceanic?) crust is dominant west of the Cordillera Central, and reasonable interpretation of the data suggests crustal thickening from approximately 16 km in the Pacific to about 35 km under the Magdalena valley. Hospers and Van Wijnen (1959) suggested thicknesses of 30–40 km under the Venezuelan Andes. The steep anomaly gradients as one proceeds from the strong positive anomaly over the Serrania Baudo across the pronounced negative belt of the San Juan–Atrato basin back to the West Colombia gravity high are interpreted in terms of horsts (+) and graben (−), respectively.

Case and MacDonald (1973) conclude that the high positive Bouguer anomalies over the topographic high of the Santa Marta massif reflect thin continental crust and pronounced lack of isostatic balance. The coincidence of positive Bouguer anomalies with terrain of high relief both in the Santa Marta area and portions of the West Colombia gravity high suggests with little doubt that both anomalies are extremely young.

The tectonic significance of the nature and distribution of earthquakes has been commented on by Case *et al.* (1971) and by Dewey (1972), among others. The concentration of epicenters along the Pacific coast of South and Central America, the greater and lesser Antilles, and in such oceanic features as the Panama and Ecuador fracture zones is well known and depicted in many standard texts. Superimposed on the overall pattern, certain details are worth noting here. The famous Benioff zone associated with the Peru trench largely disappears approximately at the Gulf of Guayaquil; to the north, intermediate to deep-focus earthquakes are far fewer, and the belt is of restricted width and is not associated with a trench, as noted by Case *et al.* (1971, Fig. 15). Some intermediate-depth shocks appear to be associated with the northern portion of the Dolores fault (near its junction with the Palestina fault) and along the southernmost portions of the Santa Marta fault. The Boconó fault shows abundant shallow focus activity. The interpretation of all these data is far from securely established. Dewey (1972) has interpreted the distribution of hypocenters and ellipses of confidence, and of first-motion patterns in terms of plate theory and horizontal motion of the Caribbean relative to South America.

He suggests that the principal interface of motion between the two plates may have moved to the Boconó fault within the last 5 m.y. Dewey makes it clear that the data are not susceptible to unique interpretation. First-motion indications of thrusting for some earthquakes and the predominance of very shallow events do not add up, in the writer's opinion, to cast-iron support for the Boconó fault as a right-lateral, strike-slip *plate-boundary* fault.

Intermediate epicenters only occur in a north–south zone over the Cordillera Oriental of Colombia and indicate deepest focal depths of approximately 200 km. It is difficult to relate these to current subduction of a lithospheric plate at the Pacific coast because such a site currently occurs some 450 km west of the epicentral region but the area between it and the Cordillera Oriental is seismically inactive. Dewey (1972, p. 1729 *et seq.*) suggests that this zone of intermediate-depth activity may be a remnant of a lithospheric slab whose surface trace of subduction occurred until about 5 m.y. ago to the west of the Cordillera Oriental but well to the east of the current zone of shallow seismic activity off the Pacific coast. First-motions on the intermediate-depth activity of the so-called Bucaramanga source are in the main not consistent with a source related to left-lateral motion on the Santa Marta–Bucaramanga fault. The data may be reconciled with a fault striking N 25 W and dipping about 60° E. This conforms with the writer's view of high-angle dip-slip movement on this fault, but the orientation of tensional axes directed down dip does not conform with the writer's view of reverse (thrust) motion. This may simply mean that seismic energy release on this fracture is a relaxation phenomenon following reverse faulting. Possibly the most significant overall inference from the seismological data is that the Mérida Andes and northern Cordillera Oriental are under compressive stress oriented approximately east–west.

The principal radiometric data have been presented in the regional descriptions in Sec. IV. The significance of some of the radiometric data was noted in Sec. V in the discussion of thermal events, and further conclusions are drawn in Sec. VI.

Paleomagnetic studies pertinent to the region under discussion have been made by Hargraves and Shagam (1969), Creer (1970), Shagam and Hargraves (1970), MacDonald and Opdyke (1972, 1974), and Shagam and Hargraves (in press). The bulk of the paleomagnetic data may be variously interpreted because of doubts concerning ages and/or magnetization history. There is excellent agreement in the results of MacDonald and Opdyke (1974), Hargraves and Shagam (1969), and Shagam and Hargraves (in press) concerning apparent primary magnetization vectors for Triassic igneous rocks. The Gondwanide fit of Africa and South America yields near identity of the respective Triassic paleopoles (see MacDonald and Opdyke, 1974, Fig. 6) and is strong evidence for post-Triassic drift between the two continents.

VI. INTEGRATED HYPOTHESES OF TECTONIC EVOLUTION OF THE NORTHERN ANDES

Useful insights into the geotectonic processes which may have occurred in this region may be gained by reviewing the regional geology in terms of a layer cake constructed of successive orogenic cycles. It is simplest and most instructive to begin with the youngest orogenic cycle. Elsewhere the writer (1972b, p. 458) has simply referred to the youngest cycle in the Venezuelan Andes as being represented by the post-Paleozoic layer of geology. In retrospect, this was too crude a definition. East of the Cordillera Central, the Triassic–Jurassic Girón–La Quinta Red Beds are best considered as piedmont deposits ("molasse" of some workers) related to erosion following orogeny at the close of the Paleozoic. As such, they are better grouped with and considered as the final event of the late Paleozoic cycle. The transition from the latter to the start of the present cycle conforms approximately with the transition from continental to marine environment, and it may be convenient to take as a datum for the current cycle the development of the late Jurassic peneplain on which the Cretaceous was deposited. West of the Cordillera Central, this would roughly conform in time with the start of basic volcanism (Dagua Group and subsequent volcanics) which extends westward from the western flanks of the Cordillera Central to the coast.

The essential features of this uppermost layer of the geologic cake may be summarized in terms of three major components:

1. Basic volcanic piles + ultramafics, showing imbricate fault slices and overfolding to the West (Cordillera Occidental and Baudo)
2. The intrusion of large batholiths with relatively minor associated basic volcanism and sedimentation, and moderately intense associated folding (Cordillera Central)
3. An extensive region to the east marked by marine transgression during the late Jurassic and Cretaceous on which were superimposed unstable basins, followed by the development of large block uplifts beginning at the close of the Cretaceous, uplifts which apparently are the surface expression of large lithospheric anticlinal flexures at depth (Cordillera Oriental, Perijá, and Venezuelan ranges)

This great range of tectonic environments may be readily interpreted within the framework of plate tectonics. Some syntheses of plate type are summarized in Fig. 19. The volcanics and structural characteristics of the western cordillera are viewed in terms of subduction (and associated obduction?) which began perhaps in what is now the San Juán–Atrato basin and then progressed steadily westwards to and off the present coast. The Cordillera Central, site of the change from oceanic to far thicker continental crust, absorbed the bulk of the mechanical reaction to impact between the eastward-moving Pacific plate and the westward-moving continental plate. Impact led, on the one hand, to a compressional environment which retarded volcanism and, on the other hand, to uplift which limited sedimentation. At depth the lower part of the continental crust of the Cordillera Central was dragged down by the subducting Pacific plate and partially melted to produce the Antioquian plutonic event. Further east, in the Cordillera Oriental and the Perijá and Mérida ranges, the initial tensional, subsiding phase characterized by marine transgression in the Cretaceous was related in part to the uplift of the Cordillera Central; but, inasmuch as the effects of transgression extended all across the northern part of the continent as far as Trinidad, it is more likely that the marine transgression was related to a down-dragging of the northern part of the continent by southward subduction of the Caribbean under the continental plate, as outlined originally by Schuchert (1935) and subsequently within the framework of modern plate theory by Stainforth (1969) and Bell (1972). The Tertiary phases of lithospheric folding and uplift in these eastern ranges are again viewed as the mechanical results of impact of crustal plates at the subduction zone(s). Because in this eastern region motion in the underlying mantle was horizontal, impact led to décollements and lithospheric anticlinal folds previously described (see Fig. 15 and Shagam, 1972b, p. 459). The depth of décollement was probably about 150 km, i.e., in or close to the mantle low-velocity zone.

The geometric interrelationships of the ranges to possible subduction zones suggest a westerly component of motion for most of Colombia and a northwesterly component for the Mérida Andes–Perijás–Santa Marta region. Resolution of east–west drift vectors into compressional northwest–southeast stresses has been suggested by Bucher (1952), among many others. Stainforth (1969) and Bell (1971, 1972) have proposed a southward vector of spreading of Caribbean crust under the continental edge of Colombia–Venezuela, and the northwest–southeast axis of compression can be reasonably integrated with Bell's brilliant synthesis of the tectonic evolution of the southern margin of the Caribbean, though with some modification. He views (Bell, 1972, Figs. 1 and 4 and text, p. 378) the Venezuelan Andes as being in effect "back-thrust" eastwards (or underthrust from the east). A similar suggestion was originally put forward by Dufour (1955). As outlined here, the regional geological and geophysical evidence strongly favors external (i.e., northwest) thrusting, which may still be integrated within the framework of underthrusting of the continent by the Caribbean plate. The writer views the underthrusting by the Caribbean as continuing through the Tertiary, whereas Bell (1972) proposed termination of such movement in the Paleocene.

As remarked in the section on structural geology in Sec. V, major fractures such as the Boconó, Oca, Santa Marta–Bucaramanga, Palestina, etc.

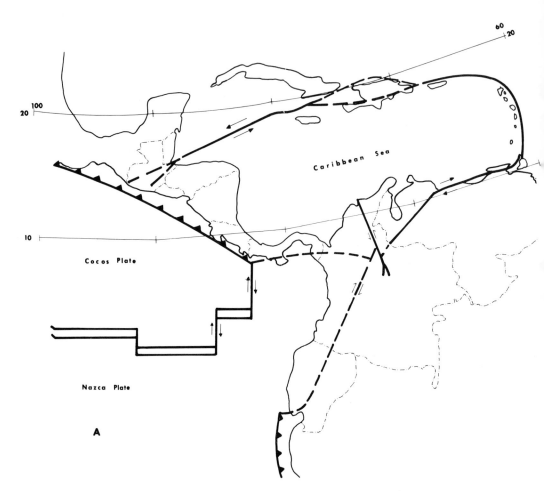

Fig. 19. Variations in plate-type tectonic syntheses. (A). Composite synthesis following
Molnar and Sykes (1969), Case *et al.* (1971), and Dewey (1972). Note participation of continental South America in Nazca and Caribbean plates and dominance of strike-slip faulting
(modified after Dewey, 1972). (B) Writer's views. South America separated from oceanic
plates. Northern boundary located seaward of off-shore island chain to coincide approximately with limit of continental crust (cf. Eardley, 1954, Fig. 6). Dominance of reverse
faulting. Lithospheric geanticlines develop on the continental plate associated with shallow

are no more than boundary faults defining shallow crustal blocks attempting
to make space for themselves. There may well have been lateral movement
on these fractures, but they are here interpreted as representing the effects of
shallow displacements uncoupled from underlying mantle motion rather than
deep fractures of strike-slip or transform type representing plate boundaries.
With the possible exception of the Sierra de Baudo, no portions of geographic
South America are considered ever to have been portions of the Pacific or

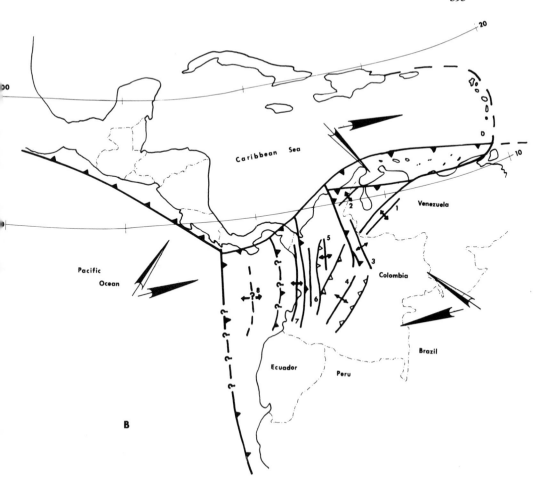

(crustal) faults and constitute partly uncoupled blocks. Subregional stress fields and pre-existing geostructural features decide spatial orientation of anticlinal blocks. Stress fields produced by forces related to principal drift directions (black arrows) or resolved therefrom (half-white arrows). Lithospheric folds shown are Mérida Andes (1), Perijás (2), Cordillera Oriental (3, 4), Cordillera Central (5), Cordillera Occidental (6), Sierra de Baudo (7), and postulated new fold in process of formation (8).

Caribbean plates as viewed by some workers, e.g., Molnar and Sykes (1969), Case *et al.* (1971), and Dewey (1972). See also Fig. 19.

Plate theory is flexible enough that the views outlined above may be modified without radical change in the basic framework. Thus, the apparent Northwest–Southeast axis of compression may be explained by southward spreading of the Caribbean, by resolution of east–west drift vectors between Caribbean and continent, or as a result of resolution of stresses produced in westward drift of the continent along a structural grain inherited from Paleozoic deformation.

Whereas the basaltic products of subduction and/or obduction are neatly piled up along and east of the Pacific Coast, they are sparse to absent along the Caribbean coast. This may be explained in terms of smearing-out of the volcanics in the course of extensive eastward movement of the Caribbean relative to the continent. Some of these volcanics (e.g., the Villa de Cura Group of northern Venezuela) appear to have been obducted away from the subduction zone onto the continent as a vast allochthonous mass, as originally suggested by Hess (Hess, in Bell, 1972). Also, the Caribbean subduction zone may have been more in the nature of a crustal down-buckle than a subduction zone involving conveyor-belt-type tectonics. As such, the volcanic products may have been of smaller volume. One may resort to similar arguments to explain the marked contrast in degree and nature of metamorphism of the volcanics in the Cordillera Occidental as compared to those in the Caribbean ranges of northern Venezuela. The seismological data provided by Dewey (1972) indicate that a compressional environment oriented east–west obtains for eastern Colombia and western Venezuela, as outlined in Sec. V. The pronounced positive Bouguer anomalies of the West Colombia high and over the Santa Marta block and the paleomagnetic data also fit the plate concept as discussed under geophysical criteria in Sec. V.

In the preceding paragraphs, I have attempted to show that the regional geology of the youngest "layer" can reasonably be interpreted in terms of plate theory. My purpose has been not so much as to be a standard bearer of plate theory as to show that when the moment of truth approaches, all aspects of the regional geology will have to fit the plate model chosen. Too many workers are selective and devise plate models to meet only one set of criteria. In the last analysis, plate theory is attractive because it readily explains the abundant evidence for intense lateral compression which obtained during virtually all of Tertiary time. The evidence includes the overall negative Bouguer anomalies, the seismological data, the structural evidence for the lithospheric anticlinal flexures of the eastern ranges and the overfolding with imbricate fault slices in the western ranges, the highly inclined paleo-erosion surfaces (see geomorphologic criteria in Sec. V), and the hint of high-pressure metamorphic assemblages at depth in the Cordillera Occidental. Possibly, a suitable mechanism to explain the Tertiary plutons in the western ranges, and the Antioquia and related plutons in the Cordillera Central at the close of the Cretaceous, would require a compressive environment too.

From Secs. IV and V, it will be clear to the reader that uncertainties remain concerning the ages of some rocks and events. The age limits of the volcanics in the western ranges are the most obvious example. Also, the K/Ar ages on the plutons in the Cordillera Central and Occidental may not correspond to original intrusive ages. In addition, the writer is at a loss to explain the Quaternary volcanism in the Cordillera Central. Many of the volcanoes fall on a major fracture which may have reached depths sufficient to allow for partial melting by reduction of pressure and subsequent explosive volcanism. But this requires a tensional environment, the reason for which is not readily apparent. On the other hand, it is not clear why the accumulation of basic volcanics along the Pacific coast should cease in the early–mid-Tertiary during the height of the latest sea-floor spreading cycle when one would expect them to be at their most voluminous, within the framework of subduction and/or obduction.

The next older "layer" of geology embraces the time span from the early Devonian to the late Jurassic. The documented story of those times is largely restricted to the Mérida, Perijá, and Oriental ranges where its format takes distinctly classical geosynclinal overtones in that a thick wedge of marine sediment is intensely deformed, intruded by granitic plutons and metamorphosed, and is finally uplifted with the consequent formation of piedmont or molasse deposits. It is true that the sedimentary accumulations are of miogeosynclinal type and that no associated eugeosynclinal facies has been recognized, but, in any event, there is a spatial association of these different components of orogeny quite distinct from the asymmetric spatial distribution required by plate theory. The opening phase of marine transgression [Devonian–Permian as elucidated by Stibane (1967, 1968) and Cediel (1972)] resembles the subsequent Cretaceous transgression in that it appears to have advanced from the north and west and to have involved unstable subsiding basins superimposed on the overall transgression in the areas of the three eastern ranges. Also, the Cordillera Central may have formed a positive borderland as it did through much of Cretaceous time, but this suggestion of Stibane's (1968) is weakened by the fossiliferous Devonian finds of Forero (Forero, in Cediel, 1972). This orogenic cycle differs from the Cretaceous–Tertiary cycle in that in the last-named the Cretaceous marine transgression was followed by mechanical lithospheric *up*-buckling devoid of thermal events, whereas the marine transgression of the older cycle was followed by *down*-buckling (as one would infer, in view of the plutonism and deformation in the climactic late Permian–early Triassic event of orogeny). The spatial distribution of the components of orogeny conjure up a tectonic environment more consonant with the early symmetrical tectogene than with the asymmetric subduction zone of plate theory, and a major question concerns what was happening to the west while the eastern ranges were undergoing the thermal and structural events of orogeny previously described. We immediately run into the problem of the age of the Cajamarca Group and related sections in the Cordillera Central. Possible large-scale tectonic interpretations of the regional relationships are shown diagrammatically in Fig. 20. If the metamorphics along the west side of the Cordillera Central are considered as of early Paleozoic age (Fig. 20A), following Nelson (1957), then it is conceivable that the late

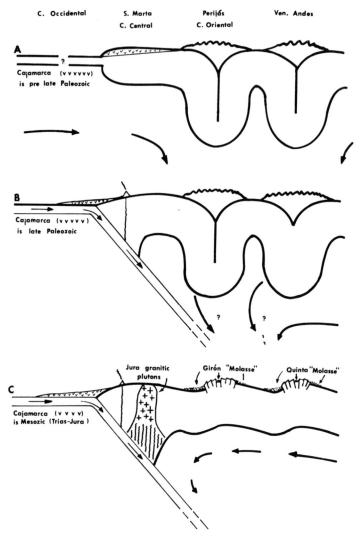

Fig. 20. Possible regional tectonic relationships (or lack of same) between oceanic and continental crust during late Paleozoic or Triassic–Jurassic time. Continental orogenies are best explained in terms of crustal down-buckles (see text). Basic volcanics of Cajamarca Group are diagrammatically shown here in terms of remelting of oceanic crust by subduction mechanism, but could just as well be obducted blocks of oceanic crust. (A) If volcanics are of pre-Devonian age, then Devonian–Permian orogenic cycle in eastern ranges may have occurred in tectonic isolation from oceanic environment. (B) If volcanics are of late Paleozoic age, then the volcanism and the continental orogeny may have been genetically related. However, down-buckling and broad regional extent of continental and orogenic belts are difficult to reconcile with collision of continental and oceanic plates in current plate theory. (C) If volcanics are of Triassic–Jurassic age (cf. Butterlin, 1972), then continent–ocean collision of plate theory provides rational explanation of coeval thermal and mechanical events. However, regional geologic/stratigraphic relationships are not consistent with this view (see text). With suitable readjustments for space and time, this geotectonic scheme does provide a rational framework for the Cretaceous–Present cycle of orogeny.

Paleozoic cycle of orogeny was restricted to the eastern ranges, and one might postulate orogeny related to subduction under thick continental crust, though the near-symmetrical geometry of the orogens in these ranges cannot easily be related to subduction. Also, there is a notable paucity of basaltic volcanics. If, following Irving (1971) and others, we consider the Cajamarca Group as of late Paleozoic age (Fig. 20B), a subduction zone located approximately along the Cauca valley might be postulated, but it would be extremely difficult to find a suitable explanation for the orogenic events in the Mérida, Perijá, and Santander Andes at the close of the Paleozoic. Those ranges occupy far too great an area for their respective orogenic phases to be explained simply in terms of downward drag of the continental crust overlying the subduction zone. If one persists in following plate theory, one is bound to inquire why processes, symmetries, etc. in this older cycle should differ so markedly from those in the Cretaceous–Tertiary cycle. Where, for example, are the corresponding lithospheric anticlinal folds? Finally, if we follow Butterlin (1972) and view the Cajamarca metamorphics as of Triassic–Jurassic age, we may resurrect the hypothesis of tectonism similar in many respects to that outlined for Cretaceous–Tertiary time. In this case (Fig. 20C), a subduction zone is postulated along the west side of the Cordillera Central. Subduction led to downward dragging of the continental crust in the overlying plate with the consequent granitic plutonism and volcanicity that occurred along the east side of the Cordillera Central in the Jurassic. This plutonism parallels in genetic setting that suggested for the Antioquia batholith and other plutons further to the west during the succeeding Cretaceous–Tertiary cycle. In turn, tectonism further east took the form of simple uplift as in the succeeding cycle, resulting in the deposition of the Girón–La Quinta red-bed "molasse" paralleling in tectonic setting the Tertiary uplifts and molasselike phases in the Tertiary. This hypothesis clearly requires redefinition of the second "layer" of geology as beginning approximately at the start of the Triassic rather than in the Devonian. Furthermore, it implies that the Girón–La Quinta red beds bear no relation to the preceding late Permian–early Triassic orogeny, as the writer has stipulated above. Workers favoring plate theory and/or continental accretion may find the hypothesis attractive for obvious reasons. However, the late Permian K/Ar ages reported by Irving (1971) on metamorphics and plutons in the Cajamarca belt of rocks and the Devonian finds near Rovira argue strongly against a Mesozoic age for the Cajamarca rocks. To isolate the Girón–La Quinta red beds from the proceeding orogeny at the close of the Paleozoic in the eastern ranges does not appear reasonable to the writer. The Cajamarca Group is best viewed as either early Paleozoic or Precambrian in age on the basis of the evidence supplied in Secs. IV and V. Although the interpretations in Fig. 20A, B, and C are not the only possible interpretations, they cover the principal range of possibilities and yet are all defective to some degree or other. Aside from the Cajamarca problem, there is also the problem of explaining the Jurassic plutonic and volcanic activity along the east side of the Cordillera Central, which appears to be partly coeval with the magmatism southeast of the Sevilla Lineament in the Santa Marta block during the Mesozoic.

Concerning the late Paleozoic "layer" of geology, one other matter arises which deserves attention. As outlined in Secs. IV and V, the evidence for orogeny at the close of the Paleozoic comes from the Sevilla metamorphic belt of the Santa Marta block, the western portion of the Cordillera Central, and portions of the Perijás, Santander massif, and Mérida Andes. The question is whether the late Paleozoic orogeny was restricted to these specific, roughly linear belts or whether they are merely the exposed remnants of an orogenic belt some 500 km wide extending approximately north–northeast through Colombia and Venezuela. The writer believes it possible that the Maracaibo platform (that region enclosed on three sides by the Perijás, Santander massif, and Mérida Andes) was *not* involved in the orogeny. We have noted (see stratigraphic and sedimentary evidence, Sec. V) that late Paleozoic deposition was probably very thin if it occurred at all in that region. For both Perijá and Mérida Andes, there is field evidence to indicate that shorelines of late Paleozoic marine basins were located within the current borders of the ranges. The writer knows of no mechanism whereby one might explain pervasive and penetrative thermal events occurring coevally throughout a belt 500 km wide. Virtually all geotectonic models would require a lateral progression of events with time in a belt of that width. The correspondence in spatial distribution of the late Permian–early Triassic orogenic belts with the Tertiary-block uplifts suggests that the latter were controlled not so much by directions of drift and subduction as by pronounced structural grain imparted in the older orogeny. This approximate coincidence in identity can only reasonably be anticipated for the Perijás, Mérida Andes, and Cordillera Oriental. Whether the current separate identities of the Cordillera Central and Occidental were similarly inherited from ancestral ranges is not known.

In turn, the late Paleozoic–early Mesozoic "layer" of geology is underlain by another representing the early Paleozoic cycle. It remains to be established whether the basal conglomeratic portion of the Devonian should be included with this cycle as a molasselike facies much as Girón–La Quinta was included with the late Paleozoic cycle. Evidence suggesting that ancestral Andes may have extended this far back in time includes the parallelism of regional structural grain of the Silgara schists in the Santander massif and of the early Paleozoic section in the Venezuelan Andes with that produced in subsequent deformations. Further evidence is the line of Devonian (and older?) granitic plutons roughly paralleling the late Permian plutons located further north in the Mérida Andes and the areal association of late Ordovician–late Silurian

plutons with those of late Permian–early Triassic age in the Santander massif
and possibly the Perijá ranges (see discussion of thermal events in Sec. V).
On the other hand, Irving (1971, Fig. 3) depicts the axis of late Ordovician
orogeny as being at a pronounced angle to the northeast-trending (Bogotá)
leg of the Cordillera Oriental, although parallel to the north-trending (Santander) leg. Burgl (1973, Fig. 9) depicts the paleogeography of Cambrian–
Ordovician in terms of geosynclines located in approximate coincidence with
the present Cordillera Oriental and Central. The writer (1972a, p. 1239)
suggested that the accumulation of thick conglomerates in the late Silurian
derived from the immediately adjacent Bella Vista schists to the Southeast
may have represented the first "Andean" event in the Venezuelan ranges.
There is hence the possibility that superimposed on the vast early Paleozoic
marine transgression which extended across to the interior plains of Colombia–
Venezuela there existed unstable basins in similar sites to those which developed
in subsequent orogenies.

The available data, though fragmentary, suggest tectonic settings which
closely resemble those for the subsequent late Paleozoic–early Mesozoic cycle.
In the Venezuelan Andes and Santander portion of the Cordillera Oriental,
we find evidence of thick sedimentary accumulations in unstable subsiding
basins succeeded by phases of granitic plutonism, regional metamorphism,
and deformation. In Secs. IV and V we debated inconclusively the possibility
of similar events in the Perijá ranges. What may have occurred in the region
from the Cordillera Central westward is again obscured by the Cajamarca
problem and the absence of known section. If Cajamarca volcanism and
sedimentation occurred during the early Paleozoic, then one might summarize the geometry of regional tectonic environments as similar to that in
Fig. 20B.

For lack of data, we are forced to include all Precambrian events in one
lowermost "layer" of geology, though it may embrace two or more orogenic
cycles. Of geotectonic interest here is the fact that the Precambrian appears
in granulite facies in the region close to its believed westerly limit (Cordillera
Central and Santa Marta block), in amphibolite facies in the Santander massif
and northern Mérida Andes, and in greenschist facies in the southern Mérida
Andes (Bella Vista schists). The oldest radiometric ages ($\pm$1300 m.y.) have
been measured on the granulites of the Santa Marta block; the youngest are
$\pm$680 m.y. (Santander massif) and $\pm$660 m.y. (Bella Vista schists). There is
thus some hint of petrologic and age zonation of the Precambrian, but the
data are too fragmentary for further comment here.

There are some indications that the tectonic environment outlined for
the Paleozoic in the eastern ranges may be traced back into the late Precambrian
in terms of the sedimentary accumulations of the Bella Vista and Bucaramanga
schists, the associated granitic plutons or migmatites, and the regional meta-

morphism and deformation with regional grain subparallel to those of subsequent cycles.

Further west the apparent association of granulite facies with the believed westerly limits of continental crust appears to be a reasonable inference, but proponents of different tectonic mechanisms will have no difficulty relating those rocks to their preferred models because available data are so scanty. The fact that Triassic–Jurassic batholiths and their cover of volcanic equivalents are preserved in the Santa Marta and Cordillera Central areas indicates that the exposure of the granulites at the surface is not the result of Tertiary or Quaternary uplift. This is further indicated by the Bouguer anomaly pattern (Fig. 18). Whichever tectonic setting is preferred, it is clear that Precambrian orogeny involved dewatering and degranitizing of the crust. We have noted for this reason that some form of subduction mechanism or large-scale overfolding is required in order to find a suitable source for the Mesozoic batholiths which occur in the granulite setting (see thermal events, Sec. V). Therein lies an interesting speculation, viz., that even deep levels of the crust need not be closed systems but may undergo chemical replenishment.

The subparallelism of structural grain in the granulite terrains with those of subsequent deformations may also be noted. This feature is particularly remarkable in view of the sharp change in regional strike from northerly in the Cordillera Central to almost easterly in the Santa Marta block. Part or all of this could be related to subsequent transposition of bedding or other S-surfaces.

Figure 21 attempts to summarize diagrammatically the pattern of orogeny in space and time as outlined in the preceding discussion. The principal inferences appear to be:

    a. A threefold division of tectonic environments: eastern-continental, western-oceanic, and the intermediate Cordillera Central which appears to show a mixture of the two

    b. A pattern of repeated orogeny in space which is particularly evident in the eastern ranges and in portions of the Santa Marta block and Cordillera Central and somewhat less so (for lack of section) in the western ranges

    c. The tentative identification of the Cordillera Central and its apparent continuation in the Santa Marta block as the fundamental element of the tectonic framework to which the eastern and western elements are in subordinate relationship

As here reviewed, we have set, if not in detail at least in general outline, a new framework of geotectonism for the region; the basis for this is discussed in the following pages.

Fig. 21. Summary of tectonic environments in space and time. In the eastern ranges (Cordillera Oriental, Perijás, and Mérida Andes) spatial relationships of sedimentation, granitic plutonism, and metamorphism (↓) and uplift (↑) best fit orogeny within a framework of symmetrical crustal down-buckling for most of geologic time (cf. Fig. 20). A–A' and B–B' represent variations in Paleozoic orogenic cycles for the Mérida Andes and the Cordillera Oriental/Perijás, respectively. In the western ranges (Cordillera Occidental/Baudo) the spatial relationships of basic volcanics, ultra-

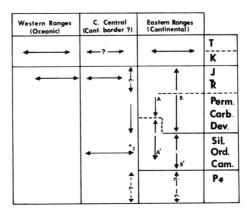

mafics, plutons, and regional structure better fit sub- or obduction mechanisms (↔). A far shorter geologic history would be postulated than for ranges to the east. In the central ranges (Cordillera Central/Santa Marta block) a mixture of orogenic environments is postulated. *1 represents a hypothetical event of subduction assuming early Paleozoic age for Cajamarca Group. Subduction in Triassic–Jurassic time preferred in order to explain occurrence of granitic plutons in Precambrian granulite terrains.

A. Problem of the Threefold Division of Tectonic Environments in Four-Dimensional Space

Alvarez (1971) noted the fact that the crystalline terrain of the Cordillera Central separates a Cretaceous–Tertiary miogeosyncline in the eastern Andes from a eugeosynclinal environment in the region of the western ranges. We have noted that Case *et al.* (1971) have indicated the likelihood of a transition from simatic to sialic crust conforming approximately with the location of the Cauca–Romeral fault system. The question is whether this asymmetry in crustal and supracrustal environments constitutes a relatively youthful development or whether it has been a long-term characteristic of the region. That the latter alternative may have obtained is suggested by the Precambrian ages (1300 m.y.) on the granulites of the Santa Marta block, the history of repeated miogeosynclinal-type accumulations going back at least to the early Paleozoic if not the Precambrian in the eastern Andes, and parallel events of granitic plutonism, metamorphism, and deformation in the same region. If the Cajamarca Group should prove to be of Paleozoic or Precambrian age, there would be a corresponding record of long-term development of eugeosynclinal environments in the region of the western ranges. If so the nature and distribution of the sedimentary environments would resemble Wells' (1949) ensialic and ensimatic geosynclines.

We note that Case *et al.* (1971) depict the transition from simatic to sialic crust as of abrupt, fault type (their Fig. 8) but that the boundary appears to diverge from the Romeral fault north of 5° N. This is not entirely consistent with the regional geology. The numerous large ultramafic masses in the middle to lower reaches of the Cauca valley suggest to the writer that continental crust is very thin or absent west of the fault at that latitude. Possibly, the continental crust "seen" by the gravity studies is compound crust consisting of a significant fraction of clastic sediment of continental origin even though resting on oceanic crust.

Proponents of plate theory may immediately object to this simple view of the geologic setting and claim that in the several cycles of drift going back to Precambrian time such vast total and relative displacements may have taken place that the present spatial relationship of orogens to continental or oceanic crust need bear absolutely no relationship to their true original spatial relationships. Thus, Tschanz *et al.* (1974, p. 283) raise the possibility of "...reassembly of microcontinental cratonic fragments along transform fault systems..." in explanation of the intrusion of the Mesozoic batholiths in the Santa Marta block. Also, many authors have suggested removal or relocation of continental crust through processes of subduction or obduction. Those who prefer geotectonic mechanisms set within a more static framework may claim, on the other hand, that orogeny may have served through time to unite isolated cratonic fragments, and the fact that there was continental crust under the Cordillera Central in the Precambrian does not preclude the existence to the east of oceanic crust in Paleozoic time. In line with such objections, one may consider the use of such terms as "oceanization" whereby continental crust is removed either by obduction onto or by subduction under the opposed plate. Gilluly (1972, p. 427) thus refers to oceanization of the Mediterranean in the Tertiary, i.e., the establishment of an oceanic environment in a region formerly of continental type. One may, in turn, infer processes of "continentalization" or "cratonization" whereby formerly oceanic environments are converted to regions of continental crust. Thus Peive *et al.* (1972) discuss "intracontinental geosynclines" but consider them to be floored by oceanic crust and to represent regions converted from oceanic to continental type by lateral compression resulting in tectonic piling-up or clustering of the original materials of the oceanic environment to a point where a granitic foundation develops and the former oceanic environment is eliminated. In addition, the terms aulacogen and taphrogeosyncline have been used to describe intracratonic fault-bound troughs, though the connotation is largely related to accumulation of thick sedimentary piles rather than specifically to orogeny.

Concerning the objections raised above, reassembly of microcratons could hardly have occurred on a significant scale without extensive accompanying basaltic volcanism for which there is very little evidence. The possibility that "oceanization" may have occurred in the region prior to the Mesozoic cannot be tested. For such a process in post-Paleozoic time, the only supporting evidence is the apparent absence of a high-pressure, low-temperature metamorphic belt west of the Cordillera Central (by inference, the metamorphic belt was subducted under the continental crustal plate). As indicated in the discussion of petrologic criteria in Sec. V, although the characteristic high-pressure mineral species such as jadeite have not been recognized, the mineral assemblages in the volcanics of the Cordillera Occidental are indicative of at least intermediate pressures, so that minor thrusting may explain the absence in outcrop of the truly high-pressure forms. From other points of view, such as sediment character, distribution, and volume, and geophysical criteria, oceanization is not required and, failing convincing evidence to the contrary, is herewith discarded.

There is little evidence to support the possibility that "cratonization" has been a significant process in the extensive region here considered to be underlain by continental crust. There is no sign in any of the ranges of an oceanic crustal floor, and we have noted the relative paucity of mafic volcanics east of the Cajamarca Group. Though we have certainly not eliminated the possibility that these various processes occurred, it appears to me a reasonable inference, and the simplest, that there has been long established continental crust extending eastward approximately from the Cauca–Romeral fracture system.

B. Problems of Orogenic Style and Superimposition

In the eastern ranges (Mérida and Perijá Andes and Cordillera Oriental), the components of orogeny have the spatial relationships associated with classical geosynclinal concepts, i.e., thick accumulations of sediment in unstable subsiding troughs were followed by terminal phases of granitic plutonism, metamorphism, and deformation in the same region. By contrast, in the Cordillera Central there is a separation and asymmetry of these components of orogeny (see Fig. 21). Thus, middle to late Mesozoic volcanic and sedimentary piles in the western piedmont are geographically well separated from the gigantic late Cretaceous Antioquia batholith and related masses. The orogenic image conveyed is that of the subduction zone of plate tectonics. Tschanz et al. (1974) have suggested a similar orogenic model for the region of the Santa Marta block southeast of the Sevilla Lineament where Mesozoic batholiths intruded Precambrian granulites but where there is no corresponding "geosynclinal" sedimentary accumulation. The subduction-type model appears to have extended westward to the Pacific coast.

Concerning the spatial distribution of orogeny in time, the writer (1972b) has noted the apparent northward migration of orogenic axes in the Venezuelan Andes. Based on the distribution of granitic plutons, orogenic axes appear to have moved from the Bella Vista belt along the southern piedmont at the

close of the Precambrian (ca. 660 m.y.) to a parallel belt passing approximately
through Valera in the northern ranges (ca. 225 m.y.), i.e., a lateral movement
(continuous or discontinuous?) of about 80 km in 430 m.y. In the Santa Marta
block there appears to be a similar external advance of orogenic axes from
Precambrian to Paleocene(?) time, though the granitic plutons appear to become progressively younger toward the southeast in the region southeast of
the Sevilla lineament. Because of uncertainties in dating metamorphic and
plutonic events, the distance–time relationships cannot be calculated. Very
approximately, one may claim ± 40 km of migration of orogenic (?) axes
northwest of the Sevilla lineament in Permian–Paleocene time (approximately
200 m.y. range in radiometric ages), which is a similar rate to that noted for
the Mérida Andes. Evidence for such lateral migrations in the Perijás is very
flimsy (there is an apparent increase in metamorphic grade southeastward,
as mentioned in the regional descriptions in Sec. IV), and there does not appear
to be any indication of such an increase in the Santander massif. In the Cordillera Central, the axes of orogenic events appear to have moved from west
(Paleozoic?) to east (Triassic–Jurassic) and back to central (late Cretaceous,
Quaternary), but the nature and timing of these events is not clear. West of
the Cordillera Central the possibility of progressive westward stepping of
orogenic events has already been noted. If this is true, the lateral migration has
been about 200 km since approximately the mid-Jurassic, i.e., faster by a factor
of at least 5 than the apparent migration rates in the Mérida Andes. From the
perspective of this review, one may consider such lateral migrations of orogenic
axes as those in the Mérida Andes as relatively insignificant against the scale
of the mountain-building processes in this portion of the continent and that,
with the exception of the ranges external to the Cordillera Central, orogeny
has effectively been superimposed in space.

It may be claimed that superimposition of orogeny is merely a matter of
inheritance of structural grain from a previous orogeny. But to claim so is to
miss the point. Involved in the complex processes of orogeny are several
energy-related gradients: temperature, pressure, and compositional gradients.
Superimposition requires remarkable changes in the slope if not direction of
these gradients. Thus, changes in pressure gradients are indicated by alternate
subsidence of basins and uplift of ranges. In the Venezuelan Andes, for example,
subsidence related to the accumulation of Bella Vista sediments is followed
by uplift of a mountain belt at the close of the Precambrian, subsidence during
the early Paleozoic, uplift of a Devonian mountain belt, deep subsidence for
accumulation of Pennsylvanian flysch, uplift of a late Permian mountain belt,
Uribante basin subsidence during the Cretaceous, and finally Tertiary uplift.
There must surely have been corresponding changes in geothermal gradients
through most of these alternations. The problem posed is suitably summarized
by a homely analogy. Consider a bicycle tire with a small linear weakness. It

is quite understandable that with rising air pressure a portion of the inner tube should protrude through the line of weakness. In this sense the protrusion is indeed mimetic after a previous deformation. Now, however, when air pressure is reduced, the protruding tube must not merely retreat back inside the tire but become a low-pressure depression! In geologic reality, most of the mass transfers necessary to accomplish subsidence may be summarized in terms of changes in the depth to the M discontinuity. There is a relatively limited number of ways in which this can be accomplished: The crust may be thinned by stretching, or it may be thinned as the result of subcrustal erosion as proposed by Gilluly (1963, p. 159), or by down-buckling in a tectogenelike structure as suggested by Hess (1938). These hypotheses may be combined to some extent and the processes further speeded up or retarded by changes in geothermal gradient. The fact that all but the youngest (Cretaceous–Tertiary) orogenic cycle in the eastern ranges involved granitic plutonism and regional metamorphism with only minor basaltic volcanism inclines the writer to favor the down-buckling mechanism. We are at a loss to explain convincingly repeated down-buckling in space. In basic theory of strength of materials, it is known that under both static and cyclic conditions energy may concentrate at the edges of minute fractures and other inhomogeneities and then propagate outwards to form fractures. Hodgson (1961) has discussed the possible relationship of such local energy concentrations to the development of joints. For the same reason, structural engineers do not find the behavior of the bicycle tire necessarily anomalous: There would be a concentration of energy at the edges of the linear weakness which could be propagated downward as well as upward depending on regional stress trajectories. This view is not precisely the same as that of inheritance of structural grain from a previous deformation. Nonetheless it has a flavor of facile explanation.

The piling up of basic volcanics in the western ranges as a result of subduction and obduction does not raise the same problem of superimposition, but the factor of time becomes crucial. If the western ranges should prove to consist of late- or post-Jurassic volcanics resting on oceanic crust, the parallelism of regional trends can reasonably be explained within the framework of sea-floor spreading as described above. Should the volcanic history prove to go back to Paleozoic time, there will be a problem similar to that outlined for the eastern ranges of explaining regional structural parallelism for such an extended period of time.

C. Problems of the Interrelationships between the Three Major Geotectonic Divisions

We have in effect defined, in the foregoing, "old" continent to the east located on crust at least 1300 m.y. old and "new" continent to the west located on oceanic crust and with a history going back to the late Jurassic, if not before, depending on the tectonic setting of the Cajamarca Group. The Cordillera Central constitutes a transition zone between the two environments, and we have now to inquire whether tectonism in the three regions has been genetically related or independent. Inasmuch as most of the sedimentary environments on old continent were of miogeosynclinal (ensialic?) type, whereas most of those on "new" continent were of eugeosynclinal (ensimatic?) type, depending on one's view of the interrelationships between the three principal component geotectonic elements, one may view these sedimentary environments in terms of paired sedimentary basins (set, according to preference, within the framework of either classical geosynclinal theory or modern plate tectonic theory) or in terms of two kinds of genetically unrelated geosynclines. How various workers have in effect taken a stand on this question may be illustrated by some examples.

Irving (1971, Fig. 10) refers to continental accretion through progressive external development of eugeosynclines beginning with the Bucaramanga gneisses in the Precambrian (Santander massif), extending to the western portion of the Cajamarca Group in the late Paleozoic (western side of the Cordillera Central), and occupying the site of the Cordillera Occidental in the Mesozoic. Burgl (1967) diagrams a similar progression but initiates the process west of the Magdalena valley in the early Paleozoic. Both writers depict in word and diagram miogeosynclinal and epicontinental environments trailing in turn behind the advancing eugeosynclinal front. Butterlin (1972), without providing a *raison d'etre*, considers Colombia the meeting place or junction of three independent Cordilleras: The Cordillera Oriental and Santa Marta are considered Andean, the Cordillera Central is considered to be a continuation of the Caribbean Island Arc environment (in so doing, the Cajamarca Group is viewed as of Mesozoic age), and the Cordillera Occidental–Sierra de Baudo are correlated with the Central American province. In contrast to the views of Irving and of Burgl, there is implied here a large measure of independence in the sedimentary environments and most other aspects of the orogenic process in the three principal tectonic subdivisions.

Case *et al.* (1971) speculate on the tectonic evolution of northern South America with particular relationship to the Caribbean–Antillean and Central American Arcs. Their model (see Fig. 19) is one of several variants within the framework of modern plate theory. From our point of view, the most notable feature is the fact that the interpretation is restricted to Mesozoic–Tertiary events and then almost entirely to the eugeosynclinal facies (i.e., Cordillera Occidental and Baudo). Reference to the eastern Andean domain is restricted to structural considerations, largely the fact that the boundaries of the postulated plates involved in drift conform with such major faults as the Guayaquil–El Pilar rather than with present coastlines. Other syntheses in a similar mold are those of Bell (1972) and Dewey (1972). Reference by the former to the Andes is again almost entirely structural; the latter's work is also concerned with major structural features and is most pertinent to late Tertiary events only.

In comment, it may be said that only Burgl (1967) and Irving (1971) among modern workers have attempted to view the tectonic evolution of these northern Andes in four-dimensional space. Their model of continental accretion implies, in line with the questions raised above, genetic association of mio- and eugeosynclinal sedimentary facies. In any event, two major criticisms of the model may be leveled. As early as 1300 m.y. ago (if not earlier) there appears to have been thick continental crust extending approximately as far west as the Cauca–Romeral–Sevilla fracture lineament, so that subsequent orogenies on that crust did not necessarily result in continental accretion in that region. Moreover, *behind* the broad sweep of Burgl's and Irving's accreting continental crust, major events of orogeny involving granitic plutonism were taking place over much the same time span in the Venezuelan Andes. Certainly a simple straight-line, time–distance view of accretion would not be indicated.

The Butterlin model recognizes the problem but tends to avoid rather than explain it; those models in the plate tectonic framework tend to ignore the basic problems by severely restricting space–time considerations.

In attempting to come to grips with the same problems, the writer is much impressed with the fact that repeatedly the answers to major questions appear to reside in the Cordillera Central and its extensions to the north in the Santa Marta block. The Precambrian granulite terrains and the Quaternary volcanics embrace the longest thermal history of all the ranges. In addition to the vast Antioquian and other granitic events, there have been extensive basic volcanic and ultramafic events strongly indicating mantle participation in the orogenic history. By contrast, the thermal events in the eastern ranges are very largely restricted to crustal events; no Alpine-type ultramafic masses have yet been identified, and basic volcanism occurs in much reduced volume. If basic volcanic and ultramafic igneous events are known from the western ranges, their history is far shorter.

It appears to me clear that the particular tectonic mechanism appealed to as a basis for explanation of the evolution of these ranges requires permanent or near-permanent belts peculiarly susceptible to repeated orogeny. Furthermore, the permanency of the eastern belt is likely to be genetically related but subordinate to that of the Cordillera Central and its extensions into the Santa Marta belt. A review of the tectonic models proposed for this region suggests that almost all fall within a tetrahedron whose apices may be labeled mechanical, classical, plate, and vertical syntheses, respectively. Mechanical syntheses are those which propose regional stress patterns sufficient to explain the linear elements of the furcate pattern of fault-bound ranges and basins in the northern Andes. Classical syntheses are set to greater or lesser degree within the framework of classical geosynclinal theory, whereas plate syntheses follow modern tectonics and the lateral movement of large lithospheric plates. Vertical syntheses propose dominantly vertical movements of lithospheric blocks set within a more static framework. On closer inspection, it is seen that a clean separation of tectonic theses into one or other model is difficult. Case *et al.* (1971) have noted that the seed of some plate hypotheses may be found in the early writings of Schuchert (1935), and Eardley (1954) proposed vertical movements in the central Caribbean leading to the horizontal compressional environment of marginal tectogenes. Syntheses illustrative of one or more of the characteristics outlined above include those of Schuchert (1935), Hess (1938), Von Estorff (1946), Bucher (1952), Lees (1952), Eardley (1954), Woodring (1954), Jacobs *et al.* (1963), Beloussov (1967), Burgl (1967), Stibane (1967, 1968), Stainforth (1969), Case *et al.* (1971), Irving (1971), Dallmus and Graves (1972), Butterlin (1972), Dewey (1972), Shagam (1972a), and Gansser (1973), among others.

Much of the variation in the tetrahedron is related to the fact that the work of the great majority of investigators has been severely limited by geologic space–time considerations. The majority of mechanical syntheses center on the major fractures, many of which were initiated or rejuvenated in the Tertiary. Stibane (1967, 1968), concentrating on the late Paleozoic history of the Cordillera Oriental, which was characterized by pronounced subsidence and marine transgression followed by equally pronounced uplift, understandably viewed the tectonic mechanism within the framework of vertical movement. We have already noted the tendency of those partial to plate theory to restrict their focus to the eugeosynclinal, late Mesozoic–Tertiary elements of the regional geology. Almost invariably, when a given tectonic mechanism is applied outside its original intended time–space focus, it proves to be deficient to a serious degree. This is particularly evident in the syntheses of mechanical type. In my opinion, the crucial test question is: Will the tectonic mechanism adequately fulfill the requirements for intense horizontal movement and compression in the mantle during the orogenic process? We have noted the compelling evidences for horizontal mantle stresses seen in the anticlinal lithospheric folds of the eastern ranges and the need for either subduction or vast overfold mechanisms to explain the Mesozoic granites in the granulite terrains of the Cordillera Central and the Santa Marta block. Following classical or modern plate models, one may readily explain such horizontal stresses. On balance, the modern plate model is the more satisfying because, in addition to explaining the horizontal motions, the conveyor-belt mechanism neatly solves the related problems of conserving volume and redistributing mass. By contrast, as Hess once noted, classical theory required *stretching* of crust on a global scale. Finally, many aspects of plate theory are susceptible to some degree or other of testing. At least for the region under discussion, plate theory provides the best basis for supplying a suitable geotectonic mechanism.

It is ironic, in view of the foregoing conclusion, that most recent workers [e.g., Burgl (1967) and Irving (1971)] who have attempted to devise an integrated hypothesis of geotectonism that matches the geology in four-dimensional space should return to the classic mold of hypothesis. The reason for this is clear: Plate theory does not adequately explain the matter of superimposed orogeny in subparallel geometry over extensive periods of time. It would be more logical to expect a more random distribution of orogenic geometries; also, one might expect extensive periods of quiescence between orogenic upheavals instead of the almost unbroken succession of cycles (at least since latest Precambrian time).

We may attempt to evade the problem of superimposed orogeny by inquiring whether the process of orogeny might not have changed with time, implying that plate theory holds only for the youngest cycle of orogeny. We noted that within that cycle the four-dimensional relationships can be reasonably related to plate-type syntheses. In so doing, we in effect arranged for a sound explanation for lateral movement and compression. The preceding cycles showed differences in tectonic style from the youngest cycle, but the basic requirement for intense lateral compression remains fundamental. It appears to me rational to persist with the plate mechanism for the earlier cycles too. We then face the problem of superimposed orogeny. Elsewhere I have suggested (Shagam, 1972b) that the explanation may lie in the fact that ocean-floor spreading is not random with respect to the distribution of the continents. The continents may constitute thermal and mechanical barriers which tend to concentrate energy and stress at their borders. Alternatively, we may have to investigate the geophysical feasibility of there being permanent or semipermanent geosutures rooted in the mantle and peculiarly susceptible to thermal plumes or other energy-focusing devices for repeated orogeny.

Inasmuch as this volume focuses basically on the Caribbean, it is appropriate to comment on the possible implications of Andean tectonic history in the interpretation of the origin and evolution of the Caribbean. From the regional descriptions (see also Fig. 3), there appears to be a reasonable probability that elements of the geology from the Cordillera Central westward to the Pacific may extend into the northwestern portion of the Santa Marta block and thence into the Guajira Peninsula and beyond (see Fig. 22), as summarized by numerous workers including Alvarez (1971), Irving (1971), Butterlin (1972), and Tschanz (1974). There are numerous possibilities of correlating the fundamental stratigraphic elements along the several component segments of this extensive trend, but two extreme cases encompass the more interesting possibilities. On the one hand, we may correlate the Mesozoic–Tertiary volcanic piles of the Cordillera Occidental and Baudo with the volcanics of northern Venezuela and the Antilles and regard what we have referred to as "new" continent as a tectonic element largely unrelated to the lengthy history of the Cordillera Central and the eastern ranges. By implication, Caribbean geology is a young tectonic domain truncating older Andean tectonism and largely independent of it aside from some mechanical interaction. On the other hand, we may, like Butterlin (1972), attempt to trace correlation through the Cajamarca Group. Accepting the correlation, we should note our conclusion that the Cajamarca is of early Paleozoic or even Precambrian age rather than of Mesozoic age, as suggested by Butterlin. There is here the almost heretical suggestion of a corresponding ancient "memory" or "heritage" of the Caribbean arc and that its evolution may have proceeded in parallel with that of the northern Andes. In this regard, it is of interest to note increasing reference to the possibility that some of the ultramafic complexes in the Andean–Caribbean zone may have been intruded originally in pre-Mesozoic times. Mégard et al. (1971, p. 11) imply a Precambrian association (and age?) for ultramafic rocks in the Peruvian Andes, and Irving (1971, p. 45) notes the possibility that some of the serpentinites in the eastern portion of the Cordillera Central may be of late Paleozoic age. Tschanz et al. (1969) depict ultramafic rocks in the Santa Marta Massif as of Mesozoic (?) age, but the masses occur mainly in metamorphic belts of probable late Paleozoic age. In the Caribbean Mountains of northern Venezuela, the only contact aureole related to ultramafic intrusions is that around the Tinaquillo Peridotite (MacKenzie, 1960) in pre-Mesozoic basement gneisses. Where the ultramafics occur in Cretaceous–Tertiary terrain, no aureoles have been reported. Rod (1967) also has implied a Paleozoic original age for the Venezuelan ultramafics. Some of the Cuban masses may be of Paleozoic age, according to Khudoley and Meyerhoff (1971). Dengo and Bohnenberger (1969) concluded that some of the Guatemalan serpentinites were intruded in the Paleozoic.

In a classic paper, Hess (1955) brought home the fundamental significance of ultramafic rocks in the orogenic concept. Dengo (1972), in summarizing the occurrence of these rocks in the Caribbean and adjoining areas, has effectively outlined the connecting thread which is also the fundamental tectonic element of this vast region. After noting the evidence for long-term superimposed orogeny in the northern Andes, the possibility of the extension of same into the Caribbean region does not seem to me to be entirely unreasonable.

At a somewhat lower order of magnitude, Andean relationships may prove to be crucial to the solution of problems concerning postulated allochthonous masses in the ranges of north-central Venezuela. Some of the gross stratigraphic relationships in those ranges have been interpreted in terms of the following:

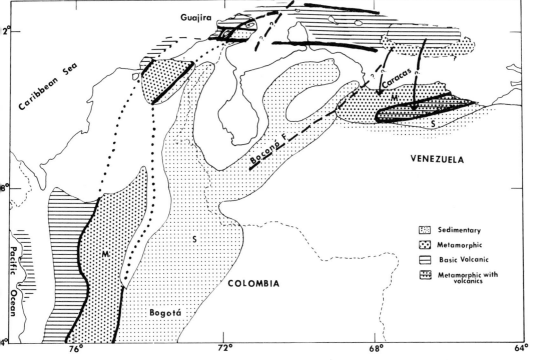

Fig. 22. Tectonic implications of regional distribution of basic volcanic, metamorphic, and sedimentary (mainly miogeosynclinal) facies in Colombia and Venezuela. Volcanic and sedimentary rocks are mainly of Mesozoic–Cenozoic age, while the metamorphics include Paleozoic and Precambrian? rocks. Broad spatial distribution at latitude of Bogotá narrows along Guajira Peninsula. In north-central Venezuela both spatial location and interrelationships of facies are disordered with respect to the rest of the region. Implied (arrows) are possible allochthonous movements of large blocks and lateral shortening by subduction or obduction (N–S) in the Guajira–Santa Marta region. Pronounced discordance between the metamorphic–volcanic complex and the Andean sedimentary belt in Venezuela has been interpreted in terms of unrelated orogenic belts, while parallelism of the facies in Colombia has been viewed by some in terms of paired geosynclinal belts and a common orogeny (see p. 408ff).

a. An allochthonous block of basic volcanics obducted onto *in situ* continental crust (Bell, 1972)

b. An allochthonous block of volcanics resting on allochthonous microcraton which is in turn resting on Paleocene flysch (Bellizzia, 1972)

c. *In situ* relationships for both volcanics and continental crust (Harvey, 1972)

It can be seen from Fig. 22 that the threefold subdivision of the geologic pattern in the northern Andes is somewhat identifiable in Venezuela, but in far more compact spatial array, suggesting that allochthonous movements may indeed have occurred.

In line with the comments rendered herein concerning the setting of the principal fractures, it is considered most unlikely that even the Oca and El Pilar faults ever constituted plate boundaries. It seems to me far more logical to set a possible Caribbean–South American plate boundary seaward of the offshore islands of the Aruba trend, especially in view of the apparent continental crustal boundary depicted by Krause (1971, Fig. 10). The several major inferred faults north of the Venezuelan coast (see Ball *et al.*, 1971), in addition to the El Pilar fracture, are considered to be probably high-angle reverse faults showing oblique-slip movement of continent north–northwest over the Caribbean. Ball *et al.* (1971) consider the offshore faults to be dominantly of normal type, but these may be shallow-seated features reflecting reaction of sedimentary cover and shallow crustal levels to the more deep-seated reverse faulting postulated here.

For the immediate future, it would seem most productive to concentrate research on the problems of the Cordillera Central and its extensions in the Santa Marta block, particularly the age, origin, and petrology of the Cajamarca Group and of the ultramafic rocks. Provided suitable basement cores can be found, a potentially valuable study in the eastern ranges would be on Rb/Sr and K/Ar ages on the metamorphic rocks underlying the Maracaibo platform. If the writer's thesis is correct, viz., that orogeny has been consistently restricted to the mobile belts of the Mérida, Santander, and Perijá ranges, this should be reflected in progressively older metamorphic ages as one moves from these ranges toward the Maracaibo basin. Available data are yet so sparse for the western ranges that virtually any study contributing to the basic outline of the stratigraphy and petrology of the volcanic piles would be a significant contribution.

ACKNOWLEDGMENTS

I am indebted to the Dirección de Geología, Ministerio de Minas e Hidrocarburos, Republic of Venezuela, for long-continued financial support of Andean studies. Among individuals who provided valuable discussion and/or logistic support were E. Araujo, A. Bellizzia, A. Espejo, E. Lavie, A. Menéndez, A. Sabater, and A. Vivas. In the course of this work, it was my fortune to be blessed with brilliant students, research associates, and colleagues who repeatedly made important contributions. Among them were P. O. Banks, L. Burkley, and F. G. Stehli of Case Western Reserve University, Cleveland, and R. F. Giegengack, R. I. Grauch, and L. Kovisars of the University of Pennsylvania, Philadelphia. A. Espejo and L. Gonzalez kindly showed me type sections in the Venezuelan Perijás. Discussion with F. Cediel, Universidad Nacional de Colombia, and T. Kassem, Director, Instituto Nacional de Investigaciones Geologico–Mineras, Republic of Colombia, were extremely useful. My colleague H. J. Kisch of Ben Gurion University in the Negev, Israel, has been a constant source of inspiration through numerous critical discussions. It would have been difficult for me to complete this work without the moral support of my wife, Livia, and both moral and material help of Dr. M. L. Cohen and family, Ramat Hasharon, Israel.

NOTE ADDED IN PROOF

One of the most important aspects in contention concerns the possible geotectonic affinities of the volcanic rocks of the western ranges. Meyerhoff (written communication, 1973) views the volcanics and associated sedimentary rocks of the Pacific coast region as belonging to a Jura–Cretaceous geosyncline that extends from Ecuador through Panama to Guatemala. Butterlin (1972) arbitrarily refers the Cordillera Occidental to the Central American domain. Alvarez (1971) notes possible continuity with either or both the Caribbean and Central America. Continuation of the Sierra de Baudo into Panama is supported by some of the geophysical evidence (see Sec. V).

REFERENCES*

Alvarez, W., 1971, Fragmented Andean belt of northern Colombia: Geol. Soc. Am. Mem. 130, p. 77–96.

Anderson, T. A., 1972, Paleogene Nonmarine Gualanday Group, Neiva Basin, Colombia, and regional development of the Colombian Andes: *Geol. Soc. Am. Bull.*, v. 83, p. 2423–2438.

Arnold, H. C., 1966, Upper Paleozoic Sabaneta–Palmarito sequence of Merida Andes, Venezuela: *Am. Assoc. Petr. Geol.*, v. 50, n. 11, p. 2366–2387.

Ball, M. M., Harrison, C. G. A., Supko, P. R., Bock, W., and Maloney, N. J., 1971, Marine Geophysical measurements on the southern boundary of the Caribbean Sea, Donnelly, T. W., ed.: Geol. Soc. Am. Mem. 130, p. 1–33.

Barrero, D., Alvarez, J., and Kassem, T., 1969, Actividad ígnea y tectónica en la Cordillera Central durante el Meso-Cenozóico: *Bol. Geól. (Colombia)*, Ingeominas, v. 17, p. 145–173.

Bass, M. N., and Shagam, R., 1960, Edades Rb/Sr de las rocas cristalinas de los Andes Merideños, Venezuela: III Cong. Geol. Venezolano, Caracas: Bol. Geol. (Venezuela), Publ. Espec. No. 3, v. 1, p. 377–381.

Bell, J. S., 1971, Tectonic evolution of the central part of the Venezuelan Coast Ranges: Geol. Soc. Am. Mem. 130, p. 107–118.

Bell, J. S., 1972, Geotectonic evolution of the southern Caribbean area: Geol. Soc. Am. Mem. 132, p. 369–386.

Bellizzia G. A., 1972, Is the entire Caribbean mountain belt of northern Venezuela allochthonous?: Geol. Soc. Am. Mem. 132, p. 363–368.

Beloussov, V. V., 1967, Esbozo de la tectónica de los Andes: *Geol. Colomb.*, N. 4, p. 5–24.

* In addition to those listed here, reference material can also be found in Irving (1971), Burgl (1973), Bowen (1972), Tschanz *et al.* (1974), and Shagam (1972a).

Botero Arango, G., 1963, Contribución al conocimiento de la geología de la zona central de Antioquia: Anales de la Facultad de Minas, n. 57, Medellín, 101 p.

Botero Restrepo, G., 1950, Reconocimiento geológico del área comprendida por los municipios de Belén, Cerinza, Corrales, Floresta, Nobsa y Santa Rosa de Viterbo, Departamento de Boyacá: Servicio Geológico Nacional, Ministerio de Minas y Petroleos, v. 8, p. 245–311, Bogotá.

Bowen, J. M., 1972, Estratigrafía del precretáceo en la parte norte de la Sierra de Perijá: IV Cong. Geól. Venezolano, v. II, p. 729–761: Bol. de Geól., Publ. Espec. No. 5, Ministerio de Minas e Hidrocarburos, Caracas.

Bucher, W. H., 1952, Geologic structure and orogenic history of Venezuela: Geol. Soc. Am. Mem. 49, 113 p.

Burgl, H., 1961, Historia geológica de Colombia: Rev. Acad. Colomb. Cien. Exactas, Fís. Nat., v. 11, n. 43, p. 137–193.

Burgl, H., 1967, The orogenesis in the Andean system of Colombia: Tectonophysics, v. 4, p. 429–443.

Burgl, H., 1973, Precambrian to middle Cretaceous stratigraphy of Colombia: Translated by Allen, C. G., and Rowlinson, N. R., Bogota, 214 p.

Butterlin, J., 1972, La posición estructural de los Andes de Colombia: IV Cong. Geól. Venezolano, v. II, p. 1185–1200: Bol. de Geól., Publ. Espec. No. 5, Ministerio de Minas e Hidrocarburos, Caracas.

Campbell, C. J., 1968, The Santa Marta wrench fault of Colombia and its regional setting: Fourth Carib. Geol. Conf. Trans., Port-of-Spain, Trinidad, W. I., 1965, p. 247–261.

Campbell, C. J., and Burgl, H., 1965, Section through the eastern cordillera of Colombia, South America: Geol. Soc. Am. Bull., v. 76, p. 567–590.

Case, J., E., Duran, S. L. G., Lopez R., Alfonso, and Moore, W. R., 1971, Tectonic investigations in western Colombia and eastern Panama: Geol. Soc. Am. Bull, v. 82, p. 2685–2712.

Case, J. E., and MacDonald, W. D., 1973, Regional gravity anomalies and crustal structure in northern Colombia: Geol. Soc. Am. Bull., v. 84, p. 2905–2916.

Cediel, F., 1968, El Grupo Girón; una molasa mesozoica de la cordillera oriental: Bol. Geológico, v. XVI, n. 1–3, p. 5–96; Servicio Geológico Nacional, Bogotá.

Cediel, F., 1972, Movimientos tectónicos en el intervalo Paleozoico-Mesozoico en Colombia y su influencia en reconstrucciones paleogeográficas: Preprint, International Symposium on Carboniferous and Permian System, Sao-Paulo, Brazil.

Compañia Shell de Venezuela and Creole Petroleum Corporation, 1964, Paleozoic rocks of Merida Andes, Venezuela: Am. Assoc. Petr. Geol. Bull., v. 48, p. 70–84.

Creer, K. M., 1970, A paleomagnetic survey of South American rock formations: Phil. Trans. Roy. Soc., London, Ser. A, v. 267, p. 457–557.

Dallmus, K. F., and Graves, G. R., 1972, Tectonic history of Maracaibo and Barinas basins: VI Conf. Geol. del Caribe, Isla de Margarita, Venezuela, Memorias, 1971, p. 214–226.

Dengo, G., 1972, Review of Caribbean serpentinites and their tectonic implications: Geol. Soc. Am. Mem. 132, p. 303–312.

Dengo, G., and Bohnenberger, O., 1969, Structural development of northern Central America: Am. Assoc. Petr. Geol. Mem. 11, p. 203–220.

Dewey, J. W., 1972, Seismicity and tectonics of western Venezuela: Bull. Seismol. Soc. Am., v. 62, n. 6, p. 1711–1751.

Dufour, J., 1955, Some oil-geological characteristics of Venezuela: IV World Petroleum Cong. Proc., Rome, v. 1, p. 19–55.

Eardley, A. J., 1954, Tectonic relations of North and South America: Am. Assoc. Petr. Geol. Bull., v. 38, n. 5, p. 707–773.

Feininger, T., Barrero, D., Castro, N., Ramírez, O., Lozano, H., Vesga, J., et al., 1970, Mapa geológico del oriente del depto. de Antioquia, Colombia (2 sheets): Instituto Nacional de Investigaciones Geológico-Mineras, Ministerio de Minas y Petróleos, Bogotá.

Feininger, T., 1970, The Palestina fault, Colombia: Geol. Soc. Am. Bull., v. 81, n. 4, p. 1201–1216.

Féo-Codecido, G., 1972, Breves ideas sobre la estructura de la falla de Oca, Venezuela: VI Conf. Geol. del Caribe, Margarita, Venezuela, Memorias, 1971, p. 184–190.

Forero, A., 1967, Notas preliminares sobre la estratigrafía del Paleozoico en el norte de los Andes de Colombia: Inst. Colombo-Alemán Investigaciones Cientificas, Santa Marta, p. 31–37.

Gansser, A., 1955, Ein Beitrag zur Geologie und Petrographie der Sierra Nevada de Santa Marta (Kolumbien, Sudamerika): Schweiz. Mineral. Petrogr. Mitt., v. 35, n. 2, p. 209–279.

Gansser, A., 1973, Facts and theories on the Andes: J. Geol. Soc. London, v. 129, p. 93–131.

García, R. J., and Campos, V., 1972, Las rocas paleozoicas en la región del Rio Mombóy, Estado Trujillo: IV Cong. Geol. Venezolano, v. II, p. 796–806: Bol. de Geól., Publ. Espec. No. 5, Ministerio de Minas e Hidrocarburos, Caracas.

Giegengack, R. F., and Grauch, R. I., 1972, Bocono Fault, Venezuelan Andes: Science, v. 175, p. 558–560.

Giegengack, R. F., and Grauch, R. I., 1973a, Quaternary geology of the central Andes, Venezuela; a preliminary assessment: Excursion no. 1, Cordillera de los Andes: II Cong. Geol. Latinoamericano, Caracas, in press.

Giegengack, R. F., and Grauch, R. I., 1973b, Late Cenozoic climatic stratigraphy of the Venezuelan Andes: Excursion no. 1, Cordillera de los Andes: II Cong. Geol. Latinoamericano, Caracas, in press.

Giegengack, R. F., Grauch, R. I., and Shagam, R., 1973, Geometry of Late Cenozoic displacement along the Bocono Fault, Venezuelan Andes: II Cong. Geol. Latinoamericano, Caracas, in press.

Gilluly, J., 1963, The tectonic evolution of the western United States: Q. J. Geol. Soc. London, v. 119, p. 133–174.

Gilluly, J., 1972, Tectonics involved in the evolution of mountain ranges, in: The Nature of the So'id Earth, Robertson, E. C., Hays, J. F., and Knopoff, L., eds.: McGraw-Hill International Series in the Earth and Planetary Sciences.

Goldsmith, R., Marvin, R. F., and Mehnert, H. H., 1971, Radiometric ages in the Santander massif, eastern cordillera, Colombian Andes: U.S. Geol. Surv. Prof. Pap. 750-D, p. D44–D49.

Grauch, R. I., 1971, Geology of the Sierra Nevada south of Mucuchíes, Venezuelan Andes: an aluminum-silicate-bearing metamorphic terrain: Ph.D. thesis, Pennsylvania University, Philadelphia, Pa. 180 p.

Grauch, R. I., 1972, Preliminary report of a Late(?) paleozoic metamorphic event in the Venezuelan Andes: Geol. Soc. Am. Mem. 132, p. 465–473.

Hargraves, R. B., and Shagam, R., 1969, Paleomagnetic study of La Quinta Formation, Venezuela: Am. Assoc. Petr. Geol. Bull., v. 53, n. 3, p. 537–552.

Harvey, S. R. M., 1972, Origin of the southern Caribbean mountains: Geol. Soc. Am. Mem. 132, p. 387–400.

Hea, J. P., and Whitman, A. B., 1960, Estratigrafía y petrología de los sedimentos precretácios de la parte norte central de la Sierra de Perijá, Estado Zulia, Venezuela: III Cong. Geol. Venezolano, Memoria, v. I, p. 351–376: Bol. de Geól., Publ. Espec. No. 3, Ministerio de Minas e Hidrocarburos, Caracas.

Hess, H. H., 1938, Gravity anomalies and island arc structure with particular reference to the West Indies: *Proc. Am. Philos. Soc.*, v. 79, p. 1–96.

Hess, H. H., 1955, Serpentines, orogeny, and epeirogeny: Geol. Soc. Am. Spec. Pap. 62, p. 391–407.

Hodgson, R. A., 1961, Regional study of jointing in Comb Ridge–Navajo Mountain area, Arizona and Utah: *Am. Assoc. Petr. Geol. Bull.*, v. 45, n. 1, p. 1–38.

Hospers, J., and van Wijnen, J. C., 1959, The gravity field of the Venezuelan Andes and adjacent basins: *K. Ned. Akad. Wet., Versl. Gewone Vergad. Afd. Natuurk.*, v. 23, n. 1, p. 1–95.

Irving, E. M., 1971, La evolución estructural de los Andes mas septentrionales de Colombia: Bol. Geológico, v. XIX, n. 2, p. i–xiv, 1–90: Instituto Nacional de Investigaciones Geologico-Mineras, Ministerio de Minas y Petróleos, Bogotá.

Jacobs C., Burgl, H., and Conley, D. L., 1963, Backbone of Colombia, in: *Backbone of the Americas*, Childs, O. E., and Beebe, B. W., eds.: Am. Assoc. Petr. Geol. Mem. 2, p. 62–72.

Kaula, W. M., 1969, A tectonic classification of the main features of the earth's gravitational field: *J. Geophys. Res.*, v. 74, n. 20, p. 4807–4826.

Kehrer, L., 1938, Some observations on the stratigraphy in the States of Táchira and Mérida, S.W. Venezuela: Bol. de Geología y Minería, v. II, n. 2, 3, 4, p. 44–55.

Khudoley, K. M., and Meyerhoff, A. A., 1971, Paleogeography and geological history of the Greater Antilles: Geol. Soc. Am. Mem. 129, 197 p.

Kovisars, L., 1971, Geology of a portion of the north-central Venezuelan Andes: *Geol. Soc. Am. Bull.*, v. 82, n. 11, p. 3111–3138.

Krause, D. C., 1971, Bathymetry, geomagnetism, and tectonics of the Caribbean Sea north of Colombia, Donnelly, T. W., ed.: Geol. Soc. Am. Mem. 130, p. 35–54.

Kündig, E., 1938, The precretaceous rocks of the central Venezuelan Andes with some remarks about the tectonics: Bol. de Geología y Minería, v. II, n. 2, 3, 4, p. 21–43.

Lees, G. M., 1952, Foreland Folding: *Q. J. Geol. Soc. London*, v. 108, p. 1–34.

MacDonald, W. D., Doolan, B. L., and Cordani, U. G., 1971, Cretaceous–early Tertiary metamorphic K-Ar age values from the south Caribbean: *Geol. Soc. Am. Bull.*, v. 82, n. 5, p. 1381–1388.

MacDonald, W. D., and Opdyke, N. D., 1972, Tectonic rotations suggested by Paleomagnetic results from Northern Colombia, South America: *J. Geophys. Res.*, v. 77, n. 29, p. 5720–5730.

MacDonald, W. D., and Opdyke, N. D., 1974, Triassic paleomagnetism of northern South America: *Am. Assoc. Petr. Geol. Bull.*, v. 58, n. 2, p. 208–215.

MacKenzie, D. B., 1960, High-temperature alpine-type peridotite from Venezuela: *Geol. Soc. Bull.*, v. 71, p. 303–318.

Martín-Bellizzia, C., 1968, Edades isotópicas de rocas Venezolanos: *Venez. Bol. Geól.*, v. 10, n. 19, p. 356–380.

McLaughlin, D. H., Jr., Arce H., M., et al., 1969: Mapa geológico del cuadrángulo K-11, "Zipaquira"-Colombia: Instituto Nacional de Investigaciones Geológico-Mineras, Ministerio de Minas y Petróleos, Bogotá.

Mégard, F., Dalmayrac, B., Laubacher, G., Marocco, R., Martinez, C., Paredes, J., and Tomasi, P., 1971, La chaine Hercynienne au Pérou et en Bolivie premiers résultats: *Cah. O.R.S.T.O.M., Sér. Géol.*, v. III, n. 1, p. 5–44.

Mencher, E., 1963, Tectonic history of Venezuela, in: *Backbone of the Americas*, Childs, O. E., and Beebe, B. W., eds.: Am. Assoc. Petr. Geol. Mem. 2, p. 73–87.

Miller, J. B., 1962, Tectonic trends in Sierra de Perija and adjacent parts of Venezuela and Colombia: *Am. Assoc. Petr. Geol. Bull.*, v. 46, n. 9, p. 1565–1595.

Miyashiro, A., 1961, Evolution of metamorphic belts: *J. Petrol.*, v. 2, part 3, p. 277–311.

Molnar, P., and Sykes, L. R., 1969, Tectonics of the Caribbean and Middle America regions from focal mechanisms and seismicity: *Geol. Soc. Am. Bull.*, v. 80, p. 1639–1684.

Morales, L. G., and The Colombian Petroleum Industry, 1958, General geology and oil occurrences of Middle Magdalena Valley, Colombia, in: *Habitat of Cil*: Am. Assoc. Petroleum Geol., p. 641–695.

Nelson, H. W., 1957, Contribution to the geology of the Central and Western Cordillera of Colombia in the sector between Ibague and Cali: *Leidse Geol. Meded.*, v. 22, p. 1–76.

Nelson, H. W., 1962, Contribución al conocimiento de la Cordillera Central de Colombia. Sección entre Ibagué y Armenia: *Colombia, Bol. Geól.*, v. 10, p. 161–201.

Nygren, W. E., 1950, Bolivar geosyncline of northwestern South America: *Am. Assoc. Petr. Geol. Bull.*, v. 34, n. 10, p. 1998–2006.

Olmeta, M. A., 1968, Determinación de edades radiométricas en rocas de Venezuela y su procedimiento por el método K/Ar: *Venez., Bol. Geol.*, v. 10, n. 19, p. 339–344.

Peive, A. V., Perfiliev, A. S., and Ruzhentsev, S. V., 1972, Problems of intracontinental geosynclines: 24th International Geol. Cong., Montreal, Sec. 3, p. 486–493.

Polson, I. I., and Henao, D., 1968, The Santa Marta wrench fault—a rebuttal: Fourth Carib. Geol. Conf. Trans., Port-of-Spain, Trinidad, 1965, p. 263–266.

Raasveldt, H. C., 1957, Las glaciaciones de la Sierra Nevada de Santa Marta: *Rev. Acad. Colomb. Cien. Exactas, Fig. Nat.*, v. 9, n. 38, p. 469–482.

Radelli, L., 1967, Géologie des Andes Colombiennes: Mem. 6, Laboratoire de Géologie de la Faculté des Sciences de Grenoble, 470 p.

Ramírez, C., and Campos, V., 1972, Geología de la región de La Grita–San Cristóbal, Estado Táchira: IV Cong. Geól. Venezolano, v. II, p. 861–897: Bol. de Geól., Publ. Espec. No. 5, Ministerio de Minas e Hidrocarburos, Caracas.

Rod, E., 1956, Strike-slip faults of northern Venezuela: *Am. Assoc. Petr. Geol. Bull.*, v. 40, n. 3, p. 457–476.

Rod, E., 1967, Paleotectonic reconstruction of the Antillean–Caribbean area for the close of the Carboniferous: Asoc. Venezolano de Geología, Minería y Petróleo, Bol. Informativo, v. 10, n. 7, p. 197–204.

Schubert, C., 1969, Geologic structure of a part of the Barinas Mountain Front, Venezuelan Andes: *Geol. Soc. Am. Bull.*, v. 80, n. 3, p. 443–458.

Schubert, C., and Sifontes, R. S., 1970, Boconó Fault, Venezuelan Andes: Evidence of postglacial movement: *Science*, v. 170, p. 66–69.

Schuchert, C., 1935, *Historical Geology of the Antillean–Caribbean Region*: John Wiley & Sons Inc., New York, 811 p.

Shagam, R., 1968, Commentary on the Caparo area: Asoc. Venezolano de Geología, Minería y Petróleo, Bol. Informativo, v. 11, n. 6, p. 171–182.

Shagam, R., 1972a, Evolución tectónica de los Andes Venezolanos: IV Cong. Geol. Venezolano, v. II, p. 1201–1261: Bol. de Geól., Publ. Espec. No. 5, Ministerio de Minas e Hidrocarburos, Caracas.

Shagam, R., 1972b, Andean research project, Venezuela: principal data and tectonic implications: Geol. Soc. Am. Mem. 132, p. 449–463.

Shagam, R., and Hargraves, R. B., 1970, Geologic and paleomagnetic study of Permo-Carboniferous red beds (Sabaneta and Merida facies) Venezuelan Andes: *Am. Assoc. Petr. Geol. Bull.*, v. 54, n. 12, p. 2336–2348.

Shagam, R., and Hargraves, R. B., in press, Additional paleomagnetic data on tuffs of the La Quinta Formation, Venezuelan Andes: II Cong. Geol. Latinoamericano, Caracas, 1973.

Stainforth, R. M., 1969, The concept of sea-floor spreading applied to Venezuela: Asoc. Venezolano de Geología, Minería y Petróleo, Bol. Informativo, v. 12, n. 8, p. 257–274.

Stibane, F. R., 1967, Paläogeographie und Tectogenese der Kolumbianischen Anden: Geol. Rundschau, v. 56, p. 629–642.

Stibane, F. R., 1968, Zur Geologie von Kolumbien, Sudamerika: Das Quetame- und Garzon-Massiv: Geotektonische Forsch., Heft 30, p. 1–85.

Stibane, F. R., and Forero, S. A., 1969, Los afloramientos del paleozoico en La Jagua (Huila) y Rio Nevado (Santander del Sur): Geol. Colomb., n. 6, p. 31–66.

Trumpy, D., 1943, Pre-Cretaceous of Colombia: Geol. Soc. Am. Bull., v. 54, n. 9, p. 1281–1304.

Tschanz, C. M., Jimeno V., Andres, Cruz B., Jaime, et al., 1969, Mapa Geológico de Reconocimiento de la Sierra Nevada de Santa Marta-Colombia: Colombia Instituto Nacional de Investigaciones Geologico-Mineras, Ministerio de Minas y Petróleos, Bogotá, scale 1 : 200,000.

Tschanz, C. M., Marvin, R. F., Cruz B. J., Mehnert, H. B., and Cebula, G. T., 1974, Geologic evolution of the Sierra Nevada de Santa Marta, Northeastern Colombia: Geol. Soc. Am. Bull., v. 85, n. 2, p. 273–284.

Useche, A., and Fierro, I., 1972, Geología de la región de Pregonero, Estados Táchira y Mérida: IV Cong. Geól. Venezolano, v. II, p. 963–998: Bol. de Geól., Publ. Espec. No. 5, Ministerio de Minas e Hidrocarburos, Caracas.

Van der Hammen, T., and Gonzáles, E., 1963, Historia de clima y vegetación del Pleistoceno superior y del Holoceno de la Sabana de Bogotá, Colombia: Colomb. Bol. Geol., v. 11, n. 1–3, p. 189–260.

Van Houten, F. B., and Travis, R. B., 1968, Cenozoic deposits, upper Magdalena Valley, Colombia: Am. Assoc. Petr. Geol. Bull., v. 52, n. 4, p. 675–702.

Vázquez, E. E., and Dickey, P. A., 1972, Major faulting in north-western Venezuela and its relation to global tectonics: VI Caribbean Geol. Conf., Trans., Margarita, Venezuela, 1971, p. 191–202.

Von Estorff, F. E., 1946, Tectonic framework of northwestern South America: Am. Assoc. Petr. Geol. Bull., v. 30, n. 4, p. 581–590.

Ward, D. E., Goldsmith, R., Cruz B., J., et al., 1969, Mapa geológico del cuadrángulo H-12 (Bucaramanga), Colombia Instituto Nacional de Investigaciones Geológico-Mineras, Ministerio de Minas y Petróleos, Bogotá.

Ward, D. E., Goldsmith, R., Cruz B., J., et al., 1970, Mapa geológico del cuadrángulo H-13 (Pamplona), Colombia Instituto Nacional de Investigaciones Geológico-Mineras, Ministerio de Minas y Petróleos, Bogotá.

Wellman, S. S., 1970, Stratigraphy and petrology of the nonmarine Honda Group (Miocene), upper Magdalena Valley, Colombia: Geol. Soc. Am. Bull., v. 81, p. 2353–2374.

Wells, F. G., 1949, Ensimatic and ensialic geosynclines (abstract): Geol. Soc. Am. Bull., v. 60, p. 1927.

Woodring, W. P., 1954, Caribbean land and sea through the ages: Geol. Soc. Am. Bull., v. 65, p. 719–732.

Zambrano, E., Vásquez, E., Duval, B., Latreille, M., and Coffinieres, B., 1971, Síntesis paleogeográfica y petrolera del occidente de Venezuela: IV Cong. Geól. Venezolano, v. I, p. 483–551: Bol. de Geología, Publ. Espec. No. 5, Ministerio de Minas e Hidrocarburos, Caracas.

Chapter 10

GEOLOGY OF THE CARIBBEAN CRUST*

Paul J. Fox

Department of Geological Sciences
State University of New York at Albany
Albany, New York

and

Bruce C. Heezen

Lamont-Doherty Geological Observatory of Columbia University
Palisades, New York

I. INTRODUCTION

The Caribbean presents several provocative geotectonic problems. Water depths are less than in the main ocean basins. Over most of the Caribbean Basin, seismic-refraction measurements reveal (Fig. 1) crustal thicknesses of 12 to 15 km, much greater than the 6 km crust generally observed beneath the main ocean basins (Officer *et al.*, 1959; Ewing *et al.*, 1960; Edgar *et al.*, 1971). In addition, the seismic-velocity sequence of the Caribbean crust is varied and contrasts sharply with the simple velocity structure generally at-

* Contribution No. 2238 of the Lamont-Doherty Geological Observatory, Palisades, New York.

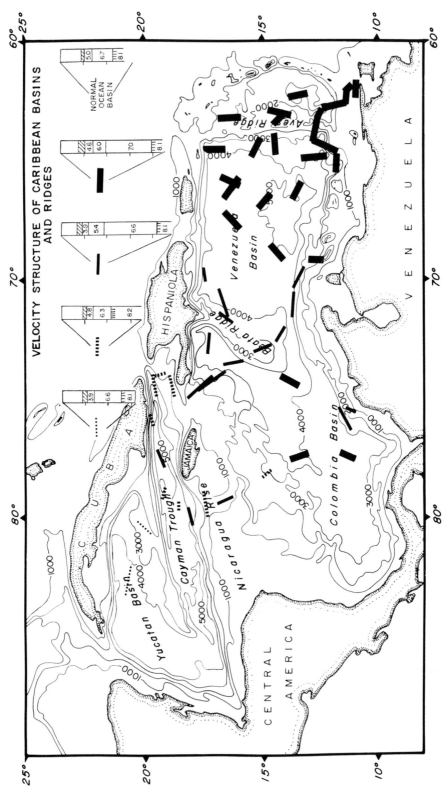

Fig. 1. A map of the Caribbean showing the location of seismic-refraction studies and the results of those studies. Data derived from Officer *et al.*, 1959; Ewing *et al.*, 1960; and Edgar *et al.*, 1971. The vertical scale for the velocity sections is in increments of 5 km; the velocity is in km/sec.

tributed to crust beneath the main ocean basins (Fig. 1). Correlatable magnetic anomalies of the type characteristic of oceanic crust generated at middle or low latitudes are absent in the Caribbean which is characterized by a relatively quiet field (Ewing *et al.*, 1960; Pitman, personal communication). Recently, however, Christofferson (1973) has attempted to correlate a series of magnetic anomalies observed in the southern Colombia Basin with the Late Cretaceous sequences established for the deep-sea crust. Differences in geophysical character of the Caribbean and the main ocean basins suggest a fundamental difference in composition, origin, and history. Some investigators have suggested that the Caribbean is a former piece of the deep-sea floor in the process of becoming a continent (Ewing *et al.*, 1957; Officer *et al.*, 1959). Others have proposed that the Caribbean is a subsided piece of continental crust well advanced in the process of becoming an ocean (Schuchert, 1935; Eardley, 1951; Butterlin, 1956; Beloussov, 1960).

Although seismic-refraction measurements provide important data on the crustal thickness and physical properties of the crust; the composition, age, and details of crustal evolution are difficult to determine from indirect geophysical measurements. Samples of Caribbean crust were needed to resolve the Caribbean controversy and further advance our knowledge. When it was generally believed that a continuous layer of sediment blanketed the ocean floor, deep crustal drilling seemed to provide the only potential method of sampling ocean crust. The layer-cake assumptions of the seismic-refraction method led to the retention of the continuous-blanket concept, and layers were shown to comformably cover even the steepest escarpments. That hard rock actually crops out on steep submarine escarpments soon became evident from broken and bent but usually empty coring tubes recovered after encounters with them. Consequently, the layer-cake interpretation was reevaluated by looking critically at what refraction measurements actually indicated. Individual seismic-refraction measurements did not always support the assumed continuous sediment cover shown in the generalized seismic-velocity cross sections. Heezen (1959) suggested that steep slopes, either tectonic or erosional, within the Caribbean basin might expose the crust lying beneath the sediment blanket. In 1966, a program to sample the Caribbean escarpments was begun aboard Duke University's *R. V. Eastward*. The original plan was to dredge and sample any outcrops that could be found. It was hoped that if crustal rocks were sampled, they could be compared with the seismic-velocity horizons of the Caribbean crust, and that if deeply buried sediments were sampled, the age and environment of deposition could be ascertained. To date, over 200 dredges have recovered material from the Beata Ridge, Aves Ridge, Lesser Antilles, Puerto Rico Trench, and Cayman Trench. In this presentation we summarize the results obtained on the Beata and Aves Ridges, Lesser Antilles, and in the Puerto Rico Trench (Fig. 2).

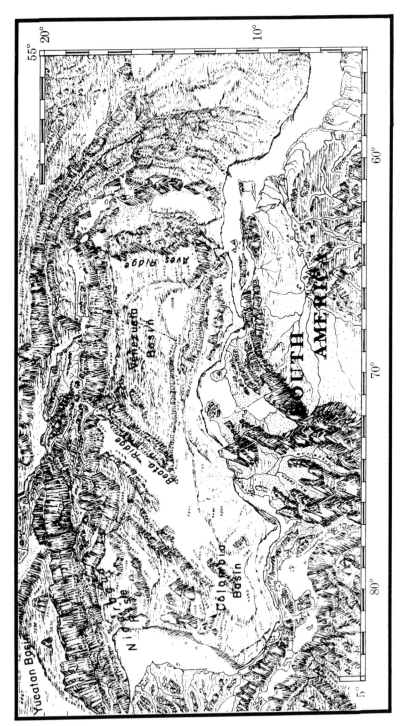

Fig. 2. Physiographic diagram of the Caribbean Basin showing major physiographic provinces (from Fox *et al.*, 1971). Reprinted with the permission of Elsevier Publishing.

II. BEATA RIDGE

The Beata Ridge extends 400 km south from Cape Beata, Hispaniola, dividing the Caribbean into the Colombia and Venezuela Basins. The ridge is bounded on the west by an escarpment with slopes exceeding 25°; its average relief above the Colombia Abyssal Plain is approximately 2500 m. Punctuated by a series of steps on the east, the ridge gently drops to the 5-km depths of the Venezuela Basin. Seismic-refraction measurements (Ewing *et al.*, 1960; Edgar *et al.*, 1971) on the Beata Ridge show that the underlying crust consists of a 5.4 km/sec layer overlying a 6.6 km/sec layer (Fig. 1). Although this velocity sequence is similar to that observed in the major ocean basins, the units here are twice as thick. This sequence of crustal velocities beneath the ridge is similar to that observed beneath the northern Colombia Basin but contrasts markedly with the crust of the southern portion of the Colombia Basin and the Venezuela Basin to the east. Refraction studies established that the crust of the Venezuela Basin and southern Colombia Basin consists of a

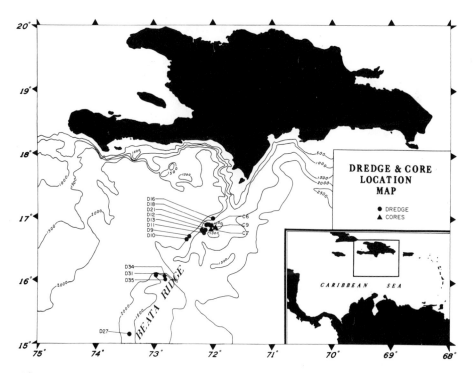

Fig. 3. Dredge and core location map. Location of dredge stations are shown by dots, with the dredge numbers keyed to Table I. Triangles locate the position of the piston cores containing the oldest (Mid-Eocene) material (from Fox *et al.*, 1970). Reprinted with the permission of Elsevier Publishing.

6.0 km/sec layer over a 7.0 km/sec layer, which in turn is underlain by a mantle of normal velocity (8.1 km/sec).

Twelve dredge stations, on the western escarpment of the Beata Ridge distributed between 4,100–2,500 m, sampled basalts and dolerite (Fox *et al.*, 1970). Nine hauls contain basaltic rocks, and although all the samples thin-sectioned are in various states of weathering, most are fresh enough for identification of primary mineral phases and textures with a petrographic microscope (Figs. 3 and 4; Table I). The textures observed are typical of shallow intrusive and/or extrusive rocks of basaltic composition with textures varying from porphyries with glassy matrix through porphyries with holocrystalline matrix to holocrystalline intergranular to subophitic intergrowths of plagioclase and pyroxene. The major mineral constituents recognized in all samples are labradorite, clinopyroxene and opaque oxides. Some samples contain vein and interstitial serpentine, some of which assume a crystal outline suggestive of olivine.

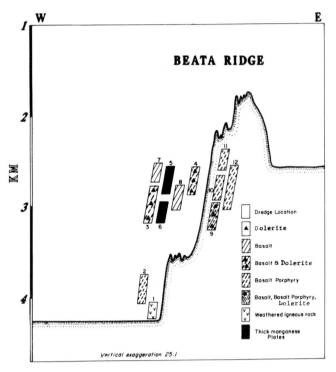

Fig. 4. Results of the 12 dredge stations projected onto a topographic profile. The length of the boxes indicates the approximate depth sampled. The number by each box is keyed to the listing of samples in Table I. The four dredge stations from the southerly localities are placed on the right-hand side of the sediment–water interface (from Fox *et al.*, 1970). Reprinted with the permission of Elsevier Publishing.

TABLE 1

Content, Position, and Depth of Dredges from the Beata Ridge

	Dredge No.	Position	Depth (m)	Rock types recovered
1	E46-67-D9	16°41' N 72°27' W	4,200 4,100	Completely weathered, manganese encrusted, igneous rock with large relict crystalline outlines
2	E46-67-D10	16°41' N 72°26' W	4,100 3,750	Partially weathered, labradorite–augite basalt porphyry
3	E46-67-D11	16°44' N 72°14' W	3,235 2,800	Partially weathered, holocrystalline, labradorite–augite basalt porphyry; partially weathered, subophitic dolerite
4	E46-67-D12	16°47' N 72°10' W	2,900 2,600	Partially weathered, holocrystalline, labradorite–augite basalt porphyry; partially weathered, subophitic dolerite; weathered, mesocrystalline, basalt porphyry
5	E46-67-D13	16°49' N 72°08' W	2,900 2,600	Manganese crust (10 cm)
6	E46-67-D16	16°60' N 71°60' W	3,200 2,980	Manganese crust (7 cm)
7	E46-67-D18	16°55' N 72°04' W	2,750 2,525	Partially weathered, holocrystalline, fine-grained basalt
8	E46-67-D21	16°54' N 72°06' W	3,100 2,800	Partially weathered, holocrystalline, fine-grained basalt
9	E50-69-D31	16°05' N 72°57' W	3,363 3,000	Partially weathered, holocrystalline, fine-grained basalt; partially weathered, holocrystalline, labradorite basalt porphyry; partially weathered, hemicrystalline, labradorite–augite basalt porphyry; partially weathered, ophitic dolerite; completely weathered volcanic rocks
10	E50-69-D34	16°04' N 72°49' W	2,988 2,702	Badly weathered, holocrystalline, plagioclase–pyroxene–olivine basalt porphyry
11	E50-69-D35	16°00' N 72°48' W	2,608 2,400	Badly weathered, plagioclase–pyroxene–olivine basalt porphyry
12	RC11-67-D27	15°10' N 73°26' W	3,146 2,500	Slightly weathered, holocrystalline, labradorite–augite basalt porphyry

The velocity of compressional waves was investigated in two holocrystal-line basalts and a dolerite. Velocity measurements were conducted first at atmospheric pressure and then at pressures up to 7 kbar. At 0.5–1 kbar, which is the *in situ* lithostatic pressure range for the upper crustal layer of the Beata Ridge, the velocities of the basalt samples ranged from 5.0–5.5 km/sec, and the velocities of the dolerite samples ranged from 5.4–5.6 km/sec. These ranges are compatible with measurements on basalts and dolerites sampled during DSDP Leg XV in the Caribbean (Fox and Schreiber, 1973) and are similar to velocities measured for basalt and dolerite sampled in the main ocean basins (e.g., Fox *et al.*, 1973).

Dredging near the ridge crest in depths shallower than 2,500 m recovered only recent foraminiferal sand. Rocks outcrop in this region, however, because several dredges were torn apart during encounters with the bottom. A 1,200-lb piston corer with a short barrel was lowered to the hard bottom 10 times. The cores recovered deep-sea pelagic carbonate oozes varying in age from late Oligocene to Recent (Fig. 5). Consolidated carbonate of Mid-Eocene age was recovered in three short cores which apparently struck outcropping rock. The

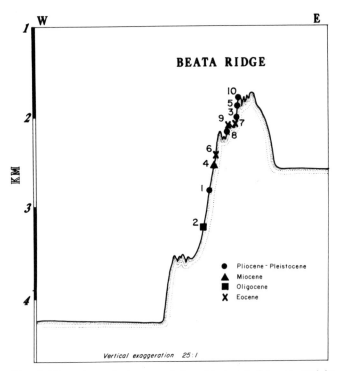

Fig. 5. The paleontologic ages of the oldest material recovered in cores projected onto a topographic profile (from Fox *et al.*, 1970). Reprinted with the permission of Elsevier Publishing.

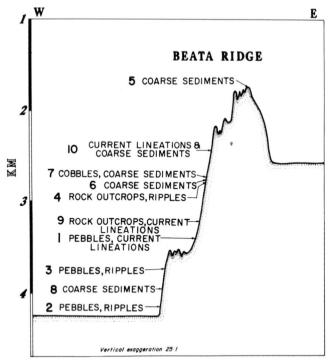

Fig. 6. Results of bottom photograph stations summarized and projected onto a topographic profile (from Fox *et al.*, 1970). Reprinted with the permission of Elsevier Publishing.

faunal assemblages of these Mid-Eocene chalks imply a neritic depositional environment.

Compass-oriented bottom photographs were obtained at 10 stations on the western escarpment in order to better understand the contemporary environment at the sediment–water interface (Fig. 6). The photographs reveal that below 2,300 m the escarpment is characterized by manganese encrusted rock outcrops, rocky pavement, current scour, and current-generated ripples. Above 2,300 m photographs generally show light-colored coarse-textured, current-swept bottom.

Seismic-refraction measurements (Ewing *et al.*, 1960; Edgar *et al.*, 1971) show that the eastern flank of the Beata Ridge is blanketed by 2 km of sediment and/or volcanics with a velocity range of 1.9 to 4.2 km/sec. Seismic-reflection profiles (Ewing *et al.*, 1967) indicate that the sediment blanket of the Caribbean is characterized by two prominent horizons (A″ and B″) which can be traced throughout most of the Caribbean. They are observed on the eastern flank of the Beata Ridge, but are absent on the western escarpment. The acoustically transparent sediments above the upper horizon (A″) and between it and the

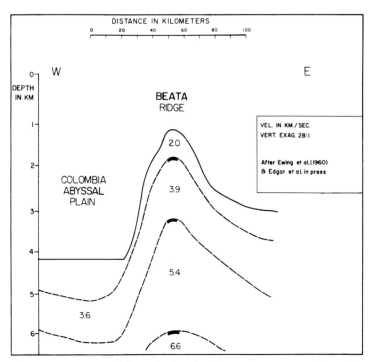

Fig. 7. Profile summarizing seismic-refraction results presented by Ewing *et al.*, 1960. Results over the ridge were obtained from a profile down the ridge axis and are shown by thick bars. These results were tied to seismic-refraction results to the east and west (not shown) (from Fox *et al.*, 1970). Reprinted with the permission of Elsevier Publishing.

lower horizon (B″) lack internal reflectors and are believed to be homogeneous. The thickness of the transparent sediment remains nearly constant across the Caribbean Basin, and this suggests that the sea floor remained fairly level throughout the depositional interval (Ewing *et al.*, 1967).

Near the southeast end where the Beata Ridge diminishes in height and begins to disappear, seismic-reflection records indicate that layer (A″) outcrops at the base of several escarpments. Piston cores obtained from the base of one escarpment recovered Early Eocene radiolarian ooze. Assuming that the upper of the two reflecting horizons (A″) is time-stratigraphic, Ewing *et al.* (1967) concluded that the sediment just above it is Early Eocene in age. The sediments at the upper reflecting horizon in the Caribbean Basin were sampled during the fourth and fifteenth legs of the Deep Sea Drilling Project (Bader *et al.*, 1970; Edgar *et al.*, 1971) and found to be chert of Early Eocene age. Although the lower reflector (B″) had not been sampled, it was generally assumed from its basin-wide distribution and nearly horizontal character to be a sedimentary horizon. During Leg 15 of the Deep Sea Drilling Project, it was sampled at five localities and shown to be Upper Cretaceous basalt and dolerite sills (Edgar *et al.*, 1971). A few seismic-refraction measurements suggest that 1–2 km of 3.9 km/sec material lie between the dolerite sills and the upper crustal layer. From the low compressional wave velocity, it seems likely that the 3.9 km/sec layer is made up of sediments and/or volcanics. The upper crustal layer under the Beata Ridge is 3–5 km thick with a compressional wave velocity of 5.4–5.9 km/sec. The lower crustal layer has a velocity of 6.6 km/sec and a thickness of 7 to 10 km.

An unreversed seismic-refraction profile, shot down the axis of the ridge crest in the area where our northern dredge hauls are located (Ewing *et al.*, 1960), was tied to two widely spaced profiles to the east and west of the ridge crest, leading to the interpretation of the ridge as an anticlinal flexure (Fig. 7).

Dredging, coring, and bottom camera results indicate that basalt and dolerite outcrop below 2,300 m on the western escarpment of the Beata Ridge (Fig. 8). The compressional wave velocities measured for representative basalt and dolerite samples in the laboratory are similar to velocities determined by seismic-refraction measurements of the upper crustal layer under the Beata Ridge, suggesting that basalt and dolerite underlie the ridge and outcrop on

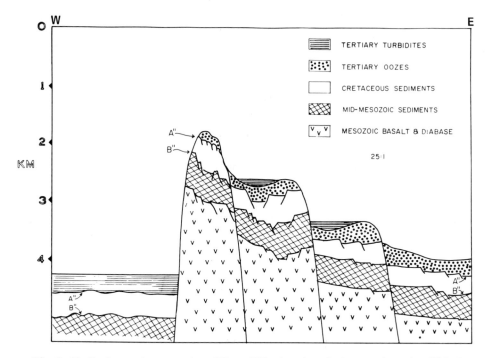

Fig. 8. Idealized crustal cross section of Beata Ridge based on dredge samples and published seismic-refraction results (Ewing *et al.*, 1970). Reprinted with the permission of Elsevier Publishing.

the western escarpment. The age and origin of the basaltic layer cannot be determined from the data so far gathered, but the results constrain possibilities. The lower acoustic layer (B″) is known to be present on the eastern flank of the Beata Ridge, and deep-sea drilling indicates that the sediments into which the basaltic sills were intruded are lowermost Mid-Upper Cretaceous in age. Several seismic-refraction stations indicate about 1 km of low-velocity material (3.9 km/sec) between the B″ reflector and crust. The evidence for a thick sedimentary sequence beneath the basalt sills suggests that the basaltic rocks which underlie the ridge and outcrop on the western escarpment could date from the Early Cretaceous or Jurassic. It is difficult to confirm without chemical analyses, but on the basis of mineralogy and textures, the basaltic rocks sampled from the Beata Ridge are similar to basalts sampled along the accreting axis of the Mid-Oceanic Ridge. If the inferred age and composition of these rocks is correct, then they could be representative of oceanic crust created during the early phases of Caribbean-basin genesis.

The time of formation of the Beata Ridge is not known. However, it apparently did not exist during Upper Cretaceous when sediments between the reflecting horizons A″ and B″ were deposited because the sediments are of uniform thickness and show no recognizable relationship to the present basement topography (Ewing et al., 1967). The shallow-water Mid-Eocene carbonates recovered in the cores from the ridge crest indicate that the Beata Ridge had formed by Mid-Eocene. Thus, the date of elevation of the ridge is pre-Mid-Eocene on the basis of the shallow-water carbonates, and post-Late Cretaceous on the basis of the acoustic stratigraphy. From Late Cretaceous to Early Eocene the margins of the Caribbean were tectonically active. The apparent synchroneity of the formation of the Beata Ridge with the tectonism of the margins of the Caribbean suggests that the interior of the Caribbean plate was locally deformed.

The inferred shallow-water depositional environment of the Mid-Eocene samples indicates that the crest of the Beata Ridge at 17° N was close to sea level in the Eocene; the Late Oligocene to present deep-sea pelagic carbonate samples indicate that by Late Oligocene the ridge crest had subsided approximately 1000 m.

III. AVES RIDGE

The Aves Ridge extends 450 km north from the continental margin of Venezuela (Fig. 9a) to the junction of the Greater and Lesser Antilles separating the Venezuela Basin (4800 m) from the Grenada Trough (3000 m). The ridge rises, with regional slopes of 5 to 10°, 2000 to 3000 m above the surrounding basins (Fig. 9b). Depths across the broad summit generally range from 500

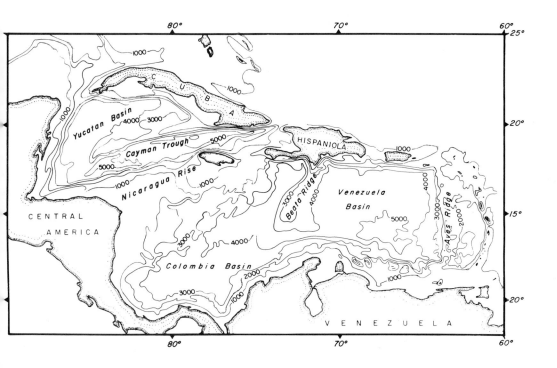

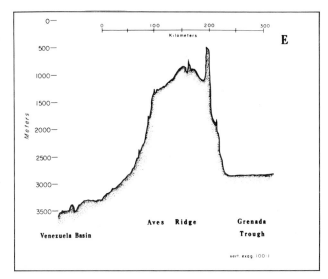

Fig. 9. (a) General contour map of the Caribbean basin showing location of Aves Ridge; (b) topographic profile of Aves Ridge (from Fox *et al.*, 1971). Reprinted with the permission of Elsevier Publishing.

to 1300 m. Locally steep-sided pedestals rise to within a few hundred meters of the surface, and one pedestal, Aves Island, breaks the surface. Dredging on pedestals at the northern end of the ridge has recovered glassy and brecciated basaltic rocks which suggest a volcanic origin for at least some pedestals (Marlowe, 1968; Chase, personal communication). Recently, andesites, basalts, dacites, and volcanic breccias have been collected from pedestals in the vicinity of Aves Island (Nagle, 1973).

Seismic-refraction results (Officer *et al.*, 1959; Edgar *et al.*, 1971) show that the Aves Ridge is capped by sediments and/or volcanics (2.0–4.8 km/sec) 5 km thick which overlie a crustal layer having a velocity of 6.2–6.3 km/sec (Fig. 10). Mantle velocities were not recorded beneath the ridge. A similar crustal layering was observed in the Grenada Trough to the east, and in the Venezuela Basin to the west. In the Venezuela Basin the 6.3 km/sec crustal layer is 2–3 km thick and is underlain by a 7.3–7.6 km/sec velocity layer 5 km thick. Normal mantle velocities (8.2 km/sec) were recorded beneath the 7.3–7.6 km/sec layer in deeper parts of the basin. Based on a comparison of the seismic-refraction profiles from the Venezuela Basin east across the Aves Ridge to the Grenada Trough, the regional topography of the ridge appears to be related to a rise or thickening of the 6.2 km/sec crustal layer.

Seismic-reflection results (Ewing *et al.*, 1968; Edgar *et al.*, 1971) show that the Aves Ridge is blanketed by 1 km of moderately stratified sediments. The characteristic acoustic reflecting horizons A″ and B″, which are recorded in the Venezuela and Colombia Basins, are not recorded over the Aves Ridge (Ewing *et al.*, 1967; Edgar *et al.*, 1971).

A dredging survey of escarpments and pedestals near the southern end of the Aves Ridge was conducted in 1969 aboard Duke University's *R/V Eastward* (Fox *et al.*, 1971).

Thirteen dredge stations, ranging in depth from 400 to 1400 m, sampled material from two pedestals and a steep scarp on the ridge flank (Figs. 11 and 12; Table II). These dredges contained limestones, marls, and cherts. The paleontologic age of the fossiliferous samples ranges from Middle Eocene to Recent. The Middle Eocene to Lower Miocene limestones were deposited in shallow, carbonate-shelf environments. The fossil assemblages in samples younger than Middle Miocene to Pleistocene were deposited in an open-ocean planktonic environment. Cobbles composed of labradorite basalt porphyry were recovered with the sedimentary rocks in one dredge (11312), which suggests that the foundations of the pedestals at the southern end of the ridge, like the ones to the north (Marlowe, 1968; Nagle, 1973), consist of volcanic rock.

Samples beneath the carbonate blanket and volcanic crust were obtained on the north wall of a precipitous, east–west trending escarpment that cuts into the southeastern flank of the ridge (Fig. 12; Table II). Two dredge locations within 10 km of each other recovered granitic rocks ranging in size from

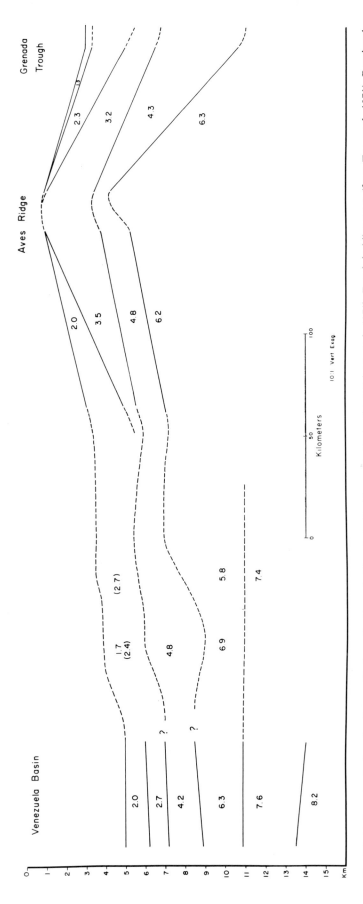

Fig. 10. East–west seismic-refraction profile from the Grenada Trough to the Venezuela Basin (from Edgar et al., 1971). Depth in kilometers (from Fox et al., 1971). Reprinted with the permission of Elsevier Publishing.

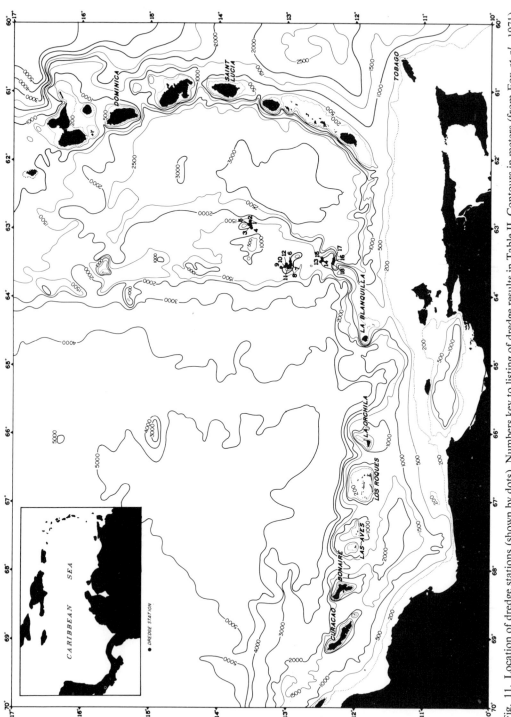

Fig. 11. Location of dredge stations (shown by dots). Numbers key to listing of dredge results in Table II. Contours in meters (from Fox *et al.*, 1971). Reprinted with the permission of Elsevier Publishing.

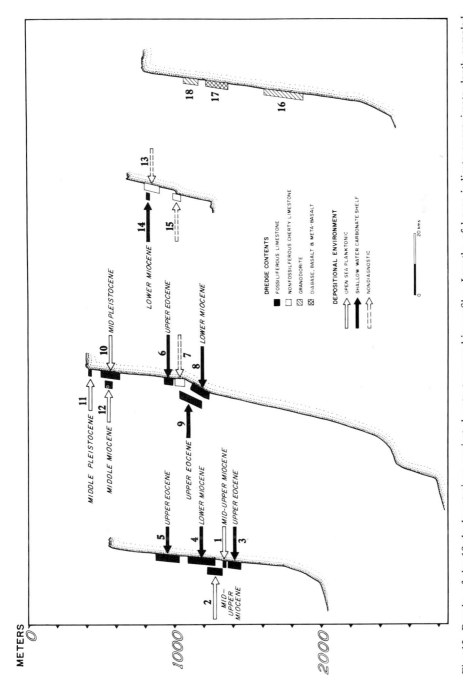

Fig. 12. Results of the 18 dredge stations projected onto topographic profiles. Lengths of boxes indicate approximate depth sampled. Number by each box is keyed to listing of samples in Table II (from Fox *et al.*, 1971). Reprinted with the permission of Elsevier Publishing.

TABLE II

Content, Position, and Depth of Dredges from the Aves Ridge

	Dredge No.	Position	Depth (m)	Rock types recovered
1	11290	13°33' N 62°56' W	1308	Consolidated, biomicrite—Middle to Late Miocene—open-sea planktonic
2	11291	13°33' N 62°57' W	1300–1200	Consolidated, biomicrite—Middle Miocene—open-sea planktonic
3	11292	13°33' N 62°58' W	1421–1336	Consolidated, biomicrite—Late Eocene—shallow carbonate shelf; nondiagnostic, cherty micrite
4	11294	13°31' N 62°59' W	1224–1077	Consolidated, biomicrite—Early Miocene—shallow carbonate shelf
5	11295	13°34' N 62°57' W	1012–845	Consolidated, biomicrite—Late Eocene—shallow carbonate shelf
6	11301	12°58' N 63°29' W	975–920	Consolidated, biomicrite—Late Eocene—shallow carbonate shelf; nondiagnostic, cherty micrite
7	11302	12°58' N 63°35' W	1049–980	Unfossiliferous, nondiagnostic, cherty micrite
8	11303	12°58' N 63°36' W	1215–1100	Consolidated, biomicrite—Early Miocene—shallow carbonate shelf
9	11306	13°07' N 63°34' W	1160–1012	Consolidated, biomicrite—Late Eocene—shallow carbonate shelf

10	11308	13°02′ N 63°32′ W	612–490	Semiconsolidated, biomicrite—Mid-Pleistocene—open-sea planktonic
11	11309	13°01′ N 63°34′ W	415	Semiconsolidated, biomicrite—Mid-Pleistocene—open-sea planktonic; nondiagnostic, cherty micrite
12	11312	13°02′ N 63°32′ W	565–528	Consolidated, biomicrite—Middle Miocene—open-sea planktonic; nondiagnostic, cherty limestone; partially weathered, labradorite basalt porphyry
13	11313	12°32′ N 63°30′ W	882–789	Nondiagnostic, cherty micrite
14	11314	12°30′ N 63°30′ W	826–817	Consolidated, biomicrite—Early Miocene—shallow carbonate shelf; nondiagnostic, cherty micrite
15	11315	12°30′ N 63°29′ N	1031–984	Nondiagnostic, micrite
16	11317	12°20′ N 63°30′ W	1839–1950	Granodiorite composed of plagioclase (An 20–30), K-feldspar, quartz, and amphibole with lesser amounts of opaque phase, chlorite, epidote, sericite, apatite, sphene, and zircon
17	11318	12°19′ N 63°27′ W	1345–1197	Subophitic diabase composed of labradorite, clinopyroxene, opaque phase, and alteration products (chlorite, epidote, and sericite); holocrystalline, labradorite–clinopyroxene basalt porphyry with groundmass alteration products (chlorite, epidote, sericite, quartz, and actinolite); metamorphosed basalt composed of chlorite, quartz, epidote, and actinolite with lesser amounts of plagioclase and pyroxene
18	11319	12°21′ N 63°32′ W	1151–1086	Granodiorite composed of plagioclase (An 20–30), K-feldspar, quartz, and amphibole with lesser amounts of opaque phase, chlorite, epidote, sericite, apatite, sphene, and zircon

600-kg angular boulders to coarse sand. The large samples are massive with no observable penetrative fabric and exhibit a random joint pattern. A few samples reveal slickensides. Some of the rocks are covered with up to 5 cm of manganese oxide. In hand specimens all samples are phaneritic and inequigranular.

Thin-section analyses of representative samples show that all samples are holocrystalline. The constituent minerals are medium-grained with grain size ranging from 0.5 mm crystals of quartz and K-feldspar to 2 mm crystals of plagioclase and amphibole. In several samples, microfractures offset crystal boundaries and plagioclase twin planes. Patches of granophyric intergrowth of K-feldspar and quartz characterize several samples. The rocks are all meso-cratic, with the color index averaging about 10% and never exceeding 20%. The major mineral constituents of the granodiorites are the following: euhedral to subhedral plagioclase, anhedral quartz, anhedral to subhedral potassium feldspar, and a light-green, pleochroic hornblende. All samples studied have been altered. If alteration is slight, fine-grained material dusts the potassium feldspar, sericite occurs in the sodic plagioclase, and epidote and chlorite partially alter the hornblende. When alteration is advanced, chlorite and epidote occur with sericite in the plagioclase phenocrysts, the hornblende is almost completely replaced by chlorite and epidote, and chlorite and epidote occur in patches and veins. Granitic rocks have also been sampled from the southern end of the Aves Ridge by other investigators (M. Bass, personal communication; G. Peter, personal communication).

A third haul (11318), weighing 40 kg and located about 40 km from the two granitic hauls, contained boulders and pebbles (Fig. 10; Table II). Thin sections of the 17 largest samples show that all are weathered and altered to varying degrees, and that dolerite predominates. Present in lesser amounts are porphyritic basalt and metamorphic rocks derived from basalts. Several of the dolerites and basalts exhibit the effects of microfracturing and breccia-tion. The medium-grained, subophitic dolerites are composed of the following: subhedral to euhedral labradorite, euhedral to subhedral clinopyroxene, opaque phase, and 5 to 10% alteration products (chlorite, epidote, and sericite). The holocrystalline basalt porphyries are composed of laths of labradorite and prisms of clinopyroxene set in a fine-grained groundmass of subhedral plagioclase, pyroxene, and opaque phase. These basalts have all experienced some degree of alteration resulting in the occurrence of patches and/or veins of chlorite, epidote, sericite, and, in some instances, quartz and actinolite. The metamorphosed, fine-grained, granular greenstones are composed pre-dominantly of chlorite, quartz, and epidote but contain, in addition, lesser amounts of Na-enriched plagioclase and pyroxene. Metavolcanics, similar to the suite of metabasalts and dolerites described above and from localities on the Aves Ridge close to our sampling location, have been recently reported (Nagle, 1973).

Four of the granitic samples and the least weathered dolerite were selected for duplicate potassium–argon age determinations. The ages in millions of years of the granitic samples are 78 and 89, 65 and 67, 57 and 58, and 18.3 and 19.3. The argon retention of potassium feldspars is reported to be good when fresh, but when altered, retention can decrease. It should be noted, however, that even fresh-appearing potassium feldspars can leak significant amounts of argon, and, therefore, these dates should be considered as minima. Of the four samples dated, that giving the youngest potassium–argon age showed the most advanced alteration (potassium feldspar dusted an opaque brown, and sericite and epidote occur in plagioclase laths) and, consequently, can be excluded on the basis of probable argon loss. The other three samples appear less altered in thin section. The ages in millions of years obtained for the dolerite sample are 57 and 60.

In an attempt to relate a rock type to defined velocity horizons in the Caribbean, the compressional wave velocity, under varying confining pressures, was measured for several granitic samples and the fresh dolerite sample. The calculated *in situ* lithostatic confining pressure of the 6.0 km/sec velocity layer under the Aves Ridge is 1 to 2 kbar. Under confining pressures of 1 to 2 kbar, the velocity of the four granitic specimens is 5.8 to 6.1 km/sec. The measurements for the fresh dolerite sample ranged from 6.5 to 6.6 km/sec at confining pressures of 1 to 2 kbar. These measured velocities are similar to those measured for continental granodiorites and dolerites (Anderson and Leibermann, 1968).

Seismic-refraction results in the Caribbean reveal a velocity structure which is anomalous when compared to the velocity structure in the main ocean basins. First, although oceanic-type velocities are found beneath the Beata Ridge and Nicaragua Plateau, the thickness of the crustal column above the mantle is $2\frac{1}{2}$ times greater than in normal ocean basins. Secondly, the majority of the Caribbean (southern Colombia Basin, Venezuela Basin, Aves Ridge, and Grenada Trough) is characterized by a Layer 2 velocity of 6.0 to 6.4 km/sec underlain by a thick (5–10 km) column of anomalous-velocity (7.0 to 7.4 km/sec) material (Fig. 1).

Based on the compilation of sound velocities in rocks by Anderson and Liebermann (1968) and the velocity measurements on oceanic rocks (Christensen and Shaw, 1970; Barrett and Aumento, 1971; Fox *et al.*, 1973), several rock types are known to have compressional wave velocities in the 6.0 to 6.4 km/sec range at 1 to 2 kbar: quartzite (5.8 to 6.2 km/sec); serpentine (4.8 to 7.0 km/sec); greenschist facies metabasalt (5.2 to 6.3 km/sec); greenschist and amphibolite facies metagabbro (5.6 to 6.5 km/sec); gneiss (5.9 to 6.4 km/sec); and granitic rock (5.8 to 6.4 km/sec). The sedimentary rocks can probably be excluded because it is unlikely that sedimentary detritus was deposited directly on 7.0 to 7.4 km/sec crust. More likely, the 6.0 to 6.4 km/sec layer is composed of igneous or metamorphic rock genetically related to the origin of the Caribbean Basin. The 6.0 to 6.4 km/sec layer of the Caribbean represents a large area and tremendous volume, and consequently the rock type making up this layer must occur in abundance in the earth's crust. Granodiorite is the most abundant rock of the continental plates (Poldervaart, 1955), and based on dredging results to date, basalt, metabasalt, gabbro, and metagabbro are the abundant rock of the oceanic crust. As already pointed out, granitic rocks have a compressional wave-velocity range, at 1 to 2 kbar, of 5.8 to 6.4 km/sec, whereas basalt (5.0 to 6.0 km/sec) and gabbro (6.5 to 7.0 km/sec) have velocities which bracket the 6.0 to 6.4 km/sec layer of the Caribbean. It is interesting to note, however, that under a confining pressure of 1 to 2 kbar the velocity range of greenschist metabasalt (characterized by epidote, Na-enriched plagioclase, chlorite, and actinolite) and metagabbro (greenschist and amphibolite facies) is compatible with the measured velocities (6.0 to 6.4 km/sec) for the upper crustal layer of the Caribbean.

The granitic rocks sampled from an escarpment in the Aves Ridge may represent the northern edge of the South American continental platform, but the similarity of the measured velocity characteristics of the dredged granitic rocks with the 6.0 to 6.3 km/sec crustal layer recorded on the Aves Ridge may indicate that the Aves Ridge is in part underlain by granitic rock. If granitic rocks characterize a portion of the crust of the Aves Ridge, then the occurrence of granite behind the Lesser Antilles island arc suggests these granitic rocks have been derived from oceanic material (crust and sediments) which was thrust beneath the Lesser Antilles. If the three K–Ar age determinations (78 and 89, 65 and 67, and 57 and 58) are correct, these dates indicate a range of granitic plutonism for the Aves Ridge region which commences in the Late Cretaceous and continues to the Late Paleocene. The diorite, like two of the granodiorite samples, has a Late Paleocene age. The granitic rocks from the Aves Ridge are of the same age as Late Cretaceous granitic plutons which characterize the margins of the Caribbean. MacGillavry (1970) has summarized the reported occurrences of granodiorite in the Caribbean and reports that granitic rock occurs in the following regions: Cuba, Jamaica, Hispaniola, Puerto Rico, St. Croix, and Virgin Gorda in the Greater Antilles; Aruba, Curaças, Los Roques, Orchilla, La Blanquilla, and Margarita in the Dutch Antilles; and Guatemala, El Salvador, Honduras, Costa Rica, and Panama in Central America. These data along with the occurrence of granodiorite under the Aves Ridge means that the Caribbean perimeter is surrounded by granodiorites of Late Cretaceous age. Hamilton (1969), in a discussion of the massive Mid-Cretaceous phase of granodiorite batholithic intrusions of the western United States, suggested that the genesis of the granodiorite could be a function of an accelerated rate of underthrusting along the west coast of North America. Recently, Larson and Pitman (1972), in a study of marine

Mesozoic magnetic anomalies, have inferred that a rapid worldwide pulse of spreading occurred from about 110 to 85 m.y. ago. Accepting the corollary that rapid spreading results in rapid subduction, Larson and Pitman suggest that the large-scale plutonism in eastern Asia, western Antarctica, New Zealand, the southern Andes, and western North America during the Late Cretaceous is directly related to the rapid phase of spreading documented for the 110 to 85 m.y. B. P. time period. As previously pointed out, the Caribbean is encircled by Late Cretaceous granodiorite which suggests that the margins of the Caribbean were all characterized by rapid underthrusting in the Late Cretaceous.

A generalized east–west crustal cross section placing the Aves Ridge in the present-day tectonic environment is shown in Fig. 13. To the east, based on the first-motion studies of Molnar and Sykes (1969) and the seismic-reflection results of Chase and Bunce (1969), oceanic crust is being underthrust, and oceanic sediment compressed, against the volcanic Lesser Antilles island arc. To the west lies the stable Venezuela Basin (Ewing *et al.*, 1968). The velocity structure of the crust beneath the Venezuela Basin and Grenada Trough is the same as the Aves Ridge, and consequently it is tempting to extrapolate thin (2 to 3 km) granodiorite crust beneath these deep-water (>4000 m) regions, but until rock of this composition is sampled elsewhere in the Caribbean Basin, a basin-wide extrapolation of granitic crust is not justified.

Carbonate samples dredged from guyot-like features on the crest of the Aves Ridge indicate that during Eocene to Early Miocene time, a shallow-water carbonate-shelf environment prevailed. The overlying deep-water pelagic samples of Middle Miocene to Recent age indicate that 400 to 1400 m of subsidence has occurred. Carbonate samples obtained from the crest of the

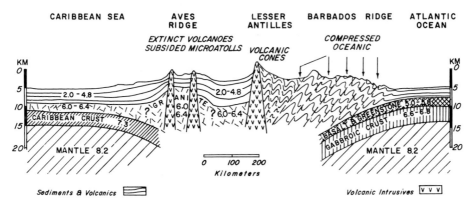

Fig. 13. An idealized crustal cross section based on our sampling results coupled with published seismic-refraction (Officer *et al.*, 1959; Edgar *et al.*, 1971) and seismic-reflection (Chase and Bunce, 1969) results. Velocity is in km/sec (from Fox *et al.*, 1971). Reprinted with the permission of Elsevier Publishing.

Beata Ridge indicate that during early Tertiary time the Beata Ridge was a shallow, carbonate platform and that by Late Oligocene it had subsided. The near synchroneity of these subsidence histories suggests that since Mid-Miocene time the Caribbean Basin as a whole may have subsided 400 to 1000 m and that within the basin vertical tectonics has played an important role.

IV. LESSER ANTILLES

The Lesser Antilles define the eastern margin of the Caribbean. This island arc complex extends almost 900 km from the continental slope of South America to the Anagada Trough at 18.5° N, a 2000-m-deep NE–SW trending feature which separates the northern end of the Lesser Antilles from the Greater Antilles. The islands that comprise the emergent Lesser Antilles can be divided, on the basis of rock type and age, into the volcanic Caribbees and the limestone Caribbees (Martin-Kaye, 1969). North of the volcanic island of Dominica, the Lesser Antilles bifurcate into two curvilinear arcs. The inner arc (consisting of Basse Terre, Monserrat, Nevis, St. Kitts, and Saba) comprises the northern limb of the volcanic Caribbees and is characterized by Late Miocene to Early Pliocene volcanics of the basalt–andesite–dacite series. The southern limb of the volcanic Caribbees (defined by Dominica, Martinique, St. Lucia, St. Vincent, Grenada) is characterized by some pre-Miocene volcanic centers as well as Late Miocene to Recent volcanism. The limestone Caribbees (consisting of Marie Galanti, La Désirade, Grande Terre, Antigua, Barbuda, St. Bartholomew, St. Martins, Anguilla, Dog, and Sombrero) lie east of the northern branch of the volcanic Caribbees and are characterized by pre-Miocene volcanics consisting of andesite, dacite, tuff, and agglomerate, which in turn have been intruded by diorite and quartz diorite. By Early Miocene the volcanic and intrusive centers were beveled and capped with shallow water limestones ranging in age from Miocene to Recent.

Seismic-refraction measurements indicate that oceanic crust dips steeply below the eastern margin of the Lesser Antilles (Officer et al., 1959). These measurements also show a thick pile (10 km) of low-velocity material (3.0 km/sec) overlying the westward-dipping oceanic crust. The Lesser Antilles is a seismically active island arc (Sykes and Ewing, 1965). Most hypocenters are confined to a seismic zone about 50 km wide that dips 60° west beneath the Lesser Antilles. Earthquake first-motion studies (Molnar and Sykes, 1969) reveal that oceanic lithosphere is underthrusting the Caribbean in a westerly direction along the Lesser Antilles. A thick accumulation of structureless sediment is observed on seismic-reflection profiles along the eastern margin of the Caribbean. The lack of discernible stratification has been attributed to complex deformation of an otherwise stratified sediment pile as the underlying oceanic crust was thrust beneath the Lesser Antilles (Chase and Bunce, 1969).

Thirteen dredge stations on the steep escarpments located along the seaward margin of the limestone Caribbees (Figs. 14 and 15; Table III) recovered material which provides important information on the evolution of the tectonically active eastern margin of the Caribbean plate. While the interior of the Caribbean basin, apart from documented vertical adjustments of the Beata and Aves Ridges, has been essentially stable since Early Tertiary, the eastern margin has been the site of extensive activity.

Four dredges (391–395) were located along an east–west escarpment located northeast of La Désirade island. Two distinct suites of rocks were

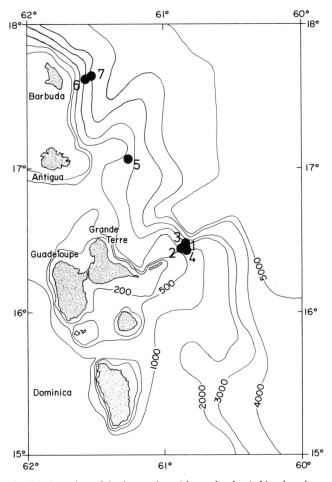

Fig. 14. Location of dredge stations (shown by dots). Numbers key to listing of dredge results in Table III.

TABLE III

Content, Position, and Depth of Dredges from the Lesser Antilles

	Dredge No.	Position	Depth (m)	Rock types recovered
1	E-391	16°26′ N 60°51′ W	1600–1525	Cataclastic metabasalt composed of Na-enriched plagioclase, epidote, quartz, chlorite, and minor actinolite; veined with quartz
2	E-392	16°26′ N 60°52′ W	900–800	Metabasalt composed of Na-enriched plagioclase, epidote, quartz, and chlorite
3	E-393	16°26′ N 60°51′ W	1700–1575	Cataclastic metagabbro composed of chlorite, pyroxene mantled by actinolite, and actinolite; metabasalt composed of Na-enriched plagioclase, chlorite, epidote, and minor actinolite; cataclastic, metamorphosed, porphyritic basalt composed of relic pyroxene phenocrysts set in a matrix of Na-enriched plagioclase, epidote, amphibole, chlorite, and quartz
4	E-395	16°26′ N 60°50′ W	800–725	Cataclastic metabasalt composed of Na-enriched plagioclase, chlorite, quartz, and epidote; cataclastic, porphyritic metabasalt with Na-enriched plagioclase set in a very fine-grained matrix; quartz diorite composed of plagioclase, quartz, epidote, and chlorite; granodiorite composed of plagioclase, K-feldspar, and quartz with minor epidote and chlorite
5	E-396	17°04′ N 61°16′ W	1450–1200	Finely laminated tuff composed of altered glass, plagioclase, pyroxene, and lesser biogenous material—Upper Cretaceous
6	E-397	17°37′ N 61°34′ W	2100–1950	Finely laminated tuff composed of altered glass, plagioclase, pyroxene quartz, and lesser biogenous material—Upper Cretaceous
7	E-399	17°38′ N 61°32′ W	3300–3100	Fine- to medium-grained, angular, feldspathic graywacke composed of plagioclase, pyroxene quartz and altered glass in a matrix of clay; fine-grained lithic graywacke composed of volcanic rock fragments (basalts and andesite), plagioclase, chlorite,

				and hornblende set in a carbonate matrix (sparry and biogenous); finely laminated tuff composed of altered glass, plagioclase, pyroxene, and lesser biogenous material—Upper Cretaceous; weathered plagioclase–pyroxene basalt porphyry
8	E-400	18°04' N 62°21' W	2500–2300	Nondiagnostic, dense, micritic limestone
9	E-402	18°29' N 62°44' W	2400–2300	Fine- to coarse-grained lithic graywacke composed of subrounded andesite fragments and angular plagioclase set in a carbonate (sparry and biogenous) matrix; fine- to medium-grained arkosic arenite composed of feldspar, rock fragments, and quartz—carbonate matrix; finely laminated tuff composed of altered glass, plagioclase, pyroxene, and lesser biogenous material—Upper Cretaceous
10	E-403	18°36' N 63°05' W	2200–2000	Fine- to medium-grained lithic graywacke composed of rounded andesite fragments and angular plagioclase and mafic grains set in a matrix of carbonate and clay; fine- to medium-grained arkosic graywacke composed of rounded andesite fragments and angular plagioclase, pyroxene, and hornblende grains set in a carbonate matrix; fine- to medium-grained, bioturbated biopelmicrite with echinoderm fragments, pelecypod fragments, and ostracod valves—skeletal, fragments abraded—shallow marine depositional environment—Paleocene–Eocene
11	E-404	18°40' N 63°06' W	4550–4000	Fine- to medium-grained arkosic graywacke composed of rounded volcanic fragments and angular plagioclase, hornblende, and augite set in a carbonate matrix; very fine-grained laminated tuff with biogenous material—Upper Cretaceous
12	E-406	18°47' N 63°46' W	5350–5200	Very fine-grained laminated tuff; medium-grained arkosic arenite composed of rounded volcanic rock fragments and angular plagioclase, hornblende, clinopyroxene, and volcanic glass set in a carbonate matrix; medium-grained lithic arenite composed of rounded volcanic fragments and angular plagioclase, hornblende, and pyroxene set in a matrix of alteration products; serpentinized peridotite composed of serpentine and scattered, partially altered orthopyroxene
13	E-408	18°47' N 63°50' W	1250–1400	Fine-grained biomicrite—nondiagnostic

recovered. The most abundant suite, characterized by primary basaltic affinities, has experienced pervasive greenschist facies metamorphism and cataclasis. Relict igneous textures range from shallow intrusive and/or extrusive basalt textures (variolitic, porphyritic) to plutonic gabbroic textures (intergranular, subophitic). The samples are characterized by abundant Na-enriched plagioclase, chlorite, epidote, quartz, and actinolite (often rimming relict primary clinopyroxene). They are often cut by veins of quartz and epidote. The mineralogy and relict textures are very similar to greenschist metabasalts and metagabbros recovered from fracture-zone escarpments of the Mid-Oceanic Ridge system. Laboratory measurements of oceanic rocks (Christensen and Shaw, 1970; Barrett and Aumento, 1971; Fox et al., 1973) show that basaltic rocks and their metamorphic equivalents have compressional wave velocities compatible with seismic-refraction determinations of oceanic basement (4.6–5.4 km/sec), and that gabbroic rocks have compressional wave velocities compatible with the oceanic layer (6.7–6.9 km/sec).

The other samples are representative of intermediate plutonic rocks and are composed predominantly of plagioclase, K-feldspar, and quartz. Minor amounts of epidote, chlorite, and opaque phase are observed. Lath-like chlorite is suggestive of pseudomorphos after hornblende. Depending on the quartz content of the samples, the rocks are either granodiorite or diorite.

A similar collection of rocks has been dredged from the Désirade escarpment (J. G. Schilling, personal communication) by the *R. V. Trident* of the University of Rhode Island.

On the island of La Désirade, which lies approximately 60 km southeast of the dredge localities, rocks, similar in mineralogy to the types recovered from the Désirade escarpment, comprise the basement complex of the island (Fink, 1970). Some of the basal metamorphic rocks (classified by Fink as a greenschist facies spilite–keratophyre suite) of the island have extrusive and intrusive basaltic textures and are often pillowed and interbedded with cherts. The greenschist facies rocks have been dated (K–Ar) and indicate an Upper Jurassic age of intrusion. These rocks are intruded by hypabyssal granodiorites. This metamorphic suite of rocks on La Désirade has been interpreted as submarine extrusions which created a volcanic ridge on the sea floor below the present-day position of the Lesser Antilles (Fink, 1970).

An alternative interpretation which seems to better account for the data is that the greenschist metabasalts and metagabbros which outcrop on the La Désirade escarpment and the pillowed metabasalts and interbedded cherts which outcrop on La Désirade are pieces of oceanic crust that have been tectonically emplaced along the subducting margin of the Lesser Antilles. The following *general* model is offered to explain the timing and emplacement of these rocks. The metabasalts and metagabbros of the La Désirade region were accreted along the axis of the Mid-Atlantic Ridge during the early phase

(Jurassic) of sea-floor accretion in the North Atlantic region as North America moved away from Africa and South America. The parcel of Jurassic sea floor, now exposed in the La Désirade region, moved passively across the early Atlantic basin until it reached the area defined today by the Lesser Antilles. The crust in the vicinity of La Désirade became involved in the initial phase of subduction of oceanic lithosphere along an incipient curvilinear line which later formed a plate boundary. The compressive stresses localized along the nascent curvilinear subduction zone resulted in the piling of slices of oceanic crust on the edge of the newly formed eastern margin of the Caribbean plate. Tectonic imbrication of the oceanic crust ceased when the oceanic crust was decoupled along the newly formed plate margin and underthrusting was initiated. When approximately 200 km of underthrusting had occurred, the calc-alkaline volcanism associated with island arcs was initiated, intruding through and building upon the imbricated pile of oceanic crust.

Nine dredges were recovered from steep escarpments located east of Antigua, Barbuda, and Anguilla and north of Sombrero (Figs. 14 and 15, Table III). Seven of the nine dredges contain sedimentary rocks rich in volcanic products. At six of these seven localities, finely laminated tuffs containing Upper Cretaceous radiolaria were recovered. These tuffs are composed of volcanic glass and lesser amounts of plagioclase, hornblende, and quartz. At five of these seven localities, sandstones, ranging from arkosic and lithic arenites to lithic and arkosic graywackes, were recovered. The sand-sized fraction of these samples is composed of volcanic rock fragments (basalts and andesites), plagioclase, hornblende, pyroxene, quartz, and volcanic glass. The occurrence of pelagic radiolaria in the finely laminated tuffs and the delicate structure of the tuffs suggests deposition in deep water. The arenites and graywackes exhibit a wide range of particle size, diverse primary structures (graded bedding, crossbedding), a varied mineralogy, and ubiquitous matrix material (10–25%). These sandstones are texturally similar to deep-sea sands found on modern abyssal plains.

The tuffs and sandstones were deposited in abyssal depths and were later uplifted when the escarpments on which they now outcrop were formed. The Upper Cretaceous age of radiolaria contained in the tuffs indicate that the latter were derived from island-arc volcanic centers active at this time. The age of the sandstones is unknown, but their constituent products (andesite, hornblende, quartz) suggest an island-arc source area. These results and observations indicate that volcanic centers were active along the Lesser Antilles in the Upper Cretaceous. The volcanoes were well developed with high relief. Erosion produced large accumulations of volcanic detritus which mantled the steep submarine volcanic slopes. Periodic slumping produced turbidity currents which created turbidite deposits in the abyssal depths (most likely the axis of a trench). Concurrent with and following deposition of the

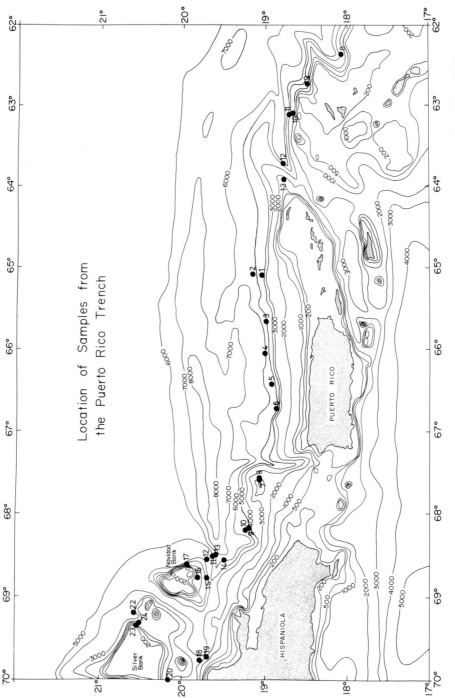

Fig. 15. Location of dredge stations (shown by dots). Numbers key to listing of dredge results in Tables III and IV.

tuffs and sandstones, the oceanic lithosphere, on which the sediments were deposited, was moving toward the subduction zone of the Lesser Antilles. Not long after deposition, the oceanic lithosphere with the overlying sediments was thrust beneath the arc. The deep-water tuffs and sandstones sampled on the Lesser Antilles escarpments were scraped off and plastered against the imbricated oceanic crust and volcanic buttress of the Lesser Antilles island arc.

Two dredges sampled nondiagnostic micritic limestone. Neither the fossil fragments nor the textures are diagnostic of age or depositional environment. The structurally chaotic nature of the compressed prism of sedimentary rocks along the Lesser Antilles front constrains these samples to neither a deep nor shallow-water origin.

In one dredge, however, biopelmicritic limestone of lowest Eocene was recovered mixed with abundant lithic and arkosic graywackes. The limestone contained echinoderm fragments, pelecypod fragments, and ostracod valves. Many of these fossil parts were abraded. Based on the fauna, the inferred depositional environment is shallow marine, indicating that shallow-water terrain, conducive to carbonate accumulation, characterized at least a portion of the northern Antilles in the Eocene.

Serpentinized peridotite mixed with tuffs and arenites was recovered in one deep dredge. The sample is composed of scattered, partially altered relict orthopyroxene and serpentine. The association of this ultramafic sample, representative of deep oceanic crust or upper mantle, with tuffs and sandstones attests to the structural complexity of the Lesser Antilles island arc front.

V. PUERTO RICO TRENCH

Paralleling the eastern portion of the Greater Antilles (Hispaniola, Puerto Rico, Virgin Islands) is the east–west trending Puerto Rico Trench which delineates the northeastern boundary of the Caribbean crust. The maximum water depth of the trench exceeds 8000 m. On the north the trench is bounded by an east–west trending outer ridge which rises approximately 5000 m above the trench. The deepest part of the trench lies at the base of the north wall at a depth of 8250 m. Towards the south the trench shallows gradually to approximately 6000 m; from 6000 m to approximately 3500, the south wall is cut by a steep escarpment. From escarpment top to the north shore of Puerto Rico the ocean floor is characterized by a broad, seaward-dipping platform.

Seismic-refraction profiles (Officer *et al.*, 1959; Talwani *et al.*, 1959; Bunce and Fahlquist, 1962) oriented perpendicular to the trench axis indicate that the ridge bounding the trench on the north is characterized by a normal, three-layer oceanic crustal velocity structure: a thin (<1000 m) layer of low-velocity material (2.1 km/sec) overlies a 2-km-thick layer (oceanic basement) with a velocity of 5.5 km/sec which in turn overlies a 5-km-thick layer (oceanic layer) with a velocity of 6.9 km/sec. The seismic-refraction results suggest that oceanic basement and overlying sediments outcrop on the north wall. Fourteen dredges have sampled material from localities along the north wall. A diverse suite of igneous and sedimentary rocks has been recovered: tuffs, cherts, and siliceous mudstones ranging from Upper Cretaceous to Pliocene; pre-Pliocene volcanic wacke; deep-water Cretaceous limestone, tholeiitic basalt and serpentinite (Bowin et al., 1966; Nalwalk, 1969). The velocity structure of the trench axis and the south wall of the trench differs from that of the northern side in that a four-layer structure is deduced and the total thickness is much greater. The axis of the trench and the south wall are characterized by a 1-km-thick layer of 2.1 km/sec material overlying a 2–3 km thick layer with a velocity range of 3.8 to 4.6 km/sec which overlies a 5-km-thick layer of 5.5 to 6.6 km/sec material which in turn overlies a 2 to 5 km thick layer of 6.9 km/sec material. The seismic-refraction results suggest that the material comprising the upper two velocity layers may outcrop along the escarpment defined by the 6000 and 3500 m contours.

The axis of the Puerto Rico Trench and the south wall are seismically active (Sykes and Ewing, 1965). Unlike the trenches of the Pacific, however, the hypocenters do not delineate a well defined, landward dipping plane of earthquakes, and first-motion studies reveal that at present the motion is not perpendicular to the trench axis. Instead, motion is parallel to the axis along nearly horizontal fault planes (Molnar and Sykes, 1969). The earthquake foci are shallow (<250 km), with most events occurring between 20–80 km.

Lying directly south of the Puerto Rico Trench are Puerto Rico and the Virgin Islands. The oldest rocks found on Puerto Rico are a structurally complex assemblage of serpentinite, basaltic amphibolite, and gneiss, which, on the basis of chemistry, have been interpreted as slices of oceanic crust (Mattson, 1960). The exact age of the assemblage is not known, but it is thought to be at least Lower Cretaceous in age because the rocks are overlain by Upper Cretaceous cherts and limestones. The Lower Cretaceous through Paleocene is characterized by predominantly calc-alkaline volcanics and volcani-clastic rocks. Upper Cretaceous plutons of granodiorite and quartz monzonite intrude Cretaceous calc-alkaline volcanics. Oligocene and younger limestones and clastic rocks flank the older sequences (Mattson, 1966).

The oldest rocks of the northern Virgin Islands are at least Lower Cretaceous in age and consist of a structurally complex assemblage of tuffs, spilites, radiolarian tuffs, and keratophyres. The entire sequence appears to have been extruded on the sea floor (Donnelly, 1966). Overlying these rocks are tuffs, volcanic breccia, radiolarian tuffaceous limestone, and volcanic wackes, all of possible Albian age. Intrusives (diorite plutons and quartz–andesine porphyry) of Late Cretaceous age intrude the older sequences. Early Tertiary sequences are not well preserved on the north Virgin Islands. The cessation of calc-alkaline volcanic activity and plutonism by Eocene along the eastern islands of the Greater Antilles suggests that large-scale subduction of oceanic lithosphere along the NE margin of the Caribbean ended at approximately this time.

The geology of the Lesser Antilles Islands, seismicity, and marine geophysical survey data suggest that this region has been and is today an active plate margin. The dredged material recovered along the seaward margin certainly indicates that the Lesser Antilles is a structurally mobile margin. The geological and geophysical data from the northeastern boundary of the Caribbean plate is not as easily resolved. The geology of the islands suggests relative quiescence since the Early Eocene, but the Puerto Rico Trench is seismically active and structurally complex. Dredging on the north wall recovered an assemblage typical of deep-ocean sedimentary environments and oceanic crust, but the south wall of the Puerto Rico Trench, the probable boundary between the Puerto Rico Trench and the eastern islands of the Greater Antilles, had not been sampled prior to the 1970 *R. V. Eastward* investigation. It was hoped that rocks, which surely must be exposed on the precipitous escarpments of the south wall and western end of the trench, would provide a better understanding of the geological history of this Caribbean region.

Twenty-four dredges were recovered from the Puerto Rico Trench region (Fig. 15; Table IV). The dredges represent two topographic provinces: the south wall of the trench and the western end of the trench around Navidad and Silver Banks (Schneidermann *et al.*, 1973). The dredges were recovered from the south wall. Six of these were located north of the Virgin Islands and Puerto Rico along a steep, topographically continuous escarpment defined by the 6000 to 3500 m contours. Two of the remaining four dredges were recovered from a steep-sided mountain located on the south wall north of Mona Passage which separates Puerto Rico from Hispaniola. The last two dredges were located on the south wall along a steep escarpment which lies directly east of Cape Samana, Hispaniola.

In all six dredges from the south wall of the trench north of Puerto Rico and the Virgin Islands, biomicrite was recovered containing ostracods, sponge spicules, benthonic foraminifera, pelecypod valves, echinoid spines, and gastropod casts. Fossils found in these samples indicate that they range in age from Upper Cretaceous to Lower Miocene. The inferred depositional environment of these carbonate samples, based on fossil assemblages and sediment textures, is a shallow marine carbonate platform. These shallow-water carbonate samples were recovered along the strike of a submarine escarpment over a distance of approximately 200 km. This distribution indicates

TABLE IV

Content, Position, and Depth of Dredges from the Puerto Rico Trench

	Dredge No.	Position	Depth (m)	Rock types recovered
1	E-409	19°03′ N 65°06′ W	4200–4000	Fine- to medium-grained biomicrite with ostracod valves, sponge spicules, gastropod casts, and benthonic foraminifera—shallow marine depositional environment—Middle Eocene
2	E-410	19°10′ N 65°05′ W	5500–5350	Fine- to medium-grained carbonate sandstone composed of biogenous and nonbiogenous carbonate with lesser plagioclase and volcanic rock fragments; fine- to medium-grained biomicrite with spicules, benthonic foraminifera, and gastropod casts—shallow marine depositional environment—Upper Cretaceous
3	E-412	19°00′ N 65°40′ W	5000–4700	Metamorphosed basalt porphyry with Na-enriched plagioclase phenocrysts set in a fine-grained matrix of altered plagioclase, epidote, and chlorite; basalt porphyry with plagioclase phenocrysts set in a fine-grained matrix of acicular plagioclase altered glass and patches of chlorite; micrite with minor amounts of plagioclase and finely laminated tuff with minor amounts of plagioclase; fine- to medium-grained biosparite with minor silicate material in matrix, biomicrite with abundant nummulites—shallow marine depositional environment—Eocene; biomicrite with abundant benthonic foraminifera—shallow marine shelf—Lower Miocene
4	E-413	19°02′ N 66°03′ W	6400–6200	Serpentinite; biopelmicrite containing ostracods, sponge spicules, pelecypod valves, echinoid spines, and benthonic foraminifera—shallow marine depositional environment—Lower Miocene; biomicrite with benthonic foraminifera—shallow marine depositional environment—Upper Cretaceous
5	E-414	18°56′ N 66°26′ W	5800–5500	Biomicrite with spicules and benthonic foraminifera—shallow marine environment—Early Miocene
6	E-415	18°53′ N 66°42′ W	5500–5350	Biomicrite containing ostracod fragments, spicules, and benthonic foraminifera—shallow marine environment—Early Miocene; recrystallized carbonate with scattered blebs of quartz

7	E-417	19°05' N 67°35' W	1300–1050	Black marble with scattered fragments of plagioclase, quartz, and minor heavy minerals
8	E-418	19°05' N 67°34' W	1150–950	Micrite incorporating angular fragments of black marble (marble similar to marble sampled in dredge haul E-417)
9	E-419	19°12' N 68°10' W	5100–4900	Recrystallized carbonate with scattered fragments of quartz and deformed bands of fibrous, white mica
10	E-420	19°15' N 68°11' W	6500–6300	Recrystallized carbonate with lineations of white mica; slightly recrystallized micrite
11	E-421	19°36' N 68°30' W	5700–5500	Biopelmicrite to biopelsparite with coral debris, skeletal fragments, algae, and benthonic foraminifera—shallow lagoon or a shelf environment of deposition—samples range from Early to Late Eocene
12	E-422	19°43' N 68°32' W	4300–4100	Bioturbated pelmicrite with no fossils—intertidal depositional environment—non-diagnostic
13	E-423	19°38' N 68°31' W	6400–6100	Pelmicrite with bird's-eye texture and scattered ostracod carapaces and miliolids—shallow lagoonal depositional environment; nondiagnostic
14	E-424	19°30' N 68°36' W	6600–6300	Fine-grained anhedral dolomite; pelmicrite with bird's-eye structure and scattered ostracod carapaces and miliolids—supratidal depositional environment—nondiagnostic
15	E-426	19°40' N 68°45' W	3500–3200	Pelmicrite with scattered ostracod carapaces, mud-cracked layers, and bird's-eye structures—intratidal to supratidal depositional environment—nondiagnostic; biopelmicrite with ostracod carapaces, benthonic foraminifera, and miliolids—shallow marine shelf environment—Upper Cretaceous
16	E-427	19°42' N 68°45' W	1150–1000	Pelmicrite with algal mat structures—intratidal depositional environment—nondiagnostic

TABLE IV (*continued*)

	Dredge No.	Position	Depth (m)	Rock types recovered
17	E-428	19°57′ N 68°38′ W	3950–3700	Stromatolitic biopelmicrite with intraclasts, miliolids, ostracod carapaces, and benthonic foraminifera—very shallow marine depositional environment—Upper Cretaceous; finely laminated algal mat with interbeds of pelmicrite—intertidal depositional environment
18	E-429	19°47′ N 69°45′ W	3800–3550	Biomicrite with coral and algal debris and benthonic foraminifera—shallow lagoon depositional environment—Early Miocene
19	E-431	19°53′ N 69°42′ W	4400–4150	Pelmicrite with scattered dolomite crystals and rare miliolids and ostracod carapaces —shallow protected lagoon depositional environment—nondiagnostic
20	E-432	20°16′ N 70°00′ W	3000–2800	Biopelmicrite with coral debris, miliolids, algal debris, and ostracod carapaces—shallow lagoon depositional environment—nondiagnostic
21	E-433	20°08′ N 71°14′ W	3400–3100	Biomicrite with coral debris and benthonic foraminifera—shallow marine environment —Late Eocene
22	E-638	20°32′ N 69°12.8′ W	3400–3200	Bioturbated pelmicrite—nondiagnostic
23	E-639	20°33.5′ N 69°25.0′ W	1850–1750	Pelmicrite—nondiagnostic
24	E-640	20°31.8′ N 69°24.3′ W	2800–2775	Bioturbated pelmicrite—nondiagnostic

that shallow water limestones, ranging in age from Upper Cretaceous to Lower Miocene, outcrop continuously along the escarpment. The escarpment from which the samples were recovered ends at approximately 3500 m, and from 3500 m to the shoreline of the Greater Antilles the submarine slopes are gentle. Therefore, although outcropping material on the escarpment below 3500 m could move downslope short distances along the steep escarpment, it is unlikely that there has been a massive downslope creep of homogeneous material across the broad, gently dipping submarine margin (0–3500 m) of Puerto Rico. If there was movement along this slope, it seems unlikely that transport would produce a similar outcrop distribution along a distance of 200 km. Consequently, these limestones show that the south wall of the Puerto Rico Trench above approximately 6000 m is underlain by shallow-water limestones of Upper Cretaceous to Lower Miocene age. These samples indicate a subsidence of at least 3500 m since the Lower Miocene.

Seismic-refraction profiles (Talwani *et al.*, 1959; Bunce *et al.*, 1962) suggest that material with a velocity range of 2.8 to 4.6 km/sec may outcrop along the escarpment from which the limestones were recovered. These limestones are dense, fine-grained carbonates. Compressional-wave-velocity measurements on limestones of this type have, at low confining pressure (0.25 to 0.75 Kb), measured velocities ranging from 3.6 to 6.0 km/sec (Anderson and Leibermann, 1968).

Three of the six dredges from the south wall contained lesser amounts of the following material: carbonate sandstone composed of biogenous and nonbiogenic carbonate with lesser plagioclase and volcanic rock fragments (dredge 410); greenschist facies metamorphic porphyritic basalt, finely laminated tuff, and carbonate micrite containing plagioclase (dredge 412); and serpentinite (dredge 413). Without ages or a better understanding of the distribution of these rock types, it is difficult to unambiguously fit them into a geologic framework of the south wall. The serpentinites were recovered in the deepest haul, which was located near the base of the escarpment. They could reflect serpentinization along shear zones and faults, or they could reflect the tectonic emplacement of hydrated ultramafic rock (upper mantle or lower oceanic crust) along fault zones at the base of the south wall. Ultramafic rocks were recovered in a similar tectonic environment from the base of the west wall of the Tonga Trench (Fisher and Engel, 1969). The metabasalt probably reflects the composition of igneous basement lying below the carbonate platform. Mineralogically and texturally, the metabasalts are similar to metabasalts found in the ocean basins. Also, the velocity measurements made on metabasalts recovered from deep-ocean escarpments record velocities in the 4.5–6.0 km/sec range (0.25–1.00 Kbar confining pressure). The axis of the Puerto Rico Trench and the south wall are underlain by a layer of this velocity range. The finely laminated tuffs (dredge 412) are probably deep-water deposits. The carbonate sandstone and micrite samples containing volcanic rock fragments and/or plagioclase (dredges 410 and 412) indicate proximity to volcanic terrain. It is difficult to infer the environment of deposition, and, therefore, it is not clear whether these mineral-bearing carbonates were deposited in shallow water with the other carbonate samples or whether they were deposited in deep water. If they were deposited in deep water, they, along with the tuffs, have probably been structurally brought into proximity with the carbonate-platform rocks by faulting.

The four remaining dredges from the western end of the south wall contrast markedly with the shallow-water carbonates recovered in dredges from north of Puerto Rico. Two dredges recovered from the sides of a large steep-sided mountain on the south wall north of Mona Passage contained black marble. The two other dredges, located east of Cape Samana, Hispaniola, recovered siltstone, recrystallized carbonate with deformed bands of white mica. Rocks of this description associated with serpentinites are reported to outcrop on steeply dipping east-west trending faults on Cape Samana (Nagle, 1970). The age of the material is unknown, but Nagle proposes that the material represents structurally dismembered and metamorphosed fragments of oceanic crust and sediments mechanically plastered along the northern boundary of the Caribbean plate. The recrystallized carbonates sampled from the south wall at four localities line up with the faults of the Samana Peninsula.

The second suite of rocks sampled from the Puerto Rico Trench area was recovered from Silver and Navidad Banks. Navidad Bank, the southernmost extension of the Blake–Bahama Platform, defines the northwestern boundary of the Puerto Rico Trench. The 14 dredges recovered from this area all contain lithified carbonate rocks with similar textures. The rocks often contained evidence of bioturbation and ranged from diverse types of micrites to sparites. Bird's-eye textures, mud cracks, and dolomite recrystallization are observed in many samples. The fossil material, when present, included rounded coral debris, skeletal fragments, algal mat structures, benthonic foraminifera, miliolids, ostracod carapaces, echinoderm spines, and annelid tubes. The samples containing diagnostic fauna range in age from Upper Cretaceous to Early Miocene. The faunal associations recognized in several dredges and the characteristic textures (bird's-eye texture, algal mats, intraclasts, dolomitization) indicate a very shallow, intertidal to supratidal environment of deposition for the carbonate samples. The shallow-water limestones record continuous carbonate-platform subsidence history from at least the Upper Cretaceous to the Middle Miocene. The subsidence history documented for the Navidad–Silver Bank area is similar to the subsidence history of the Blake Plateau (Heezen and Sheridan, 1966; Sheridan et al., 1969) and the Bahamas (Spencer, 1967). Navidad and Silver Bank are topographically continuous with the Bahama Platform to the northwest, and it is not surprising that the whole platform has been characterized by similar depositional environments and subsidence histories.

There is, however, a marked difference between the textures and faunal assemblages of the shallow-water carbonates sampled from the Blake–Bahama escarpment (Heezen and Sheridan, 1966; Sheridan *et al.*, 1969) and samples from the south side of Silver Bank (samples 427 to 431) and the southeast side of Navidad Bank (samples 421 to 426). The Blake–Bahama carbonate samples contain algal and reef debris, and the inferred depositional environment ranges from fore-reef to carbonate shelf. Deep-sea drilling to the east of the Blake–Bahama escarpment recovered oceanic sediments of Late Jurassic to Recent (Hollister *et al.*, 1972). The shallow-water carbonates dredged from the Blake–Bahama escarpment indicate 5000 m of subsidence since the Early Cretaceous. The textural and faunal assemblages of the samples from the south side of Silver and Navidad Banks suggest tidal flat-back reef-carbonate environment. The precipitous escarpments on which these samples are found rise from the floor of the Puerto Rico Trench. Apparently, unlike the Blake–Bahama escarpment, the escarpments which define the south side of Silver and Navidad banks are tectonic and were created during the formation of the Puerto Rico Trench when the carbonate platform was torn asunder. The shallow-water carbonate samples found along the south wall of the Puerto Rico Trench also have faunal assemblages indicative of a shallow, intertidal, to subtidal environment of deposition.

The occurrence of carbonates on the south wall of the Puerto Rico Trench adds a new dimension to the geology of the trench and the early history of the area. There are not sufficient data available, however, to uniquely explain the shallow-water carbonates in terms of the region's geological evolution. The limestones of the south wall can be integrated with the known geology of the area in several ways. First, the limestones could represent a carbonate platform, separate from the Blake–Bahama Platform, which rests on the northern portion of the Puerto Rico–Virgin Islands island arc foundation (slices of oceanic crust and volcanics). This reconstruction implies that the Late Mesozoic–Early Tertiary Trench of the region lay to the north of the carbonates and the island arc volcanic centers lay to the south of the carbonates. The volcanogenic products shed from the documented Cretaceous and Early Tertiary volcanic centers of Puerto Rico and the Virgin Islands were trapped in intervening tectonic troughs, and, thereby, to a large degree, island-arc erosional products were isolated from the carbonate platform. However, several carbonate sandstones, which were recovered with shallow-water carbonate samples at two localities along the south wall, contain volcanic rock fragments and plagioclase. This volcanic debris may be representative of detritus shed from Cretaceous volcanic centers to the south which, to a minor degree, mixed with and contaminated the carbonate platform sediments.

A second interpretation of the occurrence of shallow-water carbonates on the south wall of the Puerto Rico Trench is that the carbonates of the Blake–Bahama Platform and carbonates of the south wall of the Trench were continuous from Mid-Cretaceous to Early Miocene. This reconstruction implies that the Late Mesozoic–Early Tertiary trench of the Puerto Rico region lay to the south of the long NW–SE trending shallow-water carbonate platform. The trench could have been located south of the Late Mesozoic volcanic centers exposed on Puerto Rico, or it could have been positioned between the volcanic centers to the south and the carbonate platform to the north. If the Late Mesozoic trench lay to the south of the volcanic centers, then in the Mid-Tertiary the trench must have relocated along the north side of the Puerto Rico volcanic centers, disrupting the carbonate trend and forming the Puerto Rico Trench. The alternative positioning suggests that the carbonate platform now located along the south wall of the Puerto Rico Trench was positioned northward of the trench in the Mid-Cretaceous to Lower Tertiary. During this interval, as oceanic lithosphere was subducted beneath the proto-Puerto Rico trench, the shallow-water carbonate platform and the crust on which it rested approached the trench. When the carbonate platform reached the trench axis, it was decoupled from the underlying crust and tectonically emplaced against the landward wall of the trench. More data are needed to fully understand the relationship of the shallow-water carbonates to the evolution of eastern Greater Antilles and the proto-Puerto Rico Trench.

VI. CONCLUDING REMARKS

The rocks dredged from crustal outcrops provide the first direct evidence of the composition of the Caribbean crust. When these data are combined with seismic-refraction studies, the geology of the outcrop areas can be generalized over the Caribbean basin. The extrusive and shallow intrusive basaltic rocks indicate phases of volcanic activity, granodiorites indicate times of intrusion, and sedimentary rocks provide ages and environments of deposition. In addition, rocks recovered from submarine escarpments contiguous with Caribbean islands indicate the composition of the rocks which comprise the foundations of the subaerial crustal elements.

The data from the dredged rocks, however, are not sufficient by themselves to establish a detailed geological history of the Caribbean basin. But when these additional facts are combined with the geology of the islands, submarine pedestals and escarpments, the continents presently and previously adjacent to the basin, and the inferred sea-floor accretion history in the Atlantic, a general model of the evolution of the Caribbean basin emerges.

The pre-rift reconstructions of North and South America with Africa indicate that in the Triassic the Caribbean region could not have been more than one-quarter of its present size (Carey, 1958; Bullard *et al.*, 1965; Freeland and Dietz, 1971; Le Pichon and Fox, 1971). As North America moved away from Africa and South America in the Late Triassic–Early Jurassic, the oceanic parts of the Caribbean region experienced an early history similar to the contemporaneous juvenile North Atlantic. As North America moved north-westward relative to Africa and South America, a segment of the mid-Mesozoic Mid-Oceanic Ridge must have passed through the Atlantic and Caribbean regions into the Jurassic Pacific. This ridge was analogous to the present Mid-Oceanic Ridge and was characterized by a rift-transform tectonic fabric. Although complex in geometry, the ridge must have run from the vicinity of the present Azores–Gibraltar line south through the Caribbean region into the Pacific. Comparison of seismic-refraction results obtained throughout the world ocean shows that the velocity structure of the main ocean basins is very similar (Raitt, 1963). This similarity suggests a uniform oceanic composition and, to a first approximation, a uniform accretion process along the axis of the Mid-Oceanic Ridge. There is, therefore, no obvious reason why the oceanic crust created during the Jurassic in the Caribbean would not have had the same composition, thickness, and velocity structure as measured today in the main ocean basins. The basalts and dolerites from the western escarpment of the Beata Ridge may be representative of the oceanic basement layer generated during this Jurassic phase of Caribbean crustal accretion. The basaltic rocks of the Beata Ridge are overlain by Cretaceous marine sediments and volcanics.

The initial crustal accretion phase in the Caribbean came to an abrupt halt in the Early Cretaceous when the South Atlantic began to open (Maxwell *et al.*, 1970). At this time, the North Atlantic segment of the Mid-Atlantic Ridge was disengaged from the Caribbean–Pacific ridge segment and joined the newly formed ridge segment in the South Atlantic. From the Early Cretaceous on, the oceanic crust in the Caribbean region was caught between the differential movements of North and South America. The resultant compressive stresses resulted in the formation of subduction zones along much of the Caribbean perimeter. The initiating phase of subduction zone formation resulted in the tectonic emplacement of slices of oceanic crust along the axis of the Greater and Lesser Antilles and mobilization and metamorphism along the northern margin of Venezuela. The basal serpentinites, amphibolites, and pillow lavas of Hispaniola, Puerto Rico, and the Virgin Islands have been interpreted as fragments of oceanic crust (Hess, 1964; Mattson, 1966; Nagle, 1973). The occurrence of Upper Jurassic pillow lavas on La Désirade (Fink, 1970) and the dredged metabasalts and metagabbros on La Désirade escarpment suggest a similar geology for the foundation for the Lesser Antilles. Basal Lower Cretaceous gneissic basement overlying metasediments (Nagle,

1972) and ultramafic intrusives indicate the initiation of subduction to the north coast of Venezuela in the Early Cretaceous (Bell, 1972).

Partial melting of the underthrust oceanic lithosphere resulted in the Mid-Cretaceous initiation of calc-alkaline volcanics in the Greater Antilles (Nagle, 1972; Bowin, 1966; Mattson, 1966; Donnelly, 1966) and along the northern coast of Venezuela (Bell, 1972). Cretaceous volcanic tuffs recovered from escarpments off the Lesser Antilles indicate island-arc volcanism along this margin at the same time.

Recent attempts at dating the Mesozoic anomalies in the western Pacific and western North Atlantic led to the hypothesis that in the Cretaceous (110 to 85 m.y. B. P.) the Atlantic and Pacific spreading centers were characterized by rapid sea-floor accretion (Larson and Pitman, 1972). The Cretaceous granodiorite plutonism associated with inferred paleo-subduction zones around the margin of the Pacific are thought to be a consequence of rapid lithospheric accretion (Hamilton, 1969; Larson and Pitman, 1972). The previously documented (MacGillavry, 1970) coeval Mid-Cretaceous to Paleocene emplacement of granodiorite plutons around the margins of the Caribbean (Greater Antilles and Dutch Antilles) and the recovery of Upper Cretaceous to Paleocene granodiorites from the Aves Ridge may well reflect this time of rapid subduction along the Caribbean margins.

Apparently, this Cretaceous phase of rapid subduction around its margins also affected the interior of the Caribbean. Frictional heating, associated with slippage between the underthrust oceanic crust and upper mantle and the stationary material that it underrides, elevated the isotherms (McKenzie and Sclater, 1968, Oxburgh and Turcotte, 1970) within the Caribbean basin. Heat generated by this mechanism behind the Greater Antilles and Lesser Antilles subduction zones metamorphosed the normal oceanic crust of the Caribbean basin. Metamorphism altered the velocity structure of the crust and decreased the natural remanent magnetization intensities of the magnetized basalt layer. Further, peridotite diapirs, transporting heat generated by shear along the deeper parts of the Benioff zone (300–500 km), rose to shallow depths beneath the interior of the island-arc-bounded basin (Oxburgh and Turcotte, 1970; Karig, 1971). The thermal diapirs rising beneath the Caribbean basin increased the volume of the upper mantle and caused extension across the interior of the Caribbean. Extensional stress was sufficient to split the crust asunder, allowing basaltic magmas to rise through the crust up into the overlying pelagic sediments. The basaltic magma spread out laterally in the sediment, emplacing throughout much of the basin a coeval sequence of Upper Cretaceous sills. This mechanism explains the origin and timing of.the basalt and dolerite sills of Upper Cretaceous age, called "B" (Ewing et al., 1968) and sampled at five widely spaced localities during DSDP Leg XV (Edgar et al., 1971). Apart from the basaltic material accreted in the vertical feeder dikes which feed the overlying sills, this event did not create a significant amount of lithosphere within the Caribbean basin. The extrusion of basaltic rocks within the sediment pile, however, was probably accompanied by accretion of material on the underside of the Caribbean basin. This process would have thickened the Caribbean crust.

The heating of the upper mantle and crust of the Caribbean caused thermal expansion of the mineral phases which resulted in basinwide uplift. Because of differential heating, rates and amounts of uplift were not uniform throughout the basin and produced faulting in the basement. Lagging behind the Cretaceous phase of subduction around the Caribbean margin, uplift started in the Late Cretaceous, reaching a maximum in the Eocene. Extensive volcanism had ceased along the Greater Antilles and the Northern coast of Venezuela by Eocene. From Eocene to the present these margins have been relatively stable, indicating that significant plate consumption there had ceased. Shallow-water carbonate samples containing faunal assemblages of Eocene age have been recovered at the northern end of the Beata Ridge and from pedestals on the Aves Ridge. These samples show that at least portions of these ridges were elevated during Eocene times. Shallow-water samples up to Early Miocene in age were recovered from these regions, suggesting that the ridges remained high and did not sink appreciably until Mid-Miocene. After the Early Miocene both the Beata and Aves Ridges had sunk sufficiently that open-ocean pelagic deposition occurred where shallow marine and terrestrial environments had existed in the Paleogene. The synchronous subsidence history of these two ridges suggests that the whole Caribbean basin may have been uplifted in the Late Cretaceous–Early Tertiary and that its subsequent history has been one of cooling and slow subsidence.

ACKNOWLEDGMENTS

The research reported here was made possible by the Duke University Marine Laboratory which provided *R. V. Eastward* for cruises E-46-67, E-50-69, and E-52-70 under their Cooperative Oceanographic Program. This program, directed by R. T. Barber, is supported through N. S. F. Grants GB-8188, GB-8189, and GB-8190 to Duke University. We wish to note the excellent cooperation of the experienced masters, party chiefs, officers, and crew of *Eastward*. Party chiefs J. Cason, T. Piner, and G. Newton and the students who made up the scientific crew contributed greatly to the success.

This research was supported by the Office of Naval Research under contracts N00014-67-A-0108-0004 and N00014-67-A-0108-0036, and by the National Science Foundation under grant GA27281.

REFERENCES

Anderson, O., and R. Leibermann, 1968, Sound velocities in rocks and minerals: Experimental methods, extrapolations to very high pressures, and results, in: *Physical Acoustics*, Mason, W. P., ed.: Academic Press, New York, v. 4B, p. 330–472.

Bader, R. G., Gerard, R. D., Benson, W. E., Bolli, H. M., Hay, W. W., Rothwell, W. T., Ruef, M. H., Riedel, W. R., and Sayles, F. L., 1970, Initial reports of the Deep Sea Drilling Project: U.S. Govt. Printing Office, Washington, D.C., v. 4, p. 753.

Barrett, D. L., and Aumento, F., 1971, The Mid-Atlantic Ridge near 45° N. XI. Seismic velocity, density, and layering of the Crust: *Can. J. Earth Sci.*, v. 7, p. 1117–1124.

Bell, J. S., 1972, Geotectonic evolution of the southern Caribbean area: Geol. Soc. Am. Mem. 132, p. 369–386.

Beloussov, V., 1960, Development of the earth and tectogenesis: *J. Geophys. Res.*, v. 65, p. 4127–4146.

Bowin, C. O., 1966, Geology of central Dominican Republic (A case history of part of an island arc): Geol. Soc. Am. Mem. 98, p. 11–84.

Bowin, C. O., Nalwalk, A. J., and Hersey, J. B., 1966, Serpentinized peridotite from the north wall of the Puerto Rico Trench: *Geol. Soc. Am. Bull.*, v. 77, p. 257–270.

Bullard, E., Everett, J. E., and Smith, A. G., 1965, The fit of the continents around the Atlantic, in: A symposium on continental drift: Blackett, P. M. S., Bullard, E., and Runcorn, S. K., eds.: *Phil. Trans. Roy. Soc. London, Ser. A*, v. 258, p. 41–51.

Bunce, E. T., and Fahlquist, D. A., 1962, Geophysical investigation of the Puerto Rico Trench and Outer Ridge: *J. Geophys. Res.*, v. 67, p. 3955–3972.

Butterlin, J., 1956, *La constitution geologique et la structure des Antilles*: Centre National de la Recherche Scientifique, Paris, 453 p.

Carey, S. W., 1958, The tectonic approach to continental drift, in: *Continental Drift, a Symposium*, Carey, S. W., ed.: Univ. of Tasmania, Hobart, p. 177–358.

Chase, R. L., and Bunce, E. T., 1969, Underthrusting of the eastern margin of the Antilles by the floor of the western Atlantic Ocean and origin of the Barbados Ridge: *J. Geophys. Res.*, v. 74, p. 1913–1920.

Christensen, N. I., and Shaw, C. M., 1970, Elasticity of mafic rocks from the Mid-Atlantic Ridge: *Geophys. J. Roy. Astron. Soc.*, v. 20, p. 271–284.

Christofferson, E., 1973, Linear magnetic anomalies in the Colombia Basin, Central Caribbean Sea: *Geol. Soc. Am. Bull.*, v. 84, p. 3217–3230.

Donnelly, T. W., 1966, Geology of St. Thomas and St. John, U.S. Virgin Islands: Geol. Soc. Am. Mem. 98, p. 85–176.

Eardley, A., 1951, *Structural Geology of North America*: Harper and Row, New York, 624 p.

Edgar, T., Ewing, J., and Hennion, J., 1971, Seismic refraction and reflection in the Caribbean Sea: *Am. Assoc. Petr. Geol. Bull.*, v. 55, p. 833–870.

Edgar, N. T., Saunders, J. B., Donnelly, T. W., Schneidermann, N., Maurrasse, F., Bolli, H. M., Hay, W. W., Riedel, W. R., Premoli-Silva, I., Boyce, R. E., and Prell, W., 1971, Deep-sea drilling project: Leg XV: *Geotimes*, v. 12, p. 12–16.

Ewing, J., Officer, C., Johnson, M., and Edwards, R., 1957, Geophysical investigations in the eastern Caribbean: Trinidad Shelf, Tobago Trough, Barbados Ridge, Atlantic Ocean: *Geol. Soc. Am. Bull.*, v. 68, p. 897–912.

Ewing, J., Antoine, J., and Ewing, M., 1960, Geophysical measurements in the western Caribbean Sea and in the Gulf of Mexico: *J. Geophys. Res.*, v. 65, p. 4087–4126.

Ewing, J., Talwani, M., and Ewing, M., 1968, Sediment distribution in the Caribbean Sea, in: *Fourth Caribbean Geological Conference*: Caribbean Printers, Trinidad and Tobago, p. 317–325.

Ewing, J., Talwani, M., Ewing, M., and Edgar, T., 1967, Sediments of the Caribbean, in: *Symposium of Tropical Oceanography*: Univ. of Miami, v. 5, p. 88–102.

Fink, L. K., Jr., 1970, Field guide to the island of La Désirade with notes on the regional history and development of the Lesser Antilles Island Arc, in: Guidebook to the Caribbean Island-Arc System: Am. Geol. Institute, Washington, D.C.

Fisher, R. L., and Engel, C., 1969, Ultramafic and basaltic rocks dredged from the nearshore flank of the Tonga Trench: *Geol. Soc. Am. Bull.*, v. 80, p. 1373–1378.

Fox, P. J., Ruddimen, W. F., Ryan, W. B. F., and Heezen, B. C., 1970, The geology of the Caribbean crust. I. Beata Ridge: *Tectonophysics*, v. 10, p. 495–513.

Fox, P. J., Schreiber, E., and Heezen, B. C., 1971, The geology of the Caribbean crust: Tertiary sediments, granitic and basic rocks from the Aves Ridge: *Tectonophysics*, v. 12, p. 89–109.

Fox, P. J., and Schreiber, E., 1973, Compressional wave velocities in basalt and dolerite samples recovered during Leg XV. Initial reports of the Deep Sea Drilling Project: *U.S. Govt. Printing Office*, Washington, v. 15, p. 1013–1016.

Fox, P. J., Schreiber E., and Peterson, J. J., 1973, The geology of the oceanic crust: Compressional wave velocities in oceanic rocks: *J. Geophys. Res.*, v. 78, p. 5155–5172.

Freeland, G. L., and Dietz, R., 1971, Plate tectonic evolution of Caribbean–Gulf of Mexico region: *Nature*, v. 232, p. 20–23.

Hamilton, W., 1969, Mesozoic California and the underflow of Pacific mantle: *Geol. Soc. Am. Bull.*, v. 80, p. 2409–2430.

Heezen, B. C., 1959, Some problems of Caribbean submarine geology, *Trans. Second Caribbean Geol. Conf.*, p. 12–16.

Heezen, B. C., and Sheridan, R. E., 1966, Lower Cretaceous rocks (Neocomian–Albian) dredged from Blake escarpment: *Science*, v. 154, p. 1644–1647.

Hess, H. H., 1964, The oceanic crust, the upper mantle and the Mayagüez serpentinized peridotite: *Nat. Acad. Sci., Nat. Res. Council*, Publ. No. 1188, p. 168–175.

Hollister, C. D., and Ewing, J. I., 1972, Initial reports of the Deep Sea Drilling Project: *U.S. Gov. Printing Office*, Washington, v. 11, 1077 p.

Karig, D. E., 1971, Origin and development of marginal basins in the western Pacific: *J. Geophys. Res.*, v. 76, p. 2542–2561.

Larson, R. L., and Pitman, W. C., 1972, Worldwide correlation of Mesozoic magnetic anomalies, and its implications: *Geol. Soc. Am. Bull.*, v. 83, p. 3645–3662.

Le Pichon, X., and Fox, P. J., 1971, Marginal offsets, fracture zones, and the early opening of the North Atlantic: *J. Geophys. Res.*, v. 76, p. 6294–6308.

MacGillary, M. J., 1970, Geological history of the Caribbean, Proceedings: *Koninkl. Ned. Akad. Wetenschap.*, Ser. B, v. 73, p. 51–52.

Marlowe, J. I., 1968, Geological reconnaissance of parts of Aves Ridge, *Fifth Caribbean Geol. Conf.*: Univ. Puerto Rico, Mayagüez, p. 51–52.

Martin-Kaye, P. H. A., 1969, A summary of the geology of the Lesser Antilles: *Overseas Geol. Mineral Resources*, v. 10, p. 172–206.

Mattson, P. M., 1960, Geology of the Mayagüez area, Puerto Rico: *Geol. Soc. Am. Bull.*, v. 71, p. 319–362.

Mattson, P. M., 1966, Geological characteristics of Puerto Rico, in: *Continental Margins and Island Arcs*, Poole, W. M., ed.: Geol. Surv. Pap., Can., No. 66-15, p. 224–238.

Maxwell, A. E., Von Herzen, R. P., Hsu, K. J., Andrews, J. E., Saito, T., Percival, S. F., Milow, E. D., and Boyce, R. E., 1970, Deep sea drilling in the South Atlantic: *Science*, v. 168, p. 1047–1059.

McKenzie, D. P., and Sclater, J. G., 1968, Heat flow inside the island arcs of the northwestern Pacific: *J. Geophys. Res.*, v. 73, p. 3173–3179.

Molnar, P., and Sykes, L. R., 1969, Tectonics of the Caribbean and Middle American regions from focal mechanism and seismicity: *Geol. Soc. Am. Bull.*, v. 80, p. 1639–1684.

Nagle, F., 1970, Serpentinites and metamorphic rocks at Caribbean boundaries: *Geol. Soc. Am., Abstr. with Programs*, v. 2, n. 7, p. 633.

Nagle, F., 1972, Rocks from seamounts and escarpments of the Aves Ridge, in: *Sixth Caribbean Geol. Conf.*, C. Petzall, ed., p. 409–413.

Nalwalk, A. J., 1969, Geology of a portion of the north wall of the Puerto Rico Trench: *Tectonophysics*, v. 8, p. 403–425.

Officer, C., Ewing, J., Hennion, J., Harkinder D., and Miller, D., 1959, Geophysical investigations in the eastern Caribbean—Summary of the 1955 and 1956 cruises, in: *Physics and Chemistry of the Earth*, Ahrens, L. M., Press, F., Rankama, K., and Runcorn, S. K., eds.: Pergamon Press, London, v. 3, p. 17–109.

Oxburgh, E. R., and Turcotte, D. L., 1970, Thermal structure of island arcs: *Geol. Soc. Am. Bull.*, v. 81, p. 1665–1688.

Poldervaart, A., 1955, Chemistry of the earth's crust, in: *Crust of the earth*, Poldervaart, A., ed.: Geol. Soc. Am. Spec. Pap. 62, p. 119–145.

Raitt, R. W., 1963, The crustal rocks, in: *The Sea*, Hill, H. W., ed.: Wiley–Interscience, New York, v. 3, p. 85–100.

Schneidermann, M., Beckmann, J. P., and Heezen, B. C., 1972, Shallow-water carbonates from the Puerto Rico Trench Region, in: *Sixth Caribbean Geological Conference*, C. Petzall, ed., p. 423–425.

Schuchert, C., 1935, *Historical Geology of the Antillean–Caribbean Region*: John Wiley and Sons, New York, 980 p.

Sheridan, R. E., Smith, J. D., and Gardner, J., 1969, Rock dredges from Blake Escarpment near Great Abaco Canyon: *Am. Assoc. Petr. Geol. Bull.*, v. 54, p. 2032–2039.

Spencer, M., 1967, Bahama deep test: *Am. Assoc. Petr. Geol. Bull.*, v. 51, p. 263–268.

Sykes, L. R., and Ewing, M., 1965, The seismicity of the Caribbean region: *J. Geophys. Res.*, v. 70, p. 5065–5074.

Talwani, M., Sutton, G. M., and Worzel, J. L., 1959, A crustal section across the Puerto Rico Trench: *J. Geophys. Res.*, v. 64, p. 1545–1555.

Chapter 11

THE LESSER ANTILLES AND AVES RIDGE

John Frederick Tomblin

Seismic Research Unit
University of the West Indies
Trinidad

I. INTRODUCTION

A. Summary of Physiographic and Structural Units

The island-arc complex of the Lesser Antilles (Fig. 1) borders the eastern margin of the Caribbean plate where the latter meets and is underthrust by the western Atlantic Ocean floor. The continuity of the topographic features, the seismicity, and the gravity anomalies from the Greater Antilles through the Lesser Antilles and into northern Venezuela establishes the existence of an active tectonic belt through these areas. The main physiographic feature of the Lesser Antilles is the crustal upwarp which is a relatively narrow ridge from Grenada to Guadeloupe but from there northward becomes a nearly flat-topped platform. Islands along the eastern margin of this platform are composed of Eocene to Oligocene volcanics and minor intrusives capped extensively by Miocene limestones, while the western margin is bordered by a chain of younger, Plio-Pleistocene volcanoes.

The southern half of the central-arc ridge slopes gently into the Tobago Trough, while to the west it dips more steeply into the Grenada Trough. Off the northern half of the island-arc platform the bathymetric profile is reversed,

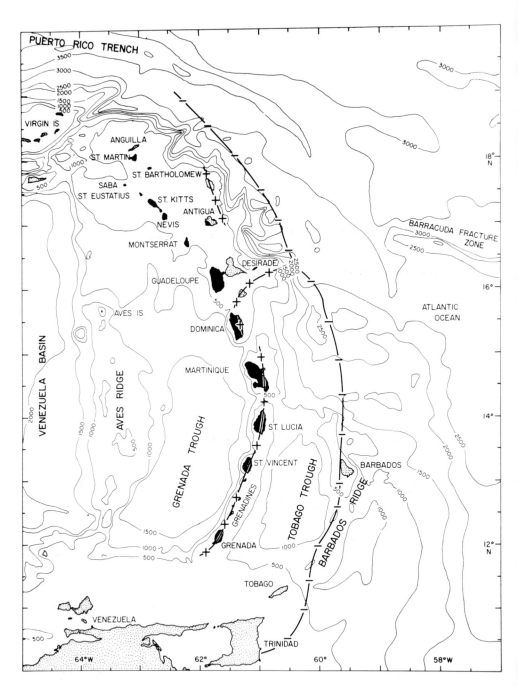

Fig. 1. Map of the Eastern Caribbean showing bathymetry (contours at 500-fathom intervals) and gravity anomaly axes. Tertiary to Recent volcanic rocks are indicated by solid shading. Heavy lines with minus and plus signs indicate axes of negative and positive gravity anomalies.

with the Atlantic floor dipping steeply eastward off the chain of older islands into the southeastern extension of the Puerto Rico Trench but descending more gently to the west of the island-arc platform into the shallower, northern part of the Grenada Trough. The outermost features of the eastern Caribbean arc complex are the Barbados Ridge in the east and the Aves Ridge in the west. While the Barbados Ridge is a relatively short feature, composed of a thick pile of low-density sediments still rising in response to isostatic readjustment, the Aves Ridge by contrast extends northward along almost the full length of the main arc ridge and apparently represents a former volcanic arc which became extinct in Lower Tertiary or earlier time and is now nearly in isostatic equilibrium. The western flank of the northern Aves Ridge slopes steeply, in a way similar to the northeast flank of the main arc.

B. Brief History of Research in the Region

The earliest geologic descriptions of the Lesser Antilles are those which refer to special phenomena such as volcanic eruptions, solfataric activity, and damaging earthquakes. These include various eyewitness reports, mostly by nongeologists, dating back to the late 1600's. The first detailed geological research carried out in the region was on the catastrophic eruptions of 1902 in St. Vincent and Martinique, in which the pyroclastic flow or nuée ardente type of eruption was first recognized, and the mechanics of the eruptive process discussed, by Anderson and Flett (1903) and Lacroix (1904). Lacroix (1904) and Flett (1908) also presented some of the earliest petrographic and geochemical data, primarily for Martinique and St. Vincent but with brief descriptions of other islands of the arc. More recent studies on the volcanic history and petrology of different islands have been carried out by MacGregor (1938), Christman (1953), Westermann and Kiel (1961), and by research groups at the University of Oxford, England (Baker, 1963; Tomblin, 1964; Lewis, 1964; Rea, 1970), the University of the West Indies (Jackson, 1970), and the Universities of Durham and Leeds, England, where broad geochemical, palaeomagnetic and isotopic dating studies are in progress.

Geophysical studies in the eastern Caribbean were initiated by Vening Meinesz (1932, 1934) whose submarine gravity expeditions here and in Indonesia established for the first time the dramatically large gravity anomalies associated with island arcs. These studies were developed and interpreted by Vening Meinesz and by Hess (1938) into the tectogene theory which was the precursor of modern notions of subduction. More recent studies of the gravity and magnetic anomalies of the Lesser Antilles, including measurements on land as well as offshore, have been made by Andrew et al. (1970).

The next major development in Caribbean geophysics was the determination of the seismic-velocity structure of the crust through refraction and reflection studies, the majority of which have been carried out by the Lamont Geological Observatory of Columbia University and the Woods Hole Oceanographic Institution (Officer *et al.*, 1957, 1959; Edgar, 1968; Chase and Bunce, 1969; Bunce *et al.*, 1970; Edgar *et al.*, 1971). Recent, and as yet unpublished, refraction surveys have been conducted by the University of Durham and cooperating institutions using large explosions at sea and recording at a network of sites, mainly on land.

The seismicity of the region was poorly observed, with only one seismograph station (Martinique) located in the eastern Caribbean until the early 1950's, when a network of stations was set up in the British Commonwealth islands with headquarters in Trinidad. Since 1953, regional hypocenters have been computed and published by the Seismic Research Unit (formerly the Volcanological Research Department) in Trinidad using arrivals from all stations in the Lesser Antilles, Greater Antilles, and Venezuela, which by 1973 numbered 21. The first synthesis of the regional seismicity using regional plus teleseismic arrivals (Sykes and Ewing, 1965) and studies of two focal mechanisms in the Lesser Antilles (Molnar and Sykes, 1969) were carried out at the Lamont Geological Observatory. Research on hypocenter distributions and focal mechanisms for more recent events are in progress at the University of the West Indies (Tomblin, in preparation).

Studies of the stratigraphy and paleontology of the marine sedimentary rocks in the Lesser Antilles have been published by numerous independent workers. Among the more important are those by Senn (1940) and Baadsgaard (1960) on Barbados, by Martin-Kaye (1959) on Antigua, and by Martin-Kaye (1969) on the Lesser Antilles as a whole.

II. VOLCANIC AND SEDIMENTARY HISTORY OF THE MAIN ARC RIDGE

A. Pre-Tertiary

The only area in the Lesser Antilles where pre-Tertiary rocks have been identified is in the small island of Désirade to the east of Guadeloupe. The Basal Volcanic Complex of this island, which had been tentatively attributed to the lower Tertiary by Barrabé (1934, 1954) and de Reynal (1966), has recently been identified by Fink (1968, 1972) from isotopic age measurements as not younger than Late Jurassic. An age of 142 ± 10 m.y. was obtained on a trondhjemite intrusive into submarine volcanics which consist of spilitic pillow lavas, cherts, and quartz–keratophyres. On the basis of the similarity between the Désirade volcanics and the pre-Middle Cretaceous Water Island Formation in the Virgin Islands, Fink (1972) has suggested that the Greater and Lesser Antilles were "a continuous feature during Late Mesozoic–Early

Tertiary time." The present writer prefers a more cautious approach in interpreting the Désirade volcanics, particularly in view of the long time interval (over 70 m.y.) between these and the next oldest rocks in the Lesser Antilles and the large horizontal movements which undoubtedly took place in the region during this interval. It is possible, for example, that the crust now exposed in Désirade represents a small block which was much closer to the Virgin Islands in Late Jurassic time and which subsequently separated from this area and moved relatively eastward. Furthermore, any reconstruction of the eastern Caribbean in pre-Tertiary time has to take account of the possibility that the Aves Ridge may have represented the main site of volcanism and subduction in the Cretaceous, and that uplift and underthrusting along the Lesser Antilles did not begin until late Cretaceous time.

B. Eocene–Oligocene

The geological history of the Lesser Antilles during the Lower Tertiary is far from complete owing to the relatively small areas of outcrop and, in numerous cases, uncertainties over the age of volcanics which have been tentatively attributed to this period. No isotopic ages are available, and dating has been achieved solely through correlation with fossiliferous marine sediments. The latter are rare on most of the islands, and their relations to nearby volcanics are often not clearly exposed.

Whatever uncertainties exist about the presence of an arc ridge at the site of the Lesser Antilles in the late Jurassic, there is no doubt that by Middle to Upper Eocene time the north–south ridge now represented by the Grenada–Guadeloupe–Anguilla chain had developed. Large volumes of pyroclasts were emitted along the entire length of the chain, with subaerial eruptions at least at the south end, while a flysch-type sedimentary sequence including epiclastic material from South America, plus volcanic ash from the main arc ridge, was being deposited in the external trough which is now Barbados.

At the northern end of the volcanic arc, the St. Bartholomew Formation, in the island of the same name, consists of andesite tuffs and breccias of submarine origin, interbedded with five different limestone formations each between 10 and 65 m thick, and of Middle to Late Eocene age (Christman, 1953). Thus, although not yet above sea level, the volcanic summits in this area were being progressively built up by repeated eruptions, while the limestones indicate that the water depth was shallow. Eocene pyroclastic formations similar to those in St. Bartholomew are exposed in the neighboring islands of St. Martin and Anguilla, indicating a considerable total extent and volume of volcanic products. In addition to these surface emissions, intrusives of quartz diorite and subordinate basalt with a late Eocene or early Oligocene age are seen in St. Bartholomew and St. Martin.

The next island southward along the arc where Lower Tertiary rock is exposed in Antigua, which displays a very thick Oligocene sequence (Martin-Kaye, 1959). This consists of the Basal Volcanic Series with an estimated thickness of about 1800 m, composed of coarse and fine pyroclasts mainly of andesitic composition, plus intrusives and minor limestones. Overlying this, the Central Plain Succession is composed of about 1200 m of pyroclasts with conglomerates, cherts, limestones, and minor intrusions. This is followed by the Antigua Formation, made up almost entirely of coral limestone with a thickness of some 500 m and an age of Middle to Upper Oligocene. The entire sequence in Antigua is tilted toward the northeast, with dips decreasing from 15 to 20° in the southwest to between 2 and 3° in the northeast of the island.

Guadeloupe contains pre-Early Miocene, i.e., Eocene or possibly Oligocene, volcanics which outcrop in two small areas of Grande Terre. In Martinique, alternating volcanic conglomerates and silicified limestones underlie limestone lenses tentatively dated as Lower Eocene (Grunevald, 1965), while the andesite–dacite series of the Diamant center in the south of the island, and the basalts to basaltic andesites of Piton Conil and Morne du Vauclin, are associated with calcareous tuffs containing *Lepidocyclina* and *Asterigerina* which have been identified as Oligocene.

Towards the southern end of the volcanic chain, in the Grenadine island of Mayreau, silicified limestone with Lower to Middle Eocene foraminifera dips eastward at 60°, while in Carriacou (Martin-Kaye, 1958; Jackson, 1970; Robinson and Jung, 1972) andesite and basalt volcanics with ages between Upper Eocene and Lower Miocene have been dated from their relationships with fossiliferous sediments which occupy most of the eastern third of the island. The latter include a reef-type limestone dated as Upper Eocene on the basis of its faunal assemblage which includes *Lepidocyclina* (*Pliolepidina*) *pustulosa tobleri* H. Douvillé, *Lepidocyclina* (*Pliolepidina*) *macdonaldi* Cushman, and *Asterocyclina minima* Cushman.

The island of Grenada is unique in the volcanic chain in that it possesses a basement of strongly folded flysch, named the Tufton Hall Formation and described by Martin-Kaye (1969). Of Upper Eocene age, this deposit is at least several hundred meters thick, consisting of volcanic graywacke and shale layers, with a foraminiferal assemblage including *Globorotalia cerro-azulensis* (Cole), *G. centralis* Cushman and Bermudez, and *G. opima nana* Bolli. The fine grain size of the volcanic component and the deep-water depositional environment of this formation suggest that it was erupted from vents not closer than a few kilometers and that the graywacke component was washed in from either a shallower-water or a terrestrial environment. The shale, representing continuous sedimentation as a background to the rapidly and intermittently deposited graywackes, was probably derived from greater distance and may represent detritus from the South American continent.

Evidence of continued volcanic activity into the Oligocene in the Grenada area is provided by the occurrence of a fossiliferous limestone associated with volcanics in Mt. Sinai in southern Grenada, while the coarser grain size of these pyroclastics suggests that an eruptive center was present in this immediate area.

The end-Eocene orogeny which affected most parts of northern South America, including Trinidad, can be recognized clearly in Grenada but apparently had little effect on the northern Lesser Antilles. In Grenada, the Tufton Hall Formation was strongly folded, mostly on east–west axes, resulting locally in near-vertical or overturned strata. Folding on comparable, northeast–southwest axes in Barbados took place earlier, between the Middle and Upper Eocene.

C. Miocene

A general submergence of the Lesser Antilles took place in the Lower Miocene, resulting in the growth of reefal limestones in many parts of the Caribbean. In some places, e.g., northwestern Guadeloupe, northern St. Lucia, and southern and central Grenada, these reefal sediments remained subordinate to, and occurred as lenticles in, the volcanics, but elsewhere the large thickness of limestone demonstrates that deposition took place over a long period free from volcanic activity. In Sombrero, Anguilla, and St. Martin, reefal limestones and marls about 50 m thick unconformably overlie the Eocene–Oligocene volcanics, while the similar and very extensive Grande Terre Formation, with a thickness of 120 m, covers almost the entire eastern half of Guadeloupe and most of Désirade.

In Martinique, a series of tuffs with limestone horizons crops out in the northeast (Caravelle) area. This formation and the Morne Jacob basaltic andesite center are tentatively attributed to the Miocene (Grunevald, 1965), while in northern St. Lucia a primarily basaltic pyroclastic formation contains lenses of reef limestone with a foraminiferal assemblage including *Operculinoides cojimarensis* (D. K. Palmer), identified as Lower Miocene (Martin-Kaye, 1969). Similar faunal assemblages occur in the Grenadine islands of Canouan and Carriacou. In the latter they are at least 30 m thick and are overlain by the Grand Bay Beds consisting of 150 m of calcareous tuffs with foraminifera. In Grenada, minor outcrops of limestone within the volcanics of the central and southern parts of the island have been tentatively identified as Miocene (Martin-Kaye, 1969).

The end of the Miocene was marked throughout South America by the spectacular earth movements of the Andean orogeny. In the Lesser Antilles, these caused the general uplift and emergence of the arc ridge, resulting in the termination of widespread limestone deposition. They also appear to have been responsible for the shift of volcanic activity in the northern half of the Lesser Antilles from the east to the west side of the arc platform, with the initiation of volcanoes along the line from Montserrat to Saba. In the islands from Guadeloupe southward, where some volcanism had continued through the Miocene, the tectonic movements appear to have caused an increase in magmatic activity in the early Pliocene, resulting in minor intrusions in the older volcanics as well as the initiation of new eruptive centers. Although there was considerable uplift on a regional scale, there is no evidence of local deformation involving anything more than gentle tilting, i.e., nothing so intense as at the end-Eocene in the southern part of the arc. In Antigua, for example, where because of the large area of well-stratified deposits the pattern of deformation can be clearly distinguished, the whole island underwent northeastward tilting, which is almost imperceptible (about 2°) in the northeast but increases progressively to a maximum of around 20° in the southwest.

D. Plio-Pleistocene

From Early Pliocene through Recent time, activity in the volcanic Lesser Antilles has remained relatively constant in character and in location, with the growth of a new branch of the volcanic chain from Guadeloupe northwestward to Saba and the building of new volcanic centers adjacent to the older ones in the islands from Guadeloupe southward. Most of the erupted material was pyroclastic, emitted in violently explosive eruptions at large stratovolcanoes. Over 90% of these pyroclasts were erupted to a great height and fell as ash showers on the sea at distances up to many hundred miles from their source. On land, the primary materials include pyroclastic flow and fall in roughly equal proportions, with subordinate massive lava (domes or flows). A large proportion of the pyroclastics on land, however, have been reworked as secondary mudflows, although volcanoes with crater lakes, such as St. Vincent has at present, have emitted primary, laharic mudflows which are generally indistinguishable from secondary mudflows. The southern islands of Grenada, the Grenadines, and St. Vincent have a considerably higher proportion of basalt to more acid rocks than in the remainder of the arc, and this is reflected in the greater abundance of lava flows and scoria conelets.

Although notorious for their very large and destructive eruptions (which so far in the present century have killed nearly 30,000 people, i.e., more than the total killed during the same period by volcanoes in all other parts of the world), the Lesser Antillean volcanoes have produced numerous smaller eruptions in historic time, including several which have been purely phreatic, with no fresh magma emitted. Two such events, for example, took place at Mt. Pelée in Martinique in 1792 and 1851, while one occurred historically in Dominica, and seven in Guadeloupe (Robson and Tomblin, 1966). However, until the 1971–1972 eruption in St. Vincent, all of the previous 23 unambiguously reported historic eruptions in the Lesser Antilles had been explosive, not quietly effusive, and although the recent explosion-free growth of the 1971–1972 dome in St. Vincent is probably not without prehistoric precedent, this quiet and very gas-poor type of eruption is clearly the exception in the West Indies.

Within the last few tens of thousands of years in the Lesser Antilles, eruptions can be recognized which must have been at least an order of magnitude larger than the largest historic ones. Such an occurrence, apparently representing a single eruptive phase and tentatively dated from radiocarbon evidence at about 14,000 years B. P., took place in St. Vincent when the upper unit of the ash-fall sequence described by Hay (1959) was deposited with a thickness of about 24 m on the lower flanks of the Soufriere and about 3 m in Kingstown at a distance of 20 km to the south (probably not downwind) of the crater. Similarly, the large pumice eruptions which immediately preceded the collapse of the Qualibou Caldera in St. Lucia (Tomblin, 1968a) emptied several cubic kilometers of magma from depth and must have approached in scale the eruptions preceding the collapse of Mt. Mazama to form Crater Lake in the Oregon Cascades.

The dates of the most recent large eruptions at the younger volcanic centers in the Lesser Antilles (Table I) show that the two islands of St. Vincent and Martinique have had major eruptions in historic time, and apparently over the last several thousand years, at mean intervals of the order of a century. Most of the other islands, however, appear to have had major outbursts at a frequency of only one or two per thousand years. This applies to Montserrat,

TABLE I

Dates of the Most Recent Large Eruptions at Lesser Antillean Volcanoes[a]

Island	Volcano	Large eruptions
St. Kitts	Mt. Misery	2158 ± 94 B.C.
Montserrat	Soufriere Hills	A.D. 1646 ± 54
Guadeloupe	La Soufrière	A.D. 1400 ± 150
Dominica	Morne Micotrin	26,900 ± 900 B.C.
Martinique	Mt. Pelée	A.D. 1902; A.D. 1929
St. Lucia	Qualibou	37500 ± 1,500 B.C.
St. Vincent	The Soufriere	A.D. 1718; A.D. 1812; A.D. 1902

[a] Dates for which errors are quoted have been determined by radiocarbon analysis. This list is complete only for the historic period (A.D. 1700 onward).

Guadeloupe, and St. Kitts, for which dates are available, and probably also to Grenada, Nevis, St. Eustatius, and Saba, although no radiocarbon dates for specific eruptions have yet been obtained for these last four islands. In the remaining two larger volcanic islands of Dominica and St. Lucia, intervals of 28,000 and 39,000 years, respectively, have passed since the youngest known large eruptions. A general conclusion can therefore be drawn that the average state of a young volcanic center in the Lesser Antilles involves one or two major eruptions per millennium plus several smaller magmatic or phreatic events, and several volcano-seismic crises with local earthquake swarms which may be accompanied by increased solfataric activity but will not necessarily end in an eruption.

III. PETROGRAPHY AND GEOCHEMISTRY OF THE IGNEOUS ROCKS

A. Mineral Assemblages and Rock Textures

The great majority of the Lesser Antillean igneous rocks belong to the calc-alkaline assemblage which is characteristic of island arcs and circum-oceanic regions. Alkaline, undersaturated rocks are present at the southern end of the arc, and basalts with tholeiitic affinities have been reported in very small volume from St. Lucia. Almost all rocks are highly porphyritic, with plagioclase as the predominant phenocryst. The larger of these phenocrysts usually have unzoned cores in the compositional range bytownite–anorthite, with strong oscillatory zoning down to labradorite or andesite at the margins.

Mafic phenocrysts in the basalts are augitic clinopyroxene, olivine and/or orthopyroxene, set in a groundmass ranging from intersertal to glassy. In the basaltic andesites, clinopyroxene tends to predominate over orthopyroxene and in some rocks is either the only mafic phenocryst or may be accompanied by hornblende.

In the dacites, amphibole (green or brown hornblende, or cummingtonite) is the predominant mafic phenocryst. It is accompanied in most rocks by ortho-pyroxene, in some by biotite, and in a few by clinopyroxene or olivine with reaction rims. The dacites of certain islands, especially St. Lucia and Martinique, are distinguished by their containing between 5 and 15% quartz phenocrysts with a texture which may vary in a single thin section from euhedral to highly embayed. The groundmasses of the andesites and dacites range from felsitic to glassy, and many of these rocks are mildly vesicular, with irregular-shaped cavities.

Rocks of rhyolitic composition ($SiO_2 > 70\%$) occur in small volume in only a few islands. Those from Antigua are strongly porphyritic, with up to

30% of quartz phenocrysts. Those from St. Lucia, by contrast, have very few phenocrysts and are conspicuously flow-banded due to the alternation of 2 to 10 mm layers of darker, more glassy and paler, more crystalline material.

Coarse-grained, xenolithic accumulates are present in certain islands, especially St. Vincent (Wager, 1962; Lewis, 1964), St. Kitts (Baker, 1968), Grenada, and the Grenadines. These consist of widely varying proportions of anorthite, hornblende, and olivine, with minor accessory augitic clino-pyroxene and titaniferous magnetite. The occurrence and composition of these accumulates and, in particular, the well-developed banding which many of them display, indicate fractionation by crystal settling in shallow magma chambers.

B. Major-Element Geochemistry

Over 340 analyses made by wet chemical methods have been published for the Lesser Antilles. A compilation of these is given by Tomblin (1968b), and a selection of typical analyses is shown in Table II. The main geochemical features of the Lesser Antillean rocks are their high mean silica content, high abundance of alumina and lime, low titania, and marked decrease in iron-to-magnesium ratio with increasing acidity. The distribution of silica is illustrated in Fig. 2. This shows histographically the frequency distribution by silica content of a total of over 1200 rocks selected to give as even an areal coverage as possible, and all were analyzed by Dr. J. G. Holland at the University of Durham using x-ray fluorescence with uniform sample preparation and calibration procedures. It can be seen from these histograms that the two islands of Grenada and St. Vincent, at the southern end of the arc, possess distinctive characteristics: Grenada contains an abnormally large proportion of basalts, including some very low-silica, normatively undersaturated types, with alkaline affinities (Sigurdsson et al., 1973). The remainder of the Grenada analyses range in silica up to 64%, with the majority of these falling in the range from 57 to 62% SiO_2, creating a second mode on the histogram which corresponds with the only or principal mode for most other islands. St. Vincent is unique among the larger islands of the Lesser Antilles for its total lack of rocks with silica above 59% and its strong mode at about 55% SiO_2. This suggests that the majority of the St. Vincent rocks were produced by the fractionation of a basaltic parental liquid. This suggestion is reinforced by the observation that almost the full extent of this chemical range in St. Vincent is represented among the products of the 1902–1903 eruption of the Soufriere, indicating that a strongly differentiated body of magma must have underlain the volcano prior to this eruption. The islands from St. Lucia northward all follow a similar pattern of silica distribution, with a single or greatly predominant mode at between 60 and 64% SiO_2. St. Lucia is unusual in that it exhibits

TABLE II

Mean Chemical Composition of Typical Lavas from the Lesser Antilles[a]

Oxide, wt %	G-B (27)	L-AB (2)	V-B (47)	V-A (17)	V-BG (10)	K-A (17)	L-D (8)	L-R (1)
SiO_2	46.1	53.0	51.9	55.6	57.1	59.3	64.2	70.7
Al_2O_3	14.8	15.7	18.4	18.8	17.4	18.4	16.7	13.4
Fe_2O_3	4.8	3.8	3.4	3.0	3.6	3.5	0.8	1.2
FeO	5.1	7.6	5.4	4.9	4.4	3.5	3.1	2.5
MgO	10.6	3.4	5.5	3.4	2.8	2.2	1.7	0.0
CaO	11.6	7.6	9.8	8.1	7.1	6.7	6.4	2.8
Na_2O	2.4	3.4	3.0	3.4	3.8	3.9	3.1	4.9
K_2O	0.8	0.6	0.5	0.7	0.9	0.6	1.5	2.0
H_2O^+	2.0	1.4	0.7	0.7	0.6	0.9	1.3	1.9
H_2O^-	0.3	0.8	0.2	0.3	0.0	0.3	0.3	0.5
TiO_2	0.9	1.0	0.9	0.9	1.0	0.6	0.5	0.3
P_2O_5	0.2	0.1	0.1	0.1	0.2	0.1	0.1	0.0
MnO_2	0.2	0.3	0.2	0.2	0.2	0.2	0.1	0.3
CO_2	0.0	1.4	0.0	0.0	0.0	0.0	0.0	0.0
Other	0.0	0.0	0.0	0.1	0.1	0.0	0.0	0.0
Total	99.8	100.1	100.0	100.1	99.2	100.2	99.8	100.5

[a] G-B = Grenada basalts; L-AB = St. Lucia aphyric basalts; V-B = St. Vincent basalts; V-A = St. Vincent andesites; V-BG = St. Vincent basalt groundmasses; K-A = St. Kitts andesites; L-D = St. Lucia dacites; L-R = St. Lucia rhyolites. Figures in brackets at the top of each column indicate the number of analyses contributing to the mean. Data are summarized from Tomblin (1968b).

a consistent trend, both in the island as a whole and within the youngest volcanic center, toward increasing acidity with decreasing age. Unlike the Soufriere of St. Vincent, the youngest center in St. Lucia has emitted large volumes of material (up to several cubic kilometers) in individual prehistoric eruptions which have shown no detectable chemical variation, suggesting either the absence of fractionation within the magma body before eruption or the presence of so large a body of magma at depth that the volume removed was insufficient to tap significantly different compositional levels.

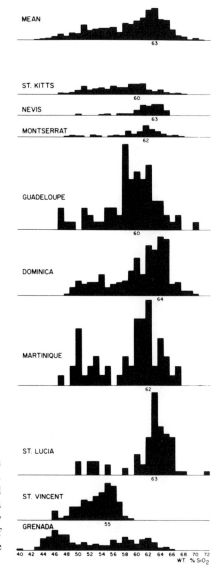

Fig. 2. Histograms showing silica distribution in the volcanic islands of the Lesser Antilles. The area of each histogram is proportional to the area of the island represented. Data are from unpublished XRF measurements by Dr. J. G. Holland made at the University of Durham on over 1200 samples. Oxides have been recalculated to 100%, water-free.

Rocks from the southern islands, including Grenada and the Grenadines, differ from those in the remainder of the Lesser Antilles not only in that they extend over a lower range of silica content, but also in that they have consistently higher potash-to-silica ratios (Fig. 3). This higher ratio is most pronounced in the silica-poor members of the sequence but remains visible throughout the full compositional range. It must reflect a significant difference either in the composition of the material partially melted to generate the magmas or in physical conditions at the original place of melting, or between this place and the surface, in the southern part of the arc. Further discussion of this question will be deferred to a later part of the chapter.

Variation in the mafic components in the Lesser Antillean volcanics is illustrated by a plot of the total iron against magnesium oxide ratios (Fig. 4). This provides, as was pointed out by Brown and Schairer (1968), a dramatic illustration of the divergence between the calc-alkaline rocks of the Lesser Antilles, which show iron impoverishment in successive differentiates, and rocks of the tholeiitic kindred (e.g., of Hawaii), in which the ratio of iron to magnesium becomes progressively higher in the residual liquids. The removal of iron in greater amounts than magnesium is attributed by Brown and Schairer (1968) to the early separation of amphibole and possibly also iron oxide under conditions of high water vapor pressure and oxygen fugacity. Evidence for the fractional crystallization of amphibole is seen in the accumulative xenoliths which are abundant is several islands, especially St. Vincent (Lewis, 1964) and St. Kitts (Baker, 1968), while the highly explosive nature of eruptions in the Lesser Antilles confirms the high water vapor content of the magma.

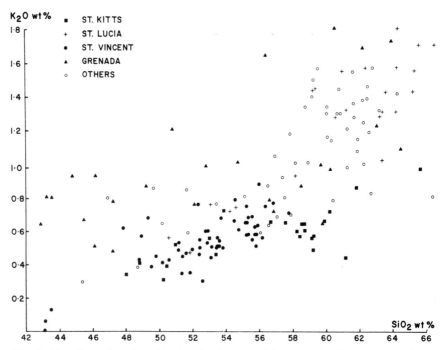

Fig. 3. Plot of potash against silica in volcanic rocks of the Lesser Antilles. Data are selected from wet chemical analyses, as received, quoted in Tomblin (1968b).

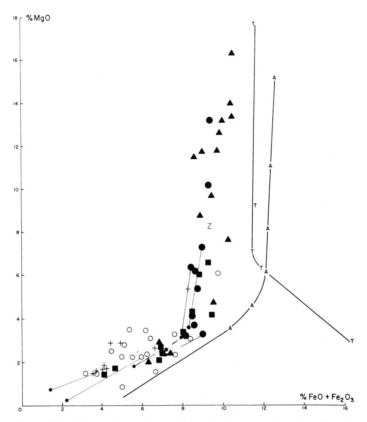

Fig. 4. Plot of magnesia against total iron in representative volcanic rocks
of the Lesser Antilles (selected from Tomblin, 1968b). Triangles, Grenada;
large solid circles, St. Vincent; crosses, St. Lucia; squares, St. Kitts; open
circles, other islands; small solid circles, groundmasses. Heavy lines through
points marked A represent Hawaiian alkali, and points marked T represent Hawaiian tholeiite series from Tilley, Yoder, and Schairer (1968). The
symbol Z indicates the mean mid-Atlantic basalt of Melson *et al.* (1968).

C. Trace Elements

Data on trace elements in the Lesser Antilles include emission-spectro-
graph measurements by Nockolds and Allen (1953), Baker (1963, 1968),
Tomblin (1964, 1968), Lewis (1964), and Rea (1970), while a summary of
determinations by x-ray fluorescence for the island of Grenada is given in
Sigurdsson *et al.* (1973). The latter represent the first published results from
a comprehensive survey of trace elements in the Lesser Antilles by x-ray
fluorescence which has recently been made at the University of Durham,
England. The abundances of most elements conform closely with those in
other calc-alkaline assemblages (Table III).

TABLE III

Average Abundance of Trace Elements in Lesser Antillean and Other Calc-alkaline Volcanic Rocks[a]

Element, p.p.m.	M-B (7)	K-B (5)	C-B (6)	K-A (12)	L-A (3)	CP-A (many)	L-D (4)	C-D (7)
Rb	10	—	40	—	51	31	112	115
Ba	108	130	250	266	383	270	243	900
Sr	380	264	975	286	270	385	277	780
Zr	83	83	80	93	97	110	88	170
Th	1.5	1.5	—	—	1.4	2.2	7.7	—
U	0.9	0.4	—	—	0.9	0.7	2.3	—
Cu	73	60	—	28	8	54	5	—
Co	28	27	35	16	9	24	5	1
Ni	13	17	100	2	6	18	5	1
Sc	24	32	20	15	15	30	7	n.d.
B	280	275	160	98	52	175	35	40
Cr	18	49	200	7	32	56	27	3
Ga	21	18	20	17	16	16	16	19
Li	11	—	25	—	21	10	34	47
K/Rb	591	—	122	—	227	430	123	151

[a] M-B = Montserrat basalts (Rea, 1970, Table 16); K-B = St. Kitts basalts (Baker, 1968, Table 7); C-B = Cascade basalts (Nockolds and Allen, 1953); K-A = St. Kitts andesites (Baker, 1968, Table 7); L-A = St. Lucia andesites (Tomblin, 1964, Table 15); CP-A = Circum-Pacific calc-alkaline andesites (Taylor, 1969, Tables 1–6, No. 4); L-D = St. Lucia dacites (Tomblin, 1964, Table 15); C-D = Cascade dacites (Nockolds and Allen, 1953, Tables 2, 3, and 7). Figures in brackets at the top of each column indicate the number of analyses included in the average. Analyses for Th and U are from Donnelly et al. (1971), and averages are based on fewer analyses than for the other elements.

The thorium–uranium relationships for Lesser Antillean rocks were reported by Donnelly et al. (1971), who found that the abundances of the elements (Th, 1.7–7.7 ppm; U, 0.7–3.3 ppm) and the Th/U ratios (basalts, 2.5; andesites, 3.1; dacite, 3.4) were both very low in comparison with continental igneous rocks. Donnelly et al. (1971) conclude that "these differences alone would suffice to rule out any appreciable involvement of continental crustal materials in the history of generation of the Lesser Antillean magma series."

D. Isotopic Studies

The study of certain isotopic abundances provides valuable information not only on the age, but also on the origin of volcanic rocks. For the latter purpose, the ratio of Sr^{87}/Sr^{86} appears to be one of the most reliable. This ratio increases primarily as a function of the amount of continental material or calcareous sediment involved in the genesis of the magma. The main studies on this ratio in the Lesser Antilles are those by Pushkar (1968), Hedge and Lewis (1971), and Pushkar et al. (1973). The 41 ratio measurements available from these papers are plotted against the silica percentage in Fig. 5. Thirty-

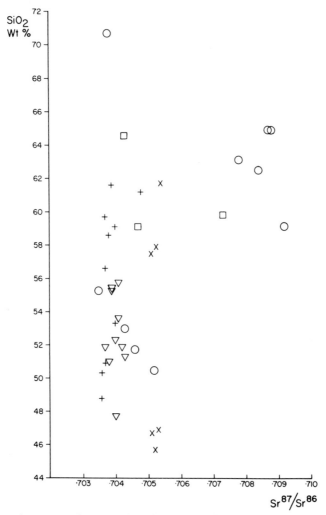

Fig. 5. Strontium isotopic ratios as a function of silica percentage in volcanic rocks of the Lesser Antilles. Strontium ratios are from Pushkar (1968), Hedge and Lewis (1971), and Pushkar et al. (1973). Silica percentages are from the wet chemical analyses quoted by Tomblin (1968).

five of these ratios fall between 0.7035 and 0.7054 and are compatible with a fairly recent, primarily mantle origin for the magmas as primitive basalts or their differentiates. The remaining six analyses, all for relatively acid rocks from St. Lucia and Dominica, have considerably higher Sr^{87}/Sr^{86} ratios of between 0.7073 and 0.7092, which cannot be explained as the result of simple differentiation of any of the analyzed basalts. Pushkar *et al.* (1973) conclude that these ratios reflect the anatexis of former ocean-floor sediments, probably including a combination of carbonate sediments and continental detritus of the type forming the Lower Tertiary sequence exposed in Barbados. It remains perplexing, however, that those of the more acid volcanics in the Lesser Antilles which appear to have incorporated a substantial proportion of ocean-floor sediment remain indistinguishable in mineralogy and major oxide composition from the others which, on the evidence of their low strontium isotopic ratios, appear to be simple differentiates from oceanic basalt or mantle. The restriction, among the strontium analyses so far published, of the high ratios to the products of large caldera-type structures (in St. Lucia and Dominica) might alternatively be attributed to assimilation at relatively shallow depth beneath the volcano. A more complete set of strontium analyses from all the islands of the arc will be necessary before these uncertainties can be resolved.

The abundances of lead isotopes in nine samples from the Lesser Antilles are presented by Donnelly *et al.* (1971), who note the high ratio of Pb^{206}/Pb^{204} in all of these samples, which are accordingly regarded as a "chemically evolved" group in contrast to a "chemically primitive" group recognized among the pre-Tertiary rocks of the eastern Greater Antilles. The Lesser Antillean lead ratios are explained either by the derivation of the magma from higher in the mantle or crust where radiogenic material has been concentrated with time or, as Donnelly *et al.* seem to prefer, by the incorporation of ocean-floor sediments into the magma.

IV. BARBADOS

The island of Barbados lies 160 km to the east and is both geographically and structurally independent of the volcanic arc. About 85% of the island is capped by reef limestone of Pleistocene and possibly Late Pliocene age. The remainder consists of a marine sedimentary sequence belonging to the Lower and Middle Tertiary which outcrops in what has been interpreted as a broad anticline with a NE–SW axis (Senn, 1940; Baadsgaard, 1960).

The oldest material belongs to the Scotland Group. This is a Paleocene to Lower Eocene flysch deposit with an estimated thickness of 1500 m (Saunders, 1968), containing epiclastic material from the South American continent which indicates the northward transport of detritus along a contemporary trench. Volcanic ash beds occur mainly in the top Eocene, showing that subaerial eruptions were already in progress from the volcanic chain to the west. The continuity of sedimentation in the Barbados area at this time contrasts with the highly interrupted growth of the volcanic islands where short periods of rapid volcanic deposition alternated with long intervals of dormancy and slow erosion. Two shallower-water limestone formations of Lower Miocene age overlie the Ocean Formation, and the whole of this sequence suffered moderately strong folding, presumably during the Late Miocene to Early Pliocene (Andean) earth movements.

Pleistocene uplifts of the island as a whole are recorded in well developed terraces in the flat-lying Coral Formation (Broecker *et al.*, 1968) of which the maximum elevation is 350 m above present sea level. The moderately large negative gravity anomaly, i.e., mass deficiency, beneath the island shows that this uplift is probably the isostatic response to an earlier period of more active tectonic downbuckling.

The significance of Barbados with regard to the evolution of the arc complex as a whole is, first, that its rocks indicate the presence of a moderately deep trench eastward of the southern volcanic arc in the early Tertiary; second, that it exhibits in its Early Tertiary sequence clastic continental and calcareous marine sediments representing the type of material which may have been carried down the subduction zone to contribute to magma genesis beneath the Lesser Antillean volcanoes; third, that it confirms the east–west axial trend of Early Tertiary fold structures which is seen to have taken place in the post-Upper Eocene in Grenada and post-Cretaceous in Trinidad and Tobago.

V. THE AVES RIDGE

The Aves Ridge is an undersea mountain chain with a north–south length of about 500 km, lying 200 km to the west of the young volcanic arc of the Lesser Antilles. It rises some 3000 m above the general level of the Venezuelan Basin and the Grenada Trough. The numerous peaks along the ridge have a local relief of up to 1500 m, and one of these projects slightly above sea level to form Aves Island. The latter in 1970 had a north–south length of 580 m and a maximum width of 150 m, with an elevation of 3 m (Herrera, 1972). The several historical maps reproduced by Herrera indicate that the island has probably decreased considerably in area since it was first mapped in 1647, although there are large uncertainties about the accuracy of early maps, and there has been no significant change in area between 1954 and 1970. Herrera notes the existence of an anticline with an east–west axis and flanks dipping at up to 25°, and indicates that the subrecent reef coral now exposed on land must be at about 12 m above its original elevation with respect to sea level.

The surface deposits forming Aves Island are calcareous gravel and sand, partly cemented and partly replaced by phosphatic material. Information on the composition and structure of the ridge as a whole comes from geophysical measurements and from dredge samples. The latter include granodiorite, dated by the K–Ar method as between mid-Cretaceous and Paleocene, from a steep escarpment on the eastern flank near the south end of the ridge at latitude 12.3° N (Fox *et al.*, 1971), which on the basis of matching compressional wave velocities (6.0–6.4 km/sec at confining pressures of 1–2 kbar) were tentatively correlated with the 6.0–6.3 km/sec layer which forms the greater part of the crustal profile beneath the Aves Ridge. From seamounts at approximately 13.0 and 13.5° N, Fox *et al.* (1971) obtained shallow-water limestones of Upper Eocene to Lower Miocene age, mostly from depths between 1000 and 1400 m below sea level, indicating subsidence of the ridge by nearly that amount since the Miocene. From the eastern flank of the Aves seamount (15.7° N) at depths between 550–1650 m, Nagle (1972) has described volcanic conglomerates and Early Miocene or older limestones. The volcanic rock types reported by Nagle from this area, and from the southern end of the Aves Ridge, range in composition from basalt through andesite to dacite, as do those of the Lesser Antilles.

The upper crustal sediments near the south end of the ridge have been sampled in the boreholes at sites 30 and 148 of the Deep Sea Drilling Project. These passed through 370 and 250 m, respectively, of Plio-Pleistocene marls and clays with volcanic ash layers. At Site 30 they continued downward in normal sequence into Middle Miocene siltstones to the hole bottom at 425 m. At Site 148, by contrast, they were separated by unconformity, with apparent subaerial weathering, from underlying volcanic sandstones including Miocene through Cretaceous fossils which are probably reworked.

From the morphology of the seamounts and the widespread distribution of calc-alkaline volcanics along the Aves Ridge, it is concluded that this feature is an extinct volcanic arc closely similar to that of the Lesser Antilles. From the capping of sediments, it seems likely, although not certain, that volcanism ended by Early Miocene at the latest. The subsidence of the ridge by over 1000 m since Early Miocene possibly represents the end of subduction along the Aves Ridge and isostatic readjustment from strongly positive to nearly normal gravity anomalies. The intrusive igneous activity, involving the emplacement of granodiorite and diabase between Middle Cretaceous and Paleocene, was probably accompanied by contemporaneous surface volcanism.

VI. GEOPHYSICAL STUDIES

A. Gravity

The main gravity features of the eastern Caribbean (Fig. 1) are the large negative-anomaly belt to the oceanward side of the arc, first described by Vening Meisnesz (1932, 1934) and Hess (1938), and the strongly positive values which coincide broadly but not exactly with the older branch of the main arc ridge. A second axis of relatively small, positive values follows the Aves Ridge.

The axis of the negative free-air-anomaly belt off the northern Lesser Antilles (Fig. 1) coincides with the foot of the eastern flank of the arc ridge and with the "outcrop" of the seismic zone. Southward from here it continues beneath the topographic high formed by the Barbados Ridge. Thus the negative-anomaly axis maintains an approximately constant distance of 150 km from the active volcanic arc. The positive axis is less regular: Andrew et al. (1970) show isostatic highs of between +120 and +150 mgal along a line running directly beneath the islands from Grenada through Martinique, corresponding with the axis of recent volcanic centers. From Martinique northward, however, a number of isolated highs occur to the east of the younger volcanic arc. The first of these is an east–west feature extending from Guadeloupe through Désirade. This coincides with a topographically elevated region which may be the result of recent tectonic uplift. Further northward, local gravity highs occur to the east of Antigua and, especially, in Barbuda where an isostatic value of +184 mgal, the highest in the eastern Caribbean, was recorded by Andrew et al. (1970).

It is clear from the distribution of these anomalies that there is a mass excess across the volcanic arc ridge and a mass deficiency across the line where the subduction zone surfaces. Areas in the Northern Lesser Antilles with the higher positive values coincide with localities where there is topographic and seismic evidence for faster recent uplift, indicating that the control of the gravity field is essentially tectonic. In the southern half of the arc, where the topographic high of Barbados lies close to the negative axis, the mass deficiency apparently results from the great thickness of low-density sediments.

The Aves Ridge is characterized by small positive anomalies, rarely exceeding +40 mgal. Free-air profiles closely follow the topography, and this together with the low values suggests that the Aves Ridge is close to isostatic equilibrium. Between Aves Island and the central Lesser Antilles, free-air anomalies are close to zero, indicating that there are no major density differences between the material of the upper crust here and beneath the Aves Ridge. Further south, in the Grenada Trough, the mildly negative free-air anomalies (−20 to −50 mgal) reflect the greater water depths and the thicker section of low-density sediments.

B. Magnetic Anomalies

Magnetic-anomaly profiles for offshore areas in the eastern Caribbean have been reported briefly by Ewing *et al.* (1960) and by Bunce *et al.* (1970). These have shown the Lesser Antilles and Aves ridges to possess high amplitude, short-wavelength anomalies which contrast with the relatively smooth field over the Grenada and Tobago basins. Both the older and the younger branches of the Lesser Antillean arc ridge are characterized by amplitudes of up to 600 γ, although the anomaly-field becomes smoother towards the south. Similarly, the magnetic anomaly-field is relatively featureless across the Barbados and southern Aves ridges. Further north along the Aves Ridge, anomalies of up to 500 γ, i.e. slightly lower than in the Lesser Antilles, are associated with topographic peaks which have been identified by seismic refraction studies as basement rocks protruding through the sedimentary cover.

C. Seismic Refraction and Reflection Studies

Extensive marine seismic-refraction studies of the Lesser Antilles were carried out by Officer *et al.* (1957, 1959) who presented several cross-profiles showing the velocity structure of the crust. More recently, Edgar (1968) and Edgar *et al.* (1971) reported refraction and reflection profiles, while Bunce *et al.* (1970) presented and discussed a sequence of reflection profiles.

In a typical refraction profile through the Eastern Caribbean arc complex (Fig. 6), the top of the mantle is lost at a depth of about 20 km as it dips eastward below the Aves Ridge and westward below the Barbados Ridge. Overlying the mantle on the Atlantic (east) side of the section is typical oceanic crust (Layer 3) with velocity of 6.7–6.8 km/sec. This has not been identified beneath the Barbados Ridge, but reappears with considerably increased thickness to form the greater part of the crustal profile beneath the Lesser Antilles and Aves ridges. Beneath the Grenada Trough and Venezuela Basin, an anomalously high (7.2–7.4 km/sec) layer with a thickness of about 5 km intervenes above the mantle, followed by a layer with velocity of 6.2 km/sec. Of the overlying, lower-velocity layers, the most extensive are the 3.0–4.0 km/sec layer which comes to the surface in the Aves and Lesser Antilles ridges and the low-velocity (1.7 km/sec) sediments which drape all but the ridge crests and reach a thickness of over 1 km in the troughs.

The seismic-refraction data show that the crust thickens notably beneath the Lesser Antilles and Aves ridges and that the Caribbean floor to the west of the Aves Ridge is closer to oceanic than to continental crust. Thus, the Lesser Antilles represent an intraoceanic island arc and were not built along the margin of a former continent.

The seismic-reflection studies of Edgar (1968), Edgar *et al.* (1971), and

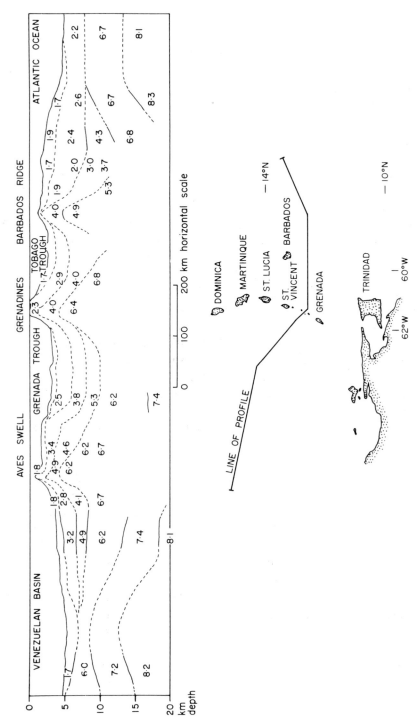

Fig. 6. Typical refraction profile through the Aves Ridge–Lesser Antilles–Barbados Ridge. Based on Officer *et al.* (1959, Fig. 17).

Bunce *et al.* (1970) have provided continuous profiles with penetration normally down to the top of the 3.0–4.0 km/sec layer (B″) and in some places, to the higher velocity layer below it. These profiles show the crumpling of the sediments on the margins of the ridges and are especially useful in defining the boundary, on the eastern flank of the Barbados Ridge, between the Caribbean and Atlantic plates.

D. Seismicity

The earliest summary of epicenters in the eastern Caribbean was that of Gutenberg and Richter (1954). This established the coincidence of major earthquakes with the margins of the Caribbean Sea, although the total number of events plotted was small, and the locations were relatively uncertain. A series of epicenter maps, one for each quarter-year, has been published in the bulletins of the Seismic Research Unit from 1953 onward, and a summary of epicenters for the period 1953–1955 has been given by Robson (1958). The regional plus teleseismic phase arrivals for the period 1950–1964 were computed by Sykes and Ewing (1965). The most reliable hypocenters for the region are those for larger events in the years 1964 onward which have been recomputed from regional and teleseismic arrivals by the International Seismological Center (Fig. 7). From this map and the corresponding cross sections (Fig. 8), it can be seen that the great majority of hypocenters fall within a layer 30 to 50 km thick with westerly dip which varies in angle and depth beneath different parts of the volcanic chain.

The northern third of the Lesser Antilles is notable for the exclusively shallow focal depths and the very low angle of dip of the seismic zone. All but four events in Fig. 8 have depths of less than 70 km and lie within a layer 30 km thick dipping at 17° towards the arc. The notable lack of activity directly beneath this part of the arc leaves the seismic zone poorly defined at greater depth. However, if extended downward to enclose the available hypocenters, this layer increases in dip to about 30°, and its top lies at a depth of 100 km beneath the recent volcanoes.

The central segment of the arc, from Guadeloupe through Martinique, contains hypocenters which in contrast with those in the northern segment are distributed almost evenly by depth down to 190 km. The hypocenters define· a seismic layer about 50 km thick which increases in dip with depth from 20° to 50°, with the result that its upper surface is at a depth of 120 km directly below the volcanic chain. The only notable irregularity in the distribution of hypocenters as a function of depth is the cluster at 150 km, which forms a bulge on the underside of the seismic zone (Fig. 8). Most of these events belong to a string of hypocenters extending east–west, normal to the island arc, and passing close to the south end of Dominica. The bathymetric map (Fig. 1)

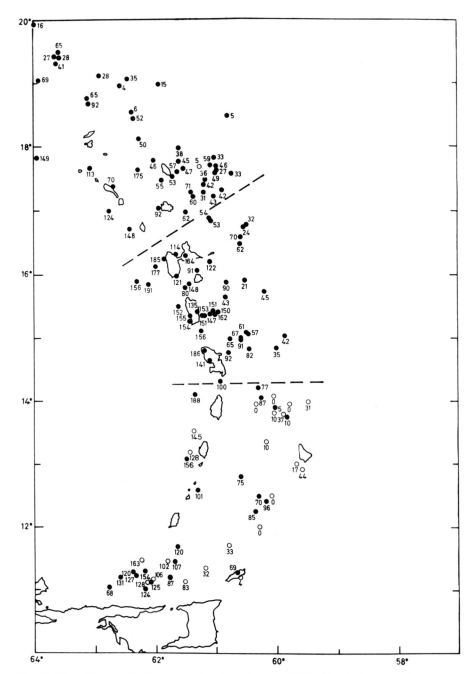

Fig. 7. Map of Eastern Caribbean earthquake epicenters, 1964–1970. Latitude and longitude are as determined by the International Seismological Centre, Edinburgh. The number beside each circle represents the focal depth in kilometers. Solid circles indicate depth determinations by the International Seismological Centre. Open circles represent depths determined by the Seismic Research Unit, Trinidad, from regional P-phase arrivals only.

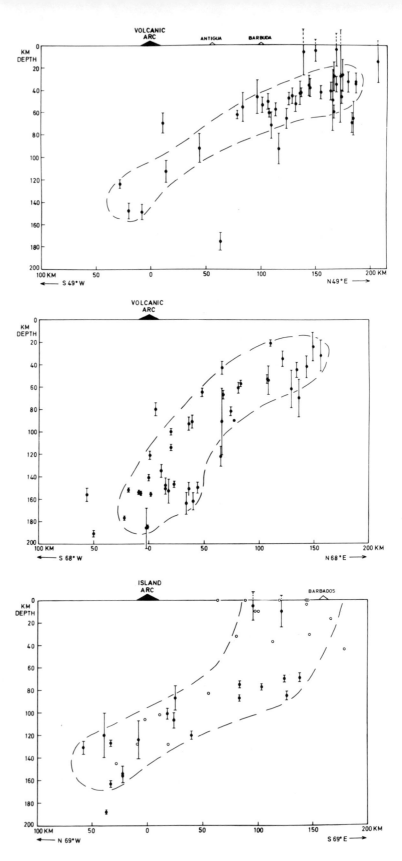

Fig. 8.

shows a relatively deep trench crossing the arc ridge at this point, which may be the surface expression of a major transverse fault. The existence of this fault and similar ones extending radially to the arc east–northeast of Guadeloupe, and from Montserrat through the south of Antigua, has been postulated by Tomblin (1972) from the distribution of all determined hypocenters, including smaller-magnitude events, during certain three-month intervals.

The southern third of the arc, from St. Lucia through Grenada, is characterized by a slightly lower level of recent seismicity. Most events fall within a layer 40 km thick which dips westward under the arc at 30°, with its upper surface at a depth of about 100 km below the volcanic axis. In contrast to the central and northern segments of the arc, the seismic zone in the southern segment appears to steepen in dip as it approaches the surface. However, there is more scatter of the shallow hypocenters in this area than elsewhere in the arc, so that the phenomenon is not very clearly established.

In addition to information which can be derived directly from the spatial distribution of earthquakes, important conclusions can also be drawn on plate movements from earthquake mechanisms. Studies of this kind have been carried out by Molnar and Sykes (1969) on two Lesser Antillean events, both of which suggested westward-dipping underthrust motion, although neither was particularly well defined.

Mechanism solutions on 14 Lesser Antillean events have recently been obtained by Tomblin (in preparation). Twelve of these are located to the east of the islands from Martinique northward, and nine of the twelve have their tension axes dipping at 40° to 50° between south and west, i.e., close to the plane of the seismic zone. The three solutions in the northern Lesser Antilles which differ from the majority all have strike-slip motions, with approximately north–south and east–west nodal planes. The largest of these events, at shallow depth to the east of Dominica, was followed by aftershocks over an area with north–south elongation, suggesting that this was the direction of the fault plane and that the movement across the fault was left-lateral. It is difficult to reconcile this solution with the general pattern of westward underthrusting of the Caribbean by the Atlantic lithosphere. The southern half of the Lesser Antilles has had fewer large-magnitude earthquakes in recent years, and consequently there are only two events which gave satisfactory focal-mechanism solutions. Both of these correspond to the predominant type of mechanism for events at the junction between the Lesser Antilles and South America

Fig. 8. Earthquake hypocenters projected into cross-profiles through the Lesser Antilles. The top, middle, and bottom profiles are for the northern, central, and southern thirds of the arc, divided along the dashed lines in Fig. 7. Solid circles with error bars indicate depth determinations and standard errors given by the International Seismological Centre, Edinburgh. Open circles are for depths determined by the Seismic Research Unit.

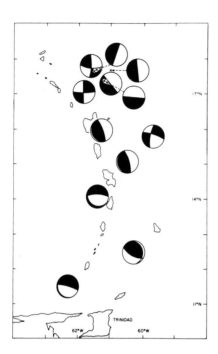

Fig. 9. Focal-mechanism solutions for earthquakes near the Lesser Antilles. Circles represent stereographic projections of rays emerg-
ing through the lower hemisphere. Shaded
quadrants represent zones of compressional
first motions.

and are tentatively interpreted as hinge faults, involving rotation on an east–
west, near-vertical plane.

The main conclusions from the distribution of hypocenters in the Eastern
Caribbean are that they define a relatively thin layer, with variable westerly
dip which represents the zone of differential movement between·the Atlantic
and Caribbean lithosphere. The mechanism solutions (Fig. 9) indicate that
the direction of movement within this zone is predominantly underthrust in
the northern Lesser Antilles and suggest that the hinge-type faulting postulated
for the southeast corner of the tectonic belt (Molnar and Sykes, 1969; Tomblin,
in preparation) may extend as far northward as the bend in the middle of the
Lesser Antillean arc, between St. Lucia and Martinique. It is of interest to
note, with regard to the thickness of the seismic zone, that this has been
progressively reduced with improvement in the accuracy of hypocenter determinations. Its present thickness corresponds approximately to the uncertainty in present focal parameters (± 25 km), suggesting that with future
improvements in instrumentation, crustal models, and computing methods,
the thickness of this zone is likely to be further reduced.

VII. SUBDUCTION AND MAGMA GENESIS

The geological and geophysical evidence point clearly to the occurrence
of crustal subduction at the present time along the Lesser Antilles. The most

convincing evidence is the high seismicity in a dipping zone which defines the plane of underthrusting of the Caribbean plate by the Atlantic lithosphere. The adjacent, strongly positive and negative gravity anomaly belts confirm that the lithosphere in this region is far out of isostatic equilibrium, while the calc-alkaline volcanism seen in the Lesser Antilles is the characteristic geological manifestation of subduction in most parts of the world. The shallower depth of the seismic zone and lower seismic energy release compared, for example, with the central Andes or the Tonga–Kermadec arcs reflect the lower rate of subduction in the Lesser Antilles.

The actual rate of subduction is not known from measurements within the region but is probably similar to the half-spreading rate in the equatorial mid-Atlantic. It seems reasonable to accept this at least until it can be demonstrated that the Caribbean plate itself is either spreading or contracting in an east–west direction. The variation in the present dip and depth of the seismic zone along its length may be related to the fact that the islands in the central section of the arc, which have the steepest and deepest seismic zone, lie almost exactly perpendicular to the direction of the underthrusting, while the northern and southern thirds of the arc diverge by 10–20° from this perpendicular and hence involve a small component of transcurrent motion. Alternatively, the different angles and depths of the seismic zone may reflect the independent behavior of different blocks of lithosphere which become separated from one another, as they approach the arc, by east–west striking faults extending perpendicularly from it.

The distribution of volcanic activity in the Lesser Antilles indicates that the subduction zone has maintained its present location at least since the early Pliocene. The westward shift of this zone in the northern half of the arc from the "Limestone Caribbees" (eastern Guadeloupe through Anguilla) to the present volcanic arc took place within, or at the end of, the Miocene. It appears that there may have been an earlier, eastward shift of the subduction zone from the Aves Ridge to the Lesser Antilles, possibly around the beginning of the Tertiary, although evidence for shifting at this time, or for the occurrence at any period of subduction beneath the Aves Ridge, is certainly not conclusive.

The occurrence of much older rocks in Désirade does not necessarily mean that these represent an early stage in the history of the Lesser Antillean subduction zone. The long time interval between these and the next oldest rocks in the Lesser Antilles suggests that they may have belonged to a separate, older orogeny and have been carried eastward on the eastern front of the Caribbean plate. Alternatively, they may have only recently been "cast up" onto the front of the Caribbean plate, off the westward-moving Atlantic floor. The occurrence of spilitic lavas in samples dredged from the Barracuda fracture zone (Bonatti, 1967) supports the latter interpretation by demonstrating the presence of more of this material in the Atlantic which has not yet reached the subduction zone. To regard the Désirade block as recently arrived from the east also avoids the issue of how, if subduction occurred beneath the Aves Ridge during the Cretaceous, the Late Jurassic or older material now in Désirade was not carried westward into the Aves subduction zone.

Another possible alternative to account for the origin of the Aves Ridge is that it moved westward away from the present site of the Lesser Antilles by the opening of the Grenada Trough along a north–south axis, in the same way as Karig (1970, 1971) has proposed for the West Mariana arc in the Philippine Sea. However, uncertainties over the operation of such a process in the Lesser Antilles are that there appear to be no median ridge or north–south basement fractures in the Grenada Trough of the kind that have been found in the Mariana Trough.

The predominant geochemical features of the Lesser Antilles are the calc-alkaline affinities of almost all the rocks and the relatively high abundance of the intermediate-silica (andesitic) members of this assemblage. Of the analyses in Fig. 2, 58% have silica contents between 54.0 and 63.9%, equivalent to andesite, while 21% have lower (basaltic) and 21% have higher (dacitic) silica contents. Although there is no way of establishing the relationship of the ratio of basic, intermediate, and acid rocks at the surface to the proportions at depth, circumstantial evidence is provided by the composition of the larger plutonic bodies in the region, e.g., the Virgin Islands Batholith, the Tobago Batholith, and the St. Martin, Désirade, Canouan, and Aves Ridge plutons, all of which have mean silica contents of not less than 60%. Hence it would appear that considerable volumes of intermediate to acid magma remain at depth, and that the high-silica volcanics represent by no means the complete intermediate-to-acid fraction from a basic-to-ultrabasic residue. On the other hand, evidence of crystal accumulation in upper crustal magma chambers is seen in certain islands, especially St. Vincent and, to a lesser extent, in St. Kitts and Grenada, from which it can be concluded that fractionation of basaltic parent magma operates at least to a certain degree. For the present we must therefore accept that either very large volumes of basalt magma fractionate to produce the observed andesites and dacites, or these more acid magmas are produced directly by partial melting. The place of melting, from the seismological evidence, is probably within or immediately above the seismic zone. If this is the case, two different types of material will be available for partial melting: first, oceanic crust which contains relatively low-melting components carried down to a high-temperature environment, and second, the mantle on the underside of the Caribbean plate which, since it is not actively descending, will already be at a temperature corresponding to the regional geothermal gradient and will receive additional heat by friction plus water by percolation upward from the upper layers of the subducted oceanic plate. This dichotomy of the possible source materials, and also of the strontium isotopic ratios, is in paradox with the homogeneity seen in the petrography and major-element geochemistry of the Lesser Antillean igneous suite.

REFERENCES

Anderson, T., and Flett, J. S., 1903, Report on the eruption of the Soufrière in St. Vincent in 1902 and on a visit to Montagne Pelée in Martinique, Part I: *Phil. Trans. Roy. Soc. London, Ser. A*, v. 200, p. 353–553.

Andrew, E. M., Masson-Smith, D., and Robson, G. R., 1970, *Gravity Anomalies in the Lesser Antilles*: H.M.S.O., London, IV, 21 p. plus maps. (Inst. Geol. Sci., Geophys. Pap. 5.)

Baadsgaard, P. H., 1960, and Barbados, W. I., Exploration results, 1950–1958: Rep. Intern. Geol. Congr. XXI, Copenhagen: Norden, Copenhagen, Pt. 18, p. 21–27.

Baker, P. E., 1963, The geology of Mt. Misery volcano, St. Kitts: Ph.D. Thesis, Univ. of Oxford, England, Unpublished, 225 p.

Baker, P. E., 1968, Petrology of Mt. Misery volcano, St. Kitts, West Indies: *Lithos*, v. 1, p. 124–150.

Barrabé, L., 1934, Sur l'affleurement du socle ancien des Petites Antilles dans l'île de la Désirade: *C. R. Acad. Sci. Paris*, v. 198, p. 487–499.

Barrabé, L., 1954, Observations sur la constitution géologique de la Désirade (Guadeloupe): *Bull. Soc. Geol. France*, v. 3, n. 6, p. 613–626.

Bonatti, E., 1967, Mechanisms of deep-sea volcanism in the South Pacific, in: *Researches in Geochemistry*, Vol. 2, Abelson, P. H., ed.: Wiley, New York, p. 453–491.

Broecker, W. S., Thurber, D. L., Goddard, J., Ku, T., Matthews, R. K., and Mesolella, K. J., 1968, Milankovitch hypothesis supported by precise dating of coral reefs and deep-sea sediments: *Science*, v. 159, n. 3812, p. 297–300.

Brown, G. M., and Schairer, J. F., 1968, Melting relations of some calc-alkaline volcanic rocks: *Carnegie Institution Yearbook*, v. 66, p. 460–463.

Bunce, E. T., Phillips, J. D., Chase, R. L., and Bowin, C. O., 1970, The Lesser Antilles Arc and the eastern margin of the Caribbean Sea, in: *The Sea*, Vol. 4, Maxwell, A. E., ed.: Wiley–Interscience, New York, p. 359–385.

Chase, R. L., and Bunce, E. T., 1969, Underthrusting of the eastern margin of the Antilles by the floor of the western North Atlantic Ocean, and origin of the Barbados Ridge, *J. Geophys. Res.*, v. 74, n. 6, p. 1413–1420.

Christman, R. A., 1953, Geology of St. Bartholomew, St. Martin, and Anguilla, Lesser Antilles: *Geol. Soc. Am. Bull.*, v. 64, p. 65–96.

De Reynal, A., 1966, Carte géologique détaille de la France, Département de la Guadeloupe, 1:50,000, Feuille de Marie Galante, La Désirade, Iles de Petit-Terre et Notice explicative: Ministère de l'Industrie, Paris.

Dickinson, W. R., and Hatherton, T., 1967, Andesitic volcanism and seismicity around the Pacific: *Science*, v. 157, p. 801–803.

Donnelly, T. W., Rogers, J. J. W., Pushkar, P., and Armstrong, R. L., 1971, Chemical evolution of the igneous rocks of the eastern West Indies: an investigation of thorium, uranium, and potassium distributions, and lead and strontium isotopic ratios, Geol. Soc. Am. Mem. 130, p. 181–224.

Edgar, N. T., 1968, Seismic refraction and reflection in the Caribbean Sea: Dissertation, Dept. of Geology, Faculty of Pure Science, Columbia Univ., unpublished.

Edgar, N. T., Ewing, J. I., and Hennion, J., 1971, Seismic refraction and reflection in the Caribbean Sea: *Am. Assoc. Petr. Geol. Bull.*, v. 55, n. 6, p. 833–870.

Edgar, N. T., Saunders, J. B., Donnelly, T. W., Schneidermann, N., Maurrasse, F., Bolli, H. M., Hay, W. W., Riedel, W. R., Premoli-Silva, I., Boyce, R. E., and Prell, W., 1971, Deep Sea Drilling project. Leg 15: *Geotimes*, v. 16, n. 4, p. 12–16.

Ewing, J., Antoine, J., and Ewing, M., 1960, Geophysical measurements in the western Caribbean Sea and in the Gulf of Mexico: *J. Geophys. Res.*, v. 65, n. 12, p. 4087–4126.

Ewing, J. I., Officer, C. B., Johnson, H. R., and Edwards, R. S., 1957, Geophysical investigations in the Eastern Caribbean: Trinidad Shelf, Tobago Trough, Barbados Ridge, Atlantic Ocean: *Geol. Soc. Am. Bull.*, v. 68, p. 897–912.

Fink, L. K., Jr., 1968, Marine geology of the Guadeloupe region, Lesser Antilles island arc: Thesis, University of Miami, Miami, Florida, unpublished, 121 p.

Fink, L. K., Jr., 1972, Bathymetric and geologic studies of the Guadeloupe region, Lesser Antilles island arc: *Mar. Geol.*, v. 12, p. 267–288.

Flett, J. S., 1908, Petrographic notes on the products of the eruptions of May 1902 at the Soufrière in St. Vincent: *Phil. Trans. Roy. Soc. London, Ser. A*, v. 208, p. 305–332.

Fox, P. J., Schreiber, E., and Heezen, B. C., 1971, The geology of the Caribbean crust: Tertiary sediments, granitic and basic rocks from the Aves ridge: *Tectonophysics*, v. 12, n. 2, p. 89–109.

Green, D. H., 1970, The origin of basaltic and nephelinitic magmas, *Leicester Lit. Phil. Soc. (England) Trans.*, v. 64, p. 26–54.

Grunevald, H., 1965, Geologie de la Martinique, Memoires pour servir à l'explication de la carte géologique détaillée de la France: Imprimerie Nationale, Paris, 144 p. plus plates.

Gutenberg, B., and Richter, C. F., 1954, *Seismicity of the Earth*: 2nd ed., Princeton Univ. Press, Princeton, New Jersey, 310 p.

Hatherton, T., and Dickinson, W. R., 1969, The relationship between andesitic volcanism and seismicity in Indonesia, the Lesser Antilles, and other island arcs: *J. Geophys. Res.*, v. 74, n. 22, p. 5301–5310.

Hay, R. L., 1959, Formation of the crystal-rich glowing avalanche deposits of St. Vincent, B.W.I.: *J. Geol.*, v. 67, p. 540–562.

Hedge, C. E., and Lewis, J. F., 1971, Isotopic composition of strontium in three basalt–andesite centers along the Lesser Antilles arc: *Contrib. Mineral. Petrol.*, v. 32, p. 39–47.

Herrera, J. P., 1972, Nuevas observaciones geologicas acerca de la Isla Aves, Venezuela, in: Caribbean Geological Conference VI, Isla de Margarita, 1971, Trans.: Impreso por Cromotip, Caracas, p. 74–78.

Hess, H. H., 1938, Gravity anomalies and island arc structure with particular reference to the West Indies: *Am. Phil. Soc. Proc.*, v. 79, n. 1, p. 71–96.

Hutton, C. O., 1968, The mineralogy and petrology of Nevis, Leeward Islands, British West Indies, A progress report, in: Caribbean Geological Conference IV, Trinidad and Tobago, 1965, Trans.: Caribbean Printers, Trinidad, p. 383–388.

Jackson, T. A., 1970, Geology and petrology of the volcanic rocks of Carriacou: M.S. Thesis, Univ. of West Indies, Jamaica, unpublished, 82 p.

Karig, D. E., 1970, Ridges and basins of the Tonga–Kermadec island arc system, *J. Geophys. Res.*, v. 75, p. 239–254.

Karig, D. E., 1971, Structural history of the Mariana island arc system: *Geol. Soc. Am. Bull.*, v. 82, p. 323–344.

Lacroix, A., 1904, *La montagne Pelée et ses éruptions*: Masson et Cie., Paris, 662 p.

Lewis, J. F., 1964, Mineralogical and petrological studies of plutonic blocks from the Soufrière volcano, St. Vincent, B.W.I.: Ph.D. Thesis, Univ. of Oxford, England, unpublished, 270 p.

MacGregor, A. G., 1938, The volcanic history and petrology of Montserrat: *Phil. Trans. Roy. Soc. London, Ser. B*, v. 229, p. 1–90.

Martin-Kaye, P. H. A., 1958, The geology of Carriacou: *Bull. Am. Palaeont.*, v. 38, n. 175, p. 395–405.

Martin-Kaye, P. H. A., 1959, *The Geology of the Leeward and British Virgin Islands*: Voice Publishing Co., St. Lucia, 117 p.

Martin-Kaye, P. H. A., 1969, A summary of the geology of the Lesser Antilles: *Overseas Geol. Miner. Resour.*, v. 10, n. 2, p. 172–206.

Melson, W. G., Thompson, G., and Van Andel, T. H., 1968, Volcanism and metamorphism in the mid-Atlantic ridge, 22°N latitude: *J. Geophys. Res.*, v. 73, p. 5925–5941.

Minear, J. W., and Toksoz, M. N., 1970, Thermal regime of a downgoing slab: *Tectonophysics*, v. 10, p. 367–390.

Molnar, P., and Sykes, L. R., 1969, Tectonics of the Caribbean and Middle America regions from focal mechanisms and seismicity: *Geol. Soc. Am. Bull.*, v. 80, p. 1639–1684.

Nagle, F., 1972, Rocks from the seamounts and escarpments on the Aves Ridge, in: Caribbean Geological Conference VI, Isla de Margarita, 1971, Trans.: Impreso por Cromotip, Caracas, p. 409–413.

Nockolds, S. R., and Allen, R., 1953, The geochemistry of some igneous rock series: *Geochim. Cosmochim. Acta*, v. 4, p. 105–42.

Officer, C. B., Ewing, J. I., Richards, R. S., and Johnson, H. R., 1957, Geophysical investigations in the eastern Caribbean: Venezuelan basin, Antilles island arc, and Puerto Rico trench: *Geol. Soc. Am. Bull.*, v. 68, p. 359–378.

Officer, C. B., Ewing, J. I., Hennion, J. F., Harkrider, D. G., and Miller, D. E., 1959, Geophysical investigations in the Eastern Caribbean: summary of 1955 and 1956 cruises, in: *Physics and Chemistry of the Earth*, vol. 3, Ahrens, L. H., Press, F., Rankama, K., and Runcorn, S. K., eds.: Pergamon, London, p. 17–109.

Pushkar, P., 1968, Strontium isotope ratios in volcanic rocks of three island arc areas: *J. Geophys. Res.*, v. 73, p. 2701–2714.

Pushkar, P., Steuber, A. M., Tomblin, J. F., and Julian, G. M., 1973, Strontium isotopic ratios in volcanic rocks from St. Vincent and St. Lucia, Lesser Antilles: *J. Geophys. Res.*, v. 78, n. 8, p. 1279–1287.

Rea, W. J., 1970, The geology of Montserrat, British West Indies: Ph.D. Thesis, Univ. of Oxford, England, unpublished, 196 p.

Robinson, E., and Jung, P., 1972, Stratigraphy and age of marine rocks, Carriacou, West Indies: *Am. Assoc. Petr. Geol. Bull.*, v. 56, n. 1, p. 114–127.

Robson, G. R., 1958, Seismological and volcanological work in the eastern Caribbean, 1952–1955, abstract and map, in: Caribbean Geological Conference, First Meeting, Antigua, 1955, Rep., p. 26: Argosy Co., Bel Air Park, E. C. Demerara [Guyana].

Robson, G. R., and Tomblin, J. F., 1966, Catalogue of active volcanoes of the world including solfatara fields, Part XX, West Indies: Intern. Assoc. Volcanology, Rome, 56 p.

Saunders, J. B., 1968, Field trip guide: Barbados, in: Caribbean Geological Conference IV, Trinidad and Tobago, 1965, Trans.: Caribbean Printers, Trinidad, p. 443–449 plus panorama.

Senn, A., 1940, Paleogene of Barbados and its bearing on history and structure of the Antillean Caribbean region: *Am. Assoc. Petr. Geol. Bull.*, v. 24, n. 9, p. 1548–1610.

Sigurdsson, H., Tomblin, J. F., Brown, G. M., Holland, J. G., and Arculus, R. J., 1973, Strongly undersaturated magmas in the Lesser Antilles island arc: *Earth Planet. Sci. Lett.*, v. 18, p. 285–295.

Sykes, L. R., and Ewing, M., 1965, The seismicity of the Caribbean region: *J. Geophys. Res.*, v. 70, n. 20, p. 5065–5074.

Taylor, S. R., 1969, Trace element chemistry of andesites and associated calc-alkaline rocks, in: Proceedings of the andesite conference, Bull. 65: State of Oregon, Department of Geology and Mineral Resources, p. 43–63.

Tilley, C. E., Yoder, H. S., and Schairer, J. F., 1965, Melting relations of volcanic tholeiite and alkali rock series: *Carneg. Instn. Year Book 64*, p. 69–89.

Tomblin, J. F., 1964, The volcanic history and petrology of the Soufrière region, St. Lucia: Ph.D. Thesis, Univ. of Oxford, England, unpublished, 213 p.

Tomblin, J. F., 1968a, The geology of the Soufrière volcanic centre, St. Lucia, in: Caribbean Geological Conference IV, Trinidad and Tobago, 1965, Trans.: Caribbean Printers, Trinidad, p. 367–376.

Tomblin, J. F., 1968b, Chemical analyses of volcanic rocks from the Lesser Antilles: Univ. West Indies Seismic Res. Unit Spec. Publ. 15, 23 p.

Tomblin, J. F., 1972, Seismicity and plate tectonics of the eastern Caribbean, in: Caribbean Geological Conference VI, Isla de Margarita, 1971, Trans.: Impreso por Cromotip, Caracas, p. 277–282.

Tomblin, J. F., in preparation, Eastern Caribbean seismicity and focal mechanisms.

Vening Meinesz, F. A., 1932, Gravity expeditions at sea, 1923–1930: Neth. Geod. Commun. Publ., Waltman, Delft, 109 p.

Vening Meinesz, F. A., 1934, Gravity expeditions at sea, 1923–1932: Neth. Geod. Commun. Publ. II, Waltman, Delft, 208 p.

Wager, L. R., 1962, Igneous cumulates from the 1902 eruption of Soufrière, St. Vincent: *Bull. Volcanol. Ser. II*, v. 24, p. 93–99.

Westermann, J. H., and Kiel, H., 1961, The geology of Saba and St. Eustatius: Uitvoerige Natuurw. Werkgrp. Ned. Antillen (Publ. Foundation Sci. Res. in Surinam and the Netherlands Antilles), v. 24, p. 1–175.

Chapter 12

THE GEOLOGY OF HISPANIOLA

Carl Bowin

Woods Hole Oceanographic Institution
Woods Hole, Massachusetts

I. INTRODUCTION

At present we have a very imperfect picture of the evolution of an island arc. Linear chains of active volcanoes presently occur in many island arcs and are associated with active subduction of an oceanic lithospheric plate beneath the arc, as are deep-sea trenches. How lithospheric underthrusting begins is much in doubt. The proposed nascent island arc in the Indian Ocean (Sykes, 1970) outward from the Java Trench, suggested by a diffuse distribution of epicenters, has as yet not been supported by further studies in the area. The development of an island arc is poorly known, and the validity of interarc spreading (Karig, 1971a, 1971b) is debated. What happens to an island arc when underthrusting ceases is also a matter of conjecture. Obviously, detailed geologic studies of island arcs can and will contribute to a better understanding of these problems. This paper attempts to summarize the present geologic knowledge of the island of Hispaniola toward that undertaking.

II. GEOGRAPHY

The Caribbean island arc lies between the continents of North and South America and comprises the Greater Antillean islands on the west and the Lesser

Antillean islands on the east. The Greater Antillean islands, as the name implies, are larger and consist of the islands of Cuba, Jamaica, Hispaniola, and Puerto Rico. These islands have virtually no recent volcanic activity. The Lesser Antillean islands, on the other hand, are smaller, and many of them have present volcanic activity. The Dominican Republic and the Republic of Haiti share the island of Hispaniola, the second largest of the Greater Antilles. Hispaniola is especially intriguing because four structural trends converge upon it: the main axis of the island arc, the southeastern part of the Bahamas, the swell (Nicaragua Rise) extending from Central America to southwestern Hispaniola, and the Beata Ridge.

The coastline of Hispaniola varies from sandy beaches to raised coral terraces with undercut cliffs facing the sea, and to steep mountainous slopes disappearing beneath the water's edge. Inland, the terrain varies from wide, nearly flat valleys to high rugged mountain ranges. Most of the valleys support only a semiarid vegetation and have the appearance of a desert. Locally, particularly where irrigated, the valleys are very fertile and support highly productive agriculture. The mountains generally support a dense forest or jungle growth which considerably inhibits geologic exploration, although in much of the country the forest is devastated by shifting cultivation and burning.

In marked contrast to the other West Indian islands, the physiography of much of Hispaniola is characterized by alternating valleys and mountain ranges (Figs. 1 and 2). These trend, in general, obliquely to the length of the island, the principal exception being the Massif de la Hotte and Massif de la Selle of the Southern Peninsula of Haiti. The grain of the topography in central and northern Haiti trends about 40° N to 50° W. This area includes the Massif du Nord, Terre-Neuve Mountains, Montagnes Noires, and the Chaine des Matheux. The southeastward continuations of these mountains show a change in trend that occurs along an imaginary line trending about N 45° E. To the east of this line the mountainous topography trends more nearly N 85° W, while farther to the east it curves to a trend of about N 45°–60° W. This curvature seems to be most pronounced in the 200-m contour on the south side of the Cibao Valley. The Sierra de Bahoruco, the Sierra Martin Garcia, and the high topography of the Cordillera Central terminate to the southeast along another imaginary line that also trends about N 45° E. This trend is coincident with the strike of the steep submarine topography along the southeast side of the Barahona Peninsula and with the crest of the western scarp of the Beata Ridge. Between the two imaginary northeast-trending lines just discussed lies almost all of the highest topography of the Greater Antillean islands, and the trend of that topographic grain is slightly concave to the southwest.

The northern margin of the Cibao Valley is very straight, in marked contrast to the southern margin. The crest of the Cordillera Septentrional

Fig. 1. Location map identifying principal topographic features and cities of Hispaniola. From Weyl (1966, Fig. 36).

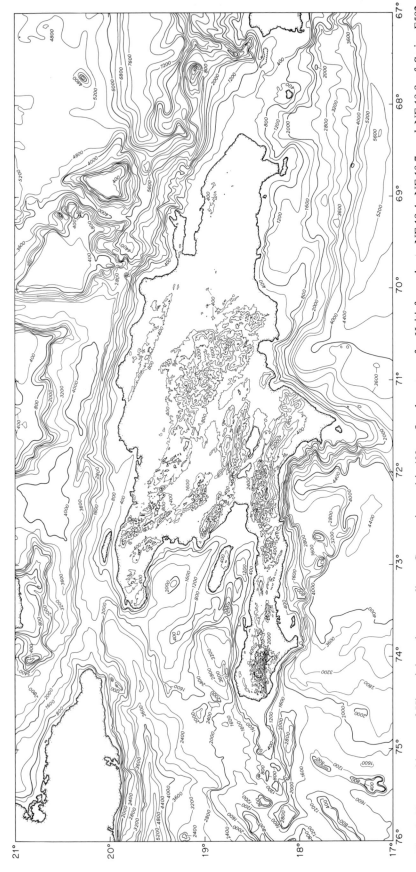

Fig. 2. Topographic map of Hispaniola and surroundings. Contour interval is 400 m. Land contours for Haiti from sheets NE 18-4, NE 18-7, and NE 18-8 of Series E502, U.S. Army Map Service (1963); for Dominican Republic provided by J. E. Matthews, U.S. Navy Oceanographic Office. Isobaths are adapted from Hydrographic Office charts BC 0804, both 1st and 2nd editions, and BC 0704; and from J. E. Matthews, U.S. Navy Oceanographic Office, for the area south and east of eastern Hispaniola. The 800-, 2000-, 4000-, and 6000-m isobaths are drawn heavier.

trends close to N 67° W over the entire 145 km length of the range. Although not well shown in Fig. 2, the trend of the topographic grain of the western and central portion of the Sierra del Seibo is approximately N 70° W. The grain in the easternmost end appears to be about N 40° W. The Sierra del Seibo, however, is not a high topographic feature, and the low topography of the eastern portion of the island differs considerably from the remainder of Hispaniola. The Sierra del Seibo borders on the north a broad raised coastal plain of Quaternary reef and back reef facies sediments.

Cucurullo (1956, p. 106) has given the average elevation for the various topographic provinces of the Dominican Republic. These are as follows: Cordillera Septentrional, 600 m; Cordillera Central, 1,800 m; Sierra de Neiba, 1,000 m; Sierra de Bahoruco, 1,100 m; Sierra de Samana, 400 m; Sierra del Seibo, 450 m; Cibao Valley, 80 m; San Juan Valley, 350 m; Enriquillo Valley, 20 m; and Coastal Plain, 60 m.

The Cordillera Central range reaches its greatest height in Pico Yunque (for a time named Pico Trujillo) with an elevation of 3,175 m (10,414 ft). This is the highest peak in the West Indies. The lowest part of the island is the surface of Lake Enriquillo, which in 1950 was 44 m below sea level. The land area of Hispaniola is approximately 80,000 km², of which approximately 30,000 km² is under the jurisdiction of the Republic of Haiti and approximately 50,000 km² is under the jurisdiction of the Dominican Republic. These measurements include adjacent islands and some lakes and lagoons.

III. GENERAL GEOLOGY AND STRUCTURE

Exposed on Hispaniola are igneous, metamorphic, and sedimentary rocks (Fig. 3). Both plutonic and volcanic igneous rocks occur, and they range in composition from ultramafics (peridotites) to leucocratic tonalites (mica trondjemites). The metamorphic rocks originally were primarily volcanic flows and tuffs and limestones. The metamorphism appears principally regional, with contact metamorphic effects only found locally around some intrusions. Sedimentary rocks are mainly of latest Cretaceous and younger age.

Exposed plutonic and hypabyssal igneous bodies appear to have been emplaced over a considerable span of time. The oldest emplacement occurred in pre-Middle Albian time and includes hornblendite and possibly augite norite and serpentinized peridotite. Tonalite (quartz diorite) plutons, some of batholithic proportions, may have been emplaced in Turonian, Santonian, Maestrichtian, and Eocene time. In Late Eocene time hypabyssal stocks of pyroxene, diorite, diabase, and gabbro were emplaced. Skarns, locally with iron deposits, were developed locally where the stocks came in contact with limestone.

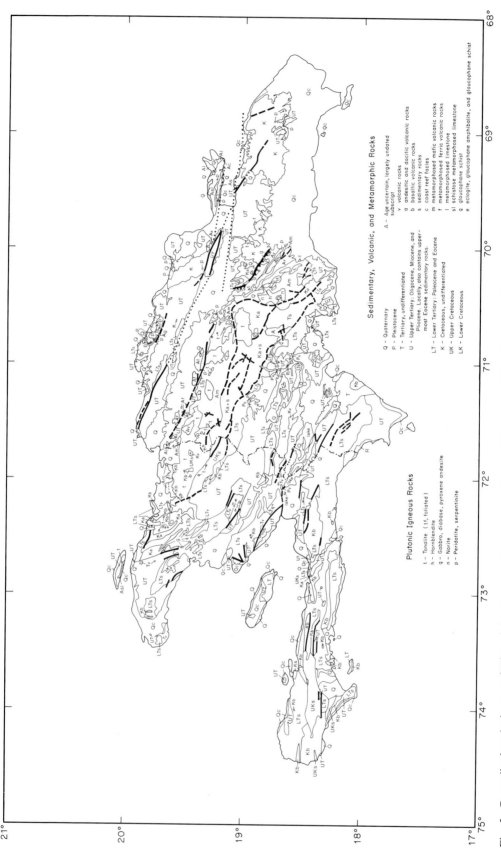

Sedimentary, Volcanic, and Metamorphic Rocks

A - Age uncertain, largely undated
subscript
   v - volcanic rocks

Q - Quaternary
   a - andesitic and dacitic volcanic rocks

P - Pleistocene
   b - basaltic volcanic rocks

T - Tertiary, undifferentiated
   s - sedimentary rocks

U - Upper Tertiary: Oligocene, Miocene, and
   Pliocene. Locally, also contains uppermost Eocene sedimentary rocks.
   c - coast reef facies

   m - metamorphosed mafic volcanic rocks

LT - Lower Tertiary: Paleocene and Eocene
   f - metamorphosed ferric volcanic rocks

K - Cretaceous, undifferentiated
   l - metamorphosed limestone

UK - Upper Cretaceous
   sl - schistose metamorphosed limestone

LK - Lower Cretaceous
   g - glaucophane schist

   e - eclogite, glaucophane amphibolite, and glaucophane schist

Plutonic Igneous Rocks

t - Tonalite (tf, foliated)
h - Hornblendite
g - Gabbro, diabase, pyroxene andesite
n - Norite
p - Peridotite, serpentinite

Fig. 3. Generalized geologic map of Hispaniola. Sources of data include: Blesch, 1966; Bowin, 1960, 1966; Butterlin, 1960; MacDonald and Melson, 1969; Kesler and Speck, in press; Nagle, 1966, in press, personal communication, 1973; Palmer, 1963; Bermudez, 1949; and G. A. Antonini, unpublished map "Surficial Geologic Map of Northwestern Dominican Republic," dated 1970.

Products of volcanic activity are very common on the island. The oldest of the unmetamorphosed volcanic rocks appear to be of Lower Cretaceous age (Middle Aptian–Middle Albian) and are predominantly flows. Upper Cretaceous volcanic rocks are composed in large part of tuff. A tectonic event which led to the formation of an unconformity over much of the island appears to have occurred in Maestrichtian time. Volcanic activity continued locally from Maestrichtian to earliest Late Eocene time. Volcanism practically ceased in Hispaniola in the Late Eocene, coincident with important structural deformation, including the development of the Hatillo Thrust (Bowin, 1966). Some Upper Tertiary volcanic deposits have been reported in Haiti (Butterlin, 1960; Woodring *et al.*, 1924), and several outpourings of Late Cenozoic limburgite, hornblende–augite, andesite porphyry, and nepheline basalt occurred in the southern part of the island. These flows appear to be approximately of Pleistocene age.

The metamorphic rocks occur primarily in two zones; one is a belt diagonally across the center of the island, and the other consists of isolated exposures near the north coast. The diagonal belt consists primarily of two distinct assemblages, one of metamorphosed mafic volcanic rock and the other characterized by the mineral assemblage albite–quartz–sericite. These two assemblages are separated by a fault zone (herein named the Hispaniola Fault Zone) along which serpentinized peridotite bodies occur, some of which are large. The Hispaniola Fault Zone includes the Loma Caribe Fault (Bowin, 1966) and the Tavera, Amina, and Inoa Faults (Palmer, 1963). Three types of metamorphosed rocks occur along the north coast, but their interrelations are obscure. The most widely known type is the metamorphosed schistose limestone and marble that occurs on the island of Tortue and on the Samana Peninsula. Only recently have glaucophane and actinolite schist been reported from the Puerto Plata area (Nagle, 1966, 1971), and the occurrence of eclogite on Samana Peninsula has just been recognized (Nagle, in press).

Sedimentary rocks occur as clastic and epiclastic deposits of conglomerate, sandstone, and shale. These rocks apparently began to accumulate in significant deposits in the latest Cretaceous (Maestrichtian). Clastic sedimentary rocks of Maestrichtian to Recent age are common. In a general way, they become more abundant through Tertiary time, and over a thousand meters of clastic sedimentary rocks were deposited during Pliocene time in the Cibao and Enriquillo basins. The occurrence of limestone ranges from small lenticular pods about 1 m long to units of limestone more than 1,000 m thick. The metamorphosed limestones are undated. Paleontologically dated limestones in Hispaniola range in age from Lower Cretaceous to Recent. The common occurrence of limestone in an island-arc setting may be surprising to those geologists who think of limestones as being principally deposited in miogeosynclinal and stable continental regions.

The structural grain of the island (Fig. 4) as shown by trends of major faults, distribution of metamorphic rocks in the central belt, the zone of tonalite plutonic bodies, and the general trend of fold axes is very similar to that shown by the grain of the topography (Fig. 2). The foliation in the metamorphic rocks west of La Vega is approximately N 70° W but changes to about N 40° W to the southeast of La Vega. These trends and their change are concordant with the previously discussed curvature in the topographic trend of the Cordillera Central. It is perhaps also noteworthy that the Hispaniola Fault Zone separating the metamorphosed mafic from femic volcanic rock assemblages of the central belt bulges northeastward (hence is concave to the south) near La Vega, departing from its trend both to the northwest in easternmost Dominican Republic and to the southeast where serpentinized peridotite has been emplaced along the fault zone. A similar curvature is shown by the northwestern part of the Hatillo Thrust and by the Bonao Fault (Bowin, 1966). This suggests that the Bonao Fault is probably a thrust.

The linear topography of the Cordillera Septentrional, the straight trend of the Septentrional Fault (Fig. 4) on the south margin of that range, the straight contact between Quaternary alluvium and Upper Tertiary sedimentary rocks along several sites in the Cibao Valley, and the straight trend of submarine scarps off northern Hispaniola all suggest that strike-slip faulting has been active in the northern Hispaniola region. Several trends are present and range between about N 65° W and N 85° W.

As noted previously, there is a contrast in the type of metamorphic rock exposed on the northern margin of Hispaniola with those of the median metamorphic belt. This contrast has been likened to paired metamorphic belts by Nagle (in press), but the age relationships of the metamorphic types is not established. The Hispaniola Fault Zone has been identified (Fig. 4) westward from the area mapped by Palmer (1963) to near the border with Haiti on the basis of the surficial rock types mapped by G. A. Antonini (Unpublished map, "Surficial Geologic Map of Northwestern Dominican Republic," dated 1970). This fault zone is inferred to continue into the northeasternmost corner of Haiti and hence offshore. The straight bathymetric contours along the north coast of Haiti (Fig. 2) suggest that they are fault controlled, and a strike-slip fault (Tortue Fault) is inferred (Fig. 4) to lie between the island of Tortue and the mainland. The trend of the southern contact between Quaternary alluvium and Upper Tertiary rocks in the western Cibao Valley (Fig. 3) is inferred to indicate the eastward continuation of the Tortue Fault into northwestern Dominican Republic. The Tortue Fault is inferred to truncate the Hispaniola Fault Zone, but this relationship does not clarify the age relations between the two metamorphic zones of Hispaniola: the north coast zone (Tortue Island, Puerto Plata–Gaspar Hernandez area, and Samana Peninsula) and the central zone (Duarte, Maimon, and Amina Formations).

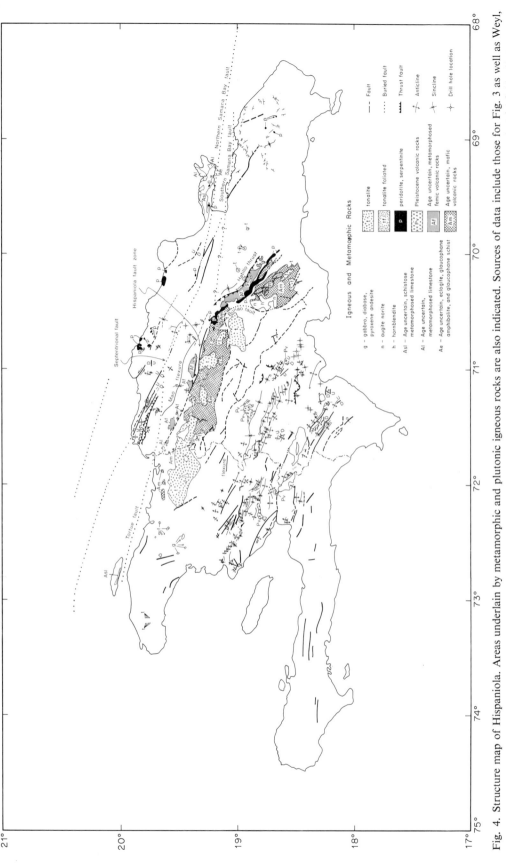

Fig. 4. Structure map of Hispaniola. Areas underlain by metamorphic and plutonic igneous rocks are also indicated. Sources of data include those for Fig. 3 as well as Weyl, 1966; Dominican Seaboard Oil Company reports by M. B. Arick, J. W. Hunter, P. J. Bermudez, R. Beall, W. M. Small, and C. F. Dohm; and Atlantic Richfield Oil Company reports by H. M. Kirk. Note that in places the structure map (this figure) and the geologic map (Fig. 3) are not compatible because different sources of information were used.

The structure of most parts of the island is not well known, particularly in the mountain portions and in eastern Dominican Republic. A compilation of structure cross sections is presented in Fig. 5. The fold axes mapped in the Upper Tertiary sedimentary rocks of central Hispaniola indicate that portion of the island has been subjected to a northeast–southwest directed compression at least from latest Eocene to Pliocene time. The foliation in the metamorphic rocks and the distribution of tonalite plutons suggest that similarly directed compressive forces were also active during the Cretaceous.

IV. STRATIGRAPHY

A. Pre-Middle Albian Metamorphic and Nonmetamorphic Rocks

A belt of metamorphic rocks extends diagonally across Hispaniola. In central Dominican Republic the metamorphic rocks occur at the eastern end of the Cordillera Central, continue westward along its northern flank, and may extend into northeastern Haiti. These metamorphic rocks are probably of early Lower Cretaceous (pre-Middle Albian; Bowin, 1966) or pre-Cretaceous age. A radiometric age (127 m.y.) on a hornblendite intrusion suggests they are earliest Cretaceous or older, although excess argon, if present, would yield an erroneous age. The metamorphic rocks of this belt occur as two distinct assemblages separated by a fault zone along which serpentinized peridotite commonly occurs. To the south of the fault zone, metamorphosed mafic volcanic rocks occur, now generally represented by actinolitic amphibolite (Duarte Formation). To the north of the fault zone, the metamorphosed volcanic rocks (Maimon Formation) are characterized by the mineral assemblage albite–quartz–sericite and appear to be predominantly metamorphosed quartz keratophyres.

The Duarte and Maimon Formations have been regionally metamorphosed to the greenschist facies where they occur in the metamorphic belt at the eastern end of the Cordillera Central (Bowin, 1966). To the west along the north flank of the Cordillera Central, the Duarte Formation decreases in metamorphic grade to the sub-greenschist facies, but the recrystallization of the Amina Formation (equivalent to the Maimon Formation) is more intense than in the Maimon Formation (Palmer, 1963).

The actinolitic epidote amphibolites (Duarte Formation) are intruded by several plutons of schistose leucocratic muscovite tonalite (trondjemite), a batholith of augite norite, and small stocks of hornblendite. These igneous types are probably of at least two ages: earliest Cretaceous or older (hornblendite) and Upper Cretaceous (tonalite). At the east end of the Cordillera Central the trend of the peridotite and of the schistosity in the metamorphic rocks is approximately N 40° W, but near La Vega it changes to N 70° W. Farther west, along the north flank of the Cordillera Central, the trend may be closer to N 60° W.

In Haiti, metamorphic rocks occur as float on the North Plain and on the Léogâne Plain and are exposed in outcrop on the island of Tortue (Woodring *et al.*, 1924, p. 84). On the North Plain east of Limonade, float of quartz, schist, and mica–schist was found. These schists are much more metamorphosed than in any of the Cretaceous volcanic rocks or sediments and, therefore, were considered of Paleozoic or early Mesozoic age by Woodring *et al.* (1924, p. 84). Farther to the east on the North Plain, a little east of Trou du Nord (formerly called Le Trou), Woodring *et al.* (1924, p. 84) reported finding float of garnetiferous quartz–schist composed essentially of quartz, chlorite, biotite, magnetite, and garnet. These schists on the North Plain may be correlative with the metamorphosed keratophyres and quartz keratophyres of the Maimon Formation of central Dominican Republic. Woodring *et al.* (1924, p. 296–297, 308–309, and 467–469) reported the occurrence of amphibolitic schists on the north flank of the Massif du Nord and in some of the low hills occurring in the North Plain. Apparently these amphibolitic schists were considered to be the result of contact metamorphism from the quartz diorite plutons exposed in the Massif du Nord. On Morne Beckly, a 10-m high swell, on the North Plain about 5 km east of Limonade, occur talcose, chloritic, and amphibolitic schists having a schistosity that strikes about N 30°–40° W and dips steeply (Woodring *et al.*, 1924, p. 468). Woodring, *et al.* (1924, p. 309) also report that outcrops of chlorite schists occur on the north flank of the Massif du Nord and (west of Grande-Riviere du Nord) and about 6 km northeast of Saint-Michel de l'Atalaye. These schists are described as being light greenish in color and spotted with dark flakes of chlorite. These rocks were inferred to be the result of chloritization of basaltic rocks along zones of thermal activity. Calcite has partially replaced chlinochore in some of these rocks.

It is here inferred that the amphibolitic and chloritic (presumably sub-greenschist) rocks exposed in northern Haiti are at least in part equivalent to the Duarte Formation of the Dominican Republic. In general, the occurrences of the quartz and mica–schists appear to be northeast of the occurrences of amphibole and chlorite schists, and this spatial relation is consistent with an eastward extension into northeastern Haiti of the boundary between the Maimon and Duarte Formations of central Dominican Republic. If so, this boundary would appear to trend offshore between Cap Haitien and the Dominican border (Fig. 4).

Mica and quartz schist, like those found in the North Plain, were found as float on the north coast of the Southern Peninsula of Haiti on the western part of the Léogâne Plain about 30 km west of Port-au-Prince (Woodring *et al.*, 1924, p. 84).

Schistose limestone occurs both on the island of Tortue and on the Samana Peninsula. On Tortue Island they occur beneath limestone of probable Upper Oligocene age (Woodring *et al.*, 1924, p. 85) and consist of recrystallized and much sheared limestones, in some places partly replaced by chlorite, epidote, and quartz. The exposures are near the center of the south coast of the island. They also occur inland at an exposure where the overlying limestone has been removed by erosion. The schistosity, in general, strikes N 80° W and dips 20° northeast according to Woodring *et al.* (1924, p. 85). Schistose limestone and quartz–calcite–chlorite–muscovite schists of unknown age are found in the Samana Peninsula. Vaughan *et al.* (1921, p. 182–183) reported that micaceous schist and schistose limestone are exposed in the mountains of the peninsula. Reconnaissance by Nagle in 1970 (in press) shows the calcareous mica–schist and marble to occupy the center of the peninsula. Gray massive limestone occurs to the south of the schist and marble, and black marble and marble breccia occur on the north side at the east end of the peninsula. Upper Tertiary white vuggy and fossiliferous limestone is exposed on much of the eastern end of the peninsula and also locally at the western end. The foliation in the schist strikes nearly east–west and has dips in the range of 40–60°. Tight, overturned folds were noted in the schist and marble assemblage.

Unmetamorphosed volcanic rocks of inferred pre-Middle Albian age (Bowin, 1966) have been mapped in central Dominican Republic. Vitric basalts of the Siete Cabezas Formation unconformably overlie the metamorphosed Duarte Formation, and pyroxene andesite and tuff of the Peravillo Formation unconformably overlie the Maimon Formation. These undated metamorphic and unmetamorphosed volcanic rocks occur within the fault-bounded belt of central Dominican Republic.

Metamorphic (?) rocks are reported to occur (oil company report) to the west of Samana Peninsula at the southeast end of the Cordillera Septentrional, and Nagle (personal communication, July 1973) confirms the occurrence there of gneissic amphibolites, serpentinite, and rocks of a quartz dike-complex. This region is virtually unexplored geologically, and very little is known of these rocks. Although no regionally metamorphosed rocks are known to occur in place in the remainder of the Cordillera Septentrional, nor in the exposures of older rocks near Puerto Plata, tectonic inclusions of actinolite and glaucophane schist occur in the peridotite bodies near the north coast of the Dominican Republic east of Imbert, and eclogite occurs in mica-schist on the Samana Peninsula (Nagle, in press).

B. Cretaceous

The oldest dated rocks of Hispaniola are Lower Cretaceous, Middle Aptian–Middle Albian. In central Dominican Republic to the northeast of the belt of metamorphic rocks they are exposed at the base of the known Cretaceous section. This Lower Cretaceous sequence (Los Ranchos Formation) began with the extrusion of quartz keratophyre, dacite, pyroxene andesite, uralite andesite, tuff, and other less common volcanic types, and deposition of very minor limestone, siltstone, and massive chert. Deposition of the Hatillo Limestone of Early Cretaceous age followed the eruption of Los Ranchos rocks. Fine tuffs of the Las Lagunas Formation (lower Cretaceous?) were deposited conformably upon the limestone.

An Aptian–Albian Cretaceous age has been reported by Reeside (1947) from about 23 km northwest of Jacmel where limey sediments containing caprinids occur interbedded with pillowed basalt. According to Mitchell (1953), a fauna ranging from Hauterivian to Barremian (Lower Cretaceous) appears to have been found in the northern part of Haiti. However, this identification has been questioned (Butterlin, 1960, p. 93).

Upper Cretaceous rocks occur in many localities in Hispaniola, in both Haiti and the Dominican Republic. They are predominantly volcanics, and in large part of pyroclastic origin, although in places clastic sedimentary rocks or limestones dominate. In Haiti, Upper Cretaceous rocks occur in the Southern Peninsula (Macaya Formation), at a few localities in the Northern Peninsula, and in the region of the Massif du Nord of northern Haiti. In western Dominican Republic they continue along the crestal region of the Cordillera Central, which is the eastern continuation of the Massif du Nord of Haiti. Upper Cretaceous rocks also occur farther eastward on both the southern and northern flank of the Cordillera Central, and they form the core of that mountain range at its eastern end. The Sierra del Seibo, in the eastern part of the Dominican Republic, is composed predominantly of fine-grained tuff and interbedded dark-gray limestone which has been dated in a few localities as Late Cretaceous. Undated rocks, probably of Late Cretaceous age, occur at the eastern end of the Barahona Range, within the Sierra de Neiba Range, in the Puerto Plata area of northern Dominican Republic, and near the eastern end of the Cordillera Septentrional.

In the Southern Peninsula of Haiti, Upper Cretaceous unmetamorphosed limestone (Macaya Formation) is exposed over a large part of the Massif de la Hotte (an area about 80 km long by 13 km wide), underlying the main part of the Massif du Macaya. A smaller area (about 30 km by 24 km) occurs to the south in the region of Morne Sinai. The Macaya Formation (Butterlin, 1960, p. 44) consists predominantly of fine-detrital clayey limestone that occurs as thick beds. It is of variable color, usually from chocolate brown to yellowish, but in places green or violet. Layers of fissile clay and radiolarite are interbedded at various levels. Veinlets of silica and calcite are common. The upper beds are massive gray or brown shales with conchoidal fracture. The formation is highly folded, probably more than 2000 m thick, and the underlying basement is not observed. On the route from Les Caye to Jeremie, in the valley of the Glace River, a fairly complete assemblage of rocks of the formation is observed to be overlain with angular discordance by Paleocene and Lower Eocene rocks. Small exposures occur in the volcanic terrain in the region of Capafou, in the valley of the river Bras á Gauche, and south of Petit Goave (Butterlin, 1960, p. 31). Microfossils (small and large foraminifera and radiolaria) of the Macaya Formation indicate a range in age from Campanian to Maestrichtian, with a major part of the formation being of Late Campanian age (Ayala, 1959, p. 120; Butterlin, 1960, p. 31). On the Massif de la Selle (Butterlin, 1960, p. 44) there are numerous small outcrops of Cretaceous clay limestones that are dark gray with siliceous zones, and which, except for the exposures near the region of Temisseau, are too small to be shown on the geologic map prepared by Butterlin (1960). These clay limestones were correlated with the Macaya Formation, and they are overlain, also with angular disconformity, by Paleocene and Lower Eocene detrital rocks.

Ammonites indicating an Early Senonian age were found by Woodring and identified by Reeside (1947) in boulders of calcareous sandstone also containing pebbles and granules of basalt. These boulders were found at two localities in Riviere Corail, a southeastward-flowing stream entering Grand Riviere about 8 miles (13 km) north of Jacmel. South of Furcy, Butterlin (1960, p. 44) found limestone with a Maestrichtian microfauna interstratified with basalt.

In the Montagnes Noires of central Haiti, slaty shales occur (Butterlin, 1960, p. 75). The shales are thin-bedded, gray to brown in color, with blue or green spots, and interlayered with black limestone and slate with small indeterminable foraminifera. These rocks are highly folded and have been metamorphosed. Woodring et al. (1924) considered these rocks to be of Early Cretaceous age because of their metamorphism. However, Butterlin reported that the metamorphism was due to a dacite intrusion. The contact metamorphosed rocks contain epidote, chlorite, quartz, and pyrite. Similar argillites were found by Butterlin on the northeast flank of Morne Bazile. He also found clasts of massive limestone containing microfossils of Campanian–Maestrichtian age occurring in andesite tuffs on the path Bois-Carré-Fiéfié.

On the Northwest Peninsula of Haiti, Butterlin (1960, p. 111) described the occurrence in the Massif de Terre-Neuve of limestone, containing radiolaria and small foraminifera identified as Cenomanian–Turonian in age, which resemble limestones of the Trois Rivieres Formation. In a more recent study in the Terre–Neuve mountains, Kesler (1971, p. 121) determined from mapping that the Upper Cretaceous (?) Colombier volcanic sequence has an apparent compositional gradation from a basal basalt (?) through andesite to dacite (?) and is unconformably overlain by Upper Cretaceous volcanic sediments and limestone which reach a maximum thickness of 500 m on Morne Miguinda

(Kesler, 1971, p. 121). The limestone is the same as that dated by Butterlin (1960) as Cenomanian–Turonian in age (Kesler, personal communication, 1973). Samples (M 322B and M 322) collected by Kesler from the upper part of the volcanic sediment and limestone sequence were found by Butterlin to contain *Globotruncana*, cf. *ventricosa* WHITE, *G.* aff. *arca* (CUSHMAN), *G.* aff. *ganseri* BOLLI, *Hedbergella* sp. or *Rugoglobigerina* sp. and were dated as Campanian–Maestrichtian (??) (Kesler, 1966; personal communication, 1973). This sequence therefore may range in age from the Cenomanian to the Maestrichtian, spanning most of the Upper Cretaceous. These Upper Cretaceous rocks are intruded by a quartz monzonite stock. A K–Ar age of 66.2 ± 1.3 m.y. on biotite from the stock was reported by Kesler and Fleck, 1967, and Kesler, 1971.

In central Dominican Republic, Bowin (1960, 1966) found limestones from several localities to contain fossils which were dated as Late Cretaceous, and three Upper Cretaceous formations were named: the Tireo Formation which includes the Constanza member, the Las Canas Limestone, and the Don Juan Formation. The Tireo Formation underlies the high rugged mountains of the eastern end of the Cordillera Central; it is comprised predominantly of unmetamorphosed volcanic rocks and includes rocks of possibly Cenomanian to Maestrichtian age, thereby spanning a large part of the Late Cretaceous. Because of dense vegetation, rugged terrain, and difficulty of access, the formation was studied only to a limited extent. Three volcanic rock types predominate: coarse to fine tuff, lapilli–tuff, and quartz keratophyre. Buff-colored fissile limestone and interbedded dark-gray limestone are exposed on some low hills in the center of the valley of Constanza in the Cordillera Central and were named the Constanza member of the Tireo Formation.

The Las Canas Limestone occurs to the northeast of the median metamorphic belt of central Dominican Republic and is considered to be upper Upper Cretaceous, probably Maestrichtian, and to unconformably overlie the Lower Cretaceous Los Ranchos and Hatillo Formations. It is thought to be overlain, with apparent conformity, by the Paleocene to middle Eocene wacke and tuff of the Loma Caballero Formation. Where the Las Canas Limestone is intruded by small pyroxene diorite stocks, it has been converted to white, medium- to coarse-grained marble, and deposits of magnetite and hematite were formed locally at the contacts.

Clastic sediments, limestone, and tuff comprise the Don Juan Formation which lies to the northeast of the Hatillo Thrust about 40 km directly north of the city of Santo Domingo. The conglomerates of this formation most commonly are composed of pebble-sized clasts, primarily of volcanic rocks, that are angular to subrounded. Sandstone, siltstone, and limestone are also present. The very fine-grained tuff or mudstone is noncalcareous and is shown by x-ray diffraction traces to be composed mainly of feldspar, quartz, and chlorite. This formation is considered to be Late Cretaceous, Campanian to Maestrichtian in age from limited fossiliferous localities.

Late Cretaceous (?) to Eocene thin-bedded siltstones, dark-gray to light brown, very fine-grained limestone, and minor interbedded pebble-conglomerate layers occur west of La Vega. These rocks are in fault contact with serpentinized peridotite, basalt, metabasalt, and metasedimentary rocks; and they are overlain with angular unconformity by Oligocene and Miocene sedimentary rocks of the Tabera and Cercado Formations. Many small stocks and sills of gabbro intrude rocks of this unit.

On the south flank of the Cordillera Central, sedimentary rocks, locally with minor volcanic rocks, are exposed between San Jose de Ocoa and Valle Neuvo along the highway to Constanza. The road cuts show that the rocks are highly and complexly deformed. Southward from Valle Neuvo, for about 15 km, mainly indurated conglomerates are exposed along the road. Most of the clasts are volcanic, but locally limestone clasts are common. A limestone clast from one locality is dated as Early Eocene. Just south of that locality there is a pronounced lithologic change, and from there to San Jose de Ocoa, shale, siltstone, sandstone, limestone, and locally minor volcanic rocks are exposed. Limestones sampled from two different localities on this part of the highway have been dated as Early Eocene and Late Cretaceous.

In the Puerto Plata area of northern Dominican Republic (Nagle, 1966, 1971), the basement rocks appear to be serpentinites, both massive and breccia, upon which the Los Canos Formation of pre-Paleocene age was deposited. The formation is probably several kilometers thick and consists of the following rock types in order of decreasing abundance: crystal and lithic andesite tuffs, hornblende andesite flows, spilites, feldspathic andesites (keratophyres?), and pyroxene andesite flows. Internal stratigraphic relations are poorly known, but spilites and keratophyres (?) are more abundant near the core on an inferred anticline, and therefore they may be lower in the section than the other types. Many of the tuffaceous rocks have relic devitrified glassy textures which Nagle inferred to have resulted from explosive submarine eruptions, and he considered the entire formation to be submarine in origin. The Los Canos Formation is unconformably overlain by Paleocene to Lower Eocene tuffs of the Imbert Formation.

Blocks of actinolite schist and glaucophane schist lie on the surface of massive serpentinites and Los Canos volcanic rocks east of Imbert and have been found nowhere else in the area. Nagle (1971, p. 80) surmised that these blocks represent either part of the oceanic crust or some fragments of metamorphic rocks, in which case they would be either from rocks older than the Los Canos Formation or from the oldest members of that formation.

Cretaceous (?) volcanic rock (basalts and andesites) are reported to occur at the east end of the Sierra de Bahoruco and on the south side of the Sierra de Neiba (Weyl, 1953; Blesch, 1966). In the Cordillera Septentrional, Vaughan *et al.* (1921, p. 54) noted that dense, hard, dark-blue banded, slightly magnesian calcareous argillite, which breaks into rectangular fragments, underlies the Eocene limestone in the front range near Damajagua, northwest of Navarréte. Fragments of this rock are included in the Eocene limestone and hence are stratigraphically lower than Upper Eocene. Vaughan *et al.* (1921) inferred that it is of Cretaceous age. The locality near Damajagua as well as a small patch of Cretaceous (?) volcanic rocks near the center of the Cordillera Septentrional, and a large area of volcanic rocks near the eastern end of the range, are indicated on the map of Blesch (1966). The small exposure indicated on the map of Blesch (1966) is shown to be crossed by the highway to Puerto Plata from Santiago but was not observed by the writer nor by Nagle (personal communication, 1973) on their travels of that road, although float of porphyritic volcanic rocks was observed in a nearby stream bed. Vaughan *et al.* (1921, p. 54) noted that Alberti (1912) found Cretaceous fossils near Guayubin at the western end of the Cibao Valley. I infer that these fossils occurred in clasts carried by rivers from either the north or south.

C. Paleocene to Eocene

At several locations in Hispaniola, Paleocene to Lower Eocene rocks unconformably overlie Upper Cretaceous or older rocks (Table I). In central Dominican Republic, Bowin (1960, 1966) concluded that the best dated unconformities occurred in the (1) Late Campanian–Middle Maestrichtian, (2) Late Eocene, and (3) latest Oligocene. The geologic mapping of the north flank of the Cordillera Central by Palmer (1963) supports this conclusion. It therefore may be that an important period of deformation occurred in the latest Cretaceous rather than coincident with the Mesozoic–Tertiary transition as proposed by Butterlin (1960).

Limestone of Paleocene to Lower Eocene age is the predominant rock type in only a few locations in Hispaniola. In the Terre-Neuve Mountains of the Northwest Peninsula of Haiti, it unconformably overlies Upper Cretaceous volcanic rocks, limestone, and quartz monzonite (Kesler, 1971; Butterlin, 1960, Table 1). A similar unconformable relation of overlying Lower Eocene limestone upon Upper Cretaceous volcanic rocks and quartz diorite occurs in the region of the North Plain (Morne Mantègue) and is inferred from float to occur on the island of Tortue (Butterlin, 1960, Table I).

At other locations, such as in the Massif du Nord and Chaine des Matheux in central Haiti, and in the Massif de la Hotte and Massif de la Selle in the Southern Peninsula, sections of Paleocene detrital sediments occur between overlying Lower Eocene massive limestone and the unconformably underlying Cretaceous volcanic rock and limestone (Butterlin, 1960, Table I). Lower

Eocene volcanic rocks (andesite and dacite) occur in the central and north-western parts of the Montagnes Noires.

In northern and central Dominican Republic, the Paleocene to Lower Eocene rocks are predominantly tuffs or tuffaceous wackes with minor inter-bedded limestone. In the Puerto Plata area, Nagle (1966, 1971) mapped the Imbert Formation; a kilometer-thick succession of Paleocene to Lower Eocene fine-grained, graded-bedded, calcareous tuffs which grade upward to vitric andesite and dacite tuffs with rare interbedded green radiolarian cherts and thin white aphanitic limestone was inferred to be of submarine origin. In central Dominican Republic east of the eastern terminus of the Cordillera Central, Bowin (1960, 1966) mapped the Loma Caballero Formation of Paleocene to Middle Eocene age composed predominantly of wacke and fine to lapilli–tuff with rare small lenses of dark gray to black limestone. Large masses of algal limestone also occur in the Formation, and appear to be algal reefs within the tuffs. The Los Banitos Formation (Bowin, 1960, 1966) of Early Eocene age also occurs in the central part of the country. It consists of fine-grained tuff or mudstone, fine tuff, limestone, and limestone pebble conglomerate. The limestones and the conglomerate show signs of structural stress—sheared, faulted, and veined by calcite—but not of thermal meta-morphism.

Lower Eocene rocks were mapped by Bowin (1966, p. 49–51) in an area about 30 km west of Santo Domingo. There a basal Lower Eocene cream-colored fine-grained limestone about 230 m thick rests with angular un-conformity upon Upper Cretaceous rocks of the Tireo Formation and upon a quartz diorite pluton. The clayey limestone at the top grades into calcareous mudstone, and the mudstone is followed up-section by bedded siltstones. Approximately 500–600 m stratigraphically above the top of the basal lime-stone occurs the first of several pebble conglomerates in which the rounded clasts are all cream-colored fine-grained limestone, which appears identical with that lower in the section. This Eocene section dips moderately (15–50°) southward.

On the south flank of the Cordillera Central shale, siltstone, sandstone, and limestone are exposed northwest from San Jose de Ocoa along the highway to Constanza. Fossils from two localities about 37 km northwest of San Jose de Ocoa have been dated as Early Eocene (Bowin, 1960, p. 109; 1966), but Paleocene and Upper Cretaceous sedimentary rocks are probably also included in that section.

The Middle and Upper Eocene rocks are predominantly massive lime-stones in the Massif de la Hotte and Massif de la Selle in southern Haiti. North of the Enriquillo–Cul de Sac Basin, in the Chaine des Matheux, rocks of this age consist of limestones, both massive and chalky, and farther east-ward in the Montagnes du Trou d'Eau the Middle and Upper Eocene rocks are massive limestones. In southern Dominican Republic the Middle and Upper Eocene rocks occurring in the Sierra de Bahoruco and Sierra de Neiba are mainly massive fine-grained limestones.

Farther north in central Haiti, however, limestone is not the most prevalent lithologic type. In the Montagnes Noires, the Middle Eocene consists of three units which, stratigraphically from lower to the upper, are the following: the Abuillot Formation, consisting of about 1000 m of layered calcareous mudstone; the Perodin Formation, consisting of about 1000 m of andesite tuff and lava with interstratified limestone; and overlying chalky limestone and massive limestone. These rocks are intruded by granodiorite and dolerite, considered of Upper Eocene age by Butterlin (1960, Table I). The intrusions were followed by deposition of Upper Eocene chalky and siliceous limestone and massive limestone which is now exposed in the southeast part of the mountains.

To the northwest of the Montagnes Noires, on the Northwest Peninsula, Middle Eocene massive and chalky limestone (exposed in the Massif de Terre-Neuve and Montagnes du Nord-Ouest) are overlain by Middle Eocene tuff and basalt flows. Microdiorites of inferred Upper Eocene age (Butterlin, 1960, Table 1) occur in the Massif de Terre-Neuve. To the east, in the Massif du Nord, the Middle Eocene Crete Sale Formation of impure, sandy, and marly limestone is overlain by Upper Eocene basalt and dolerite (exposed in the regions of Ennery and Borgne) and Upper Eocene massive limestone and chalky and siliceous limestone. On Tortue Island, north of Haiti, rocks of Upper Eocene age are also massive limestone.

As mentioned previously, the wacke and tuff of the Loma Caballero Formation of central Dominican Republic range in age from Paleocene to late Middle Eocene (Bowin, 1966, p. 42–43), and are intruded by stocks of gabbro. As also previously mentioned, indurated sedimentary rocks west of La Vega, considered to be of Late Cretaceous (?) to Eocene age, are also intruded by several gabbro sills and small stocks. Lithic conglomerate, conglomeratic volcanic breccia, limestone, and minor sandstone and mudstone of the Magua Formation (Palmer, 1963) on the north flank of the Cordillera Central similarly range in age from Maestrichtian (?) to Middle Eocene and include a basalt flow at least 100 m thick. The Los Caguelles Limestone Breccia with minor volcanic rock fragments of Late Eocene (?) age was inferred by Palmer to be younger than the Magua Formation and older than the Tabera Formation.

Uppermost Upper Eocene sedimentary rocks (Bowin, 1966, p. 54) occur west of La Vega and were included in the Tabera Formation (dated as Early and Middle Oligocene by Bermudez, 1949) by Bowin because of the similarity of degree of induration and appearance to rocks of that formation. Thus, a major unconformity appears to be very closely dated in north central

Dominican Republic as between upper Middle Eocene and uppermost Upper Eocene.

On the north coast of the Dominican Republic, Middle Eocene tuffaceous limestones occur only as exotic blocks in the San Marcos Olistostrome (Nagle, 1971, p. 80). The olistostrome is a several-hundred-meter-thick allochthonous, chaotic, clay unit, containing boulders of various lithologies ranging from a few millimeters to 2 km in greatest dimension. The matrix material, consisting of quartz and kaolinite with minor montmorillonite and illite, has been dated with foraminifera as Paleocene–Early Eocene. Nagle (1971, p. 80) inferred that sometime in the Middle Eocene an earthquake shocked a water-charged tuffaceous unit and caused the unit to slide down a submarine slope, rupturing overlying strata and scraping off older rocks protruding from irregularities in the slope. Portions of the Puerto Plata area were apparently above sea level by early Late Eocene because the older Imbert and Los Canos Formations supplied streamworn detritus to a basal conglomerate of the Upper Eocene Luperon Formation. The formation is estimated to be 1,000 m thick, and above the basal conglomerate occurs an alternating sequence of calcareous, tuffaceous, graded-bedded sandstones and shales with occasional interbedded bioclastic limestones (Nagle, 1966, 1971).

D. Oligocene

During the past decade and a half, the delineation of the boundary between the Oligocene and Miocene has varied considerably from earlier assignments. Thus, most of the formations assigned by Bermudez (1949) to the Oligocene would now be assigned to the Miocene. Accordingly, the formations assigned to the Middle and Lower Miocene by Bermudez would now be distributed differently within the Miocene, and the Upper Miocene formations are now considered most likely to be of Pliocene age.

In Haiti, according to Butterlin (1960) the Oligocene formations are limestone: coralline in the Northwest Peninsula, massive in the Massif du Nord, chalky and massive in the Montagnes Noires, chalky and cherty in the Chaine des Matheux, and chalky with detrital fragments in the Massif de la Hotte. Butterlin (1960) inferred the Oligocene to rest with unconformity upon Eocene formations at all the above sites with the exception of the Chaine des Matheux where conformable relations were inferred. Bermudez (1949) indicates that in the Central Plain of Haiti, clastic fossiliferous limestone underlies the Madame Joie Formation (of Oligo-Miocene age) and hence is probably Oligocene. Oligo-Miocene formations are massive limestones in the Massif de la Hotte, Massif de la Selle, Montagnes du Trou d'Eau, and on Tortue Island. They are chalky and cherty limestones in the Chaine des Matheux, and chalky limestones intermixed with argillaceous sediments in the Montagnes Noires, in the Central Plateau, and in the Northwest Peninsula.

In southeastern Dominican Republic, probable Oligocene clastic fossiliferous limestone occurs in the Sierra Neiba and in the Commendador region of the San Juan Valley. North of the Cordillera Central, however, the Tabera Formation of uppermost Upper Eocene, Oligocene, Oligo-Miocene, and Miocene age consists of shales, sandy shales, sandstone, and limestone, with conglomerate at the base. The Tabera Formation is exposed on the south side of the Cibao Valley and in the Cordillera Septentrional. Nagle (1971) infers that the Puerto Plata area on the north coast of the Dominican Republic was above sea level during the Oligocene. To the south, east, and west of the immediate Puerto Plata area, Oligo-Miocene sediments were deposited in a marine basin now uplifted to form part of the Cordillera Septentrional.

E. Miocene

In Haiti, Butterlin (1960) indicates the occurrence of Miocene rocks in many parts of the Republic. Lower Miocene limestone occurs in the Massif du Nord, Montagnes du Trou d'Eau, and on the island of Gonave. Lower Miocene rocks, however, are detrital sedimentary rocks (flysch facies) in the Massif de la Selle, Chaine des Matheux, and Montagnes Noires; and sandstones, shales, and limestones occur in the Massif de la Hotte. The Middle and Upper Miocene rocks are largely sandstones, and shales in the Northwest Peninsula, Central Plateau, Montagnes Noires, Chaine des Matheux, Cul de Sac Plain, Massif de la Selle, and Massif de la Hotte. Limestone and lignite occur locally in the Massif de la Hotte, limestone in part of the section in the Cul de Sac Plain, gypsum in the Chaine des Matheux, and lignite and gypsum in the Central Plateau. Butterlin (1960) reports Lower Miocene basalt to occur in the Massif du Nord, and Upper Miocene basalt to occur in the Chaine des Matheux.

As indicated in the discussion of Oligocene rocks, most of the formations assigned by Bermudez (1949) to the Oligocene are now considered to be of Miocene age. A review of the foraminifera identified by Bermudez (1949) for the Sombrerito, Trinchera, and Gurabo formations by R. C. Tjalsma (personal communication, May 1973) resulted in the following tabulation and age assignments.

In the Sombrerito Formation the following planktonic foraminiferal species were found: *Globigerinita (G.) dissimilis*, *Globorotalia (Gr.) fohsi*, *Hastigerinella digitata*, *Globoquadrina (Gq.) conglomerata* (= *altispira*?), *Orbulina universa*, and *Turborotalia mayeri* (= *siakensis*); and hence the formation may range in age from Lower Miocene to upper Middle Miocene. The Trinchera Formation contains *O. universa*, *O. bilobata*, *Gr. lobata*, *Sphaeroidinel-*

lopsis (*S.*) *grimsdalei* (= *seminulina*), and *S. rutchi* (= *seminulina*); it is assigned a Middle Miocene age. Although Bermudez (1949, p. 18) considers the Trinchera Formation to overlie the Sombrerito Formation, the ages just given suggest that they are in part age equivalent. The Gurabo Formation contains *Gr. menardii*, *Gr. cibaoensis*, *Sphaeroidinella* (*Sa.*) *dehiscens*, *Candeina nitida*, *Globigerinoides* (*Gs.*) *sacculiferus*, *Gq. quadraria* var. *advena*, and *Gq. altispira*; it ranges in age from the uppermost Middle or Upper Miocene to the Lower Pliocene. From this limited information the following formations of Bermudez (1949) for the southern Dominican Republic are herein assigned to the Miocene: Sombrerito, Trinchera, Gaspar, Bao, Quita Coraza, Lemba, Higuerito, Arroyo Blanco, and Arroyo Seco Formations and the Florentino Limestone. These formations consist of conglomerate, shales, sandy shales, sandstone, marl, coral limestone, and some massive and chalky limestone. A fence diagram of the thickness of these formations in the Enriquillo and Azua valleys is shown in Fig. 6. In northern Dominican Republic, north of the Cordillera Central, this report considers the following formations to be of Miocene age: Villa Trina, Bulla, Baitoa, and Cercado Formations and the Cevicos Limestone. The Villa Trina Formation in the eastern Cibao region is composed of marls and shaley limestone. The Bulla Formation is largely conglomeratic with some sandstone, and the Baitoa and Cercado Formations consist of sandstone, sandy clays, and conglomerate. The Cevicos Limestone, named by Vaughan *et al.* (1921, p. 63), is yellow or cream-colored, more or less argillaceous, and in places nodular. Wherever bedding was observed by Bowin (1966, p. 44), it is very nearly horizontal. The Cevicos Limestone has a well developed karst topography that shows up clearly on air photographs. The formation lies to the south of the west end of the bay of Samana. As suggested by Vaughan *et al.*, the limestone may dip gently northward or northeastward. Vaughan *et al.* (1921, p. 64) report a massive reefal limestone to cover the Cevicos Limestone near its northern margin. Along its southern margin the Cevicos Limestone unconformably overlies the amygdaloidal vitric volcanic rock and hornblende tonalite (Bowin, 1960, p. 117). A sample of the Cevicos Limestone collected by Bowin was assigned a Miocene age by P. Bronnimann (in Bowin, 1966, p. 44).

F. Pliocene–Quaternary

In Haiti, Butterlin (1960) reports that about 100 m of gravels and sandstone (Formation de Hinche) of Pliocene age occur in the Central Plateau; about 250 m of Pliocene detrital rocks (Rivière Gauche Formation) occur in the Massif de la Selle in the valley of the rivers Lavagne and Gauche; and Pliocene coralline and sandy limestones occur in the Massif de la Hotte. Butterlin (1960) indicates that an unconformity separates older rocks from

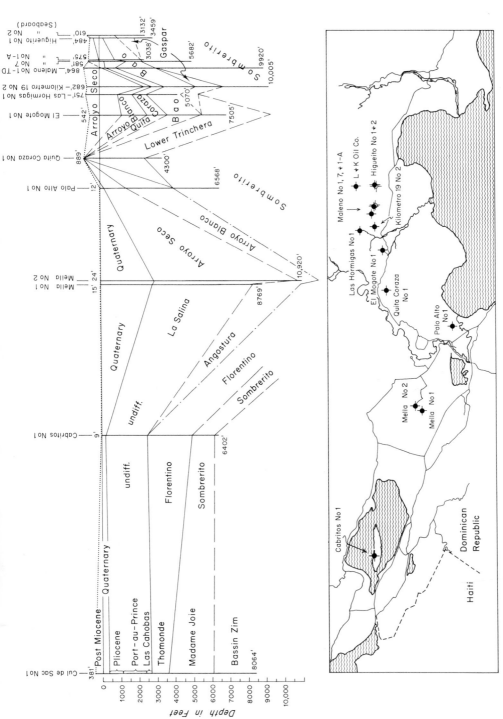

Fig. 6. Fence diagram of thickness of Tertiary and Quaternary formations in the Enriquillo and Azua Valley from drill-hole information. Modified from figure prepared by A. Weymouth and J. F. Jerez of Petrolera Dominicana Corporation in 1961. Elevation of ground at well head and total Depth in feet are annotated for each well. Map shows location of well holes.

Pleistocene and Recent sediments. Pleistocene coastal reef limestones occur along much of the coastline of Haiti and the islands of Gonave and Tortue. Pleistocene and Recent alluvium occur in river valleys and on the plains.

In northern Dominican Republic, the Mao Formation, to which Bermudez (1949) assigned an Upper Miocene age, is now assigned to the Pliocene. The fauna indicated by Bermudez (1949) to occur in the Mao Formation—*Gr. miocenica* (= *crassula*), *Pulleniatina obliquiloculata*, *Globigerinella* (*Gg.*) *siphonifera*, *Gs. conglobatus*, *S. seminulina*, and *Gr. cibaoensis*—is interpreted by R. C. Tjalsma (personal communication, May 1973) to indicate a Lower to Middle Pliocene age. The Mao Formation is composed of clay, sandy clay, shale, sandstone, conglomerate, and coral limestone. Over 3,600 m (12,000 ft) of the upper part of this formation has been encountered in a drill hole in the Cibao Valley near Santiago (P. F. Martin, personal communication, 1960).

By analogy with the Mao Formation of northern Dominican Republic, the Via Formation of the Azua and San Juan Valleys of southern Dominican Republic, which were also indicated by Bermudez (1949) to be of Upper Miocene age, are herein assigned to the Pliocene. The Via Formation consists of sandy clay, sandstone, and conglomerate. Unconformably overlying the Via Formation in the Azua and San Juan Valley area is the Las Matas Formation consisting of sandy clay, sandstone, and conglomerate. The Las Matas Formation is probably of Pliocene to Pleistocene age. The upper Miocene Angostura and Las Salinas formations of Bermudez (1949), occurring in the Enriquillo Basin, are herein also tentatively assigned to the Pliocene. The Angostura Formation is composed of gypsum, black shale, sandy clay, and limestone. The Las Salinas Formation is composed of clay, silt, conglomerate, chalky limestone, and coquina.

The Jimani Formation, of probable Pliocene to Pleistocene age, overlies the Las Salinas Formation and consists of chalky, fossiliferous reef limestone and some thin layers of conglomerate. In the Enriquillo Basin these three formations (Angostura, Las Salinas, and Jimani) and the underlying Arroyo Seco Formation are formations of the Cerros de Sal Group. These formations are deformed, and the salt layers with vertical bedding may be diapiric structures.

Coastal reef limestones are common along the northern and southern peninsulas of Haiti, and they are particularly extensive in the southern half of eastern Dominican Republic where, together with back-reef facies, they underly a considerable area inland from the coast. They occur locally elsewhere as well. Alluvium occurs in most stream valleys and covers much of the floor of the Enriquillo and Cibao Valleys. A post-Miocene fault breccia is reported by Nagle (1966, p. 91) to occur along a fault scarp in the Puerto Plata area of northern Dominican Republic. The breccia is composed of angular cobbles, boulders and pebbles of serpentinite, altered gabbro, and volcanic rocks in a calcite cement.

V. IGNEOUS ROCKS

A wide variety of igneous rocks occurs on the island of Hispaniola, though very few areas have been mapped in detail. Spilite, basalt, andesite, keratophyre, and quartz keratophyre are common pre-Tertiary volcanic types. Late Cenozoic volcanism of limburgite and hornblende–augite andesite porphyry in the central western Dominican Republic (MacDonald and Melson, 1969), and of nepheline basalt north of the Cul de Sac in Haiti (Woodring *et al.*, 1924; Butterlin, 1960), occurs approximately in an east–west trending zone. This zone may continue eastward into the Cordillera Central to near south of the valley of Constanza (Weyl, 1941, p. 27–28; MacDonald and Melson, 1969, p. 81; Table II, Nos. 18, 20, 21, and 22). Many of the occurrences are of Pleistocene age (MacDonald and Melson, 1969).

Chemical analyses for 36 igneous and metamorphic rocks from Hispaniola are tabulated in Table II. They are arranged approximately by their silica content. The metamorphic rocks (amphibolites) are probably metamorphosed mafic volcanic rocks.

Hypabyssal stocks of pyroxene diorite and gabbro, of probable Late Eocene age, occur on the north and northwest side of the median metamorphic belt in central Dominican Republic. Deposits of magnetite and hematite were formed in places at the contacts between marble and pyroxene diorite and have been the source of approximately 700,000 tons of iron ore (Bowin, 1966, p. 65). The basaltic hypabyssal intrusive in the Terre-Neuve Mountains of the Northwest Peninsula of Haiti has a similar setting. This intrusive cuts Lower Eocene limestone, and a 5- to 15-m-thick skarn and marble is formed at the upper contact (Kesler, 1971, p. 124). Similar intrusives occur in the l'Hormand River valley toward Gros Morne (Woodring *et al.*, 1924; Butterlin, 1960; Kesler, 1971). All these bodies are herein inferred to be of Late Eocene age. Some stocks of hornblendite occur within the metamorphosed Duarte Formation and are probably of pre-Middle Albian age.

Plutons of ultrabasic, basic, and acidic igneous rocks occur on Hispaniola, and some are of batholithic proportions. Serpentinized peridotite masses constitute the ultramafic bodies, a batholith of augite norite being the basic pluton, and tonalite occurs in many plutons, of which at least two are of batholithic size.

Serpentinized peridotite bodies occur at many sites within the Dominican Republic (Figs. 3 and 4) and possibly define two parallel belts. One belt occurs in the Hispaniola Fault Zone which parallels the trend of the Cordillera Central and at its eastern end curves to the southeast. The other belt lies along the north coast and extends from the Puerto Plata region to Samana Peninsula where a curve to the southeast could link the north coast belt with the small occurrence near Higuey at the eastern end of the island. These belts are at

TABLE II

Chemical Analyses of Hispaniola Igneous and Metamorphic Rocks[a]

Rock type	Location	Chemical analyses				
		SiO_2	Al_2O_3	Fe_2O_3	FeO	MgO
1. Nepheline basalt	Near Maneville, Haiti	38.64	11.14	5.35	5.31	13.04
2. Limburgite	NW of San Juan de la Maguana, D.R.	37.2	9.8	10.2*		12.8
3. Limburgite	NW of San Juan de la Maguana, D.R.	39.5	9.8	10.4*		13.5
4. Limburgite	NW of San Juan de la Maguana, D.R.	43.3	10.7	10.3*		10.4
5. Limburgite	NW of San Juan de la Maguana, D.R.	43.5	11.5	11.3*		9.6
6. Limburgite	NW of San Juan de la Maguana, D.R.	45.9	11.4	10.0*		9.3
7. Limburgite	NW of San Juan de la Maguana, D.R.	47.0	12.1	10.2*		9.3
8. Limburgite	NW of San Juan de la Maguana, D.R.	47.0	11.6	10.1*		8.8
9. Hornblendite	Loma Angola, Cordillera Central, D.R.	40.0	8.0	11.3	12.2	12.1
10. Olivine gabbro	Polo Arriba, Sierra de Bahoruco, D.R.	46.20	13.36	4.09	6.40	11.07
11. Amphibolite	Duarte formation, on road to Jautia at E end of Cordillera Central, D.R.	48.0	13.78	3.35	6.64	11.37
12. Amphibolite	Below summit of Pico del Yaque, Cordillera Central, D.R.	46.2	13.0	6.4	6.2	13.6
13. Amphibolite	SWW of Monte entre los Rios, Cordillera Central, D.R.	48.2	13.1	4.2	5.1	13.2
14. Amphibolite	Duarte formation, Carretera Duarte near Madrigal, D.R.	49.8	9.5	5.7	8.6	12.3
15. Basalt	Massif de la Selle, Haiti	48.97	14.90	0.96	10.27	7.09
16. Aphanitic vitric rock	Siete Cabezas formation, Carretera Duarte, 8 km S of Madrigal, D.R.	46.9	13.97	4.58	9.24	6.51
17. Aphanitic vitric rock	Siete Cabezas formation, about 5 km S of Madrigal, D.R.	50.9	13.8	10.9*		7.85
18. Olivine andesine trachybasalt	Cordillera Central on road between Valle Nuevo and Constanza, D.R.	52.21	13.93	1.05	7.79	8.99
19. Quartz diorite	Cordillera Central near Dajabon, D.R.	53.58	17.23	5.23	5.47	3.99
20. Phonolitic-leucite, tephyritic tuff	Cordillera Central, in pass S of highland of Valle Nuevo on the old trail, D.R.	54.42	15.27	5.82	1.15	6.74

[a] Notation used: * indicates total iron expressed as Fe^2O^3; † indicates total H^2O; NR means not reported.

Chemical analyses									Total	Ref.	Sample no.
CaO	Na$_2$O	K$_2$O	MnO	TiO$_2$	P$_2$O$_5$	H$_2$O$^+$	H$_2$O$^-$	CO$_2$			
14.40	3.43	1.90	0.14	2.85	0.71	3.01	0.29	NR	100.21	1	p. 316
16.9	1.0	0.4	0.19	2.22	NR	6.05	0.61	NR	97.37	2	16
13.7	3.7	1.2	0.18	2.18	NR	2.81	0.55	NR	97.52	2	16B
13.0	2.4	1.2	0.17	2.41	NR	4.61	0.70	NR	99.29	2	39
12.5	1.9	2.1	0.18	3.60	NR	3.86	0.22	NR	100.26	2	26
11.7	1.8	2.3	0.15	2.55	NR	2.66	0.11	NR	97.87	2	36B
10.7	2.3	1.8	0.16	2.71	NR	0.91	0.63	NR	97.81	2	15
11.6	2.4	1.3	0.16	2.65	NR	1.97	0.66	NR	98.24	2	36A
13.8	2.1	0.8	0.1	0.6	0.1	0.13†		NR	101.23	3	9
11.57	1.77	0.03	0.18	0.91	0.10	3.17	0.80	NR	99.65	3	8
9.78	2.30	0.10	0.16	0.79	0.00	1.92	NR	00.00	98.19	4	70-149
8.9	1.0	1.5	0.17	3.5	0.5	0.8†		NR	101.77	3	5
11.1	2.8	1.2	0.2	1.2	0.1	0.2†		NR	100.6	3	4
8.4	2.8	0.8	0.1	1.3	0.06	0.2†		NR	99.56	3	3
11.72	2.06	0.33	0.15	2.05	0.24	1.35	0.28	NR	100.37	1	p. 325
10.33	1.05	0.22	0.23	1.35	0.10	5.78	NR	NR	100.43	4	70-117
12.6	2.8	0.1	NR	0.85	NR	NR	NR	NR	99.80	4	Bowin-431b
10.56	2.73	1.96	0.15	0.93	0.25	0.10	0.07	NR	100.72	3	15
8.65	2.65	0.63	0.17	0.50	0.09	1.62	0.04	NR	99.85	3	11
7.70	2.85	3.24	0.10	1.03	0.43	1.06	0.18	NR	99.99	3	18

TABLE II (*continued*)

Rock type	Location	Chemical analyses				
		SiO$_2$	Al$_2$O$_3$	Fe$_2$O$_3$	FeO	MgO
21. Trachyandesite	Cordillera Central above Valle Nuevo, D.R.	54.95	16.38	5.80	1.80	4.72
22. Trachyandesite	Cordillera Central, highland S of Valle Nuevo, D.R.	57.86	14.45	4.50	1.80	6.13
23. Diorite	Cordillera Central, Monte entre los Rios, D.R.	54.15	10.65	3.77	8.49	11.03
24. Syenodiorite	Terre-Neuve Stock, Haiti	53.1	14.0	11.3*		6.5
25. Granodiorite	Terre-Neuve Stock, Haiti	59.2	14.3	8.4*		3.4
26. Granodiorite	Memé Valley, Haiti	59.37	13.23	3.45	4.25	3.76
27. Quartz diorite	N slope of Morne Madeleine between Les Perches and Valliere, Haiti	59.66	13.33	5.12	5.53	4.50
28. Pyroxene andesite	Colombier Volcanic Sequence, Terre-Neuve Mountains, Haiti	58.06	15.12	6.31	1.06	3.99
29. Pyroxene andesite	Colombier Volcanic Sequence, Terre-Neuve Mountains, Haiti	59.99	13.52	3.68	3.96	3.36
30. Pyroxene andesite	S slope of Morne Dumuraille, Commune of Terre-Neuve, Haiti	61.41	14.00	2.56	2.69	3.01
31. Late quartz monzonite	Terre-Neuve Stock, Haiti	62.6	14.6	6.8*		2.6
32. Late quartz monzonite	Terre-Neuve Stock, Haiti	63.97	13.62	2.09	4.35	2.89
33. Early quartz monzonite	Terre-Neuve Stock, Haiti	65.0	14.3	5.6*		1.7
34. Early quartz monzonite	Terre-Neuve Stock, Haiti	65.79	15.08	2.05	3.04	1.68
35. Tonalite	Cordillera Central, on trail San Jose de las Matas to San Juan, D.R.	67.32	14.36	2.37	2.37	1.46
36. Plagioclase granite	Sierra del Seibo, between Monte Plata and Sabana Grande, D.R.	71.98	13.29	1.57	2.01	0.82

References: 1. Woodring, W. P., Brown, J. S., and Burbank, W. S., 1924, *Geology of the Republic of Haiti*: The Lord Baltimore Press, 631 p. - 2. MacDonald, W. D., and Melson, W. G., 1969, A Late Cenozoic volcanic province in Hispaniola: *Caribb. J. Sci.*, v. 9, n. 3–4, p. 81–91, Table I. - 3. Weyl, R., 1966, *Geologie der Antillen, Band 4. Beiträge zur Regionalen Geologie der Erde*, H. J. Martini, ed.; Gebrüder Borntraeger, Berlin, 410 p., Table 14. - 4. Donnelly, T., 1972, personal communication. Analyst for 70-149 and 70-117

Chemical analyses									Total	Ref.	Sample no.
CaO	Na$_2$O	K$_2$O	MnO	TiO$_2$	P$_2$O$_5$	H$_2$O$^+$	H$_2$O$^-$	CO$_2$			
7.69	3.40	2.84	0.12	0.85	0.37	0.61	0.07	NR	99.60	3	16
6.65	4.07	2.12	0.11	0.72	0.34	0.84	0.18	NR	99.77	3	17
7.93	1.52	0.34	0.22	0.57	0.04	1.51	0.06	NR	100.28	3	12
8.5	3.7	1.3	0.1	1.4	NR	NR	NR	NR	99.9	5	1500-30
5.5	3.8	2.8	0.1	1.0	NR	NR	NR	NR	98.5	5	M-1
5.63	3.69	2.62	0.17	2.14	0.22	1.08	0.06	NR	99.67	1	p. 304
5.96	3.14	0.97	0.12	1.55	0.18	0.47	0.08	NR	100.61	1	p. 292
4.90	3.03	2.63	0.05	1.09	0.55	3.73[†]		0.04	100.56	6	M-185
4.95	2.95	1.60	0.12	1.01	0.33	3.58[†]		0.25	99.30	6	M-173
5.29	3.39	1.56	0.14	1.17	0.37	3.93	0.55	NR	100.07	1	p. 276
5.0	3.7	3.4	0.1	0.9	NR	NR	NR	NR	99.7	6	M-2
2.58	3.58	2.65	0.07	0.51	0.40	1.43[†]		2.20	100.14	6	M-111
3.9	3.8	3.6	0.1	0.7	NR	NR	NR	NR	98.7	6	M-150
3.29	3.33	2.75	0.06	0.51	0.15	1.16[†]		0.98	99.87	6	SK-5
5.09	3.48	1.09	0.12	0.67	0.04	0.76	NR	NR	99.13	3	10
2.85	4.41	0.71	0.08	0.36	0.12	0.97	0.11	NR	99.28	3	13

is M. Budd. Analyst for Bowin-431b is T. Donnelly. - 5. Kesler, S. E., 1968, Mechanisms of magmatic assimilation at a marble contact, northern Haiti: *Lithos*, v. 1, p. 219–229, Table 2. - 6. Kesler, S. E., 1971, Petrology of the Terre-Neuve igneous province, northern Haiti: *Geol. Soc. Am. Mem.* 130, p. 119–137, Table 3.

least in part coincident with the twin metamorphic belts suggested by Nagle (in press).

The only occurrence of peridotite from Haiti has been reported by Woodring *et al.* (1924, p. 288, 308, and 466), and it was found only as float in the valley of Las Lomas northeast of Saint-Michel de l'Atalaye. This augite peridotite is dissimilar to those of the Dominican Republic on two accounts. Firstly, plagioclase, probably bytownite, constitutes about 10% of the Haitian rock, whereas plagioclase is very rare in those in the Dominican Republic. However, the dunite reported by Kesler and Speck (in press) from western Dominican Republic also is plagioclase bearing. This occurrence is south of the large tonalite batholith at the border with Haiti. Secondly, the Haitian sample is only incipiently serpentinized, whereas those of the Dominican Republic are highly serpentinized. Thus, it is possible that the Haitian and western Dominican Republic occurrences are not genetically related to the other Dominican Republic occurrences. Instead, they may be more related to differentiated mafic bodies, perhaps similar to the occurrence of dunite in the augite norite batholith (Bowin, 1966, p. 63).

An analysis of the O^{18}/O^{16} fractionations for coexisting serpentine and magnetite in a serpentinized peridotite sample from the large body in central Dominican Republic by Wenner and Taylor (1971) showed it to belong to the normal lizardite–chrysotile type of serpentinite which they concluded to have formed at very low temperatures: no higher than 175°C and more likely in the range 85° to 115°C. They further concluded that meteoric ground waters and/or "connate" formation waters (brines) appear to be important in the formation of continental and island arc lizardite–chrysotile serpentinite bodies, but that such waters do not seem to be involved in any significant way with antigorite formation which appears to occur at higher temperatures of 220° to 460°C and presumably results from deep-seated metamorphic waters.

There is a notable contrast in the occurrence of serpentinized peridotite between that on Hispaniola and on the adjacent islands of Cuba and Puerto Rico. Whereas, in Hispaniola, serpentinized peridotite appears generally to occur in close association with faults and hence to be mainly tectonic slivers, those in eastern Cuba and southwestern Puerto Rico (Mattson, 1973) are more circular or elliptical in plan and may be exposed through the erosion of domical and anticlinal uplifts.

A batholith of augite norite and two satellitic bodies were found by Bowin (1966) in central Dominican Republic. On meager evidence, Bowin surmised that the batholith may be a differentiated and stratified intrusive rock. A narrow zone of dunite occurs along the eastern border, and quartz–augite norite is common in the western part of the batholith. The apparently conformable contacts of the batholith and of a satellite pluton suggest mesozonal emplacement of the augite norite.

More than 25 tonalite plutons have been found in Hispaniola. Many parts of the island remain geologically unexplored, and hence more may yet be discovered. Although plutons of tonalite are numerous, they are restricted to a belt that trends roughly about S 75° E from the Northwest Peninsula of Haiti across the island. The tonalite intrusions of Hispaniola fall into two groups: foliated and unfoliated. The foliated tonalites occur only in the metamorphosed mafic volcanic rocks of the Duarte Formation in central Dominican Republic (Bowin, 1966). These rocks are moderately to strongly foliated, primarily due to alignment of muscovite and biotite flakes, and show cataclastic texture. All modal analyses showed quartz to be in excess of plagioclase, and thus they are mica trondjemites.

The unfoliated tonalites do not show cataclastic textures, and the major varietal mineral is hornblende; however, very little hornblende is observed in some plutons. Less than 2% K_2O was found in the unfoliated tonalite samples from central Dominican Republic analyzed by Bowin (1966, p. 68), and modal analyses of only three samples out of nine indicated the presence of potash feldspar following staining. The maximum potash feldspar percentage found is 1.3%. Lewis and Kesler (1973) report that some of the tonalite bodies of the Cordillera Central are dominantly hornblende tonalite, but that some are also complex and consist of rocks ranging from diorite to quartz monzonite. They also state that the tonalitic rocks from the Cordillera Central resemble similar rocks from the Virgin Islands batholith, but are distinct from the comparatively potash-rich rocks of the Terre-Neuve stock in the Northwest Peninsula of Haiti, the northern Utuado pluton in Puerto Rico, and the Above Rocks in Jamaica.

Boulders of hornblende tonalite are found in the conglomerates of the Magua Formation (Palmer, 1963, p. 10), which occurs on the north side of the Cordillera Central and which may range in age from Late Maestrichtian to Middle Eocene (Palmer, 1963, p. 121). Cobbles of unfoliated tonalite are also found in the uppermost Upper Eocene conglomerate a short distance north of the pre-Tertiary rocks also on the north side of the Cordillera Central (Bowin, 1960, p. 160).

VI. RADIOMETRIC AGES

Eight radiometric age dates are available from Hispaniola. Seven of these measurements for the Dominican Republic are tabulated in Table III. The upper four samples listed in Table III were collected by C. Bowin, and the lower three by S. Kesler (Personal Communication, 1973). In addition to those seven measurements, a radiometric age of 66 ± 1 m.y. has been reported for a quartz monzonite from the Meme Mine 1600-ft level in the Terre-Neuve

TABLE III

Radiometric Age Dates for Dominican Republic

Sample	Rock type	Location	Method	Analyst	Mineral	Purity	K, %	% radiogenic	Age, m.y.
450	Muscovite trondjemite	From foliated tonalite pluton west of El Puerto, Dominican Republic, 18°46′ N, 70°16′ W on Bowin, 1966 map	K–Ar	L. T. Aldrich, Carnegie Institute of Washington	Muscovite	—	—	—	68
328a	Tonalite	Near center of El Rio batholith, Dominican Republic, 19°01′ N, 70°37′ W on Bowin, 1966 map	K–Ar	S. R. Hart, Carnegie Institute of Washington	Hornblende	99%	0.359	49	86 ± 3%
B92	Amphibolite schist	Duarte formation, in contact metamorphic aureole around foliated tonalite pluton, Dominican Republic, 18°47′ N, 70°19′ W on Bowin, 1966 map	K–Ar	S. R. Hart, Carnegie Institute of Washington	Hornblende	99%	0.0537	16	91 ± 10%
B27f	Hornblendite	Hornblendite stock, Dominican Republic, 18°51′ N, 70°20′ W on Bowin, 1966 map	K–Ar	S. R. Hart, Carnegie Institute of Washington	Hornblende	99%	0.160	36	127 ± 5%
RD-72-2	Tonalite	Medina stock, Dominican Republic, 18°36′10′′, 70° 9′36′′ on Bowin, 1966 map	K–Ar	J. L. Barton, University of Montreal	Hornblende	—	0.54	76	79 ± 2
RD-72-2	Tonalite	Medina stock, Dominican Republic, 18°36′10′′, 70°9′36′′ on Bowin, 1966 map	K–Ar	J. L. Barton, University of Montreal	Biotite	—	5.71	65	79 ± 2
RD-72-10	Tonalite	Medina stock, Dominican Republic, 18°36′10′′, 70°9′36′′ on Bowin, 1966 map	K–Ar	J. L. Barton, University of Montreal	Biotite	—	5.17	54	79 ± 2

Mountains of the Northwest Peninsula of Haiti (Kesler and Fleck, 1967; Kesler, 1968).

The unmetamorphosed hornblendite (sample B27f) intrusion into the Duarte Formation probably provides the best minimum age date (127 m.y.) for the Duarte Formation and its metamorphism. Sample B92 (91 m.y.) of the Duarte Formation is from the contact aureole around a foliated tonalite body. This amphibole schist had been reheated and thus probably provides an age closer to that of the time of intrusion of the foliated tonalite. A radiometric age of 68 m.y. has been determined for the muscovite of sample 450 from the foliated tonalite. This age is close to that of 66 m.y. that was reported by Kesler and Fleck (1967) for biotite from a quartz monzonite in northern Haiti which intrudes the Colombier volcanics of Cenomanian, Lower Cretaceous, or older age. These dates, however, are less than that obtained (86 m.y.) from hornblende separated from a sample (328a) of unfoliated tonalite of the El Rio batholith. The retention of argon by hornblende is probably superior to that of muscovite or biotite (Hart, 1966). Kesler et al. (1972) cite K–Ar ages of 48 to 35 m.y. on phases of the Loma de Cabrera Batholith (the large batholith that appears to straddle the border between the Dominican Republic and Haiti in Figs. 3 and 4) and of the El Bao Batholith (Las Placetas–Carrizal Batholith of Palmer, 1963). Kesler (personal communication, 1973) also reports K–Ar ages of about 79 m.y. on the Medina tonalite pluton (Table III) and on a quartz diorite phase of the Jautia Augite Norite Batholith (Bowin, 1966).

Owing to uncertainties in the interpretation of the radiometric dates and in the paleontological dating of the host rocks for the plutonic intrusions, the history of plutonism remains speculative. Assuming that all the reported radiometric ages (except that for sample 450) indicate the time of intrusion, then four episodes of tonalite emplacement would appear to have occurred in Hispaniola: about 90 m.y. ago in the Turonian, about 80 m.y. ago in the Santonian, about 66 m.y. ago in the Maestrichtian, and about 40 to 45 m.y. ago in the Eocene. The two older episodes may be confined to exposures in central Dominican Republic and appear to define a northwest–southeast trending zone, whereas the two younger episodes define an east–west trending belt from the north flank of the Cordillera Central south of Santiago westward to the Northwest Peninsula of Haiti. The latter zone may also continue eastward and include the epizonal tonalite plutons near the villages of Cotui, Cevicos, and Sabana Grande de Boya on the east side of the Hatillo Thrust.

VII. GRAVITY ANOMALIES

Gravity measurements in the Haitian part of Hispaniola have been made by Bowin (1968), and in the Dominican Republic part, by the Inter-American Geodetic Survey, Cartographic Institute, and Bowin (1971). Free-air gravity anomaly and simple Bouguer anomaly maps for the island and the surrounding area are presented in Figs. 7 and 8, respectively. High positive free-air anomaly values occur over most of Hispaniola, whereas negative values generally occur offshore. Large negative free-air anomaly values occur over the Puerto Rico Trench and its westward continuation between northern Hispaniola and the southeastern Bahamas, over the Oriente deep in the Cayman Trough south of eastern Cuba, south of Haiti, and in the Muertos Trough south of eastern Dominican Republic. In the marine area between the two peninsulas of Haiti, the island of Gonave occurs at the eastern end of a gravity high that separates two areas of negative free-air anomalies.

Positive free-air anomalies are usually associated with mountains, and low free-air anomaly values with valleys. In the central portion of Hispaniola the free-air anomaly contours trend generally about N 75° W approximately parallel to the ridges and valleys. In the western and eastern portions they trend more nearly east–west, as does the coastline of those portions of the island. Although Hispaniola has the highest mountains in the Greater Antilles, its highest free-air anomaly value ($+239$ mgal) is not very much greater than the maximum values for Jamaica and Puerto Rico (Bowin, in press).

Bouguer anomalies on land are an approximate attempt to remove the mass relating to the topography above sea level, thereby revealing mass distributions below a datum plane, which on land is sea level, and for the ocean areas is normally the sea floor. Bouguer anomalies generally have large negative values over mountain ranges, and large positive values over deep ocean areas.

The simple Bouguer anomaly map of Hispaniola (Fig. 8) shows a pattern not easily discernible from an examination of free-air anomaly and topographic maps. The Bouguer anomaly high associated with the central axis of the Cayman Trough (Bowin, 1968) does not follow the axis of negative free-air anomalies between Cuba and northwestern Haiti, but instead turns eastward into the area between the two peninsulas of Haiti. An apparent eastward continuation of that high occurs in the eastern portion of the Northwest Peninsula and eastward along the Massif du Nord and along the northern part of the western portion of the Cordillera Central. Near north central Dominican Republic (19°10′ N, 70°30′ W) the Bouguer anomaly high diminishes in value and merges with the east–west-trending Bouguer anomaly high associated with the Sierra del Seibo of eastern Hispaniola, and a northwest–southeast-trending Bouguer anomaly (and free-air) high associated with metamorphic rocks and a serpentinized peridotite mass occurring in a belt trending about N 40° W north of the capital city of Santo Domingo. A poorly defined Bouguer anomaly high trends north of east into the Mona Passage from easternmost Hispaniola.

An axis of Bouguer anomaly lows extends from offshore easternmost Cuba to the east, north of Haiti and western Dominican Republic. This low comes onto the land in central Hispaniola and appears to continue eastward

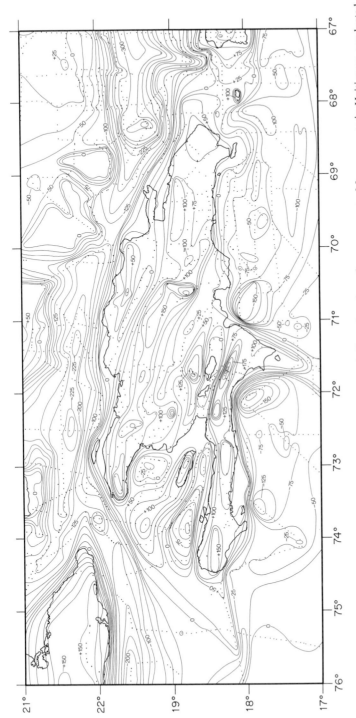

Fig. 7. Free-air gravity anomaly map. Contour interval is 25 mgal. Dots indicate location of measurements. Measurements in Haiti were conducted by C. Bowin and R. Geller in 1963, in the Dominican Republic by the Inter-American Geodetic Survey and Cartographic Institute in 1955–1957, and by C. Bowin and R. Geller in 1961. Marine measurements are from observations conducted aboard the *R/V Chain* of the Woods Hole Oceanographic Institution using La Coste and Romberg Sea Gravity meter S-13 during cruises 46 and 55. Measurements in Cuba are from Shurbet and Worzel (1957), and those in western Puerto Rico are from Shurbet and Ewing (1956) and from observations conducted by the United States Geological Survey in 1964.

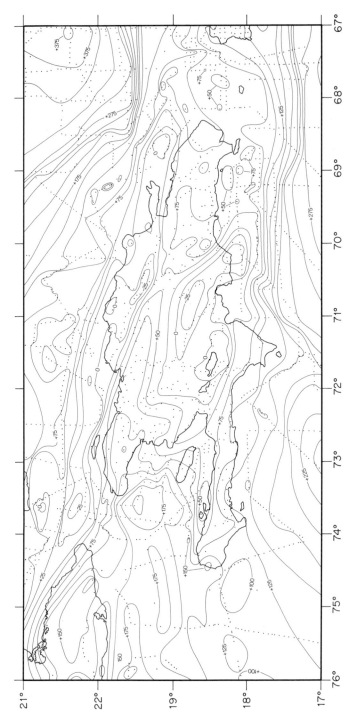

Fig. 8. Simple Bouguer gravity anomaly map. Contour interval is 25 mgal. The assumed crustal density is 2.67 g/cm³, and a water density of 1.03 g/cm³ was assumed for reduction of the marine measurements. Sources of data are given in caption for Fig. 6.

across the Samana Peninsula and into the Atlantic Ocean, and may extend eastward as far as the north coast of central Puerto Rico (Bowin, in press). Other simple Bouguer anomaly lows of lesser extent occur over the Cibao Valley, over the south of the Cordillera Central and Massif du Nord, over the Enriquillo Valley, and over the marine area north of the Southern Peninsula of Haiti. The low over the Enriquillo–Cul de Sac Valley apparently curves to a northwest trend to the west and approximately follows the coastline of Haiti there. The Bouguer anomaly low associated with the southern flank of the Cordillera Central extends southeastward onto the shelf off southern Dominican Republic. The Massif de la Hotte, Massif de la Selle, and the Sierra de Bahoruco have neither simple Bouguer anomaly highs nor lows associated with them, but instead are crossed by a Bouguer anomaly gradient. Although control is sparse, the Chaine des Matheux, the Sierra de Neiba, and the Montagnes Noires appear to have only weak Bouguer anomaly signatures.

VIII. SEISMICITY

Earthquake epicenters occurring in the region of Hispaniola over a 73-year period since 1900 are indicated in Fig. 9. A majority of the epicenters are offshore from the islands. They cluster in several areas, and the greater activity east of longitude 70°30′ W is rather marked. Intermediate-depth earthquakes occur only in the vicinity of eastern Hispaniola. A few hypocenters from depths greater than 150 km occur near southeasternmost Hispaniola. These relationships are but little different from those shown on the epicenter map for the shorter time period 1950 through 1964 presented by Sykes and Ewing (1965).

West of longitude 70°30′ W, except for a single pre-1961 earthquake, the entire region of the Cordillera Central, Massif du Nord, Sierra de Neiba, Montagnes Noires, and the Terre-Neuve Mountains is devoid of teleseismic activity. This is of note because the highest elevation in the Caribbean island arc lies in this area, and stream-terrace surfaces high above the present river level on the south side of the Cordillera Central indicate continued recent uplifting (Bowin, 1966, p. 83).

Slip mechanisms for several earthquakes in the Hispaniola region were determined by Molnar and Sykes (1969). They interpreted those mechanisms to indicate thrust faulting, with mechanisms 112 (southeasternmost Hispaniola) and 104 (in the Hispaniola channel southeast of Great Inagua Island) as also having components of strike-slip faulting. The choice of fault plane and auxiliary plane could not be made for mechanism 104.

Bracey and Vogt (1970) proposed that a section of the Atlantic plate is underthrusting beneath eastern Hispaniola because of the thrust-earthquake-

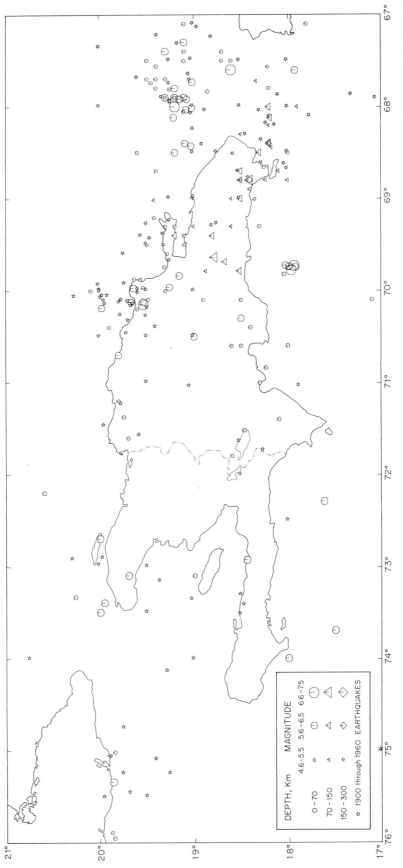

Fig. 9. Earthquake epicenters in the Hispaniola region for the period January 1, 1900 through January 31, 1973. Plotted from a compilation on magnetic tape provided by the Environmental Data Service, National Oceanic and Atmospheric Administration.

mechanism solutions there and the clustering of epicenters at both ends of the inferred underthrusting section. The clustering they inferred is characteristic of hinge faults which would occur at such locations. Other locations of epicenter clusters such as near Samana Peninsula and Mona Canyon were not accounted for. They presumed the Hatillo Thrust, which appears to be of Late Eocene age, to support their underthrusting hypothesis, although they favored a Late Miocene time for initiation of the underthrusting section. The Pleistocene volcanism which they also associated with the underthrusting segment lies rather far to the west in central and western Hispaniola, and the outpourings lie more on an east–west trend than parallel to the proposed underthrusting zone. However, the occurrence of intermediate-depth earthquakes in a southward-dipping zone beneath eastern Dominican Republic and the thrust-slip mechanisms do strongly suggest that a portion of the Atlantic plate is underthrusting beneath that part of Hispaniola.

IX. EVOLUTION OF HISPANIOLA

Within the last decade the concepts of sea-floor spreading and plate tectonics have proved to be quite powerful for understanding much of the geologic record and in deciphering the evolution of many geologic structures. It is within that framework that the following discussion attempts to interpret the geologic evolution of the island of Hispaniola. The geologic record in Hispaniola is imperfect, and our knowledge of that record is even more imperfect. The ages, lithologies, and structural relations of the various rock types need much more study. With these shortcomings in mind, we speculate on a possible tectonic history of Hispaniola in the following paragraphs.

First we will recapitulate the general characteristics of the stratigraphic succession found on the island. Metamorphic rocks occur in a belt diagonally across the middle of the island and in small exposures near the north coast. In the central belt, a 127 m.y. radiometric age for a hornblendite intrusive probably provides the best minimum date for the Duarte Formation and hence suggests that it is earliest Cretaceous or pre-Cretaceous. The age of the Maimon and Amina Formations is known only as pre-middle Albian. They are probably similar in age to the Duarte Formation. The age of the gneissic amphibolites, glaucophane schist, eclogite, micaceous marble, and massive marble found near the north coast is known only as pre-Tertiary. The metamorphic rocks of the central belt are predominantly volcanic flows (Bowin, 1960, 1966; Palmer, 1963), as is the Lower Cretaceous section (the oldest paleontologically dated rocks). Upper Cretaceous rocks are also predominantly volcanic rocks, but they are in large part of pyroclastic origin consisting of coarse to fine tuff and lapilli–tuff. Limestones of Upper Cretaceous age are locally prominent, particularly in Haiti. During the Maestrichtian and earliest Tertiary, pyroclastic material was locally deposited, but clastic sedimentation was becoming more common, although deposition of limestone continued being prevalent in portions of Haiti. Late Tertiary time was virtually devoid of volcanic activity, being overwhelmingly a period of deposition of clastic sediments. This succession leads to the view that the island has become increasingly emergent with time.

The greenschist of the Duarte Formation is mainly metamorphosed mafic volcanic flows. These rocks are similar in composition (Table II) to greenschist rocks dredged from the base of the rift scarp on the Mid-Atlantic Ridge (Melson *et al.*, 1966) and the Carlsberg Ridge in the Indian Ocean (Matthews, Vine, and Cann, 1965). Thus, it seems probable that the Duarte Formation may be metamorphosed oceanic crust that formed by sea-floor spreading. This sea floor may have been formed during the Jurassic when there is evidence from the reconstructions of North America and Africa, and South America and Africa, that the North American and South American plates separated rapidly through the Jurassic period (Phillips and Forsyth, 1972; Francheteau, 1970; Bowin, in press). The metamorphism may have occurred in part within a few million years of the formation of the basaltic ocean crust under conditions of a few kilometers of burial beneath the sea floor. Such conditions would presumably be similar to those that formed the greenstones exposed on the rift scarps of the Mid-Atlantic Ridge and Carlsberg Ridge. However, the foliation in the Duarte Formation, exclusive of the contact aureoles around intrusive bodies, suggests that a later metamorphism was probably responsible for the foliation since the greenstones from the central rift of the Mid-Atlantic Ridge have no well-defined fabric (Melson *et al.*, 1966).

The metamorphosed quartz keratophyres and keratophyres and other lavas of the Maimon Formation and of the apparently equivalent but more highly recrystallized Amina Formation range in composition from andesitic to dacitic to keratophyric, and are not typical of ocean basins. These rocks are more characteristic of island arcs. Keratophyres are reported by Donnelly (1964) to be an early deep-water volcanic phase in the development of the Virgin Islands. They also occur in the Lower Cretaceous sequence in central Dominican Republic (Bowin, 1966). A major fault (the Loma Caribe Fault, part of the Hispaniola Fault Zone), along which serpentinized peridotite has been emplaced, appears to separate the Maimon and Amina Formations from the metamorphosed mafic volcanic rocks of the Duarte Formation. This fault may be evidence of a former Benioff zone which separated a developing island arc from oceanic crust underthrust beneath the arc. If the Loma Caribe Fault has been exposed through erosion following simple regional uplift, then it would seem probable that the Benioff zone had originally dipped northward rather than southward in order to account for the oceanic crustal plate being on the south side of the zone. However, structural complications have probably occurred, and hence the original dip direction of the presumed Benioff zone is uncertain from the present outcrop pattern.

Several intrusive bodies, including hornblendite and a batholith of augite norite, have been found in the Duarte Formation. Were these intrusives emplaced while the Duarte Formation was young oceanic crust? For example, gravitationally stratified gabbro and norite rocks appear to occur in the Atlantic Ocean basin (Melson and Thompson, 1970). Or, were they emplaced much later as the island-arc structure was developing? More reliable radiometric age dating of the metamorphosed formations and of the intrusions which cut them is needed. Because of the foliation in the Duarte Formation, and the fact that the contact aureoles around the norite batholith and tonalite intrusives are better developed than in the regionally metamorphosed rocks, I infer that the intrusives were emplaced as the Greater Antillean island arc was developing. The radiometric dates previously described suggest that the norite and tonalite intrusions in the Duarte Formation between La Vega and Santo Domingo are about 80 to 90 m.y. old.

Lower Cretaceous, Aptian to Albian (about 112 to 100 m.y. ago; Van Hinte, 1972) rocks are known from central Dominican Republic and from near Jacmel in southern Haiti. The paleontologic dating is obtained from fossiliferous limestones and limey sediments interbedded with volcanic rocks. The volcanic rocks are pillow basalt in Haiti and more acidic types, in large part quartz keratophyre, in the Dominican Republic. Caprinids are found in the limey sediments interbedded with the pillow basalt and suggest that the pillows were formed in shallow water and, therefore, most probably not at a normal site of sea-floor spreading. The keratophyre, pyroxene andesite, uralite andesite, and other volcanic types in central Dominican Republic indicate island-arc volcanism and hence are presumably related to lithospheric underthrusting. The Lower Cretaceous volcanics are predominantly flows, and the island-arc ridge crest is inferred to have been largely submerged below sea level at that time (Bowin, 1966), perhaps because the arc was still in a youthful stage of development.

The predominantly pyroclastic nature of the volcanic rocks of Upper Cretaceous age suggests that at that time the volcanism was principally subaerial. Although Upper Cretaceous volcanic rocks are known to occur in many places in Hispaniola, their principal occurrences are restricted to a roughly east–west trending zone along the length of the island. The zone extends approximately across the center of Dominican Republic and westward into northern Haiti to the Northwest Peninsula. This main zone comprises the apparently widespread exposures of Upper Cretaceous tuff in eastern Dominican Republic in the Sierra del Seibo and its foothills north of the wide coastal terrace, and the tuffs and quartz keratophyre of the Tireo Formation at the east end of the Cordillera Central. Farther west along the crest of the Cordillera Central, the distribution of Upper Cretaceous volcanic rocks and Upper Cretaceous sedimentary rocks on the south flank, and metamorphosed mafic volcanic rocks (Duarte Formation) on the north flank, remains obscure. Access to the area is very limited, and the most recent geologic exploration of the region was conducted by Weyl (1941) in 1938 and 1939. Near the Dominican/ Haitian border the rocks exposed across the Cordillera Central appear to be predominantly quartz diorite in the northern part, and Upper Cretaceous(?) sedimentary rocks in the southern part. The topography of the Cordillera is much subdued in that region.

Upper Cretaceous volcanic rocks are exposed farther west in the western-most part of the Massif du Nord in north central Haiti, and also southwest of there as the Colombier volcanic sequence (Kesler, 1971) in the Terre-Neuve mountains. In the Terre-Neuve mountains, the Colombier volcanic sequence of Upper Cretaceous(?) age is unconformably overlain by volcanic sediments and limestone (Kesler, 1971), which appear to range in age from Cenomanian–Turonian to Campanian–Maestrichtian(??). Thus, the volcanic sediments and limestone may be age-equivalent to the Tireo Formation of Central Dominican Republic, and the Colombier volcanics are Cenomanian or Lower Cretaceous, or older in age.

Pre-Paleocene andesite tuff, andesite flows, spilite, and keratophyre(?) of the Los Canos Formation occur in the Puerto Plata area in northern Dominican Republic (Nagle, 1966, 1971) and are interpreted to be of submarine origin. The blocks of actinolite schist and glaucophane schist that presently lie on the surface of the Los Canos volcanic rocks and the underlying serpentinite basement are inferred by Nagle (1966, 1971) to be metamorphosed oceanic crust or fragments of the oldest members of the Los Canos Formation or an older unexposed formation. Thus, it seems probable that a subduction zone existed in the vicinity of the Puerto Plata area during at least a part of the Cretaceous period.

The dating of the volcanic and deformational events during Cretaceous time is meager. From the evidence available, I infer that from at least late Early Cretaceous time to early Maestrichtian time, lithospheric underthrusting was taking place beneath Hispaniola, and presumably from the north towards the south. The former island-arc-trench edge may now be exposed in the Puerto Plata region of northern Dominican Republic. The main volcanic arc may now be exposed as the Cretaceous volcanic belt across the length of the island.

Probing beyond these facile relationships is more speculative. Several fundamental problems concerning the evolution of island arcs are involved. For example, what is the nature of the structural deformation that may occur at depth beneath a volcanic arc while underthrusting is taking place? If the metamorphic belts of northern and central Hispaniola are exposures of the same zone of underthrusting, and both the Duarte Formation and the actinolite and glaucophane schists and eclogite of the north coast are metamorphosed oceanic crust, then what accounts for their different temperature and pressure regimes of formations, and why would not the rocks of the presumed deeper part of the underthrust crust have first been metamorphosed to glaucophane schist or eclogite? Could it be that the high pressures involved in the formation of glaucophane occur during rapid uplift of the underthrust zone by buoyancy forces following cessation of underthrusting, rather than during the time of underthrusting? Perhaps, the active zone of underthrusting shifts to a deeper zone and underthrusts not only the island-arc crust but portions of previously underthrust oceanic crust. If an island arc undergoes concomitant uplift, then the dip of the underthrusting surface might actually not vary much, but metamorphosed oceanic crust would continually be raised to shallower levels. However, and here considered more likely, the Duarte Formation and the glaucophane schist and eclogite of the north coast may not be part of the same lithospheric slab. The Duarte Formation probably is part of the lithosphere plate on which the island arc devoped during the Cretaceous, whereas the metamorphosed oceanic crustal rocks of the north coast probably are fragments of the underthrusting Cretaceous lithospheric slab. The Hispaniola Fault Zone, separating the Duarte Formation from the Maimon and Amina Formations, is interpreted as being a pre-Middle Albian, possibly Jurassic, Benioff zone that formed very early in the development of the Caribbean Island arc. Faulting along the Hispaniola Fault Zone also occurred in the Oligocene on the north flank of the Cordillera Central, bringing conglomerates of the lower part of the Tabera Group into fault contact with the Duarte and Amina Formations (Palmer, 1963).

Another fundamental question concerns the relationship of tonalite intrusions to underthrusting. Do the intrusions occur at the same time as underthrusting, only after cessation of underthrusting, or both during and after? If the range in radiometric ages for the tonalites of Hispaniola are accepted, then it is highly likely that tonalite intrusions are emplaced while underthrusting is active. If the 35 m.y. radiometric age for some tonalite bodies is verified, then some intrusions probably continued after cessation of underthrusting, which in central Hispaniola is herein inferred to have happened in late Eocene, about 40 m.y. ago. If the 35 m.y. age is not verified, then it would be likely that the acidic intrusions require active underthrusting for their formation and emplacement.

The texture and radiometric ages of the older tonalite intrusions provide evidence that uplift of the crust beneath an island arc does occur during underthrusting. The tonalite intrusions that were apparently emplaced in the Turonian, about 90 m.y. ago, occurred at epizonal depths in the Tireo Formation and at mesozonal depths in the Duarte Formation. But, following this emplacement, the Duarte Formation underwent uplift, as evidenced by the Medina tonalite intrusion in the Duarte Formation, which is unfoliated and hence of epizonal depth. Radiometric dating of this intrusion (Table III; Kesler, personal communication, 1973) suggests an age of about 79 m.y.

In Maestrichtian time a major period of uplift and deformation is inferred to have taken place. Emplacement of the Terre-Neuve quartz monzonite may also have occurred at this time. Volcanism continued, largely resulting in tuff deposits, but substantial clastic sedimentary deposits began accumulating in basins adjacent to the area of uplift. These accumulations are particularly noteworthy on the south side of the Cordillera Central.

Clastic and epiclastic Upper Cretaceous sedimentary rocks occur in the Montagnes Noires, as the Trois Rivières Formation in the Massif du Nord (Butterlin, 1960), on the south flank of the Cordillera Central (Ocoa Formation) in the Dominican Republic, and as the Don Juan Formation occurring about 40 km due north of Santo Domingo. The extensive exposures of conglomerate with primarily volcanic clasts on the south flank of the Cordillera Central south of Valle Neuvo may be in part of Late Cretaceous age, and the volcanic rock pebble conglomerate of the Don Juan Formation is apparently Late Cretaceous, Campanian to Maestrichtian in age. The Upper Cretaceous clastic sedimentary rocks occur principally on the south flank of the Cordillera Central and the Massif du Nord, on the flank opposite the metamorphic belt. This suggests that in the Late Cretaceous, probably principally in the late Maestrichtian, the northern portion of the Cordillera Central was undergoing uplift and the southern portion was a basin of deposition. This tectonic setting apparently continued into the Paleocene and Eocene. Some limestone is interbedded in the predominantly clastic section. Late Cretaceous(?) to Eocene clastic sedimentary rocks are found locally north of the Cordillera Central to the west of La Vega, and their preservation suggests that the Late Cretaceous and Early Tertiary uplift of the Cordillera Central was restricted to a zone about 50 km wide. Farther west from La Vega, along the south edge of the Cibao Valley, except for the limited occurrences of the Magua Formation of Maestrichtian(?) to Middle Eocene age (Palmer, 1963), Oligocene clastic sediments of the Tabera Formation lie directly upon the Cretaceous and older(?) basement.

In contrast to the rest of the island, Upper Cretaceous massive limestone and marl is the dominant rock type in the Southern Peninsula of Haiti. This evidence argues against the suggestion of Malfait and Dinkelman (1972, p. 262) that west of the Beata Ridge, crust of a Cretaceous to Eocene "Caribbean" Plate underthrust western Hispaniola on the south. Upper Cretaceous emplacement of basalt and dolerite apparently followed the limestone and shale deposition of the Macaya Formation (Campanian to Maestrichtian, but mainly of Upper Campanian age). The basalt and dolerite thus appears to be younger than the basalt encountered in the bottom of Deep Sea Drilling Project (DSDP) holes 151 and 152 to the south on the Beata Ridge and in the Colombian Basin, respectively (Edgar *et al.*, 1971). The basalt in hole 151 appears to be Santonian in age, and that of hole 152 was dated as Campanian. Cretaceous basalt and dolerite of the Southern Peninsula, as well as the limestone and shale, are unconformably overlain by Paleocene detrital sediments. It is problematical whether the Southern Peninsula of Haiti had greater limestone deposition than other parts of the island because it was isolated from clastic deposition in part by the inferred Late Cretaceous basin on the south side of the Cordillera Central/Massif du Nord, or because it was a greater distance away and has been displaced to its present position by a Maestrichtian translation. The translation may have occurred in a manner similar to that inferred by Hess and Maxwell (1953). The emplacement of the basalt and dolerite may be associated with rifting accompanying such translation. The limestone deposition of the Maestrichtian Las Canas Limestone northeast of the median metamorphic belt in central Dominican Republic appears to have occurred during a quiescent period of volcanic activity.

As just mentioned above, deposition of volcanic tuff and epiclastic and clastic deposition continued into the Paleocene and Lower Eocene, apparently following the pattern of the latest Upper Cretaceous rocks in the region of the Cordillera Central and Massif du Nord. Lower Eocene volcanic rocks occur in the Montagnes Noires, and Paleocene detrital sediments unconformably overlie Cretaceous volcanic rocks and limestone. Paleocene detrital sediments also unconformably overlie Upper Cretaceous limestone and basalt and dolerite [pre-late Maestrichtian(?)] in the Massif de la Hotte and Massif de la Selle on the Southern Peninsula. These occurrences suggest three possibilities: one, that the Southern Peninsula had been translated to near its present proximity to the Cordillera Central, and hence received Paleocene sediments from the Cordillera Central. Two, that the Paleocene clastic sedimentation simply spread farther south of the Cordillera Central/Massif du Nord than in the latest Cretaceous. Or, three, that uplift of the Southern Peninsula resulted in local clastic deposition.

Limestone of Paleocene to Lower Eocene age overlies Upper Cretaceous volcanic rocks and limestone, and quartz monzonite in the Terre-Neuve Mountains of the Northwest Peninsula, and suggests that, following the Maestrichtian deformation, uplift, and erosion to expose the Upper Cretaceous rocks, the area was isolated from the clastic sedimentation occurring to the east on the flank of the Massif du Nord and Cordillera Central. A further restriction of the areal extent of clastic sedimentation apparently occurred in the Lower Eocene because limestone of that age was deposited in the Massif du Nord, Plaine du Nord (Morne Mantegue), Island of Tortue, Chaine des

Matheux, and the Massif de la Hotte and Massif de la Selle in the Southern Peninsula. This restriction in extent of clastic deposits may have resulted because of a lessening rate of uplift of the Cordillera Central.

Lower Eocene clastic sedimentary rocks and limestone unconformably overlie Cretaceous(?) volcanic rocks at the south end of the metamorphic belt west of Santo Domingo (Bowin, 1966, p. 49). The Eocene rocks strike N 70°–85° E and dip moderately (15°–50°) southward. This occurrence may indicate that this site was near the eastern terminus of the Cordillera Central uplifted zone in Early Eocene time. This condition appears also to have been the situation in Oligocene and Miocene time since sedimentary rocks of that age there strike about N 25° E and dip 5°–30° southeastward (Bowin, 1966, p. 49). It is probably significant that this location of the apparent eastward termination of the Cordillera Central uplifted region from Lower Eocene into the Miocene is near a northwestward extension of the Beata Ridge lineament.

Except for the volcanic tuff and clastic sedimentary rocks of the south flank of the Cordillera Central, the limited exposures near La Vega on the north side of the Cordillera Central, and the Loma Caballero Formation to the east of the eastern terminus of the Cordillera Central, the Middle Eocene was elsewhere in Hispaniola predominantly a time of massive limestone deposition. The Middle Eocene limestone found in the Puerto Plata area of northern Dominican Republic is tuffaceous, perhaps because of the volcanic activity to the south that deposited the Loma Caballero rocks. Sometime in the Middle Eocene, an earthquake apparently caused a water-charged tuffaceous unit of Paleocene to Early Eocene age to slide down a submarine slope and become the San Marcos Olistostrome now exposed in the Puerto Plata area. Information for eastern Dominican Republic is sparse.

In the Late Eocene, volcanism ceased to be an important contributor to the rocks deposited in Hispaniola. The Hatillo Thrust (Bowin, 1966) carried rocks of the median metamorphic belt over Lower Cretaceous, Upper Cretaceous, and Lower Tertiary rocks, and hypabyssal stocks of pyroxene diorite, gabbro, and dolerite were intruded on the north side of the median metamorphic belt in central Dominican Republic and near La Vega, and in Haiti on the south side of the Massif du Nord in the Montagnes Noires, Ennery, and Terre-Neuve Mountain areas. This was a time of profound change in tectonism in Hispaniola and over much of the Greater Antilles. It presumably marks the time of cessation of active lithospheric underthrusting beneath Hispaniola and Puerto Rico that had resulted in the volcanism occurring in the central belt along the length of Hispaniola and in Puerto Rico.

The change in tectonism in the latest Eocene appears to have had an important effect on the trend of the structural grain in Puerto Rico and presumably in eastern Dominican Republic. There structural trends, in terms of present geographic orientation, were about N 70° W prior to latest Eocene time and nearly east–west afterwards. However, such a large change in trend does not appear to have occurred in central and western Hispaniola, since axes of folds and flexures, and trends of faults, in Oligocene, Miocene, and Pliocene sedimentary rocks in that area have a general north-of-west trend (about N 65° W). The change in Upper Tertiary deformation pattern may have taken place near where the structural trend of the metamorphic belt changes trend in central Dominican Republic (Bowin, 1966, p. 83) from about N 70° W to N 40° W near the town of La Vega. The formation of the Muertos Trough by lithosphere underthrusting south of eastern Dominican Republic (J. E. Matthews, personal communication, 1973) may have begun following the change in tectonism in the latest Eocene. That may also have been the time at which sea-floor spreading may have begun forming the Cayman Trough (Holcombe *et al.*, 1973; Heezen *et al.*, 1973).

The change in Tertiary deformation pattern may take place across the Beata Ridge lineament, which is defined by the crest of the Beata Ridge, and the straight eastern coastline of Barahona Peninsula (about N 34° E). The inland projection of this lineament extends to near the location of the bend in the structural trend of the metamorphic belt and of the Hispaniola Fault Zone, and to near the apparent eastern terminus of Lower Eocene to Miocene uplift of the Cordillera Central. The mountains and structures of the Barahona Peninsula and Enriquillo Basin are truncated at the coast. Were it not for the continuity of the metamorphic belt across the northern end of this lineament, large transcurrent displacement would be a likely assumption. It may be that the Beata lineament separates two crustal behaviors under compression. The region of Hispaniola to the west of the lineament deformed by folding, faulting, and uplift, whereas eastern Hispaniola deformed principally by underthrusting of Caribbean Sea crust beneath the Muertos Trough and eastern Hispaniola (J. E. Matthews, July 1973, personal communication), and perhaps more recently, also, by underthrusting of a portion of the Atlantic plate beneath eastern Dominican Republic as indicated by earthquake slip mechanisms. The trend of the Beata lineament is close to being normal to the general trend of the fold axes and fault traces in central and western Hispaniola, supporting the view that the lineament may be the location of a transform fault.

In the Oligocene, following the major deformation in the latest Eocene, limestone was the principal sediment deposited over almost all of Haiti and over the southwest part of the Dominican Republic south of the Cordillera Central. North of the Cordillera Central, clastic sedimentation predominated, and the Cordillera Central appears to have been the primary source of the clastic sediments, judging from the conglomerates at the base of the section on the south side of the Cibao Valley. Nagle (1971) infers that the Puerto Plata area was above sea level during the Oligocene, while in the Oligo-Miocene, areas to the south, east, and west were basins and accumulated marine sediments. Oligo-Miocene sediments in Haiti, where they are recognized, are massive or chalky limestone, and in the northern part of the country (Northwest Peninsula, Central Plateau, and Montagnes Noires) they are interbedded with argillaceous sediments.

An angular unconformity is reported by Butterlin (1960) to separate the Miocene formations of Haiti from the underlying Oligo-Miocene and Oligocene sedimentary rocks. Both limestone and clastic sedimentary rocks of Miocene age are reported from Hispaniola, but detrital sediments seem to be more abundant. Limestone appears to be somewhat more common in the Lower Miocene in places. The Miocene Cevicos Limestone occurs at the southeast end of the Cibao Valley and has very nearly horizontal bedding. Gypsum and lignite occur locally in Miocene sedimentary rocks.

A few hundred meters of Pliocene gravel and sandstone occur in the Central Plateau and in the valleys of the Lavagne and Gauche rivers in the Massif de la Selle. However, over 3,600 m (12,000 feet) of Pliocene shale and sandy shale with some sand layers were drilled in the Cibao Valley near the town of Santiago. At its total depth, the well is considered to be above the top of the Mao Adentro Limestone which is the lower member of the Mao Formation. A thick section of Pliocene sediments (the Angostura, Las Salinas, and Jimani Formations; Fig. 6) appears to have also accumulated in the Enriquillo–Cul de Sac Basin. Thus, both the Cibao and Enriquillo basins were sites of very active subsidence in the Pliocene and probably also in the Miocene. An inferred fault (Southern Samana Bay Fault, Fig. 4) separates the flat-lying Miocene Cevicos Limestone from the Cibao Basin in which several thousand feet of Pliocene sediments were deposited. Presumably, this fault is principally one of strike-slip motion.

The distribution of Quaternary alluvium in the Cibao Valley (Fig. 3) suggests that important subsidence in the central and western part of this basin has probably ceased, and the distribution of older sedimentary rocks in the center of the valley suggests that uplift of the basin may be beginning. The occurrence of Pleistocene coral reefs on the edge of Lake Enriquillo indicates that the lake had previously been open to the Caribbean Sea. The lake is presently below sea level, and that part of the basin is isolated from the sea by a ridge near the coast. Information from oil-well drilling in the area indicated the presence of a buried structural ridge across the end of the basin near the coast (structure on which Palo Alto No. 1 was drilled; A. A. Weymouth, 1961, personal communication; Llinas, 1972, Part 2, Figs. 11 and 12; Fig. 6). Presumably this ridge is still rising and has recently isolated the Enriquillo Valley floor from the sea.

The alkaline mafic volcanic flows (nepheline basalt and limburgite) north of the Cul de Sac–Enriquillo Valley are approximately of Pleistocene age

(MacDonald and Melson, 1969). Raised Pleistocene coastal reef limestone along much of the coast of the island may indicate recent uplift.

I infer that the strike-slip faulting along the northern margin of Hispaniola dates back only to about Pliocene time, and that during Oligocene and Miocene time the northern margin was principally being downwarped and a thick section of clastic sediments accumulated there. The locally very thick section (possibly 4 km or more) of Pliocene sediments in the Cibao Basin probably accumulated in a depression that developed as the Cordillera Septentrional was being uplifted by a component of thrusting that accompanied strike-slip motion on the Septentrional Fault at the northern border of the Cibao Valley. About 6 km of vertical displacement appears to have occurred along that fault (Bowin, 1960, p. 178), exposing Eocene rocks along the south edge of the Cordillera Septentrional. Hence, the long straight faults in northern Hispaniola are inferred actually to be oblique-slip faults. The several trends (between N 65° W and N 85° W) of the presumed predominantly strike-slip faults suggest there has been a complex adjustment along northern Hispaniola between the approximate N 81° E motion on the transform fault in the eastern Cayman Trough (Holcombe et al., 1973; Molnar and Sykes, 1969) and the nearly east-west motion north of Puerto Rico on the south side of the Puerto Rico Trench (Molnar and Sykes, 1969). The initiation of strike-slip faulting in the Pliocene would indicate that a change in the tectonic pattern in this region has occurred in the last few million years. Such a recent change might explain why the trend of the Bouguer gravity anomaly high along northern Hispaniola cuts across the structural grain that includes rocks of Miocene age. The southwestward underthrusting of the Atlantic Plate beneath eastern Dominican Republic (Molnar and Sykes, 1969; Bracey and Vogt, 1970) may also be a Pliocene to Recent development. Eastern Dominican Republic thus would be a wedge-shaped crustal block lying between underthrusting faults on the north and south. Seismic activity may alternate with time between these faults.

ACKNOWLEDGMENTS

The wholehearted cooperation and assistance received from many individuals interested in the geology of Hispaniola has been inspiring. After having been away from active studies of the area for several years, this help was most welcome and needed. I would particularly like to express my appreciation to F. Nagle, J. Butterlin, R. Weyl, S. Kesler, G. Antonini, J. E. Matthews, T. Donnelly, W. MacDonald, S. R. Hart, L. T. Aldrich, R. C. Tjaslma, R. Geller, B. Bornhold, J. J. Hungria, P. A. Brouwer, O. Cucurullo, Jr., and A. A. Weymouth. I would like again to express my indebtedness to Harry Hess who made the initial arrangements for my studies in Hispaniola and supported my gravity measurements in the Dominican Republic from a National Science Foundation grant. E. Pawley of Petrolera Dominicana, C. Dohm of Pan American International Oil Company, Lee Lamar and W. E. Wallis of Standard Oil Company, and H. M. Kirk of the Atlantic Refining Company were generous in providing data on Hispaniola. C. Dean and L. Gove helped in the preparation of the manuscript. The gravity measurements in Haiti were conducted with support from National Science Foundation Grant 24142. This study was supported by the Office of Naval Research, Department of the Navy.

REFERENCES

Alberti y Bosch, N., 1912, *Apuntes Para la Prehistoria de Quisqueya*, Tomo 1, *Geologica y Partes Descriptivas*, La Vega, 148 p.

Ayala, A., 1959, Estudio de Algunos microfosiles Planctonicos de las Calizas del Cretacio Superior de la Republica de Haiti: *Paleontol. Mex.*, n. 4, 42 p.

Berggren, William A., 1972, A Cenozoic Time Scale—Some Implications for Regional Geology and Paleobiography: *Lethaia*, v. 5, n. 2, p. 195–215.

Bermudez, P. J., 1949, *Tertiary Smaller Foraminifera of the Dominican Republic*: Cushman Laboratory for Foraminiferal Research, Spec. Publ. No. 25, 322 p.

Blesch, R. R., 1966, Mapa Geologico Preliminar, Republica Dominicana, 1 : 250,000, from: Mapas, Vol. II, Reconocimiento y Evaluacion de Los Recursos Naturales de la Republica Dominicana, Union Panamericana.

Bowin, C., 1960, Geology of Central Dominican Republic: Ph.D. Dissertation, Princeton University, 211 p.

Bowin, C., 1966, Geology of Central Dominican Republic: a Case History of Part of an Island Arc: Geol. Soc. Am. Mem. 98, p. 11–84.

Bowin, C., 1968, Geophysical Study of the Cayman Trough: *J. Geophys. Res.*, v. 73, p. 11–84.

Bowin, C., 1971, Some Aspects of the Gravity Field and Tectonics of the Northern Caribbean Region: Trans. Fifth Carib. Geol. Conf., Geol. Bull. No. 5, Queens College Press, p. 1–6.

Bowin, C., in press, The Caribbean: Gravity Field and Plate Tectonics: Geol. Soc. Am. Spec. Pap. 169.

Bracey, D. R., and Vogt, P. R., 1970, Plate Tectonics in the Hispaniola Area: *Geol. Soc. Am. Bull.*, v. 81, p. 2855–2860.

Butterlin, J., 1960, Geologie Generale et Regionale de la Republique d'Haiti, Vol. VI of *Travaux et Memoires de l'Institut des Hautes Etudes de l'Amerique Latine*: Université de Paris, 194 p.

Cucurullo, O., Jr., 1956, *Geografia de Santo Domingo*, Vol. 1: Editora Montalvo, Ciudad Trujillo, 118 p.

Donnelly, T. W., 1964, Evolution of Eastern Greater Antillean Island Arc: *Bull. Am. Assoc. Petr. Geol.*, v. 48, p. 680–696.

Edgar, N. T., Saunders, J. B., Donnelly, T. W., Schneidermann, N., Maurrasse, F., Bolli, H. M., Hay, W. W., Riedel, W. R., Premoli-Silva, I., Boyce, R. E., Prell, W., Broecker, W., Gieskes, J., Horowitz, R., and Waterman, L., 1971, Deep Sea Drilling Project, Leg 15: *Geotimes*, p. 12–16.

Francheteau, J., 1970, Paleomagnetism and Plate Tectonics: Ph.D. Dissertation, University of California, San Diego, 305 p.

Hart, S. R., 1966, Current Status of Radioactive Age Determination Methods: *Trans. Am. Geophys. Union*, v. 47, n. 1, p. 280–286.

Heezen, B. C., Perfit, M. R., and Dreyfus, M., 1973, The Cayman Ridge: Geol. Soc. Am. Abstr. with Programs, v. 5, n. 7, 1973 Annual Meetings, p. 663.

Hess, H. H., and Maxwell, J. C., 1953, Caribbean Research Project: *Geol. Soc. Am. Bull.*, v. 64, p. 1–6.

Holcombe, T. L., Vogt, P. R., Matthews, J. E., and Murchison, R. R., 1973, Sea-Floor Spreading in the Cayman Trough (Abstract): *Trans. Am. Geophys. Union*, v. 54, n. 4, p. 327.

Karig, D. E., 1971a, Origin and Development of Marginal Basins in the Western Pacific: *J. Geophys. Res.*, v. 76, p. 2542–2561.

Karig, D. E., 1971b, Structural History of the Mariana Island Arc System: *Geol. Soc. Am. Bull.*, v. 82, p. 323–344.

Kesler, S. E., 1966, Geology and Ore Deposits of the Memé-Casseus District, Haiti: Ph.D. Dissertation, Stanford University, 159 p.

Kesler, S. E., 1968, Mechanisms of Magmatic Assimilation at a Marble Contact, Northern Haiti: *Lithos*, v. 1, p. 219–229.

Kesler, S. E., 1971, Petrology of the Terre-Neuve Igneous Province, Northern Haiti: *Geol. Soc. Am. Mem.* 130, p. 119–137.

Kesler, S. E., and Fleck, R. J., 1967, Age and Possible Origin of a Granitic Intrusion in the Great Antilles Island Arc (Abstract): Geol. Soc. Am. Spec. Pap. 115, p. 482.

Kesler, S. E., Barton, J. M., Jones, L. M., and Walker, R. L., 1972, Eocene Porphyry Copper Province, Greater Antilles (Abstract): Geol. Soc. Am. Abstr. with Programs, v. 5, n. 7, p. 603.

Kesler, S. E., and Speck, R. C., in press, Plagioclase-Bearing Dunite from the Western Cordillera Central, Dominican Republic: *Carib. J. Sci.*

Lewis, J. F., and Kesler, S. E., 1973, Tonalites (Quartz Diorites) from the Cordillera Central, Dominican Republic—A Relatively Minor Plutonic Rock Type in the Greater Antilles and Nicaraguan Rise (Abstract), EOS: *Trans. Am. Geophys. Union*, v. 54, n. 4, p. 490.

Llinas, R. A., 1972, Geologia del Area Polo-Duverge, Cuenca de Enriquillo, Codia: Publication of Colegio Dominicano de Ingenieros, Arquitectosa y Agrimensores, Part 1 in No. 31, p. 55–65, and Part 2 in No. 32, p. 40–53.

MacDonald, W. D., and Melson, W. G., 1969, A Late Cenozoic Volcanic Province in Hispaniola: *Carib. J. Sci.*, v. 9, n. 3–4, p. 81–90.

Malfait, B. T., and Dinkelman, M. G., 1972, Circum-Caribbean Tectonic and Igneous Activity and the Evolution of the Caribbean Plate: *Geol. Soc. Am. Bull.*, v. 83, p. 251–272.

Matthews, D. H., Vine, F. J., and Cann, J. R., 1965, Geology of an Area of the Carlsberg Ridge, Indian Ocean: *Geol. Soc. Am. Bull.*, v. 76, p. 675–682.

Mattson, P. H., 1973, Middle Cretaceous Nappe Structures in Puerto Rican Ophiolites and Their Relation to the Tectonic History of the Greater Antilles: *Geol. Soc. Am. Bull.*, v. 84, p. 21–38.

Melson, W. G., and Thompson, G., 1970, Layered Basic Complex in Oceanic Crust, Romanche Fracture, Equatorial Atlantic Ocean: *Science*, v. 168, p. 817–820.

Melson, W. G., Bowen, V. T., van Andel, T. H., and Siever, R., 1966, Greenstones from the Central Valley of the Mid-Atlantic Ridge: *Nature*, v. 209, n. 5023, p. 604–605.

Mitchell, R. C., 1953, New Data Regarding the Dioritic Rocks of the West Indies: *Geol. Mijnbouw, N.S.*, 15ᵉ Jaarg., p. 285–295.

Molnar, P., and Sykes, L. R., 1969, Tectonics of the Caribbean and Middle America Regions from Focal Mechanisms and Seismicity: *Geol. Soc. Am. Bull.*, v. 80, p. 1639–1684.

Nagle, F., 1966, Geology of the Puerto Plata Area, Dominican Republic: Ph.D. Dissertation, Princeton University, 171 p.

Nagle, F., 1971, Geology of the Puerto Plata Area, Dominican Republic, Relative to the Puerto Rico Trench: Trans. Fifth Carib. Geol. Conf., Geol. Bull. No. 5, Queens College Press, p. 79–84.

Nagle, F., 1974, Blueschist, Eclogite, Paired Metamorphic Belts and the Early Tectonics of Hispaniola: *Geol. Soc. Am. Bull.*, v. 85, p. 1461–1466.

Palmer, H. C., 1963, Geology of the Moncion–Jarabacoa Area, Dominican Republic: Ph.D. Dissertation, Princeton University, 256 p.

Phillips, J. D., and Forsyth, D., 1972, Plate Tectonics, Paleomagnetism, and the Opening of the Atlantic: *Geol. Soc. Am. Bull.*, v. 83, p. 1579–1600.

Reeside, J. B., Jr., 1947, Upper Cretaceous Ammonites from Haiti: *U.S. Geol. Surv. Spec. Pap.* 214A, p. 1–11.

Shurbet, G. L., and Ewing, M., 1956, Gravity Reconnaissance Survey of Puerto Rico: *Bull. Geol. Soc. Am.*, v. 67, p. 511–534.

Shurbet, G. L., and Worzel, J. L., 1957, Gravity Measurements in Oriente Province, Cuba: *Bull. Geol. Soc. Am.*, v. 68, p. 119–124.

Sykes, L. R., 1970, Seismicity of the Indian Ocean and a Possible Nascent Island Arc between Ceylon and Australia: *J. Geophys. Res.*, v. 75, p. 5041–5055.

Sykes, L. R., and Ewing, M., 1965, The Seismicity of the Caribbean Region: *J. Geophys. Res.*, v. 70, p. 5065–5070.

Van Hinte, J. E., 1972, The Cretaceous Time Scale and Planktonic–Foraminiferal Zones: *Proc. Koninkl. Nederl. Akad. Wetenschap.*, Ser. B, v. 75, n. 1, p. 1–8.

Vaughan, T. W., Cooke, W., Condit, D. D., Ross, C. P., Woodring, W. P., and Calkins, F. C., 1921, *A Geological Reconnaissance of the Dominican Republic*: Gibson Brothers, Inc., Washington, 268 p.

Wenner, D. B., and Taylor, H. P., Jr., 1971, Temperatures of Serpentinization of Ultramafic Rocks Based on O^{18}/O^{16} Fractionation between Coexisting Serpentine and Magnetite: *Contrib. Mineral. Petrol.*, v. 32, p. 165–185.

Weyl, R., 1941, *Bau und Geschichte der Cordillera Central von Santo Domingo*, Vol. 2 of *Veröffentlichungen des Deutsch-Dominikanischen Tropenforschungsinstituts*, A. Meyer-Abich, ed.: Gustav Fischer, Jena, 70 p.

Weyl, R., 1953, Die Sierra de Bahoruco von Santo Domingo und ihre Stellung in Antillen-bogen: *Neues Jahrb. Geol. Palaeontol., Abhand*, v. 98, p. 1–27.

Weyl, R., 1966, Geologie der Antillen, Band 4, Beiträge zur Regionalen Geologie der Erde, H. J. Martini, ed.: Gebrüder der Borntraeger, Berlin, 410 p.

Woodring, W. P., Brown, J. S., and Burbank, W. S., 1924, *Geology of the Republic of Haiti*: The Lord Baltimore Press, 631 p.

Chapter 13

GEOLOGY OF CUBA

Georges Pardo

Gulf Oil Corporation
P. O. Box 1166
Pittsburgh, Pa. 15230

I. INTRODUCTION

For many years the geology of Cuba has been a challenge to every geologist who has come in contact with any of its aspects, and many were attracted by some of the exceptional features of the island, for example, the extensive development of ultrabasic rocks; the numerous oil and gas seeps; the rich Jurassic ammonite faunas, the variety of its Tertiary foraminifera, and the presence of some of the largest Upper Cretaceous rudistids ever discovered, to name only a few. The greatest geological attraction of the island, however, has been the challenge of understanding the stratigraphy and the structures, not to mention the difficulty of reconstructing the geologic history in a logical and probable way.

This chapter is a departure from the traditional format in that no references will be given in the text because the bulk of the information presented is derived from Gulf Oil Corporation files and has not been previously published. However, all available published literature was studied and taken into consideration; a bibliography is provided at the end of the chapter. Furthermore, for the sake of clarity, no references will be given within the text, and there will be no attempt made to present all existing conflicting opinions on the subject. The interpretation presented is strongly influenced by the author's personal experience in Cuba during the period from 1951 to 1955. In that period of time, Gulf Oil Corporation undertook to establish precise stratigraphic correlations and to map in detail the central part of the island with the purpose of resolving, if possible, the geological puzzle. (The reasons for selecting the central part of Cuba for this study are discussed later in the text.) This effort was under the direction of the author who wishes to give credit to P. B. Truitt and H. Wassall who were responsible for most of the field work and the general structural-stratigraphic studies; to P. Bronnimann who was in charge of the stratigraphic laboratory and was assisted by N. K. Brown in paleontology and K. K. Dickson in petrography. It is worth mentioning at this point that P. Bronnimann was the first man to make systematic use of nannoplankton to solve a large-scale, very difficult problem; his studies were one of the keys to the understanding of the Cretaceous and Upper Jurassic stratigraphy of the island. Credit is also due M. T. Kozary who was closely associated with this project. Although this detailed study of central Cuba has provided the core of the information contained herein, many data were also provided by the reconnaissance work done by Gulf geologists on other parts of the island.

In the following text, formation and other geologic names are given only when these have been previously published and are considered valid by the author. Names of stratigraphic belts, however, whether previously published or unpublished, will be used, and their original definition will be given. On the other hand, many formation names that were established and defined during Gulf's project will not be used as these have not been previously reported in the geologic literature and their definitions are beyond the scope of this chapter.

This paper has been prepared with the help of Gulf Oil Corporation, and permission to publish it is acknowledged with gratitude.

II. REGIONAL SETTING

Cuba, the largest of the Caribbean islands, has an arcuate shape, concave to the south; it separates the Gulf of Mexico to the west, and the Florida–Bahama platform to the east from the Caribbean (Fig. 1). Over most of its length, Cuba is the dividing line between extremely stable geologic conditions to the north and unbelievably complex ones to the south. Generally speaking, much of the north coast of Cuba, extending from central Matanzas to western Oriente, belongs to the Florida–Bahama carbonate province. On the other hand, most of the southern part of the island (the Isle of Pines, the Trinidad Mountains, south-central Camaguey, and the Sierra Maestra) consists of metamorphosed sedimentary and acidic igneous rocks. In between is a relatively

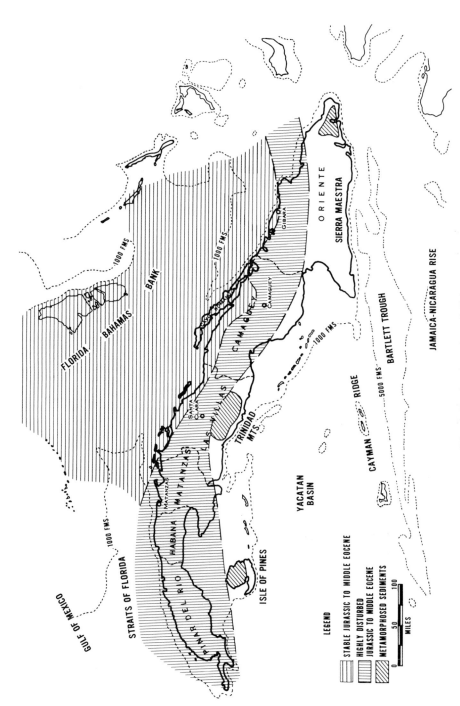

Fig. 1. Regional setting of Cuba.

narrow belt (from less than 30 up to 100 miles in width) of extremely folded and faulted material that consists of almost detrital-free platform carbonates, pelagic carbonates, cherts, locally great thicknesses of quartz-derived sandstones and shales, ultrabasic igneous rocks, and many types of volcanics and volcanic-derived sediments. South of, and parallel to, the Sierra Maestra (which is the eastward extension of the Cayman Island ridge) lies the Bartlett Trough separating Cuba from the Jamaica–Nicaragua rise. With the exception of the Trinidad and Sierra Maestra Mountains, Cuba has a generally low elevation but with many areas of rugged topography due to recent uplift and erosion.

III. MAJOR GEOPHYSICAL FEATURES

A. Gravity

The Bouger gravity map of Cuba (Fig. 2A) shows intermediate values; with one exception, they range between $+80$ and -30 mgal. A west–northwest-trending minimum extends from Gibara to Matanzas and follows approximately parallel and slightly to the south of the coast of Cuba. The minimum values range from -30 mgal in Matanzas to -20 mgal in Camaguey and northern Oriente. South of this belt, near the southern coast, are areas showing relatively high values of $+60$ to $+80$ mgal and over $+100$ mgal in southern Oriente. These gravity highs are separated from each other by narrow gravity minima. Two very pronounced transverse lows are worth noticing. One extends in an east–northeast direction from the southern end of the Trinidad Mountains to Cayo Romano, and the other, with the same trend, runs south of the axis of the Pinar del Rio Province.

B. Magnetics

The aeromagnetic survey (Fig. 2B) shows two featureless areas along the north coast, from Gibara to Matanzas, and along the northwestern half of Pinar del Rio Province, indicating deep magnetic basement. Depth estimates show as much as 30,000 ft to that basement.

South of southern Matanzas and along the southeastern two-thirds of the island, magnetics indicate the presence of many shallow igneous bodies. The trend of this complex area is interrupted by an east–northeast trending discontinuity that coincides with one of the previously described transverse gravity minima.

Matanzas and Habana Provinces, as well as southeastern Pinar del Rio Province, are characterized by a relatively featureless magnetic expression showing, in places, sharp anomalies suggesting isolated shallow igneous bodies over a deep magnetic basement.

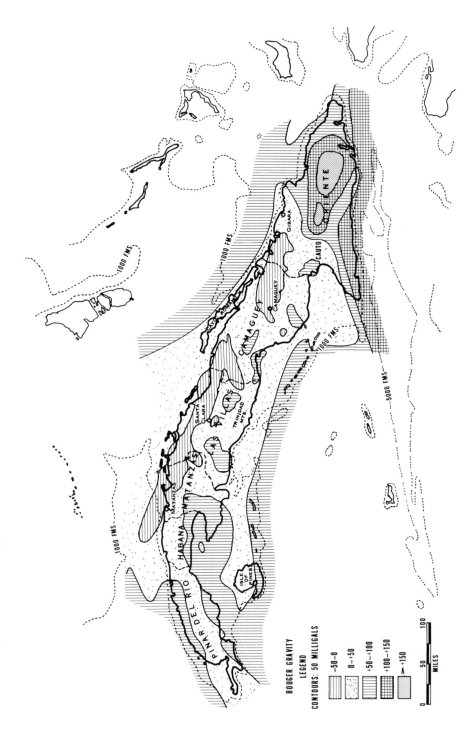

Fig. 2A. Bouguer gravity map of Cuba.

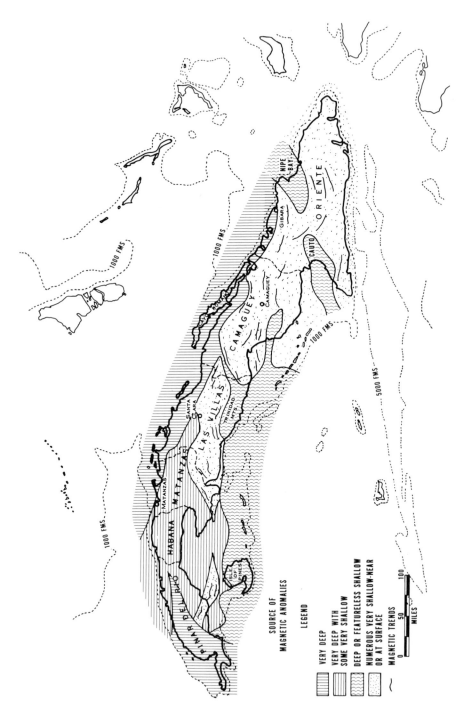

Fig. 2B. Magnetic interpretation map of Cuba.

The magnetic trends along Pinar del Rio Province are parallel to the gravity trends. Between southwestern Pinar del Rio and the Isle of Pines many shallow magnetic anomalies are present and possibly represent the continuation of the Las Villas Province shallow anomaly belt.

The south-central part of the island from the Isle of Pines to southeast of the Trinidad Mountains has a somewhat featureless magnetic expression which appears to be characteristic of the metamorphosed sediments.

It is worthwhile to point out that in many areas characterized by Bouger gravity lows the magnetic survey shows shallow igneous material, such as south and east of the town of Santa Clara or near the town of Camaguey.

IV. STRATIGRAPHY

A. General Description

The geologic map of Cuba (Fig. 3) shows that the island is segmented into six general areas, each one having its own structural and stratigraphic character. From west to east these areas are: (1) Pinar del Rio, (2) the Isle of Pines, (3) Habana–western Matanzas, (4) eastern Matanzas–Las Villas–western Camaguey, (5) Camaguey–northern Oriente, and (6) southern Oriente.

These areas are present-day structural highs surrounded by a relatively thin and undisturbed cover that ranges in age from the late Lower Eocene to the Pleistocene. These large-scale uplifts are mostly post-Eocene and are responsible for the present topographic relief of the island. The Lower to Middle Eocene and older rocks cropping out in the core of these uplifts are strongly deformed, and their general character can be described as follows.

In Pinar del Rio, sediments from Middle Jurassic (and possibly older age) to Lower to Middle Eocene are present in complexly faulted blocks, most of them dipping to the north. Along the north coast, ultrabasic and Cretaceous volcanics are present. The general strike is northeast–southwest.

The Isle of Pines consists mostly of metamorphics. Some Cretaceous volcanics outcrop in the northwest of the island.

In the Habana–western Matanzas area only Cretaceous volcanics and volcanic-derived sediments and Lower Eocene sediments are present in an extremely complex series of fault blocks. There are scattered ultrabasic bodies. Dips are highly variable from horizontal to vertical, and the faults appear to be high angle. The general strike is west–northwest.

The eastern Matanzas–Las Villas–western Camaguey area is similar to the Pinar del Rio Province in the sense that sediments from Upper Jurassic to Lower to Middle Eocene are well represented in a large number of facies.

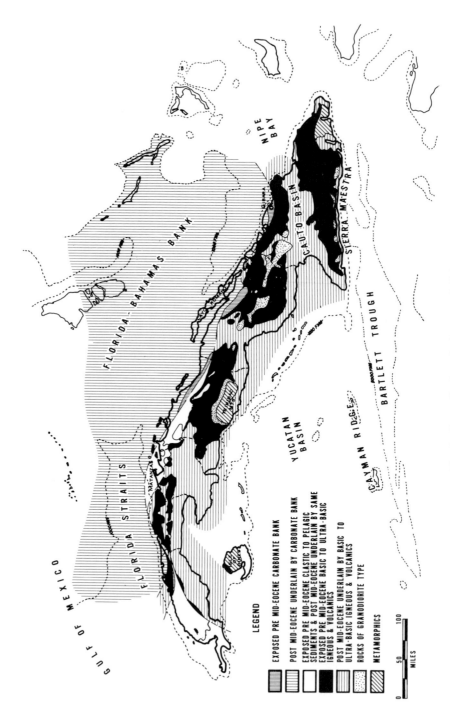

Fig. 3. Generalized geologic map of Cuba.

However, although the area is highly fragmented by vertical faults and is complexly folded, the general appearance is that of south-dipping blocks. The ultrabasics, volcanics, and volcanoclastics are found to the south. Finally, further to the south the metamorphic and igneous complex of the Trinidad Mountains is present. The general strike is northwest–southeast. This area has the most complete sequence of sedimentary, volcanic, and igneous rocks on the island and exhibits a greater variety of observable relationships than any other. For this reason, it will be considered as a "type geologic province" from which many of the opinions expressed in this chapter have been deduced and extended to other parts of the island.

The Camaguey–northern Oriente area is similar to the preceding one; however, most of it is covered by serpentines, volcanics, and igneous rocks. Except for outcrops of restricted lithologies, the sedimentary facies are poorly exposed. The general strike is northwest–southeast in the western part of the area and swings in an east–west direction in northern Oriente where the sedimentary facies, as well as the ultrabasic bodies, point out to sea in the area of Gibara.

In southern Oriente only ultrabasic, volcanic, acid igneous, and metamorphic rocks are present.

In the following description of the stratigraphy, the representative sections for each area or subdivision will be described from bottom to top. Measured thicknesses will be representative of areas and therefore approximate for any specific locality. Estimated (est.) thicknesses will be indicated as such. Only pre-Upper Eocene rocks will be described.

B. Central Cuba

The eastern Matanzas–Las Villas–western Camaguey area will be covered in the following description of central Cuba (Fig. 4).

The central part of Cuba can be divided into a series of narrow linear and roughly parallel belts that extend along the north-central part of the island in a northwest–southeast direction. Each one of these belts is characterized by a diagnostic stratigraphy and structural style. These belts have been named according to their geographic position; however, the relationships between the belts are sometimes complex, and one has to abandon the geographic definition and resort to a stratigraphic definition of the belt; that is, it will be defined as an association of several lithologies which occur invariably together. This definition can be carried even further in a paleogeographic and paleotectonic sense; every part of a belt will have had an identical succession of tectonic, sedimentary, and igneous events during geologic time. Furthermore, and of major importance, is the fact that each one of the belts represents an orderly sampling, in proper relative position, of the facies and structural

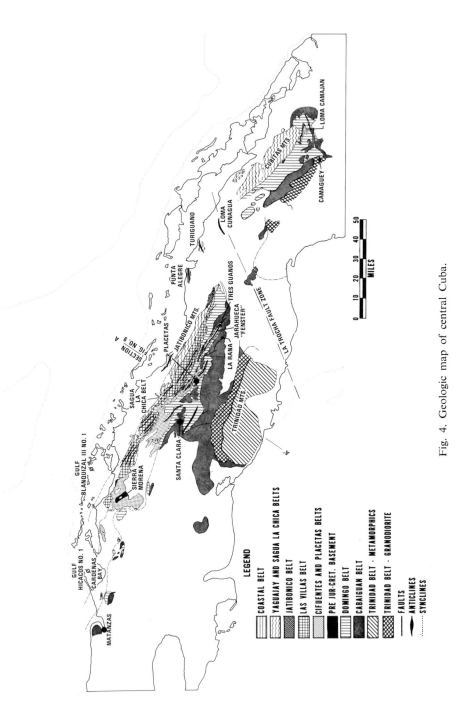

Fig. 4. Geologic map of central Cuba.

character that one would expect across a "classical" orthogeosyncline—the stable miogeosyncline being to the northeast in the Bahamas, and the mobile, thermally active eugeosyncline to the south toward the Caribbean.

There are, however, important differences between the Cuban situation and the classical concepts. For instance, the miogeosyncline is not marginal to a stable craton but to a rapidly subsiding area—perhaps more than 20,000 ft during Upper Jurassic and Cretaceous. Remarkably the eugeosynclinal material is completely unmetamorphosed. Finally, during much of the geosynclinal evolution there was essentially no terrigenous contribution from the continent.

The following is a description of the stratigraphy and general structural features of the belts of central Cuba (Fig. 5).

1. *Platform*

Cayo Coco Belt. This belt is mostly known through drilling along the north coast of Cuba and has only two small incomplete outcrop areas near Punta Alegre. The base of the section is unknown. The total thickness observed is approximately 14,000 ft

1. 3000 ft plus of interbedded massive limestones, secondary dolomites, and anhydrite of Jurassic age (Punta Alegre Formation).
2. 6500 ft of massive secondary dolomites, shallow-water platform-type limestones with abundant foraminifera (miliolids) ranging in age from Neocomian through Aptian.
3. 3200 ft of dense limestones, marls with chert nodules, and marly limestones of Albian through Coniacian age. Pelagic organisms are abundant.
4. 650 ft of fragmental limestone, in places coarsely conglomeratic, of Maestrichtian age.
5. Several hundred feet of interbedded marly limestones and marls overlain in turn by fragmental limestones of Lower to Middle Eocene age.

Fossils of Paleocene age have not been recognized. This belt represents the southern limit of the Bahama platform, and its total thickness, according to magnetic depth estimates, could reach 30,000 to 40,000 ft. The structure of this belt is gentle except in its southern edge where gypsum and salt diapirs are present along faults. The gypsum of the diapirs contains, among other exotics, fragments of quartz–sandstone and shales that have never been observed *in situ* in the Cayo Coco or other nearby belts.

Yaguajay Belt. The Yaguajay Belt is essentially similar to the Cayo Coco Belt in the sense that it belongs to the same carbonate province. The total thickness exposed is approximately 14,000 ft. No base is known.

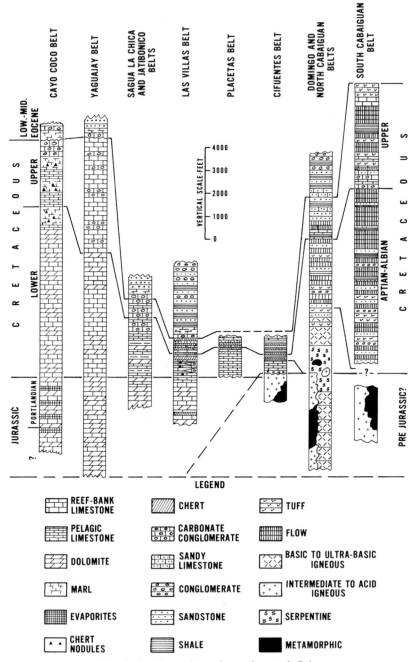

Fig. 5. Stratigraphic sections of central Cuba.

1. 4200 ft plus of massive secondary dolomites and shallow-water lime-stones that are thought to be equivalent to the Jurassic anhydrite–dolomite section of the Cayo Coco Belt.

2. 5400 ft of interbedded shallow-water massive limestone with abundant fossils (miliolids), secondary crystalline dolomites, thin-bedded fine-grained dolomites, and occasional intraformational breccias that suggest collapse features. The age ranges from Neocomian to Cenomanian. It should be noted that the upper part of this section is equivalent to pelagic sediments to the north.

3. 5000 ft of detrital and shallow-water limestones ranging in age from the Turonian through the Maestrichtian.

4. 200 ft of carbonate conglomerate of Lower to Middle Eocene age containing clasts of all underlying units.

5. Several hundred feet of marls and fragmental limestones that contain an increasing amount of igneous-derived material including brown shales and some sandstone with basic igneous detritals. The age is Lower to Middle Eocene.

The Paleocene is not recognized.

The Yaguajay Belt consists of southward-dipping blocks (40° to 60°) cut and dislocated by high-angle faults. It is bounded to the north by a south-dipping high-angle fault system and to the south by a relatively high-angle southward-dipping thrust fault whose sole rests on the Lower to Middle Eocene clastics.

2. Slope or Scarp

Sagua la Chica Belt. This belt has a fairly restricted outcrop area but is thought to be of great paleogeographic importance.

1. 1500 ft plus of carbonate conglomerates with components ranging from one-sixteenth of an inch up to tens of feet. There are occasional to abundant chert fragments. Interestingly enough, there is no ground-mass, and all the fragments have suture contact with each other in such a way that smaller fragments perfectly fill the voids between the larger ones. The carbonates contained in this conglomerate consist of all the types found in the Yaguajay and Cayo Coco Belts, including fragments of a reefoidal origin. The age of the components varies from Aptian–Albian at the base to Lower to Middle Eocene at the top. Interrupting the coarse clastic deposition are a few fine-grained lime-stone beds containing pelagic microfossils of Maestrichtian age. The lower part of the Upper Cretaceous has not been identified; nor has the Paleocene. However, no break in the section is observable.

2. Several hundred feet of marls and fragmental limestones that show upward an increasing contribution of basic igneous components. Their age is Lower to Middle Eocene.

The size of the components of these carbonate conglomerates, the very poor sorting, as well as their position between a Cretaceous, shallow-water carbonate bank, and (as will be seen later) deeper-water pelagic sediments of the same age, strongly suggest that the deposition of the Cretaceous part of these deposits was as submarine talus derived from the steep flanks and rim of the banks. The Lower to Middle Eocene part, as indicated by the nature of the fragments, suggests an orogenic origin.

Jatibonico Belt. This belt, as the Sagua la Chica Belt, is also of limited extent but of stratigraphic importance.

1. Several hundred feet of massive secondary dolomites like those found at the base of the Yaguajay Belt.
2. 2300 ft of interbedded dolomites, limestones, and clays of an intermediate facies between the Yaguajay Belt lithologies and the pelagic ones found further south. This unit ranges from Portlandian to Aptian age.
3. 900 ft of carbonate detrital conglomerate similar to the one present in the Sagua la Chica Belt but in which only Aptian–Albian faunas have been recognized.

This belt and the Sagua la Chica Belt are probably fault-isolated remnants of a much more extensive belt.

3. *Pelagic*

Las Villas Belt. This is unquestionably the one in which the most complete and better exposed sections are present. The oldest exposures occur in the southern part of the belt.

1. 2000 ft of dark gray limestones and brown-to-black secondary dolomites with occasional shale partings. In the middle of the unit there are well-developed characteristic oölitic limestones containing shallow-water microorganisms. This unit contains occasional ammonites and is of Upper Jurassic (Upper Kimeridgian and Portlandian) age. It should be noted that the appearance, as well as the faunal content of the oölitic beds, is typically Tethyan.
2. 800 ft of gray, buff weathering, thick-bedded, dense, fine-grained limestone with abundant thin yellow-orange wavy laminae. There are abundant Radiolaria, and calcareous nannoplankton are often rock-forming. Aptychi are common. Siliceous nodules are frequent but

secondary in origin. These limestones are interbedded with a number of intraformational conglomerates, thickening and increasing in abundance toward the south, and containing fragments of carbonates of Jurassic and lowermost Cretaceous age. The age is Neocomian to Aptian. To the north the conglomerates disappear, and only the dense limestones are present and grade upward into a thinly-bedded vari-colored succession of limestones and paper-thin marly shales of some 350 ft in thickness and of Aptian age.

3. 500 ft of thinly and regularly interbedded gray limestone and black primary chert beds. Radiolaria and other pelagic organisms are abundant. There are occasional fragmental limestones and limestone conglomerates mainly derived from equivalent age and older formations. An interesting aspect of these fragmental limestones is their content of shallow-water organisms such as *Orbitolina*; they occur in beds, at times no more than 1 in. thick, in between thin radiolarian and globigerinan limestones. The abrupt and repeated changes from coarse, shallow-water, fragmental limestone to dense, pelagic limestone excludes the possibility of rapidly changing sea level and points toward an origin by submarine slides or turbidity currents in a relatively deep sea. The age of these beds ranges from Upper Aptian through Turonian, and the contact with the overlying unit shows some degree of weathering.

4. 400 ft of sugary, fragmental limestone; red and brown jasper; and coarse, fragmental-to-conglomeratic limestones. The sugary appearance of the limestone is due to very small fragments of broken rudistid shells forming the bulk of the rock. These limestones are manganese stained at the base of the section. This unit is of Maestrichtian age and appears to have been deposited in shallower marine water. There is a possibility of a hiatus during the Coniacian. No Paleocene has been recognized.

5. 200 ft of carbonate conglomerate with chert pebbles that increase in abundance to the south. It contains components representing the entire section of the Yaguajay, as well as of the Las Villas Belt, and is of Lower to Middle Eocene age. Above this conglomerate, the percentage of igneous detrital material increases, and some impure marls and fragmental limestones are present.

6. 3300 ft plus of gray, weathering-to-brown shales and siltstones (sometimes calcareous in the lower part) interbedded with coarse-grained sandstones derived from basic igneous and volcanic rocks. The percentage of coarse detritals increases toward the top of the section, and there is an increasing amount of conglomerate. The matrix of the conglomerates is argillaceous, and the components range up to 10 ft

in size and consist of many types of crystalline igneous rocks, volcanics, and volcanic-derived sediments. The basic and ultrabasic components are dominant, but occasional fragments of Cretaceous and Jurassic pelagic carbonates are present. Sorting is very poor. This suggests that, during the Lower to Middle Eocene, a nearby source of volcanic and igneous material was being actively eroded. The upper part of this section is barren of organisms, but the lower part is rich in Radiolaria, Discoasters, and Coccoliths, suggesting that the initiation of clastic sedimentation took place in fairly deep water. It is worthwhile noticing that the igneous-derived clastic material is invariably found in synclines and along fault zones such as those separating the Las Villas Belt from the Yaguajay Belt or from the Placetas Belt to the south. It therefore appears that the Lower to Middle Eocene detrital sediments were the result of violent tectonic activity and at the same time provided some of the necessary lubrication for thrusting.

The structure of the Las Villas Belt consists of tightly folded anticlinoria in contrast with the monoclinal blocks of the Yaguajay Belt. However, in spite of the complex folding, the Las Villas Belt units can be relatively easily followed and mapped.

Stratigraphically, it is worth noticing the differences in thickness—approximately 1500 ft for the pre-Maestrichtian Cretaceous of the Las Villas Belt versus 7000 ft for the equivalent section in the Yaguajay Belt. The paleo-relationships between the Yaguajay Belt and the Las Villas Belt have certain similarities to the present relationship between the Bahamas and the sequence encountered in Joides Nos. 99, 100, and 101 where pelagic sediments, not more than a few thousand feet thick, are equivalent to more than 15,000 ft of shallow-water carbonates present some 50 miles away.

Placetas Belt. This consists of the following:

1. 1000 ft (est.) of brown, pelagic limestones interbedded with dark gray shales that increase in abundance toward the south. The limestones themselves consist almost entirely of Radiolaria and nannoplankton. This unit, of Neocomian age, can be distinguished from its age equivalent in the Las Villas Belt by the dense aspect of the limestones and the presence of layers of clay. However, the distinction is not always sharp as, depending on the position along the belt, a number of transitional types can be found. The base has not been observed.

2. 50 ft of brown, sandy, micaceous limestones and calcareous quartz and mica sandstones interbedded with tan shales. This represents the first observed influx of material derived from an "acid" source. Abundant pelagic organisms of Aptian age are present.

3. 500 ft (est.) of an interbedding of fine, fragmental limestones; dense, argillaceous limestones with abundant secondary cherts; thin-bedded, brown, primary cherts; brown-to-gray noncalcareous shales; and occasional fine-grained quartz sandstones. This unit is usually thin and even bedded and contains abundant pelagic organisms such as Radiolaria, *Globigerina*, and *Globotruncana*. Some of the limestones are black. The age ranges from Albian through Turonian. This section shows that there was a definite influx of fine-grained noncalcareous material during the lower part of the Upper Cretaceous. Long intervals of thin-bedded, primary cherts and siliceous shales and the presence of swelling clays are suggestive of a volcanic origin for most of the noncalcareous material.

4. 200 ft (est.) of a group of associated lithologies consisting of gray-to-cream, medium-to-coarse, fragmental limestones; gray, dense-to-fine, fragmental, argillaceous limestones; and pink, dense, massive-bedded, intensely fractured limestones. All of the above units are of Campanian through Maestrichtian age; part of the Coniacian and Santonian appear to be missing. The fragmental beds contain larger foraminifera such as *Sulcoperculina*, *Sulcorbitoides*, *Vaughanina*, and *Orbitoides*. Rudistid fragments are exceedingly abundant, indicating a shallow-water reefoidal origin. The finer-grained beds, on the other hand, contain a very rich pelagic fauna. No sediments younger than the Maestrichtian are present in this belt.

The deformation of the Placetas Belt is much more intense than that of Las Villas Belt. The lower Cretaceous carbonates and shales are sometimes intensely pleated with a fold periodicity of a few tens of feet. The Upper Cretaceous lithologies are represented by discontinuous chaotic outcrops which, however, form mappable units. It should also be noted that many of the fault contacts between the Placetas Belt and the Las Villas Belt are characterized by highly deformed Lower to Middle Eocene clastics caught in the fault planes.

Cifuentes Belt. The Cifuentes Belt is a fairly complex unit; however, its association of lithologies can be recognized and mapped in a number of places in spite of the structural complications.

1. The Cifuentes Belt shows the only known base of the sedimentary section. In a number of places a pink, plutonic rock that grades from granite to a labradorite–quartz–amphibole rock showing cataclastic alteration has been observed. This basement rock of general granodioritic composition and texture shows locally a number of inclusions of coarse crystalline marble and other metamorphics. Patches of ser-

pentine are usually present, but the association is believed to be structural. This basement is overlain in places by an old weathered zone (regolith), and in its turn overlain by up to 50 ft of a granule- to pebble-sized conglomerate characterized by quartz and other igneous fragments as well as limestone detritus. Since this basal conglomerate is overlain by Neocomian limestones and contains pebbles of the same limestone, it must be of Neocomian age. The three best exposures of basement are allochthonous and are located near La Rana, Tres Guanos, and Sierra Morena. It should be noted that these granodiorites, as well as others that will be discussed later, are very poor in muscovite and therefore cannot be considered as the source for the quartz–muscovite-bearing sandstones of Aptian age found in the Placetas and Cifuentes Belts.

2. 450 ft (est.) of pelagic limestones of late Neocomian to early Aptian age similar to those of the same age in the Placetas Belt but black in color when fresh and weathering to light gray. These carbonates are richly organic and often give a bituminous odor when broken. They are interbedded with yellowish clay intervals. Toward the top, muscovite-rich quartz sandstones similar to those of the Placetas Belt are found.

3. 500 to 1000 ft (est.) of a monotonous section of thin-bedded black, brown, red, yellow, and gray cherts and clays. Some gray, dense limestone is occasionally present. The age is Albian through Turonian. The thickness cannot be measured due to tectonic complications. This section is almost certainly of volcanic origin and was deposited under pelagic conditions, as indicated by the faunal content. The gradation between this unit and its time equivalent in the Placetas Belt can be observed; it shows a decrease in chert and swelling clays northward accompanied by an increase in limestone.

4. An unknown thickness of shallow-water fragmental limestones with occasional pelagic influx similar to the Campanian–Maestrichtian section found in the Placetas Belt.

The Cifuentes Belt also shows some evidence for a pre-Campanian post-Turonian unconformity. No Paleocene or Eocene rocks are known in this belt.

The Cifuentes Belt shows the most complex structure of all the sedimentary belts. Outcrops of its different units are found in apparently disconnected blocks; however, detailed mapping of these units permits a series of plates stacked on the top of each other to be delineated. The lowermost plates are, facieswise, more similar to the Placetas Belt, while the uppermost ones are the most argillaceous and siliceous and have the highest content of silicate clastics. Although, as will be discussed later, the basement outcrop of La Rana is completely out of place, the ones at Sierra Morena and Tres Guanos are considered to be part of the topmost observable plate of this belt.

4. *Basic Igneous-Volcanic (Zaza)*

The igneous-volcanic sequence has been subdivided on the basis of lithologic character. The Domingo Belt to the north consists mainly of basic to ultrabasic igneous rocks, while the Cabaiguan Belt consists of volcanics and associated sediments. These two belts are intimately related, and it is believed that the Domingo Belt at one time was the basement upon which the volcanic pile of the Cabaiguan Belt was deposited.

Domingo Belt. This belt consists of an association of intermediate to ultrabasic igneous rocks having a definite layered distribution. These rocks will be described according to their apparent stratigraphic sequence. In addition, some sediments and metamorphic rocks are invariably associated with rocks of the Domingo Belt. Although the association is unquestionably tectonic, this material will be described under this heading.

1. The lowermost unit crops out in the eastern and northeastern part of this belt. It is a basic to ultrabasic igneous complex that consists of fine to very coarsely crystalline uralite gabbro, olivine gabbro, hornblende gabbro, and epidiorite. The coarsest crystalline development appears to be restricted to the top of the unit. The thickness of this body is unknown but must be on the order of several thousand feet. North of the Placetas–Cifuentes Belt exposures and also caught along the fault zone that separates the Las Villas Belt from the Placetas–Cifuentes Belt, southeast of the town of Cifuentes, the above-described gabbros are associated with diorites and some rock types which are almost a mechanical mixture between diorites and basic igneous. The diorites are coarse-grained quartz diorites with biotites or hornblende. The feldspar is labradorite or andesine. This rock is very similar in composition and texture to the granite and granodiorite which form the basement of the southern plate of the Cifuentes Belt. They show a similar crushing of the grains.

2. Serpentine, which can be divided (in ascending order) into the three following types: (1) waxy serpentine, which has a scaly, highly sheared, glossy aspect. The shearing is of tectonic origin as it is usually found near fault zones or in areas of intense deformation. This type of serpentine contains, in places, exotic blocks of pegmatite, vein quartz and metamorphic rocks sometimes reaching many hundreds of feet in size. The metamorphics consist of anthophyllite schist, chlorite schist,

phyllite, muscovite schist, graphite schist, garnet schist, etc. These metamorphics are totally foreign to the ultrabasic complex and appear to be more related to the metamorphics found in the Trinidad Mountains to the south and the Isle of Pines further to the west. Some of these metamorphic blocks show a definite tear-drop shape, indicating the intensity of the deformation; (2) "reticulate" serpentine, which is relatively massive with numerous thin criss-crossing bands of dark green serpentine; and (3) porphyritic serpentine, which is similar to the reticulate serpentine but contains large lighter-colored bastite crystals. The above three types of serpentine were derived from pyroxene-bearing peridotites. The total thickness of this body is unknown but appears to vary considerably. North of the Placetas Belt the thickness of the serpentine might not exceed a few hundred feet, and in some places it might have been absent altogether, while south of Santa Clara the serpentine might be several thousand feet thick.

3. Hornblende dolerites with augite and uralite. These dolerites or micro-gabbros are very similar to the finer facies of the unit that underlies the serpentine. They seem, in general, to indicate more hypabyssal conditions. Quite often, the contact between the serpentine and dolerites shows interbedding of both types of rock. These dolerites are quite often cut by thick quartz veins mineralized with copper, magnetite, etc.

4. Uralite basalts interbedded with dolerites and crystal tuffs with abundant amygdules. This unit shows flow structure and considerable spilitization that suggest submarine volcanic origin. This unit is believed to be well over a thousand feet thick and is always overlain by the volcanics of the Cabaiguan Belt.

Two main types of dykes which usually cut the Domingo Belt are black and white sparkling fine-grained diorites and hornblende trachyte and andesite porphyries similar to some of the Upper Cretaceous flows found in the Cabaiguan Belt.

It is worthwhile mentioning that the basic igneous rocks of this belt are sometimes cut by asphalt dykes. In addition, the great majority of the hydrocarbon seeps, as well as the small local oil production, come from or are closely related to rocks of this belt.

The Domingo Belt shows, therefore, a very interesting association of basic to ultrabasic igneous rocks which, according to the grain size and texture, must have been formed under hypabyssal conditions with effusive rocks of similar composition but which were unquestionably deposited on the surface of the sea floor. The consistent succession of layers of the above-described types for hundreds of miles along the strike is believed to be a very strong argument against a deep-seated intrusive origin. Rather, the evidence points toward a combination of submarine volcanic flows and associated ultrabasic sills which took place above some continuous basement. No evidence has been found of intrusion of the limestones of the Cifuentes, Placetas, and other carbonate belts by any kind of igneous rock; and, although the serpentine contains many exotics of foreign origin, not a single one of the lithologies belonging to the sedimentary belts has been found.

Whether this basement was continental or oceanic in nature is a fundamental question. However, the similarity between the diorites of the Domingo Belt and the diorites of the Trinidad and Cifuentes Belts, the abundance of intermediate types of igneous rocks, and the similarity between the metamorphics included in the serpentine and the metamorphics of the Trinidad Belt strongly suggest that prior to the formation of the Domingo Belt rocks, a large area of complex metamorphic basement with associated diorites was in close proximity to an area where basic to ultrabasic material was being extruded. As a matter of fact, the ultrabasic extrusion could well have taken place through the metamorphic basement.

The age of the rocks of the Domingo Belt cannot be definitely ascertained. As will be seen later, the oldest fossils found in the overlying volcanic Cabaiguan Belt are of Cenomanian age, the igneous and metamorphic exotics found in the serpentine are of possible pre-Upper Jurassic age, and the first indications of volcanic activity in the Cifuentes Belt is in the Albian. This would suggest a Lower Cretaceous age of formation. However, the age of formation of the rocks of the Domingo Belt is quite different from the age of emplacement to their present position which, as will be discussed later in this chapter, is considered to be Lower to Middle Eocene.

Rocks of the Domingo Belt in the Santa Clara–Placetas area are associated with outcrops of red, olive brown, noncalcareous shales, mudstones, and fragmental limestones with abundant igneous grains. They contain a rich pelagic fauna of Maestrichtian age. These outcrops are always associated with fault zones. Lithologically speaking, these sediments do not belong to any previously described belt and show influx of igneous material much greater than in beds of equivalent age in the northern belts; their position within the igneous rocks of the Domingo Belt is certainly tectonic. The fact that no older sediments are found in association might indicate that originally this formation was laid directly on a basement high.

The structure of the Domingo Belt varies from chaotic toward the front to apparently simple broad anticlines and synclines. Faults are common, and in places the sections appear to be repeated by northward-directed thrusts.

Cabaiguan Belt. This belt is characterized by a large development of volcanic and volcanic-derived sediments that can hardly be separated from the Domingo Belt as both are intimately associated geographically and strati-graphically. The Cabaiguan Belt shows significant changes in stratigraphy and apparent thickness, depending on the area.

The thickest and most complete section of volcanics is found along the southern part of the Cabaiguan Belt along the southern flank of a broad syncline north of the Trinidad Mountains. There the base of the volcanic sequence is not observable; the entire section dips to the north with the oldest beds in fault contact with the igneous-metamorphic complex of the Trinidad Mountains.

1. Unknown thickness (probably many thousands of feet) of porphyritic flows; sandstones; conglomerates; yellow-brown, massive, spherulitic porphyries and dolerite flows. The age is unknown. Exposures of these "old volcanics" make up the bulk of the southeastern Cabaiguan Belt (good outcrops are seen along the Zaza River). Toward the north they are always in contact with the Domingo Belt.

2. 2000 ft plus of basalt flows, amygdular basalts, and porphyritic basalts interbedded with thin-bedded, brown, siliceous shales and slightly calcareous sandstones and conglomerates. This unit is almost barren of organisms except for some questionable Globigerinidae.

3. 2500 ft of olivine and augite dolerites in thick, massive flows, sometimes porphyritic, and frequently showing pillow structures. These flows are interbedded with a minor amount of conglomerate, shale, and noncalcareous tuff. This formation is barren of fossil organisms. The above succession of volcanics is considered to be upper Lower Creta-ceous to lower Upper Cretaceous.

4. 600 ft of an interbedding of fine- to medium-bedded oölitic, dense, and fragmental limestones, and thin-bedded shales. It contains pelagic foraminifera and Radiolaria and is considered Cenomanian in age. Some of the fragmental limestones contain oölites definitely derived from the type of Upper Jurassic limestones present in the Las Villas Belt. It is therefore an important evidence that the shallow-water carbonates of the Upper Jurassic must have extended as far south as the volcanic province and must have been exposed during Cenomanian time.

5. 500 ft of brown, dense-to-fragmental limestones and thinly bedded shales. Toward the top some tuffaceous sandstones and conglomerates are present as well as rudistid reefs. A porphyritic flow is present in the middle of the succession. Most of this unit is of Cenomanian age and was deposited under pelagic conditions, as indicated by abundant pelagic foraminifera and Radiolaria.

6. 600 ft of coarse- to fine-grained tuffs, sandstones, shales, and occasional limestones of Turonian age that were deposited under pelagic condi-

tions, as indicated by a planktonic assemblage similar to that of the underlying formation.

7. 2000 ft of a group of lithologies consisting, in the lower part, of porphyritic quartz–andesite flow interbedded with siliceous, tuffaceous, radiolarian shales and volcanic agglomerates containing characteristic glass bombs. The flows show abundant pillows coated with a glass layer. The middle of this group consists of thin, even-bedded, brown-to-yellow, radiolarian shales; and the upper part consists of hard, brown, siliceous shales with massive, black cherts and flows of augite basalt and andesite porphyry. It contains *Globigerina* and Radiolaria, thus indicating deposition under pelagic conditions. The age is considered Turonian–Senonian because of its stratigraphic position.

8. 300 ft of orange and buff, mottled, sandy limestones and marls that were deposited under fairly shallow-water conditions as, in addition to the usual planktonic assemblage, *Pseudorbitoides*, *Sulcoperculina*, and algae are abundant. The age is Maestrichtian. It is usually associated with local massive rudistid and other mollusk reefs and amygdular basalt porphyry.

9. 300 ft of thin-bedded, green and brown, volcanic-derived sandstones; crystal tuffs; shales; and occasional sandy-to-argillaceous limestones. The fauna is pelagic and of Maestrichtian age.

10. 400 ft of medium- to massive-bedded green tuffs.

Toward the north, the base of the Cabaiguan volcanic sequence is invariably in contact with the igneous rocks of the Domingo Belt. This contact is usually tectonically disturbed but is believed to be close to a normal stratigraphic contact. The pre-Cenomanian sequence of basic flows thins considerably from more than 7000 ft in the south to possibly less than 5000 ft near the Domingo Belt, to 1500 ft within the Domingo Belt proper, and to probably less than 1000 ft in the klippe separating the Placetas Belt from the Las Villas Belt. The Cenomanian–Turonian carbonate section also has an equivalent of reduced thickness toward the north. The thickness of the post-Turonian pre-Maestrichtian flows, interbedded tuffs, and volcanic-derived sediments decreases rapidly northward (from nearly 3000 ft to 0) through truncation by the pre-Maestrichtian unconformity. On the other hand, the Maestrichtian thickens northward from 1200 ft to approximately 2000 ft with a full development of a basal, coarsely crystalline hornblende–biotite porphyry and with local white massive rudistid and orbitoidal limestones that indicate very definite reefoidal conditions. In this unit the very large Rudistids, *Hippurites*, and *Radiolites* are found. At the top, shallow-water fragmental-to-dense limestones with marly, coarse-grained, igneous-derived sandstones and conglomerates are present. Toward the north the Maestrichtian is overlain by some 300 ft of pseudoölitic medium-bedded limestones containing igneous grains and marls of possible Paleocene age.

The youngest formations found are of Lower to Middle Eocene age and occur in the northeastern part of the Cabaiguan Belt. They consist of inter-bedded igneous-derived sandstones and shales characterized by the presence of large, well-rounded boulders of granite and basalt. Occasional beds of argillaceous limestones, marls, and reef limestones are present. To the north-west the possible Paleocene section is overlain by a sequence of basic igneous-derived sandstones, shales, and conglomerates with occasional limestones and marls. The thickness of the Lower Tertiary is at least 2000 ft.

The Cabaiguan Belt is, therefore, characterized by three main periods of volcanic activity. The oldest one, of possible Lower Cretaceous to early Cenomanian age, is typified by either dolerites or basalts containing augite and often olivine. The second period of volcanic activity is of Turonian–Senonian age and is characterized by more acidic flows of andesite or quartz–andesite porphyries. The third and last period of volcanic activity, of late Maestrichtian age, is characterized by basalts with hypersthene phenocrysts. These periods of volcanic activity are separated by two periods of rest; one during the Cenomanian, and the other, during the early Maestrichtian. Judging by the volume of volcanic flows, the earliest period of activity was by far the most important. The bulk of the volcanism, as well as sedimentation, appear to have been submarine throughout, and, except during the Maestrichtian, there are practically no evidences of shallow-water conditions. Most of the indigenous fauna is planktonic, and in the volcanic-derived sediments fragile minerals, such as plagioclases and ferromagnesians, are perfectly fresh, in-dicating lack of exposure to the atmosphere.

The structure of this belt, as that of the Domingo Belt, is highly disturbed and chaotic to the north and northeast. There, as in the klippe between the Las Villas and Placetas Belts, fragments of the various Domingo and Cabaiguan Belt lithologies are found as somewhat unrelated blocks. To the south, and especially south of the Domingo Belt, the Cabaiguan Belt forms a broad asymmetric syncline cut by a number of transverse high-angle faults.

5. *Metamorphic—Trinidad Belt*

The Trinidad Belt, named after the Trinidad Mountains, has a general dumbbell outcrop shape and appears to be surrounded on the west, north, and possibly east by rocks of the Cabaiguan Belt. The bulk of the Trinidad Belt consists of folded and faulted metasediments. Some of them show medium-grade metamorphism and consist of marbles, dolomites, schists, graphitic schists, quartzite, gneiss, etc. Along the north flank of the mountains are low-grade metamorphic rocks consisting of quartzites, quartzitic phyllites, phyllites, and marbles. Patches of serpentinite are locally common. The age of these metamorphic rocks has not been definitely established; however, it is believed to be Jurassic or older. Similarly, the age of metamorphism is unknown but could be as young as uppermost Cretaceous.

Along the northern part of the Trinidad Belt there is an extensive body of hornblende–diorite and microdiorite which is found in intrusive relationship with amphibolite and relatively high-grade metamorphic rocks, some of them possibly metamorphosed volcanics that appear to be the result of an earlier metamorphic phase. This diorite–metamorphic complex, in many places in fault contact with the Trinidad metamorphics, in composition is related to diorites and granodiorites considered basement in the northern belts. It is also related to granodiorites intermingled with ultrabasics in the northern front of the Domingo Belt. It might be important to mention that one sample of the granodiorite has been given a radiometric age of 180 million years. If this date can be confirmed, it would support the correlation between parts of the Trinidad Belt and the igneous and metamorphic rocks that are considered to be basement in the Cifuentes Belt. The Trinidad Belt contact with the Cabaiguan Belt, wherever observable, is by fault. Almost invariably, the fault appears to dip away from the Trinidad Belt and tends to be roughly parallel to the dips observed in the Cabaiguan Belt. This fault system has therefore a semicircular shape concave to the south. The Trinidad Belt is thought to represent an autochthonous massif.

Of great importance is the fact that along the faults that separate the northern part of the Trinidad Belt from the Cabaiguan Belt completely un-altered and unmetamorphosed outcrops of characteristic Upper Jurassic oölitic limestones and Neocomian pelagic limestones that occur in the Las Villas and Cifuentes Belts have been found. This definitely suggests that the granodiorite–metamorphic complex is a pre-Upper Jurassic basement.

C. Western Cuba

1. Pinar del Rio

In Pinar del Rio the concept of belts as developed in central Cuba holds true (Fig. 6). Although detailed maps are not available, the sedimentary belts show a succession roughly similar to that of central Cuba; i.e., during upper-most Jurassic and Cretaceous there was a general tendency to go from platform to the north to deeper water sediments to the south. However, there are some major differences: the shape of the belts is not as linear; the equivalent of the Domingo–Cabaiguan Belts is found to the north, along the coast, instead of south of the deeper water facies; and there are visible facies changes along the strike of the belts.

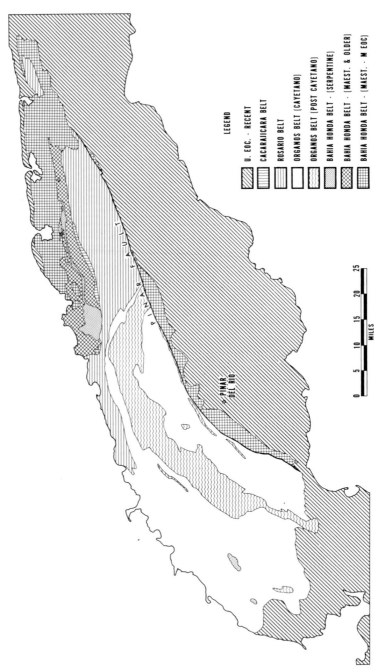

Fig. 6. Geologic map of Pinar del Rio Province.

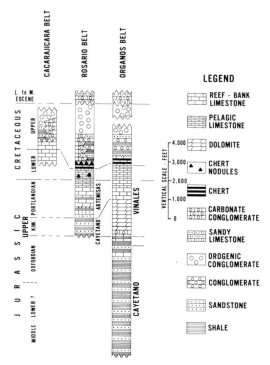

Fig. 7. Stratigraphic sections of Pinar del Rio Province.

Following is a description of the stratigraphy of the Pinar del Rio belts (Fig. 7).

Bahia Honda Belt. In its lower part, this belt consists of the same serpentine, gabbro, and spilite that has been described in the Domingo Belt. The serpentine is in fault contact to the south with volcanics of younger age and contains abundant volcanic blocks at its base. Overlying the basic and ultrabasic complex is a section of massive, weathered, and highly sheared, basic volcanic flows, cherts, and volcanic-derived shales and sandstones. Occasional impure limestones are present and permit paleontological dating of the middle part of the section as Cenomanian. It is of approximately the same age and lithologically similar to the lowermost carbonates present in the Cabaiguan Belt.

Above this unit, but in a complexly faulted relationship, are green and white tuffs, tuffaceous sandstones interbedded with light-colored, siliceous flow rocks, and occasional amygdaloidal porphyries. These volcanics are also very similar to some of the Maestrichtian volcanics in central Cuba.

Overlying the volcanics with unconformity and possibly in part equivalent to the previous unit is a section consisting of basic igneous-derived conglomerates, sandstones, siltstones, and shales. Argillaceous red and white

limestone, coarse- to medium-fragmental, heterogeneous, orbitoidal limestone, and limestone conglomerate are present. Dark brown and black thin-
bedded cherts are present near the base as well as rare basalt porphyry. Near
the top are enormous tectonically jumbled blocks of actinolite schist, serpentine, diabase, spilite, tuff, and assorted flow rocks mixed with the sediments.
This unit is of Maestrichtian and possibly of Campanian age. Similar conglomerates and breccias of Maestrichtian age are found in all the belts of Pinar
del Rio; they rest unconformably on all previous lithologies, and their composition reflects immediate nearby sources.

The youngest observed lithologic unit consists of sandstones, shales,
marls, and conglomerates with some sandy and fragmental limestones. The
lower part is conglomeratic with rudistid reefs and orbitoidal limestones and
is of possible Maestrichtian age. The Lower to Middle Eocene part is usually
sandy.

Thicknesses of the lithologic units of this belt are impossible to estimate
because of the complex structural condition but are certainly in the thousands
of feet. The outcrops dip steeply toward the north, as do the numerous faults.

Cacarajicara Belt. This belt, only a mile wide, extends for some 20 miles
along the southern edge of the previous belt. It consists of approximately a
thousand feet of massive shallow-water limestones of Aptian to Coniacian age
identical to those found in the Yaguajay Belt. There is no terrigenous material
interbedded with or included in the limestones. They are unconformably
overlain by coarse conglomerates similar to those described in the previous
belt, but the clasts consist mostly of underlying limestones. Although no
Maestrichtian-Yaguajay-type carbonates are identified in place, they must
have been present nearby at one time because they are well represented as
clasts in the conglomerates. This belt is essentially a fault sliver, and all the
beds are steeply dipping to the north.

Rosario Belt. The Rosario Belt extends south of the previous two belts
and continues westward along the north coast in the central part of Pinar del
Rio. An inlier 10 miles long is present north of Guanajay, near Habana.

1. A few hundred feet of white-to-red quartz sandstone and siltstone
   with limonite cement; mica-rich, light gray-to-purple, clayey shale with
   occasional carbonaceous plant material; medium- to coarse-grained
   friable arkose (Cayetano Formation).

2. 2000 ft of thin- to medium-bedded, dense, dark gray-to-purple, wavy-
   banded limestone with occasional red-brown chert stringers and
   papery, brown shale. The limestone is pseudoölitic and oölitic toward
   the top, contains abundant ammonites and aptychi, and is very similar
   to the Upper Jurassic of the Las Villas Belt. The lower part is barren

and grades into the Cayetano sandstones. In the southern part of the
Rosario Belt only the basal part of the formation contains sandstones
which weather to a characteristic red color. To the northwest the
upper part of the formation also contains abundant quartz sandstones
and shales which weather a yellow brown. This unit wedges out to the
northeast. The age is Portlandian (Artemisa Formation).

3. 600 ft of dense, white-to-cream limestone with abundant Radiolaria:
   black cherts; occasional medium fragmental limestone with rare green,
   volcanic grains; friable quartz sandstone and siltstone with limonite
   cement. At the top of the unit there is an interval approximately 100 ft
   thick of massive sandstone with some shales. The age of this unit is
   Neocomian to Aptian and is therefore equivalent to the pelagic carbonates found in the Las Villas, Placetas, and Cifuentes Belts to which
   it also shows some lithologic similarities.

4. 150 ft of thin-bedded radiolarian chert and dark shales with occasional
   light brown, dull, dense, slightly argillaceous, radiolarian limestone.
   This unit is of Cenomanian to Turonian age and has a slight unconformity at the base. Its carbonate content decreases southward and is
   lithologically identical to units of the same age found in the Placetas
   Belt.

5. 3000 ft of limestone and chert conglomerates and coarse fragmental
   limestones of Maestrichtian age. Some of the components are clasts
   derived from the Cacarajicara Belt, but others are derived from the
   immediately underlying rocks. These conglomerates are found caught
   along fault planes, indicating that violent tectonic activity was taking
   place during the Maestrichtian time. This is in contrast with central
   Cuba where such a situation is not recognized until Lower to Middle
   Eocene time.

6. Several hundred feet of red limestones, limestone conglomerates, and
   quartz sandstones of Lower to Middle Eocene age.

The structure of this belt consists of north-dipping fault blocks separated
by apparently north-dipping faults. Most of the southern boundary of the
Rosario Belt consists of a large down-to-the-south normal fault (the Pinar
Fault) that in many places brings the Miocene into contact with the Jurassic.
It is worth noticing that, to the southwest, many of the outcrops are made of
Upper Jurassic rocks while, to the northeast, the Upper Jurassic is essentially
nonexistent, and the Cretaceous forms the bulk of the outcrops. The north-
western extension of the Rosario Belt is not well subdivided but consists of a
sandy and shaley equivalent of the previously described Jurassic and pre-
Maestrichtian Cretaceous sediments and is hardly distinguishable from the
Cayetano Formation. The southwesternmost belt, Organos Belt, is separated

from the Rosario Belt by a fault zone across which there is a marked change in the lithologic character of equivalent sedimentary sections.

Organos Belt. This consists of the following:

1. 6000 ft plus of white-to-red, quartz sandstone and mica-rich, light gray-to-purple shale. Near the top are interbeds of black oyster hash limestone. All the lithologies show some diagenetic changes, and in some areas the rocks are slightly metamorphosed to phyllites and semi-quartzites. Along the Pinar Fault extreme metamorphism to schist, gneiss, and marble has taken place in local areas of concentrated faulting. The base has not been observed; however, in general, this formation appears to get thicker toward the southwest. The age of this unit is mostly Oxfordian and could well extend into the Lower Jurassic (Cayetano Formation).

2. 300 ft of thick-bedded, dark gray, oyster hash limestone at the base; purple-black calcareous shale with nodular limestone concretions containing ammonites, fish, and reptile bones in the middle; and medium-bedded, barren, black, carbonaceous limestone at the top. It is of Upper Oxfordian age. It grades downward into, and is a partial lateral equivalent of, the Cayetano sandstones.

3. 4000 ft of an unbroken sequence of thick-bedded, dark gray, carbonaceous limestone and dolomite with occasional nodular chert stringers and a basal intraformational conglomeratic limestone (it contains fragments of both underlying and overlying units). The upper part of the formation is often oölitic and pseudoölitic and contains a fauna of ammonites. The age of this formation is definitely Portlandian to Neocomian. The Kimeridgian has not been specifically identified but could very well be represented by the unfossiliferous dolomitized lower part of the unit (Viñales Formation).

4. 150 ft of thin-bedded, dense-to-porcellaneous, gray limestones with thin, black, chert stringers. It contains nannoplankton and is of Aptian age.

5. 100 ft of red, argillaceous limestone with an occasional microconglomerate at the contact with the underlying unit. It was deposited during Cenomanian through Turonian time. In the northern part of the Organos Belt this formation changes facies to a section of thin-bedded, white and gray cherts and shales. It is similar to the varicolored chert development of the same age in the Cifuentes Belt in Las Villas Province. The pre-Maestrichtian Upper Cretaceous is often absent, and its scarcity is believed to be due to an extended period of non-deposition. In places though, the pre-Maestrichtian unconformity has cut the section deeply into the Lower Cretaceous.

6. An unknown thickness of a sequence of basic volcanic-derived sandstones and conglomerates, orbitoidal fragmental limestones, porcellaneous white limestone, fine fragmental argillaceous limestones, shales, and occasional volcanic flows of Maestrichtian age. In addition to these bedded sediments, the upper part of the section contains tectonically mixed, large masses of serpentine, diabase, gabbro, schists, and assorted volcanics. This material is of orogenic origin and is similar to sediments of the same age in the Bahia Honda Belt.

7. An unknown thickness of red, argillaceous, fragmental limestones with occasional limestone conglomerates locally rich in manganese. It is tectonically mixed with the underlying "wildflysch" and is present below faults. It is of Lower to Middle Eocene age.

The Organos Belt gives the appearance of a large anticlinorium dipping to the southeast in the southeastern part of the belt near the Pinar Fault and north to northwestward in the northwesternmost two-thirds of the belt. The Cayetano Formation shows a higher degree of folding and faulting than the overlying massive carbonates, and this is believed to be due to the difference in competence. Northward and northwestward there are a number of faults apparently dipping in the same direction as the sediments and repeating the entire section a number of times. Most dragfold axial planes dip northward. In detail the structural complexities are such that it is difficult to ascertain, by field observation, whether these faults are high-angle reverse faults thrusted toward the south, or whether they separate northward-plunging nappes that were originally being thrusted northward. In places the section can be completely overturned; however, overturning is rather the exception than the rule. At least one sill of porphyry has been observed in the Cayetano quartzites. A few outcrops of serpentine have also been mapped in the southwestern part of the Organos Belt.

To the south the Organos Belt, like the Rosario Belt, is bounded by the Pinar Fault which is a normal fault with large displacement downthrown to the south. In places, along the fault trace, there are linear outcroppings of serpentine.

South of the Pinar Fault there are occasional outcrops of igneous and volcanic-derived Maestrichtian and Lower to Middle Eocene sediments. These beds are steeply dipping to the south near the fault. They are overlapped by younger Tertiary. Further south, as indicated by drilling, the dip reverses northward.

2. Isle of Pines

The Isle of Pines is almost entirely made of metamorphic rocks. Among these, the dominant ones consist of quartzites, quartzitic phyllites, and schists.

Some marble also crops out. Many authors have noted the similarity between these metamorphics and the sediments exposed in the Organos Belt. Unfortunately, no fossils have been found to confirm the correlation. These metamorphosed sediments also show a strong similarity to some of the ones exposed in the Trinidad Mountains. The Isle of Pines section is intruded by peridotite, gabbro, and granodiorite. Some serpentine is present. Three potassium–argon age determinations were made on muscovite and muscovite schists from the Isle of Pines and gave dates from 73 to 78 million years. Along the northwest coast, there are some volcanics believed to be of upper-Lower to lower-Upper Cretaceous age.

3. Habana–Western Matanzas

In the Habana and western Matanzas area only rocks belonging to the Domingo and Cabaiguan Belts are exposed. There is one exception: A large outcrop of Neocomian pelagic limestone, of the type found in the sedimentary belts of Las Villas Province, is found in fault contact with serpentine and Cabaiguan Belt material. Rocks of the Rosario Belt that occur in western Habana have been discussed under Pinar del Rio. The main and more varied outcrop of pre-Upper Eocene sediments occurs along a large late Tertiary anticline that extends from the city of Habana to Matanzas and shows complexly folded and faulted older rocks. South of this anticline, there are also a number of exposures showing, through the upper Tertiary cover, intensely deformed Lower Eocene sediments.

The rocks of the Domingo Belt consist of large isolated outcrops of serpentines in which small- to large-sized fragments of marble are found. There are also large outcrops of gabbro that are only occasionally associated with the serpentine. The serpentines are found in greater abundance in the northern half of the Habana–Matanzas anticline.

Rocks of the Cabaiguan Belt are represented by (Fig. 8):

1. 60 ft plus of andesite porphyries and acid trachyte that are thought to be the oldest representative of the Cabaiguan Belt type of rocks. They are associated with silicified limestones, siliceous shales, radiolarites, and graywacke siltstones of questionable Cenomanian to definitely Turonian age (Pre-Via Blanca beds).

2. 1500 ft of volcanic-derived sediments with limestone lenses that carry a Campanian to Lower Maestrichtian planktonic fauna. Conglomerates containing intermediate igneous rocks and graywacke sandstones are present. The youngest members of this section are thought to be green-to-tan siliceous tuffs and volcanic detritals that are lithologically identical to some of the Maestrichtian tuffs of the Cabaiguan Belt

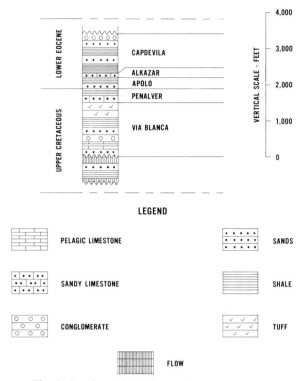

Fig. 8. Stratigraphic sections of Havana Province.

in Las Villas Province. There appears to be an unconformity at the base as no Coniacian to Santonian is recognized (Via Blanca Formation).

3. 60 to 400 ft of calcareous clastics ranging from coarse-grained at the base to lutitic at the top and of Upper Maestrichtian age (Peñalver Formation).

4. 300 ft of red-to-brown clays, dirty silt, and sands with some igneous-derived conglomerates of Lower Eocene age. A major stratigraphic break could be present between this unit and the preceding one in view that fossils of Paleocene age have not been recognized in many parts of the island (Apolo Formation).

5. 30 to 120 ft of interbedded greenish marls, chalks, and locally silicified, fragmental limestones of Lower Eocene age (Alkazar Formation).

6. 1100 ft of well-bedded, brownish shales, dirty silts, and sands. In the lower part of the section the shales and the silts are better developed than the sands. In the upper part of this unit the coarser beds are better developed, and conglomerates are present. The fragments of the

conglomerates are of many sources—igneous, volcanic, and carbonate—and it is of great significance that lithologies present in the Yaguajay, Las Villas, Cifuentes, and Cabaiguan Belts of the Las Villas Province can be recognized. It is of interest that volcanic fragments in some of the conglomerates are highly angular and totally unweathered, suggesting little exposure to atmospheric agents. Some marly beds are present in the uppermost part of this unit (Capdevilla Formation).

Overlying this unit with unconformity is the relatively undisturbed section that consists of dominantly calcareous sediments such as chalks, silicified limestones, fragmental limestones, etc. (Marianao Group); the age of the base of this section is considered upper-Lower to Middle Eocene (Universidad Formation).

Toward the eastern end of the Habana–Matanzas anticline, there are several large outcrops of gypsum in contact with volcanic and volcanic-derived sediments of Upper Cretaceous through Lower to Middle Eocene age of the Cabaiguan Belt. There are serpentine outcrops nearby, and one of them is in direct contact with the gypsum. Within the gypsum are a very large number of fragments of rocks that do not crop out in the Matanzas or Habana Provinces. They consist of silty, micaceous, gray shale and coarse-grained sandstone that suggest the Cayetano Formation of Pinar del Rio; fine-grained, beige limestone and fine-grained, gray-to-dark-gray limestone containing *Nannoconus* and Radiolaria that suggest the Neocomian pelagic carbonates of the Las Villas, Placetas and Cifuentes Belts; fine-grained sandy limestone and dolomitic limestone resembling some of the types found in the Yaguajay Belt. In addition, fragments of quartz mica schist, marble, and bluish sedimentary quartzite have been identified. One inclusion of serpentine has also been reported. Although there is no direct evidence for the age of the gypsum, it probably belongs to the evaporite section, Punta Alegre Formation, of the Cayo Coco Belt. There is little doubt that these outcroppings of gypsum are diapiric structures.

The outcrops of the pre-Middle Eocene sediments in Matanzas and Habana Provinces are highly disturbed and cut by numerous faults. Dips are very high, near vertical, and the strike is west–northwest.

D. Eastern Cuba

1. *Central and Eastern Camaguey*

In north-central Camaguey the Yaguajay Belt is very well represented in the Cubitas Range; most of the lithologic types recognized in Las Villas are also found there.

The deeper-water nonvolcanic equivalent facies are very poorly represented as the serpentines, ultrabasics, and associated Cretaceous volcanics cover a large portion of the state. One window through the Domingo–Cabaiguan Belts at Loma Camajan, 15 miles northeast of the town of Camaguey, shows exposures of the Las Villas Belt to the north in fault contact with representatives of the Cifuentes Belt to the south. Two other small windows next to the Cubitas front show Las Villas Belt to the northeast and Cifuentes Belt to the southwest.

The igneous lithologies of the Domingo Belt are well represented in north-central Camaguey and are apparently overlain with very complex relationships by members of the Cabaiguan Belt.

To the south there is a large intrusive body that appears to extend almost uninterruptedly in a east–southeast direction across the entire province of Camaguey into Oriente Province. It consists of diorites to granodiorites very similar to those found north of the Trinidad Mountains. The relationship to the country rock is uncertain but in many places appears to be faulted.

South of the igneous material, Cretaceous volcanics and volcanic-derived sediments are reported, but no detailed map of the area is available. However, an area of exposures of Lower to Middle Eocene volcanics and associated sediments is present near the south coast.

Lower to Middle Eocene chaotic conglomerates containing enormous blocks of metamorphic, basic igneous and volcanic rocks are well developed in the area southeast of the Cubitas Range, north of Loma Camajan.

The pre-Upper Eocene exposures from this province are separated from those in Las Villas Province by what appears to be a large east–northeast trending left-lateral strike-slip fault (the Trocha Fault).

A domal structure along this fault, Loma Cunagua, has been drilled and found to be a northward recumbent salt diapir separating Cabaiguan Belt facies to the south from those of the Cayo Coco Belt to the north.

The structures in the Yaguajay Belt are somewhat more complex than those of the same belt in Las Villas Province showing definite internal folding and faulting. The windows that show the Las Villas and Cifuentes Belts exhibit very strongly deformed and sheared sediments. The structures within the Cabaiguan and Domingo Belts have not been mapped in detail but are complex.

2. Northern Oriente

In northern Oriente there is, west of the town of Gibara, a relatively small area (10 by 15 miles) where carbonates typical of the Yaguajay Belt crop out. The general strike is east–west with nearly vertical dips. There are a number of normal, as well as south-dipping, thrust faults. This is the easternmost exposure in Cuba of the platform carbonate facies.

Exposures of the Las Villas, Placetas, and Cifuentes Belts are essentially absent in this area.

To the south, there is a large area of exposures of a mixture of Domingo and Cabaiguan Belt lithologies. The ultrabasics form a series of arcuate, linear parallel outcrops concave to the north that strike in an east–northeast direction along the coast. These ultrabasic belts are separated by volcanic-derived sandstones and conglomerates with a small amount of volcanic flows and pyroclastics.

The ultrabasics consist mostly of slightly serpentinized peridotite, gabbro, and serpentine with pillow-like structures suggesting ultrabasic submarine flows. Within the ultrabasic complex are exotics consisting of blocks of white marble and thin-bedded black limestone with Radiolaria and *Globigerina* of possible Upper Cretaceous age. There are also blocks of reefoidal limestones of possibly Cenomanian to Turonian age; dense, thinly laminated limestone with pelagic Campanian fauna; and orbitoidal, reefoidal, fragmental limestones of early Campanian through Maestrichtian age.

Two conglomeratic units believed to be of Lower to Middle Eocene age are present; one is finer grained and appears to contain only elements from a nonvolcanic sequence, while the other contains fragments of basic igneous rocks and of various types of carbonates in a fine-grained matrix.

Although the elements of these miscellaneous carbonate facies are discontinuous, the different types are persistent along the strike. It is difficult at this point to ascertain whether the different carbonates belong to non-volcanic belts such as Las Villas and Cifuentes or belong to the Cabaiguan Belt sequence where both pelagic and reefoidal limestones are present.

The sequence of Domingo and Cabaiguan Belts dips steeply to the south in what appear to be isoclinal, relatively thin, thrust sheets.

3. *Southern Oriente*

In southern Oriente, south of the Cauto Tertiary basin, is an area where rocks of the Domingo Belt dominate. There the ultrabasics have the appearance of being relatively flat-lying and undisturbed. Some of the elements of the Cabaiguan Belt are associated with them. To the southwest of these igneous exposures is a large area covered by Lower to Middle Eocene sediments and volcanics. These are the fill-in of the "old Cauto Basin." There, as much as 12,000 ft of basalt, andesite, dacite, rhyolite flows, and volcanic breccias have been estimated. Also there are tuffs, conglomerates, and limestones (Cobre Formation). The Cobre Formation is present also north of the ultrabasic complex, on the flank of the Nipe Basin and along the south coast of the island in the Guantanamo Basin.

The Sierra Maestra consists of volcanics, volcanic-derived clastics, and sediments of the Cabaiguan Belt type in which Turonian limestones have been found. These volcanics and associated sediments are intruded by diorites and granodiorites that have been dated as 46 to 58 million years old.

Toward the southeastern extremity of the island, east of Guantanamo, is an area of metamorphic outcrops which, according to some authors, resembles the Cayetano Formation of western Cuba.

The structure of the Lower to Middle Eocene rocks of the "old Cauto Basin" is relatively simple although cut by numerous faults. It is also worth noting that the extreme degree of orogenic activity present everywhere else on the island in Lower to Middle Eocene time is not apparent here; no major unconformity is present, and along the flanks of the Cauto Basin the Cobre Formation is overlain disconformably by the late Eocene clastics and carbonates.

V. METAMORPHISM AND INTRUSIONS

Regional metamorphism of sedimentary material is encountered only along the backbone of the island—the Isle of Pines, the Trinidad Mountains, and southeastern Oriente Province. In all three areas low-grade metamorphics are present, and these have been correlated on a lithologic similarity basis only to the Jurassic Cayetano Formation. A few reported K–Ar dates indicate an Upper Cretaceous age for the formation of muscovite schists in the Isle of Pines. This date therefore corresponds with the period of intense volcanic activity found everywhere on the island. On the other hand, a pegmatite dike cutting similar metamorphics in the Oriente Province was dated as late Jurassic to early Cretaceous; the significance of this is difficult to evaluate at the present time except that it gives support to the pre-Cretaceous age of the metamorphic country rock. The only sediments from the sedimentary belts that show the effect of metamorphism are those of the Cayetano Formation in the Organos Belt of Pinar del Rio where, along and north of the Pinar Fault, they grade into schists and quartzites. In at least one area the quartzites are intruded by a sill of granodiorite.

In the Trinidad Mountains, as previously mentioned, there are medium- to high-grade metamorphic rocks that could probably be older than the previously discussed low-grade metamorphics. An older age is suggested by the fact that some of them are intruded by a large dioritic to granodioritic body that (1) has given a K–Ar date of early Jurassic, and (2) is essentially identical to the three allochthonous pre-Neocomian basement outcrops at Sierra Morena, Tres Guanos, and La Rana. These basement outcrops also show inclusions of white marble and other high-grade metamorphics.

The origin and mode of emplacement of the ultrabasic complex of the Domingo Belt presents an interesting problem. There is only one reported occurrence of intrusion of rocks of the sedimentary belts by ultrabasics. It is in eastern Pinar del Rio Province where the Cayetano Formation appears to be altered around a serpentine body; otherwise the contact is always faulted. On the other hand, blocks of metamorphic rocks resembling the "older metamorphics" of the Trinidad Mountains are extremely common in the serpentine and frequently show streamlining along the serpentine flow lines. In Las Villas Province, along the northern edge of the Domingo Belt and immediately south of the Las Villas Belt, there are extensive exposures of a dioritic rock very similar to that of the allochthonous basement outcrops which appear to have been intruded by ultrabasics. Although the serpentine itself varies greatly in thickness, the ultrabasic complex shows a definite stratification, with the serpentine generally occurring in the middle of it. In view of its position under a thick section of submarine basalt flows, this ultrabasic sequence is considered to represent a part of the oceanic basement, with its original magmatic differentiation, that was subsequently mobilized under the orogenic forces. How the ultrabasic material was exposed to the sea floor is conjectural; however, a mechanism of tension fractures through rifting, such as that presently evident in the Yucatan Basin and the Bartlett Trough, can be visualized. The great abundance of metamorphic fragments suggests that the fracturing took place on a metamorphic terrain with granodioritic intrusives formed during an earlier phase of orogenic activity. However, it should be remembered that the present position of the ultrabasics is allochthonous, and that the exotic blocks present in the serpentine could have been incorporated during the late mobilization of the ultrabasics along large thrusts. The fact that only metamorphics are present in the serpentine would tend to negate this latter possibility. It is obvious, from the stratigraphic record, that the ultrabasic complex came to its present position, according to the geographic location, between the late Cretaceous and the Middle Eocene. There is also evidence that, to a much lesser degree, the serpentine has been mobilized along faults even during the recent Tertiary. On the other hand, the time of its original exposure as sea-floor basement is more difficult to ascertain. The oldest dated sediments within the volcanic sequence of the Cabaiguan Belt are of Cenomanian age, and they are separated from the ultrabasic complex by many thousands of feet of pillow basalt. However, (1) blocks of Upper Jurassic oölitic limestones and Lower Cretaceous pelagic limestones are found along the faults that separate the Cabaiguan Belt from the Trinidad metamorphics; (2) in the southern part of the Cabaiguan Belt the Cenomanian sediments interbedded with the volcanics contain large amounts of clasts of the above limestones, and these are in every respect identical to the carbonates of the same age present in the Las Villas Belt; (3) younger Tertiary conglomerates northeast of the Trinidad Mountains contain abundant components of these limestones; and (4) there is no evidence in the sedimentary belt of proximity to volcanic activity until late Lower

Cretaceous. These facts suggest that shallow-water conditions existed during Upper Jurassic south of the present and former position of the Cabaiguan and Domingo Belts, that deepening of the sea took place simultaneously to the north and south of the future volcanics in early Cretaceous, and that the rifting responsible for the formation of an ultrabasic sea floor, as well as the following volcanic activity, took place during the middle part of the Lower Cretaceous. Finally, the geographic position of the area where the rifting responsible for the ultrabasic and volcanic outpouring took place is much more difficult to establish, and this problem will be discussed later under the description and interpretation of the structure of Cuba.

The Domingo and Cabaiguan Belts show numerous Upper Cretaceous intrusives which are related to the intense volcanic activity. These intrusives, in the form of dikes and possibly sills, caused only contact metamorphism, leaving the Cabaiguan Belt as a beautiful example of an unmetamorphosed orogenic volcanic sequence. In southern Oriente, some diorites have been dated as Lower to Middle Eocene by the K–Ar method; they are undoubtedly related to the period of early Tertiary volcanism evidenced by the Cobre Formation in the Cauto Basin. This is the youngest period of volcanism recorded in Cuba and might be related to geophysical evidences of submarine volcanism that have been recorded along the Cayman Ridge.

It is of great importance that, with the exception of the metamorphism and intrusions in the Cayetano Formation in southwestern Pinar del Rio, there has not been any observed indication of intrusion or metamorphism in any of the sedimentary belts. Even under the most complex structural conditions, the sedimentary components are always in fault contact with the igneous rocks and are often separated from them by thin slivers of Lower to Middle Eocene clastics.

In conclusion, one can postulate the existence of an older metamorphic and granodioritic basement of possible late Paleozoic to early Mesozoic age over which the clastics of the Cayetano Formation were deposited south of a line extending through eastern Pinar del Rio, the northern Trinidad Mountains, and southeastern Oriente. North of this line, platform carbonates and evaporites were being deposited. During the middle part of the Lower Cretaceous, rifting took place to the south and is associated with the formation of the "old" Yucatan Basin and the Bartlett Trough. This rifting was accompanied by strong volcanic activity (first ultrabasic to basic and becoming acidic during the Upper Cretaceous) that was at least in part responsible for the lower-grade metamorphism of the Cayetano Formation. This volcanic activity was terminated to the north of the Yucatan Basin in the Maestrichtian, but continued into the Lower to Middle Eocene to the southeast of the Basin. The hypothesis for the existence of a Paleozoic basement is supported by the presence of, in the Maya Mountains of British Honduras, Permo-carboniferous sediments intruded by Late Paleozoic and Triassic granite which separate a section of Cretaceous platform carbonates (similar to the Yaguajay Belt) to the north from thin-bedded Neocomian limestones with aptychi (identical to those of the same age in the Las Villas Belt) to the south (G. Flores, personal communication).

VI. STRUCTURES

There have been endless arguments as to the true nature of Cuban tectonics, and this is understandable when one considers the structural and stratigraphic complexities. The multiplicity of interpretations is due to the facts that: (1) well-developed fold belts are narrow and relatively scarce; (2) the existing folds are very tight, suggesting drag folding of incompetent materials; (3) simple clear-cut faulting is seldom present; (4) many of the faults must be inferred as the only evidence is two nearby outcrops of completely disparate orientation and composition; and finally (5) the great majority of the observable faults are high-angle, some of them exhibiting a strike-slip component, while the apparent shuffling of the previously described belt succession cannot be satisfactorily explained without the existence of large-scale low-angle thrusting.

Another source of disagreement among students of Cuban geology is that detailed geological maps of the most complex parts of the island are not publicly available, and for a number of reasons most of the principal workers have never had a chance to discuss the problems over the outcrop. The structures are so complex that a valid observation of some relationship at one particular point can be totally opposite to one observed a few tens of feet away.

The better documented features of central Cuba will be described (Figs. 4 and 9) and an attempt made to demonstrate that the same structural principles can be applied elsewhere on the island.

A. Central Cuba

1. *Cayo Coco Belt*

This belt is characterized by a relatively low structural relief. There are, however, three remarkable structures—the Punta Alegre and Isla de Turiguano gypsum diapirs and the Loma Cunagua salt diapir.

By drilling, the Punta Alegre diapir was found to be a mass of Jurassic evaporites that overlie Oligocene, Lower to Middle Eocene, and possibly Upper Jurassic sediments. Normal and reverse faults are associated with the diapir, and it is therefore believed that the extrusion is in part related to fairly recent movement along the faults.

The Turiguano diapir has not been drilled but is likely to be similar in nature to Punta Alegre. The two features appear to be en-echelon.

The Loma Cunagua diapir has been found to overlie Cayo Coco Belt sediments to the north and underlie Domingo–Cabaiguan Belt lithologies to the south. It is related to a major northeast–southwest fault, La Trocha, that shows a left-lateral component of displacement.

In the Cayo Coco Belt folding, normal and reverse faulting are unquestionably common; however, this is known only through very sparse drilling, and therefore nothing can be said about regional trends. When approaching the Yaguajay Belt to the south, there are some fairly tight folds, some of which are isoclinal, with axial planes dipping either to the south or to the north. They are believed to be part of a system of drag folds related to the major fault separating the Cayo Coco Belt from the Yaguajay Belt.

2. *Yaguajay Belt*

This belt is characterized by fairly uniform south dips and a west–north-west strike. It is cut by a large number of high-angle faults striking to the northeast. This belt is limited to the north by a fault zone which consists of a set of en-echelon faults. Within the fault zone itself, which shows intense brecciation, there is a series of northward-tilted blocks showing relatively little displacement in relation to the Cayo Coco Belt but a much larger one (in excess of 10,000 ft) in relation to the main part of the Yaguajay Belt. The dip of the fault is unknown but appears to be vertical. The fault system seems to have a left-lateral component. The development of such a fault is difficult to understand, and the outcrops suggest the near-surface expression of a high-angle, south-dipping, reverse fault. It is believed that since the time of faulting in early Eocene, there have not been great changes in the elevation of the Cayo Coco Belt.

3. *Sagua la Chica Belt*

Because of the limited exposures of this belt, little can be said about its structure; however, the belt appears to consist of a fault sliver along the major fault that separates the Yaguajay from the Las Villas Belt.

4. *Jatibonico Belt*

This belt represents a monocline with high dips to the south. There is some sharp folding toward the center of the belt. It is bounded to the northeast by a high-angle fault with some 5,000 ft or more of throw, upthrown to the south. The Jatibonico Belt is a southeastward continuation of the Yaguajay Belt but is separated from it by a fault that brings Lower to Middle Eocene igneous-derived clastics against it. The Jatibonico Belt is in contact to the north and to the south with typical Upper Cretaceous to Lower Eocene Las Villas Belt lithologies, which in their turn are overlain by gabbros and other lithologies of the Domingo Belt. This situation will be discussed later.

5. *Las Villas Belt*

This belt is bounded to the north by a reverse fault that extends the entire length of the exposures of the Yaguajay and Jatibonico Belts. Nearly everywhere the fault overrides the relatively incompetent Lower to Middle Eocene igneous-derived shales, sandstones, and conglomerates that were deposited as a flysch over the sediments of the Yaguajay Belt. In some areas only thin slivers of Eocene are preserved; in others, one can observe the normal Yaguajay Belt succession going from carbonates up into several hundred feet of clastics.

In the Jatibonico Mountains, at the southeastern end of the Yaguajay Belt and in the area of exposures of the Jatibonico Belt, Las Villas Belt lithologies can be seen rimming Yaguajay Belt carbonates to the southwest, southeast, and northeast. It is separated from the Yaguajay carbonates by several hundred feet of Lower to Middle Eocene igneous-derived clastics belonging to the Yaguajay Belt. The dips in the Las Villas Belt can be seen progressively changing from South 60°, southwest of the Yaguajay Belt, to North 10–20° to the northeast of it. Slickensides and dragfolds indicate a northeastward movement of the Las Villas over the Yaguajay Belt. This area shows the only indisputable evidence of a folded overthrust in central Cuba. The Las Villas

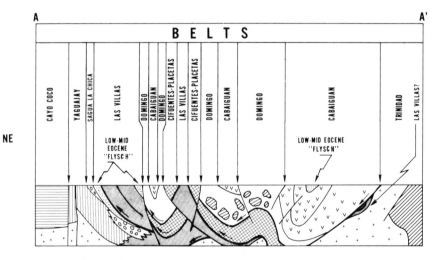

Fig. 9. Diagrammatic structural sections through central Cuba.

Belt is overlain to the northeast and to the southwest by rocks of the Domingo Belt, indicating a second, superimposed overthrust. As previously mentioned, the Jatibonico Belt appears as a faulted window showing through Las Villas and Domingo Belt lithologies. It is difficult to estimate the magnitude of the overthrusting of this major fault. The normal rapidity of change from the shallow- to the deep-water carbonates, as presently observed on the margin of the Bahama Banks, does not require a very large displacement to bring such disparate facies in superposition. However, the displacement must have been at least several miles because the deep-water facies had to "climb up the slope" to come in contact with the youngest part of the carbonate platform section. It should also be noted that, although the southward dip of the fault is approximately 60°, the Yaguajay Belt is in itself a southward-dipping block which was probably tilted simultaneously with the northward push of the Las Villas Belt, thereby steepening the angle of the fault separating the two belts.

The Las Villas Belt, along its length, consists of a tightly folded anticlinorium which is subdivided into three major areas of outcrop that suggest broad regional highs separated by lows showing northward encroachment of the southern belts. To the north the structures consist of several long (up to 20 miles) and narrow (approximately one mile) anticlines exposing Upper Jurassic in the core. In the intervening synclines, rocks as young as the Lower to Middle Eocene carbonate and chert conglomerates are preserved. The relatively competent Jurassic to Lower Cretaceous carbonates show fairly gentle dips and are broken by a large number of faults. On the other hand, the Upper Cretaceous part of the section, which is less competent, is very intensely folded and shows evidence of flowage and bedding plane slippage. Due to the fact that the total structural relief is small compared to the size of the Las Villas Belt, it is believed that the individual folds are not deep seated but are instead drag folds on a relatively thin plate caught between two competent masses sliding horizontally in relation to each other.

From the central part of the Las Villas Belt to the southeast, two parallel, tightly folded, anticlinoria are present, with the younger beds showing greater deformation than the older ones. The northern structure is not as strongly deformed as the southern one, and the oldest beds exposed in the core are of Lower to Upper Cretaceous age. In the more intensely deformed southern structure the Upper Jurassic is present in the core of the anticline. These two features are separated by a south-dipping reverse fault, and, as to the northwest, the folds do not appear to be deep seated.

Further to the southeast, the Las Villas Belt appears to "plunge" under rocks of the Domingo Belt until it becomes a monocline, showing first a succession of repeated sedimentary sections; then these repeated sections are imbricated with rocks of the Domingo Belt; and finally only one is left extending from the Upper Jurassic into the Lower Eocene.

The Las Villas Belt has a maximum width of approximately 6 miles, which must have represented at least 10 miles prior to folding. However, 6 miles from the southwestern edge of the main Las Villas Belt exists an area where exposures of characteristic Las Villas Belt lithologies, ranging in age from Lower to Upper Cretaceous, are completely surrounded by rocks of the Cifuentes Belt. Along the southwestern edge of the Las Villas Belt is a long, linear series of exposures where Las Villas lithologies and, in the Maestrichtian, lithologies transitional between the types found in the Las Villas and the Placetas Belts form a very tightly folded and crushed syncline which is separated from the Las Villas Belt proper by a steep southward-dipping fault. These Las Villas–Placetas Belt lithologies can be seen again some 3.5 miles to the southwest of the Las Villas Belt forming an elongated dome some 4 miles long by 1.5 miles wide; it is completely surrounded by Placetas Belt lithologies which are separated from the main Las Villas Belt by a band of Domingo and Cabaiguan Belt lithologies.

6. *Placetas and Cifuentes Belts*

For the purpose of describing the structures, it is impossible to separate the Placetas from the Cifuentes Belt since they are intimately associated. However, several areas are distinguishable by their characteristic tectonic style. They are the Cifuentes Belt proper, the northwest Placetas Belt, and the southeast Placetas Belt (Jarahueca Fenster).

(a) The Cifuentes Belt proper is characterized by the almost exclusive presence of Cifuentes Belt lithologies. In detail the structural condition is extremely complex with intense fracturing and, in places, very tight isoclinal folding. In many cases several lithologic units are mixed together in what appears to be a gigantic breccia the components of which vary in size from a few feet to over a thousand feet across. Detailed mapping of this area suggests a broad syncline–anticline combination in a superposition of perhaps as many as three thrust plates. Lithologic types more similar to those of the Las Villas Belt are found in the core of the anticline and around the previously mentioned Las Villas Belt window, while the types which appear more remote from a facies point of view (terrigeous detrital influence in the Lower Cretaceous and volcanic influence in the Upper Cretaceous) occur in the troughs of the synclines. The granodioritic pre-Neocomian basement of Sierra Morena is surrounded also by the younger components of the Cifuentes Belt lithologies and is interpreted as the highest plate of the belt.

The northern boundary of the Cifuentes Belt shows an area of varying width, but usually very narrow, where igneous rocks of the Domingo Belt, including granodiorite and serpentine, are tectonically mixed with the Lower to Middle Eocene igneous-derived clastics belonging to the Las Villas Belt.

The dip of the fault that separates the Cifuentes Belt from the igneous rocks is usually steep, over 60° S, but can be as low as 35° S. Observable along this fault zone, toward the Las Villas Belt, are clear evidences of intense deformation such as the squeezing out of entire formations and "boudinage" phenomena. It should also be mentioned that in the low separating the northwestern from the central segment of the Las Villas Belt, there is a large isolated remnant of Cifuentes Belt some 1.5 miles north of the belt proper and separated from it by Domingo Belt lithologies. Toward the east the Cifuentes Belt is in fault contact with the Placetas Belt.

(b) In the northwestern Placetas Belt area, the Placetas and Cifuentes Belt lithologies occur tectonically intermingled. In general, the structure appears to be that of an intensely crushed anticline. The crest of the anticline is at the previously mentioned window of Placetas–Las Villas Belt lithologies. There is a strong contrast between the relatively coherent and less disturbed nature of the window itself and the chaotic nature of the surrounding Placetas Belt sediments. The northwestern Placetas Belt area is bounded to the north by a reverse fault dipping 50° to 75° S. This fault appears to be the continuation of the fault zone separating the Cifuentes Belt from the Las Villas Belt; however, toward the southeast the fault zone becomes wider, and the amount of Domingo Belt material increases and becomes intermingled with volcanics, volcanoclastics, and sedimentary rocks that lithologically and agewise are identical to those found in the Cabaiguan Belt proper. This zone of Domingo–Cabaiguan lithologies has an average width of approximately 4 miles and extends for 50 miles between sediments of the same age of the Las Villas and the Placetas Belts that contain essentially no trace of igneous or igneous-derived material. It is a northwestward finger-like extension of the bulk of the Domingo and Cabaiguan Belts bounded to the north by a high-angle fault showing a strong left-lateral slip component.

(c) The southeast Placetas Belt area (Jarahueca Fenster) consists of a complexly faulted and folded anticlinorium 16 miles long by 2 miles wide, where mostly Placetas Belt lithologies from Lower to Upper Cretaceous age are found in the core and are surrounded by Cifuentes Belt lithologies. Here again, at the southeastern end of the structure (Tres Guanos), the Neocomian–Cifuentes Belt carbonates can be seen resting on a pre-Cretaceous igneous basement of granodioritic and metamorphic composition. This anticlinorium is completely surrounded by Domingo and Cabaiguan Belt lithologies. It is bounded to the north by a high-angle south-dipping fault that appears to be the continuation of the one making the northern boundary of the northwest portion of the Placetas Belt. This feature is considered additional evidence for the superposition of Cifuentes over the Placetas Belt and Domingo over both of them, and it should be noted that the basement outcrop is allochthonous.

Two additional occurrences of Cifuentes–Placetas lithologies are worth mentioning: (1) A window northeast of the city of Santa Clara which shows a semicircular dome of Placetas Belt lithologies with a partial rim of Cifuentes Belt lithologies. It is surrounded on all sides by Domingo Belt lithologies consisting of a tectonic breccia characterized by a mixture of blocks of igneous-derived Maestrichtian clastics, metamorphics, and serpentine containing meta-morphic–exotic blocks. The window itself is highly sheared and shows evidence of flowage. It appears to be the southeastern continuation of the Cifuentes Belt proper. (2) The La Rana Klippe that consists of a small area of Neo-comian–Cifuentes Belt lithologies on granodioritic basement that lies in the trough of a syncline in the Cabaiguan Belt. It is surrounded by Maestrichtian, but appears to be partially overlain by young Maestrichtian sediments and by Lower to Middle Eocene igneous-derived clastics that contain large well-rounded boulders of basalt and granite.

In general, it can be said that the structural complexity of the sedimentary belts increases with proximity to the Domingo Belt, and the internal structure of the Cifuentes and Placetas Belts is by far the most complex.

An accurate determination of the northward displacement of these belts is impossible, but an attempt at some rough estimates can be made. The present maximum width of the two belts is 12 miles; however, there is a mini-mum of 11 miles of displacement in relation to the Las Villas Belt. Considering that they might consist of up to 4 superimposed plates, a northward displace-ment of up to 40 miles for the southernmost exposures of Cifuentes Belt lithologies in relation to the southernmost exposures of Las Villas Belt is not unlikely. If one adds the possible 20 miles of displacement of the southernmost Las Villas Belt in relation to the Yaguajay Belt, one comes up with 60 miles of northward movement for the southernmost exposures of Cifuentes Belt. This is further south than the southern shore of the island near the Trinidad Mountains.

In conclusion, these two belts have a tectonic style characterized by a superimposed succession of relatively flat-lying "thrust" sheets forming folds of relatively great amplitude. The detailed structure of these belts is very chaotic and in no way resembles Alpine nappes.

7. Domingo and Cabaiguan Belts

As in the previous case, the Domingo Belt cannot be separated structurally from the Cabaiguan Belt. However, there are marked differences in structural style depending on the geographic position. These differences will be described as follows.

Southeastern Yaguajay–Jatibonico Belt Area. In this area, Domingo Belt lithologies consisting of gabbros, uralite basalts to dolerites, and bedded crystalline tuffs are found north of the massive carbonates of the Yaguajay and Jatibonico Belts. As previously described, they appear to be overlying, above a north-dipping fault contact, Upper Cretaceous pelagic carbonates of the Las Villas Belt. This is very important evidence because here the relationships are the same as those between the Bahia Honda Belt and the Cacarajicara Belt of Pinar del Rio.

Area between the Las Villas and the Cifuentes–Placetas Belts. Here, as already mentioned, Domingo and Cabaiguan Belt lithologies are found in what can be considered a very complexly faulted synclinorium the axis of which strikes N 35° W and plunges toward the southeast. There is evidence of high compression and squeezing with common strike faults. This body of Domingo–Cabaiguan Belt rocks can be traced for 50 miles, becoming narrower and structurally more disturbed toward the northwest until it becomes infaulted together with Lower to Middle Eocene igneous-derived clastics in the fault zone that for over 60 miles separates Las Villas Belt from Cifuentes Belt lithologies; there the outcrops of Domingo material become discontinuous and decrease in abundance toward the northwest. It is believed that this relationship between the sedimentary and the igneous-volcanic belts is strong evidence for the existence, at one time, of an extensive low-angle thrust sheet of Cabaiguan and Domingo material which overlaid the sedimentary belts and became infolded and caught in faults as subsequent northward thrusting took place.

Northeastern Domingo–Cabaiguan Belt Area. This area extends from the southeastern Placetas Belt area eastward. To the west the southeast Placetas Belt (Jarahueca Fenster) forms the core of a breached anticline in the Domingo–Cabaiguan Belts. Serpentine is nearly everywhere in contact with the sediments of the Placetas and Cifuentes Belts; and although deformation is intense with much faulting, a normal section from the serpentine through the Upper Cretaceous volcanics can invariably be mapped. Between the southeastern Placetas Belt and the southeastern extension of the Las Villas Belt, there is a broadening synclinorium, asymmetric toward the south, showing gabbros toward the north overlain by serpentine and Cretaceous volcanics in the axis. It is the southeastward continuation of the area between the Las Villas and the Cifuentes–Placetas Belts; here, as previously mentioned, the structures appear to plunge to the southeast, and the sedimentary belts are progressively covered, in a southeastward direction, by Cabaiguan and Domingo Belt material.

Area between the Towns of Santa Clara and Placetas. This is an area with a large proportion of Domingo Belt outcrops whose front almost invariably consists of serpentine in contact with Cifuentes or Placetas Belt lithologies.

The contact itself consists of a sinuous fault zone dipping steeply to the south; occasionally, a calcareous mesh is found along the contact. In this area a number of anticlines and synclines can be mapped; the anticlines bring to the surface highly sheared serpentine containing abundant metamorphic blocks, while the synclines show Cretaceous volcanics in the trough rimmed by ultra-basics and serpentines. It is of interest that all along the northern edge of the Cabaiguan Belt the synclines are found separating outcrops of the Placetas and Las Villas Belts. This is also true southeast of the town of Placetas where a syncline of Domingo and Cabaiguan Belt lithologies separates the northwest from the southeast Placetas Belt areas discussed previously. This is considered to be further strong evidence of extensive thrusting of the Domingo Belt over the sedimentary facies. Throughout the whole area deformation is intense, and there is evidence of northward movement. There is also evidence of thrusting within the Domingo–Cabaiguan Belts themselves. West of Placetas a syncline of Domingo and Cabaiguan Belt lithologies lies within the trough of another similar syncline, the uppermost Cabaiguan plate being in places in contact with Lower to Middle Eocene detritus. Toward the south, Domingo Belt lithologies are in fault contact with the Cabaiguan Belt volcanics, the fault being a high-angle thrust upthrown toward the south. The area between Santa Clara and Placetas shows the most impressive development of serpentines of central Cuba. They form the largest percentage of ultrabasic outcrops, and in this area they contain a very large number of exotic metamorphic blocks.

Southern Domingo–Cabaiguan Belt Area. This area extends south of the Domingo Belt proper to the Trinidad Mountains and contains the bulk of the Cabaiguan Belt volcanics. The structure consists of a broad syncline some 8 to 10 miles wide whose axial plane dips toward the north. Although this feature is cut by numerous high-angle faults and some southward directed thrusts, the deformation of the Cabaiguan Belt proper is relatively gentle and comparable to that of the Yaguajay Belt.

In conclusion, the northern portion of the Domingo–Cabaiguan Belts, where the bulk of the Domingo-type lithologies occurs and which is in contact with the sedimentary belts, shows the most intense deformation. The Cabaiguan Belt proper to the south is much less deformed. It has a characteristic magnetic expression with numerous anomalies. Wherever observed, the Cabaiguan Belt volcanics are found to be in fault contact with the Trinidad Belt; the fault system separating the two belts curves around the western end of the Trinidad Mountains and appears to be parallel to dips of the Cabaiguan volcanics. The presence along this fault zone of outcrops of Upper Jurassic to Lower Cretaceous carbonates which definitely belong to the northern sedimentary belts is possible evidence of the tectonic superposition of the volcanics over the northern facies.

There is, therefore, ample evidence that the Domingo and Cabaiguan Belts are allochthonous, and that the ultrabasics, especially the serpentines, served as the lubricating material over which the volcanic pile of the Cabaiguan Belt moved northward. Diapiric structures might be the cause for the thickness variations of the serpentine. The order of displacement of the Domingo Belt material must be at least greater than the order of displacement of the Cifuentes Belt in relation to the Yaguajay Belt, i.e., greater than 60 miles northward. Although large, this figure is not unusual if one considers an unstable volume of material sliding under the action of gravity. Such a slide would therefore have been responsible for the near chaotic structures of the Cifuentes and Placetas Belts, the mechanical mixing of these belts with the northern Domingo and Cabaiguan Belts, and the intense, drag-like folding of the Las Villas Belt. It would also explain the relatively mild deformation of the southern part of the Cabaiguan Belt. The sliding was probably rendered easier by the continuous erosion and sedimentation of the volcanic pile during Lower to Middle Eocene time; this clay-rich detritus was deposited along the front of the slide, thereby increasing the lubrication along the fault planes. If one accepts this type of mechanism and the magnitude of the horizontal displacements as explaining the structural deformation of central Cuba, it becomes apparent that the Domingo and Cabaiguan Belts originated south of the Trinidad Mountains in the area presently occupied by the Yucatan Basin.

8. *Trinidad Belt*

No detailed geologic map of the Trinidad Mountains is available; however, a large anticlinorium with locally intense faulting and folding has been reported. In places the metamorphics appear relatively undisturbed, and there is no evidence that they were thrust over the Cabaiguan Belt. On the contrary, the fault separating the bulk of the metamorphics from the body of grano-diorite and amphibolite is north-dipping, as is the fault that separates the granodiorite from the Cabaiguan volcanics. In the northern part of the mountains, the metamorphics show abundant shearing and fracturing that indicate approximately horizontal movements. The Trinidad Mountains are therefore considered an essentially autochthonous massif.

9. *Paleotectonic Stratigraphic Implications*

From the previously described stratigraphy, the following relationships can be inferred about some structural events that took place in Cuba during the late Mesozoic and early Tertiary.

(a) The southern Las Villas Belt contains abundant conglomerates in the Neocomian and Aptian carbonates that are not present in the northern Las

Villas Belt. These carbonate conglomerates contain clasts of Portlandian oölitic limestones as well as of Neocomian pelagic limestones.

The allochthonous basement outcrops are overlain by a fossil regolith which is overlain by Neocomian pelagic limestones containing abundant clasts of the same type of limestone. In addition, there are interbeds of quartz sandstones rich in muscovite.

In the Aptian of the Cifuentes and Placetas Belts, there are thin quartz-derived sandstones also rich in muscovite which apparently came from the same source as the underlying Neocomian ones. It should be mentioned that none of the observed igneous outcrops (specifically, the basement outcrops) contain muscovite in any quantity, and therefore the source of this mineral has to be the reworking of Cayetano-like sediments or muscovite-rich metamorphics such as those found in the Trinidad Mountains.

The above observations indicate that during lowermost Cretaceous there was an uplift of basement with erosion of the Jurassic as well as part of the Neocomian. The fact that the basement outcrops are directly overlain by pelagic carbonates also indicates that these basement uplifts were of a local nature and short duration, suggesting that there were violent vertical movements, such as are associated with rifting, during the Neocomian and Aptian.

(b) Attention is called to the fact that from the Cenomanian through the Coniacian, pelagic shales and marls are present in the Cayo Coco Belt and are equivalent to shallow-water platform carbonates further to the south. This indicates that at least during the Upper Cretaceous the carbonate banks might have had a bathymetry similar to the present-day Bahama Banks, i.e., platforms separated by deep-water tongues. There is no reason why such a situation might not have existed during the Lower Cretaceous.

(c) In the Albian through Coniacian the Las Villas, Placetas, and Cifuentes Belts show an interbedding of cherts, fragmental limestones containing occasional shallow-water organisms, and dense pelagic carbonates. The section is believed to have been entirely deposited in deep water, the fragmental material being the result of turbidity currents. These fragmental limestones decrease in abundance toward the Cifuentes Belt and are replaced by varicolored clays believed to have a volcanic origin. This is, therefore, the first indication of "nearby" volcanic activity in the Cifuentes Belt, and it happens at the close of the Aptian.

(d) There is a possible widespread disconformity during the Lower Senonian. The reason for it is unknown. It might be due to a period of nondeposition or to a lack of diagnostic fossils. It is believed not to be due to uplift and erosion of the Las Villas, Placetas, and Cifuentes Belts.

(e) In the Campanian through Maestrichtian, deep-water conditions continue with periodic influx of shallow-water coarse material from the north in the Las Villas Belt, but with an influx of fine-grained igneous material derived from the Domingo and Cabaiguan Belts in the Placetas and Cifuentes Belts. There are also abundant reworked shallow-water faunas, indicating that to the south active erosion was taking place.

(f) The Paleocene appears to be missing throughout central Cuba, and it is not clear whether this is due to a true disconformity or whether diagnostic fossils have not been recognized.

(g) The lower part of the Lower to Middle Eocene shows chert and carbonate conglomerates, free of igneous detritus, overlying the Yaguajay, Sagua la Chica, Jatibonico, and Las Villas Belts. Although some of the clasts come from adjacent belts, the majority of the components appear to have been derived from the immediately underlying formations, suggesting that these units were being deformed and eroded in Lower to Middle Eocene time but were out of the reach of the igneous detrital influence.

(h) During the latter part of the Lower to Middle Eocene the amount of detritus increases and the sections consist entirely of shales, sandstones, and conglomerates derived from the Domingo and Cabaiguan Belts. It is also worth mentioning that the size of the fragments increases upward in the section, indicating decreasing distance from the source. These clastics are found wedged in every fault zone north of the Placetas Belt and south of the Yaguajay Belt.

(i) At the contact between the Domingo and the Cifuentes Belts, there are a number of chaotic blocks of igneous-derived sandstone of Maestrichtian age that have never been seen associated with any other sedimentary or volcanic belts. Only metamorphic-type blocks are found in the vicinity. This is interpreted as the possible displaced remnant of an igneous-metamorphic ridge over which only detrital clastics were deposited before it was overrun by the slide of Domingo and Cabaiguan Belt rocks.

(j) The only evidences of shallow-water conditions are found in the carbonates of all ages of the Cayo Coco and Yaguajay Belts, in the Portlandian oölitic limestones, and in the Upper Cretaceous carbonates found occasionally interbedded with the volcanics of the Cabaiguan Belt.

B. Western Cuba

From central Cuba westward, the succession of belts found in Las Villas Province continues west–northwest toward the sea. Along the keys, Gulf Blanquizal III No. 1 was drilled in the Yaguajay Belt, while further to the west Gulf Hicacos No. 1 in Cardenas Bay was drilled in carbonates of the Las Villas Belt. Further to the west, along the coast near the city of Matanzas, only rocks belonging to the Domingo and Cabaiguan Belts are exposed.

However, there are three important evidences supporting the possibility that the igneous and volcanic belts are riding over the sedimentary belts. (1) The gypsum diapir, in the Habana–Matanzas anticline, contains fragments that appear to come from lithologies found in the Yaguajay, Las Villas, Placetas, and Cifuentes Belts; this suggests that these belts, in addition to the Cayo Coco Belt, are present at depth underlying the Domingo–Cabaiguan Belt. (2) South of Habana on the south flank of the same anticline, in a fault block, there is a large outcrop of Neocomian *Nannoconus* limestone definitely belonging to the Las Villas, Placetas, or Cifuentes Belt. This again suggests the presence of these belts at depth. (3) Near Habana the conglomerates of the Lower Eocene Capdevilla Formation contain, mixed in with volcanic and igneous detritus, fragments that unmistakenly belong to the Yaguajay, Las Villas, Cifuentes, and Cabaiguan Belts. The magnetic expression of Matanzas and Habana Provinces does not contradict this possibility as it shows a deep and/or featureless basement that is overlain by very shallow material giving sharp magnetic anomalies.

Further to the west, in Pinar del Rio, conditions appear to change drastically. With the exception of the southern part of the Organos Belt, all dips are steeper toward the north, and the section appears to be repeated by faults that nearly parallel the dips. In the majority of the fault blocks, the younger part of the section is to the north. The dips of the beds and faults increase northward until they become nearly vertical. Drag folds are exceptionally well developed on the north limb of the Organos Belt, and their axial planes dip to the north, indicating that movement of the faults was downward to the north. Locally, structural complications are extreme, and exposed sections are completely overturned. Along many of the faults Maestrichtian conglomerates are found in which the clasts belong to immediately underlying lithologies, as well as lithologies from nearby belts, indicating that the deformation started earlier than in central Cuba.

The Organos and Rosario Belts have a very different make-up in comparison to their counterparts in central Cuba. The Upper Cretaceous is generally thin to absent, while the lowermost Cretaceous and Jurassic consist of many thousands of feet of sandstones, shales, and massive carbonates. In central Cuba the Upper Jurassic and Lower Cretaceous involved in the deformation always appear to be relatively thin and consist exclusively of carbonates. On the other hand, to the north of the Rosario Belt the Cacarajicara Belt, which appears to be a north-dipping fault block, contains lithologies identical to those encountered in the Yaguajay Belt. It is overlain by the Bahia Honda Belt, where many of the lithologies originally observed in the Domingo and Cabaiguan Belts can be recognized. The Bahia Honda Belt is considered to be allochthonous. Its position north of the Cacarajicara Belt is analogous to that of the Domingo Belt north of the Yaguajay and Jatibonico

Belts in eastern Las Villas Province. In addition to the similarity with Las Villas Province, the Maestrichtian and Lower to Middle Eocene conglomerates contain increasing amounts of igneous, volcanic, and metamorphic components in a southerly direction, and serpentine is found in a number of localities along the Pinar Fault. South of this fault, magnetic surveys show an area similar in character to central Las Villas and Camaguey, and wells drilled in this region have encountered Cabaiguan volcanics at depth. It appears, therefore, that the structure of Pinar del Rio can be explained in the same manner as that of Las Villas; i.e., a mass of ultrabasic and volcanic rocks slid north to northwestward and caused large-scale imbrications in the thick section of Jurassic and Lower Cretaceous clastics and carbonates. Unfortunately, no detailed study of the origin of the sediments of the Organos and Rosario Belts is available, making it difficult to reconstruct the relative movements of individual fault blocks. At the present time, however, the original width of these belts appears to have been shortened without major shuffling of lithologies. It should be noted that although the Organos Belt with its greater content of Cayetano clastics lies to the southwest of the Rosario Belt, both belts appear to increase in Cayetano-like clastics in a westward direction. There is, therefore, the definite possibility that there existed in this area a pre-Upper Cretaceous basin with isoliths at an angle with the present structural trends.

C. Eastern Cuba

Eastern Cuba is separated from central Cuba by the large east–northeast La Trocha fault zone that has an apparent left-lateral displacement of some 30 miles. According to gravity, magnetics, and subsurface evidence, this fault zone, which probably extends into the Yucatan Basin, is responsible for a deep basin filled with younger Tertiary sediments. In Camaguey Province, the outcrop distribution is very similar to the southeastern part of central Cuba. Volcanics and ultrabasics of the Domingo and Cabaiguan Belts are nearly everywhere in contact with the carbonates of the Yaguajay Belt in the Cubitas Range and appear to drape around its southeastern end. The general appearance is that the Domingo Belt continues the southeastern plunge already apparent in the eastern part of central Cuba. There is a body of granodiorite, apparently a continuation of the one north of the Trinidad Mountains, which extends along the center of Camaguey and northwestern Oriente. However, there are no associated metamorphics, and it is surrounded by Cabaiguan volcanics. Toward northern Oriente Province ultrabasic rocks mixed in with Cabaiguan Belt volcanics are seen almost standing on end and draping around the area of Yaguajay Belt carbonate exposures near the coast. Further to the south, the relatively undisturbed Cauto Basin trends across the island through the Bay of Nipe and appears to be the eastward continuation of the Yucatan Basin. In southern Oriente the ultrabasics of the Domingo Belt cover large areas and appear to be relatively undisturbed. They are overlain by Cretaceous volcanics, which in turn are overlain by the Eocene Cobre Formation. Along the southern coast, surrounded mostly by Cretaceous volcanics, is a number of Eocene granodiorites. With the exception of a strip along the north coast and the deeper parts of the Cauto Basin, the provinces of Camaguey and Oriente exhibit the complex magnetic pattern observed over the exposures of the Domingo and Cabaiguan Belts in Las Villas Province.

In conclusion, the overall structure of Cuba can be described well by representing the base of the Domingo–Cabaiguan sequence as a twisting surface dipping to the northwest in Pinar del Rio exposing the sediments to the south of it, nearly flat-lying in Habana and Matanzas Provinces, gently dipping to the south in Las Villas and western Camaguey Provinces showing the sedimentary belts to the north of it, and finally dipping steeply to the south in Camaguey and northern Oriente Provinces showing only the northernmost sedimentary belts. Local imbrications and deformation of that surface could describe nearly all the major structural traits of the island.

At this point, it is appropriate to reiterate the fact that in this chapter the Bartlett Trough is not considered a wrench fault of great lateral displacement but a tensional rift, bounded by faults, with a left-lateral component of movement of a few tens of miles at most. From the geophysical evidence, the same origin is postulated for the Yucatan Basin.

VII. STRATIGRAPHIC AND STRUCTURAL DEVELOPMENT

An attempt now will be made to combine the previously described facts and opinions into a theory concerning the events leading to the creation of Cuba as it exists today (Figs. 10 and 11).

A. Pre-Middle Jurassic

As has already been indicated, very little is known about the geology of the general Cuban–Bahamas area prior to this time. There are some indications, however, that an area existed during the Lower Jurassic where various types of Paleozoic sediments and metamorphics were present. There were also various types of granitic to dioritic intrusives, some possibly as young as Triassic to early Jurassic. This basement for Jurassic and later sedimentation was probably an extension of the North American craton and might have been present as far south as the Jamaican–Nicaraguan rise. In early to Middle Jurassic, the entire Cuban region must have been close to sea level.

B. Middle Jurassic to Neocomian

During this time, active sedimentation began in western Cuba and possibly along the southern coast by the deposition of the Cayetano Formation clastics. The origin of these sediments is unknown, but the fact that they are in part the northward equivalent to marine carbonates in Pinar del Rio suggests a northern to western source. The base of the sedimentary section in the northern part of Cuba is also unknown; however, as suggested by exotics in the gypsum diapirs, Cayetano clastics could be present. During much of the Upper Jurassic the entire northern Cuba–Bahama region must have been isolated from free-water circulation and became part of a large evaporite basin covering the entire Gulf of Mexico. Toward the close of the Jurassic, subsidence continued over the entire Cuban area, but the rate was higher in Pinar del Rio; there a large thickness of pelagic carbonates, rich in organic matter, replaced clastic sedimentation, indicating a considerable deepening of the water. In central and northern Cuba, and possibly as far south as the Yucatan Basin, water

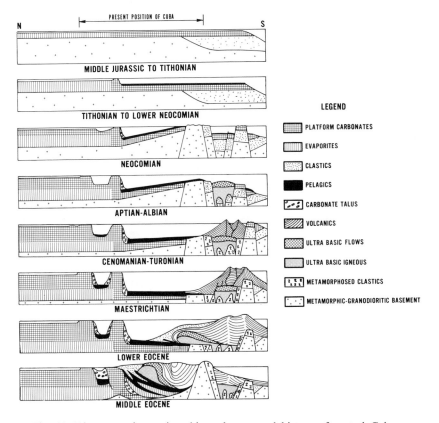

Fig. 10. Diagrammatic stratigraphic and structural history of central Cuba.

conditions remained fairly shallow, and widespread dolomites and oölitic limestones were deposited. Along the northern shore of Cuba, evaporite and carbonate bank sedimentation was taking place. During the Upper Portlandian and into the lower part of the Neocomian, there was a marked deepening of the basement. To the north, construction of shallow-water carbonate banks kept up with subsidence; while to the south, sedimentation became entirely pelagic, with the formation of relatively thin carbonate deposits made up almost exclusively of nannoplankton and Radiolaria.

C. Neocomian to Santonian

Deepening of the basement continued. However, south of central Cuba, and possibly even south of the Trinidad Mountains, large-scale rifting began to take place, with the formation of horsts and grabens corresponding to the ancestral Yucatan Basin. Some of the horsts must have been emergent for a period of time during which the Upper Jurassic and lowermost Neocomian were eroded and the resulting detritus dumped northward as evidenced by the coarse, carbonate clastics interbedded with the Neocomian pelagic limestones of the southern Las Villas Belt. Some of the horsts further south must have rapidly collapsed to great depth, as the rapid resumption of pelagic conditions over a granodioritic basement indicates. Other blocks, still further south, must have remained uplifted through much of the Cretaceous, as indicated by the quartz and mica detritus from a southern source found interbedded in the Aptian pelagic carbonates. Little can be said about eastern Cuba because of the lack of exposure of pelagic facies; however, conditions must have been similar to those south of central Cuba. There is the very definite possibility that large faults trending in a general east–northeast direction in relation to Cuba came closer to the shallow-water bank carbonates along the north coast. It is possible too that during that time faulting along the major fault zones of El Pinar and La Trocha might have been initiated. It is also likely that during this period of time associated with faulting, ultrabasic material came to the sea floor, and the submarine volcanism of the Domingo and Cabaiguan Belts was initiated, so that great outpourings of basalt took place in the troughs possibly in Albian time. From the Albian through the Coniacian, subsidence continued north of these fracture zones, as indicated by the abundance of radiolarian cherts and the smaller amount of calcareous material in the Las Villas and Placetas Belts. Further north, the building of shallow-water carbonate banks continued, and their talus contributed to the coarse, carbonate clastics of the Sagua la Chica and Jatibonico Belts, as well as to the numerous fragmental limestones found in the Las Villas and Placetas Belts. The Cifuentes Belt, which was closest to the volcanic activity, received only chert and volcanic-derived clay deposits. These conditions seem to have extended from Pinar del

Rio at least as far as western Camaguey. To the south, volcanic activity slowed down at the base of the Upper Cretaceous, and there was a period of deposition of clastic carbonates, still containing the Upper Jurassic oölitic limestones, from a source even further south. Volcanism resumed during the Upper Cretaceous but had a more acidic character. At times, volcanism must have been of an explosive nature, as evidenced by the layers of ash and glass bombs.

D. Campanian and Maestrichtian

During the Campanian and Maestrichtian, subsidence continued in the north accompanied by continuous growth of the platform carbonates; conditions remained somewhat stable in the area occupied by the Las Villas Belt. In the Placetas and Cifuentes Belts, the first evidence of coarser, clastic grains derived from volcanic and other igneous sources appeared. The volcanism continued with widespread deposition of tuffs and acidic to basaltic flows; common, shallow-water carbonate reefs, interbedded with the volcanics, indicate the existence of emergent volcanic islands. This situation must have persisted along the southern part of the island. However, toward the west, there is the first evidence of major deformation taking place in the Maestrichtian. It is believed that in Pinar del Rio, probably due to the northward tilting of the area and possibly triggered by movements along the faults, the volcanic mass started to move northward. During this sliding, imbrication of the sediments in the Organos and Rosario Belts took place. As the front of the thrusts advanced, material was incorporated into the Maestrichtian breccias and caught in the slide planes. The slide overran the platform carbonates of the Cacarajicara Belt and continued into the Straits of Florida. There, at the southwestern edge of the Cretaceous Bahama Platform, the banks were probably discontinuous and not as well developed as the ones in the Yaguajay Belt and did not provide a very effective stopping place for the volcanic slide. Unfortunately, it is impossible at this time to estimate the amount of northward displacement of the Organos Belt. There is a possibility that the fracture system responsible for the Pinar del Rio volcanism was not continuous with the one responsible for the Cabaiguan Belt; it might have been an en-echelon and parallel system perhaps present between Cuba and the Isle of Pines. This is suggested by the incipient metamorphism and dioritic intrusions in the southern part of the Organos Belt. There is also evidence that the sliding originated in Maestrichtian time in Habana Province.

E. Paleocene and Middle Eocene

The interval of time from the Paleocene through the Middle Eocene witnessed most of central Cuba's deformation. Due to possible basement

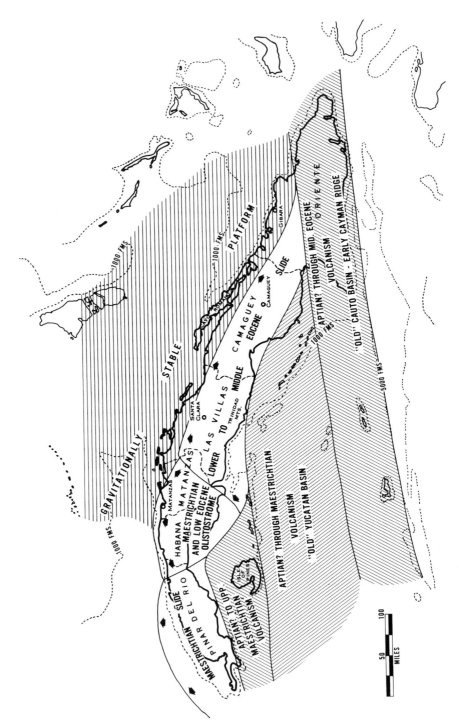

Fig. 11. Diagrammatic area of distribution of the main tectonic events in Cuba.

uplifts in the south, an unstable situation resulted, and the volcanic mass began to slide "downhill" on the sea floor. To what extent crustal compression was responsible for the initial thrusting of the ultrabasics over oceanic sediments is unknown; however, thrusting is considered to be unnecessary because, in an active rifting situation, the material filling the basins could become exposed along fault scarps and collapse under the action of gravity. This mass of volcanics with a front of serpentine diapirs (possibly very similar to the salt and shale diapirs that are being presently observed at the bases of continental slopes throughout the world) kept moving. It crushed the deep-water, oceanic-type sediments of the Cifuentes and Placetas Belts in its path, and possibly pushed ahead of it the Las Villas Belt, partially exposing it, so that its own detritus mixed with that of the Yaguajay Belt in the Lower to Middle Eocene carbonate conglomerates. As the volcanic mass moved north-ward, the Las Villas Belt was finally thrust over a depressed Yaguajay Belt and, in places, overridden by the Domingo Belt front. Erosion products of the exposed igneous and volcanic rocks kept filling the basin between the slide front and the Cayo Coco Belt, and this sedimentary fill became involved in, and probably facilitated, the sliding. This gravity slide was unquestionably a complex feature with numerous imbrications that could therefore explain the observed field relationships between sediments and igneous and volcanic rocks. Lack of coherent structures in Domingo and Cabaiguan Belt lithologies in Matanzas and Habana Provinces, as well as the lack of the typical magnetic signature of these belts, as seen in Las Villas, Camaguey, and southern Pinar del Rio, could be due to the fact that in this area the slide became a true olistostrome with complete loss of structural continuity. Toward the east, as previously mentioned, the Yaguajay Belt might not have been a continuous linear feature, and it is quite possible, in some of the places where the igneous rocks and volcanics are seen draping around the Yaguajay Belt, as in the Cama-guey Province, that they reflect the original flow of slide material up the deep-water tongues between the banks. It is also possible that the fracture system responsible for the main volcanic events of southern Las Villas inter-sected the island through central Oriente parallel to and north of the Cauto Basin. In central Cuba all volcanic activity terminated with the Cretaceous; however, a southern and parallel rift system that included the Cauto Basin, the Cayman Ridge, and possibly an earlier form of the Bartlett Trough re-mained active, and volcanism continued through the Lower and Middle Eocene. It should also be mentioned that while over most of Cuba tectonic activity was prevalent until the end of the Middle Eocene, this activity had already ceased in the Lower Eocene in the Habana and Pinar del Rio Provinces. The fact that the deformation occurred over an area where active rifting was taking place explains the large number of high-angle faults that modify the original geometry of the structures. Furthermore, the many faults exhibiting a left-lateral movement are compatible with the left-lateral component of movement along the Bartlett Trough graben. The general shape of present-day Cuba must have been established toward the close of the Middle Eocene, which was a time of strong erosion and peneplanation. During that time, as a consequence, thick deposits of clastic material were deposited in local basins such as the one south of the Pinar Fault. There must also have been erosion of some source areas that have since disappeared, as evidenced by the numerous boulders of granite and gabbro that are found in the Lower to Middle Eocene clastics that overlie the Cabaiguan Belt. It is difficult at this time to attach significance to the lack of recognized Paleocene fauna throughout most of the island. It might be due, in part, to a problem of paleontologic definitions, but could be related to the fact that the Paleocene was a time of maximum change of conditions.

F. Post-Middle Eocene

By the Upper Eocene the island had been much eroded and began a slow subsidence. The present areas of pre-Upper Eocene exposure remain as topographic highs that were onlapped by successively younger sediments. These sediments are mostly of shallow-water origin and consist of marls, fragmental limestones, reef banks, conglomerates, and clastics. There are occasional deeper-water marly deposits in some of the basins. Some of the larger normal faults remained active. It is possible that the higher massifs of Cuba were never submerged after the Middle Eocene. From the Miocene on, the island began to emerge, accentuating the already existing uplifts. During the late Miocene to Pliocene, Cuba had essentially the same expression as it has today. Throughout the Tertiary the carbonate banks along the northern shore of Cuba continued their growth; and for all practical purposes, the keys along the north shore of Cuba can be considered part of the Bahama banks.

VIII. CONCLUSIONS

This brief and greatly oversimplified history of Cuba is not intended as a final treatment of the subject. On the contrary, it is a first approximation since many problems remain to be solved before a more definitive statement can be made. However, it seems appropriate to point out that there is a very definite relationship between the geology exposed on the surface of Cuba and the formation of trenches and ridges that separate the Bahama Platform from the Jamaican–Nicaraguan rise. As a matter of fact, Cuba can be considered as the "outcrop" of the north and east rim of the Yucatan Basin, as well as the outcrop of part of the northern flank of the Bartlett Trough. Present-day geophysical evidence, as well as deductions made from the geology of Cuba, lead to the conclusion that the entire area was probably under tension beginning in the early part of the Cretaceous. This did not require large-scale spreading of the crust—only enough so that tension fractures could occur, forming the horsts and grabens, with associated volcanism, typical of rift situations. From the evidence available, it appears that prior to rifting there was a "continental" crust in the sense that it was not basaltic but had been the product of former orogenies at the close of the Paleozoic or earlier. It also appears that the area encompassing Cuba and the Yucatan Basin was under shallow-water conditions during late Jurassic. However, as a consequence of rifting, typical oceanic crust material came to the surface and covered the former crust. While this was taking place, the area north of the rifting was sinking rapidly, and sediments typical of present-day oceanic sedimentation were being deposited. In other words, during much of the process the only emerged parts were the carbonate banks of the Bahamas to the north where, from a structural point of view, subsidence was occurring at a fairly high rate. Cuba therefore reflects a process that might well have been independent from a continental margin and might well defy the definitions of "continental" or "oceanic." Yet, Cuba has been and is being called a classical geosynclinal belt, and it is one. It has an eugeosyncline and a miogeosyncline. It has belts of "ophiolites" and the typical gravity sliding that are characteristic of alpine and other mountain systems. It is therefore clear that a complete understanding of the island of Cuba might well bridge the gap between continental and marine geology. Perhaps there is no fundamental difference between the Middle Atlantic Ridge, the Andes, the Tonga Trench, or the great African rift. All could turn out to be the result of one universal process, and differences in appearance might depend entirely on the environment, the nature of the crust and whether this process is taking place in an area undergoing active erosion, active sedimentation such as the continental margin, or slow sedimentation such as the ocean basin. At any rate, if Cuba is a sample of "marine" geology, the thought of what might be hidden under the "acoustic basement" of the oceans of the world is frightening.

REFERENCES

Baie, L. F., 1970, Possible structural link between Yucatan and Cuba: *Am. Assoc. Petr. Geol. Bull.*, v. 54, p. 2204–2207.

Bateson, J. H., 1972, New interpretation of geology of Maya Mountains, British Honduras: *Am. Assoc. Petr. Geol. Bull.*, v. 56, p. 956–963.

Bermudez, P. J., 1961, Las formaciones geológicas de Cuba: *La Habana, Ministerio de Industrias, Inst. Cubano Recursos Minerales, Geologia Cubana*, No. 1, 177 p.

Bowin, C. O., Nalwalk, A. J., and Hersey, J. B., 1966, Serpentinized peridotite from the north wall of the Puerto Rico Trench: *Geol. Soc. Am. Bull.*, v. 77, p. 257–269.

Bronnimann, P., 1953, On the occurrence of calpionellids in Cuba: *Eclogae Geol. Helv.*, v. 46, n. 2, p. 263–268.

Bronnimann, P., 1955, Microfossils *incertae sedis* from the Upper Jurassic and Lower Cretaceous of Cuba: *Micropaleont.*, v. 1, n. 1, p. 28–49.

Bronnimann, P., and Rigassi, D., 1963, Contribution to the geology and paleontology of the area of the city of La Habana, Cuba, and its surroundings: *Eclogae Geol. Helv.*, v. 56, n. 1, p. 193–480.

Bucher, W. H., 1947, Problems of earth deformation illustrated by the Caribbean Sea basin: *Trans. N.Y. Acad. Sci., Ser. 2*, v. 9, n. 3, p. 98–116.

Butterlin, J., 1956, La constitution géologique et la structure des Antilles: *Paris, Centre National Recherche Scientifique*, 453 p.

Dengo, Gabriel, 1969, Problems of tectonic relations between Central America and the Caribbean: *Trans. Gulf Coast Assoc. Geol. Soc.*, v. 19, p. 311–320.

Dillon, William P., and Vedder, John G., 1973, Structure and development of the continental margin of British Honduras: *Geol. Soc. Am. Bull.*, v. 84, p. 2713–2732.

Ducloz, C., 1960, Apuntes sobre el yeso del Valle de Yumurí, Matanzas: *Soc. Cubana Hist. Nat. Mem.*, v. 25, n. 1, p. 1–9.

Ducloz, C., and Vaugnat, M., 1962, À propos de l'âge des serpentinites de Cuba: *Arch. Sci., Soc. Phys. et d'Hist. Nat. Génève*, v. 15, fasc. 2, p. 309–332.

Engel, R. L., 1962, Geology and tectonics of Trinidad Mountains, Las Villas Province, Cuba (abstract): *Am. Assoc. Petr. Geol. Bull.*, v. 46, n. 2, p. 266.

Erikson, A. J., Helsley, C. E., and Simmons, G., 1972, Heat flow and continuous seismic profiles in the Cayman Trough and Yucatan Basin: *Geol. Soc. Am. Bull.*, v. 83, p. 1241–1260.

Flint, D. E., Albear, J. F. de, and Guild, P. W., 1948, Geology and chromite deposits of the Camaguey district, Cuba: U.S. Geol. Surv. Bull. 954-B, p. 39–63.

Furrazola-Bermúdez, G., Judoley, C. M., Mijailóvskaya, M. S., Miroliúbov, Yu. S., and Solsona, J. B., 1964, Geología de Cuba: *La Habana, Ministerio de Industrias, Inst. Cubano Recursos Minerales*, 239 p. + folio volume.

Guild, P. W., 1947, Petrology and structure of the Moa chromite district, Orient Province, Cuba: *Am. Geophys. Union Trans.*, v. 28, n. 2, p. 218–246.

Hatten, C. W., 1967, Principal features of Cuban geology: Discussion: *Am. Assoc. Petr. Geol. Bull.*, v. 51, p. 780–789.

Hatten, C. W., and Meyerhoff, A. A., 1965, Pre-Portlandian rocks of western and central Cuba (abstract): *Geol. Soc. Am. Spec. Pap. 82*, p. 301.

Hill, P. A., 1959, Geology and structure of the northwest Trinidad Mountains, Las Villas Province, Cuba: *Geol. Soc. Am. Bull.*, v. 70, p. 1459–1478.

Imlay, R. W., 1942, Late Jurassic fossils from Cuba and their economic significance: *Geol. Soc. Am. Bull.*, v. 53, p. 1417–1477.

Iturralde-Vinent, M. A., 1969, Principal characteristics of Cuban Neogene stratigraphy: *Am. Assoc. Petr. Geol. Bull.*, v. 53, p. 1938–1955.

Keijzer, F. C., 1945, Outline of the geology of the eastern part of the Province of Oriente, Cuba (E. of 76° W.L.) with notes on the geology of other parts of the island: *Geogr. Geol. Mede., Physiogr. Geol. Reeks, Geogr. Inst., Utrecht*, ser. 2, n. 6, 239 p.

Khudoley, K. M., 1967a, Principal features of Cuban geology: *Am. Assoc. Petr. Geol. Bull.*, v. 51, p. 668–677.

Khudoley, K. M., 1967b, Reply: *Am. Assoc. Petr. Geol. Bull.*, v. 51, p. 789–791.

Khudoley, K. M., and Meyerhoff, A. A., 1971, Paleogeography and Geological History of Greater Antilles: Geol. Soc. Am. Mem. 129.

Kozary, M. T., 1953, Conglomerates associated with the Cubitas plateau, Cuba (abstract): *Geol. Soc. Am. Bull.*, v. 64, p. 1446.

Kozary, M. T., 1956, Ultramafics in the thrust zones in northwestern Oriente, Cuba (abstract): *20th Intern. Geol. Congr., Mexico, Resumenes*, p. 138–139.

Kozary, M. T., 1968, Ultramafic rocks in thrust zones of northwestern Oriente Province, Cuba: *Am. Assoc. Petr. Geol. Bull.*, v. 52, p. 2298–2317.

MacGillavry, H. J., 1937, Geology of the Province of Camaguey, Cuba, with revisional studies in rudistid paleontology: *Geogr. Geol. Meded., Physiogr. Geol. Reeks, Geogr. Inst. Utrecht*, ser. 2, n. 14, 168 p.

Malfait, B. T., and Dinkelman, M. G., 1972, Circum-Caribbean tectonic and igneous activity and the evolution of the Caribbean Plate: *Geol. Soc. Am. Bull.*, v. 83, p. 251–272.

Meyerhoff, A. A., and Hatten, C. W., 1968, Diapiric structures in central Cuba: Am. Assoc. Petr. Geol. Mem. 8, p. 315–357.

Meyerhoff, A. A., Khudoley, K. M., and Hatten, C. W., 1969, Geologic significance of radiometric dates from Cuba: *Am. Assoc. Petr. Geol. Bull.*, v. 53, p. 2494–2500.

Pardo, G., 1966, Stratigraphy and structure of central Cuba (abstract): *New Orleans Geol. Soc. Log.*, v. 6, n. 12, p. 1, 3.

Pyle, T. E., Antoine, J. W., Fahlquist, D. A., and Bryant, W. R., 1969, Magnetic anomalies in Straits of Florida: *Am. Assoc. Petr. Geol. Bull.*, v. 53, p. 2501–2505.

Pusharowski, Yu. M., Knipper, A. L., and Puig-Rifa, M., 1966, Mapa Tectonico de Cuba: Instituto Geologico de la Academia de Ciencias de USSR, Moscu.

Rutten, L. M. R., 1934, Geology of Isla de Pinos, Cuba: *Akad. Wet. Amsterdam Proc.*, v. 37, n. 7, p. 401–406.

Rutten, M. G., 1936, Geology of the northern part of the Province of Santa Clara (Las Villas), Cuba: *Geogr. Geol. Meded., Physiogr. Geol. Reeks, Geogr. Inst. Utrecht*, ser. 2, n. 11, 59 p.

Soloviev, O. N., Skidan, S. A., Skidan, I. K., Pankratov, A. P., and Judoley, C. M., 1964a, Comentarios sobre el mapa gravimétrico de la Isla de Cuba: *La Habana, Ministerio de Industrias, Revista Tecnológica*, v. 2, n. 2, p. 8–19.

Soloviev, O. N., Skidan, S. A., Pankratov, A. P., and Skidan, I. K., 1964b, Comentarios sobre el mapa magnétometrico de Cuba: *La Habana, Ministerio de Industrias, Revista Tecnológica*, v. 2, n. 4, p. 5–23.

Thayer, T. P., 1942, Chrome resources of Cuba: *U.S. Geol. Surv. Bull.* 935-A, p. 1–74.

Thiadens, A. A., 1937, Geology of the southern part of the Province of Santa Clara (Las Villas) Cuba: *Geogr. Geol. Meded., Physiogr. Geol. Reeks, Geogr. Inst. Utrecht*, ser. 2, n. 12, 69 p.

Uchupi, Elazar, 1973, Eastern Yucatan continental margin and western Caribbean tectonics: *Am. Assoc. Petr. Geol. Bull.*, v. 57, p. 1075–1085.

Chapter 14

GEOLOGY OF JAMAICA AND THE NICARAGUA RISE

Daniel D. Arden, Jr.

Georgia Southwestern College
Americus, Georgia

I. REGIONAL SETTING

A. Geographic Names

Topographic features of submerged oceanic areas have received less attention from geographers than have inhabited islands or economically important fishing banks, with the result that geographic names are slow to become established and some areas may have several designations. This has been the case in the western Caribbean where the suboceanic ridge extending eastward from Central America to Jamaica and beyond—the Nicaragua Rise—is called by different names on recently published maps. Current practice seems to favor the direct use of a nearby land name for the oceanic feature without an adjective modification. Examples are *Colombia Basin*, *Barbados Ridge*, *Grenada Trough*, and *Puerto Rico Trench*. Where a name has not already become firmly established, this practice will be followed here.

B. Location and Physical Geography

1. *Nicaragua Rise*

The Nicaragua Rise is a western extension of the Greater Antilles arc. It is a continuation of the Mesozoic and younger rocks exposed in Honduras and Nicaragua and extends eastward to Hispaniola. It is separated from the Cuban arc by the Yucatan Basin and the younger Cayman Trough which now forms the northern boundary of the rise. The present crest of the Nicaragua Rise is marked by the island of Jamaica and a broad, shallow-water platform 250 km wide with many islets and carbonate banks. It drops off steeply on the northern side into the Cayman Trough, whereas toward the southeast there is a regular gradient for about 200 km, then a steeper decline into the Colombia Basin. A northern extension of the Colombia Basin separates the Nicaragua Rise from the Beata Ridge. It is postulated that the geographic boundaries are faults and that formerly the Beata and Cayman Ridges were attached to the Nicaragua Rise. In its greatest dimension (northeast–southwest) the rise is about 1000 km long, and the total area under consideration in this paper is approximately 500,000 km².

Only a small portion of this area is accessible for direct geologic observation. The banks and most islands are composed of late Cenozoic carbonates, exceptions being found only in Jamaica and in some minor volcanic material on Providencia and the adjacent islet of Santa Catalina. Information comes from geologic work in Jamaica and Central America, the stratigraphic sections encountered in drilling for oil, and from a variety of geophysical projects.

About a fourth of the Nicaragua Rise area is above the 100-fathom line, of which nearly all is held under petroleum exploration licenses from various governments.

2. *Landforms of Jamaica*

Jamaica is the third largest island of the Caribbean and has a rugged topography, with about half of its 11,424 km² above 300 m. The highest point is Blue Mountain Peak which has an elevation of 2256 m. There are three main types of land forms on the island that result mainly from bedrock control.

Interior Mountain Ranges. The interior mountains expose early Tertiary and Cretaceous rocks which include sediments of many types plus metamorphic and igneous rocks.

Dissected Limestone Plateaus and Fault Blocks of Tertiary Age. Karst topography is developed in regions where the thickness of the carbonate strata may total 2000 m. The "Cockpit Country" in the north-central part of the island is a typical mature karst topography on which there is almost no surface drainage. The area is so intricately dissected that there are no roads into the center of the Cockpit Country and no settlements. This is the only wilderness portion of Jamaica except for the very highest parts of the Blue Mountains.

Coastal Plains and Interior Valleys. Coastal plains are narrow in the north but elsewhere extend inland to the limestone foothills. In down-faulted areas within the limestone country are "poljes," which are flat-bottomed interior valleys floored with *terra rossa* clays or alluvium. An interesting example is in the Linstead area where it is suggested that a large Pleistocene lake existed when the narrow gorge of the Rio Cobre became dammed near Bog Walk.

C. Geological Investigations

A substantial literature on Jamaican geology has accumulated, most of it representing work done since 1960. In the mid-1960's renewed interest in the petroleum possibilities of Jamaica and the Nicaragua Rise resulted in considerable geophysical work and at least 20 exploratory wells. In addition, there is information from earlier drilling. Four wells were drilled on Jamaica in the late 1950's (Fig. 1), two each in the coastal regions of Nicaragua and Honduras, and one in offshore Nicaragua.

I wish to thank the director and staff of the Geological Survey of Jamaica for their assistance. Appreciation is also expressed for stratigraphic and geophysical information received from the petroleum industry. Active exploration is still going on, and it will probably be many years before complete results of the work can be synthesized and published.

II. THE CRUST OF THE NICARAGUA RISE

A. Geophysical Evidence

Published results of seismic refraction studies in the western Caribbean have provided data from which certain generalizations can be made about the thickness and composition of the crust (Edgar *et al.*, 1971; Ewing *et al.*, 1960; Officer *et al.*, 1957). A reflection profile across the Nicaragua Rise and one extending northwestward from the rise to the Yucatan Basin are described by Edgar *et al.* (1971). Molnar and Sykes (1969) discussed the significance of earthquake distribution in the Caribbean and Central America to plate-tectonics theories. The Cayman Trough has been the subject of several geophysical studies which provide peripheral data, although these investigations generally do not extend far onto the Nicaragua Rise itself (Banks and Richards,

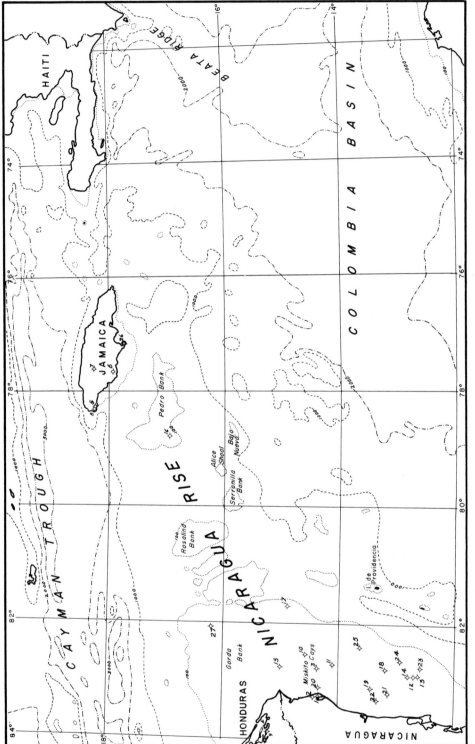

Fig. 1. Exploratory drilling, Nicaragua Rise. Locations identified in Table I. Water depths in fathoms.

1969; Bowin, 1968; Erickson *et al.*, 1972; Pinet, 1971, 1972). Much work throughout the area has been done since 1966 in conjunction with petroleum exploration, including reflection and refraction seismic, marine and airborne magnetics, and gravity. This work was primarily concerned with details of the sedimentary section or in defining the top of "magnetic basement," but the results were also useful for crustal studies in that they confirmed the general configuration and velocities of the upper crustal layers as interpreted from the refraction lines.

The overall shape of the Nicaragua Rise is that of a thick, somewhat elongated crustal prism that has been sheared off along either flank. This is illustrated in an isopach map of total crustal thickness (Fig. 2) and a section across the rise (Fig. 3). Maximum crustal thickness is about 22 km. Although the line of maximum thickness parallels the present topographic crest of the rise, it lies about 200 km south of that crest. This probably resulted from Tertiary movement along the Cayman Trough, which in turn caused a southward tilting of the rise.

Mantle velocity beneath the rise is about 8.1 km/sec, which agrees with average values elsewhere (Gutenberg, 1955; Edgar *et al.*, 1971). Above the mantle is a crustal zone with a maximum thickness of 19.5 km and averaging 13 km in thickness. Seismic velocities in the zone range from 6.2 to 7.0 km/sec and average about 6.7 km/sec. This is "Layer 3," the oceanic layer, described by Raitt (1963). Vogt *et al.* (1969) have summarized research on crustal layering and illustrate ocean crustal sections from various areas of the world, including areas in the Caribbean. Layer 3 has been characterized as serpentinite (Hess, 1962), amphibolite (Cann, 1968), and basalt (Oxburgh and Turcotte, 1968; and Raitt, 1963).

Above Layer 3 is a well-defined layer, correlated with Raitt's (1963) Layer 2, which is also known as the "volcanic layer" (Vogt *et al.*, 1969). Under the Nicaragua Rise it reaches a maximum thickness of 4.5 km. It has not been identified in the Cayman Trough crustal block, but appears to be about 2.5 km thick at the northern boundary of the rise. Velocities range from 5.2 to 5.8 km/sec. Along the section (Fig. 3) it appears that Layer 2 thins against an upward bulge of oceanic crust near 14° N latitude and 79° W longitude. The data are too incomplete to suggest whether this is a case of stratigraphic overlap against a previous high area, the result of subsequent intrusion, or an incorrect interpretation.

Layer 2 is overlain by a highly variable sequence, ranging from less than 1.0 km to more than 5.0 km in thickness, with a seismic velocity range of 3.9 to 4.8 km/sec, and interpreted as Cretaceous and Tertiary sediments and volcanics. An uppermost layer of late Tertiary and younger material, with a velocity of 1.7 to 2.0 km/sec, has a maximum thickness of 1000 m along the present crest of the rise.

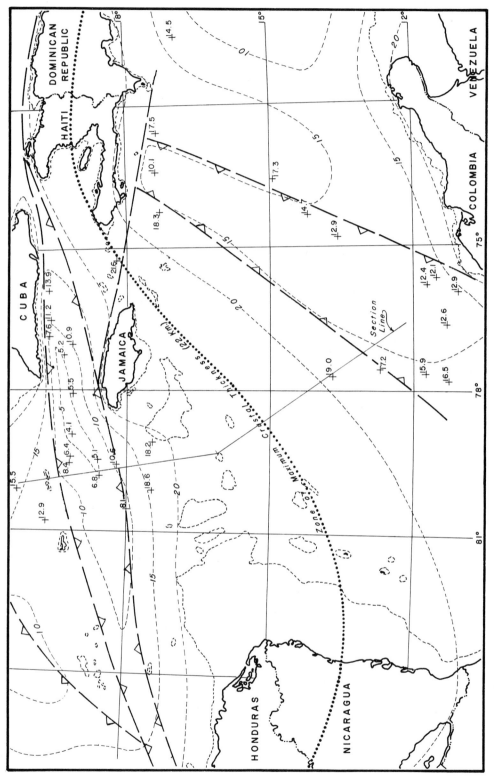

Fig. 2. Western Caribbean. Isopach map of total crustal thickness, not including water depth. Thickness in kilometers.

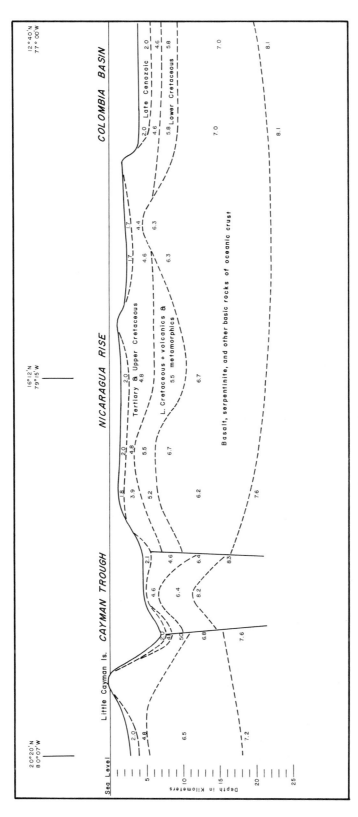

Fig. 3. Section across the Nicaragua Rise. Vertical exaggeration times 10. *P*-wave velocities shown in kilometers per second. Seismic data from published sources, chiefly Ewing *et al.*, 1960.

The Cretaceous–Tertiary succession exhibits greater thickness in two areas (Fig. 3). One area is a zone along the southeastern margin of the Nicaragua Rise; the other is a parallel belt along the southern edge of the present crest, i.e., along the edge of such features as Pedro Bank, Alice Shoal, and Serranilla Bank. These belts of thick section may represent former reef edges during different orientations of the Nicaragua Rise. The belt bordering the Colombia Basin may have marked the edge of the rise crest prior to the last large-scale movement along the Cayman Trough during which time the rise was tilted to the south, whereas the second belt represents a reef front related to the present bathymetric configuration.

B. Structural Framework of Jamaica

The oldest rocks known in Jamaica are Lower Cretaceous sediments and volcanics, dated by fossils. Metamorphic rocks in the Blue Mountains are probably Upper Cretaceous sedimentary and volcanic rocks altered during orogenic episodes occurring between Late Cretaceous and Middle Eocene time (Fig. 4). Thus, an ancient "basement" is unknown in Jamaica. As interpreted from seismic results published by Ewing *et al.* (1960), the basaltic crustal layer beneath the island is 12 to 15 km thick, above which is a 5.4 km/sec Layer 2 about 4.0 km in thickness, neither of which is exposed.

Jamaica consists of three structural blocks separated by faulted troughs (Fig. 5). The central Clarendon Block is separated on the west by the Montpelier–Newmarket Trough from a relatively small Hanover Block, and on the east it is separated from the Blue Mountain Block by the Wagwater Trough.

The dominant feature of the eastern block is the Blue Mountains massif (Fig. 6), the least known region of Jamaica. Although mapped mainly as a mass of metamorphosed Cretaceous rocks, the metamorphosed portion is much smaller than shown on the 1958 geologic map (Jamaica Geological Survey, 1958) and appears to make up only a minor part of the inlier. An arcuate fault system bounds the Blue Mountains on the west and south. Its northwestern portion is the Yallahs fault zone which is continuous with the Plantain Garden fault to the east. The Yallahs zone consists of at least three fault wedges, along which the latest movement was apparently normal. The Plantain Garden portion is clearly visible on the surface today and seems to be a steeply dipping system confined to a rather narrow zone and with uplift on the northern side accompanied by right lateral movement. Parallel to the Yallahs faults 10 km to the southwest is the Wagwater fault zone, which conforms to the general stress pattern of the eastern block. Movement along this fracture system has apparently been right lateral with a vertical component, resulting in a graben bounded by normal faults.

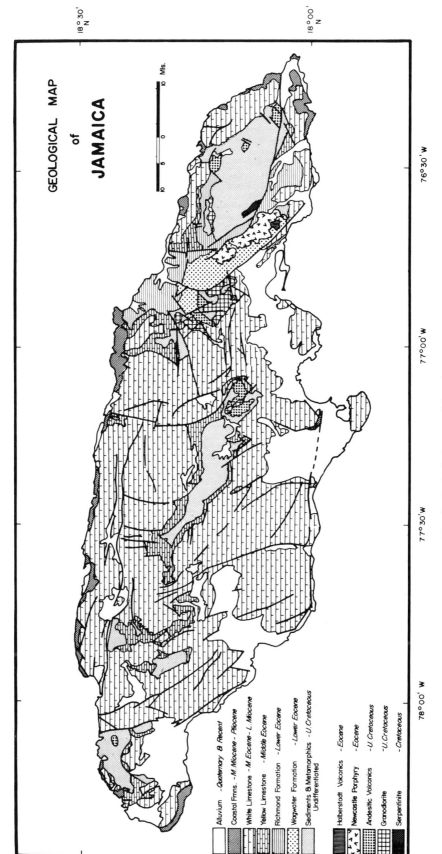

Fig. 4. Geological map of Jamaica.

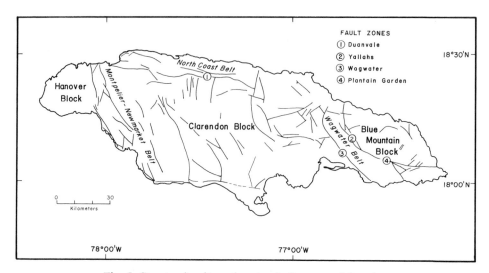

Fig. 5. Structural units and major fault zones of Jamaica.

The outstanding structural features of the Clarendon Block are several inliers of Cretaceous and Eocene rocks and numerous tilted fault blocks bounded by steep scarp slopes. The dominant fault pattern varies from a north–south to a northwest–southeast orientation. A strip 9 to 13 km wide along the north coast is the North Coast Belt, set off from the Clarendon Block by faults trending east–west called the Duanvale fault system. It is parallel to the coast and probably is related to the formation of the Cayman Trough and the late Tertiary tilt of the Jamaican side of the rift.

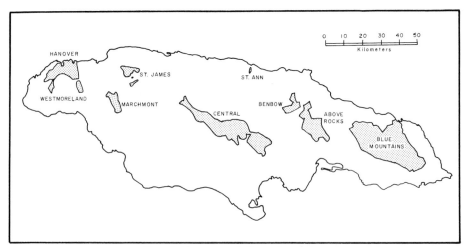

Fig. 6. Cretaceous inliers of Jamaica.

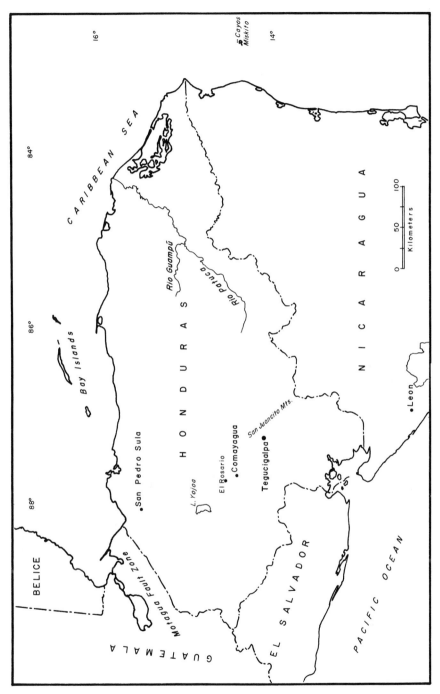

Fig. 7. Locality index, Honduras and surrounding area.

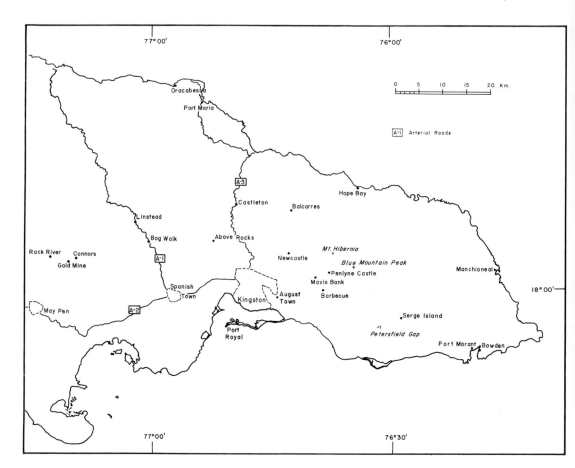

Fig. 8. Locality index of eastern Jamaica.

Although the general fault pattern may have been impressed as early as the first phase of the Laramide orogeny, when the Wagwater Trough was formed, later episodes renewed the early Laramide faults and also fractured the brittle Tertiary White Limestone. There are numerous examples throughout the island of relatively recent movement. Displacement on the fault forming the west-facing scarp of the Don Figuerero Mountains* may be more than 1500 m, and vertical movement in the Santa Cruz Mountains is at least 900 m. The east–west fault that shapes the coastline between Great Pedro Bluff and Alligator Pond is either very young or has had recurrent movement recently. Here the scarp is steep, and several large landslides have occurred so recently that weathering and erosion have not begun to attack them.

* See Figs. 7–9 for locality indexes.

III. PRE-CRETACEOUS HISTORY

A. Origin of the Nicaragua Rise—Modern Theories

The core of the Nicaragua Rise is nowhere exposed, so its complete history cannot be interpreted from the record of the rocks. Indirect evidence must be employed, and there are few areas of the earth where the evidence has been interpreted with so many contradictory results. Some investigators have considered the rise to be the site of a Paleozoic or Mesozoic geosyncline fringing a landmass to the south in the present location of the Colombia and Venezuela Basins (e.g., Eardley, 1954; Chubb, 1960; Khudoley, in Khudoley and Meyerhoff, 1971), or as a geosyncline related to a northern land area (Meyerhoff, in Khudoley and Meyerhoff, 1971).

Since the introduction of the plate-tectonics concept, several hypotheses have been proposed relating to the origin of the Caribbean and its various crustal features. Typical of these is the conclusion of Edgar et al. (1971)

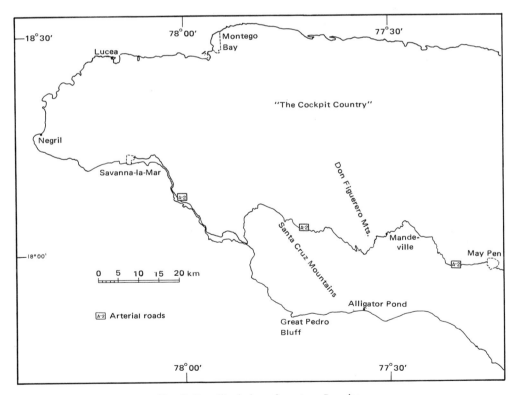

Fig. 9. Locality index of western Jamaica.

that the Caribbean was originally Pacific crust caught between North and South America during the Mesozoic opening of the Atlantic. The interaction of the Caribbean plate with the surrounding land areas then resulted in the formation of geosynclines. A variation of this hypothesis was presented by Malfait and Dinkelman (1972) who suggested that the Caribbean plate decoupled from the East Pacific plate near the end of the Laramide orogeny and that Caribbean lithosphere was transferred to the North American plate as the Cayman Trough developed from west to east. A series of transform faults migrated eastward, accompanied by underthrusting of the North American beneath the Caribbean plate. A further modification, by Walper and Rowett (1972), proposes that a sphenochasm formed in the Gulf of Honduras in early Tertiary time as a result of clockwise rotation of the North and South American plates, forming the Cayman Trough and Yucatan Basin. Rotational movements of the two American plates created tension along the northern margin of the Caribbean plate and compression along the southern margin.

Arden (1969) suggested that the Nicaragua Rise originated as a mobile belt between crustal plates and based an explanation for the mechanics upon a concept presented by Donnelly (1964). Briefly, this concept holds that a thickening root of oceanic crustal material develops along the boundary between plates having contrasting physical properties. Magmas originate through partial melting of the somewhat hydrated upper mantle material along the axis of deformation, and the thickened crustal zone grows into an island arc.

B. Pre-Cretaceous Geology

Drilling on the rise appears not to have penetrated below Cretaceous rocks, and nowhere on Jamaica have rocks older than Cretaceous been identified. The western buttress of the rise in Central America may provide clues to its early history.

Pre-Mesozoic rocks are widely distributed in Honduras and northern Nicaragua. They have been described by Zoppis Bracci (1957), Dengo (1969), McBirney and Bass (1969), Williams and McBirney (1969), and Fakundiny (1970). Dengo (1969) has designated the basement rocks south of the Motagua fault zone the Palacagüina Metamorphics. Generally, they consist of phyllites and schists of the greenschist facies, although Williams and McBirney (1969) report quartzite and boulders of metamorphosed chert, diabase, and basalt in addition. Fakundiny (1970) found chiefly metasediments, including marble, in El Rosario Quadrangle. He has named the formation Cacaguapa Schist and believes it is a structurally deeper facies of the Palacagüina Schist.

It has been stated as a generalization that the strike of the basement rocks in this region is parallel to the east–west trend of the Nicaragua Rise (e.g., Khudoley and Meyerhoff, 1971, p. 14). I believe much of this orientation is more anticipated than real because of map trends associated with young topographic and structural features. Prominent structural trends of the islands along the northern coast of Honduras parallel those of the Cayman Trough. McBirney and Bass (1969) find that folding in the metamorphic rocks is also parallel with the long axes of the islands and conclude that this orientation can best be explained by compressive forces acting in a direction almost at right angles to the present Cayman Trough axis. They believe subsequent movement has resulted in a general southward tilting up to 30°.

In Honduras proper, various alignments can be found. For example, grabens trending N to N 15° E form a prominent pattern (Mills *et al.*, 1967), and Fakundiny (1970) points out a geographic trend which is apparent on the topographic maps. The latter is a series of belts trending N 60° W, parallel to the Middle America Trench. Fakundiny (1970) says that the largest pre-Mesozoic fold in the El Rosario Quadrangle plunges S 15° W with a direction of maximum shortening oriented WNW–ESE. The significance of these examples is that the structural trends in the older, presumably Paleozoic, rocks have been obscured by younger tectonic movements to the extent that paleogeographic interpretation based on the apparent trends is of questionable value. More detailed field investigation of the older rocks, such as the recent work by Dupré (1970), Everett (1970), and Fakundiny (1970), should help in better defining the trends and age of origin.

No offshore well has been drilled deep enough to test for the presence of pre-Mesozoic rocks, except that probable Paleozoic "basement" was reached at a depth of 1900 m in the Tuara-1 well, located on land near the eastern coast of Nicaragua.

An important study of the Mesozoic rocks of Honduras was made by Mills *et al.* (1967). They recognize five major lithologic units: (1) El Plan Formation (Jurassic), a thin-bedded dark shale sequence, overlain by (2) Todos Santos Formation (Jurassic and Lower Cretaceous), a red-bed section containing thin limestone and some tuffs; (3) the Yojoa Group (primarily Aptian and Albian) is a thick carbonate series followed by (4) the Valle de Angeles Group (Late Cretaceous), which in some places contains thick volcanics; (5) the Esquias Formation (Maestrichtian and Eocene) is a brown, marly shale and limestone formation, generally very fossiliferous, which overlies and interfingers with the upper red beds.

The El Plan Formation was named by Carpenter (1954) for sedimentary rocks occurring in the Rosario mining district of the San Juancito Mountains. The formation overlies the metamorphics unconformably and consists predominantly of dark shale and siltstone with thin sandstone beds and lenses of conglomerate. The overlying Todos Santos beds are terrestrial sediments without fossils diagnostic of age. The El Plan beds have been called Triassic or Jurassic on the basis of plant remains identified as Rhaetic by Newberry (1888). It should be pointed out that Newberry did not collect the specimens but received them from a mining engineer, and although San Juancito is given as the collecting locality, no other such fossils have been found in the El Plan beds. Several workers, myself included, have searched the outcrop area without success. Newberry says (1888, p. 342) that below the "hydromica schists" containing the plant beds there are said to be limestones with Carboniferous fossils. While limestones, or marbles, certainly exist in the Palacagüina–Cacaguapa sequence, no fossils have been reported. In fact, no limestones are known from these rocks in the Rosario area, which makes one wonder if there is not some confusion regarding the locality from which the Newberry collection was made.

Carpenter (1954) described siltstone and sandstone beds near Rosario from which he collected *Trigonia* cf. *quadrangularis* and other fossils identified as Jurassic in age. The inference is that the El Plan and undefined Jurassic beds are part of the same sequence. These sediments appear to have been deposited under flood plain, paludal, and shallow-water marine conditions.

C. Interpretation of Early History

It seems reasonably well established that rocks representing a long Paleozoic history exist in Central America and that the western end of the Nicaragua Rise somehow connects with this segment of continental crust. The shape and thickness of the crust composing the rise, however, suggests that it is not a direct continuation of these old rocks, for its axial thickness of 22 km and the relative thickness of the major crustal zones is closer to "subcontinental" than to any type of continental crust (Ronov and Yaroshevsky, 1969).

Several authors, whose works are cited above, have suggested that a plate boundary exists to the north of the Nicaragua Rise. It is not my purpose here to go into a history of the Caribbean, so it will suffice to say that such a situation would be compatible with the suggested origin of the rise, that is, the rise originated as a belt of thickened oceanic crust because of its location along the boundary of two active crustal plates. The eventual evolution of this belt into a volcanic island chain can be visualized, but there is no clear evidence as to when the volcanic activity began.

A previous section pointed out that the crust forming the Nicaragua Rise is essentially the same as ocean crust except that the individual layers tend to be thicker. Edgar *et al.* (1971) report a similar situation for the Beata Ridge. The isopach map of total crustal thickness (Fig. 2) shows the central mass of the rise to be a pod-shaped feature, thickening gradually to a maximum along a central axis. Between the Beata Ridge and the Nicaragua Rise there is a segment in which the crust is distinctly thinner. I have interpreted this as a possible rift. The bathymetry supports this interpretation. The western edge of the Beata Ridge is probably a fault scarp (Ewing *et al.*, 1967; Fox *et al.*, 1968; Edgar *et al.*, 1971). If the postulated rift were closed, the regular thinning of the crust making up the southeastern flank of the Nicaragua Rise would be uninterrupted.

Both the cross section (Fig. 3) and the isopach map (Fig. 2) show a sudden and drastic thinning of the crust within the Cayman Trough. This also is probably a rift.

Jurassic clastics containing marine fossils in Honduras indicate that marine conditions were present in the Caribbean before the end of Jurassic time. Additional evidence for a Jurassic sea is, of course, found in Cuba and the Gulf of Mexico region. I would point out, however, that evidence for a *pre-Jurassic* Caribbean Sea has not been demonstrated. The Jurassic beds of Honduras are so limited in areal extent that few paleogeographic inferences can be made. A coastal environment undoubtedly existed in the San Juancito Mountain area, but the configuration of the coastline is unknown, and there is no direct evidence as to what terrain, if any, lay to the east. We can only suggest that the thickened welt that formed the embryonic Nicaragua Rise grew in an easterly direction, and it may have been late Jurassic or even early Cretaceous time before islands developed.

IV. LOWER CRETACEOUS HISTORY

A. Lower Cretaceous Rocks of Jamaica

There has been a misconception regarding the abundance and age of metamorphic rocks in Jamaica. The geologic map (Jamaica Geol. Survey, 1958) shows a large portion of the Blue Mountains as metamorphics intruded by ultrabasics and Cretaceous granodiorite or andesite. Perhaps it is this map and Matley's (1951; *also* Matley and Higham, 1929) description that have been chiefly responsible for the widely held idea that ancient—perhaps Precambrian—rocks occur there along with an extensive ophiolite zone. Recent work in the Blue Mountains, such as that of Green and Holliday (1970) and Kemp (1971), indicates that metamorphism is largely confined to a belt a few kilometers wide paralleling the southwestern fault zone.

Fossiliferous Upper Cretaceous beds have been sampled and a variety of sediments and volcanics of this age described from the massif, including all the possible parent rocks for the various metamorphic types. There is no direct evidence at this time, either structural or radiometric, for assigning a Lower Cretaceous or older date to any of the rocks making up the Blue Mountain massif.

The only Lower Cretaceous section in Jamaica for which the age designation is supported by fossil evidence is in the Benbow Inlier, descriptions of which have been published by Burke *et al.* (1968) and Robinson *et al.* (1971). A review of the volcanic rocks in the section was presented by Roobol (1971). Two formations have been designated: the Devils Racecourse Formation, Barremian(?) through Aptian in age, and the Rio Nuevo Formation, of Albian age. Dating is based entirely on rudist species that occur in limestones intercalated with volcanic and clastic beds.

The Devils Racecourse Formation is dominantly a volcanic mélange, about 1000 m thick, within which three horizons of limestone have been identified. The base of the formation is not exposed, but beneath the lowest limestone level is about 400 m of volcanics, the lower 300 m of which consists largely of lava flows. The rocks are hydrothermally altered, but preserved structures indicate that the greenstones were originally andesite and basalt. Flow banding is well developed, and pillow structures are present near the top of the sequence. The upper 100 m is a poorly bedded pyroclastic breccia and conglomerate.

The lower limestone consists of the persistent Copper Member and the lenticular Phillipsburg Member. The Phillipsburg may be somewhat older or in part equivalent to the Copper Member. The limestones are primarily clastic, ranging in size from calcareous mud to cobble-sized rudist fragments. Oolites are present but not abundant. Chubb (1967) described a new species of rudist, *Monopleura diaboli*, from these limestones and considers them to be Lower Cretaceous.

The middle section of the Devils Racecourse volcanics is similar to the conglomeratic material that occurs beneath the Copper limestone. There is a mixture of subangular porphyritic andesite and rudist fragments. Some feldspathic sandstone and at least two basaltic flows are present. The carbonate content is variable and ranges from none to the nearly pure limestone of the Jubilee Member. This member reaches a thickness of 100 m but is not present throughout the outcrop area. Fossils consist of numerous Nerineid gastropods and a new species of rudist, *Pachytraga jubilensis* CHUBB.

Overlying the middle volcanic sequence is the Benbow Limestone Member, 400 m thick, which is persistent throughout the inlier. No fossils have been identified. Above the Benbow layer, the Devils Racecourse is mainly volcanic conglomerate that contains a hematite and chert horizon with possible radiolarian structures and a variolitic lava with zeolite-filled amygdales. Roobol (1971) says that a basaltic pillow lava approximately 300 m in thickness occurs in the upper unit.

The Rio Nuevo Formation overlies the Devils Racecourse Formation. It is composed largely of well-bedded feldspathic sandstone and shale. In the lower part of the formation is a conglomerate composed of basalt, andesite, and limestone fragments. Limestone interbeds have been assigned to the Seafield Limestone Member, which consists of rudist and gastropod fragments within a micritic matrix. Burke *et al.* (1968) report that Chubb identified *Caprinuloidea perfecta* from the limestone and *Sabinia totiseptata* plus new species of *Sphaerucaprina* and *Tepeyacia* from shale immediately overlying the limestone. The age is considered to be Albian.

Adjoining the Benbow exposure to the southeast with fault contact are formations of the Above Rocks area. They have been mapped and described by members of the Jamaica Geological Survey (1972). A granodiorite which has been dated at 65 ± 5 m.y. (Chubb and Burke, 1963; Harland *et al.*, 1964) intrudes the Border Volcanic Formation and the underlying Mount Charles Formation. The Mount Charles is primarily bedded chert and hornfels–mudstone, with occasional arkosic and conglomeratic beds. The Border Volcanic Formation contains andesitic lava, tuff, agglomerate, and interbedded conglomerate. Except for establishing their age as pre-granodiorite, these rocks are not precisely dated. Reed (1966) has assigned them with uncertainty to the Lower Cretaceous on the evidence of north dip, which if continued across the fault and into the Benbow Inlier would place them beneath the Devils Racecourse Formation.

In the eastern part of the Central Inlier, about 20 km southwest of the Benbow exposures, there are some poorly known volcanics, consisting of basic flows, dikes, and breccia, that may be equivalent to the Border Volcanics or part of the Devils Racecourse Formation.

B. Lower Cretaceous Section in Honduras

Above the dark clastics of the Jurassic El Plan Formation is the Todos Santos Formation, a sequence of terrestrial clastics and volcanics that may exceed 1000 m in thickness. Todos Santos beds are unconformable wherever the lower contact is seen and lie upon metamorphics as well as on the El Plan Formation. It is a red-bed succession with conglomerates forming a significant percentage of the volume.

Fakundiny (1970) says that four genetically different lithologies make up the Todos Santos Formation: (1) conglomerates with metamorphic clasts, interbedded with reddish-brown sandstone, shale, and spotted arkose; (2) volcanic clast conglomerate and spotted arkose; (3) devitrified crystal-poor vitric tuff; and (4) light-colored, bimodal arkose. No fossils have been found in these rocks. In the Rio Guampú drainage are about 150 m of dark red andesitic flows (Mills *et al.*, 1967) which are unconformable upon El Plan beds and are also unconformable beneath the Todos Santos strata, although the andesites were included with that formation for mapping.

Unconformably overlying the Todos Santos Formation is a carbonate and shale section, the Yojoa Group of Mills *et al.* (1967). The four named formations of the Yojoa are more indicative of facies than age sequence. The Cantarranas Formation is a back-reef facies containing thin-bedded, dark-colored limestone and shale. The Atima Formation is massive limestone that originated as patches of oyster and rudist reefs. The Guare is a fore-reef formation of thin-bedded, black shale and limestone. The Ilama Formation is a limestone conglomerate interbedded with Atima and Cantarranas lime-stones. Fossils have been found in each formation but are most abundant in the Atima. Primarily, Albian age is indicated, although Upper Aptian to Turonian fossils have been identified.

On the basis of stratigraphic position and similarity with Todos Santos beds elsewhere in Central America, Mills *et al.* (1967) have considered the formation in Honduras to be uppermost Jurassic and lowermost Cretaceous in age. They suggest the Yojoa Group may represent Upper Neocomian through Turonian time, with a marine transgression reaching its greatest extent during the Albian.

C. The Nicaragua Rise during Jurassic–Lower Cretaceous Time

The rocks thought to be Jurassic or Lower Cretaceous occurring at either end of the Nicaragua Rise probably exceed 2000 m in thickness and represent a variety of terrestrial and shallow-water marine environments. Fossils, however, are not common, and reliable dating is seldom possible. It appears that the Devils Rececourse beds of Jamaica are pre-Albian, and the same can be said for the lower portion of the Yojoa and older formations in Honduras.

The amount of volcanic material in the section available for examination is appreciable. It cannot be stated with certainty when volcanism was first manifest, but it seems to have occurred in Honduras in latest Jurassic or early Cretaceous because there are flows beneath the Todos Santos red beds, and some conglomerates in the Todos Santos contain angular fragments of primarily volcanic origin that could not have traveled far. The thickening of the crustal welt was accompanied by volcanoes and possibly fissure eruptions that helped to build a ridge along the mobile belt. Ash beds and terrestrial ag-glomerates verify the presence of subaerial volcanic cones.

One can visualize the environmental setting as a belt of active volcanic islands with rudist or oyster reefs developing in favored locations on their flanks. Lavas were dominantly andesitic or dacitic, but basalts are also known.

V. UPPER CRETACEOUS CYCLE OF DEPOSITION

A. Distribution of Upper Cretaceous Beds

A depositional cycle, with the maximum marine advance occurring in Albian time and ending around late Turonian, is indicated in Honduras by the sediments of the Yojoa Group. The overlying Valle de Angeles Group is not diagnostic of a normal cycle. It consists of a thick accumulation of red-bed sandstone, shale, and conglomerate, with lesser calcareous beds, all of which is indicative of a terrestrial and shoreline environment. The age of this group ranges from Late Cretaceous to Middle Miocene. Marine beds younger than about Middle Eocene are not known.

In Jamaica the outcrops of Lower Cretaceous and earliest Upper Cretaceous sediments are not sufficiently extensive to put together a detailed chronology. Upper Cretaceous beds, with a total thickness of at least 2000 m, have been recognized in seven major inliers plus several smaller related areas of exposure and in exploratory wells. The beginning of a cycle of deposition is not clearly defined, but its development and terminal phase can be followed quite well.

B. Upper Cretaceous Rocks of Jamaica

1. *Blue ‑Mountain Inlier*

It has long been known that a variety of lithologies occur in the Blue Mountains, including intrusive and extrusive igneous rock, fossiliferous Maestrichtian beds, and metamorphosed sedimentary and volcanic rocks. Until the late 1960's, however, a lack of adequate base maps and difficulties of access, combined with scarcity of bedrock exposures, effectively deterred systematic mapping. Work related to utilization of the region's water resources has stimulated geologic investigations, so that the relations and distribution of rock types are becoming clearer. Robinson *et al.* (1971) have reported on some of this work, which has been carried out by Green and Holliday (1970), Acker and Willis (1971), Kemp (1971), and members of the Jamaica Geological Survey Department.

Most of the Blue Mountain massif exposed beneath the Tertiary carbonates consists of a thick, broadly folded and faulted mass of Upper Cretaceous sediments with some interbedded volcanics and minor intrusives.

The most severely deformed rocks belong to the Westphalia Schists which outcrop on the east side of the Yallahs fault as a block about 7 km long and 1 km wide, located between Penlyne Castle and Mavis Bank in the southwestern portion of the Blue Mountains. They are quartzo-feldspathic and greenschist rocks that are transitional to amphibolite. They weather to a rusty-brown color, and only in fresh outcrop is schistosity apparent. East of the Westphalia Schist, and in fault contact, is a belt of rocks with a maximum width of 1.8 km, within which there is a variety of lithologies. Robinson *et al.* (1971, p. 28) called these "metamorphosed eugeosynclinal rocks (greenschist belt) of presumed Cretaceous age.... . Rock types include serpentinite, metavolcanics, marble, and calc- and magnesian schists." A strip of grano-diorite occurs between the two metamorphic belts.

Outcrops of the marble occur near the town of Serge Island and extend northwestward along the metamorphic belt in discontinuous patches up to 30 m thick as far as Mt. Hibernia (Zans and Bailey, 1961). Rudist fragments, reportedly *Monopleura* (Chubb, 1962), a genus that ranges from Valanginian to Maestrichtian, occur in a slightly metamorphosed brecciated limestone at Whitfield Hall, north of Mt. Hibernia.

Apparently unconformable upon the metamorphics is a sequence of dark shale, red beds, volcanics, and siltstone from which Maestrichtian (Robinson *et al.*, 1971) and perhaps Campanian (Kauffman, 1966) fossils have been identified. Green and Holliday (1970) described the section along the summit of Blue Mountain Peak, which is on the axis of an anticline. Along the trail on the western limb of the fold the sequence appears to be conformable through 425 m of exposure. It consists of red tuffaceous arkose and arkosic sandstone at the center of the fold, overlain by thick alternations of red and purple shale, tuff and conglomerate, and fossiliferous siltstone and mudstone. The latter is the Blue Mountain Shale of Chubb (1961). A more detailed description of formations of the "Blue Mountain Complex" is presented in an Annual Report of the Jamaica Geological Survey (1962, p. 4–7).

On the northern slopes of the mountains are brown marine shales containing Maestrichtian fossils, including *Titanosarcolites*. Farther into the mountains on the north side are some slightly metamorphosed limestones containing *Barrettia monilifera*, which is Campanian in age (Chubb, 1962).

Three K–Ar age determinations have been made by Lewis *et al.* (1973) on rocks from the Westphalia Schist. Hornblende gives an Upper Cretaceous age of 76.5 ± 2.5 m.y. Biotite provides ages of about 50 m.y., which are interpreted as having been reset during Early Eocene Laramide activity.

2. Central Inlier

The Central Inlier is about 310 km² in area. Although it represents the crest of a large Tertiary anticline, it also reflects earlier structure developed in Cretaceous beds. In the latter case, the anticline plunges westward, so that the oldest Cretaceous beds are exposed at the eastern end of the inlier. The stratigraphic section in the Central Inlier is shown in Fig. 10.

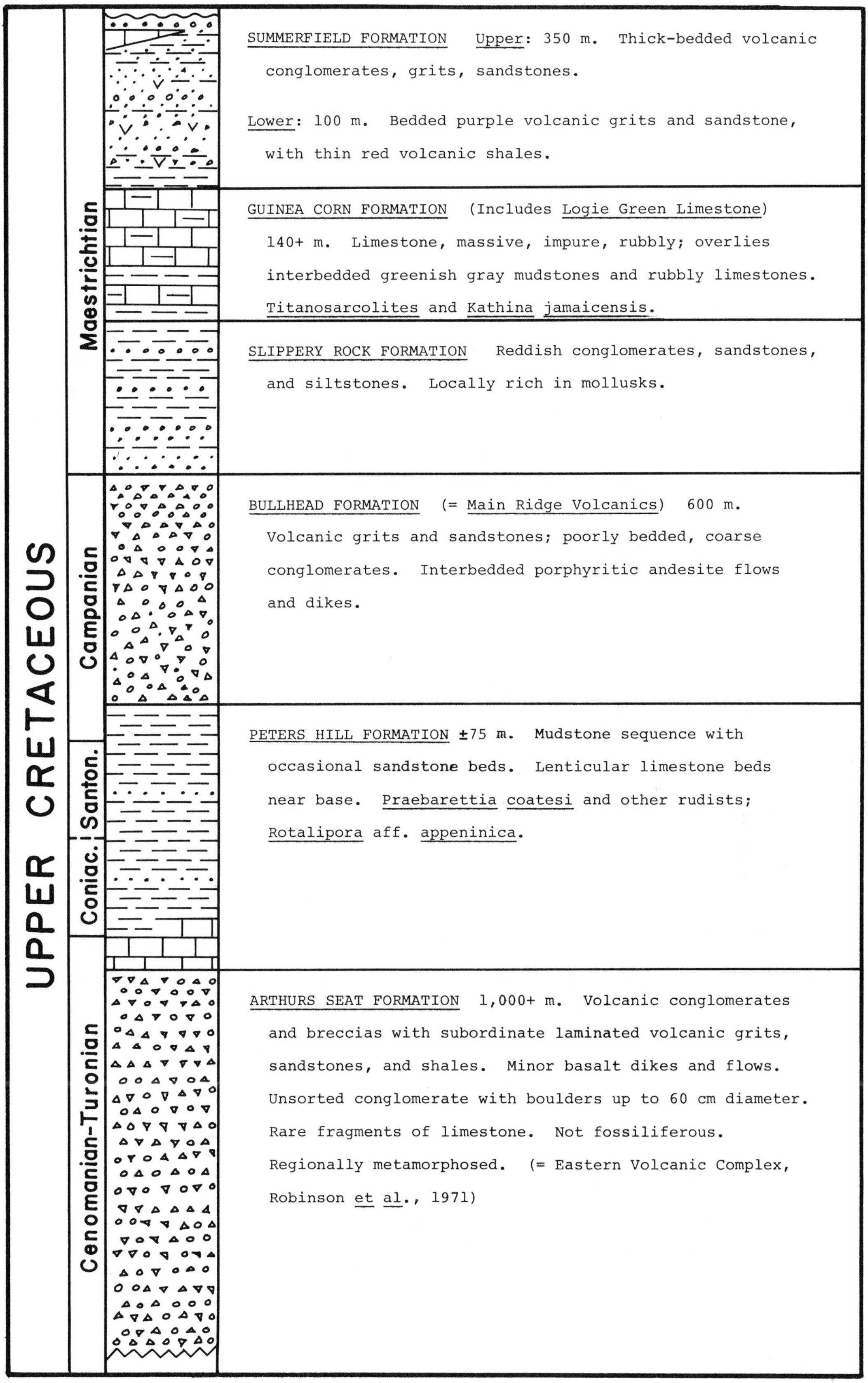

UPPER CRETACEOUS	**Maestrichtian**		**SUMMERFIELD FORMATION** Upper: 350 m. Thick-bedded volcanic conglomerates, grits, sandstones. Lower: 100 m. Bedded purple volcanic grits and sandstone, with thin red volcanic shales.
			GUINEA CORN FORMATION (Includes Logie Green Limestone) 140+ m. Limestone, massive, impure, rubbly; overlies interbedded greenish gray mudstones and rubbly limestones. Titanosarcolites and Kathina jamaicensis.
			SLIPPERY ROCK FORMATION Reddish conglomerates, sandstones, and siltstones. Locally rich in mollusks.
	Campanian		**BULLHEAD FORMATION** (= Main Ridge Volcanics) 600 m. Volcanic grits and sandstones; poorly bedded, coarse conglomerates. Interbedded porphyritic andesite flows and dikes.
	Coniac. Santon.		**PETERS HILL FORMATION** ±75 m. Mudstone sequence with occasional sandstone beds. Lenticular limestone beds near base. Praebarettia coatesi and other rudists; Rotalipora aff. appeninica.
	Cenomanian-Turonian		**ARTHURS SEAT FORMATION** 1,000+ m. Volcanic conglomerates and breccias with subordinate laminated volcanic grits, sandstones, and shales. Minor basalt dikes and flows. Unsorted conglomerate with boulders up to 60 cm diameter. Rare fragments of limestone. Not fossiliferous. Regionally metamorphosed. (= Eastern Volcanic Complex, Robinson et al., 1971)

Fig. 10. Upper Cretaceous section, Central Inlier of Jamaica.

Robinson *et al.* (1971) believe that the oldest rocks exposed are basic flows, dikes, and shallow intrusives in the area of Rock River, between Connors and Gold Mine. Even though these rocks have been intruded by granodiorite and have been variously affected by hydrothermal alteration, Roobol (1971) suggests that the eastern volcanics are equivalent to Arthurs Seat Volcanic Formation, the oldest formation recognized by Coates (1968) in the central portion of the inlier. Lewis *et al.* (1973) have calculated a K–Ar isochron age of 85 ± 9 m.y. for the time of granodiorite emplacement.

Overlying Arthurs Seat Volcanics is the Peters Hill Formation, a mudstone sequence with some sandstone and limestone. Fossils from different localities within the formation suggest an age range of Turonian to Campanian.

Unconformable on Peters Hill beds is a volcanic sequence called the Bull-head Formation and Main Ridge Volcanics by Robinson *et al.* (1971). The Bullhead Formation consists of poorly bedded volcanic clastics interspersed with andesite flows and dikes. Main Ridge is composed of andesite flows and breccias intruded by basaltic and andesitic dikes.

Three Maestrichtian formations have been named. The Slippery Rock Formation (Robinson *et al.*, 1971) is a series of reddish clastics. The Guinea Corn Formation, about 186 m thick, was designated by Coates (1965) to include the *Titanosarcolites*-bearing limestones and interbedded greenish-gray siltstones. The Summerfield Formation consists of 100 m of bedded and often graded volcanic sandstone overlain by massive volcanic conglomerate and sandstone. Clasts are dominantly hornblende–plagioclase andesite with minor pyroxene andesite and quartz dacite.

3. *St. Ann Inlier*

This inlier, located near the north-central coast, exposes about 4 km² underlain by Cretaceous sediments. Chubb (1958, 1962) reports that the oldest rocks, the *Inoceramus* Beds, and the succeeding thick conglomerate are Turo-nian to Coniacian in age. There is presumed to be an unconformity between the conglomerate and the overlying *Barrettia*-bearing limestone, which is regarded as basal Campanian. This is followed in sequence by three formations that are also believed to be Campanian: the *Diozoptyxis* Shale, New Ground Conglomerate, and Windsor Shale. A succession of *Globotruncana* species occurs throughout the section and has served to date the beds.

4. *St. James Inliers*

Three Cretaceous inliers in St. James Parish cover about 38 km². Chubb (1962) estimates that at least 2300 m of section is exposed in the Sunderland Inlier where a constant southward dip of 30° to 40° is recorded. There is a covered interval of 2.5 km; then the Cretaceous beds reappear in the small

Calton Hill Inlier, and about 800 m farther south again in the Maldon Inlier where southerly dips of 10° to 20° are reported and a thickness of 425 m is calculated.

Beds exposed in the Sunderland Inlier begin with the John's Hall Conglomerate, which has a *Globotruncana* fauna indicating a probable Lower Campanian age. Succeeding formations, also Campanian, are the Sunderland and Newman Hall Shales, with a combined thickness of 1370 m; a *Barrettia*-bearing limestone, about 7.5 m thick; and the Shepherds Hall Formation, composed of conglomerate and tuff, about 850 m in thickness. Although the Shepherds Hall contains no fossils, it is tentatively included in the Campanian.

In the soil at the southern end of the inlier one finds a large number of fossils which are probably weathering from a limestone. *Titanosarcolites* and *Praebarrettia* are present, and it is assumed that the lowest Maestrichtian beds underlie the soils.

The *Titanosarcolites* fauna reappears in the Calton Hill and Maldon Inliers, where four limestone formations are separated by mudstone and siltstone. The shaly beds contain a molluscan fauna and a characteristic Maestrichtian foraminifer, *Kathina jamaicensis*.

5. *Marchmont Inlier*

This Maestrichtian exposure is comparable in many ways with the section found in the Maldon Inlier. There are three rudist-bearing limestones with intervening shale or siltstone sequences. The limestones are all dominated by *Titanosarcolites*, and the shaly beds contain *Kathina jamaicensis*.

6. *Hanover and Westmoreland Inliers*

This group of exposures is located at the northwestern end of the island. Part of the area is within the Negril Sheet (Geological Sheet 1—Provisional; 1:50,000 scale) published in 1972 by the Jamaica Geological Survey Department. Two sequences are recognized: the Hanover Shale, of Campanian and possibly older age; and a Maestrichtian group including red beds and limestone.

The Hanover Shale has been divided into Upper and Lower units by Chubb (1962). The Lower unit consists of calcareous mudstone with thin intercalated limestone and calcareous sandstone. At the top of the unit is the Clifton Limestone, which forms a prominent topographic escarpment on the west side of the Lucea East River valley. It is of reef origin and is composed largely of corals and rudists. Among the rudists are species of *Barrettia*, indicative of early Campanian age. Kauffman (1966) has collected two subspecies of *Inoceramus balticus* at levels above the Clifton Limestone. He believes that the fossils indicate a low horizon in the Campanian and suggests the Clifton Limestone must be very early Campanian, thus intimating that the Lower Hanover

Shale may be older. The Upper unit is more arenaceous and contains conglomerate and shaly lenses. I have collected *Barrettia gigas* and other Campanian fossils from limestones in this unit. W. F. Scott (reported on Negril Geological Sheet) estimated the total thickness of the Hanover Shale at about 3650 m, with half the thickness represented by shaly beds.

The Maestrichtian succession is about 1000 m thick and includes reddish brown tuffaceous sandstone with interbedded shale and conglomerate. Limestone containing *Titanosarcolites* and oysters are the basis for dating this section.

C. Historical Synthesis

Unquestionably, one of the most important areas of investigation for understanding the early geologic history of the Nicaragua Rise is that of absolute dating. In Jamaica we have built the entire Lower Cretaceous stratigraphy around the identification and assumed time span of a few rudist genera occurring in the Benbow Inlier and have tied in other sections by such tenuous threads as dip directions and degree of metamorphism. I am not suggesting that either the dating, the correlations, or the logic by which they were established is faulty, but I would feel far more confident of the geologic history if there were a few isotopic dates to substantiate it. Even though the pattern of paleomagnetic analyses by Steinhauser *et al.* (1972) is not conclusive, it does not support a Lower Cretaceous age interpretation for the Devils Racecourse volcanics.

There is better fossil control in the Upper Cretaceous strata. Clearly there was a volcanic island chain in the western Caribbean, even though the details of its geography are still somewhat vague. Roobol (1971) has pointed out that there are no extensive lava formations exposed in Jamaica, and he suggests that dacite and andesite lavas seldom travel more than 8 km from their source. Assuming that an occurrence of more than 5 different flows in an area of 2 km² is regarded as marking a volcanic center, he has plotted several Cretaceous volcanoes in the eastern half of the island. He concludes that the picture is one of small scattered volcanic centers.

Pyroclastic material could have traveled farther than lava, but its abundance in the Upper Cretaceous section and the fact that it often occurs as coarse-grained, angular fragments require a substantial, nearby source. The axis of the Nicaragua Rise runs south of Jamaica, close to the southern edge of the Pedro Bank. The Occidental–Signal Pedro Banks-1 exploratory well (Table I), located in the area, penetrated Tertiary limestone to 1920 m, and then a coarse conglomerate for 52 m, 90% of the fragments of which were granodiorite and the remainder dolomite; the final 5.8 m was granodiorite. The bottom formation has not been dated radiometrically but may well be

TABLE I

Exploratory Wells: Jamaica and Nicaragua Rise

Map No. (Fig. 1)	Well name	Company	N. latitude	W. longitude	Completion year	Total depth (m)
1	Tuara-1	Gulf	14°21.5'	83°17.5'	1944	1910
2	Punta Gorda-1	Gulf	14°24.0'	83°17.0'	1944	2053
3	Touche-1	Waterford	14°22.5'	82°47.0'	1957	4419
4	Lagunado Caratasca	Mecom	15°17.0'	83°47.5'	1957	2117
5	Negril Spots-1	Base Metals Min.	18°15.0'	78°16.9'	1955	1925
6	Santa Cruz-1	Jam. Stanolind	17°55.2'	77°40.4'	1957	2662
7	Cockpit-1	Pan-Jamaican	18°15.7'	77°38.5'	1957	2373
8	West Negril-1	Pan-Jamaican	18°15.0'	78°19.4'	1957	2818
9	Mosquitia-1	Pure	15°12.1'	83°49.8'	1963	4236
10	Martinez Reef-1	Union of Calif.	14°34.0'	82°32.2'	1968	3688
11	Zelaya-1	Chevron	14°05.0'	82°42.9'	1969	2741
12	Perlas-1	Shell	12°41.9'	82°53.5'	1969	3810
13	Perlas-2	Shell	12°38.0'	82°53.5'	1970	3794
14	Perlas-3	Shell	12°47.6'	82°53.5'	1970	3802
15	Coco Marina-1	Union of Calif.	15°00.0'	82°43.5'	1969	3047
16	Pedro Banks-1	Occidental–Signal	16°56.2'	78°48.0'	1970	1978
17	Miskito-1	Occidental–Signal	14°52.4'	81°41.2'	1970	2051
18	Atlantico-1	Shell	13°10.2'	82°47.7'	1970	3253
19	Prinzapolka-1	Chevron	13°25.9'	83°06.4'	1970	2252
20	Toro Cay-1	Chevron	14°20.2'	83°06.9'	1970	2292
21	Rama-1	Shell	13°08.3'	83°12.7'	1970	2209
22	Escondido-1	Chevron	13°20.5'	83°17.2'	1970	2448
23	Tinkham-1	Mobil–Esso	13°38.3'	82°25.1'	1971	1993
24	Tyra-1	Mobil–Esso	12°54.5'	82°38.3'	1971	2646
25	Nica-1	Mobil–Esso	12°35.9'	82°41.9'	1971	3379
26	Portland Ridge-1	Occidental–Signal	17°44'	77°11'	1971	2262
27	Berta-1	Columbia	16°13'	82°04'	1971	2265

related to the episode of granodiorite intrusion associated with the Laramide orogeny. Indeed, the magnetic and seismic work in the Pedro Bank area support such an idea (Arden, 1969), but there also seems to have been a volcanic ridge in this area. The suggestion is that the main volcanic chain was along the crest of the Nicaragua Rise as it existed in Cretaceous time, and Jamaica was a subsidiary chain located 100 to 200 km to the north.

The metamorphics and presumed eugeosynclinal beds of the southwestern Blue Mountains area appear to represent two phases of metamorphism, as suggested by Robinson et al. (1971) and Kemp (1971), with the Westphalia Schist being the older. The Mt. Hibernia trend contains serpentinite and has been variously metamorphosed, but it is a relatively narrow belt, not more than 2 km wide. As more data are reported, it is becoming clear that the Blue Mountains Cretaceous section does not differ much from that of the Clarendon Block, and the amount of metamorphics and fault-controlled intrusives is not large in comparison with the normal sequence of Upper Cretaceous rocks.

The Upper Cretaceous cycle of deposition was brought to a close by the first phase of the Laramide orogeny. The western end of the Nicaragua Rise was largely emergent long before the end of the period. Jamaica became emergent by the end of the Maestrichtian, a condition that persisted generally throughout Paleocene time. In Jamaica the uplift was accompanied by widespread intrusions, principally of granodiorite and related rocks. Dating by Rb–Sr ratios from biotite and U–Pb from sphene, collected at Zion Hill Bridge in Above Rocks area, gave an age of 65 ± 5 m.y. (Chubb and Burke, 1963; Harland et al., 1964). The first phase of metamorphism may have been coincident with the intrusions.

One of the most outstanding effects of the Laramide movements was the development of the Wagwater Trough in Jamaica, probably as a result of right lateral movement. I believe that at approximately the same time there was even greater right-lateral shift along a fault passing north of Jamaica and south of Hispaniola, and that it was this movement that caused the rift along the south flank of the Nicaragua Rise, separating the rise from the Beata Ridge.

VI. PALEOCENE–LOWER EOCENE EVENTS

A. Nicaragua Rise Sections

Paleocene sediments have been reported from wells drilled on the Nicaragua Rise only in the Touche-1 well (Table I). It encountered calcareous shale at 4145 m, which, on the basis of foraminifers, was considered to be of Midway age. The well bottomed at 4571 m in the same formation. This is the deepest well to date and seems to have been drilled in a graben.

To the west 56 km, and inland from the Nicaraguan coast about 16 km, the Tuara-1 well reached "magnetic basement," presumably a metamorphic rock, at 1900 m after penetrating 710 m of Middle Eocene marine beds overlying about 180 m of undated coarse-grained clastics. In the other direction along the rise, the Miskito-1 well, about 135 km ENE of the Touche-1, encountered a hydrothermally altered porphyritic granodiorite at 2021 m beneath a Middle Eocene shale and shaly limestone sequence.

To the north, on the Honduras coast, the Mosquitia-1 found Middle Eocene shale above possibly Cenomanian limestone. At the Berta-1 location, near the crest of the rise, hard, dark-green to gray, blocky andesite was drilled from 2073 m to total depth at 2265 m, again with no Paleocene or Lower Eocene reported. Overlying the andesite was 396 m of brown and red sandstone with streaks of hard, black claystone, believed to be Middle Eocene. The Pedro Banks-1 well encountered granodiorite at 1973 m, overlain by 53 m of conglomerate, succeeded by Middle Eocene limestone.

There are no isotopic dates available from any of the igneous rocks within which many of the wells bottomed. Stratigraphic position and extrapolation from similar sections on land suggest shallow Upper Cretaceous or Laramide intrusives that were unroofed soon after emplacement. The condition of these rocks indicates exposure, weathering, and some alteration by circulating waters. The occurrence of Paleocene and Lower Eocene only in what appears to be a fault trough is interesting because there is a similar pattern in Jamaica.

B. Jamaican Lower Tertiary Rocks

1. Clastic and Volcanic Units

In eastern Jamaica there is a succession of clastics with minor limestone inclusions that attain a thickness of 4500 m in the Wagwater Trough. It has been customary to call the coarse sandstone and conglomerate deposits Wagwater Beds, or some variation of this term (Fig. 11). The finer-grained, thinly bedded clastics are often referred to as Richmond Formation, or Group, or Shale, or simply as Richmond Beds. The limestone units have been given local names wherever they are prominent enough to be noticeable.

Wagwater Beds are recognized only in the Wagwater Trough, a graben bounded on the east by the Yallahs–Blue Mountain fault zones and on the west by the Wagwater fault zone. A description of the sequence in the Buff Bay Quadrangle is included in the Jamaica Geological Survey Department Annual Report for 1966 (p. 9). There is a 3350-m-thick basal unit composed of a poorly sorted boulder conglomerate lacking granodiorite clasts. This is distinguished from an upper 600-m-thick conglomerate by an abundance of

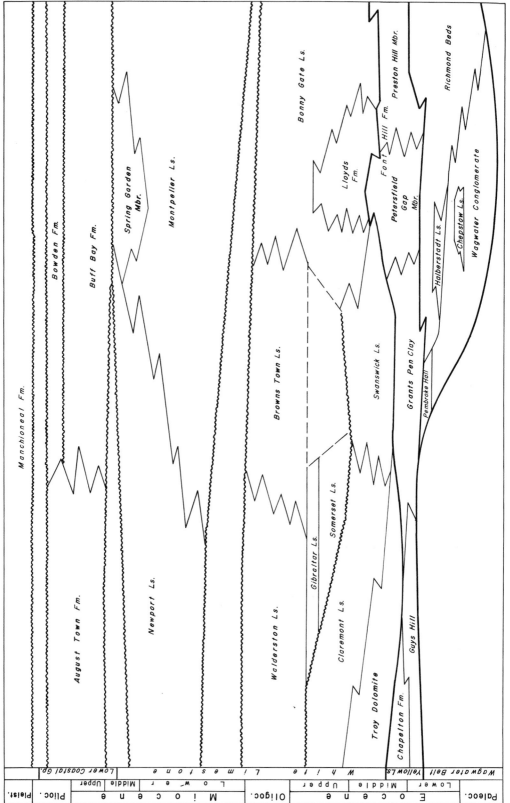

Fig. 11. Tertiary formations of southeastern Jamaica. Based on an unpublished diagram by A. Holliger and D. D. Arden, Jr.

granodiorite clasts in the upper unit. The implication is that the upper unit represents sedimentation after the early Laramide intrusions were exposed. Both units are characteristically colored dark-red to purple by hematite stain and cement. They are separated by 600 m of green to gray arkosic grit and sandstone. Near the top of the upper conglomerate are some basalt flows and green tuffs. Fossils within the Wagwater lithology are rare and are usually fragments of thick-shelled bivalves or poorly preserved gastropods (Matley, 1951).

Greiner (1965) reviewed the data on file with the Department of Mines, Kingston, relating to four exploratory wells drilled in Jamaica between 1955 and 1957 in a search for petroleum. I have suggested (unpublished report, 1969) that the "Cretaceous Santa Cruz Volcanics," reported from the Jamaican Stanolind Santa Cruz-1 and Pan Jamaican Cockpit-1 wells, are actually Wagwater-type beds, and that the cuttings of "andesite" and "lavas" described in the well logs are from boulders in a conglomerate. Some confirmation comes from Roobol (1971; and personal conversation, 1972) who expressed doubt that andesite and dacite lavas would be as thin as reported from the wells (less than 10 m). He has recovered the original cuttings and cores and says that they may represent boulders of Summerfield volcanics.

In the Barbecue area, Black and Bailey (1968) described two Wagwater units separated by Newcastle Volcanics. Roobol (1971) has reviewed the recent work on these volcanics. He reports that they consist of five andesite and dacite flows up to 600 m thick, broken by faulting. They outcrop over an area of about 240 km² and are interbedded with Wagwater and Richmond lithology. Copper mineralization in porphyritic hornblende–andesite flows has been examined in the Castleton area (Black et al., 1972), where the flows reach 450 m thick and are contained within the Wagwater Beds.

Other volcanics occur along the Wagwater Trough, including basalt pillow lava. Southwest of the Newcastle exposures is a mass of basaltic pillow lava, 200 to 300 m thick, known as the Halberstadt Lavas (Matley, 1951; Roobol, 1971). Vesicles in the lower layers suggest water depths of 200 to 300 m (Roobol, 1971). Parts of the flow are intimately associated with gypsum beds, so the lava appears to have flowed from a volcano located near coastal embayments.

At the northern end of the Wagwater belt are the Nutfield Volcanics, which consist of a thin basalt pillow lava overlain by a dacite flow, with the sequence contained in Richmond lithology, which is laterally equivalent to the Upper Wagwater.

The Richmond Beds overlie the Lower Wagwater Beds but intertongue with the Upper Wagwater sequence. Richmond rocks are thin-bedded brown mudstone and sandstone, plus conglomeratic layers and minor limestone lenses. Graded bedding and slump structures are common and are especially well displayed in the road cuts north from Petersfield Gap. A thickness of at least 660 m has been measured at the northeastern end of the Wagwater belt between Port Maria and Annotto Bay.

Anhydrite and gypsum occur at the southern end of the Wagwater belt at a stratigraphic position between marine beds represented by Richmond lithology and overlying nonmarine Wagwater lithology. Holliday (1971) suggests the evaporites originated in an arid supratidal environment. The gypsum is at the same horizon as the Halberstadt Lavas and a lens of reefal limestone known as Halberstadt Limestone.

Wright and Dickinson (1972) have studied the provenance of sandstone in the Richmond Beds and conclude that the clasts were derived from an andesite–dacite source with secondary contribution from granitic areas. These are the sources one would anticipate in late Lower Eocene time. Andesite or dacite was widespread, and the granodiorite intrusives were just being exposed to erosion.

Near the north coast, between Balcarres and Hope Bay District, a red and purple conglomerate is associated with dark-gray shaly limestone and brown shale. The limestone is known as the Chepstow Limestone; the clastics are considered to be Wagwater-Richmond. The Chepstow contains algae similar to some in the Halberstadt Limestone plus the foraminifer *Ranikothalia catenula*, which is considered to be a Lower Eocene marker (Robinson, 1962). "*Globorotalia*" *palmerae*, also a Lower Eocene foraminifer, occurs in the Langley Mudstone Member near the top of the Richmond Beds (Robinson, 1968b).

C. Geologic History

The Laramide movements that began near the end of Maestrichtian time are clearly recorded throughout the Nicaragua Rise. Generally, the result was to elevate the region, but rifts accompanying lateral fault displacements created depositional troughs. The Wagwater Trough is most accessible to examination; therefore, much of the interpretation of the history of the entire region is based on observations from that area. There is evidence that other rifts also developed at this time, including one east of the present coastline of Nicaragua in the Miskito Cays area. A reinterpretation of volcanics in exploratory wells in Jamaica indicates that other troughs are present under later cover. In summary, the tectonic pattern is one of lateral stresses, crustal fragmentation, and general elevation of the rise, but with deep, narrow grabens occurring in some of the main fracture zones.

Shallow intrusions of granodiorite were widespread. By the beginning of Eocene time some had probably been exposed to weathering and erosion. Most wells that reached "basement" penetrated Upper or Middle Eocene carbonates and shales and went through a basal clastic sequence or went directly into igneous rocks, commonly granodiorite. Sometimes this igneous material showed signs of exposure and weathering.

The only reported fossiliferous Paleocene beds are hard, brown, calcareous shales from the deep Touche-1 well, off Nicaragua in the Miskito Cays Trough. Although there is no fossil confirmation, it is reasonable to assume that the Lower Wagwater Beds, which lack granodiorite clasts, were deposited before the Laramide intrusives were exposed. The Upper Wagwater is the age equivalent of the Richmond Beds, in which a full Lower Eocene section has been recognized.

Rubbly limestones and thick-shelled marine mollusks attest the occasional presence of the sea in the Wagwater Trough. The rapid changes in lithology, cross-bedding, coarse conglomerates, and evaporites all indicate near-shore conditions and rapid deposition. Lava flows occur throughout the trough. Pillow structures and vesicles show submarine flows into water as deep as 200 to 300 m, but association of some lavas with gypsum and reefal limestone demonstrates that the vents were probably on land.

Examples of exotic blocks of Wagwater conglomerate within Richmond Beds may be seen north of Yallahs Hill. The blocks are up to 100 m long and 25 m thick and probably represent the collapse of steep sea cliffs. Cambray and Jung (1970) have shown that the Richmond Beds contain sole marks indicating turbidity currents flowing down the sides of the trough and then along the axis. Graded bedding and slump structures support this picture.

Subsequent phases of the Laramide orogeny have been recorded throughout the Greater Antilles at different intervals during the Eocene (Weyl, 1966; Khudoley and Meyerhoff, 1971). Chubb (1962) envisioned a second phase of Laramide activity in Jamaica and cited evidence which he interpreted as indicating thrusting of the Wagwater belt sediments towards the west. Robinson *et al.* (1971) do not believe that there were either compressive stresses or uplift associated with tectonic activity at this time, and I am inclined to agree with them. Chubb's observations can be reconciled with a different pattern of movements from that which he postulated. There is no widespread unconformity in Jamaica at this time. Some lateral fault movements accompanied by rifting and volcanism occurred during Lower Eocene time, and in early Middle Eocene a general subsidence began so that deposition of the Yellow Limestone Group spread across the island.

No Wagwater or Richmond beds are exposed west of the Wagwater belt, so there is no direct evidence for what happened during this time interval in the Clarendon Block. Perhaps the northwest–southeast-trending fault system which can be seen today throughout southern Jamaica had its inception then, and rejuvenation along these lines resulted in the fault-controlled topography of the Santa Cruz, Don Figuerero, and other mountainous blocks. It will be pointed out in the next section that facies of the Yellow and White Limestones were directly related to a horst and graben type of depositional surface.

VII. MIDDLE EOCENE–MIDDLE MIOCENE EVENTS

A. Nicaragua Rise Section

Well control is too sparse to make satisfactory paleogeographic maps for specific time horizons over the Nicaragua Rise. Tertiary deposition in Jamaica was influenced by the presence of fault blocks and troughs, and it seems likely that a similar situation existed throughout the rise.

At the western end of the rise near Nicaragua, the Eocene section is dominantly normal limestones, sometimes reefal and porous, occasionally containing streaks of gray to reddish brown, calcareous shale. Foraminifers include *Fabularia*, *Dictyoconus*, *Coskinolina*, and *Robulus*. Lignitic traces are also known, so that the assemblage suggests an environment much like the present-day banks of the area.

Middle and Upper Eocene are probably present throughout the area, but Oligocene beds are missing or very thin. The unconformity is even more pronounced off Honduras, where it may extend from Middle Eocene to Miocene. Lower Tertiary evaporites in the form of anhydrite exist in the subsurface near the present coastline of Honduras and are probably somewhat higher stratigraphically than the Lower Eocene gypsum of Jamaica, although the depositional environment may have been similar. More widespread evaporites, such as were postulated by Pinet (1972), have not been verified by drilling, and no evaporites of significance have been found offshore.

B. Jamaican Stratigraphy

1. *Yellow Limestone Group*

The carbonate deposits which began with the Eocene submergence were separated by Sawkins (1869) into an older Yellow Limestone and a younger White Limestone. These terms have persisted and are now used in the sense of stratigraphic groups. The first serious attempt to establish subdivisions based on lithological and paleontological considerations was by Hose and Versey (1956). Further work has been done by Robinson (1968a; 1969a), especially in eastern Jamaica.

Where the topmost units of the Wagwater–Richmond beds are present, there is no unconformity at the base of the Yellow Limestone formations. Figure 11 shows diagrammatically the relationship of some of the Yellow and

White Limestone subdivisions. Essentially, the units reflect facies differences rather than time divisions.

In the Blue Mountain Block of eastern Jamaica, the 600-m-thick Font Hill Formation is gradational above Richmond beds or overlaps onto older rocks. It ranges in age from uppermost Lower Eocene through lower Middle Eocene. Three members include: Preston Hill Member, consisting of planktonic foraminiferal marl; Palmetto Grove Member, evenly bedded shaly marl alternating with dense structureless limestone; and the Petersfield Gap Member, with interbedded marl, bioclastic limestone, and olistostrome masses.

In central Jamaica, the Yellow Limestone Group begins with the Chapelton Formation, which contains basal grits and conglomerate overlain by cross-bedded arkose and siltstone with plant fossils. This is the Guys Hill Member. The coarse basal beds have received various local names and may be equivalent to the Pembroke Hall Conglomerate, which grades eastward into typical Richmond Beds. The Ham Walk Limestone Member is made up of brown-weathering, silty and sandy, concretionary limestone. The Grants Pen Clay Member is a series of reddish to brown clay, sandy clay, and conglomerate.

In Trelawny, on the north side of the Central Inlier, there is a lower limestone (Cascade Limestone) containing an echinoid and mollusk fauna, followed by clays and tuffs (Versey, 1962) that may reach 300 m in thickness. This is overlain by a second limestone containing *Lepidocyclina antillea*, *Yaberinella jamaicensis*, and *Fabiana cubensis*.

In the western part of the island, Chapelton Formation equivalent has been identified. The lower contact with the "Red Beds" of the Hanover area is not well defined because lower Chapelton beds are similar in lithology. An upper unit is composed of orange-brown marl and calcarenite (Negril Sheet, Geological Map Sheet No. 1).

2. *White Limestone Group*

This is the most widespread sequence of formations in Jamaica, covering about two-thirds of the island. In addition to the work listed in the preceeding section, Wright (1966) and Robinson (1967) have made special studies of White Limestone areas.

The group ranges in age from Middle Eocene to Middle Miocene. It is a remarkably pure limestone, with a carbonate content of about 98% (Robinson, 1967). There is no general unconformity between White and Yellow Limestone formations nor a clear time boundary. A dozen facies have been described and named (Fig. 11). During the Laramide movements, Jamaica west of the Wagwater Trough fractured into three regions: a rigid mass, the Clarendon Block, made up the eastern portion; a smaller positive area, the Hanover Block, constituted the western segment; and a trough, the Montpelier–Newmarket Belt, separated the blocks. In Tertiary time the Clarendon Block formed a shallow submarine bank. East of it was the newly submerged Wagwater Belt and the relatively stable Blue Mountain Block. On the west, a scarp separated the Clarendon Block from the Montpelier–Newmarket Trough, in which was deposited deeper-water chalk which contains diagenetic flint nodules. This facies occurs in other areas where deeper water surrounded the stable blocks.

Versey (1962) has pointed out that a facies pattern can be recognized, which includes a Middle Eocene dolomite or dolomitic limestone (Troy Dolomite), a Middle to Upper Eocene bioclastic facies (Swanswick Limestone), a chalk with flint (Bonny Gate), and several less widespread foraminiferal facies, the various units being characterized by peneroplids, miliolids, or orbitoids. The conclusion is that the dolomitic and peneroplid facies are characteristic of a shallow-water bank environment and that the miliolids indicate a transitional environment between the shallow-water peneroplid and deeper-water orbitoid facies. Depths here are relative evaluations, and it is unlikely that even the basin environment was much deeper than 200 m.

There are two unconformities, one in Upper Eocene time and the other at the top of Middle Oligocene. Submarine slumps, calcarenites, and intraformational conglomerates attest the emergences, as do gaps in the faunal record. However, no terrigenous deposits are known, and no detritus other than intraformational carbonate has been found, indicating that erosion did not reach to older formations.

The Lower and Middle Oligocene interval is marked by the Walderston Limestone, a comparatively soft limestone, composed largely of miliolids; and the Brown's Town Limestone, loose and nodular, with a large amount of interstitial powdery lime on fresh exposure but developing a durable "case hardening" upon weathering.

The Oligo-Miocene section is characterized by the Newport Limestone, a moderately well-bedded and compact limestone which reaches a thickness of at least 800 m. The Montpelier Limestone, composed of chalky micrite, chalk, and some basal cherty layers, is the deeper-water equivalent of the Newport. The Spring Garden Member is within the upper portion of the Montpelier and consists of massive white chalk beds with abundant planktonic foraminifera and sponge spicules.

C. Depositional History

Submergence seems to have been the rule in Middle Eocene time, and emergence was probably at a minimum throughout the Nicaragua Rise until late Oligocene or Miocene. By the start of White Limestone deposition in Jamaica, submergence was complete, evidence for which is found in the widespread presence of carbonates. The White Limestone beds have no more than 2% terrigenous matter, and clays or other clastics are minor constituents everywhere. Occasional streaks of gray or brown shale found in wells on the rise contain traces of lignite. This, in association with reef limestones, suggests an environment of carbonate banks and cays, much like the present conditions in the region. The shales and plant material probably represent thin paleosols and former vegetation on low-lying islands.

An unconformity occurs in Upper Eocene beds of Jamaica, but the uplift was not of sufficient intensity or duration for erosion to reach pre-White Limestone beds. If the uplift was effective beyond Jamaica, it has not yet been recognized. A second White Limestone unconformity is found above Middle Oligocene beds. It was followed by further submergence until about the start of Middle Miocene time.

It was in Middle Miocene that the Alpine orogeny began to affect the Antillean region. Robinson *et al.* (1971) conclude that there was gradual southward tilting in Jamaica, indicated by the southward thickening of the Newport Limestone. It is my belief that eventually all of Jamaica was uplifted and there was rejuvenation of old Laramide fault trends. The Blue Mountains were elevated and various fault blocks in the Clarendon Block were tilted, with some being raised as much as 1000 m. The well sections everywhere on the Nicaragua Rise reveal the unconformity. At most, there is only a remnant of Oligocene, and frequently none is left. Late Miocene beds may rest on Upper or even late Middle Eocene limestone.

I believe that it was during the latter half of the Miocene that the last major lateral movement along the Cayman Trough rift began. Before movement on the rift had stopped, the Nicaragua Rise, including Jamaica, was tilted southward. The crest of the rise shifted northward fairly rapidly and has probably occupied its present location since Upper Miocene.

Faulting parallel with the rift shapes the northern coastal region and offshore topography of Jamaica as well as the entire northern flank of the Nicaragua Rise. The faulting in Jamaica resulted in a minor volcanic episode along the north coast near Orange Bay, in Portland, where the Low Layton Volcanics form an east–west ridge about 2.5 km long and 190 m high, probably the result of a fissure flow (Williams, 1962).

VIII. LATE TERTIARY AND QUATERNARY DEPOSITS

A. Nicaragua Rise Section

Upper Miocene beds commonly lie upon Upper Eocene or, rarely, upon an eroded Oligocene section. In the Honduras sector, i.e., in the northwestern portion of the rise, the Miocene beds are dominantly shale and red beds.

Marine stringers are marly and rich in selenite. The red beds are replaced by carbonates towards the east and southeast.

Along the Nicaraguan coast there is a dominantly continental and near-shore section that quickly grades eastward into a shallow-water marine clastic facies. This, in turn, rapidly increases in limestone content, so that between the Touche-1 and Miskito-1 wells, the Miocene changes from dominantly fine-grained sandstone to biogenetic carbonate.

The Pliocene and Quaternary beds tend to be highly porous carbonates of mixed origin. Reefal zones are cavernous, and some caverns are breccia-filled. Drilling through this zone may yield only occasional samples because of lost circulation. Layers of carbonaceous lime mud are typical of deposition in lagoons floored by sediments rich in organic matter and constantly worked by burrowing animals. Turtle grass (*Thalassia*) and eel grass (*Zostera*) were the probable sources for most of the carbonaceous debris.

B. Jamaican Deposits

The late Tertiary rocks of Jamaica have been collectively mapped and described as the Coastal Formations. They never spread completely across the island, but overlap around the edges or occur as alluvial and lacustrine deposits. Thus, the individual formations are of limited extent and have received several local names. The marine beds are divisible into two age groups, which Robinson (1969*b*) has called Lower Coastal Group and Upper Coastal Group.

The Lower Coastal Group is late Middle Miocene and Pliocene in age. The August Town Formation is present on the south coast in the August Town area, east of Kingston. About 260 m of beds are exposed. A basal marl rests unconformably on White Limestone and is succeeded by yellowish lime-stone, marl, coral reefs, and rubbly coraline limestone.

Robinson (1969*b*) has redescribed the section near Buff Bay, on the north coast, which lies unconformably upon the Spring Garden Member of the Montpelier Formation and is probably generally comparable with the August Town Formation. The Buff Bay Formation consists of bluish white, semi-indurated marl, with occasional pea-sized pebbles scattered through it. At the top is a section correlated with the Bowden Beds of the south coast and now considered by Robinson (1969*b*) to form part of the Bowden Formation. The type section of the Bowden Formation, near Port Morant, starts at the base of the classical Shell Bed from which Woodring (1925, 1928) collected over 600 species of mollusks. The section consists of coarse, sandy and con-glomeratic layers occurring within silty, planktonic foraminiferal marl. The coarse beds containing the megafossils are thought to be allochthonous masses that slid into deeper water. Included in the slumped beds are some strata of Miocene age, so a mixed Miocene–Pliocene fossil hash has resulted.

Pleistocene beds along the eastern coast of Jamaica have been named Manchioneal Formation. It is chiefly composed of rubbly limestone and reef debris. Pleistocene coastal terrace deposits have also been identified. Cant (1971) described four main terrace levels from the north coast between Ora-cabessa and Port Maria.

Alluvial plains and interior valley deposits are important aquifers. The Liguanea Formation, upon which the city of Kingston is located, is a wedge of sand, gravel, and clay more than 200 m thick.

C. Late Tertiary and Quaternary History

Differential movement between fault blocks was probably a continuous feature of the Nicaragua Rise following the Miocene emergence. There is not much detailed information on the environment of this interval available from the oil-exploration effort because the thickness of the interval may be well under 100 m and drilling returns, if any, often are only hard nodules that have been badly abraded. Lost circulation and the problem of setting the surface casing may result in no returns at all for 200 or 300 m. Coring in 610 m of water between Walton Bank and Jamaica revealed the Pliocene–Pleistocene boundary at 23.5 m (C. Emiliani, personal communication, 1968). The sediment was primarily pelagic ooze with graded and nongraded turbidites containing echinoderm and shell fragments as well as remains of plankton. The sample is from a basinal area between the Clarendon Block and carbonate banks, and the seismic sections reveal that the fine-grained sediments are thicker here than over the shallow banks.

In Jamaica, instability is reflected in the discontinuous pattern of fringing reefs and associated sediments. A general unconformity is recognized during Late Pliocene time, which probably represents the last island-wide uplift, when the Blue Mountains and other mountainous blocks reached their maximum elevation.

Earthquakes are frequently recorded in Jamaica. The city of Kingston was badly shaken in 1907, and the destruction of Port Royal in 1692 is a well-known story. Recent fault scarps can be recognized throughout the island, and hot mineral springs emerge along the trace of several faults.

IX. SUMMARY

Geological and geophysical data from Jamaica, Central America, and the Nicaragua Rise are compatible with an interpretation that the Nicaragua Rise is a continuation of the Greater Antilles arc and originated in Jurassic time as a belt of thickened crust along the boundary of two oceanic crustal plates. A study of refraction-seismic velocities reveals that the general configuration of the Nicaragua Rise is a pod-shaped feature about 500 km wide, 1000 km long, and a maximum of about 22 km thick along its axis. The line of maximum thickness is today 150 to 200 km south of the topographic crest of the rise. There is evidence that northward migration of the crest was the result of southward tilt in late Tertiary time.

The Beata Ridge once formed the southern flank of the Nicaragua Rise and was separated from it by a rift during a Laramide orogenic episode at the end of the Cretaceous. Crustal zones in the two areas are alike, and evidence for the rift are the steep scarps along the southeastern side of the Nicaragua Rise and the northwestern side of the Beata Ridge and the thinner crust between them. Other fault movements occurring about the same time caused grabens such as the Wagwater Trough of Jamaica. The north flank of the rise is also faulted and is bounded by the Cayman Trough, which originated from left-lateral movement in late Middle or Upper Miocene time.

The Nicaragua Rise is believed to have begun as a line of crustal thickening. Volcanic clasts are found in the Todos Santos Formation (Upper Jurassic–Lower Cretaceous) in Honduras, and lava flows occur immediately beneath some Todos Santos beds, thus indicating that volcanic activity began in the western part of the rise at least by the end of Jurassic time.

Jamaica may have been a subsidiary ridge located on the northern side of the Nicaragua Rise. The oldest rocks in Jamaica are dated on the basis of fossil rudists as Lower Cretaceous. These rocks are primarily shallow-water marine limestone and interbedded volcanics. Upper Cretaceous strata are exposed in numerous inliers throughout the island, the largest of which is the Blue Mountain Inlier. The Blue Mountains are bounded on the southwest by fault zones. Along part of this boundary is a belt of metamorphosed rocks up to 3 km wide and intermittently observable for a length of 25 km. The Westphalia Schist consists of quartzo-feldspathic and greenschist rocks transitional to amphibolite. Serpentinite and granodiorite occur between these and some less strongly metamorphosed rocks in which fossils of *Monopleura* have been reported. K–Ar age determinations from the Westphalia Schist give an Upper Cretaceous age of 76.5 $\pm$ 2.5 m.y., which is compatible with the age of *Monopleura*. Most of the Blue Mountain Inlier consists of dark shale, red beds, volcanics, and siltstone. Fossils from these rocks indicate Campanian and Maestrichtian age.

Laramide movements affected the Nicaragua Rise during at least two episodes between latest Cretaceous and late Lower Eocene time. Intrusion of granodiorite was accompanied (or closely followed) by general uplift and the formation of grabens bounded by faults having both normal and lateral slip components. Rapid erosion dumped clastics into the grabens and soon exposed the granodiorite intrusions.

Subsidence was widespread by Middle Eocene time, and clastic deposition gave way to carbonate accumulation as the land areas became submerged beneath a shallow sea. Uplift is again recorded about Middle Oligocene time, which resulted in an important unconformity in the western part of the Nicaragua Rise where Upper Miocene may rest on Middle Eocene beds. In Jamaica the emergence was not of sufficient duration for erosion to reach below the Eocene carbonates.

The Alpine orogeny, beginning in Middle Miocene time, affected the western Caribbean in several ways. The last major movement of the Cayman Trough rift occurred then, which also caused southward tilting of the Nicaragua Rise. Thus the northern side of the rise was uplifted, and the topographic crest migrated northward. Jamaica was exposed so that late Miocene and younger sediments occur only in coastal areas of small extent.

X. FUTURE INVESTIGATIONS

Petroleum exploration has stimulated interest in the western Caribbean, and much basic information has been collected as a result of geophysical and drilling programs. No commercial discovery has yet been announced, but exploratory efforts are continuing, and there is still a certain amount of competition for acreage as concession agreements expire and contractual relinquishments come due. When full information can be disclosed, details of Cretaceous and Tertiary stratigraphy will be much better defined.

Seismic investigations supported by gravity and magnetics are valuable for structural interpretation. This is especially true where gravity and magnetics can be adapted for analysis by computerized modeling programs. One of the most significant unsolved problems is the relation between the Paleozoic rocks of Central America and the structure of the Nicaragua Rise. The answer must come from geophysical studies. There is a paucity of seismic-refraction data in the western Caribbean. A line of refraction surveys across the western end of the Nicaragua Rise would be valuable. Marine gravity and magnetics in the same region could be designed to extend work already completed in the Cayman Trough and Gulf of Honduras areas.

Jamaica is more accessible to geologic observation than any other part of the rise. The Geological Survey Department there has begun publication of geologic maps at a scale of 1:50,000. We look forward to the completion of this project. Radiometric dating has helped in understanding the metamorphics of the southwestern Blue Mountains, but additional dates are needed from there and elsewhere. If the volcanic material in the Devils Racecourse Formation could be accurately dated, doubt regarding its Lower Cretaceous age could be dispelled. Other useful ages could be obtained from the igneous rocks encountered in exploratory wells. More careful examination of Jamaican volcanic rocks should be made to locate the volcanic centers and determine when they were active.

REFERENCES

Acker, R. C., and Willis, C. L., 1971, Section *C*, Geology, of Prefeasibility report: the Blue Mountain Water Supply Project: Harza Engineering Co. and Hue Lyew Chin (for the Government of Jamaica and the Water Commission).

Arden, D. D., Jr., 1969, Geologic history of the Nicaraguan Rise: *Trans. Gulf Coast Assoc. Geol. Soc.*, v. 19, p. 295–309.

Banks, N. G., and Richards, M. L., 1969, Structure and bathymetry of western end of Bartlett Trough, Caribbean Sea, in: Tectonic relations of northern Central America and the western Caribbean—the Bonacca Expedition (A. R. McBirney, ed.): Am. Assoc. Petr. Geol. Mem. 11, p. 221–228.

Black, C. D. G., and Bailey, B. V., 1968, Investigation of the Barbecue Copper Prospect: Jam. Geol. Surv. Econ. Geol. Rep. 2.

Black, C. D. G., Green, G. W., and Nawrocki, P. E., 1972, Preliminary investigation of the Castleton copper prospect: Jam. Geol. Surv. Econ. Geol. Rep. 4.

Bowin, C. O., 1968, Geophysical study of the Cayman Trough: *J. Geophys. Res.*, v. 73, p. 5159–5173.

Burke, K., and Robinson, E., 1965, Sedimentary structures in the Wagwater belt, eastern Jamaica: *J. Geol. Soc. Jam.* (*Geonotes*), v. 7, p. 1–10.

Burke, K., Coates, A. G., and Robinson, E., 1968, Geology of the Benbow Inlier and surrounding areas, *Trans. 4th Caribbean Geol. Conf., Trinidad*, 1965, p. 299–307.

Cambray, F. W., and Jung, F., 1970, Provenance of the Richmond Formation from sole marks: *J. Geol. Soc. Jam.*, v. 11, p. 13–18.

Cann, J. R., 1968, Geological processes at mid-ocean ridge crests: *Geophys. J.*, v. 5, p. 331–341.

Cant, R. V., 1971, Pleistocene terraces near Oracabessa, in: Field guide to aspects of the geology of Jamaica (E. Robinson, ed.), *International Field Institute Guidebook to the Caribbean Island-Arc System, 1970*: Am. Geol. Inst., Washington, p. 40–43.

Carpenter, R. H., 1954, Geology and ore deposits of the Rosario Mining District and the San Juancito Mountains, Honduras, Central America: *Bull. Geol. Soc. Am.*: v. 65, p. 23–38.

Chubb, L. J., 1958, The Cretaceous inlier of St. Ann's Great River: *J. Geol. Soc. Jam.* (*Geonotes*), v. 1, p. 148–152.

Chubb, L. J., 1960, The Antillean Cretaceous geosyncline, *Trans. 2nd Caribbean Geol. Conf., Mayaguez, P. R.*, 1959, p. 17–26.

Chubb, L. J., 1961, Blue Mountain shale: *J. Geol. Soc. Jam.* (*Geonotes*), v. 4, p. 1–7.

Chubb, L. J., 1962, Cretaceous formations, in: Synopsis of the geology of Jamaica (V. A. Zans et al., eds.): Jam. Geol. Surv. Bull. 4, p. 6–20.

Chubb, L. J., 1967, New rudist species from the Cretaceous rocks of Jamaica: *J. Geol. Soc. Jam.*, v. 9, p. 24–31.

Chubb, L. J., 1971, Rudists of Jamaica: *Palaeontogr. Am.*, v. 7, p. 161–257.

Chubb, L. J., and Burke, K., 1963, Age of the Jamaican granodiorite: *Geol. Mag.*, v. 100, p. 524–532.

Coates, A. G., 1965, A new section in the Maestrichtian Guinea Corn Formation near Crawle River, Clarendon: *J. Geol. Soc. Jam.* (*Geonotes*), v. 7, p. 28–33.

Coates, A. G., 1968, The geology of the Cretaceous Central Inlier around Arthurs Seat, Trans. *4th Caribbean Geol. Conf., Trinidad*, 1965, p. 309–315.

Dengo, G., 1969, Problems of tectonic relations between Central America and the Caribbean: *Trans. Gulf Coast Assoc. Geol. Soc.*, v. 19, p. 311–320.

Donnelly, T. W., 1964, Evolution of eastern Greater Antillean island arc: *Bull. Am. Assoc. Petr. Geol.*, v. 48, p. 680–696.

Dupré, W. R., 1970, Geology of the Zambrano Quadrangle, Honduras: M. A. Thesis, Univ. Texas, Austin, 128 p.

Eardley, A. J., 1954, Tectonic relations of North and South America: *Bull. Am. Assoc. Petr. Geol.*, v. 38, p. 707–773.

Edgar, N. T., 1968, Seismic refraction and reflection in the Caribbean Sea: Ph.D. Dissertation, Columbia Univ., New York, 159 p.

Edgar, N. T., Ewing, J. I., and Hennion, J., 1971, Seismic refraction and reflection in the Caribbean Sea: *Bull. Am. Assoc. Petr. Geol.*, v. 55, p. 833–870.

Erickson, A. J., Helsley, C. E., and Simmons, G., 1972, Heat flow and continuous seismic profiles in the Cayman Trough and Yucatan Basin: *Bull. Geol. Soc. Am.*, v. 83, p. 1241–1260.

Everett, J. R., 1970, Geology of the Comayagua Quadrangle, Honduras, Central America: Ph.D. Dissertation, Univ. Texas, Austin, 152 p.

Ewing, J. I., Antoine, J. W., and Ewing, W. M., 1960, Geophysical measurements in the western Caribbean and in the Gulf of Mexico: *J. Geophys. Res.*, v. 65, p. 4087–4126.

Ewing, J. I., Talwani, M., Ewing, M., and Edgar, N. T., 1967, Sediments of the Caribbean, in: *Studies in Tropical Oceanography*: Miami (Florida) Univ. 5, p. 88–102.

Fakundiny, R. H., 1970, Geology of the El Rosario Quadrangle, Honduras, Central America: Ph.D. Dissertation, Univ. Texas, Austin, 234 p.

Fox, P. J., Heezen, B. C., Ruddiman, W. F., and Ryan, W. B. F., 1968, Igneous rocks from the Beata Ridge (abstract), in: *Abstracts of Papers, 5th Caribbean Geol. Conf., St. Thomas, V. I.*, p. 26–27.

Green, G. W., and Holliday, D. W., 1970, The Cretaceous rocks of Blue Mountain Peak, Jamaica: *J. Geol. Soc. Jam.*, v. 11, p. 19–23.

Greiner, H. R., 1965, The oil and gas potential of Jamaica: Jam. Geol. Surv. Bull. 5, 24 p.

Gutenberg, B., 1955, Wave velocities in the earth's crust, in: Crust of the Earth (A. Poldervaart, ed.): Geol. Soc. Am. Spec. Pap. 62, p. 19–34.

Harland, W. B., Smith, A. G., and Wilcock, B., eds., 1964, *The Phanerozoic Time-Scale, a Symposium*: Geol. Soc. London, 458 p.

Hess, H. H., 1962, History of ocean basins, in: *Petrologic Studies: a Volume in Honor of A. F. Buddington*: Geol. Soc. Am., p. 599–620.

Holliday, D. W., 1971, Origin of Lower Eocene gypsum–anhydrite rocks, southeast St. Andrew, Jamaica: *Trans. Inst. Mining Met. (Sec. B)*, v. 80, p. B305–B315.

Hose, H. R., and Versey, H. R., 1956, Palaeontological and lithological divisions of the Lower Tertiary limestones of Jamaica: *Colon. Geol. Miner. Resour.*, v. 6, p. 19–39.

Jamaica Geological Survey, 1958, Geologic map of Jamaica, 1:250,000 scale.

Jamaica Geological Survey, 1962, Annual report of the Geological Survey Department for the year ended 31st March, 1961: Kingston, 25 p.

Jamaica Geological Survey, 1972, Above Rocks, Geol. Sheet 22, 1:50,000 scale.

Kauffman, E. G., 1966, Notes on Cretaceous Inoceramidae (Bivalvia) of Jamaica: *J. Geol. Soc. Jam. (Geonotes)*, v. 8, p. 32–40.

Kemp, A. W., 1971, The geology of the southwestern flank of the Blue Mountains, Jamaica: Ph.D. Dissertation, Univ. West Indies, Kingston.

Khudoley, K. M., and Meyerhoff, A. A., 1971, Paleogeography and geological history of Greater Antilles: Geol. Soc. Am. Mem. 129.

Lewis, J. F., Harper, C. T., Kemp, A. W., and Stipp, J. J., 1973, Potassium–argon retention ages of some Cretaceous rocks from Jamaica: *Bull. Geol. Soc. Am.*, v. 84, p. 335–340.

McBirney, A. R., and Bass, M. N., 1969, Geology of Bay Islands, Gulf of Honduras, in: Tectonic relations of northern Central America and the western Caribbean—the Bonacca Expedition (McBirney, A. R., ed.): Am. Assoc. Petr. Geol. Mem. 11, p. 229–243.

Malfait, B. T., and Dinkelman, M. G., 1972, Circum-Caribbean tectonic and igneous activity and the evolution of the Caribbean plate: *Bull. Geol. Soc. Am.*, v. 83, p. 251–272.

Matley, C. A., 1951, Geology and physiography of the Kingston District, Jamaica: Inst. Jamaica, Kingston, 139 p.

Matley, C. A., and Higham, F., 1929, The Basal Complex of Jamaica, with special reference to the Kingston District: *Q. J. Geol. Soc. London*, v. 85, p. 440–492.

Mills, R. A., Hugh, K. E., Feray, D. E., and Swolfs, H. C., 1967, Mesozoic stratigraphy of Honduras: *Bull. Am. Assoc. Petr. Geol.*, v. 51, p. 1711–1786.

Molnar, P., and Sykes, L. R., 1969, Tectonics of the Caribbean and Middle America regions from focal mechanisms and seismicity: *Bull. Geol. Soc. Am.*, v. 80, p. 1639–1684.

Newberry, J. S., 1888, Rhaetic plants from Honduras: *Am. J. Sci., 3rd Ser.*, v. 36, p. 342–351.

Officer, C. B., Ewing, J. I., Edwards, R. S., and Johnson, H. R., 1957, Geophysical investigations in the eastern Caribbean: Venezuelan Basin, Antilles island arc, and Puerto Rico Trench: *Bull. Geol. Soc. Am.*, v. 68, p. 359–378.

Oxburgh, E. R., and Turcotte, D. L., 1968, Mid-ocean ridges and geothermal distribution during mantle convection: *J. Geophys. Res.*, v. 73, p. 2643–2661.

Pinet, P. R., 1971, Structural configuration of the northwestern Caribbean plate boundary: *Bull. Geol. Soc. Am.*, v. 82, p. 2027–2032.

Pinet, P. R., 1972, Diapirlike features offshore Honduras: implications regarding tectonic evolution of Cayman Trough and Central America: *Bull. Geol. Soc. Am.*, v. 83, p. 1911–1922.

Raitt, R. W., 1963, The crustal rocks, in: *The Sea* (M. N. Hill, ed.): Wiley–Interscience, New York, Vol. 3, p. 85–102.

Reed, A. J., 1966, Geology of the Bog Walk Quadrangle: Jam. Geol. Surv. Bull. 4, 54 p.

Robinson, E., 1962, Lower Tertiary conglomerates and shales, in: Synopsis of the geology of Jamaica (V. A. Zans, L. J. Chubb, H. R. Versey, J. B. Williams, E. Robinson, and D. L. Cooke): Jam. Geol. Surv. Bull. 4, p. 21–24.

Robinson, E., 1967, Submarine slides in White Limestone Group, Jamaica: *Bu'l. Am. Assoc. Petr. Geol.*, v. 51, p. 569–578.

Robinson, E., 1968a, Biostratigraphic position of late Caenozoic rocks in Jamaica: *J. Geol. Soc. Jam. (Geonotes)*, v. 9, p. 32–41.

Robinson, E., 1968b, Stratigraphic ranges of some larger foraminifera, Jamaica, *Trans. 4th Caribbean Geol. Conf., Trinidad*, 1965, p. 189–194.

Robinson, E., 1969a, Studies in the Tertiary stratigraphy of eastern Jamaica: Ph.D. Dissertation, Univ. London.

Robinson, E., 1969b, Geological field guide to Neogene sections in Jamaica, West Indies: *J. Geol. Soc. Jam.*, v. 10, p. 1–24.

Robinson, E., Lewis, J. F., and Cant, R. V., 1971, Field guide to aspects of the geology of Jamaica, in: *International Field Institute Guidebook to the Caribbean Island-Arc System*, 1970: Am. Geol. Inst., Washington.

Roobol, M. J., 1971, The volcanic geology of Jamaica (preprint), 6th Caribbean Geol. Conf., Margarita, Venezuela, 1971.

Ronov, A. B., and Yaroshevsky, A. A., 1969, Chemical composition of the earth's crust, in: The earth's crust and upper mantle (P. J. Hart, ed.): Am. Geophys. Union Geophys. Monogr. 13, p. 37–57.

Sawkins, J. G., ed., 1869, Reports on the geology of Jamaica, Geol. Surv. Mem.: H. M. Stationery Office, London, 340 p.

Steinhauser, P., Vincenz, S. A., and Dasgupta, S. N., 1972, Paleomagnetism of some Lower Cretaceous lavas on Jamaica (abstract): *Eos Trans. Am. Geophys. Union*, v. 53, p. 356–357.

Versey, H. R., 1962, Older Tertiary limestones, in: Synopsis of the geology of Jamaica (V. A. Zans, L. J. Chubb, H. R. Versey, J. B. Williams, E. Robinson, and D. L. Cooke): Jam. Geol. Surv. Bull. 4, p. 26–43.

Vogt, P. R., Schneider, E. D., and Johnson, G. L., 1969, The crust and upper mantle beneath the sea, in: The earth's crust and upper mantle (P. J. Hart, ed.): Am. Geophys. Union Geophys. Mon. 13, p. 556–617.

Walper, J. L., and Rowett, C. L., 1972, Plate tectonics and the origin of the Caribbean Sea and the Gulf of Mexico: *Trans. Gulf Coast Assoc. Geol. Soc.*, v. 22, p. 105–116.

Weyl, R., 1966, *Geologie der Antillen*: Gebrueder Borntraeger, Berlin, 410 p.

Williams, H., and McBirney, A. R., 1969, Volcanic history of Honduras: *Univ. Calif. Publ. Geol. Sci.*, v. 85, p. 1–101.

Williams, J. B., 1962, Igneous and metamorphic rocks, in: Synopsis of the geology of Jamaica (V. A. Zans, L. J. Chubb, H. R. Versey, J. B. Williams, E. Robinson, and D. L. Cooke): Jam. Geol. Surv. Bull. 4, p. 58–66.

Woodring, W. P., 1925, Miocene mollusks from Bowden, Jamaica, Pt. 1: Carnegie Inst. Wash. Publ. 336.

Woodring, W. P., 1928, Miocene mollusks from Bowden, Jamaica, Pt. 2: Carnegie Inst. Wash. Publ. 385.

Wright, R. M., 1966, Biostratigraphical studies on the Tertiary White Limestone in parts of Trelawny and St. Ann, Jamaica: M.S. Thesis, Univ. London.

Wright, R. M., and Dickinson, W. R., 1972, Provenance of Eocene volcanic sandstones in eastern Jamaica—a preliminary note: *Caribb. J. Sci.*, v. 12, p. 107–113.

Zans, V. A., and Bailey, B. V., 1961, Marble deposits of Mount Hibernia and adjacent areas in north-western St. Thomas: *J. Geol. Soc. Jam. (Geonotes)*, v. 4, p. 8–11.

Zoppis Bracci, L., 1957, Estudio geológico de la región de Palacagüina y su depósito de antimonio: *Bol. Serv. Geol. Nac. Nicaragua*, v. 3, p. 29–34.

Chapter 15

THE GEOLOGICAL EVOLUTION OF THE CARIBBEAN AND GULF OF MEXICO—SOME CRITICAL PROBLEMS AND AREAS

Thomas W. Donnelly

Department of Geological Sciences
State University of New York
Binghamton, New York

I. INTRODUCTION

The fascinating and complex area of Middle America—roughly the Caribbean Sea, Gulf of Mexico, and adjacent land areas—has always attracted the attention of geologists. Geographically and politically fragmented, this dominantly oceanic area displays a rich panoply of geologic phenomena: earthquakes, volcanoes, deformed and metamorphosed sedimentary rocks of various ages, and a wide spectrum of igneous rock types. Because of the wide variety of phenomena displayed and because of the limited area of outcropping rock, geological interpretations of this region have largely been framed to fit prevailing contemporary geological fancies and to reflect one side or another of a geological debate centering around more distant areas. Thus, the Caribbean has figured prominently in debates such as "where do the Appalachians go?". In some cases the disputants have not been Caribbean geologists; in fact, the earliest summarizer of the Caribbean and Gulf of Mexico—the estimable C.

Schuchert (1935)—did not even visit the area and based his conclusions solely on his admirable summary of the existing literature.

In recent years the Caribbean region has found itself squarely in the middle of a new debate centering around the sea-floor-spreading concept— and once again much of the debate has centered on inferred local effects of distant phenomena. Two schools of thought have arisen which might be called the mobilist and the stabilist views. The former seeks to explain the geological evolution of the Caribbean region by extensive reference to sea-floor spreading. This group's philosophy stems from the work of Carey (1958), but also includes the more recent contributions of Mattson (1969), Malfait and Dinkelman (1972, 1973), Walper and Rowett (1972), Freeland and Dietz (1971, 1972), as well as several others. The stabilist viewpoint is espoused most eloquently by the Meyerhoffs (1972, 1973) but receives additional support from observations in summary papers by MacGillavry (1970), Barr (1974), and Weyl (1964). The reader is invited to examine the above-cited papers, as well as the very useful summaries of Nagle (1971), A. Meyerhoff (1967), Edgar *et al.* (1971), Ewing *et al.* (1971), the indispensable volumes 10 (Worzel, Bryant, *et al.*, 1973) and 15 of the Deep Sea Drilling Project (Edgar, Saunders, *et al.*, 1973), and papers on the seismicity of the region (Sykes and Ewing, 1965; Molnar and Sykes, 1969). Having done so, he will find that the problem is not as simple as either school contends. Further, and more disturbing, the reader will find an approximate negative correlation between Caribbean field geology experience and readiness to apply the recent concepts of sea-floor spreading to this area. Stated another way, many geologists with extensive field experience in the Caribbean region find important discrepancies between the geology of the region and various postulated models for its origin based on the sea-floor-spreading theory.

On the other hand, the sea-floor-spreading proponents have marshalled an impressive array of arguments in support of the stepwise opening of the Atlantic Ocean in Mesozoic time. Although many details remain to be resolved, there appears to be little doubt that the early Mesozoic opening of the North Atlantic was followed by a later Mesozoic opening of the tropical South Atlantic (Pitman and Talwani, 1972; Le Pichon and Fox, 1971; Le Pichon and Hayes, 1971). This evidence, together with the "Bullard fit" of continents around the pre-Mesozoic Atlantic (Bullard *et al.*, 1965), has two significant implications: 1) that the geological evolution of the Caribbean region cannot be considered without reference to the geology of the Old World, and 2) that the opening of the Atlantic in stages implies a differential movement between North and South America during much of the Mesozoic. It appears inescapable that the evolution of the Caribbean must be reconciled to a differential movement between the two major parts of the "Americas Plate." Ladd (1973) made a recent attempt to investigate this differential movement.

II. GEOLOGY OF THE PRE-MESOZOIC OF THE CARIBBEAN

Useful summaries of geological information relevant to the distribution and significance of older rocks in the Caribbean area are found in Banks (this volume), King (this volume), López-Ramos (this volume; 1969), Dengo (this volume; 1969), Helwig (this volume), Shagam (this volume; 1972a; 1972b), Walper and Rowett (1972), Woods and Addington (1973), de Cserna (1969; 1971a; 1971b), Kesler (1971), and Bass (1969). Possibly relevant occurrences of older rocks in northwest Africa are summarized by Hurley *et al.* (1974) and Schenck (1971).

In essence, a well-defined eastern North America Paleozoic structure (the Appalachians) encounters in Alabama an apparently distinct structure (Ouachita), which is traceable at least into northern Mexico. A more poorly defined Paleozoic structure (northern Andes) appears to terminate northward in Colombia or branch northeastward into Venezuela. A similarly poorly defined Paleozoic structure in northwest Africa extends southwestward beneath younger cover near the coast. A reconstructed pre-Mesozoic Atlantic requires that some attempt be made to link some or all of these trends through the Caribbean region.

The identification of old rocks does not by itself necessarily carry with it the possibility of finding old structures. For example, the essential difference between the Appalachian and Ouachita structures is that the former are the locus of thick, early Paleozoic geosynclinal accumulations, and the latter have noteworthy late Paleozoic accumulations on a normal (platform mainly) early and middle Paleozoic section. What little is known of the Paleozoic of Mexico and Central America is either referrable to the Ouachita structural "style" or is problematical; certainly, no thick geosynclinal accumulations correlative with the Appalachian "style" have been identified, although de Cserna (1971a) has postulated deformation of Taconian age of several older rock units of eastern Mexico. The southernmost old rock units of Central America are the Chuacús and Las Ovejas gneiss groups of Guatemala and the variously named schist–phyllite unit of northern Nicaragua (Palacagüina), Honduras (Cacaguapa), and southeastern Guatemala ("Santa Rosa"). At present neither the age nor the structural orientation of these units is well known; Kesler's (1971) analysis is in conflict with our own (unpublished) experience in this regard. The contrast in structural–stratigraphic styles at the tectonic nexus of central Alabama (King, this volume), though amenable to differing interpretations, at least seems to negate Walper and Rowett's suggestion that the Appalachian–Ouachita belts could be bent back into a collinear trend by a plastic reconstruction. Unfortunately, too little is known of the Paleozoic rocks of eastern Mexico to shed much light on the existence or nonexistence of a Paleozoic Gulf of Mexico.

The extensions northeastward of the Paleozoic Andes and southwestward of the Paleozoic Atlas are equally poorly known. Although both areas show early Paleozoic structural trends (as evidenced by thick Ordovician sections, mid-Paleozoic deformations, etc.), there has been a sufficiently strong overprint in both areas of Mesozoic deformation as to make the earlier development nearly undecipherable. However, there is sufficient similarity to postulate that these belts could well have been collinear Paleozoic marginal belts of deformation facing a Paleozoic ocean to the west and northwest. Such a scheme would remove the possibility of an Andean–Appalachian link. Either the Appalachians simply end in Alabama by passage into a different sort of structure or possibly they swing westward into the Wichita–Arbuckle belt, with a later Ouachita tectonic overprint obscuring the earlier structural bend in the subsurface of the Mississippi embayment. Uncertainties about the pre-Mesozoic structural configuration around the northern, western, and southern margins of the Caribbean region are likely to remain one of the most unsatisfactory areas of Caribbean geology in view of the very limited available exposures of older rocks.

In the island part of the Caribbean, pre-Mesozoic rocks are presently considered to occur only in Cuba, but this occurrence is by no means universally accepted (Khudoley and Meyerhoff, 1971). The description by Kuman and Gavilán (1965) of sillimanite-bearing gneisses on the Isle of Pines remains one of the most convincing arguments for pre-Mesozoic rocks in Cuba, and the recent dredging of metamorphic rocks between the Yucatán Peninsula and Cuba (Pyle et al., 1973) strengthens the argument that some older rocks extend east of the mainland. However, the relationships of these rocks and any older structural trends are still unknown.

III. THE AGE OF THE GULF OF MEXICO

The division of Caribbean geologists into mobilists and stabilists is paralleled by a division of Gulf of Mexico geologists into a group that believes that the Gulf of Mexico is very old and a group that believes that the Gulf is Mesozoic (or, at most, late Paleozoic) in age. Martin and Case (this volume) summarize the arguments.

Much of the argument for the age of the Gulf of Mexico centers on the existence or nonexistence of a Ouachita ocean, on the nature of post-Ouachita sediments, and on the distribution and nature (deep or shallow water) of the Jurassic salt deposits. The Ouachita ocean problem has no easy solution, but there is no direct evidence in the available rocks for the existence of a continental-ocean boundary in the Gulf of Mexico area at this time. Serpentines and volcanic rocks of the island arc or continental border type are wholly absent.

Post-Ouachita sediments (Woods and Addington, 1973), which are known mainly from some deep wells in the southern United States and northeastern Mexico, appear to reflect sedimentation in a region increasingly faulted in a tensional environment (rifted) with the passage of time. The Triassic Eagle Mills Formation, which is an analog of the Newark Series of the eastern United States, shows this structural facies quite clearly. The Jurassic (Louann) Salt of the Gulf of Mexico perhaps has been the focus of the most intensive debate on the origin of the Gulf. If the Gulf of Mexico is very old, then the salt, much of which underlies the Sigsbee Deep, must be, in large part, of deep-water origin, or formed in stages during the desiccation of a Gulf of Mexico isolated from the main world ocean. The latter possibility would appear to require some renewal of water periodically in order to supply adequate salt. The relevance of the salt to the main problem is underscored by the identification (by Wilhelm and Ewing, 1972) of a high-velocity layer (3.3 to 4.3 km/sec) at about 5 km depth beneath the Gulf as the salt. Direct identification of this layer is not possible. Although the higher velocity could reflect salt, it could as easily reflect compaction and lithification of oceanic sediments. On the other hand, arguments against the existence of a continuous salt layer across the Sigsbee Abyssal Plain are 1) that the salt is observed to occur in diapirs only in well-defined belts (Martin and Case, this volume), and that diapirs should occur throughout its area of occurrence, and 2) a salinity anomaly in the pore-fluids was not found at Deep Sea Drilling Project Sites 90 and 91 (Leg 10), whereas such anomalies were found at other sites (92, 2, 3, 85, 88, 89) known to be underlain by salt (Manheim et al., 1973). The second point is weaker, but it should be stressed that there is no direct evidence for a continuous salt layer beneath the Gulf of Mexico. The point is critical, not so much for the deep or shallow water origin of the salt as for the age assignment for the bulk of the sediment beneath the Sigsbee Abyssal Plain. Clearly a Jurassic salt layer halfway down in a thick (10–12 km) sedimentary section implies a late Paleozoic, or older, age for the base of this sediment. However, the alternative is that the *base* of the sediment section is Early Mesozoic and that the entire Sigsbee Plain section represents a high rate of pelagic and turbidite sediment accumulation since that time. The accumulation rates of the pre-Pleistocene sediments at Sites 90 and 91 of Leg 10, Deep Sea Drilling Project (Worzel, Bryant, et al., 1973) would, if extrapolated backward to the early Mesozoic, be responsible for 6–7 km of sediment accumulation since that time. If these rates had been higher, due to a greater abundance of turbidites in the early Tertiary and Mesozoic, then the entire 10–12-km sedimentary section could have accumulated since that time. The persistence of the Mississippi drainage system since the Mesozoic would appear to support this idea.

Another stratigraphic point relevant to the antiquity of the Gulf of Mexico is the absence of marine Permian deposits south of nuclear Central America, the absence of marine Triassic deposits throughout Middle America, and the restriction of normal marine Jurassic deposits to limited areas adjacent to the Gulf of Mexico, on the north, and to northeastern South America. Within most of the area of Middle America the Triassic–Jurassic appears to be represented by continental facies (although recent work suggests that much of the so-called Todos Santos of Guatemala and Honduras is more normal marine and early Cretaceous in age—see Burkart et al., 1973).

IV. THE AGE OF THE CARIBBEAN BASINS

The age of the basinal areas of the Caribbean presents a less tractable problem even than that of the Gulf of Mexico. In several land areas marginal to the basins occur sediments or volcanic rocks of Jurassic age—including the clastic sediments of northern Trinidad (Hutchinson, 1938; Potter, 1968), clastics, limestones, and evaporites of western Cuba (Khudoley and Meyerhoff, 1971), and the Jurassic volcanic rocks of La Désirade (Fink, 1972). The last of these, as will be discussed below, appears to be definitively oceanic, the other two more doubtfully so, at least from the standpoint of the occurrence of oceanic crust; only the great clastic sediment thicknesses in Cuba and Trinidad are suggestive of a continental margin situation. These occurrences of putative oceanic (or ocean-margin) Jurassic suggest a similar age for the Caribbean Basins. Direct evidence has not been forthcoming.

One of the objectives of Leg 15 of the Deep Sea Drilling Project (Edgar, Saunders, et al., 1973) had been to find, through drilling, the age of the basement of the Venezuelan and/or Colombian Basins. The failure of the project to do so is well known,* and the age is still unknown. At Site 146 in the central Venezuelan Basin the drilling stopped in dolerite intruding latest Turonian limestones less than 0.8 km beneath the sea floor. Older seismic-refraction studies (Officer et al., 1959) and more recent sonobuoy measurements (Ludwig

---

* The reactions of geologists to the reports of the Leg 15 drilling have provided a valuable insight into the ways in which information and misinformation are circulated in the geological community. The scientific staff of Leg 15 published nothing which suggested that drilling during the Leg had reached the crust of the Caribbean, or that this crust was younger than had been anticipated (Edgar, Saunders, et al., 1971, 1973; Donnelly et al., 1971; Donnelly, 1973). However, Meyerhoff and Meyerhoff (1972) dwelt extensively on an article in the popular press purporting to show that the scientific staff "did an abrupt about face" and "abandoned" our "earlier insertion hypothesis." Uchupi (this volume) cites a press release for a similar, if less emphatic, purpose. Neither reference is valid and neither fairly represented the views of the scientific staff. In other cases (Barr, 1974; Mattinson et al., 1973) authors apparently felt that their conclusions that the basalt was younger than the crust somehow flew in the face of the views of the scientific staff of Leg 15, which was not the case.

et al., 1974) show that the distance penetrated by the drill is a fraction of the total distance to the crust (velocities greater than 6 km/sec). The latter reference, in oral presentation, showed that the distance to crust beneath horizon B″, where the drilling terminated, varies from 2.5 to 3.5 km, with the sub-B″ material having a velocity from 4.5 to 5.3 km/sec. Thus, the age and character of the older materials above the crust of the Venezuelan Basin are unknown. Information of any sort regarding the Colombian Basin is far more spotty, and information regarding the Yucatán Basin is virtually nonexistent. It is well known that the total crustal thickness (defining the crust as that material with velocities of between about 6 and 8 km/sec) of the Venezuelan and Colombian Basins is large compared with the normal oceanic crust, but the Yucatán Basin and Gulf of Mexico have a more normal oceanic crust (Officer *et al.*, 1959; Ewing *et al.*, 1960). The significance of this abnormally thick crust in the Colombian and Venezuelan Basins will be discussed below; at this point it should simply be emphasized that its age is quite unknown.

V. MESOZOIC OROGENIC EVOLUTION—THE NORTHWESTERN AREA

The Mesozoic stratigraphy of Cuba (Pardo, this volume), Jamaica (Arden, this volume) and western Hispaniola (Bowin, this volume) show important similarities and are here grouped as a probable tectonic unit. Little is known of the subsurface Bahamas (Uchupi *et al.*, 1971; Paulus, 1972), and it is included in this area with some hesitation. Stratigraphically this area is characterized by large amounts of carbonate sediment throughout the Cretaceous sections. Important thicknesses of clastic sediments occur in Cuba, and volcanic rocks occur throughout; however, both of these rock types are less significant than elsewhere in the Caribbean. The volcanic rocks include important thicknesses of basalt in the Early and Middle Cretaceous; this rock is less common elsewhere, except off the northern coast of Venezuela. In Cuba, Pardo (this volume) has emphasized the evolution of the island in terms of vertical tectonics with a later, rather violent compressive phase. The basement rocks on either side of the Las Villas belt appear to be continental, though those to the north, presumably the subsurface of the Bahamas and Florida, are not exposed. Classical island-arc evolution is difficult to identify in this area; instead, analogies with a rifted continental margin undergoing a relatively late (Early Tertiary) compressive pulse appear to be more compatible with the geology. The subsidence that much of the Bahamas platform is inferred to have undergone during the Mesozoic is compatible with a history of extensional tectonics, rather than a dominantly compressional history.

VI. MESOZOIC OROGENIC EVOLUTION—NUCLEAR CENTRAL AMERICA

In the present discussion of Nuclear Central America, I am forced to rely heavily on abundant unpublished field and laboratory data stemming from a 12-year-old project in the critically important Motagua Valley zone of Guatemala. Some of the conclusions presented here are tentative and contradict to a slight extent those of Dengo (this volume).

The Mesozoic evolutionary picture as seen especially in Guatemala and Honduras (Dengo, this volume) shows important similarities to that of the northwestern Caribbean islands. A still incompletely understood late Paleozoic orogenic pulse, which is later (Leonard–Wolfcamp) than Ouachita deformation, produced block faulting and deposition of a shale–limestone sequence (Tactic–Chochal) following a generally undated, but possibly syn-Ouachita, pulse producing thick, immature clastic wedge deposits in much of Nuclear Central America.

A part of the Permian extensional pulse apparently involved the eruption of local basalts within a rifted block (the El Tambor basalts, whose Permian age* has been tentatively determined through unpublished lithologic correlations of our project in that area). Following this episode, a compressive phase closed this basin and severely deformed the rocks on either side of the zone (the Motagua Valley fault zone of Guatemala). Later sandstones (the Todos Santos, which are Middle Cretaceous in this area) are deformed to a lesser extent. The emplacement of serpentine appears to be latest Cretaceous, based on appearance of debris in the Sepur and Subinal Formations, which are assigned to this age. Compressive deformation continued in the Tertiary and is continuing, on a much subdued level, to the present day.

Analogies with Cuba, including the serpentine tectonic zone between two continental blocks, are impressive. The serpentine emplacement is placed slightly earlier in Guatemala than in Cuba, but the tectonic environments of the two areas are very similar. Further elucidation of the evolution of this area, including the key questions as to the significance of the structure of the western Yucatán Basin (Case, this volume), the dredged metamorphic rocks along this border (Pyle *et al.*, 1973), and the age of the Basin itself, must await the resolution to a question posed by Arden (this volume): What is the nature of the Nicaragua Rise? Dissimilarities between Jamaica and Honduras are baffling; the former is far more volcanic in the early Cretaceous. However, a transition from an almost nonvolcanic early Cretaceous in Guatemala to a slightly volcanic early Cretaceous of the Mosquitia region seems to suggest a more complete transition to a more typical island arc or similar tectonic facies to the east. The early or middle Tertiary opening of the Cayman Trough

---

* See note added in proof, p. 684.

(Bowin, 1968; Holcombe *et al.*, 1973) then represents a wholly new tectonic style whose spectacular topography has tended to obscure an older and still enigmatic evolution.

VII. MESOZOIC OROGENIC EVOLUTION—THE SOUTH AMERICAN BORDERLAND

The all-important borderland of Venezuela and Trinidad is not treated elsewhere in this volume; the reader is referred to summaries by Barr and Saunders (1965), Bell (1971, 1972), Maresch (1974), Bellizia (1972), Menéndez (1967), and Beets (1972). The Mesozoic history begins with the deposition of a thick section of clastic sediments in northern Venezuela and Trinidad in the late Jurassic (the Caracas Group and its correlatives). The dominantly volcanic Villa de Cura Group, which is poorly dated but tentatively placed in the Early Cretaceous, was apparently formed north of this; however a Late Cretaceous–Early Tertiary tectonic event caused this entire group to slide southward as a huge allochthonous mass, so that its original position is not known with certainty. The oldest rock unit of the offshore islands is a tholeiitic basalt of Middle Cretaceous age (Santamaría and Schubert, 1974).

The Caracas and Villa de Cura Groups appear to represent clastic sedimentation and volcanism along a compressive continental margin. Whether an offshore island arc was present is uncertain, although this inference is made by most of the workers cited above. A climactic phase of compression produced extensive blueschist metamorphics and eclogites (Maresch, 1974) along the apparent front of this belt.

Attempts to find the northward roots of the allochthonous Villa de Cura Group have not been successful; the volcanic northern islands appear to represent a different, younger event, and the roots of the earlier island arc may lie beneath these exposed volcanics or even more seaward beneath the Curaçao Ridge.

As Hedberg (1974) has pointed out, there is a striking parallelism throughout northern South America in the stratigraphic units of Mesozoic age. Thus, the Triassic–Jurassic red beds, the transgressive sandstone–shale–limestone of the Early Cretaceous, the thick Aptian–Albian reefal limestones, and the dark, organic limestones of the Turonian attest to a tectonic unity of this entire region. However, most of these units are absent to the north, in Central America and the West Indies. The red beds are duplicated in the northern Gulf of Mexico area; their occurrence in Nuclear Central America as the Todos Santos beds may not be a true correlation, as noted above. The thick Aptian–Albian limestones occur throughout Middle America, though they are less conspicuous in the northeastern islands.) The most direct interpretation of the Mesozoic history of the entire northern South American borderland is of a compressive orogenic continental-oceanic boundary.

VIII. MESOZOIC OROGENIC EVOLUTION—THE NORTHEASTERN ISLANDS

The most complete stratigraphic-tectonic picture available for the northeastern islands is seen in the Puerto Rico–Virgin Islands area (Berryhill *et al.*, 1960; Glover, 1971; Pease, 1968; Seiders, 1971; Berryhill, 1965; Donnelly, 1964, 1966; Donnelly *et al.*, 1971; Mattson, 1973). Bowin (1966; this volume) shows that the eastern part of Hispaniola, though not thoroughly known, is very similar. Fink (1972) has shown that a part of the island of La Désirade is lithologically identical with the Water Island Formation of the Virgin Islands. Further correlatives westward of eastern Hispaniola and southward of La Désirade [with the possible exception of Tobago, which has many lithological similarities to Puerto Rico (Maxwell, 1948) in the character of the volcanic rocks] are unknown to me.

The oldest rocks in this area are of two types: cherts above the Bermeja complex of southwest Puerto Rico contain Tithonian radiolaria (Mattson and Pessagno, 1974), and radiometric dates from the keratophyre–trondjhemite complex of La Désirade (Fink, 1972; Mattinson, *et al.*, 1973) are of virtually the same age. The chert in Puerto Rico overlies a serpentine–amphibolite complex identified as oceanic crust (Hess, 1964; Donnelly *et al.*, 1971; Donnelly *et al.*, 1973*b*; Mattson, 1973; Lee *et al.*, 1974). The serpentine–amphibolite complex appears to have been deformed in two stages—at least one apparently before the deposition of the chert (Tobisch, 1968; Mattson, 1973), suggesting a Jurassic or earlier age of oceanic crust beneath the evolving island arc.

Of equal antiquity as the chert, but with uncertain tectonic-stratigraphic relationship, is the spilite–keratophyre–basaltic andesite complex seen most clearly in the Virgin Islands, dated most precisely in La Désirade, and having correlatives in eastern Puerto Rico and Hispaniola. Donnelly (1964, 1966, 1972) has presented arguments for the initial eruption of this magma in an ocean of abyssal depth (based largely on the inferred nonexplosive eruptive behavior of a hydrated magma), with a later shallowing and emergence of the eruptive centers. The volcanic history seems to span the entire history of the island arc from its beginnings on deep-ocean floor to its present emergent condition. The earlier (Jurassic to Albian) magma series is followed in Puerto Rico by a more typical calc-alkaline volcanic series (Donnelly *et al.*, 1971), but with no implied change in tectonic style. With the passage of time (Donnelly, 1964) the topography of Puerto Rico became more differentiated, with development of adjacent high areas and basins of deposition. The last record of igneous activity is in the Eocene, during which both abundant volcanic and plutonic rocks were emplaced. Since the Eocene the geologic history of the northeastern island area has consisted solely of deposition of coastal-plain sedimentary series, continued uplift, and mild deformation. The volcanic history of the Lesser Antilles (Tomblin, this volume) proper begins at almost the same time as that of the Greater Antilles ceases; La Désirade, then, is interpreted as a fragment of an older history exposed by chance beneath a younger island-arc volcanic accumulation. Results of Leg 15 of the Deep Sea Drilling Project (Donnelly, 1973a) show that volcanic activity of the Lesser Antilles waned during most of the Miocene but increased dramatically in the Plio-Pleistocene.

In summary, the stratigraphy of the northeastern islands contrasts with the northwestern islands by having only very limited amounts of carbonate sediment, and with the southern areas in having very limited clastic sediment prior to the local accumulations of Late Cretaceous and Early Tertiary age. Thus, the northwestern island area suggests subsidence of a rifted continental margin with a mixed group of volcanics and a later compressive crisis, the western (Central America) reflects rifting and a later compressive event, the southern area reflects the evolution of a compressive volcanic-sedimentary orogen along a continental boundary, and the northeastern island area reflects the evolution of a compressive island arc on oceanic crust. The further problem is to link these areas in a meaningful regional tectonic picture.

IX. THE PROBLEM OF THE EARLY VOLCANIC ROCKS

The tectonic significance of the spilite–keratophyre association of the Virgin Islands and its correlatives in La Désirade is currently a subject of considerable debate. My co-workers and I have argued (Donnelly et al., 1971) that the early volcanic association (Water Island Formation; Louisenhoj Formation) belong to the island-arc tectonic-magmatic facies. We termed these early volcanics "primitive," noting especially their low content of Th. Contemporary work on the volcanic rocks of Western Pacific island arcs (Jåkeš and White, 1972) showed the existence of a comparable early series there, which has been termed "island-arc tholeiite." I now have extensive additional unpublished trace-element (especially rare earth) data which shows clearly that the early volcanic rock associations in the two areas are highly comparable. However, the tectonic significance of this early association remains unclear. Pending further publication of the differing viewpoints, suffice it to say that one school of thought holds that the early volcanics have affinities to those of mid-ocean ridges and represent the same tectonic environment (rifting). The other school, to which I belong, holds that the early volcanics of island arcs represent the same compressive environment as the later (calc-alkaline) eruptions, but with different materials being fused. The principal distinction between "primitive" island-arc volcanic associations and those of mid-ocean ridges is the great abundance of highly siliceous igneous rocks in the former and their virtual absence in the latter. Important distinctions in minor elements (especially Th, Zr, Ti, rare earth elements) also can be found. A further point is that the "primitive" igneous rocks of island arcs evidently crystallize in many or most instances from intrinsically hydrated magmas rather than dry magmas; basalt flows are generally scarce but fragmental (pyroclastic) basaltic andesites relatively common. A discussion of the inferred deep-water eruptive environment of the early, hydrated magma of the Virgin Islands is given in Donnelly (1972).

More recently, Mattinson *et al.* (1973) have taken issue with these con- clusions with regard to the La Désirade volcanic rocks, which they term "ophiolites." This argument is a fundamental one for Caribbean geology; if the La Désirade (and by correlation, Virgin Islands) volcanic series are indeed fragments of oceanic crust, then the inferred evolution of the island arc is far different from that suggested by Donnelly (1964, 1972). It is also con- siderably younger than suspected—certainly post-Jurassic. A complication not fully brought out by Mattinson *et al.* (1973) is that *two* volcanic associations occur on La Désirade. The volcanics at the west end of the island are domi- nantly siliceous, have been dated as Jurassic, and are clearly correlative with the Virgin Islands. Those at the east end of the island are undated pillow basalts with cherty interflow sediments. Each end of the island has distinctive lithol- ogies not found at the other. Unfortunately, the relationship between the two ends is unknown. It would appear to me that the pillow basalt complex at the eastern end is probably representative of oceanic basalts (in part, then, ophiolite) and that the keratophyre–trondjhemite complex at the western end represents the oldest volcanic rocks of the island-arc association. If the age relations are shown to be reversed, then the pillow basalt complex must be interpreted as a later magmatic event. Because of the pivotal age and position of this volcanic complex, the resolution of the age and tectonic relationship between its two parts in La Désirade is a matter of the utmost significance in Caribbean geologic history. The island of La Désirade has far greater signifi- cance in the implications of its tectonic history than merely in its age. Schemes such as those of Freeland and Dietz (1971 and 1972) and Malfait and Dinkel- man (1972) simply have no place presently for an island-arc complex of Jurassic age.

X. THE AVES RIDGE

The Aves Ridge (Fox and Heezen, this volume) has long been an enigma to Caribbean geologists. A topographic analogy with the "second ridges" of Pacific island arcs, notably the Marianas islands, has led to speculation about these features having a common origin and reflecting some underlying, fundamental principle in island-arc evolution. Geophysical information about the ridge is scanty; Edgar *et al.* (1971) showed that the crustal velocities are similar to those of the Venezuelan Basin (though the layer thicknesses are, of course, greater) and dissimilar to those of the Beata Ridge. Donnelly (1973*b*) showed that there was a sharp boundary between a magnetically relatively "quiet" zone of the Venezuelan Basin and a "disturbed" zone beginning near the western foot of the ridge. More recently Watkins (1974) confirmed the boundary of the magnetically disturbed (short wavelength) zone, but he was unable to add meaningful information on the nature of the Aves Ridge magnetic sources. Seismic profiling in the Venezuelan Basin (Edgar *et al.*, 1971) shows that horizon B'' is not traceable beneath the Aves Ridge. Edgar *et al.* (1973) showed that horizon B'' deepens to the east in the Venezuelan Basin. Recent unpublished data (J. Matthews, personal communication) suggest that this deepening continues eastward to the foot of the slope, where the reflector disappears.

Dredged rocks from the Aves Ridge have received a great deal of attention because of the report of granodiorite by Fox *et al.* (1971). These authors and Fox and Heezen (this volume) have stressed the possibility that this granodiorite might be a sample of a layer beneath the Aves Ridge and, indeed, of the entire Venezuelan Basin. However, Nagle (1972) found only volcanic rocks (resembling those from Puerto Rico, based on my own examination) and sediments further north. In fact, granitic rocks have been found by several investigators, but only at the southernmost end of the ridge. The resemblance (chemical, petrographic, ages) between the dredged granodiorites and granitic rocks from La Blanquilla and other Venezuelan islands (Schubert and Motiska, 1973; Santamaría and Schubert, 1974) and my own unpublished chemical data is sufficiently strong to relate these dredged rocks directly to those of the Venezuelan borderland. Whether there are similar rocks elsewhere on the Ridge is still unknown.

Further information about the Aves Ridge has come from dredging and drilling. Nagle (1972), Bock (1972), and Edgar *et al.* (1973) have all given strong evidence for Late Tertiary vertical movements of considerable magnitude for the ridge. There is no evidence that the ridge has been a young volcanic center, but dredged and drilled rocks cited above lead irresistibly to the conclusion of Cretaceous (possibly Early Tertiary) calc-alkaline volcanic activity.

The sum of the data suggests that the Aves Ridge is an inactive island-arc segment. The polarity of subduction is not determinate but is commonly assumed to be the same as that of the present Lesser Antilles, i.e., a slab descending from the east. The deepening of horizon B'', the magnetic boundary on the westward side, and postulated subduction from within the Venezuelan Basin along the north (Matthews and Holcombe, 1974) and south side (Edgar,

Saunders, *et al.*, 1973) lead to the conclusion that the present inactivity of the ridge could result from the cessation of subduction from the west. If this is the case, subduction must have continued to at least post-horizon B″ (approximately 80 m.y.).

XI. THE GREATER ANTILLEAN "ORTHOGEOSYNCLINE"

One issue on which mobilists and stabilists alike have tended to agree is that the Greater Antilles has acted as a reasonably coherent tectonic element during its evolution. The discovery of the old rocks on La Désirade has caused most workers to extend this unit at least that far, and there is some feeling that the Aves Ridge should be included here. The contrast in tectonic and stratigraphic styles between the eastern and western ends, however, raises the question whether they have always been one unit. The stratigraphic contrast between the two ends is seen most strikingly in eastern Hispaniola, where the ultramafic belt associated with the Hatillo Thrust (Bowin, this volume) appears to separate two dissimilar units. Significantly, there is a zone of Late Cretaceous granitic plutons along the southwest side of this zone but virtually no post-Middle Cretaceous igneous activity on the northeast side. We might speculate, then, that this zone could represent the suturing (at some unspecified time in the Late Cretaceous) of two originally dissimilar tectonic elements. In this sense, the ultramafic zone of Hispaniola is analogous to those of Cuba and Guatemala, though in no case is extensive movement necessarily indicated.

Related closely to this question is that of Cretaceous polarity of the portions of the Greater Antilles. Khudoley and Meyerhoff (1971) have argued strongly for an ocean-to-the-south polarity for Cuba; Pardo (this volume) has a similar placement, but his interpretive cross sections do not resemble normal island-arc polarity. For Hispaniola and Jamaica, there is little good evidence for polarity. Puerto Rico and the Virgin Islands remain elusive. Although I (Donnelly, 1966) argued for a Cretaceous trench north of the Virgin Islands, I now feel that my evidence showed only the direction in which the water deepened and not the existence of a trench itself.

The chemistry of the volcanic rocks of Puerto Rico (Lidiak, 1972) is also unhelpful; potassium contents have little consistent variation across the present island. However, Jolly (1970) found an increase in the potassium/sodium ratio from south to north for one middle Cretaceous unit—the Lapa Lava. A recent structural study of the Bermeja Complex by Mattson (1973) has suggested thrusting of this complex, followed by a later thrust of younger rocks, from south to north. Thus a number of observations, none of them overwhelming, suggest a polarity during much of the active earlier period for the eastern Greater Antilles in a reverse sense to the present polarity.

XII. IGNEOUS PETROLOGY AND THE CARIBBEAN; THE GREAT FLOOD BASALT EVENT

The most significant single result of Leg 15, Deep Sea Drilling Project (Edgar, Saunders, *et al.*, 1973) was the discovery of basalt and dolerite at each of five central basin sites in the Colombian and Venezuelan Basins. These occurrences were very nearly of the same age, though slightly younger to the west. The fact that there appeared to be no age pattern to these basalt occurrences (and certainly no recognizable magnetic signature), as well as the fact that we have penetrated only a fraction of the distance to crustal velocities, led us to recognize immediately that we had not found the age of basement but had been prevented from doing so by a vast basaltic event. The event is significant both for its apparent size (perhaps 1 million km² ; Donnelly, 1973*a*) and for its significance in Caribbean evolution.

Basaltic magmatism is divided in contemporary petrological thinking into two tectonic environments: compressive, in which a subducted slab melts, either through frictional heating or introduction of water via hydrated materials into hot but anhydrous rock at depth; and tensional, in which melting accompanies tension or extension. Because this latter process necessarily leads to upward movement of material to fill the opening surficial gap, it is not conceptually easy to differentiate it from surficial spreading in response to a "plume" of hot material being forced upward from below. Thus, the origin of tectonic movements in a tensional environment leads to a "chicken vs. egg" argument over what is the ultimate causative agent.

Nevertheless, the two magmatic environments are tectonically useful in differentiating extensional or compressive regimes. Criteria for the separation of the two magma suites formed in the two regimes are debatable, but the consensus is that basaltic suites (MORB, Hawaiian-type basalts, alkalic basalts of a variety of oceanic and continental environments, and continental tholeiites) are characteristic of extensional regimes and andesitic suites (including the calc-alkalic suite of variable silica content; the "primitive" basaltic andesites, spilites, and keratophyres) belong here. There is some debate about the "primitive" basalts (or "island-arc tholeiite"), and there is considerable uncertainty about the tectonic significance of high-K basalts (or shoshonites) in island-arc environments, but the principal dichotomy is generally accepted.

In the Caribbean this dichotomy creates a problem in Cretaceous tectonics which is only just now being grasped. The problem is that, during the Late Cretaceous, island-arc areas such as Puerto Rico and Hispaniola erupted quantities of magma of the "compressive" type, while the basin immediately behind the arc erupted a vast flood of "extensional" basalt. The age limits of the flood are not well known. Edgar, Saunders, *et al.* (1973) originally dated the flood at about 80 m.y. on the basis of its correlation with horizon

B″, but this is obviously only the end of the event. Certain circum-Caribbean basalt occurrences [Nicoya Complex of Costa Rica and Panama, possible basalt occurrences of Campanian age in Colombia (J. Case, personal communication), basalts of southern Haiti (Butterlin, 1954), Curaçao (Beets, 1972)] were suggested as belonging to this province. However, recent radiometric age dating (Santamaría and Schubert, 1974) has shown that the northern Venezuelan–Curaçao occurrences are older (Aptian–Albian) and correlate with such occurrences as the Sans Souci basalt of Trinidad (Barr, 1963). However, the occurrence of Turonian ammonites on Aruba (Dodd and MacDonald, 1974) suggests that this basaltic event either spanned about 30 m.y. of time or occurred in two stages, at about 110 and 80 m.y. before present.

On Puerto Rico during this interval occurred basalts and shoshonites of the Robles Group. These have been grouped with the island-arc series by Donnelly *et al.* (1971), but with some *caveats*. Very possibly their occurrence represents the incursion of an extensional tectonic event into the normally compressive island-arc regime of the northeastern islands. This problem is not resolved, but the lack of occurrence of basalt with undoubted MORB chemical characteristics in the islands during this time period is noteworthy. The separation of basalt types could argue for the tectonic decoupling of the Venezuelan Basin from Puerto Rico during this time, which becomes another argument in favor of a reversed polarity for the island arc at that time.

The contributions of igneous petrology to Caribbean tectonics are potentially powerful. Certainly the distinction between compressive and extensional regimes is very useful. Thus the "South Central American Orogen" (Nicoya Complex) of Meyerhoff and Meyerhoff (1972) is now interpreted as a fragment of an extensional event that not only does not prevent Malfait and Dinkelman's (1972) Pacific plate from sliding into the Caribbean but could represent the plate itself! I suggest this not as support for the mobilist view but as an illustration of how some stabilist objections to spreading might be found to be completely invalid.

North of the Nicaragua Rise the flood event is not recognized. A subsurface dolerite in Jamaica (Donnelly *et al.*, 1973*b*) has been tentatively placed here, but exposed volcanic rocks from the Benbow inlier of Jamaica (the only sufficiently ancient rocks of the island) do not support the occurrence there (Roobol, 1972). Basalts from Cuba (Pardo, this volume), which are not known to me personally, could belong to this or a similar event. There is no record of this event known to me in Nuclear Central America.

XIII. THE CONTRIBUTION OF THE DEEP SEA DRILLING PROJECT

The five legs of the Deep Sea Drilling Project (1, 4, 10, 11, and 15) that have been undertaken in this area have profoundly altered concepts of Caribbean and Gulf of Mexico geology. Leg 1 (Ewing, Worzel, *et al.*, 1969) identified the Sigsbee Knolls as salt diapiric structures. Leg 10 (Worzel, Bryant, *et al.*, 1973) showed that much of the Gulf of Mexico margin originated as a carbonate bank area subsiding slowly since the Late Cretaceous. Legs 4 and 15 (Bader, Gerard, *et al.*, 1970; Edgar, Saunders, *et al.*, 1973) showed that the central Caribbean had been pelagic since the Late Cretaceous and had been the site of a mammoth basalt eruption at about 80 m.y. before present. In addition, the facies of sediments drilled during the Late Cretaceous showed a siliceous content and an organic nature which seemed to suggest a connection with the siliceous Pacific surface water during this time, coupled with a restriction in bottom-water circulation to allow stagnation. This was also seen north of Brazil at Site 144 of Leg 14; regrettably it was not seen in the Atlantic at the same time in Legs 1 and 11 (Hollister, Ewing, *et al.*, 1972) owing to the general absence of Turonian–Santonian deep-water sediments recovered in these legs. A problem, however, with ascribing the stagnation seen so vividly in the Turonian–Santonian of Leg 15 sediments to bottom-water conditions is that near-shore correlative sediments of northern South America (such as the Querequal, La Luna, and San Antonio Formations; Hedberg, 1974) show the same organic facies.

The siliceous sediments persistent in the Caribbean to the early Miocene have counterparts in the adjacent Gulf of Mexico (Worzel, Bryant, *et al.*, 1973), Atlantic (Hollister, Ewing, *et al.*, 1972; Bader, Gerard, *et al.*, 1970), on the Blake Plateau (Hathaway *et al.*, 1970), and on the coastal plain (Wise and Weaver, 1974). Thus, the Antillean island chain formed at this time no effective barrier to Pacific–Caribbean siliceous surface waters. The disappearance of a siliceous biota from the pelagic fauna in the middle Miocene can be ascribed to a gradual elevation of the Central American isthmus and an introduction of terrigenous sediment with a probable Amazonian source in the later Tertiary period.

The contribution of the Deep Sea Drilling Project has, hopefully, not ended for this area. Perhaps the most important future contribution would be the drilling of a site in the western Atlantic outside the Antillean arc to recover a good stratigraphic section of at least Santonian age in order to allow a careful assessment of the geography of the Antillean island chain and its effect on oceanic circulation. Such an assessment is difficult now owing to the incomplete sections presently available from Legs 1, 4, 11, and 14 outside the Antillean arc.

XIV. SOME BIOGEOGRAPHIC CONSIDERATIONS

One of the most important, and best known, biogeographic observations relevant to Middle America is the Tertiary isolation of South America and an evolution there of an endemic mammalian fauna (Keast, 1972). The connection in the Pliocene with North America and the supppression of the South American fauna by a flood of northern immigrants has been known since Charles Darwin studied Argentine vertebrate remains while on the voyage of the *Beagle*. There are, however, some observations which are less well known and potentially troublesome for the "mobilists." One is that in the Late Cretaceous there are similarities in the dinosaur faunas of North America and South America. Another is that in the Cretaceous of North America and in approximately coeval beds of South America marsupial remains are found (Keast, 1972). Whatever the ultimate origin of the marsupials, the suggested connection is difficult to understand from the contemporary mobilist point of view. Another observation more consistent with the stabilist view is the New World endemism of the Caprinuloidinae (rudists) of Albian age, which developed in parallel with the Old World Caprininae (MacGillavry, 1970). None of these observations should be given more weight than it deserves, but the stabilists will derive more comfort from them than the mobilists.

XV. THE CARIBBEAN REGION IN THE LIGHT OF CONTINENTAL DRIFT

Early in this discussion, reference was made to the current debate between the stabilists and mobilists. The former attempt to explain the origin of Middle America with no reference to large horizontal movements or rotations; the latter make such movements central to their varied schemes. The conflict is that basic elements of sea-floor spreading and plate tectonics, at least as they apply to the opening of the Atlantic Ocean, seem to be essentially incontrovertible. The origin of the Caribbean would appear to have to be explained in terms of the separation, first of North America and later South America, from Africa.

One important potential clue, the occurrence of pre-Mesozoic rocks, is of little help. The occurrences in Middle America are limited to the subsurface of Florida and the Yucatán, to limited exposures in eastern Mexico, and to more widespread exposures in nuclear Central America. Unfortunately, these do not provide firm evidence for or against a Mesozoic opening of the Gulf of Mexico or for a pre-drift linkage of structural elements among the major craton and Paleozoic fold belt areas. My prejudice is for a complete Late Paleozoic disappearance of all oceanic areas, but the strongest evidence to be marshalled for this view is that the post-Ouachita pre-Louann sedimentary facies in the Gulf of Mexico appear to reflect rifting tectonics and to parallel the rifted eastern margin of North America. A complete closure also solves the problem of the provenance of the Cuban Cayetano clastic debris (Jurassic), and opening of the Atlantic provides an environment for an Early Mesozoic shallow-water salt basin. Later subsidence of the margin of a widening, rifted Gulf of Mexico carried the salt beds and impressive Middle Cretaceous reefal accumulation (Tampico, Campeche, west coast of Florida) to depth.

One of the important unknowns is the age of the Caribbean basins. If they are coeval with the North Atlantic, they could be early Mesozoic in age. In this case, the North Atlantic might have opened between North and South America—much in the position of the present Venezuelan and Colombian Basin. The Motagua Fault Zone Permian basalts might represent a slightly earlier abortive opening which was closed shortly after it began. An obvious objection to this scheme is the geography of the Greater Antilles, which would appear to form a barrier to a single ocean separating North America from South America–Africa. However, the inference of noncontinuity between the eastern and western ends of the Greater Antilles allows these two ends to have evolved in the Jurassic and Early Cretaceous on opposite sides of the Atlantic. The eastern Greater Antilles, then, evolved off the coast of Africa and was probably collinear with the belt off the northern coast of South America. Both show important similarities in that they both formed in a compressive environment—but one end was truly oceanic and the other adjacent to a cratonic sediment source. The western end of the Greater Antilles, meanwhile, developed more as a rifted and subsided continental margin (of which the Bahamas is the purest example) modified by some subduction. The consequences of this scheme are interesting and will be strengthened if further examples of Jurassic–Early Cretaceous stratigraphic dissimilarity across the putative suture zone (central Hispaniola) can be found or inferred indirectly.

The presumed Early Cretaceous opening of the tropical South Atlantic (Le Pichon and Hayes, 1971) would have to have begun behind (east of) the eastern Greater Antillean islands and to have moved these islands westward—giving them a counterclockwise rotation at the same time. The paleomagnetic observations on Middle Cretaceous lavas from Puerto Rico (Fink and Harrison, 1972) support this; indeed, a widespread counterclockwise rotation of tectonic blocks within the Caribbean area has been summarized by MacDonald and Opdyke (1972). A further consequence of this rotation of the eastern Greater Antilles is that some of the earlier Venezuelan Basin must have been consumed. I would suggest that a reverse polarity allowing consumption of the Venezuelan Basin plate from the west would allow this rotation quite easily. The Aves Ridge, then, would be interpreted as a presently inactive island-arc segment, but with the subduction zone on the west. A goal of future dredging expeditions should be to look for serpentine on the western margin of this ridge. The present Lesser Antilles island arc then becomes a late feature built on the fragments of an earlier arc and facing the Eocene

Atlantic Ocean. Thus, we might not expect to find an equivalent of La Désirade beneath the southern half of the Lesser Antilles.

The flood basalt event remains unexplained by this (or by a stabilist) scheme. One possibility is that the abandonment of an Early Mesozoic spreading ridge in the Caribbean, as the newly formed South Atlantic mid-ocean ridge "captured" (to borrow a term from fluvial geomorphology) the North Atlantic Ridge at about 20° north latitude, left a large mass of hot upper mantle beneath the Caribbean with no well-defined eruptive center (plate separation). This mass, after a suitable delay, simply appeared as a flood basalt roughly in the vicinity of the old ridge path. Another possibility is that this basalt event records an extensional plate breakup over a large area such that little plate movement occurred. One of the future goals of deep-sea drilling should be to penetrate the horizon B'' and ascertain the distribution of basalt with depth (and age) in the Caribbean. The sonobuoy results of Ludwig *et al.* (1974) imply 2.5 to 3.5 km of moderate-velocity material— possibly largely basalt—above the crust. If this is the case, or even if the amount of basalt is relatively small, the thermal implications will be staggering.

A consequence of the flood basalt event which should be investigated is the tectonic evolution of northern South America. A still unexplained event of major significance is the slide of the allochthonous Villa de Cura block many tens of kilometers southward from the Caribbean coast. The magnitude of the Albian–Turonian thermal event as seen at the coast suggests that the Venezuelan borderland might simply have been thermally elevated so as to allow the Villa de Cura block to slide off to the south. (MacGillavry, in an unabstracted talk at the 1974 Guadeloupe Caribbean Geological Conference, showed that on Bonaire the majority of a 5-km section consisted of basalt and dolerite.)

A further consequence of the flood basalt is that the anomalously thick Caribbean crust might be explained by the event. Murauchi *et al.* (1973) show a similar thick crust beneath the Ontong Java Plateau of the southwest Pacific; Kroenke (1974) relates this to a similar Middle Cretaceous flood basalt event.

The precise delineation of the past and present fractures outlining the Caribbean Plate might be a self-defeating exercise. Seismicity gives a believable general pattern (Sykes and Ewing, 1965; Molnar and Sykes, 1969), but with many gaps. Some areas whose morphology and structure suggest neotectonic activity (Anegada Passage, Enriquillo Basin) have not been seismic in the last few decades. Very likely both the northern and southern boundaries are the loci of complex surficial manifestations of more or less simple structural accommodation to plate movement at depth. Thus, it is no surprise that we have a young spreading center in the Cayman Trough (Holcombe *et al.*, 1973), evidently young calc-alkaline volcanic activity in central Hispaniola (Donnelly, unpublished), evidently young sinistral strike-slip movement in the Polochic–Chixoy zone, but only young vertical movements in the morphologically spectacular adjacent Motagua Valley of Guatemala.

In summary, it appears that the evolution of the Caribbean can best be explained by the growth of Mesozoic ocean basins between the Americas as they separated at different rates from Africa. Certain older elements have been torn apart and rotated, and at least two elements (the eastern and western Greater Antilles) may have been fused as late as the end of the Mesozoic during transcurrent movement. Critical periods in the Caribbean correspond closely to critical periods in Atlantic Ocean opening (Pitman and Talwani, 1972). Although much of the critical evidence is still lacking—and much may never be found—arguments in support of impressive differential plate movement, such as the sliding into the Caribbean of a Pacific Plate, have not been backed by any positive evidence. A fascinating final speculation is what the next major geological surprise is going to be—who would have predicted La Désirade or the Caribbean flood basalts?

ACKNOWLEDGMENTS

In arranging the index cards for the bibliography of this chapter, I find that I know two-thirds of the authors—and half of these I have been in the field with! I am deeply indebted to all of them, and to the many other Caribbean associates from whom I have learned so much. I owe a special debt of gratitude to E. Pessagno for the unpublished information concerning the age of the Mariquita Chert, to M. Perfit for showing me dredged rocks from the Cayman Trough, to J. Watkins, T. Holcombe, and J. E. Matthews for discussing unpublished geophysical information, to my co-worker J. J. W. Rogers for his stimulation and efforts in our continuing geochemical studies, and to J. Case, J. Matthews, and H. J. MacGillavry for recent discussions (cited above). E. Silver provided me with the manuscript of a most interesting paper on geophysics of the Venezuela borderland. Minor-element work (most still unpublished) relevant to the Caribbean was done at Australian National University during a visiting Fellowship there in 1972–1973. My work in the Caribbean has been supported and/or sponsored by the National Science Foundation, Research Foundation of the State University of New York, the Center for Solid Earth Geology of the State University of New York at Binghamton, the Deep Sea Drilling Project, and the Instituto Geográfico Nacional of Guatemala. Finally, my wife and companion on numerous trips in this area once again stepped into the breach with her irreplaceable help in the assembly of this chapter.

NOTE ADDED IN PROOF

Subsequent to the submission of the manuscript for this chapter, an important paper contradicting one of my conclusions was published (H. H. Wilson, 1974, *Am. Assoc. Petr. Geol. Bull.*, v. 58, p. 1348–1396). Wilson had the good fortune to find recognizable rudist fragments in the highly deformed limestone unit that forms the top of the El Tambor sequence. The entire sequence now appears to be Late Cretaceous (specifically probably Cenomanian–Campanian) and not latest Paleozoic. Thus the El Tambor basalt–chert–shale–limestone unit becomes a highly deformed member of the "Great Flood Basalt Event." Further implications of this startling find are currently under investigation.

REFERENCES

Bader, R. G., Gerard, R. D., *et al.* 1970, *Initial Reports of the Deep Sea Drilling Project*, Vol. 4: U.S. Government Printing Office, Washington.

Barr, K. W., 1963, The geology of the Toco District, Trinidad, West Indies: *Overseas Geol. Miner. Resour.*, v. 8, p. 379–415; v. 9, p. 1–29.

Barr, K. W., 1974, The Caribbean and plate tectonics—some aspects of the problem. Verhand 1: Naturf. Ges. Basel, v. 84, p. 45–67.

Barr, K. W., and Saunders, J. B., 1965, An outline of the geology of Trinidad, *Trans. 4th Caribbean Geol. Conference, Port-of-Spain, Trinidad*, p. 1–13.

Bass, M. N., 1969, Petrography and ages of crystalline basement rocks of Florida—some extrapolations: Am. Assoc. Petr. Geol. Mem. 11, p. 283–310.

Beets, D. J., 1972, Lithology and stratigraphy of the Cretaceous and Danian succession of Curaçao. Utrecht: Natuurwetenschappelijke studiekring voor Suriname en de Nederlandse Antillen, n. 70, 153 p.

Bell, J. S., 1971, Tectonic evolution of the central part of the Venezuelan coast ranges, in: Donnelly, T. W., ed., Caribbean Geophysical, Tectonic, and Petrologic Studies: Geol. Soc. Am. Mem. 130, p. 107–118.

Bell, J. S., 1972, Geotectonic evolution of the southern Caribbean area, in: Shagam, R., ed., Studies in Earth and Space Science: Geol. Soc. Am. Mem. 132, p. 369–386.

Bellizia G., A., 1972, Is the entire Caribbean mountain belt of northern Venezuela allochthonous?, in: Shagam, R., ed., Studies in Earth and Space Science: Geol. Soc. Am. Mem. 132, p. 363–368.

Berryhill, H. L., Jr., 1965, Geology of the Ciales Quadrangle, Puerto Rico: U.S. Geol. Surv. Bull. 1184, 116 p.

Berryhill, H. L., Jr., Briggs, R. P., and Glover, L., III, 1960, Stratigraphy, sedimentation, and structure of Late Cretaceous rocks in eastern Puerto Rico—preliminary report: *Am. Assoc. Petr. Geol. Bull.*, v. 44, p. 137–155.

Bock, W. D., 1972, The use of foraminifera as indicators of subsidence in the Caribbean, *Memoria VI Conferencia Geológica del Caribe, Margarita, Venezuela*, p. 439–440.

Bowin, C. O., 1966, Geology of Central Dominican Republic (A case history of part of an island arc), in: Hess, H. H., ed., Caribbean Geological Investigations: Geol. Soc. Am. Mem. 98, p. 11–84.

Bowin, C. O., 1968, Geophysical study of the Cayman Trough: *J. Geophys. Res.*, v. 73, p. 5159–5173.

Bullard, E. C., Everett, J. E., and Smith, A. G., 1965, The fit of continents around the Atlantic: *Philos. Trans. R. Soc. London, Ser. A*, v. 258, p. 41–51.

Burkart, B., Clemons, R. E., and Crane, D. C., 1973, Mesozoic and Cenozoic stratigraphy of southeastern Guatemala: *Am. Assoc. Petr. Geol. Bull.*, v. 57, p. 63–73.

Butterlin, J., 1954, La géologie de la République d'Haiti et ses rapports avec celle des regions voisines: Publ. Com. 150ᵉ Anniv. Indép. Républ. Haïti, Secr. Prés. Port-au-Prince, 446 p.

Butterlin, J., 1956, La constitution géologique et la structure des Antilles: Centre Nat. pour la Recherche Scientifique, Paris, 453 p.

Carey, S. W., 1958, A tectonic approach to continental drift, in: Carey, S. W., ed., *Continental Drift, A Symposium*: Geol. Dept., Univ. of Tasmania, Hobart, p. 177–355.

de Cserna, Z., 1969, Tectonic framework of southern Mexico and its bearing on the problem of continental drift: *Bol. Soc. Geol. Mex.*, v. 30, p. 159–168.

de Cserna, Z., 1971a, Taconian (early Caledonian) deformation in the Huasteca structural belt of eastern Mexico: *Am. J. Sci.*, v. 271, p. 544–550.

de Cserna, Z., 1971b, Precambrian sedimentation, tectonics, and magmatism in Mexico: *Geol. Rundschau*, v. 60, p. 1488–1513.

Dengo, G., 1969, Problems of tectonic relations between Central America and the Caribbean: *Trans. Gulf Coast Assoc. Geol. Soc.*, v. 19, p. 311–320.

Dodd, P. A., and MacDonald, W. D., 1974, Structural and petrologic study of pre-Tertiary rocks of Aruba (abstract), *VII Conference Géologique des Caraïbes (Guadeloupe)*.

Donnelly, T. W., 1964, Evolution of eastern Greater Antillean island arc: *Am. Assoc. Petr. Geol. Bull.*, v. 48, p. 680–696.

Donnelly, T. W., 1966, Geology of St. Thomas and St. John, U.S. Virgin Islands, in: H. H. Hess, ed., Caribbean Geological Investigations: Geol. Soc. Am. Mem. 98, p. 85–176.

Donnelly, T. W., 1972, Deep-water, shallow-water, and subaerial island-arc vulcanism: an example from the Virgin Islands, in: R. Shagam, ed., Studies in Earth and Space Sciences: Geol. Soc. Am. Mem. 132, p. 401–414.

Donnelly, T. W., 1973a, Circum-Caribbean explosive volcanic activity: evidence from Leg 15 sediments, in: Edgar, N. T., Saunders, J. B., *et al.*, *Initial Reports of the Deep Sea Drilling Project*, Vol. 15: U.S. Government Printing Office, Washington, p. 969–988.

Donnelly, T. W., 1973b, Magnetic anomaly observations in the eastern Caribbean Sea, in: Edgar, N. T., Saunders, J. B., *et al.*, *Initial Reports of the Deep Sea Drilling Project*, Vol. 15: U.S. Government Printing Office, Washington, p. 1023–1030.

Donnelly, T. W., 1973c Evolution of island-arc magmas: the Caribbean and south-west Pacific compared (abstract): *Geol. Soc. Am. Abstr.*, v. 5, p. 602.

Donnelly, T. W., 1973d, Late Cretaceous basalts from the Caribbean: a possible flood basalt province of vast size (abstract): *Trans. Am. Geophys. Union (EOS)*, v. 54, p. 1004.

Donnelly, T. W., Kay, R., and Rogers, J. J. W., 1973a, Chemical petrology of Caribbean basalts and dolerites; Leg 15, Deep Sea Drilling Project (abstract): *Trans. Am. Geophys. Union (EOS)*, v. 54, p. 1002–1004.

Donnelly, T. W., Rogers, J. J. W., Kay, R., and Melson, W., 1971, Basalt and dolerite from the central Caribbean: Preliminary results from Deep Sea Drilling Project, Leg XV (abstract): Geol. Soc. Am. Ann. Mtg., p. 548.

Donnelly, T. W., Melson, W., Kay, R., and Rogers, J. J. W., 1973b, Basalts and dolerite of late Cretaceous age from the central Caribbean, in: Edgar, N. T., Saunders, J. B., *et al.*, *Initial Reports of the Deep Sea Drilling Project*, Vol. 15: U.S. Government Printing Office, Washington, p. 989–1012.

Donnelly, T. W., Rogers, J. J. W., Pushkar, P., and Armstrong, R. L., 1971, Chemical evolution of the igneous rocks of the eastern West Indies: an investigation of thorium, uranium, and potassium distributions, and lead and strontium isotopic ratios, in: Donnelly, T. W., ed., Caribbean Geophysical, Tectonic, and Petrologic Studies: Geol. Soc. Am. Mem. 130, p. 181–224.

Edgar, N. T., Ewing, J. I., and Hennion, J., 1971, Seismic refraction and reflection in Caribbean Sea: *Am. Assoc. Petr. Geol. Bull.*, v. 55, p. 833–870.

Edgar, N. T., Holcombe, T., Ewing, J. I., and Johnson, W., 1973, Sedimentary hiatuses in the Venezuelan Basin, in: Edgar, N. T., Saunders, J. B., *et al.*, *Initial Reports of the Deep Sea Drilling Project*, Vol. 15: U.S. Government Printing Office, Washington, p. 1051–1062.

Edgar, N. T., Saunders, J. B., *et al.*, 1971, Deep Sea Drilling Project—Leg 15: *Geotimes*, v. 16, n. 4, p. 12–16.

Edgar, N. T., Saunders, J. B., *et al.*, 1973, Initial Reports of the Deep Sea Drilling Project, Vol. 15: U.S. Government Printing Office, Washington, 1137 p.

Ewing, J. I., Antoine, J., and Ewing, M., 1960, Geophysical measurements in the Western Caribbean Sea and in the Gulf of Mexico: *J. Geophys. Res.*, v. 65, p. 4087–4126.

Ewing, J. I., Edgar, N. T., and Antoine, J. W., 1971, Structure of the Gulf of Mexico and Caribbean Sea, in: Maxwell, J., ed., *The Sea*, Vol. 4, Part II, Chap. 10: John Wiley & Sons, New York, p. 321–358.

Ewing, M. W., Worzel, J. L., *et al.*, 1969, *Initial Reports of the Deep Sea Drilling Project*, Vol. 1: U.S. Government Printing Office, Washington, 672 p.

Fink, L. K., Jr., 1972, Bathymetric and geologic studies of the Guadeloupe region, Lesser Antilles island arc: *Mar. Geol.*, v. 12, p. 267–288.

Fink, L. K., Jr., and Harrison, C. G. A., 1972, Paleomagnetic observations of selected lava units on Puerto Rico, *Memoria 6th Conferencia Geológica del Caribe, Margarita, Venezuela*, p. 379.

Fox, P. J., Schreiber, E., and Heezen, B. C., 1971, The geology of the Caribbean crust: Tertiary sediments, granitic and basic rocks from the Aves Ridge: *Tectonophysics*, v. 12, p. 89–109.

Freeland, G. L., and Dietz, R. S., 1971, Plate tectonic evolution of Caribbean–Gulf of Mexico region: *Nature*, v. 232, p. 20–23.

Freeland, G. L., and Dietz, R. S., 1972, Plate tectonic evolution of the Caribbean–Gulf of Mexico region, *Memoria 6th Conferencia Geológica del Caribe, Margarita, Venezuela*, p. 259–264.

Glover, L., III, 1971, Geology of the Coamo area, Puerto Rico, and its relation to the volcanic arc–trench association: U.S. Geol. Surv. Prof. Pap. 636, 102 p.

Hathaway, J. C., McFarlin, P. F., and Ross, D. A., 1970, Mineralogy and origin of sediments from drill holes on the continental margin off Florida: U.S. Geol. Surv. Prof. Pap. 581-E, 26 p.

Hedberg, H. D., 1974, Some distinctive lithostratigraphic units of the Caribbean region (abstract), *VII Conference Géologiques des Caraïbes* (*Guadeloupe*).

Hess, H. H., 1964, The oceanic crust, the upper mantle, and the Mayaguez serpentinized peridotite, in: Burk, C., ed., *A Study of Serpentinite*: Nat. Acad. Sci.–Nat. Res. Council Publ. 1188, p. 169–175.

Holcombe, T. L., Vogt, P. R., Matthews, J. E., and Murchison, R. R., 1973, Sea-floor spreading in the Cayman Trough? (abstract): *Trans. Am. Geophys. Union* (*EOS*), v. 54, p. 327.

Hollister, C. D., Ewing, J. I., *et al.*, 1972, *Initial Reports of the Deep Sea Drilling Project*, Vol. 11: U.S. Government Printing Office, Washington, 1077 p.

Hurley, P. M., Boudda, A., Kanes, W. H., and Nairn, A. E. M., 1974, A plate tectonics origin for late Pre-Cambrian–Paleozoic orogenic belt in Morocco: *Geology*, v. 2, p. 343–344.

Hutchinson, A. G., 1938, Jurassic ammonites in the northern range of Trinidad: Bol. Geol. Miner., Caracas, Ven. v. 2.

Jakeš, P., and White, A. J. R., 1972, Major and trace element abundances in volcanic rocks of orogenic areas: *Bull. Geol. Soc. Am.*, v. 83, p. 29–40.

Jolly, W. T., 1970, Petrologic studies of the Robles Formation, South Central Puerto Rico: Ph.D. Thesis, State Univ. of New York at Binghamton.

Keast, A., 1972, Continental drift and the evolution of the biota on southern continents, in: Keast, A., *et al.*, eds., *Evolution, Mammals, and Southern Continents*: State Univ. of New York Press, Albany, p. 23–88.

Kesler, S. E., 1971, Nature of ancestral orogenic zone in nuclear Central America: *Bull. Am. Assoc. Petr. Geol.*, v. 55, p. 2116–2129.

Khudoley, K. M., and Meyerhoff, A. A., 1971, Paleogeography and geological history of Greater Antilles: Geol. Soc. Am. Mem. 128, 199 p.

Kroenke, L. W., 1974, Origin of continents through development and coalescence of oceanic flood basalt plateaus (abstract): *Trans. Am. Geophys. Union (EOS)*, v. 55, p. 443.

Kuman, V. E., and Gavilán, R. R., 1965, Geología de Isla de Pinos: Revista Tecnológica (Ministerio de Industrias, La Habana), v. 3, p. 20–38.

Ladd, J. W., 1973, Relative motion between North and South America and the evolution of the Caribbean (abstract): Geol. Soc. Am. Abstr. with Programs, v. 5, p. 705.

Lee, V., Mattson, P. H., and Miyashiro, A., 1974, Metamorphosed oceanic crust or early volcanic products in Puerto Rico basement rock association (abstract), *VII Conference Géologique des Caraïbes (Guadeloupe)*.

Le Pichon, X., and Fox, P. J., 1971, Marginal offsets, fracture zones, and the early opening of the North Atlantic: *J. Geophys. Res.*, v. 76, p. 6294–6308.

Le Pichon, X., and Hayes, D. E., 1971, Marginal offsets, fracture zones, and the early opening of the South Atlantic: *J. Geophys. Res.*, v. 76, p. 6283–6293.

Lidiak, E. G., 1972, Spatial and temporal variations of potassium in the volcanic rocks of Puerto Rico, *Memoria VI Conferencia Geológica del Caribe, Margarita, Venezuela*, p. 203–209.

López-Ramos, E., 1969, Marine Paleozoic rocks of Mexico: *Bull. Am. Assoc. Petr. Geol.*, v. 53, p. 2399–2417.

Ludwig, W. J., Houtz, R. C., and Ewing, J. I., 1974, Profiler–sonobuoy measurements of Colombia and Venezuela Basins, Caribbean Sea (abstract), *VII Conference Géologique des Caraïbes (Guadeloupe)*.

MacDonald, W. D., and Opdyke, N. D., 1972, Tectonic rotations suggested by paleomagnetic results from northern Colombia, South America: *J. Geophys. Res.*, v. 77, p. 5720–5730.

MacGillavry, H. J., 1970, Geological history of the Caribbean: *Proc. Koninkl. Ned. Akad. Wetenschap.*, Ser. B, v. 73, p. 64–96.

Malfait, B. T., and Dinkelman, M. G., 1972, Circum-Caribbean tectonic and igneous activity and the evolution of the Caribbean plate: *Bull. Geol. Soc. Am.*, v. 83, p. 251–272.

Malfait, B. T., and Dinkelman, M. G., 1973, Circum-Caribbean tectonic and igneous activity and the evolution of the Caribbean plate: reply: *Bull. Geol. Soc. Am.*, v. 84, p. 1105–1108.

Manheim, F. T., Sayles, F. L., and Waterman, L. S., 1973, Interstitial water studies on small core samples, Deep Sea Drilling Project; Leg 10, in: Worzel, J. L., Bryant, W., *et al.*, *Initial Reports on the Deep Sea Drilling Project*, Vol. 10, U.S. Government Printing Office, Washington, p. 615–624.

Maresch, W. V., 1974, Plate-tectonics origin of the Caribbean mountain system of northern South America: discussion and proposal: *Bull. Geol. Soc. Am.*, v. 85, p. 669–682.

Matthews, J. E., and Holcombe, T. L., 1974, Possible Caribbean underthrusting of the Greater Antilles along the Muertos Trough (abstract), *VII Conference Géologique des Caraïbes (Guadeloupe)*.

Mattinson, J. M., Fink, L. K., Jr., and Hopson, C. A., 1973, Age and origin of ophiolitic rocks on La Désirade Island, Lesser Antilles Island Arc: Ann. Rpt. Director, Geophysical Laboratory, Carnegie Institution Year Book 72, p. 616–623.

Mattson, P. H., 1969, The Caribbean: a detached relic of the Darwin Rise (abstract): *Trans. Am. Geophys. Union (EOS)*, v. 51, p. 317.

Mattson, P. H., 1973, Middle Cretaceous nappe structure in Puerto Rican ophiolites and their relation to the tectonic history of the Greater Antilles: *Bull. Geol. Soc. Am.*, v. 84, p. 21–38.

Mattson, P. H., and Pessagno, E. A., Jr., 1974, Tectonic significance of Late Jurassic–Early Cretaceous radiolarian chert from Puerto Rican ophiolite (abstract): Geol. Soc. Am. Abstr. with Programs, v. 6, p. 859.

Maxwell, J. C., 1948, The geology of Tobago, B. W. I.: *Geol. Soc. Am. Bull.*, v. 59, p. 801–854.

Menéndez, A., 1967, Tectonics of the central part of the Western Caribbean mountains, Venezuela, *Intern. Conf. Tropical Oceanography, Proc., Studies in Tropical Oceanography*, Vol. 5, p. 103–130.

Meyerhoff, A. A., 1967, Future hydrocarbon provinces of Gulf of Mexico—Caribbean region: *Trans. Gulf Coast Assoc. Geol. Soc.*, v. 17, p. 217–260.

Meyerhoff, A. A., and Meyerhoff, H. A., 1972, Continental drift, IV: the Caribbean "Plate": *J. Geol.*, v. 80, p. 34–60.

Meyerhoff, A. A., and Meyerhoff, H. A., 1973, Circum-Caribbean tectonic and igneous activity and the evolution of the Caribbean plate: discussion: *Bull. Geol. Soc. Am.*, v. 84, p. 1101–1104.

Molnar, P., and Sykes, L. R., 1969, Tectonics of the Caribbean and Middle America regions from focal mechanisms and seismicity: *Bull. Geol. Soc. Am.*, v. 80, p. 1639–1684.

Murauchi, S., Ludwig, W. J., Den, N., Hotta, H., Asanuma, T., Hoshii, T., Kubotera, A., and Hagiwara, K., 1973, Seismic refraction measurements on the Ontong Java Plateau northeast of New Ireland: *J. Geophys. Res.*, v. 78, p. 8653–8603.

Nagle, F., 1971, Caribbean geology, 1970: *Bull. Mar. Sci.*, v. 21, p. 375–439.

Nagle, F., 1972, Rocks from the seamounts and escarpments of the Aves Ridge, *Memoria VI Conferencia Geológica del Caribe, Margarita, Venezuela*, p. 409–413.

Officer, C. B., Ewing, J. I., Hennion, J. F., Harkrider, D. G., and Miller, D. E., 1959, Geophysical investigations in the eastern Caribbean: summary of the 1955 and 1956 cruises, in: Ahrens, L. H., Press, F., and Runkorn, S. K., eds., *Physics and Chemistry of the Earth*, Vol. 3: Pergamon Press, London, p. 17–109.

Olson, W. S., and Leyden, R. J., 1973, North Atlantic rifting in relation to Permian–Triassic salt deposition, in: *International Permian–Triassic Conference, 1971*: Univ. of Calgary and Geol. Survey of Canada, p. 720–732.

Paulus, F. J., 1972, The geology of Site 98 and the Bahama Platform, in: Hollister, C. D., Ewing, J. I., *et al.*, 1972, *Initial Reports of the Deep Sea Drilling Project*, Vol. 11: U.S. Government Printing Office, Washington, p. 877–897.

Pease, M. H., Jr., 1968, Cretaceous and Lower Tertiary Stratigraphy of the Naranjito and Aguas Buenas Quadrangles and ajacent areas, Puerto Rico: U.S. Geol. Surv. Bull. 1253, 57 p.

Pitman, W. C., III, and Talwani, M., 1972, Sea floor spreading in the North Atlantic: *Bull. Geol. Soc. Am.*, v. 83, p. 619–646.

Potter, H. C., 1968, A preliminary account of the stratigraphy and structure of the eastern part of the Northern Range, Trinidad, *Trans. Fourth Caribbean Geol. Conference, Port-of-Spain, Trinidad*, p. 15–20.

Pyle, T. E., Meyerhoff, A. A., Fahlquist, D. A., Antoine, J. W., McCrevey, J. A., and Jones,

P. C., 1973, Metamorphic rocks from northwestern Caribbean Sea. *Earth Planet. Sci. Lett.*, v. 18, p. 339–344.

Roobol, M. J., 1972, The volcanic geology of Jamaica, *VI Conferencia Geológica del Caribe, Margarita, Venezuela*, p. 100–107.

Santamaría, F., and Schubert, C., 1974, Geochemistry and geochronology of the southern Caribbean–northern Venezuela plate boundary: *Bull. Geol. Soc. Am.*, v. 85, p. 1085–1098.

Schenck, P. E., 1971, Southeastern Atlantic Canada, Northeastern Africa, and continental drift: *Can. J. Earth Sci.*, v. 8, p. 1218–1251.

Schubert, C., and Motiska, P., 1973, Reconocimiento geológico de las islas Venezolanas en el Mar Caribe, entre Los Roques y Los Testigos (Dependencias Federales). II. Islas Orientales y conclusiones: *Acta Cien. Venez.*, v. 24, p. 19–31.

Schuchert, C., 1935, *Historical Geology of the Antillean–Caribbean Region*: John Wiley & Sons, New York, 811 p.

Seiders, V. M., 1971, Cretaceous and Lower Tertiary stratigraphy of the Gurabo and El Yunque Quadrangles, Puerto Rico: U.S. Geol. Surv. Bull. 1294-F, 58 p.

Shagam, R., 1972a, Evolución tectónica de los Andes Venezolanos, *Memoria IV Congreso Geológico Venezolano (Ministerio de Minas e Hidrocarburos, Caracas)*, Vol. 2, p. 1201–1261.

Shagam, R., 1972b, Andean research project, Venezuela: principal data and tectonic implications, in: Shagam, R., *et al.*, eds., Studies in Earth and Space Science: Geol. Soc. Am. Mem. 132, p. 449–463.

Silver, E. A., Case, J. E., and MacGillavry, H. J., Geophysical study of the Venezuelan borderlands (unpublished).

Sykes, L. R., and Ewing, M., 1965, The seismicity of the Caribbean region: *J. Geophys. Res.*, v. 70, p. 5065–5074.

Tobisch, O. T., 1968, Gneissic amphibolite at Las Palmas, Puerto Rico, and its significance in the early history of the Greater Antilles island arc: *Bull. Geol. Soc. Am.*, v. 79, p. 557–574.

Uchupi, E., Milliman, J. D., Luyendyk, B. P., Bowin, C. O., and Emery, K. O., 1971, Structure and origin of southeastern Bahamas: *Bull. Am. Assoc. Petr. Geol.*, v. 55, p. 687–704.

Walper, J. L., and Rowett, C. L., 1972, Plate tectonics and the origin of the Caribbean Sea and the Gulf of Mexico: *Trans. Gulf Coast Assoc. Geol. Soc.*, v. 22, p. 105–116.

Watkins, J. S., 1974, Implications of magnetic anomalies in the northern Venezuelan Basin (abstract), *VII Conference Géologique des Caraïbes (Guadeloupe)*.

Weyl, R., 1961, *Die Geologie mittelamerikas*: Gebrüder Bornträger, Berlin, 226 p.

Weyl, R., 1964, Die palaeogeographische Entwicklung des mittelamerikanisch-westindischen Raumes: *Geol. Rundschau*, v. 54, p. 1213–1240.

Weyl, R., 1966, *Geologie der Antillen*: Gebrüder Bornträger, Berlin, 410 p.

Wilhelm, O., and Ewing, M., 1972, Geology and history of the Gulf of Mexico: *Bull. Geol. Soc. Am.*, v. 83, p. 575–600.

Wise, S. W., and Weaver, F. M., 1974, Origin of cristobalite-rich Tertiary sediments in the Atlantic and Gulf coastal plains: *Trans. Gulf Coast Assoc. Geol. Soc.*, v. 23, p. 305–323.

Woods, R. D., and Addington, J. W., 1973, Pre-Jurassic geologic framework. Northern Gulf Basin: *Trans. Gulf Coast Assoc. Geol. Soc.*, v. 23, p. 92–108.

Worzel, J. L., Bryant, W., et al., 1973, *Initial Reports of the Deep Sea Drilling Project*, Vol. 10: U.S. Government Printing Office, Washington, 748 p.

INDEX